Microscale Testing in Aquatic Toxicology

Advances, Techniques, and Practice

Microscale Testing in Aquatic Toxicology

Advances, Techniques, and Practice

edited by
Peter G. Wells, Ph.D.
Kenneth Lee, Ph.D
Christian Blaise, D. Sc.

with the assistance of
Johanne Gauthier, M.Sc.

CRC Press
Boca Raton   Boston   London   New York   Washington, D.C.

Library of Congress Cataloging-in-Publication Data

Microscale testing in aquatic toxicology :   advances, techniques, and
   practice /   edited by Peter G. Wells, Kenneth Lee, Christian Blaise;
   with the editorial assistance of Johanne Gauthier.
       p. cm.
   Includes bibliographical references and index.
   ISBN 0-8493-2626-5 (alk, paper)
   1. Water quality bioassay.   I. Wells, P. G. II. Lee, Kenneth,
1953–   .   III. Blaise, Christian. IV. Gauthier, Johanne, 1961–   .
QH90.57.B5M535   1997
628.1′61--dc21   97-18407
                                                                    CIP

No claim to original U.S. Government works
International Standard Book Number 0-8493-2626-5
Library of Congress Card Number   97-18407
Printed in the United States of America 1 2 3 4 5 6 7 8 9 0
Printed on acid-free paper

About the Editors

Peter G. Wells, Ph.D., is a Senior Coastal Research Scientist with Environment Canada in Dartmouth, Nova Scotia, a federal agency with whom he has worked since 1974. He has academic appointments at the School for Resource and Environmental Studies (SRES) and the Marine Affairs Program, Dalhousie University, Halifax; the Acadia Centre for Estuarine Research, Acadia University, Wolfville, N.S; and the Bermuda Biological Station for Research. Since 1988, he has chaired several working groups for the United Nations GESAMP (Joint Group of Experts on Scientific Aspects of Marine Environmental Protection), of which he is now Vice-Chair.

Dr. Wells attended schools in England, Calgary, and Montreal; received degrees at McGill University (B.Sc., Zoology, 1967), the University of Toronto (M.Sc., Marine Zoology, 1969), and the University of Guelph (Ph.D., Zoology, 1976). He has worked at the Biological Station, St. Andrews, New Brunswick, (marine fisheries biologist and doctoral candidate, 1969–74); Environment Canada, Bedford Institute of Oceanography, Dartmouth, N.S. (aquatic toxicologist, Head–Aquatic Toxicology and principal research scientist, 1974–83); the University of Toronto, Institute for Environmental Studies (Research Associate and principal research scientist, 1980–83); Environment Canada, Ottawa (Toxic Chemicals Assessment Advisor, 1983–85); Environment Canada, Dartmouth, N.S. (Senior Advisor and Research Manager, Marine Environmental Quality, 1985–94); and Dalhousie University (SRES, Associate Professor-Research, 1991–94). Primary research interests are in the areas of aquatic and marine ecotoxicology — the ecotoxicology of oil-derived hydrocarbons, spill control agents, contaminated sediments and other complex mixtures; the development and application of microscale marine toxicity tests; the role of marine ecotoxicology in identifying, assessing, managing, and monitoring marine pollution problems; and the role of science in integrated coastal management. Current research is largely in the science and application of marine microscale bioassays in hazard assessments of contaminated sediments, focusing on coastal inlets and the mud flats of the Bay of Fundy.

Since 1970, Dr. Wells has written, coauthored, or edited well over 100 primary and technical publications, dealing largely with contaminants and pollutants in aquatic and marine environments. Recent contributions as writer and/or editor include: *Proceedings of the Canadian Conference on Marine Environmental Quality* (1988); the National Academy of Sciences (NRC) review, *Using Oil Spill Dispersants on the Sea,* (1989); *Controlling Chemical Hazards* (Unwin-Hyman, 1991); *Health of Our Oceans: Status of Canadian Marine Environmental Quality* (Environment Canada, 1991); *Progress and Trends in Marine Environmental Protection* (Marine Pollution Bulletin, Vol. 25(1–4), 1992); the United Nations GESAMP Reports and Studies No. 50 *Impacts of Oil Pollution and Related Chemicals and Wastes on the Sea* (IMO London 1993); *Proceedings of the Coastal Zone Canada '94 Conference* (1994, 1996); *Exxon Valdez Oil Spill: Fate and Effects in Alaskan Waters* (ASTM STP 1219, 1995); *The Rio de la Plata: An Environmental Overview. An Ecoplata Project Background Report* (1st edition, 1996); GESAMP Reports and Studies No. 58 *Opportunistic Settlers and the Problem of the Ctenophore Mnemiopsis leidyi Invasion in the Black Sea* (1997); and *Fundy Issues. Scientific Overview and Workshop Proceedings* (1997).

Dr. Wells chairs the Fundy Marine Ecosystem Science Project, a new Canadian initiative in coastal marine science and management. He is the past Chair and an active member of the Canadian Intergovernmental Aquatic Toxicity Group. He is a founding member of the National Steering Committee for the Canadian National Aquatic Toxicity Workshops and a charter member of the Society for Environmental Toxicology and Chemistry (SETAC). He has chaired many sessions on microscale aquatic toxicity testing. He has served on several editorial boards, currently those of

Ecotoxicology and the *Water Pollution Research Journal* of Canada. At Dalhousie University, he offers three graduate courses — Fundamentals of Environmental Toxicology, Aquatic Toxicology and Water Quality Assessment, and Readings in Marine Ecotoxicology — and supervises numerous graduate and honors students in marine ecotoxicology.

Kenneth Lee, Ph.D. is a research scientist with the Maurice Lamontagne Institute, Fisheries and Oceans — Canada, where he heads the Microbiology and Hydrocarbons Section, Marine Environmental Sciences Division. He is responsible for research programs in microbial ecology, toxicology, and chemical oceanography relating to the transformation of contaminants in estuarine and coastal environments. He holds academic appointments as Adjunct Professor at the University of Toronto (Department of Botany) and the University of Québec at Rimouski (Department of Oceanography).

Dr. Lee received his B.Sc. degree in biology from Dalhousie University in 1975. He obtained his M.Sc. degree in limnology and his Ph.D. degree in microbial ecology from the University of Toronto in 1977 and 1982, respectively. He was an NSERC Visiting Scientist at the Institute of Ocean Sciences, Ocean Chemistry Division, Fisheries and Oceans Canada, from 1982 to 1984.

Dr. Lee is currently a principal investigator in several multinational research programs focused on oil spill counter-measure technologies, such as bioremediation, and environmental impact assessment techniques. He has published over 80 articles in scientific journals, technical reports, and book chapters, and is a member of the editorial board of the journal, *Environmental Technology*. Dr. Lee frequently provides expert advice to scientific groups, industry, and government on R&D policy decisions.

Christian Blaise, D.Sc. is a senior research scientist at the Saint-Lawrence Centre, Environment Canada, Québec Region, where he heads the Bioanalytical Research Unit (BRU), Ecotoxicology and Environmental Chemistry Section. He also holds adjunct professor status at UQAR (Université du Québec à Rimouski), where he contributes to teaching and (co)directs graduate students in the field of ecotoxicology.

Dr. Blaise obtained university diplomas from the University of Montréal (B.A., 1967: biology and chemistry), University of Ottawa (B.Sc., 1970: cell biology; M.Sc., 1973: environmental microbiology) and the University of Metz (D.Sc., 1984: ecotoxicology). He is a member of the editorial board of the journal *Environmental Toxicology and Water Quality* and holds membership in both the biologists' (Association des Biologistes du Québec) and microbiologists' (Association des Microbiologistes du Québec) associations of the province of Québec. He regularly attends and makes presentations during major venues held in the field of ecotoxicology (SETAC: Society of Environmental Toxicology and Chemistry; SECOTOX: Society of Ecotoxicology and Environmental Safety; ATW — Canada: Aquatic Toxicity Workshop — Canada; ISTA: International Symposium on Toxicity Assessment).

Under Dr. Blaise's leadership, BRU strives to develop, validate, standardize, modernize (and promote the commercialization of) bioanalytical techniques, making use of new instrumental technologies whenever possible, in order to determine the potential (geno)toxicity of chemicals and various types of environmental matrices (e.g., effluents, sediments, pore/surface waters). BRU research output provides practical tools and approaches which facilitate decision-making for aquatic environmental management. BRU also provides (inter)national technology transfer to interested professionals and agencies and promotes graduate student training by codirecting applied research projects with university collaborators. Dr. Blaise has (co)authored over 60 scientific articles in internationally refereed journals, as well as having written several book chapters, and various government technical reports.

Johanne Gauthier, M. Sc., is a biologist in the Habitat Management and Environmental Science Division at the Maurice Lamontagne Institute, Department of Fisheries and Oceans Canada.

Ms. Gauthier received her Bachelor (1983) and her Master (1987) of Science degrees in biology from Laval University. She performed her experimentation for her M.Sc. degree at the Cancer Research Center of L'Hôtel-Dieu de Québec hospital, where she studied the antigenic expression of bladder cancer cells using flow cytometry.

Ms. Gauthier's work activities during the past 10 years have included research on bladder cancer with the objective of using specific monoclonal antibodies as prognostic and diagnostic tools in the treatment of the disease. Ms. Gauthier is now involved in the assessment of toxicity in oil spill bioremediation studies. This recent work aims at identifying the best strategies to accelerate biodegradation and reduce the toxicity of stranded oil.

Contributors

S. Alexandre
Centre des Sciences de l'Environnement
Université de Metz
Metz, France

Brian S. Anderson, M.A.
Institute of Marine Sciences
University of California–Santa Cruz
Marine Pollution Studies Laboratory
Monterey, CA, USA

Shimshon Belkin, Ph.D.
Environmental Sciences
The Fredy and Nadine Herrmann Graduate
  School of Applied Science
The Hebrew University
Jerusalem, Israel

Sharon G. Berk, Ph.D.
Center for the Management,
  Utilization, and Protection of Water Resources
Tennessee Technological University
Cookeville, TN, USA

H. Bessi, Ph.D.
Faculté des Sciences et Techniques
Mohammedia-Maroc

Gabriel Bitton, Ph.D.
Department of Environmental Engineering
  Sciences
University of Florida
Gainesville, FL, USA

Christian Blaise, D.Sc.
Centre Saint-Laurent
Environnement Canada
Montréal, QUE, Canada

Nancy J. Bowers, Ph.D.
Environmental Science and Resources (ESR)
Portland State University
Portland, OR, USA

Margaret Branton, M.E.S.
School for Resource and Environmental Studies
Dalhousie University
Halifax, NS, Canada
and
JSA Environmental
Long Beach, CA

Scott F. Briscoe
McGill University
Department of Microbiology and Immunology
Montréal, QUE, Canada

Anthony A. Bulich, Ph.D.
AZUR Environmental
Carlsbad, CA, USA

Jie Cai, Ph.D.
McGill University
Department of Microbiology and Immunology
Montréal, QUE, Canada

John Cairns, Jr., Ph.D.
Department of Biology
Virginia Polytechnic Institute
  and State University
Blacksburg, VA, USA

R. Scott Carr, Ph.D.
U.S. Geological Survey
Marine Ecotoxicology Research Station
TAMU-CC, Center for Coastal Studies
Natural Resources Center
Corpus Christi, TX, USA

Peter M. Chapman, Ph.D.
EVS Environment Consultants
North Vancouver, BC, Canada

Larry D. Claxton, Ph.D.
Environmental Carcinogenesis Division
National Health and Environmental Effects
  Laboratory
U.S. Environmental Protection Agency
Research Triangle Park, NC, USA

Chantale Côté
Group Leader, Québec Region
Beak International Inc.
Dorval, QUE, Canada

V. Cruciani, Ph.D.
Centre des Sciences de l'Environnement
Université de Metz
Metz, France

Andrea Dankwardt, Ph.D.
Department of Botany
Technical University of Muenchen
  at Weihenstephan
Freising, Germany

Francine Denizeau, Ph.D.
TOXEN
Université du Québec à Montréal
Montréal, QUE, Canada

Kenneth G. Doe, B.Sc.
Environment Canada
Environmental Conservation Branch
Moncton, NB, Canada

Michael S. DuBow, Ph.D.
McGill University
Department of Microbiology and
  Immunology
Montréal, QUE, Canada

Marie-Josée Durand, Ph.D.
Centre des Sciences de l'Environnement
Université de Metz
Metz, France

Jean-François Férard
Centre des Sciences de l'Environnement
Université de Metz
Metz, France

Vincent Ferrier
Centre de Biologie du Développement
Université Paul Sabatier
Toulouse, France

L. Gauthier
Centre de Biologie du Développement
Université Paul Sabatier
Toulouse, France

Guy L. Gilron, M.Sc., R.P.Bio.
BEAK International Incorporated
Brampton, ON, Canada

Karl E. Gustavson
Water Resources Center and Environmental
  Toxicology Center
University of Wisconsin–Madison
Madison, WI, USA

Peter-Diedrich Hansen, Prof. Dr.
Berlin University of Technology
Institute for Ecological Research and Biology
Department of Ecotoxicology
Berlin, Germany

John M. Harkin
Water Resources Center and Environmental
  Toxicology Center
University of Wisconsin–Madison
Madison, WI, USA

Armin Herbert, Dr.
Berlin University of Technology
Institute for Ecological Research and Biology
Department of Ecotoxicology
Berlin, Germany

Bertold Hock, Prof., Ph.D.
Department of Botany
Technical University of Muenchen
  at Weihenstephan
Freising, Germany

Peter V. Hodson, Ph.D.
Director, School of Environmental Studies
Queen's University,
Kingston, ON, Canada

J.C. Hoflack, Ph.D.
Centre des Sciences de l'Environnement
Université de Metz
Metz, France

John W. Hunt, M.S.
Institute of Marine Sciences
University of California–Santa Cruz
Marine Pollution Studies Laboratory
Monterey, CA, USA

Don L. Isenberg, Ph.D.
AZUR Environmental
Carlsbad, CA, USA

Colin R. Janssen
Laboratory for Biological Research
in Aquatic Pollution
University of Ghent
Ghent, Belgium

B. Thomas Johnson, Ph.D.
Environmental Microbiology and Applied
Toxicology Department
Biological Resources Division
U.S. Geological Survey
Environmental and Contaminants Research
Center
Columbia, MO, USA

E. Marshall Johnson, Ph.D.
Anatomy and Developmental Biology
Jefferson Medical College
Philadelphia, PA, USA

Naomasa Kobayashi
Laboratory of Environmental Biology
Hiroshima Jogakuin University
Hiroshima, Japan

Kenneth Lee, Ph.D.
Habitat Management and Environmental
Science Division
Maurice Lamontagne Institute
Fisheries and Oceans Canada
Mont-Joli, QUE, Canada

J. L'Haridon
Centre de Biologie du Développement,
Université Paul Sabatier,
Toulouse, France

Denis H. Lynn, Ph.D.
Department of Zoology
University of Guelph
Guelph, ON, Canada

Adrian J. MacDonald, B.Sc.
Environment Canada, Atlantic Region
Dartmouth, NS, Canada

Charles L. McKenney, Jr., Ph.D.
U.S. Environmental Protection Agency
National Health and Environmental Effects
Research Laboratory
Gulf Ecology Division
Gulf Breeze, FL, USA

Alan J. Mearns, Ph.D.
Hazardous Materials Response and
Assessments Division
National Oceanic and Atmospheric
Administration
Seattle, WA, USA

Daria G. Mochan, M.S.
Environmental Science and Resources (ESR)
Portland State University
Portland, OR, USA

Jean Louis Morel
Laboratoire Sols et Environnement
E.N.S.A.I.A.
Vandoeuvre, France

Isabel M. Müller
Institut für Physiologische Chemie, Universität
Abteilung Angewandte Molekularbiologie
Mainz, Germany

Werner E. G. Müller, Prof. Dr.
Institut für Physiologische Chemie, Universität
Abteilung Angewandte Molekularbiologie
Mainz, Germany

Czesia Nalewajko, Ph.D.
Life Sciences Division
Scarborough College
University of Toronto
Scarborough, ON, Canada

B. R. Niederlehner, M.S.
Department of Biology
Virginia Polytechnic Institute and State
University
Blacksburg, VA, USA

Marion G. Nipper, Ph.D.
NIWA–National Institute for Water and
Atmospheric Research Ltd.
Hamilton, New Zealand
and
Texas A&M University — Corpus Christi
Center for Coastal Studies
Natural Resources Center
Corpus Christi, TX, USA

Ursula Obst, Dr.
WFM Wasserforschung
Mainz GmbH,
Mainz, Germany

Mary Olaveson
Life Sciences Division,
Scarborough College
University of Toronto
Scarborough, ON, Canada

Joanne Parrott, Ph.D.
National Water Research Institute
Canada Centre for Inland Waters
Burlington, ON, Canada

Guido Persoone, Ph.D.
Laboratory for Biological Research
  in Aquatic Pollution
University of Ghent
Ghent, Belgium

Witold Piekarski
Institute of Marine Sciences
University of California–Santa Cruz
Marine Pollution Studies Laboratory
Monterey, CA, USA

Bryn M. Phillips
Institute of Marine Sciences
University of California–Santa Cruz
Marine Pollution Studies Laboratory
Monterey, CA, USA

Patricia Pocklington, B.Sc.
Bermuda Biological Station for Research, Inc.
St. George, Bermuda

James R. Pratt, Ph.D.
Environmental Science and Resources (ESR)
Portland State University
Portland, OR, USA

Sabine Pullen, Ph.D.
Department of Botany
Technical University of Muenchen
  at Weihenstephan
Freising, Germany

Ansar A. Qureshi, Ph.D.
AZUR Environmental
Carlsbad, CA, USA

Harry W. Read, Ph.D.
Water Resources Center
University of Wisconsin–Madison
Madison, WI, USA

Donald J. Reish, Ph.D.
Department of Biology
California State University–Long Beach
Long Beach, CA, USA

Robert O. Roberts, M.S.
Knoxville, TN, USA

Philippe Ross, Ph.D.
Department of Biology
The Citadel
Charleston, SC, USA

Jim Sherry, Ph.D.
National Water Research Institute
Canada Centre for Inland Waters
Burlington, ON, Canada

Eric P. Smith, Ph.D.
Department of Statistics
Virginia Polytechnic Institute
  and State University
Blacksburg, VA, USA

Terry W. Snell, Ph.D.
School of Biology
Georgia Institute of Technology
Atlanta, GA, USA

Kok-Leng Tay, Ph.D.
Environment Canada
Atlantic Region
Dartmouth, NS, Canada

Paule Vasseur, Ph.D.
Centre des Sciences de l'Environnement
Université de Metz
Metz, France

M. Wiegand-Rosinus, Dr.
Stadtwerke Mainz AG
Mainz, Germany

Judith S. Weis, Ph.D.
Department of Biological Sciences
Rutgers University
Newark, NJ, USA

Peddrick Weis, D.D.S.
Department of Anatomy, Cell Biology and
 Injury Sciences
University of Medicine and Dentistry
 of New Jersey
NJ Medical School
Newark, NJ, USA

Peter G. Wells, Ph.D.
Environment Canada
Environmental Conservation Branch
Dartmouth, NS, Canada

A. Wessler, Dr.
WFM Wasserforschung
Mainz GmbH
Mainz, Germany

Paul A. White, Ph.D.
Visiting Research Scientist
Atlantic Ecology Division
United States Environmental Protection
 Agency
Narragansett, RI, USA

Hao Howard Xu, Ph.D.
MicroGenomics, Inc.
La Jolla, CA, USA

C. Zoll-Moreux
Centre de Biologie du Développement
Université Paul Sabatier
Toulouse, France

Reviewers

Khrosrow Adeli
Department of Biochemistry
University of Windsor
Ontario

Jorma T. Ahokas
Key Center for Applied and Nutritional
  Toxicology
Melbourne, Australia

Brian S. Anderson
Institute of Marine Sciences
University of California, Santa Cruz
Marine Pollution Studies Laboratory
Monterey, CA

Jack W. Anderson
Columbia Analytical Sciences
Carlsbad, CA

Anders W. Andren
Water Chemistry Program
University of Wisconsin
Madison

Céline Audet
INRS-Océanologie
Université du Québec à Rimouski
QUE

Marc Babut
Agence de L'Eau Rhin-Meuse
Département Études et Conseil
Metz, France

Eros Bacci
Department of Environmental Biology
University of Siena
Italy

Susan T. Bagley
Michigan Technological University
Department of Biological Sciences
Environmental Microbiology Laboratory
Houghton

Arthur P. Barton
School of Biomedical Sciences
Curtin Institute of Technology
Perth, Western Australia

Steven M. Bay
Southern California Coastal Water
  Research Project
Westminster, CA

Christian Bélanger
Maurice Lamontagne Institute
Fisheries and Oceans Canada
Mont-Joli, QUE

Sharon G. Berk
Center for Water Resources
Tennessee Technological University
Cookeville

Alan Bettermann
BioRenewal Technologies, Inc.
Madison, WI

Gabriel Bitton
University of Florida
Dept. of Environmental Engineering Sciences
Gainesville

Sandra Blenkinsopp
Environment Canada
Western Regional Office
Edmonton, Alberta

Radovan Borojevic
Department of Biochemistry
Institute of Chemistry
Federal University of Rio de Janeiro
Brazil

Alain Boudou
Laboratoire d'Ecologie Fondamentale
  et Ecotoxicologie
Université de Bordeaux
Talence, France

Barbara Brown
Department of Marine Sciences
 and Coastal Management
University of Newcastle upon Tyne
UK

Allen G. Burton
Wright State University
Department of Biological Sciences
Dayton, OH

R. Scott Carr
NBS, Gulf Coast Research Station
TAMU-CC
Corpus Christi, TX

Carl E. Cerniglia
Department of Health & Human Services
Public Health Service
Food and Drug Administration
National Center for Toxicological Research
Jefferson, AR

Peter M. Chapman
E.V.S Environmental Consultants, Ltd.
North Vancouver, BC

Mildred S. Christian
Argus Research Laboratories
Horsham, PA

Nick Christofi
Pollution Research Unit
Department of Biological Sciences
Napier University
Edinburgh, Scotland

Nancy Cook
Department of Chemistry
Mount Saint Vincent University
Halifax, NS

Kristin E. Day
Environment Canada
Burlington, ON

Tom A. Dean
Coastal Resources and Associates
Vista, CA

Frans J. deBruijn
Center for Microbial Ecology
Michigan State University
East Lansing

Daniel Dive
INSERM
Villeneuve d'Ascq, France

Kenneth G. Doe
Environment Canada
Dartmouth, NS

Barney J. Dutka
AEPB, NWRI
Burlington, ON

François Gagné
St-Lawrence Center
Montreal, QUE

John P. Giesy
Michigan State University
Department of Fisheries and Wildlife
East Lansing

Mel Greene
AZUR Environmental
Carlsbad, CA

André Guillouzo
Hépathologiques
INSERM
Hôpital de Pontchaillou
Rennes, France

Peter-Diedrich Hansen
Institute for Ecological Research and
 Technology
Department of Ecotoxicology
Berlin, Germany

R. Hayes
Howard University
Washington, DC

Michael Henebry
Illinois EPA
Springfield, IL

David E. Hinton
Department of Anatomy
School of Veterinary Medicine
University of California at Davis

Kay T. Ho
U.S. Environmental Protection Agency
Narragansett, RI

Bertold Hock
Technische Universität Müenchen
Freising, Germany

Virginia Houk
Genetic Toxicology Division
US Environmental Protection Agency
Research Triangle Park, NC

B. Thomas Johnson
U.S. Geological Survey
Columbia, MO

Guy R. Lanza
Environmental Sciences Program
University of Massachusetts
Amherst

R.A. LaRossa
DuPont CR&D, Experimental Station
Wilmington, DE

Dickson Liu
Rivers Research Branch
National Waters Research Institute
Canada Center for Inland Waters
Burlington, ON

Denis H. Lynn
Department of Zoology
University of Guelph
ON

Paul V. McCormick
Everglades Systems Research Division
South Florida Water Management
  District
West Palm Beach

Judith McDowell
Woods Hole Oceanographic Institution
MA

Blahoslav Marsalek
Academy of Sciences of Czech Republic
Institute of Botany
Department of Experimental Phycology
  and Ecotoxicology
Brno, Czech Republic

Linda Medlin
Alfred-Wegener-Institut
Bremerhaven, Germany

George Morrison
US Environmental Protection Agency
Environmental Research Laboratory
Narragansett, RI

Royal Nadeau
U.S. Environmental Protection Agency
Edison, NJ

Jean-Francois Narbonne
Université de Bordeaux I
Laboratoire Toxicologie Alimentaire
Talence, France

Michael Newman
University of Georgia
Savannah River Ecology Laboratory
Aiken, SC

Giao Nguyen-Ba
Laboratoire Pharmacologie Cellulaire
Faculté de Pharmacie
Chatenay-Malabry, France

Marion G. Nipper
NIWA–National Institute for Water and
  Atmospheric Research Ltd.
Hamilton, New Zealand
and
Texas A&M University — Corpus Christi
Center for Coastal Studies
Natural Resources Center
Corpus Christi, TX, USA

Niels Nyholm
Laboratory of Environmental Sciences and
  Ecology
Technical University of Denmark
Lyngby

Mary Olaveson
University of Toronto
Scarborough, ON

Michel Pardos
Bioanalytical Research Unit
Ecotoxicology and Environmental Chemistry
Centre Saint-Laurent, Environment Canada
Montreal, QUE

Jocelyne Pellerin
Université du Québec à Rimouski
Groupe de Recherche en Environnement
Rimouski, QUE

James R. Pratt
Environmental Science and Resources
Portland State University
OR

Peter D. Premdas
Queen's University
School of Environmental Studies
Kingston, ON

Michael J. Prival
Food and Drug Administration
Washington, DC

Philippe Quillardet
Unité de Programmation Moléculaire et
 Toxicologie Génétique
Institut Pasteur
Paris

L.C. Rai
Laboratory of Algal Biology
Department of Botany
Banaras Hindu University
Varanasi, India

K. Ranga Rao
Biology Department
University of West Florida,
Pensacola

Salem Rao
Aquatic Ecosystem Protection Branch
National Water Research Institute
Environment Canada
Burlington, ON

Donald J. Reish
Department of Biology
California State University
Long Beach

Stephen H. Safe
Department of Veterinary Physiology and
 Pharmacology
College of Veterinary Medicine
Texas A&M University
College Station

Patricia L. Seyfried
Department of Microbiology
University of Toronto
Toronto, ON

Terry Snell
School of Biology
Georgia Institute of Technology
Atlanta

Richard E. Sparks
Illinois Natural History Survey
River Research Laboratory of the Forbes
 Biological Station
Havana, IL

Richard Swartz
U.S. EPA
Newport, OR

K. L. Tay
Environment Canada
Dartmouth, NS

Carmella I. Tellone
Valent USA Corporation
Walnut Creek, CA

E. Michael Thurman
US Geological Survey
Water Research Division
Lawrence, KS

Glen Thursby
U.S. Environmental Protection Agency
National Health & Environmental Effects
 Research Laboratory
Narragansett, RI

Paule Vasseur
Centre des Sciences de l'Environnement
Université de Metz
Metz, France

Donald J. Versteeg
Environmental Safety Department
The Proctor and Gamble Company
Cincinnati, OH

Torgny J. Vigerstad
Bio-Response Systems Limited
Halifax, NS

Jack Word
Battelle Marine Sciences Laboratory
Sequim, WA

Preface

Special sessions on microscale aquatic toxicity testing were held during Canada's 17th Annual Aquatic Toxicity Workshop (ATW) in Vancouver in 1990, and at the 20th Annual ATW in Québec City in 1993. They collectively identified the field's recent developments and advances and relevant applications for aquatic toxicology. The session chairs in 1993 decided that it was time to produce a comprehensive book on the topic.

As a result, a collaboration began between Environment Canada, Fisheries and Oceans Canada, and the publishers, CRC Press. Support for the production of this book was provided by Environment Canada (Centre St. Laurent, Québec Region; Environmental Conservation Branch, Atlantic Region) and Fisheries and Oceans Canada (Green Plan Toxic Chemicals Program). Ms. Johanne Gauthier was employed by Fisheries and Oceans as book coordinator, a position she filled very capably and with patience and humor. Authors were chosen and work commenced, culminating in another special session of the ATW in St. Andrews, N.B., in October 1995. Many of the talks given at the ATW were based on draft chapters written for this book, enabling authors to meet one another and the editors, discuss the subject thoroughly, and critique the chapters themselves. The submitted chapters were each reviewed by two or more anonymous reviewers, and reviewed and edited by one of the editors. The book comprises those chapters that survived this rigorous process and timely revisions.

Hence, this volume is the outcome of several years of collaboration, and with the contributors and reviewers, many careers of experience. It covers state-of-the-art knowledge generated in the area of microscale testing in aquatic toxicology by the mid- to late 1990s. This unique set of contributions contains over 40 chapters provided by internationally recognized scientists. The book is organized to provide information on relevant principles, novel techniques, and their applications to research scientists, environmental managers, academics, and the private sector. It is hoped that it will stimulate much discussion and many new advances in what is proving to be one of the most exciting and environmentally beneficial areas of aquatic toxicology. We also hope that the book and its methods contribute to the broader goal of conserving and protecting our aquatic ecosystems and their biota for future generations.

Peter G. Wells
Kenneth Lee
Christian Blaise

March 1997

Foreword

Over the past few years, monitoring agencies have become increasingly aware that to analyze environmental samples such as water, effluent, sediment, and other solid-phase samples for every suspected chemical could be exorbitantly expensive with no assurance that the detection procedures would be sensitive enough to reveal all of the potential "toxic chemicals." Moreover, chemical analyses alone cannot be used to predict joint toxicity (e.g., synergistic) responses and bioavailability. Although sediments and water bodies may contain high concentrations of toxicants, the phenomenon of toxicity or increased toxicity to organisms living in these environments may not be observed. This is due in part to the dependence of bioavailability on the trophic position of an organism, and of toxicity to benthic organisms on their relative sensitivity to interstitial and particle-bound chemicals.

Our research, and that of many others, has clearly demonstrated that one of the most cost-effective ways to address these problems is first to use various short-term, inexpensive bioassays to screen the samples for indications of toxic, genotoxic, or chronic effects, then to prioritize the samples or sampled areas for chemical analyses and/or more intensive studies. Application of such bioassays to environmental samples soon reveals, not surprisingly, that there is no single test (bioassay) which responds equally well to all toxicants or mixtures of toxicants, e.g., similar to the analogy that there is no single antibiotic for all bacterial infections. Organisms and people are similar in that they all have different resistances to toxicants, infections, and other form of stress. This realization led to the concept of using several bioassays to assess environmental water quality and the ecological impacts of effluents, discharges, and emissions on the total environment. This soon became known as the "battery-of-tests" approach. Short-term bioassays, making up various batteries, were often based on different trophic levels and included different biological endpoints, such as acute toxicity, chronic toxicity, and genotoxicity/mutagenicity.

As these short-term screening bioassays were developed and perfected, it was found that they could be used to assist chemists, engineers, and environmental protection specialists to target specific point sources, in-plant toxic stream flows, and extraction fractions of various industrial processes as sources of toxicants/genotoxicants and estrogenic chemicals. This broad application could result in significant savings in analytical costs.

Short-term bioassays, in partnership with chemical, biogeochemical, and microbiological analyses, are believed by many environmental scientists to be essential for providing appropriate information to help pinpoint the exact causes of biological responses. We believe that the use of a mixture of chemical/physical and biological measurements is essential for hazard assessment and environmental quality control and monitoring.

Environmental chemical data and procedures have become well established over the years, while bioassay procedures have only been used widely for the past two decades. As a result, bioassay methods are still being developed globally, but often used only locally. Others are used almost exclusively within a specific country. Because of this fragmentation of usage, it becomes essential that workers in the field of environmental toxicological assessment have access to a comprehensive reference to the bioassays presently being used internationally. This book contributes to filling this need by providing an excellent overview of the bioassays which use smaller life forms, enzyme systems, tissue cultures, and young life stages to estimate potential environmental hazards.

The editors of *Microscale Testing in Aquatic Toxicology: Advances, Techniques, and Practice* have literally searched the world to produce this comprehensive resource book for microbioassay techniques and applications. The microbioassay techniques are presented in several multi-sectioned parts: Biochemical assays; Tissue culture assays; Bacteria; Algae and Microalgae; Protozoa; Invertebrates; and Teleosts and Amphibians. The other chapters provide interesting and useful information on historical perspectives of microscale toxicity testing, multitrophic level testing, validation of results, bioassay selection for field studies, and "crystal ball" gazing into the future.

Within these chapters, workers in the field of ecotoxicology will find their own favorite techniques as well as others which are new and may stimulate the reader to investigate their potential. Many of the bioassays described are available commercially and are well standardized, e.g., Microtox®, Metpad®, Toxkits, and immunoassays, while the rest are laboratory based and well-utilized. Thus, developing laboratories will find an excellent shopping basket of techniques which can be applied to their specific budgets and problems.

This book is unequivocally recommended without bias for all laboratories, scientists, and managers who are involved with environmental pollution, hazard assessment of toxic chemicals, and the principles and practice of aquatic toxicology.

B.J. Dutka
Research Scientist
National Water Research Institute
Environment Canada
Burlington, Ontario, Canada

Contents

PART ONE
Introduction

Microscale Testing in Aquatic Toxicology: Introduction, Historical Perspective, and Context

Christian Blaise, Peter G. Wells, and Kenneth Lee

CONTENTS

I. INTRODUCTION

While this century's industrial boom has unquestionably contributed countless chemical-derived benefits for mankind, it has also created unwanted and, in some cases, extremely damaging pollution problems at local, regional, and global levels. In response, the field of ecotoxicology has developed enormously since the 1950s (Table 1.1).

Microscale testing in aquatic toxicology, the *raison d' être* of this book, is a rapidly-expanding field involving numerous bioanalytical techniques developed and applied at subcellular to multicellular levels of biological organization, in the laboratory or *in situ*. A small-scale test or any of its synonyms (microbiotest, microbioassay, microscale test, microtest, second-generation biotest) was recently defined as involving the exposure of a unicellular or small multicellular organism to a liquid sample in order to measure a specific effect.[1] With advances in analytical techniques, our view has expanded in that the field also includes subcellular tests involving enzymatic measurements, as well as immunoassay and biosensor techniques. Furthermore, since recent microbiotests (so-called sediment direct-contact tests) have been developed to assess the toxic potential of solid matrices,[2-7] such tests now apply to both liquid and solid matrices.

The primary purpose of this book is to discuss the nature and role of microscale tests in the broader field of aquatic toxicology,[8,9] not to unnecessarily "invent" a new branch of aquatic toxicology. However, the subject deserves separate attention. Microscale testing and microscale

Table 1.1   Some Highlights in the Evolution of Aquatic Toxicology in the 20th Century

Period	Highlights
1950s and before	Ecotoxicity problems stemming from the industrial revolution have appeared. However, environmental response is limited due to the lack of adequate assessment techniques and strategies.
1960s	Fish bioassays, in particular, confirm damage by industrial effluents and specific xenobiotics to biota of receiving waters.
1970s	Thanks in part to bioanalytical data and knowledge generated in the '60s, many industrialized countries create environmental agencies. The emphasis is placed on "banning flagrant insults to the environment."
1980s	Decade of "holistic thought" in which integrated biological/chemical ecotoxicological strategies and hazard assessment schemes are employed. Tests at various levels of biological organization are developed to address insidious (chronic) insults to the environment.
1990s	Marked upsurge in the development and application of cost-effective microscale aquatic toxicity tests. Because of their attractive characteristics (e.g., miniaturization, rapidity, automation potential), they constitute useful tools for monitoring and improving environmental quality.

approaches at different levels of biological organization and with different phyla are an important area of aquatic toxicology and hazard assessment. Recently, intense interest in the field has been demonstrated by the increasing numbers of symposia and publications on the topic, the increased acceptance of microscale tests by regulators, and the marked ongoing effervescence in experimental and field-based microbiotest studies.

Dedicating a book to the subject of aquatic microscale tests is long overdue. We hope that readers will sense the diversity of techniques, knowledge, and potential applications of this field to the management of toxic substances in aquatic environments. Our goals in preparing this book are therefore to: 1) advance knowledge in a bioanalytical sphere, and 2) promote microbiotests as useful and practical diagnostic tools in aquatic toxicology and applied ecology.

II. A BRIEF HISTORY OF MICROSCALE AQUATIC TOXICOLOGY

New areas of research generally have slow starts (a lag phase) followed by a period of rapid growth (a logarithmic phase) when upsurging interest and needs draw in more and more research groups. Microbiotesting is one such area in which modest beginnings have quickly expanded into bustling and varied activities. The number and diversity of contributions to the present book certainly offer undeniable evidence in this respect. The very first example of small-scale toxicity testing may well go back as far as 1672 when "shiny meat" sections, colonized by luminescent bacteria, were observed to rapidly lose light when rinsed in water containing various chemicals.[10] Much later, but prior to the concerted work which has ensued since the late 1970s, pioneering work provided some procedures on small-scale approaches. Examples include work performed with bacteria (the development of the Microtox® test system), daphnids[11] (classic toxicity studies), culture of fish embryos (see Chapter 33), and marine embryo and larval testing.

Fish and macroinvertebrate bioassays of the 1950s and 1960s were the forerunners of microbiotests.[12,13] They demonstrated the acute toxicity of chemicals and effluents, and often predicted effects on receiving water biota and habitats. These "macrobiotests" have been used extensively since that time, especially for screening chemicals and regulatory compliance monitoring, and have proved crucial in heightening environmental awareness on the potential and real hazards of xenobiotics and point source discharges. Environmental agencies created in several industrialized countries in the 1970s (e.g., U.S. Environmental Protection Agency, Environment Canada, National Swedish Environmental Protection Board) spent 10 to 15 years establishing basic guidelines and regulations to control specific effluent/chemical discharges. International groups such as the International Maritime Organization (London) endorsed the use of aquatic toxicity data for chemical evaluation control, under specific conditions. Using chemical characterizations and standard toxicity tests with fish or macroinvertebrates, the agencies soon began to approve the application of combined biological (detecting effects through bioassays) and chemical (cause identification of effects) strategies to assess the hazards of industrial wastes to aquatic systems.[13] A key OECD (Organization for Economic Cooperation and Development) meeting in Duluth, Minnesota, in September 1984 gave crucial policy support to this approach.[14] The general acceptance of toxicity test data, and of the combined bioassay–chemical approach, proved important to the advancement and development of new micro-bioanalytical tools in the years that followed.

Ensuring validation and standardization of bioassays can be a lengthy process demanding sustained efforts on the part of individuals and organizations. Test development, internal verification for reproducibility by the laboratory of origin, round robin exercises, building QA/QC into the experimental protocol, as well as the approval and publication of a standardized operating procedure, are necessary steps to final acceptance and recognition by the scientific community of a reliable toxicity test. In this respect, the major role that standards organizations have played at regional, national, and international levels is worthy of acknowledgment. *Standard Methods for the Examination of Water and Wastewater* (jointly published by the American Public Health Association, American Water Works Association, and the Water Pollution Control Federation), for instance, first incorporated a fish toxicity assay in its 11th (1960) edition and added additional toxicity tests in its 14th (1976) edition. These initiatives eventually led to the standardization of several microbiotests using (micro)organisms at various levels of biological organization. ISO (International Standards Organization), ASTM (American Society for Testing and Materials), OECD (Organization for Economic Cooperation and Development), DIN (Deutsches Institut Normen), and AFNOR (Association Française de Normalisation) are other examples of major (inter)national organizations which have markedly contributed to the promotion of useful microbiotests.

III. NEEDS FOR SMALL-SCALE APPROACHES IN AQUATIC TOXICOLOGY

The requirement for multitrophic and cost-effective toxicity tests has driven the evolution of biotesting from the earlier exclusive use of macrobiotests to increased use of microbiotests. Toxicity was soon found to be trophic level-specific and specifically affected by characteristics of chemicals, endpoints, test exposures, and test matrices.[8,9] Hence, it was realized that protection of aquatic resources could not be ensured by conducting bioassays solely at the macro-organism (e.g., fish) level of biological organization. Hence, microscale procedures involving bacterial, protozoan, microalgal, and micro-invertebrate indicators, for example, began to appear frequently during the 1970s.[15-19] Including an array of tests (often small-scale by design) in toxicity and impact assessment studies would contribute more effectively to the protection of all trophic levels in an affected aquatic system. Furthermore, since some microassays were carried out with early sensitive life stages,[8,19-21] the possibility of correctly evaluating chemical hazards at their source and in receiving waters was greatly enhanced.

The search for cost-effective and high performance tests was a second major factor that favored the development and application of small-scale assay procedures. In an article stressing the urgent need for cost-effective, high output environmental assessment, Cram[22] wrote: "Productivity of laboratories … is critical for [any environmental agency] to meet its legislative responsibilities. The backlog and cost of [environmental] programs alone in terms of analytical requirements is staggering. Therefore, laboratory productivity has to continue to increase, turnaround time must decrease, the cost per sample must decrease…." Because of the attractive features of microbiotests,[1] investigators and applied ecotoxicologists can achieve increased bioanalytical throughput in their laboratories. As a result, microbiotesting (ranging from bacterial assays to fish tissue tests) is now contributing to more cost-efficient delivery of environmental programs and better levels of environmental protection and conservation.

Toxicity (pre)screening for an ever-increasing number of existing and new man-made chemicals (e.g., over 2000 new chemicals submitted to U.S. EPA premanufacturing notification on an annual basis) is one example of a pressing need which can only be addressed by employing cost-effective multitrophic biotesting. To this day, the great majority of chemicals still lack toxicological information which can only be realistically provided, in part, by applying relevant arrays of microbiotests. This issue was clearly recognized in the 1980s by research groups who proposed the use of suites of cost-effective bioassays within tiered or simultaneous hazard assessment schemes.[23-27] Applying comprehensive microbiotesting should ensure that problematic substances will not filter through a "leaky sieve," in terms of toxic assessment, with potential consequences for environmental or human health. Use of small-scale tests for this purpose is expanding quickly.

IV. THE ERA OF SMALL-SCALE ECOTOXICOLOGY

The growth of ecotoxicology, a discipline of the aquatic sciences which essentially began in the 1970s,[28-30] was paralleled by equal growth of what can be termed "small-scale ecotoxicology." Indeed, the 1980s saw many advances in this field.[31,32] Strengthened by an ever-increasing array (in quantity and quality) of microtests, work during this decade demonstrated how conducting environmental studies with suites of such tests could augment their cost-effectiveness and diagnostic potential. Innovative applications using integrated ecotoxicological strategies offered fresh ways of undertaking comprehensive toxicity evaluations of xenobiotics and varied environmental matrices such as effluents, leachates, sediments, soils, and interstitial waters. Hazard assessment schemes,[33,34] test battery approaches,[35-38] multitrophic assessments,[39,40] and multispecies evaluations[41] included some of the new concepts that microbiotests helped to inspire or contribute to.

Comparative testing, coupling microassays with ambient and community responses, also confirmed their capacity to predict potential hazards toward receiving-water biota.[42-44] Again, the 1980s witnessed the development of sensitive chronic microtoxicity procedures (algae: Blaise et al., 1986[45]; freshwater micro-invertebrates: Mount and Norberg, 1984[46]; marine micro-invertebrates: Persoone et al. 1984[20]), as well as the introduction of commercialized microtoxicity test kits (Artoxkit®[47]; SOS Chromotest®[48]; Rotoxkit[49]). Since 1983, a new scientific forum, the International Symposium on Toxicity Assessment, has brought together scientists on a biennial basis (Burlington, Canada, 1983; Banff, Canada, 1985; Valencia, Spain, 1987; Las Vegas, Nevada, 1989; Kurashiki, Japan, 1991; Berlin, Germany, 1993; Santiago, Chile, 1995; Perth, Australia, 1997). This symposium has greatly contributed to promoting the use of microbiotests and to improving environmental protection by advancing toxicological understanding and knowledge. Likewise, the annual SETAC (Society of Environmental Toxicology and Chemistry) meetings are fertile grounds for the presentation and discussion of new toxicity concepts and data.

V. MICROBIOTESTING IN THE 1990s

Small-scale testing continues to expand on various fronts during this decade. The high demand for simple, rapid, and practical microtoxicological procedures has successfully created a "testing industry" built around an increasing number of commercially available bioanalytical products (e.g., cyst-based TOXKITs: Bio-International, The Netherlands; Mutatox® and Chronic Microtox® tests: AZUR Environmental Inc., USA; MetPlate®, a metal-detection test: Group 206 Technologies, Inc., USA; direct-contact Toxi-Chromotest® for solid matrix assessment: EBPI-Beak, Canada; bacterial stress gene assays for rapid cytotoxic and genotoxic screening: Xenometrix, USA). Development of new products is a continuous activity and many more commercial initiatives are expected before the end of the century.

Table 1.2 Rising Interest in Microscale Testing Applications Represented by Time-Related Publication Numbers for a) Microtox®/Mutatox® Technology and b) Toxkit Technology

Bioassay Technology	1979–83	1984–89	1990–95	Total Publications
a) Microtox/Mutatox	29	201	279	507

Bioassay Technology	1984–88	1989–92	1993–96	Total Publications
b) Toxkits	3	16	36	55

a) Time-related publications adapted from data (numerical bibliography index, January 14, 1997) provided by AZUR Environmental, CA, U.S.A.
b) Time-related publications adapted from data provided by the Laboratory for Biological Research in Environmental Pollution, State University of Ghent, Belgium.

Scientific progress in fields such as molecular biology, genetics, and biotechnology, coupled with new instrumental technologies, is quickly giving rise to a new generation of microbiotests. Luciferase gene fusions (see Chapter 11) or inclusion of promoter/reporter gene constructs (see Chapter 12) in *Escherichia coli*, allowing for rapid (<2 h) and sensitive detection of specific classes of bioavailable xenobiotics, are examples of recently developed microtests that have the potential to revolutionize the bioanalytical world.

New knowledge and technology also enable modernization and automation of bioanalytical procedures, thereby conferring better cost-efficiency and analytical throughput to previously developed microbiotests (e.g., microplate-based Ames test with luminescent *Salmonella typhimurium*[50]; post-exposure automation of microplate-based phytotoxicity tests[51,52]; robotic initiation of SOS Chromotest®[53]). An increasing frequency of publications is a useful index of the growing popularity of microbiotests, particularly during the 1990s. As an example, Table 1.2 confirms this wave of support for the Microtox/Mutatox tests and for cyst-based TOXKITs. Time-related microplate-based phytotoxicity publications also display similar trends (see Figure 18.1 in Chapter 18).

Proliferation of microbiotesting activities in the 1990s has contributed markedly to the development and application of a myriad of alternative techniques giving diagnostic information commensurate with that of more costly and older traditional tests. Immunoassays now provide cost-efficient quantitative detection of an ever-increasing number of toxic compounds (see Chapter 2). Again, rapid toxicity procedures conducted at the subcellular level with enzymes offer useful prescreening that may compete well with established microbiotests (see Chapter 5). Cell cultures, fish cell lines in particular, are becoming more frequently employed as surrogates to whole-animal testing to measure toxicity of aquatic pollutants with relevant (sub)lethal endpoints[54,55] (also Chapters 4 and 7). The recent creation of ECVAM (European Centre for the Validation of Alternative Methods) by the Commission of European Communities is a clear indication of the importance that international bodies are now attaching to the validation and use of alternative methods in toxicology and of the key role that microbiotests can play in this respect.[56]

Other contemporary activities include the tremendous proliferation of approaches with gametes, embryos, and larvae of marine invertebrates, due to their culturability, economic value, ecological value (marine biodiversity as shown by having 43 of 57 phyla in marine systems), and their generally greater sensitivity compared to adult organisms and teleosts. For example, model organisms such as sea urchin embryos and crustacean larvae have found new uses (e.g., hazard assessment of mixtures). Many laboratories and regulatory agencies, including the PARCOM (Paris) Commission, ICES (International Council for the Exploration of the Sea), IOC (Intergovernmental Oceanographic Commission), and IMO (International Maritime Organization), that have responsibilities for chemical safety, spill response, and ocean health indicators, have accepted their role. Standards organizations such as ASTM (American Society for Testing and Materials) and ISO (International Standards Organization) have protocols for many of these tests, and interest is growing in continuing this trend. Recently, early life stage assays with amphibian larvae have been developed as relevant test systems to assess the genotoxic potential of aqueous samples, enabled by post-exposure enumeration of erythrocytes displaying micronuclei (see Chapter 35). This and other cellular end-points measured in higher organisms are proving increasingly useful in investigating toxicity issues with both field and laboratory approaches.[57-60]

With the recognition of the ecological significance of microorganisms in the marine environment based on the development of new analytical procedures to measure their activity, there has been a corresponding expansion in the development of microbial biotests in the '90s (note Chapters 9 through 23 on bacteria, microalgae, and protozoa). Many of these tests can be developed into kits that will support automated procedure developments. To improve site-specific relevance, microbial biotests based on the activity of indigenous microbiota from the area of concern are now under development and evaluation (see Chapter 15).

Prospects for microbiotests appear very promising. There are clear indications that some micro-biotests will become desirable adjuncts or even alternatives to a chemical-specific approach for assessing the quality of various environmental matrices (e.g., effluents, sediments) and be incorporated into regulatory standards. Additionally, selected microscale tests are starting to replace more traditional higher organism bioassays and play a more important role, within hazard assessment schemes, for the overall protection of the aquatic environment.

VI. THE BOOK: CONTENTS, ORGANIZATION, AND CONTRIBUTORS

This volume is organized to reflect advances in microscale toxicity procedures across different levels of biological organization, different phyla, and activities of a directly practical bent (validation, applications, and training). Each chapter reviews the context, specific methodologies, and the development and use of specific types of small-scale assays. The contributors are from eight countries, largely North America and Europe, and are recognized experts in their particular disciplines.

The book attempts to be comprehensive and to acknowledge the large number of scientists making significant contributions in this area of aquatic toxicology. Indeed, some specialists who wished to contribute did not do so because of the pressure of other commitments. Hence, the reader will note gaps, and perhaps this will stimulate periodic treatises on microscale testing, which would be a signal that the book and the efforts of all of the contributors have been successful. If these initiatives together lead to cleaner environments for future generations and for our fellow creatures, we can judge our collective efforts worthwhile.

REFERENCES

1. Blaise, C., Microbiotests in aquatic ecotoxicology: characteristics, utility and prospects, *Toxicity Assessment*, 6, 145, 1991.
2. Brouwer, H., Murphy, T.P., and McArdle, L., A sediment-contact bioassay with *Photobacterium phosphoreum*, *Environmental Toxicology and Chemistry*, 9, 1353, 1990.
3. Kwan, K.K., Direct toxicity assessment of solid phase samples using the Toxi-Chromotest kit, *Environmental Toxicology and Water Quality*, 8, 223-230, 1993.
4. Dutka, B., Teichgräber, K., and Lifshitz, R., A modified SOS Chromotest procedure to test for genotoxicity and cytotoxicity in sediments directly without extraction, National Water Research Institute, Environment Canada, Burlington, Ontario, NWRI Contribution No. 95-53, 1995.
5. Bitton, G., Garland, E., Kong, I.C., Morel, J.L., and Koopman, B., A direct solid phase assay for heavy metal toxicity. I. Methodology, *Journal of Soil Contamination*, 5, 385, 1996.
6. Boularbah, A., Morel, J.L., Bitton, G., and Mench, M., A direct solid phase assay for heavy metal toxicity. II. Assessment of heavy metal immobilization in soils and bioavailability to plants, *Journal of Soil Contamination*, 5, 395, 1996.

7. Cook, N.H. and Wells, P.G., Toxicity of Halifax Harbour sediments: an evaluation of the Microtox® solid-phase test, *Water Quality Research Journal of Canada*, 31, 673, 1996.

8. Rand, G.M. and Petrocelli, S.R., Fundamentals of aquatic toxicology: methods and applications, Hemisphere Publishing Corporation, New York, 1985.

9. Rand, G. M., Fundamentals of aquatic toxicology: effects, environmental fate, and risk assessment, 2nd edition, Taylor & Francis, Washington, DC, 1995.

10. Beckman, Instruments Inc., Microtox system operating manual 015555879, Carlsbad, CA, USA, 1982.

11. Anderson, B.G., Aquatic invertebrates in tolerance investigations from Aristotle to Naumann, in *Aquatic Invertebrate Bioassays*, A.L. Buikema, Jr., and J. Cairns, Jr., Eds., American Society for Testing and Materials, 3, 1980.

12. Sprague, J.B., Review paper. Measurement of pollutant toxicity to fish. I. Bioassay methods for acute toxicity, *Water Research*, 3, 793, 1969.

13. Blaise, C., Sergy, G., Wells, P.G., Bermingham, N., and Van Coillie, R., Biological testing — development, application and trends in Canadian environmental protection laboratories, *Toxicity Assessment*, 3, 385, 1988.

14. Environment Canada, Proceedings of the OECD Workshop on the biological testing of effluents (and related receiving waters), Duluth, MN, September 1984, Environment Canada, Ottawa, October, 1984.

15. Cabridenc, R. and Lundhal, P., Intérêt et limites du test daphnie pour l'étude des nuissances des polluants, *Techniques et Sciences Municipales de l'Eau*, 6, 340, 1974.

16. Dive, D. and Leclerc, H., Standardized test method using protozoa for measuring water pollutant toxicity, *Progress in Water Technology*, 7, 67, 1975.

17. Chiaudani, G. and Vighi, M., The use of *Selenastrum capricornutum* batch cultures in toxicity studies, *Verhandl. Intern. Ver. Limnol.*, 21, 316, 1978.

18. Bulich, A. A., Use of luminescent bacteria for determining toxicity in aquatic environments, in *Aquatic Toxicology*, Marking, L.L. and Kimerle, R.A., Eds., American Society for Testing and Materials, ASTM 667, 1979, 98.

19. Buikema, A. L., Jr., Niederlehner, B.R., and Cairns, J., Jr., Toxicant effects on reproduction and disruption of the egg-length relationship in grass shrimp, *Bulletin of Environmental Contamination and Toxicology*, 24, 31, 1980.

20. Persoone, G., Jaspers, E., and Claus, C., Eds., Ecotoxicological testing for the marine environment, State University of Ghent and Institute for Marine Scientific Research, Belgium, 1984, Volumes I and II.

21. White, H.H., *Concepts in Marine Pollution Measurement*, Maryland Sea Grant Publication, University of Maryland, College Park, 1984.

22. Cram, S.P., Challenges and opportunities of environmental-analytical measurements, *American Environmental Laboratory*, September, 19, 1989.

23. Lloyd, R., Toxicity testing with aquatic organisms: a framework for hazard assessment and pollution control, *Rapport des réunions du conseil international d'exploration de la mer*, 179, 339, 1980.

24. Cairns, J. Jr., The case for simultaneous toxicity testing at different levels of biological organization, in *Aquatic Toxicology and Hazard Assessment*, Sixth Symposium of the American Society for Testing and Materials, STP 802, 1983, 111.

25. Chapman P. M. and Long, R., The use of bioassays as part of a comprehensive approach to marine pollution assessment, *Marine Pollution Bulletin*, 14, 81, 1983.

26. Slooff, W. and Canton, J.H., Comparison of the susceptibility of 11 freshwater species to 8 chemical compounds. Part II. (Semi)chronic toxicity tests, *Aquatic Toxicology*, 4, 271, 1983.

27. Johnson, E. M., A tier system for developmental toxicity evaluations based on considerations of exposure and effect relationships, *Teratology*, 35, 405, 1987.

28. Butler, G.C., Principles of ecotoxicology, SCOPE Volume 12, Wiley, Chichester, 1978.

29. Ramade, F., Ecotoxicologie, *Collection d'écologie*, No. 9, Masson éditeur, Paris, 1979.

30. Moriarty, F., Ecotoxicology: the study of pollutants in ecosystems, Academic Press, London, 1983.

31. Bitton, G. and Dutka, B., *Toxicity Testing Using Microorganisms*, Volume 1, CRC Press, Boca Raton, FL, USA, 1986.

32. Dutka, B. and Bitton, G., *Toxicity Testing Using Microorganisms*, Volume 2, CRC Press, Boca Raton, FL, USA, 1986.

33. Blaise, C., Bermingham, N., and van Coillie, R., The integrated ecotoxicological approach to assessment of ecotoxicity, *Water Quality Bulletin*, 10, 3, 1985.

34. van Coillie, R., Bermingham, N., Blaise, C., Vezeau, R., and Lakshminarayana, J.S.S., Integrated ecotoxicological evaluation of effluents from dumpsites, in *Advances in Environmental Science and Technology*, Volume 22, Lakshminarayana, J.S.S., Ed., John Wiley & Sons, New York, 1988, 161.

35. Thomas, J.M., Skalski, J.R., Cline, J.F., McShane, M.C., Simpson, J.C., Miller, W.E., Peterson, S.A., Callahan, C.A., and Greene, J.C., Chemical characterization of chemical wastesite contamination and determination of its extent using bioassays, *Environmental Toxicology and Chemistry*, 5, 487, 1986.

36. Dutka, B., Priority setting of hazards in waters and sediments by proposed ranking scheme and battery of tests approach, *German Journal of Applied Zoology*, 75, 303, 1988.

37. Couture, P., Blaise, C., Cluis, D., and Bastien, C., Zirconium toxicity assessment using bacteria, algae and fish assays, *Water, Air, and Soil Pollution*, 47, 87, 1989.

38. Chapman, P. M. Sediment quality criteria from the Sediment Quality Triad — an example. *Environ. Toxicol. Chem.*, 5, 957, 1986.

39. Taub, F.B., and White, H.H. Measurement of pollution in standardized aquatic microcosms, in *Concepts in Marine Pollution Measurements*, White, H.H., Ed., Tech. Rep. Md. Univ. Sea Grant Program, 1984, 159.

40. Taub, F.B., Standardized aquatic microcosms — development and testing, in *Aquatic Ecotoxicology: Fundamental Concepts and Methodologies*, Vol. II, Boudou, A. and Ribeyre, F., Eds., CRC Press, Boca Raton, FL, 1989, 47.

41. Blanck, H., Wallin, G., and Wängberg, S. A., Species dependent variation in algal sensitivity to chemical compounds, *Ecotoxicology and Environmental Safety*, 8, 339, 1984.

42. Mount, D.I., Thomas, N., Norberg, T., Barbour, M., Roush, T., and Brandes, W., Effluent and ambient toxicity testing and instream community responses on the Ottawa River, Lima, Ohio, US Environmental Protection Agency Report No. EPA-600/3-84-080, Duluth, MN, 1984.

43. Mount, D.I., Steen, A., and Norberg-King, T., Validity of effluent and ambient toxicity testing for predicting biological impact on Five Mile Creek, Birmingham, Alabama, US Environmental Protection Agency Report No. EPA/600/8-85/015, 1985.

44. Persoone, G., Calamari, D., and Wells, P.G., Short-term toxicity tests for non-genotoxic effects, in *SCOPE 41 IPCS Joint Symposia 8, SGOMSEC 4*, P. Bourdeau, E. Somers, G.M. Richardson, and J.R. Hickman, Eds., John Wiley & Sons, New York, 1990, 301.

45. Blaise, C., Legault, R., Bermingham, N., van Coillie, R., and Vasseur, P., A simple microplate algal assay technique for aquatic toxicity assessment, *Toxicity Assessment*, 1, 261, 1986.

46. Mount, D.I. and Norberg, T.J., A seven-day life-cycle cladoceran toxicity test. *Environmental Toxicology and Chemistry*, 3, 425, 1984.

47. Vanhaecke, P. and Persoone, G., The ARC-test: a standardized short-term routine toxicity test with *Artemia* nauplii, Methodology and evaluation, in *Ecotoxicological Testing for the Marine Environment*, Persoone, G., Jaspers, E., and Claus, C., Eds., State University of Ghent and Institute for Marine Scientific Research, Ghent, Belgium, Volume 2, 1984, 143.

48. Fish, F., Lampert, I., Halachmi, A., Riesenfeld, G., and Herzberg, M., The SOS Chromotest kit: a rapid method for the detection of genotoxicity, *Toxicity Assessment*, 2, 135, 1987.

49. Snell, T.W. and Persoone, G., Acute toxicity bioassays using rotifers. II. A freshwater test with *Brachionus rubens*, *Aquatic Toxicology*, 14, 81, 1989.

50. Côté, C., Blaise, C., Delisle, C., Meighen, E.A. and Hansen, D., A miniaturized Ames mutagenicity assay employing bioluminescent strains of *Salmonella typhimurium*, *Mutation Research*, 345, 137, 1996.

51. SRC (Saskatchewan Research Council), Development of aquatic plant bioassays for rapid screening and interpretive risk assessments of metal mining wastewaters, SRC Publication No. E-2100-2-C-95, Saskatchewan, Canada, 1995, 191.

52. Galgani, F., Cadiou, Y., and Gilbert, F., Simultaneous and iterative weighted regression analysis of toxicity tests using a microplate reader, *Ecotoxicology and Environmental Safety*, 23, 237, 1992.

53. White, P., Rasmussen, J., and Blaise, C., A semi-automated microplate version of the SOS Chromotest for the analysis of complex environmental extracts, *Mutation Research*, 360, 51, 1996.

54. Babich, H. and Borenfreund, E., Cytotoxicity and genotoxicity assays with cultured fish cells: a review, *Toxicology in vitro*, 5, 91, 1991.

55. Gagné, F., Intérêt de l'application de tests alternatifs en écotoxicologie: cas des hépatocytes primaires de truite arc-en-ciel (*Onchorhynchus mykiss*), Doctorate thesis, University of Metz, France, 1996.

56. Marafante, E. and Balls, M., The European Centre for the Validation of Alternative Methods (ECVAM), in *Alternatives to Animal Testing. New Ways in the Biomedical Sciences, Trends and Progress*, Reinhardt, C.A., Ed., VCH, Weinheim, 1994, Chapter 3.

57. Alvarez, M.R., Freidl, F.E., and Hudson, C.M., Effect of a commercial fungicide on the viability and phagocytosis of hemocytes of the American oyster, *Crassostrea virginica*, *Journal of Invertebrate Pathology*, 57, 395, 1991.

58. Lamb, T., Bickham, J.W., Whitfield-Gibbons, J., Smolen, M.J., and McDowell, S., Genetic damage in a population of slider turtles (*Trachemys scripta*) inhabiting a radioactive reservoir, *Archives of Environmental Contamination and Toxicology*, 20, 138, 1991.

59. Wan, C.P., Park, C.S., and Lau, B.H.S., A rapid and simple microfluorometric phagocytosis assay, *Journal of Immunological Methods*, 162, 1, 1993.

60. Dopp, E., Barker, C.M., Schiffmann, D., and Reinish, C.L., Detection of micronuclei in hemocytes of *Mya arenaria*: association with leukemia and induction with an alkylating agent, *Aquatic Toxicology*, 34, 31, 1996.

PART TWO

Microscale Testing at Various Biological and Phylogenetic Levels

Biochemical Assays

CHAPTER 2

Immunoassays: Applications for the Aquatic Environment

Andrea Dankwardt, Sabine Pullen, and Bertold Hock

CONTENTS

I. INTRODUCTION

Contaminants play an important role in the aquatic environment if they interfere with the ecosystem and endanger aquatic resources, such as drinking water. Several sources such as industry, urban wastewater plants, and agriculture contribute to the pollution of water bodies with a broad range of toxic substances. These consist of organic and inorganic contaminants, e.g., pesticides and heavy metals. The compounds can directly reach the surface water by drainage or erosion. Contaminants may also be transferred to the atmosphere, e.g., by evaporation during or after the application of pesticides.[1] They can be transported over long distances via the atmosphere, as in the case of the well-known pesticide DDT,[2] and may then be disposed by rainwater.[3] Water-soluble contaminants can leach from the soil to the ground water. Substances which are drained into water can be adsorbed to plants, animals, sediment, and organic matter,[4] or taken up by aquatic organisms.[5] In some cases, covalent bonds may also be formed with humic substances dissolved in the water.[6]

These residues are not available and effective for the ecosystem, but they can be released when organisms die or when the environmental conditions, e.g., pH or temperature, change. Stable compounds concentrate in the food chain, which leads to the bioaccumulation of certain contaminants.[7] The occurrence of several contaminants in the water may lead to increased stability of some compounds by the change of physicochemical parameters. This can lead to prolonged effects of a certain substance.[8]

Therefore it is necessary to monitor for the presence of pollutants in aquatic systems. This especially holds true for drinking water. The European Union drinking water regulation limits the residue concentration for pesticides and toxic metabolites in drinking water to 0.1 µg/L for a single substance and to 0.5 µg/L for the sum of all pesticides and their major metabolites. Presently, there are no governmental regulations or limitations for the concentration of contaminants in water bodies besides the drinking water regulations. But it is obvious that the maintenance of intact aquatic ecosystems and drinking water of high quality is not possible without monitoring the water bodies. Analytical techniques are required for single harmful substances and for groups of contaminants. This monitoring can be carried out by conventional analytical methods like GC, GC/MS, or HPLC. Electrophoretic methods such as isotachophoresis and capillary electrophoresis are also gaining importance. However, large-scale screening requires simpler and less expensive approaches. For this purpose, immunochemical methods are valuable supplements. They were originally developed in the medical field by Yalow and Berson for the quantification of insulin in serum.[9] Since then they have obtained broad application in medical diagnostics, especially in bacteriology, virology, or endocrinology. During the last 10 years immunochemical methods have become increasingly important for environmental analysis. They are used for the detection of pesticides and related contaminants in water, soil, food, and body fluids.[10-17]

Immunoassays (IA) offer distinct advantages for the detection of pollutants in the aquatic environment. Aquatic samples can usually be analyzed without clean-up procedures or solvent extraction, and the analytes can be detected at very low concentrations even in low sample volumes. Many samples can be analyzed in a short period of time. Conventional methods, such as GC or GC/MS, are expensive, time-consuming, and require a well-equipped laboratory. Compared to this, IA are cost-effective, easy to handle, and only need simple laboratory equipment such as balance, pipette, photometer, or reflectometer. Ready-to-use test kits offer the possibility to carry out on-site analyses.

II. IMMUNOCHEMICAL ANALYSIS

A. Antibody Production

Immunochemical analysis is based upon the specific reaction between an antibody and its corresponding antigen or hapten. Antibodies are part of the vertebrate defense system. They are serum glycoproteins of the immunoglobulin (Ig) class produced by the immune system against foreign material such as pathogens or xenobiotics and bind the target substance with high selectivity and affinity. Although there are five distinct classes of antibodies in most higher mammals, most immunoassays rely upon IgG as the major immunoglobulin. Antibodies may be raised against defined substances, even against low-molecular-mass pesticides if those are coupled to an immunogenic carrier, such as bovine serum albumin, before immunization.[11,18] Antibody production is conveniently carried out in warm-blooded animals, e.g., rabbits, sheep, or mice,[19] by immunization with suitable antigens. Polyclonal antibodies are obtained from the serum and comprise a mixture of different antibody populations. Monoclonal antibodies consist of a single monospecific antibody population. These antibodies are produced in cell culture by a single hybridoma cell line after the fusion of B-lymphocytes with myeloma cells. The hybridoma technology was introduced by Köhler

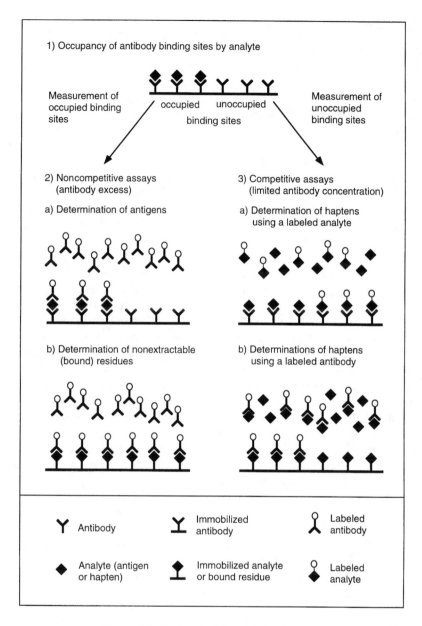

1) Occupancy of antibody binding sites by analyte

Measurement of occupied binding sites

occupied   unoccupied

binding sites

Measurement of unoccupied binding sites

2) Noncompetitive assays (antibody excess)

a) Determination of antigens

3) Competitive assays (limited antibody concentration)

a) Determination of haptens using a labeled analyte

b) Determination of nonextractable (bound) residues

b) Determinations of haptens using a labeled antibody

Y   Antibody

  Immobilized antibody

  Labeled antibody

◆   Analyte (antigen or hapten)

  Immobilized analyte or bound residue

  Labeled analyte

Figure 2.1 Basic principles of immunoassays

and Milstein.[20] Major progress is expected from recombinant techniques, i.e., the cloning of immunoglobulin genes into bacterial expression systems.[21]

B. Basic Principles of Immunoassays

IA presently belong to the most common methodology in the field of immunoanalysis. The principle is based upon the measurement of antibody binding site occupancy by the analyte (Figure 2.1). This reflects the analyte concentration in the sample. Since the binding reaction does not produce a signal that can be detected by simple means, various markers, e.g., radioactivity, enzymes, or fluorescence, are employed for the detection of the immunoreaction. Depending on the label, the assays are classified as radio immunoassay (RIA), enzyme-linked immunosorbent assay (ELISA) or enzyme immunoassay (EIA), fluorescence immunoassay (FIA), or polarization fluoroimmunoassay (PFIA). The most common methods in environmental immunoanalysis are ELISA and EIA. They are often carried out in a polystyrene 96-well microtiter plate, but other solid phases such as test tubes or membranes are available. The test can be performed as a homogeneous assay without separation of the reactants,[22] but more common are heterogeneous tests where unreacted reagents are removed before evaluation. The two basic approaches are competitive and noncompetitive assays. Noncompetitive assays can only be applied for high-molecular-mass analytes with more than one antigenic determinant (i.e., antigens) or low-molecular-mass analytes (haptens) bound to a solid phase, exposing the antigenic determinant (Figure 2.1(2)). They work using an antibody excess. For low-molecular-mass analytes (haptens) in solution, competitive tests have to be employed (Figure 2.1(3)), using limiting antibody concentrations. Two different formats are available, the direct and the indirect. In a direct test, the antibodies are bound to the solid phase. The analyte and a labeled analyte (tracer) compete for the free antibody binding sites. After removal of unbound reactants, the bound tracer yields a signal that is inversely proportional to the analyte concentration. The indirect test format employs an immobilized hapten-carrier-conjugate on the solid phase to which analyte and antibody are added. The antibody binds to the free analyte or to the immobilized hapten according to the concentration of the reactants. If a labeled antibody is used, the amount of antibody bound to the solid phase can be directly determined after a washing step. Alternatively, a secondary labeled antibody may be used to detect the antibodies which have bound to the solid phase. The signal is inversely proportional to the amount of free analyte in the solution. Very sensitive competitive immunoassays have been developed with detection limits between 1 and 50 ng/L, for example for the triazine and urea herbicides.[23-25]

Another solid-phase support for IA are membranes. They can be used for dipsticks, which are incubated for a short time in the solutions,[26] or for dot-blots and immunofiltration tests. Here the reactants are filtered through the membrane.[27,28] The test principle is the same as for the microtiter plate tests, but the reaction time is much shorter because of the high surface area of the membrane and the short distance between reaction partners. Application of remission measurements yields a proportional relationship between analyte and remitted light. By using a pocket reflectometer, this setup is ideally suited for field-monitoring purposes.[15]

Column chromatographic methods like Abicap (antibody immuno column for analytical processes) provide the opportunity to couple antibodies or haptens to a gel.[29] The immunoreaction takes place while the sample and other reagents pass the column in a very short time. Immunolocalization tests offer the possibility to detect contaminants *in situ,* such as soil or plant material.[30,31] Immunosensors combine the immunochemical detection with a transducer in order to produce a signal. The transducers can be electrochemical, optical, or gravimetric devices. The measurement of the antibody reaction is carried out by analytical instruments.[32,33]

C. Application of Immunoassays

So far, IA for approximately 100 pesticides and other compounds of environmental concern have been published. Most of them are suitable for detection of relevant concentrations of these substances in the environment. A list of the IA with information on the test format, antibody sources, test range or detection limit, and references is available from the authors. IA are commercially available for a broad range of compounds (Table 2.1). Usually a competitive EIA in the direct format is used, but they differ in their embodiments. The Rapid Assay kits (Ohmicron), for example, are based on antibody-coated magnetizable particles, and the separation step is carried out with a strong magnet. Most other commercial kits use antibody-coated microtiter plates or tubes. Emphasis has also been placed on providing field-portable immunoassay kits, packaging antibodies, reagents, standards, and substrates in field portable units that are ready to use.[34]

Table 2.1 Commercial Immunoassays for Pesticides and Industrial Contaminants

Compound	Detection Limit (D.L.)/ Measuring Range	Manufacturer
Herbicides		
Alachlor	0.1–4 ppb	Strategic Diagnostics (P)
	0.1–2.5 ppb	Millipore (P, T)
	0.05–5 ppb	Ohmicron (T)
Atrazine (Triazines)	0.02–2 ppb	Strategic Diagnostics (P)
	0.025–0.5 ppb	Millipore (P, T)
	0.015–1 ppb	Ohmicron (T)
	0.01–0.81 ppb	R-Biopharm (P)
	0.01–0.5 ppb	Riedel de Haen (P)
	0.05–2 ppb	Transia (T)
Chlorsulfuron	0.04–0.8 ppb	Millipore (P)
Cyanazine	0.04–3 ppb	Ohmicron (T)
2,4-D	0.5–100 ppb	Millipore (P, T)
	0.7–50 ppb	Ohmicron (T)
Imazapyr	0.3–30 ppb	Millipore (P)
Imazaquin	1–80 ppb	Strategic Diagnostics (P)
Isoproturon	0.05–0.5 ppb	Ohmicron (T)
	0.05–0.5 ppb	Millipore (P)
Metolachlor	0.1–4 ppb	Strategic Diagnostics (P)
	0.1–3 ppb	Millipore (P)
	0.05–5 ppb	Ohmicron (T)
Metribuzin	0.1–3 ppb	Ohmicron (T)
Metsulfuron	0.025–0.5 ppb	Millipore (P)
Paraquat	0.025–0.2 ppb	Millipore (P)
	0.020–0.5 ppb	Ohmicron (T)
Triasulfuron	0.05–1 ppb	Millipore (P)
2,4,5-T	1.4–250 ppb	Ohmicron (T)
Triclopyr	0.03–3 ppb	Ohmicron (T)
Trifluralin	1–40 ppb	Strategic Diagnostics (P)
Urea Herbicides	0.05–2 ppb	Millipore (P)
Insecticides		
Aldicarb	1–20 ppb	Millipore (P, T)
	0.25–100 ppb	Ohmicron (T)
Bioresmethrin	0.1–2 ppm	Millipore (P)
Carbaryl	0.25–5 ppb	Ohmicron (T)
Carbofuran	0.06–5 ppb	Ohmicron (T)
Chlordane	0.05–500 ppb	Millipore (T)*
Chlorpyrifos	0.1–3 ppb	Ohmicron (T)
Chlorpyrifos-ethyl	0.1–1 ppb	Millipore (P)
Chlorpyrifos-methyl	0.5–5 ppm	Millipore (P, T)*
Cyclodienes	5–20 ppb	Millipore (P, T)
	0.6–26.6 ppb	Ohmicron (T)
DDT	0.04 ppm D.L.	Millipore (T)*
Diazinon	30–500 ppt	Millipore (P)
Fenitrothion	0.5–10 ppm	Millipore (P, T)
Lindane and BHC	1 ppm D.L.	Millipore (T)*
Methomyl	0.45–15 ppb	Ohmicron (T)
Methoprene	3 ppm D.L.	Millipore (T)*
Nicotine	10–1000 ppb	Millipore (P)*
Parathion	0.04–0.4 ppb	Millipore (P)
Pentachlorophenol	0.5 ppm D.L.	Dräger (T)*
	5 ppb D.L.	Strategic Diagnostics (P)
	5 ppb D.L.	Millipore (T)
	0.06–10 ppb	Ohmicron (T)

Table 2.1 (continued)   Commercial Immunoassays for Pesticides and Industrial Contaminants

Compound	Detection Limit (D.L.)/ Measuring Range	Manufacturer
Pirimiphos-Methyl	0.05–5 ppm	Millipore (P)*
Trichloropyridinol	0.25–6 ppb	Ohmicron (T)

Fungicides

Compound	Detection Limit (D.L.)/ Measuring Range	Manufacturer
Benomyl (Carbendazim, MBC)	0.05–1.6 ppb	Strategic Diagnostics (P)
	0.4–10 ppb	Millipore (P, T)
	0.1–5 ppb	Ohmicron (T)
Captan	0.01–3 ppm	Ohmicron (T)
Carbofuran	0.1 ppb D.L.	Millipore (T)
	0.06–5 ppb	Ohmicron (T)
Chlorothalonil	0.5–50 ppb	Strategic Diagnostics (P)
	0.07–5 ppb	Ohmicron (T)
Metalaxyl	0.1–25 ppb	Millipore (P)
Procymidone	2 ppb D.L.	Millipore (T)*
	0.8–100 ppb	Ohmicron (T)
Toxaphene	0.2 ppm D.L.	Millipore (T)*

Industrial Contaminants

Compound	Detection Limit (D.L.)/ Measuring Range	Manufacturer
Benzene	5 ppb D.L.	Strategic Diagnostics (T)
Total BTEX (Xylene, Benzene,	0.6–10 ppm	Merck (M)
Naphthalene etc.)	0.1 ppm D.L.	Millipore (T)
	0.02–3 ppm	Ohmicron (T)
Dioxin (2,3,7,8-TCDD)	ppt range	Strategic Diagnostics (T)*
Gasoline	165 ppb D.L.	Strategic Diagnostics (T)
Hydrocarbons	10 ppm	Dräger (T)*
PAHs (Phenanthrene)	1 ppm D.L.	Strategic Diagnostics (P)*
	0.2 ppm	Millipore (T)*
	0.93–67 ppb	Ohmicron (T)
PAHs (Benzo[a]pyrene)	0.08–10 ppb	Ohmicron (T)
PCB (Aroclor 1254, 1260 etc.)	1 ppm D.L.	Dräger (T)*
	0.4 ppm D.L.	Strategic Diagnostics (P)*
	0.5–50 ppm	Merck (M)*
	1 ppm D.L.	Millipore (T)*
	0.2–10 ppb	Ohmicron (T)
TNT	0.7 ppm D.L	Strategic Diagnostics (P)
	5–45 ppb	Merck (M)
	0.5–50 ppb	Millipore (P)
	0.07–5 ppb	Ohmicron (T)
	0.03–30 ppb	R-Biopharm (P)
Mercury	0.25 ppb D.L.	Strategic Diagnostics (T)

Note: P = Plate kit, T = Tube kit, M = Membrane kit.
Immunoassay manufacturers (more suppliers can be found in van Emon and Gerlach[34]):
Dräger AG, Moislinger Allee 53-55, D-23542 Lübeck, Germany
Strategic Diagnostics Inc., 128 Sandy Dr., Newark, DE 19713, USA
Merck, Frankfurter Straße 250, D-64271 Darmstadt, Germany
Millipore Corporation, 50 Ashbury Road, Bedford, MA 01730, USA
Ohmicron Corporation, 375 Pheasant Run, Newton, PA 18940, USA
R-Biopharm GmbH, Rösslerstaße 94, D-64293 Darmstadt, Germany
Riedel-de Haën AG, Wunstorfer Straße 40, D-30926 Seelze, Germany
Transia GmbH, Dieselstraße 20A, D-61239 Ober-Mörlen, Germany

* Test kits have been developed for matrices other than water, e.g., soil, grain, tobacco, and wine.

D. Quality Control of Immunoassays

Precision and accuracy of IAs are important properties which deserve special attention. The quality and stability of the employed material, like microtiter plates, pipettes, and reagents, e.g., antibodies, enzyme tracer, or buffers, play a crucial role.[18] Microtiter plates from the same lot should be used; accurate pipettes are essential; and a proper protocol has to be supplied. The long-term stability of reagents has to be ensured, e.g., by freeze-drying antibodies and, if necessary, adding stabilizing components to the test reagents. Correct preparation of the buffers and proper storage are mandatory. As most tests are carried out at 20 to 25°C, it is important to allow adequate time for reagents stored in the refrigerator to reach ambient temperature. If sequential pipetting steps are used, one should be aware of possible timing differences between the start of the immunoreaction with the first and the last sample. To minimize a "drift," rapid reagent addition and an exact timing of reagent addition is necessary. The other approach is to use longer incubation times to minimize the impact of variation. Sometimes decreasing absorbencies of EIA are observed in the course of time. In many cases this is due to activity loss of the enzyme tracer, e.g., by microbial degradation or inactivation by high temperatures, especially during the summer months. Then, a new batch of enzyme tracer should be used, as sufficiently high absorbencies (>0.8) are necessary to calculate sample concentrations. Addition of stabilizing compounds to the enzyme tracer greatly reduces the risk of inactivation.[35] Also deteriorating substrate or chromogen solutions may be responsible for decreasing absorbencies. The quality of the water used to prepare standard and buffer solutions is also very important. It should be of highest purity. Problems with contaminated water may occur, especially during the summer months in areas with heavy pesticide application.

In spite of the simple handling of the assays, expert knowledge is required, especially to recognize and remove incident errors. Therefore, immunoassays should be performed by trained personnel. The development of simple and rapid assays, e.g., dipstick assays, immunofiltration tests,[15,26,28] reduces the requirement for trained users, but one has still to be aware of potential problems such as interferences from the sample matrix.

The investigation of the variation of an IA also gives valuable information about the consistency of the test. Coefficients of variation (CV) of IA measurements are usually between 10 and 20% (see references 36 and 37) for an optimized assay, although more precise results can be obtained.[38,39] Usually standards show lower CV of 2 to 10%, while samples from sources with more complex matrices, such as rain or surface water, exhibit higher CV, sometimes up to 20 to 30%.[40,41] Same-day and day-to-day CV of samples containing different matrices should be determined.[37] Dankwardt et al.[42] carried out reproducibility studies for a rainwater monitoring program. For the same-day reproducibility, samples were measured on five plates on the same day; for day-to-day reproducibility, the same samples were investigated on five different days. Usually, the same-day reproducibility was higher, showing CV mostly <10%, while for day-to-day reproducibility CV were up to 13%. Interlaboratory tests of the same immunoassay as it was carried out by Hock and the Immunoassay Study Group[43] for the standardization of triazine immunoassays help to evaluate the general applicability of a test. However, several conditions, like exact description of the assay including calibration curves, detection limits, cross-reactivities, a working range close to the middle of the test, enough parallel measurements, etc., must be met (see also AOAC criteria). Recently, a prenorm has been established for a standardized procedure for immunoassays in water in Germany.[44]

A validation of the results obtained by EIA should be carried out using statistical evaluation. To a limited extent this can be done by IA itself. Dilution of the samples as well as spiking of the authentic sample with known amounts of the contaminant can be used to check whether the matrix interferes with the IA.[45] However, spiked samples do not completely mimic real unknown samples. They do not contain potential metabolites of the contaminant nor residues from other compounds that may be present in real samples. Therefore, an IA should also be validated by a different established method like HPLC, GC, or GC/MS. During the last few years many groups have used this approach and usually obtained correlation coefficients of >0.9.[40,42,46,47] Often a slight overestimation of the immunoassay in comparison with HPLC or GC is observed due to cross-reactivities of the antibody or matrix effects.

E. Cross-Reactivity

Depending on the conjugate used for immunization and the class of chemicals under investigation, cross-reactivities of the antibodies with haptens similar to the analyte are frequently observed (see references 18 and 48). Therefore, it should be checked which compounds cross-react to which degree with the antibody. This is usually done by comparing the standard curves of the analyte under investigation with similar haptens, using the analyte concentrations at 50% of the inhibition curve as the reference.[18] However, cross-reactivities may be different at the beginning or the end of the inhibition curve. This should be taken into account if concentrations close to the detection limit are determined.

The strong cross-reactivity of an alachlor antibody to the sulfonic acid metabolite, for example, produced frequently false positive values using an alachlor screening kit.[49] This problem could be solved, however, by using solid-phase extraction (SPE) prior to immunoassay and sequential elution of the two compounds with different organic solvents.

If an antibody is selective for a single compound, it is regarded as monospecific.[50] An antibody that recognizes several compounds to the same extent, e.g., a group of s-triazines, can be used for the screening of a class of herbicides (group specific antibody).[51] If cross-reacting compounds are not expected in the samples because the compounds are not licensed, e.g., propazine in most European countries, a group-specific antibody can also be used for quantitative measurements of one compound.[41]

F. Matrix Effects

Natural water samples can contain compounds in addition to the target analyte that may interfere with the test. These may be, for example, ions or humic substances. Especially water samples from forest stands, bog water, or soil water contain a high content of organic compounds such as humic and fulvic acids which are responsible for the yellow color of some samples. Ruppert et al.[52] investigated the influence of ions on an EIA for triazine herbicides and observed an inhibition by several anions like azide, which inhibits the peroxidase by binding to the heme group of the enzyme. Most cations did not have an effect, except for Ca^{2+}, which led to an activation of the peroxidase. Some problems with interfering ions reacting with buffer components and thus leading to precipitates can be solved by changing the buffer of the assay system so that no precipitation occurs.[52] While ions may inhibit the enzyme used as a label or react with the buffer components, humic substances may bind nonspecifically to the antibody and thereby interfere with the specific binding of the analyte. These reactions usually lead to false-positive results. The addition of fulvic acids from soil water to atrazine standard solutions reduced the absorbencies of an EIA by about one fourth.[53] Therefore, an assay should be checked for its sensitivity to these substances. Matrix effects may also be detected by diluting the samples or spiking the samples with the analyte. If humic substances are present, low recoveries of the added pesticide may be observed. Matrix effects can be reduced by dilution of the sample if the concentrations of the analytes are high enough. Also, validation of the sample concentrations by other methods, such as HPLC and GC, should be carried out. Addition of bovine serum albumin to the plates prior to the addition of the standard and sample solutions[54] or to the enzyme tracer[53] greatly reduced the influence of humic and fulvic acids on the EIA. It may also be helpful to switch to a different batch of antibodies or a different assay kit, because different antibodies may show different sensitivities to humic substances. The buffering capacity of the assay buffer should also be checked, as some water samples, e.g., rain water from forest stands, may show relatively low pH values. No effects were observed between pH of 3 to 10 by different investigators.[42,51,55]

Although water samples are usually directly investigated by IA, sometimes an extraction step with organic solvent is necessary.[49] IA are to a certain degree tolerant of a variety of solvents, but each system must be tested to determine which and how much solvent can be tolerated. Nugent[56] found 10% propanol to work best in an IA for chlorpyrifos. An IA using monoclonal antibodies against s-triazines tolerated up to 10% methanol.[57] Recently, an IA for parathion was developed, in which the analyte dissolved in hexane could be measured directly in the IA without prior removal of the hexane by using antibodies encapsulated in reverse micelles composed of Aerosol T with aqueous centers.[58] However, a 10^4-fold decrease in sensitivity was observed.

G. Immunoanalysis of Natural Water Samples

IAs have been used intensively for the determination of pesticides and other environmental contaminants in rainwater (see references 42, 59, and 60), surface water (see references 40, 49, 61 through 67), and ground water (see references 46, 49, 61, 68, and 69). Selected examples are described below.

A substantial part of these studies was carried out for triazine herbicides.[40,42,46,59,64,66-69] This illustrates the widespread occurrence of these herbicides. Furthermore, atrazine is used as a model substance for pesticides in the environment, because much information is available on its behavior and persistence. Procedures for immunoconjugate synthesis and antibody production are known, facilitating the development of new assay systems or the introduction of new immunochemical techniques such as immunosensors. Methods for validation of the assays are available. It can be expected that in the coming years publications on the investigation of other pesticides and environmental contaminants by IA will increase.

Many investigators have used commercial test kits, which allow the investigation of samples without time-consuming antibody production. Thurman et al.,[46] for example, used a Res-I-Mune kit (ImmunoSystems) for the investigation of triazines in surface and ground water. The EIA was compared to GC/MS results obtained from samples that were extracted by SPE. Correlation coefficients between 0.91 and 0.95 were observed after introducing cross-reactivity factors for each of the herbicides in order to calculate a sum parameter for the GC. The majority of the samples contained only atrazine (up to 3 µg/L), and the EIA results corresponded well with the atrazine concentrations obtained by GC/MS.

Very high concentrations of triazines of up to 12.7 mg/L were found by Bushway et al.,[40] who investigated water samples from the former Czechoslovakia with a commercial test kit (Immuno-Systems). A close agreement of EIA and HPLC results with a coefficient of correlation of 0.99 was found, with a slight overestimation by the EIA, which was due to cross-reactivities with other triazines or triazine metabolites.

Mouvet et al.[70] compared four commercially available test kits and one assay developed in-house for the determination of triazines in surface and ground water. Operational characteristics, cross-reactivities, sensitivity, CV, and agreement with GC–LC measurements were investigated. Detection limits were determined between 0.003 and 0.07 µg/L. Intra-assay CV were below 7% for all tests; inter-assay CV were below 20%. Correlation studies between the EIA kits and GC–LC were carried out for samples from different water matrices. Depending on the water source, different levels of significance were observed with different tests. The best results were obtained for surface water, while not all kits showed a good agreement for lysimeter samples.

Interesting work was carried out by Thurman's group for the measurement of triazines in surface water by immunoassay.[66,67] Atrazine was measured in different layers of a lake used as a drinking water reservoir. The three-dimensional distribution of the triazine concentrations was established at different times of the year. After atrazine application, up to 10 µg/L atrazine were determined in the reservoir. In June the highest concentrations were observed at the entrance of the reservoir, while in August the concentrations moved toward the middle, with lower concentrations following behind.[66] In another study the effect of spring flush on the herbicide concentrations in surface waters was investigated. High concentrations of herbicides (up to 100 µg/L) were flushed from cropland and transported through the surface water system. The metabolite-to-parent compound ratio (deethylatrazine to atrazine ratio, DAR) was established, which may be used as a tracer of ground water movement into rivers.[67]

The use of test kits saves time-consuming antibody production, but there is often a lack of information on the immunoconjugate used for immunization or on antibody characteristics, such as their sensitivity toward interfering substances. We have produced polyclonal antibodies in rabbits and sheep to investigate surface[42,61] and rainwater[42] for atrazine. The latter study dealt with the atmospheric transportation of atrazine. We investigated more than 250 rainwater samples from open fields and forest stands and about 50 surface water samples by EIA without any cleanup or concentration steps prior to the test. The antibodies were tested for their sensitivity against humic substances and pH. No interference of humic substances up to 10 mg/L and no influence of the pH between 3 and 10 were found. Atrazine concentrations of up to 4 µg/L were observed in rain water, and up to 14 µg/L in surface water. Validation measurements by GC and GC/MS showed correlation coefficients of 0.95. The atrazine data were used to calculate annual deposition rates in open fields and forest stands, showing up to 180 µg per m^2 and year in open fields and up to 150 µg per m^2 and year in forest stands, where no atrazine had been applied, indicating atmospheric transport of the herbicide.

Alachlor was determined in ground and surface water using commercial tests.[60] SPE was carried out prior to EIA to remove interfering substances and to concentrate the analyte. Concentrations of up to 0.8 µg/L were observed, and a comparison with GC/MS showed a correlation coefficient of 0.95 with a slight underestimation by EIA. In a follow-up study, alachlor and its sulfonated metabolite were determined in rivers and reservoirs of the midwestern United States.[71] The high cross-reactivity of the ELISA for alachlor to the ethane sulfonic acid metabolite allowed the determination of both compounds after separation of the two by SPE. The concentration of the metabolite (up to 1.5 µg/L in the Mississippi river) exceeded the concentration of alachlor. The occurrence of carbaryl was determined by Marco et al.[72] in well-water from Spain with their own assay and compared with a commercial test kit. Both IAs yielded a good agreement with conventional methods. Concentrations of 0.08 to 1.37 µg/L were observed. Two commercial test kits were used in the Netherlands to determine 2,4-D concentrations in the rivers Rhine and Meuse.[65] A modification of the commercial kits significantly lowered the detection limit as originally indicated by the manufacturer. The water matrix substantially affected the recovery of 2,4-D with one assay kit, yielding unexpectedly low recoveries in demineralized and tap water. However, similar results were obtained by EIA and GC/MS for spiked samples from the river Rhine (with a slope of about 1 and r = 0.99). Routine samples were also analyzed and yielded analyte concentrations mostly below the detection limit (0.03 to 0.05 µg/L).

These examples show that EIAs are well suited for screening analyses. Substantial evidence has been collected for the reliability and robustness of this approach.

H. Integration of Liquid Chromatography with Immunoassay

Up to now, IAs are not yet available for all relevant compounds. The most interesting compounds are polar ones, because they are difficult to analyze with chromatographic methods.

If samples containing several contaminants are analyzed by IA, only the target substances are recognized, whereas unexpected compounds are detected by chromatographic methods through unusual peaks. Several compounds can usually be investigated with chromatographic methods in one run (multiresidue analysis), although different extraction methods and chromatographic parameters often have to be chosen for different compound classes.[73] While different antibodies can also be used to detect different compounds in one sample, the strength of the IA lies in the fast screening of many samples for the same compound, especially if many negative samples are expected.

More recently, antibodies have also been used in conjunction with liquid chromatography (LC), e.g., to concentrate an analyte from a large volume of sample and separate it from an interfering matrix.[74-76] In this case an immunoadsorbent column is used before analysis by LC. The immunoadsorbent column contains immobilized specific antibodies which bind the analyte, while interfering substances pass through. The analyte can be eluted by using a pH gradient[77] or an organic solvent.[75] Therefore, large sample volumes with low concentrations of the analyte can be reduced to small volumes with sufficiently high concentrations without coextracting interfering substances like humic and fulvic acids. This raises the effective sensitivity of the analysis. Antibody mixtures can be used to bind substances from different compound classes, e.g., the phenyl urea herbicides and the triazines.[78] In this case, the eluted compounds were injected into the LC, yielding a detection limit of 0.03 to 0.5 µg/L from sample volumes as low as 25 or 50 mL.

When cross-reacting antibodies are applied in IA, the obtained signal is not only related to the analyte, but also to related compounds. This problem can be circumvented by the use of LC prior to the IA. LC–IA was applied by Krämer et al.[74] to determine 4-nitrophenols. The nitrophenols were separated with different LC systems and determined by IA. LC–IA was about 8 to 10 times more sensitive compared to LC with UV detection. Therefore, the integration of LC with IA combines the high separation quality of the LC and the sensitivity of an IA.[79,80]

I. Determination of Residues Bound to Dissolved Humic Material

Compounds of environmental concern occur not only as free residues in the aquatic environment, but they can also bind to mostly low-molecular-weight, water-soluble humic substances.[81,82] The determination of these residues bound to dissolved humic and fulvic acids is of increasing relevance because humic material may alter the environmental fate and the effects of these xenobiotic compounds. The water solubility of lipophilic organic pollutants — accompanied by their mobility — may be enhanced by complexation or binding to dissolved humic material.[83] Kinetics of degradation reactions may be altered,[84] and consequently bioavailability and toxicity may change.[85,86] There is also a distinct possibility that these compounds may be released again, e.g., by microbial activities, and become available and effective to aquatic organisms.[87]

Immunochemical methods can also be useful for the detection of nonextractable (bound) residues. The application of antibodies for the detection of bound residues in soil or plants has been reported by Hahn et al.[30] and Giersch et al.[31] A competitive EIA, which has originally been designed for the determination of free atrazine residues,[42,88] was used to assay atrazine residues bound to aromatic structures like aminoaryl-s-triazines.[53] The aminoaryl-s-triazines are considered model compounds for nonextractable residues of atrazine in dissolved humic and fulvic acids. Andreux et al.[89] proposed an aminoaryl-s-triazine structure of their nonextractable atrazine residues just as was found by Adrian et al.[90] for chloroanilines.

Most of the aminoaryl-s-triazines were detected with the same affinity as free atrazine.[53] They could be determined in a range of about 0.02 to 2 µg/L. Determination of aminoaryl-s-triazine concentrations in samples from a photodegradation experiment yielded similar values when determined by EIA and HPLC. The anilines of the aryl side chain without bound triazine residues were checked for their cross-reactivities. No reactions with the antibodies were found. Work is now in progress to assay atrazine bound to aquatic humic and fulvic acids.

III. CONCLUSIONS AND OUTLOOK

Substantial evidence has been presented for the widespread and reliable use of IA in water analysis. The strength of the method lies in the screening of a large number of samples within a short time at low cost. Therefore, it can be a valuable supplement to conventional analytical methods. Important applications are seen in the analytics of ground and drinking water, where matrix effects are seldom observed.

Some restrictions are imposed by the fact that immunoassays are *de facto* single analyte methods. However, new approaches are being undertaken, such as the integration of IA with LC. Furthermore, multianalyte systems are under development. One concept is the microspot IA,[91] which uses many microspots with fluorescence-labeled antibodies of different selectivity immobilized on a chip. After incubation with the analyte (antigen or hapten) a fluorescence-labeled tracer antibody is added. The tracer antibody is either directed against the antigen or consists of an anti-idiotype antibody directed against the binding site of the capture antibody. Sensor and tracer antibody carry different fluorescence labels. Therefore, it is possible to determine the amount of analyte bound to the sensor antibodies with optical scanning methods by measuring the signal ratio (ratiometric assay).

Another possibility is the use of cross-reacting antibodies for multianalyte detection.[92] Known cross-reactivities of different antibodies can be used to calculate the different concentrations of different analytes in a sample containing several contaminants. The estimation of the individual concentrations is carried out by complex calculating procedures, e.g., by neuronal networks.[93]

Immunochemical analysis is a fast developing field with numerous possibilities for further improvements and developments. Two main directions can be observed. Much effort is put into the development of continuous measurements, such as flow injection immunoanalysis (FIIA) and immunosensors. A quasicontinuous FIIA of pesticides was developed by Krämer and Schmid[94] on the basis of a competitive IA. Here, the antibodies are immobilized on a membrane. The reaction takes place in the membrane reactor, the central part of the flow injection system. All reagents are sequentially added to the reactor, and the product is assayed with the aid of a flow fluorimeter. The measuring range of the flow injection analysis almost equals that of the EIA.

Important progress is to be expected in the field of immunosensors in which the sensing units are antibodies. Some relatively simple devices are immunoassay-based dipsticks.[26,27,95] The antibodies are immobilized on a membrane and the dipstick is then introduced into the sample. If reflectance detection is used, a quantitative signal is produced. A dipstick assay for atrazine was used to investigate atrazine-spiked water and liquid food samples.[95] No cleanup or enrichment of the samples was necessary, yielding a satisfactory agreement between the results of the dipstick immunoassay and HPLC or GC measurements. Immunofiltration, also a membrane-based test, was used to screen rain and surface water, but only visual detection was used.[28] In more complicated systems, the immunological recognition system is immobilized in the direct vicinity of a transducer, an electrochemical, optical, or gravimetric device. They respond to chemical compounds or ions and yield electrical signals which depend on the concentration of the analyte. Immunosensors with piezoelectric crystals as physical sensors are in a relatively advanced state of development.[96] They function as microbalances onto which antibodies are immobilized. Other physical sensors use optical systems, such as surface plasmon resonance (SPR), interferometry, or grating couplers.[97,98] A biosensor employing SPR was used for the determination of atrazine.[99] A detection limit of 0.05 μg/L of atrazine in water was reached with an analysis time of 15 minutes. Bier and Schmid[100] used a grating coupler immunosensor for the determination of terbutryn, a triazine herbicide. A detection limit of 15 nmol/L (about 3.6 μg/L) was established. Interesting developments are also to be expected from antibody electrodes.[101]

Also, new strategies for antibody production are being developed. Genetically engineered monoclonal antibodies appear very attractive because their selectivity and affinity can be tailored by site-directed mutations without requiring new immunizations.[102] Methods are now provided to rapidly isolate desired clones from antibody libraries and to manipulate individual recombinant antibodies to match specific demands of environmental analysis. Binding proteins derived from antibodies but consisting only of a part of their light- or heavy-chain and recombinant antibody fragments (Fabs) directed against different s-triazines have been produced.[21,103] The ultimate goal is the completely synthetic production of binding proteins or other synthetic receptors which are fitted to the structure of the analyte by molecular design. The use of libraries is guaranteed to close the bottleneck in antibody production.

ACKNOWLEDGMENTS

We are grateful to Dr. Burton Blais (Agriculture and Agri-Food Canada, Ottawa) for reading the English manuscript.

REFERENCES

1. Boehncke, A., Siebers, J., and Nolting, H.-G., Investigations of the evaporation of selected pesticides from natural and model surfaces in the field and laboratory, *Chemosphere,* 21, 1109, 1990.
2. Junge, C. E., Transport mechanisms for pesticides in the atmosphere, *Pure and Applied Chemistry,* 42, 95, 1976.
3. Buser, H.-R., Atrazine and other s-triazine herbicides in lakes and in rain in Switzerland, *Environmental Science & Technology,* 24, 1049, 1991.
4. Sundaram, K. M. S., Fate and short-term persistence of permethrin insecticide injected in a Northern Ontario (Canada) headwater stream, *Pesticide Science,* 31, 281, 1991.
5. Huber, W., Ecotoxicological relevance of atrazine in aquatic systems, *Environmental Toxicology and Chemistry,* 12, 1865, 1993.
6. Andreux, F., Scheunert, I., and Adrian, P., The binding of pesticide residues to natural organic matter and their bioavailability, *Preprinted Extended Abstract, Division of Environmental Chemistry, American Chemical Society,* 30, 501, 1990.
7. Freitag, D., Ballhorn, L., Geyer, H., and Korte, F., Environmental hazard profile of organic chemicals — an experimental method for the assessment of the behavior of organic chemicals in the ecosphere by means of simple laboratory tests with ^{14}C labeled chemicals, *Chemosphere,* 14, 1589, 1985.
8. Hindelang, D., Wirkungen von Herbizid- und Herbizid-Insektizid-Kombinationsbelastungen in aquatischen Modellökosystemen unter besonderer Berücksichtigung des Pestizidnachweises mit ELISA, Ph.D. Thesis, Technical University of Muenchen at Weihenstephan, Department of Botany, 1993.
9. Yalow, R. S. and Berson, S. A., Assay of plasma insulin in human subjects by immunological methods, *Nature,* 184, 1648, 1959.
10. Hammock, B. D., Gee, S. J., Cheung, P. Y. K., Miyamoto, T., Goodrow, M. H., Van Emon, J., and Seiber, J. N., in *Utility of Immunoassay in Pesticide Trace Analysis. Pesticide Science and Technology,* Greenhalgh, R. and Roberts, T., Eds., Blackwell Scientific Publications, Oxford, 1987, 309.
11. Vanderlaan, M., Watkins, B. E., and Stanker, L., Environmental monitoring by immunoassay, *Environmental Science & Technology,* 22, 247, 1988.
12. Hock, B., Enzymimmunoassays zur Bestimmung von Pflanzenschutzmitteln im Wasser, *Zeitschrift für Wasser- und Abwasser-Forschung,* 22, 78, 1989.
13. Sherry, J. P., Environmental chemistry: The immunoassay option, *Critical Reviews in Analytical Chemistry,* 23, 217, 1992.
14. Van Emon, J. M. and Lopez-Avila, V., Immunochemical methods for environmental analysis, *Analytical Chemistry,* 64, 78A, 1992.
15. Niessner, R., Immunoassays in environmental analytical chemistry: Some thoughts on trends and status, *Analytical Methods and Instrumentation,* 1, 134, 1994.
16. Meulenberg, E. P., Mulder, W. H., and Stoks, P. G., Immunoassays for pesticides, *Environmental Science & Technology,* 29, 553, 1995.
17. Knopp, D., Application of immunological methods for the determination of environmental pollutants in human biomonitoring. A review, *Analytica Chimica Acta,* 311, 383, 1995.

18. Hock, B., Enzyme immunoassay for pesticide analysis, *Acta Hydrochimica et Hydrobiologica*, 21, 71, 1993.
19. Hurn, B. A. L. and Chantler, M., Production of reagent antibodies, *Methods in Enzymology*, 70, 104, 1980.
20. Köhler, G. and Milstein, C., Continuous culture of fused cells secreting antibody of predefined specificity, *Nature*, 256, 496, 1975.
21. Kramer, K. and Hock, B., Recombinant antibodies for pesticide analysis, in *Residue Analysis in Food Safety*, Beier, R.C., Ed., ACS, Washington, DC, 1996, Chapter 38, 471.
22. Crowl, C. P., Gibbons, I., and Schneider, R. S., Recent advances in homogeneous enzyme immunoassays for haptens and proteins, in *Immunoassays: Clinical Laboratory Techniques for the 1980s*, Nakamura, R.M., Dito, W.R., and Tucker III., E.S., Eds., Alan R. Liss Inc., New York, 1980, 89.
23. Wittmann, C. and Hock, B., Improved enzyme immunoassay for the analysis of s-triazines in water samples. *Food and Agricultural Immunology*, 1, 211, 1989.
24. Lawruk, T. S., Lachmann, C. E., Jourdan, S. W., Fleeker, J. R., Herzog, D. P. and Rubio, F. M., Quantification of cyanazine in water and soil by a magnetic particle-based ELISA, *Journal of Agricultural and Food Chemistry*, 41, 747, 1993.
25. Schneider, P., Goodrow, M. H., Gee, S. J., and Hammock, B. D., A highly sensitive and rapid ELISA for the arylurea herbicides diuron, monuron and liuron, *Journal of Agricultural and Food Chemistry*, 42, 413, 1994.
26. Giersch, T., A new monoclonal antibody for the sensitive detection of atrazine with immunoassay in microtiter plate and dipstick format, *Journal of Agricultural and Food Chemistry*, 41, 1006, 1993.
27. Ploum, M. E., Haasnoot, W., Paulussen, R. J. A., Van Bruchem, G. D., Hames, A. R. M., Schilt, R., and Hug, F. A., Test strip enzyme immunoassays and the fast screening of nortesterone and clenbuterol residues in urine samples at the parts per billion level, *Journal of Chromatography*, 546, 413, 1991.
28. Dankwardt, A. and Hock, B., Rapid immunofiltration assay for the detection of atrazine in water and soil samples, *Biosensors & Bioelectronics*, 8, XX, 1993.
29. Erhard, U., Schnellanalytik in der Antikörperproduktion, *Biotec*, 10, 1990.
30. Hahn, A., Frimmel, F., Haisch, A., Henkelmann, G., and Hock, B., Immunolabeling of atrazine residues in soil, *Zeitschrift für Pflanzenernährung und Bodenkunde*, 155, 203, 1991.
31. Giersch, T., Sohn, G., and Hock, B., Monoclonal antibody-based immunolocalization of bound triazine residues in two aquatic macrophytes (*Elodea canadensis* and *Myriophyllum spicatum*), *Analytical Letters*, 26, 1831, 1993.
32. Cammann, K., Lemke, U., Rohen, A., Sander, J., Wilken, H., and Winter, B., Chemical sensors and biosensors — principles and applications, *Angewandte Chemie, International Edition in English*, 30, 516, 1991.
33. Schmid, R. D., Trends in Biosensors, *Biofutur*, 3, 37, 1988.
34. Van Emon, J. M. and Gerlach, C. L., A status report on field portable immunoassay, *Environmental Science & Technology*, 29, 312, 1995.
35. Reichle, J., Stabilisierung von Enzymimmunoassays, Master's Thesis, Technical University of München, Department of Botany, 1996.
36. Huber, S. J., Improved solid-phase enzyme immunoassay systems in the ppt range for atrazine in fresh water, *Chemosphere*, 14, 1795, 1985.
37. Bushway, R. J., Perkins, L. B., and Hurst, H. L., Determination of atrazine in milk by immunoassay, *Food Chemistry*, 43, 283, 1992.
38. Van Emon, J., Hammock, B. D., and Seiber, J. N., Enzyme-linked immunosorbent assay for paraquat and its application to exposure analysis, *Analytical Chemistry*, 58, 1866, 1986.
39. Manclus, J. J. and Montoya, A., Development of an enzyme immunoassay for the analysis of chlorpyrifos and its major metabolite 3,5,6-trichloro-2-pyridinol in the aquatic environment, *Analytica Chimica Acta*, 311, 341, 1995.
40. Bushway, R. J., Perkins, B., Fukal, L., Harrison, R. O., and Ferguson, B. S., Comparison of enzyme-linked immunosorbent assay and high-performance liquid chromatography for the analysis of atrazine in water from Czechoslovakia, *Archives of Environmental Contamination and Toxicology*, 21, 365, 1991.

41. Dankwardt, A., Seifert, J., and Hock, B., Magnetpartikel-Enzymimmunoassay als schnelle Screening-Methode zur Bestimmung von Atrazin in Umweltproben. *Acta hydrochimica et hydrobiologica*, 21, 110, 1993.

42. Dankwardt, A., Wüst, S., Elling, W., Thurman, E. M., and Hock, B., Determination of atrazine in rainfall and surface water by enzyme immunoassay, *Environmental Science & Pollution Research*, 1, 196, 1994.

43. Hock, B. and The Immunoassay Study Group, Enzyme immunoassays for the determination of s-triazines in water samples: Two interlaboratory tests, *Analytical Letters*, 24, 529, 1991.

44. German standard methods for the examination of water, waste water and sludge — Sub-animal testing (group T), — Part 2: Guideline for selective immunotest methods (immunoassays) for the determination of plant treatment and pesticide agents (T2), DIN V 38415-2, 1995.

45. Pengelley, W. L., Validation of Immunoassays, in *Plant Growth Substances*, Bopp, M., Ed., Springer Verlag, Berlin Heidelberg, 1985, 35.

46. Thurman, E. M., Meyer, M., Pomes, M., Perry, C. A., and Schwab, P. A., Enzyme-linked immunosorbent assay compared with gas chromatography/mass spectrometry for the determination of triazine herbicides in water, *Analytical Chemistry*, 62, 2043, 1990.

47. Franek, M., Kolar, V., and Eremin, S. A., Enzyme immunoassays for s-triazine herbicides and their application in environmental and food analysis, *Analytica Chimica Acta*, 311, 349, 1995.

48. Harrison, R. O., Goodrow, M. H., and Hammock, B. D., Competitive inhibition ELISA for the s-triazine herbicides: assay optimization and antibody characterization, *Journal of Agricultural and Food Chemistry*, 39, 122, 1991.

49. Aga, D. S., Thurman, E. M., and Pomes, M. L., Determination of alachlor and its sulfonic acid metabolite in water by solid-phase extraction and enzyme-linked immunosorbent assay, *Analytical Chemistry*, 66, 1495, 1994.

50. Wittmann, C. and Hock, B., Development of an enzyme linked immunoassay for the analysis of the atrazine metabolite hydroxyatrazine, *Acta Hydrochimica et Hydrobiologica*, 22, 60, 1994.

51. Seifert, J., Baker RaPID Assay — ein ELISA zur Bestimmung von Pestiziden mit magnetischen Partikeln als fester Phase, *Gewässerschutz-Wasser-Abwasser*, 134, 129, 1992.

52. Ruppert, T., Weil. L., and Niessner, R., Influence of water contents on an enzyme immunoassay for triazine herbicides, *Vom Wasser*, 78, 387, 1992.

53. Dankwardt, A., Hock, B., Simon, R., Freitag, D., and Kettrup, A., Determination of non-extractable triazine residues by enzyme immunoassay — Investigation of model compounds and soil fulvic and humic acids. *Environmental Science & Technology*, 30, 3493, 1996.

54. Keuchel, C., Weil, L., and Niessner, R., Enzyme-linked immunosorbent assay for the determination of 2,4,6-trinitrotoluene and related nitroaromatic compounds, *Analytical Sciences*, 8, 9, 1992.

55. Itak, J. A., Olson, E. G., Fleeker, J. R. and Herzog, D. P., Validation of a paramagnetic particle-based ELISA for the quantitative determination of carbaryl in water, *Bulletin of Environmental Contamination and Toxicology*, 51, 260, 1993.

56. Nugent, P., Enzyme-linked competitive immunoassay, in *Emerging Strategies for Pesticide Analysis*, Cairns, T. and Sherma, J., Eds., CRC Press, Boca Raton, FL, 1992, 247.

57. Schneider, P. and Hammock, B. D., Influence of the ELISA format and the hapten enzyme conjugate on the sensitivity of an immunoassay for s-triazine herbicides using monoclonal antibodies, *Journal of Agricultural and Food Chemistry*, 40, 525, 1992.

58. Francis, J. M. and Craston, D. H., Immunoassay for parathion without its prior removal from solution in hexane, *Analyst*, 119, 1801, 1994.

59. Huber, K. J., Weil, L., and Niessner, R., Total deposition monitoring of the triazine herbicide atrazine by use of an enzyme-linked immunosorbent assay (ELISA). *Fresenius Journal of Analytical Chemistry*, 343, 146, 1992.

60. Aga, D. S. and Thurman, E. M., Coupling solid-phase extraction and enzyme-linked immunosorbent assay for ultratrace determination of herbicides in pristine water, *Analytical Chemistry*, 65, 2894, 1993.

61. Wittmann, C. and Hock, B., Evaluation and performance characteristics of a novel ELISA for the quantitative analysis of atrazine in water, plants and soil. *Food and Agricultural Immunology*, 2, 65, 1990.

62. Thurman, E. M., Goolsby, D. A., Meyer, M. T., Mills, M. S., Pomes, M. L., and Kolpin, D. W., A reconnaissance study of herbicides and their metabolites in surface water of the midwestern United States using immunoassay and gas chromatography mass spectrometry, *Environmental Science & Technology*, 26, 2440, 1992.

63. Gascon, J., Durand, G., and Barcelo, D., Pilot survey for atrazine and total chlorotriazines in estuarine waters using magnetic particle based immunoassay and gas chromatography nitrogen/phosphorus detection, *Environmental Science & Technology*, 29, 1551, 1995.

64. Gruessner, B., Shambaugh, N. C., and Watzin, M. C., Comparison of an enzyme immunoassay and gas chromatography/mass spectrometry for the detection of atrazine in surface waters, *Environmental Science & Technology*, 28, 251, 1995.

65. Meulenberg, E. P. and Stoks, P. G., Water quality control in the production of drinking water from river water. The application of immunological techniques for the detection of chlorphenoxy acid herbicides, *Analytica Chimica Acta*, 311, 407, 1995.

66. Fallon, J. D., Determining the 3-dimensional distribution, transport, and relative age of atrazine and selected metabolites in Perry, Lake, Kansas, Master's Thesis, University of Kansas, Department of Civil Engineering, 1994.

67. Thurman, E. M., Herbicides in surface waters of the midwestern United States: The effect of spring flush, *Environmental Science & Technology*, 25, 1794, 1991.

68. Sherry, J. P. and Borgmann, A., Enzyme-immunoassay techniques for the detection of atrazine in water samples — Evaluation of a commercial tube based assay, *Chemosphere*, 26, 2173, 1993.

69. Brady, J. F., Lemasters, G. S., Williams, R. K., Pittman, J. H., Daubert, J. P., Cheung, M. W., Skinner, D. H., Turner, J., Rowland, M. A., Lange, J., and Sobek, S. M., Immunoassay analysis and gas chromatographic confirmation of atrazine residues in water samples from a field study conducted in the state of Wisconsin, *Journal of Agricultural and Food Chemistry*, 43, 268, 1995.

70. Mouvet, C., Broussard, S., Jeannot, R., Maciag, C., Abuknesha, R., and Ismail, G., Validation of commercially available ELISA microtiter plates for triazines in water samples, *Analytica Chimica Acta*, 311, 331, 1995.

71. Thurman, E.M., Goolsby, D.A., Aga, D.S., Pomes, M.L., and Meyer, M.T., Occurrence of alachlor and its sulfonated metabolite in rivers and reservoirs of the midwestern United States: The importance of sulfonation in the transport of chloroacetanilide herbicides, *Environmental Science & Technology*, 30, 569, 1996.

72. Marco, M.-P., Chiron, S., Gascon, J., Hammock, B. D., and Barcelo, D., Validation of two immunoassay methods for environmental monitoring of carbaryl and 1-naphthol in ground water samples, *Analytica Chimica Acta*, 311, 319, 1995.

73. Scharf, J., Wiesiollek, R. and Bächmann, K., Pesticides in the atmosphere, *Fresenius Journal of Analytical Chemistry*, 342, 813, 1992.

74. Krämer, P. M., Li, Q. X., and Hammock, B. D., Integration of liquid chromatography with immunoassay: An approach combining the strength of both methods, *Journal of the AOAC International*, 77, 1275, 1994.

75. Pichon, V., Chen, L., and Hennion, M.-C., On-line preconcentration and liquid chromatographic analysis of phenylurea herbicides in environmental water using a silica-based immunosorbent, *Analytica Chimica Acta*, 311, 429, 1995.

76. Thomas, D. H., Beck-Westermeyer, M., and Hage, D. S., Determination of atrazine in water using tandem high-performance immunoaffinity chromatography and reversed-phase liquid chromatography, *Analytical Chemistry*, 66, 3823, 1994.

77. Marx, A., Giersch, T., and Hock, B., Immunoaffinity chromatography of s-triazines, *Analytical Letters*, 28, 267, 1995.

78. Pichon, V., Chen, L., Hennion, M.-C., Durand, N., and Le Goffic, F., Selective multiresidue analysis of pesticides in surface water using immunosorbents, Abstract L-9, presented at the 5th Symposium on chemistry and fate of modern pesticides, Paris, September 6-8, 1995.

79. de Frutos, M. and Regnier, F. E., Tandem chromatographic-immunological analyses, *Analytical Chemistry*, 65, 17A, 1993.

80. Lucas, A. D., Gee, S. J., and Hammock, B. D., Integration of immunochemical methods with other analytical techniques for pesticide residue determination, *Journal of the AOAC International*, 78, 585, 1995.

81. McCarthy, J. F., Jimenez, B. D., and Barbee, T., Effect of dissolved humic material on accumulation of polycyclic aromatic hydrocarbons: Structure-activity relationships, *Aquatic Toxicology,* 7, 15, 1985.
82. Robinson, K. G. and Novak, J. T., Fate of 2,4,6-Trichloro-(^{14}C)-phenol bound to dissolved humic acid, *Water Research,* 28, 445, 1994.
83. Chiou, C. T., Malcolm, R. L., Brinton, T. I., and Kile, D. E., Water solubility enhancement of some organic pollutants and pesticides by dissolved humic and fulvic acids, *Environmental Science & Technology,* 20, 502, 1986.
84. Perdue, E. M. and Wolfe, N. L., Modification of pollutant hydrolysis kinetics in the presence of humic substances, *Environmental Science & Technology,* 16, 847, 1982.
85. Steinberg, C. E. W., Sturm, A., Kelbel, J., Kyu Lee, S., Hertkorn, N., Freitag, D., and Kettrup, A. A., Changes of acute toxicity of organic chemicals to Daphnia magna in the presence of dissolved humic material (DHM). *Acta Hydrochimica et Hydrobiologica,* 20, 326, 1992.
86. Steinberg, C. E. W., Xu, Y., Kyu Lee, S., Freitag, D., and Kettrup, A. A., Effect of dissolved humic material (DHM) on bioavailability of some organic xenobiotics to *Daphnia magna. Chemical Speciation and Bioavailability,* 5, 1, 1993.
87. Khan, S. U. and Dupont, S., Bound pesticide residues and their bioavailability, in *Pesticide Science and Biotechnology,* Greenhalgh, R. and Roberts, T.R., Eds., Blackwell Scientific Publications, Oxford, 1987, 417.
88. Dankwardt, A., Pullen, S., Rauchalles, S., Kramer, K., Just, F., Hock, B., Hofmann, R., Schewes, R., and Maidl, F. X., Atrazine residues in soil two years after the atrazine ban — A comparison of enzyme immunoassay with HPLC, *Analytical Letters,* 28, 621, 1995.
89. Andreux, F. G., Portal, J. M., Schiavon, M., and Bertin, G., The binding of atrazine and its dealkylated derivatives to humic-like polymers derived from catechol, *The Science of the Total Environment,* 117/118, 207, 1992.
90. Adrian, P., Lahaniatis, E. S., Andreux, F., Mansour, M., Scheunert, I., and Korte, F., Reaction of the soil pollutant 4-chloroaniline with the humic acid monomer catechol, *Chemosphere,* 18, 1599, 1989.
91. Ekins, R., Chu, F. W., and Biggart, E. M., Multianalyte immunoassay: The immunological "compact disk" of the future, *Journal of Clinical Immunoassay,* 13, 169, 1990.
92. Hock, B., Immunchemische Verfahren in der Umweltanalytik und ihre Perspektiven, *Zeitschrift für Umweltchemie und Ökotoxikologie,* 5, 309, 1993.
93. Vertosick, F. T. and Rehn, T., Predicting behavior of an enzyme-linked immunoassay model by using commercially available neural network software, *Clinical Chemistry,* 39, 2478, 1993.
94. Krämer, P. M. and Schmid, R. D., Automated quasi-continuous immunoanalysis of pesticides with a flow injection system, *Pesticide Science,* 32, 451, 1991.
95. Wittmann, C., Bilitewski, U., Giersch, T., Kettling, U., Schmid, R.D., Development and evaluation of a dipstick immunoassay format for the determination of atrazine residues on-site, *Analyst,* 121, 863, 1996.
96. Minunni, M., Mascini, M., Guilbaut, G. G., and Hock, B., The quartz crystal microbalance as biosensor. A status report on its future, *Analytical Letters,* 28, 749, 1995.
97. Lukosz, W., Principles and sensitivities of integrated optical and surface plasmon sensors for direct affinity sensing and immunosensing, *Biosensors & Bioelectronics,* 6, 215, 1991.
98. Brecht, A. and Gauglitz, G., Optical probes and transducers, *Biosensors & Bioelectronics,* 10, 923, 1995.
99. Minunni, M. and Mascini, M., Detection of pesticide in drinking water using real time biospecific interaction analysis (BIA), *Analytical Letters,* 26, 1441, 1993.
100. Bier, F.F. and Schmid, R.D., Real time analysis of competitive binding using grating coupler immunosensors for pesticide detection, *Biosensors & Bioelectronics,* 9, 125, 1994.
101. Cammann, K., Selectivity modulation of ion-selective membranes — the general principle of a new class of immunosensors, *Biosensors '90,* Elsevier Scientific Publications LTD, London, 1991.
102. Skerra, A. and Plückthun, A., Assembly of a functional immunoglobulin Fv fragment in *Escherichia coli, Science,* 240, 1038, 1988.
103. Ward, V. K., Schneider, P. G., Kreissig, S. B., Hammock, B. D., and Choudary, P. V., Cloning, sequencing and expression of the Fab fragment of a monoclonal antibody to the herbicide atrazine, *Protein Engineering,* 6, 981, 1993.

CHAPTER 3

Environmental Applications with Submitochondrial Particles

Harry W. Read, John M. Harkin, and Karl E. Gustavson

CONTENTS

I. INTRODUCTION

Mitochondria are distinct organelles which can occupy 2 to 22% of the volume of the cytoplasm of all aerobic eukaryotic (nucleus-containing) cells.[1,2] They are perhaps the best understood cell constituents in terms of structure, molecular organization, and function in cell metabolism. The principal role of mitochondria is to provide energy to the cell through utilization of an array of closely synchronized enzymatic activities to effect stepwise oxidative degradation of a variety of energy-supplying substrates. For these reasons, they are extremely vulnerable to toxic insult from extraneous substances intruding into the cell. Mitochondria are closely associated with myofibrils in tissues, to which they supply ATP (adenosine triphosphate), the molecule containing an energy-rich phosphate bond which activates muscular movement.

Tests based on suspensions of submitochondrial particles (or SMP) — micelle-shaped fragments of the inner convoluted enzyme-laden cristae membrane of mitochondria — occupy a special niche in the spectrum of *in vitro* bioassays for toxicants in water or aqueous solutions.[3,4] Among the many advantages of SMP as biochemical targets for *in vitro* tests is the fact that the particles are so small that they do not interfere appreciably with the passage of light in spectroscopic determinations, and because of their large surface area, they act like a chemical or biochemical reagent in solution. Because SMP comprise an integrated battery of enzymes embedded in a bilipid membrane matrix, they are sensitive to a wide range of toxic materials. SMP tests respond to substances that directly inhibit one or more of the enzymes, interfere with concerted enzyme activity, or alter the properties of the membrane in which the enzymes are located. SMP respond to either individual substances or mixtures of materials of different types.

SMP are therefore superb candidates for use in screening tests for general toxicity, i.e., broad-spectrum tests for a variety of directly acting toxicants (i.e., those not requiring metabolic activation), whether singly or in combination, that can be completed in minutes. Generally, the tests are robust in the sense that matrix waters from aquatic environments normally do not cause significant interferences, so they have many potential applications in aquatic toxicology and water-quality monitoring. Response is rapid and sensitive because the particles have no external membrane to deny or inhibit access of the toxicants to the enzymes. In this respect, SMP are preferable to unicellular organisms such as bacteria, algae, or cell cultures used in other microscale *in vitro* bioassays, because whole cells can have strong, impermeable outer membranes that may prevent some genuinely harmful chemicals from entering.

II. RATIONALE FOR USE OF BEEF-HEART SMP IN *IN VITRO* BIOASSAYS

Mitochondria and submitochondrial particles from beef hearts were selected as biologicals to develop broad-spectrum screening tests for toxicants in the environment for the following reasons:

a. Large amounts of mitochondria are easily obtained from beef hearts. Mitochondria are readily isolated from cell debris in tissue homogenates in appropriate supporting or preservation media (e.g., buffered isotonic Ca^{++}-free sucrose solutions) by differential centrifugation, but their yields, activity, and durability vary greatly with the organ and species from which they are obtained. Mitochondria from the most common source, rat livers, normally have a useful lifetime of only a few hours vs. up to three days for heart mitochondria. Fresh beef hearts are readily available from cattle slaughtered for food, avoiding the need to breed and sacrifice laboratory animals. The heart muscle has greater densities of mitochondria than liver tissue and is usually less fatty — a beneficial feature for the workup of the mitochondria. Other investigators have isolated mitochondria from soft (liver, kidney, brain, adrenal cortex, white or brown adipose tissues) or hard vertebrate tissues (heart or skeletal muscles), from insect flight muscles, plants (mung beans, potatoes, spinach), yeasts, algae, filamentous fungi, and spermatozoa;[2] cell fractionation by sedimentation or freezing/thawing has been used for production of SMP preparations.[2] However, large-scale production from beef hearts using specialized equipment is the best way to obtain large batches of uniform material.[5,6] (Such SMP are now commercially available from BioRenewal Technologies Inc., The Faraday Center, 2800 Fish Hatchery Road, Madison, WI 53711, USA.)
b. As separate organelles of primordial origin with their own genetic material, mitochondria are highly conserved within and between species.[1,2] Mitochondria are thought to be descendants of primordial bacteria which invaded and formed an endosymbiotic relationship with protoeukaryotic cells.[1,2] The outer membrane of mitochondria, unlike that of bacteria, is readily penetrated by both organic and inorganic substances. Access of these substances into the living cell is restricted, so mitochondria are normally protected from toxic concentrations of these substances. Isolated mitochondria lack this protection. As the main chemical-processing constituent of cells, they encounter and interact with most chemicals (benign or toxic, nutritive or inadvertent, organic or inorganic) which enter the cell. Consequently, the effects of any of these chemicals or their combinations on mitochondria are similar, regardless of the position of the host organism in the evolutionary chain. In other words, toxicity, as registered by the response of mitochondria, is not expected to be highly species-specific; interspecies differences in other aspects of organism adaptation (uptake, metabolic modification, and excretion) are the features which can lead to differential toxic responses among species.

c. The delicate balance of their interdependent enzyme functions makes mitochondria particularly vulnerable to toxic insult. Because of their central role in energy production in cell metabolism, mitochondria are replete with enzyme complexes acting in concert to degrade oxidizable substrates with simultaneous production of ATP, the key coenzyme in the transfer of phosphate bond energy for muscle activation and other energy-linked cell biochemical processes. This function explains why they occur in such high concentrations in heart muscle, the leanest, most active muscle in any higher organism, and why the enzyme systems must act efficiently as a team to maintain a steady supply of energy to the cell, and by extension, to the whole organism. For this reason, breakdown in mitochondrial function is thought to be a major mechanism of action of many toxicants in whole-animal toxicity. Any toxicant interfering with the smooth sequence of events in normal mitochondrion operation will produce a change likely to upset the whole process and may be discernible as a physiological phenomenon, a clinical manifestation on the whole organism, or in a laboratory test of some biochemical or physical endpoint.

d. Unlike systems of soluble enzymes, many functions of whole mitochondria and SMP depend upon membrane integrity. Both whole mitochondria and SMP are membrane vesicles, and in SMP, most of the enzymes are membrane-bound. Coupling between the respiratory chain of enzymes and the ATPase requires selective permeability to ions, just as with neurons, and will fail if the membrane is not intact. Furthermore, many of the enzymes lose functionality in the presence of certain lipid-soluble compounds, suggesting disruption of the positioning of membrane-bound, interacting enzyme complexes within the lipid bilayer. This broadens the range of toxicants to which SMP respond.

e. Variations in mitochondrial test protocols can delineate mode of action. The enzymes in isolated whole mitochondria continue to function normally, if provided with extraneous sources of the substrates and biochemicals which they utilize for their normal function. By employing different mixtures of substrates and selected mitochondrial inhibitors, individual combinations of enzymes can be segregated and scrutinized. This approach can be used to yield information about the mechanism of action and, if testing an unknown, clues about a toxicant's identity. Relatively unskilled technicians can rely on objective instrumental measurements to detect and quantify toxicity responses in these tests.

f. Unlike whole mitochondria, SMP are fully capable of coupling the respiratory chain of enzymes to the ATP-synthesizing complex even after freezing and thawing. Our first tests for toxicants were based on large-amplitude swelling of isolated mitochondria, an energy-driven process occurring in suitable iso-osmotic media as a result of normal metabolism. Light scattering of suspensions at 520 nm provided a measure of the swelling. Inhibition of swelling could be quantitatively related to the toxicity of the medium.[7] Although beef-heart mitochondria remain active for several days and can be produced quickly, they still have a restricted lifetime and cannot be readily preserved. The relatively large membranes of whole mitochondria are disrupted by freezing and thawing so that electrochemical gradients, believed to couple respiratory chain enzymes to mitochondrial ATPase and other enzymes, can no longer be maintained. Consequently, tests based on whole mitochondria were deemed unsatisfactory for routine use. However, many of the integrated enzyme functions in whole mitochondria are fully maintained in SMP, produced by sonication of whole mitochondria and purification by ultracentrifugation. The coordinated chain of enzymes involved in oxidative phosphorylation in intact mitochondria or cells is maintained with minimal loss of function in SMP and can be stored for months or years in preservative medium at $-80°C$ without significant loss of enzymatic activity.

III. UTILITY OF COMPARISONS BETWEEN SMP AND WHOLE-ORGANISM TESTS

As with all biochemical tests based on single cells, cell organelles or fragments, or cell constituents, SMP tests are still only a surrogate model of restricted validity for whole-organism toxicity. This is because the processes of absorption (uptake), metabolism (detoxification or activation), interorgan transport, and excretion (elimination) which modify overall toxicity *in vivo* in complete higher organisms cannot be accurately duplicated or modeled in the *in vitro* protocols. Some refinements of SMP tests, e.g., activation of otherwise benign substrates by liver microsomal enzymes or fractions, can usually be made more readily than with enzyme- or unicellular-organism tests.

Evaluation of the utility of any of these tests must rely on the closeness of correlations of their results with toxicities measured in whole organisms. The higher organisms which provide data for these comparisons are usually species that are the goal of toxicant-avoidance programs, e.g., humans, fish, or wildlife, where information is obtained from mortality or morbidity resulting from occupational, accidental, or environmental exposures to the toxicants of concern. In such cases, exposure levels are difficult to determine or are inferred from clinical measurements of body burdens through analysis of blood or tissue samples. Alternatively, toxicity data are derived from controlled exposures using standardized protocols of laboratory test species such as rodents, fathead minnows, or daphnids. The rodents serve as surrogates for higher species. Because the aquatic species are widely occurring, trophically important, and reasonably sensitive, they are selected as representatives for the entire range of organisms inhabiting the particular environment in question, e.g., temperate-zone freshwater rivers and lakes.

Toxicities of a large number of individual inorganic ions or organic compounds measured in tests with SMP correlate closely with toxicities reported for the same materials in cell lines, rodents, fish, and humans.[8-10] Thus, SMP bioassays are perhaps the best available substitutes for whole organisms in rapid, inexpensive *in vitro* prescreening tests for chemical toxicity. The nature of mitochondria, their enzymes, and functions provides a reasonable explanation for the expectation that SMP should reflect whole-organism toxicity well, regardless of species. In general, higher precision and closer reproducibility can be achieved with SMP tests because the underlying variability is relatively low, the techniques are simple (minimizing interlaboratory and cross-operator variability), and, because they are fast and inexpensive, large numbers of replicate tests can be completed. In whole-organism bioassays, the experimental techniques are more complicated, more expensive to perform and more difficult to duplicate, so the range of inter-investigator error is high, few cases of multiple determinations of substances are available, and the range of substances that can be tested is constrained by cost. These tend to weaken correlations with the SMP — or any other — *in vitro* toxicity bioassay and do not necessarily reflect failings in the microscale tests.

Even with good-quality data sets, poor correlations can result because toxic responses in whole organisms are modulated by higher levels of biological organization to varying degrees by exclusion, preferential uptake, bioaccumulation, metabolism, or excretion. These also explain much interspecies variability. Also, many toxic responses are mediated through specific receptors absent from *in vitro* systems, e.g., acetylcholinesterase inhibition by organophosphate or carbamate insecticides. The biologicals used in SMP and other *in vitro* tests are often sensitive to these chemicals despite their lack of the receptor, but usually to a lesser magnitude. The fact that SMP tests are reasonably good predictors of whole organism responses may be rooted in an underlying similarity of biological systems at all levels of organization: proteins must work together, and membranes must maintain selectivity. SMP are a complex enough system to share this similarity, yet remain simple enough to work in fast and convenient tests.

These factors call for judicious application and interpretation of all standardized toxicity tests, whether with whole organisms, single cells, or cell organelles or constituents. Employment of combinations of *in vitro* tests offers an efficient tool for discovering toxicant mechanism of action or for matching test capabilities with contaminants expected in a given environmental situation. SMP and other *in vitro* tests can be completed much more quickly than whole-organism tests, viz. in less than an hour vs. four to seven days for fish tests. This enables users to make decisions quickly and is a distinct advantage in applications such as spill-site evaluations, on-line effluent monitoring, and toxicity identification/reduction evaluations. Perhaps most powerfully, SMP tests can be used to serve as an extension and complement to a testing program based on whole organisms.

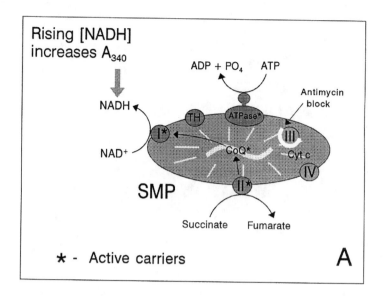

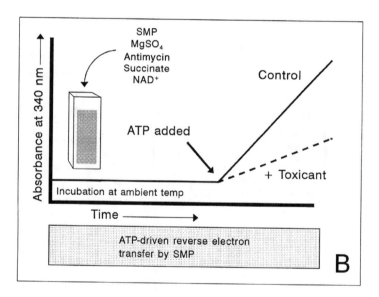

Figure 3.1 A) ATP-driven reverse electron transfer in SMP. I, II, III, and IV are mitochondrial electron transfer enzyme complexes; CoQ is coenzyme Q; Cyt c is cytochrome c; TH is mitochondrial transhydrogenase; ~ designates energy coupling between electron transfer chain and ATPase enzymes. B) Necessary reagents and generalized time course of absorbance changes in the reverse electron transfer (RET) test.

IV. OPTIONS FOR TOXICANT BIOASSAYS BASED ON SMP RESPONSES

Thanks to the broad diversity of some 50 enzymatic activities present in SMP, many of them interlinked, several choices of the target for toxicants can be chosen and suitably sensitive test protocols elaborated. Several different bioassays are described below, including: 1) the *reverse-electron transport (RET)* bioassay (Figure 3.1),[11] which segregates a portion of the mitochondrial

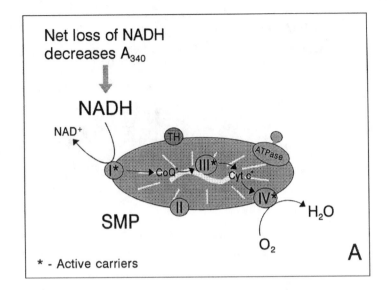

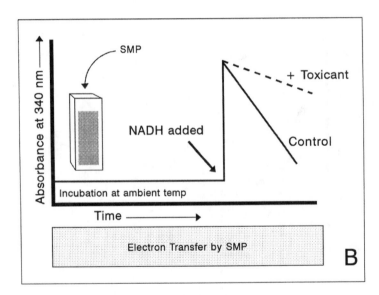

Figure 3.2 A) Electron transfer from NADH to oxygen in SMP. I, II, III, and IV are mitochondrial electron transfer enzyme complexes; CoQ is coenzyme Q; Cyt c is cytochrome c; TH is mitochondrial transhydrogenase. B) Necessary reagents and generalized time course of absorbance changes in the electron transfer (ETr) test.

respiratory chain and causes a backward flow of the electrons along the oxidation pathway; 2) the *electron transport (ETr)* bioassay (Figure 3.2),[12] which allows normal forward electron flow along a portion of the oxidation pathway; 3) the *facilitated electron withdrawal (FEW)* bioassay (Figure 3.3),[13] which uses special superoxide-dismutase-depleted SMP preparations and an electron transfer inhibitor to block the normal flow of electrons, shunting them to electron-scavenging toxicants. The RET and ETr tests give direct measures of acute toxicity to SMP, data which can be correlated with acute or chronic toxicities in whole organisms.[4,8,9] The FEW assay provides a measure of the formation of highly active oxygen free-radicals (pro-oxidant states) formed through redox cycling, species thought to be involved in cancer promotion.[14] Xenobiotics which fail to exert full toxicity unless activated by xenobiotic-activating enzymes present in the liver, lungs, kidneys, or

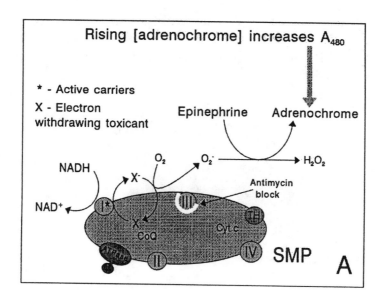

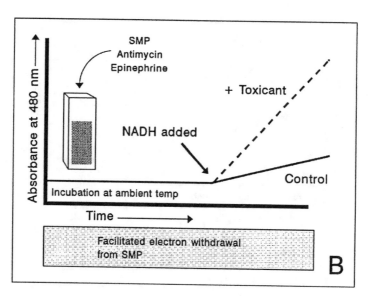

Figure 3.3 A) Electron withdrawal by toxicants from SMP in the presence of antimycin and transfer via oxygen to epinephrine. I, II, III, and IV are mitochondrial electron transfer enzyme complexes; CoQ is coenzyme Q; Cyt c is cytochrome c; TH is mitochondrial transhydrogenase. B) Necessary reagents and generalized time course of absorbance changes in the facilitated electron withdrawal (FEW) test.

other tissues are the targets of hybrid assays (RET-P450 and ETr-P450 protocols), in which RET and ETr tests are performed after preincubating the SMP with the toxicant and cytochrome P-450 enzymes.

All of these tests deploy separate sources of energy and electrons coupled through selected contiguous enzyme complexes which may be vulnerable at one or more points to toxic insult. These assays are all monitored spectrophotometrically: light absorption properties of the nicotinamide adenine dinucleotide redox couple (NAD$^+$ $\rightleftharpoons$NADH) are used to detect endpoints in the RET and ETr tests.[11,12] NADH, the reduced form of this respiratory coenzyme, exhibits an absorbance maximum at 340 nm, while NAD$^+$ is translucent at this wavelength. Absorbance can therefore be related to the rates of enzymatic conversion of measured amounts of NAD$^+$ or NADH added to

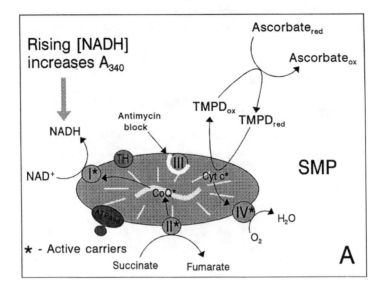

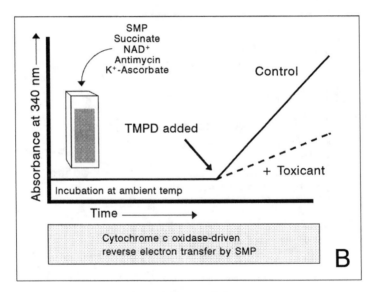

Figure 3.4 A) Ascorbate-TMPD-driven reverse electron transfer in submitochondrial particles. I, II, III, and IV are mitochondrial electron transfer enzyme complexes; CoQ is coenzyme Q; Cyt c is cytochrome c; TH is mitochondrial transhydrogenase; TMPD is tetramethylphenazine diamine; ~ designates energy coupling between complexes IV and I. B) Necessary reagents and generalized time course of absorbance changes in cytochrome oxidase (COx) test.

bioassay incubations and used to detect retardation of the reaction caused by toxicants. The conversion of the colorless catecholamine hormone epinephrine by activated oxygen species to the orange iminoquinone pigment adrenochrome is used to monitor the FEW assay by measuring absorbance at 480 nm.[13]

Additional bioassays based on other enzyme reactions have also been conceived but not as completely developed and tested; these include the *cytochrome oxidase (COx)* test (Figure 3.4), involving NAD^+ reduction to NADH by reverse electron transfer driven by ascorbate through intercession of tetramethylphenylenediamine (TMPD), and an *oxidative phosphorylation (OP)* test (Figure 3.5)[15] involving ADP phosphorylation driven by succinate oxidation. ATP is cycled back

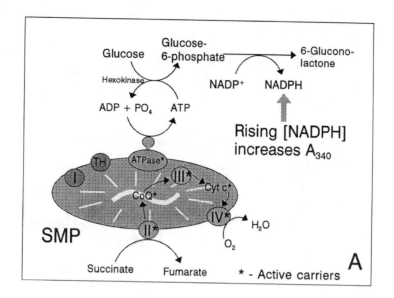

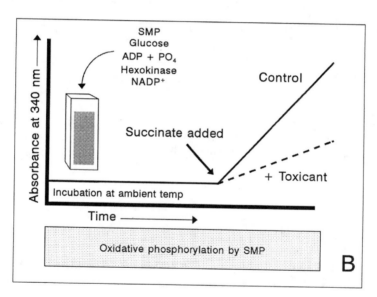

Figure 3.5 A) Succinate-driven oxidative phosphorylation by SMP. I, II, III, and IV are mitochondrial electron transfer enzyme complexes; CoQ is coenzyme Q; Cyt c is cytochrome c; TH is mitochondrial transhydrogenase; ~ designates energy coupling between electron transfer chain and ATPase enzymes. B) Necessary reagents and generalized time course of absorbance changes in the oxidative phosphorylation (OP) test.

to ADP by including glucose and hexokinase, and its production is monitored by using glucose 6-phosphate dehydrogenase to reduce NADP$^+$ to NADPH, which like NADH, absorbs at 340 nm.[16,17] ATP production can also be monitored luminometrically by substituting luciferin and luciferase (EC 1.13.12.7) for the aforementioned system.[18]

Figure 3.6 shows a generic scheme illustrating how some substances, e.g., polynuclear aromatic hydrocarbons (PAHs) or chloroalkane solvents, are transformed by cytochrome P-450 enzymes to metabolites which inflict damage on SMP detectable in one or more of the standard tests. This approach has been successfully applied with both the RET and ETr tests, although this work is ongoing and protocols have not been optimized. As an example, Figure 3.7 shows dose-response

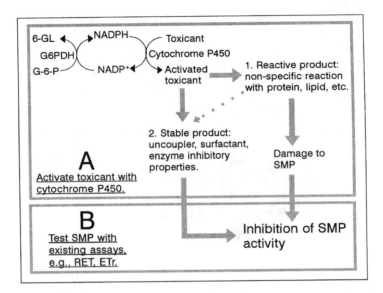

Figure 3.6   Generalized scheme for (A) exposing SMP to toxic metabolites produced by cytochrome P-450 enzymes, and (B) subsequent testing of SMP activity in electron transfer (ETr) or reverse electron transfer (RET) tests. G-6-P = glucose 6-phosphate; G6PDH = glucose 6-phosphate dehydrogenase.

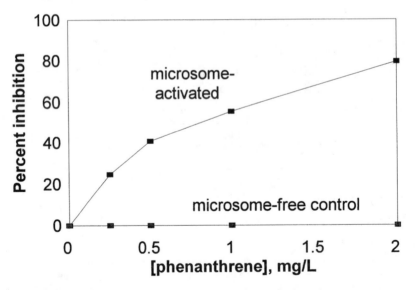

Figure 3.7   Inhibition of reverse electron transfer in SMP by metabolites of a polynuclear aromatic hydrocarbon (phenanthrene) generated by cytochrome P-450 enzymes in liver microsomes from male Fisher rats induced with β-naphthoflavone.

curves for the PAH phenanthrene comparing results in the conventional RET test with those in which phenanthrene and SMP were preincubated with liver microsomes for 20 minutes at 37°C.

In each of the illustrations described so far, the tests are conducted in a 2 to 3 mL incubation mixture in a quartz or disposable polystyrene cuvette. While simple and straightforward, a batch of tests with increasing concentrations of toxicant must be conducted in parallel to obtain the best dose-response curve from which an EC_{50} — the concentration of toxicant causing 50% inhibition of the respective reaction and the standard relative measure of toxicity — can be obtained. However, tests can now be conducted far more efficiently and dose-response curves obtained with a single set of incubations by using disposable 96-well polystyrene microtiter plates, which are read automatically in an automated plate reader. Procedures are described below.

V. TEST PROTOCOLS

Detailed practical procedures have been previously described for tests with SMP in cuvettes using the RET,[4,8,9,11] ETr,[8-10,12] COx,[4] and FEW assays.[13] While these procedures are still reliable, it has now been found unnecessary to include sucrose in the RET, ETr, and COx test media, and to keep cuvettes or microtiter plates at exactly 25°C, provided controls and test samples are processed simultaneously. Also, the amount of SMP protein prescribed is lowered, and longer intervals can be used between absorbance readings, a convenience when many tests are being run simultaneously in cuvettes by a single operator. Finally, it has been shown that cyanide can be substituted for antimycin in the FEW test.[19]

For the OP test, SMP (0.2 mg) are incubated at 25°C in 3 mL of 50 mM K$^+$-HEPES buffer (pH 7.5), 6 mM MgSO$_4$, 5 mM glucose, 1 mM K$^+$-phosphate, 1 mM NADP$^+$, 1 mM ADP, 5 mM K$^+$-succinate, and 1 unit each of myokinase-free hexokinase (EC 2.7.1.1) and glucose 6-phosphate (G-6-P) dehydrogenase (EC 1.1.1.49) with a vehicle control and various concentrations of test chemical or sample in the same medium, usually distilled deionized water (DDW) or DDW plus ethanol or dimethyl sulfoxide (DMSO). After readings of the baseline absorption at 340 nm for three 5-min intervals, the reaction is initiated with 50 µL 300 mM K$^+$-succinate, and the change in absorbance recorded for six further 5-min intervals.

VI. PROTOCOLS FOR SMP TESTS WITH TOXICANT ACTIVATION
BY CYTOCHROME P-450

RET-P450 — A stock microsome-SMP mixture is prepared by adding 500 µL SMP (3 mg/mL protein), 100 µL microsomes (approximately 40 mg/mL protein),[20] and 400 µL 0.1M K$^+$-phosphate buffer (pH 7.5) to a 1.5 mL microcentrifuge tube. A 100-µL aliquot of this mixture is added to a 1 cm pathlength cuvette containing 200 µL of an NADPH-generating system [70.6 mM K$^+$-phosphate buffer (pH 7.5), 9 mM MgSO$_4$, 1.8 mM NADP$^+$, 21.6 mM G-6-P, and 1.8 units G-6-P dehydrogenase],[21] 200 µL 37.5 mM K$^+$-phosphate buffer (pH 7.5), and 100 µL DDW including the toxicant or appropriate vehicle control. This 600-µL volume is incubated for 20 min at 37°C. The incubate is then diluted with 2.35 mL diluent [62.5 mM EPPS buffer (pH 8.0), 6.25 mM K$^+$-succinate, 1.25 mM NAD$^+$, 0.25 µg/mL antimycin A, and 30 mM MgSO$_4$] and the baseline absorbance at 340 nm read at room temperature for 15 min; the RET reaction is initiated with 50 µL 100 mM ATP and the change in absorbance at 340 nm monitored for 30 min longer. The higher pH and higher Mg^{++} content are required to inhibit SMP's endogenous transhydrogenase,[22] which interferes in the RET test by reducing NAD$^+$ to NADH by NADPH still present in the medium.[23]

ETr-P450 — In this test the microsome-SMP stock mixture contains 250 µL SMP (1 mg/mL protein), 100 µL microsomes (approximately 40 mg/mL protein),[20] and 650 µL 0.1 M K$^+$-phosphate buffer (pH 7.5). A 100-µL aliquot of stock microsome-SMP is added to a 1 cm pathlength cuvette containing 200 µL 0.15 M K$^+$-phosphate buffer (pH 7.5), 200 µL NADPH-generating system,[21] and 100 µL DDW or appropriate vehicle containing the toxicant or control. Following a 20-min incubation at 37°C the incubate is diluted with 2.35 mL 0.1 M K$^+$-phosphate buffer (pH 7.5) and, following baseline readings for 15 min at room temperature, the reaction is initiated with 50 µL 8.2 mM NADH and the absorbance read at 5-min intervals for 30 min. The high phosphate buffer concentrations help maintain SMP activity in this protocol. Lower SMP levels are preferable here because the ETr reaction rate is faster than the RET reaction, and higher sensitivities to low toxicant concentrations can be achieved in this way.

VII. RET/ETr MICROPLATE PROTOCOL

Protocols for the RET and ETr procedures have been conducted successfully in 96-well micro-titer plates using a spectrophotometric endpoint. For these assays, each microwell contains 50 mM K$^+$-HEPES (pH 7.5), 6 mM MgSO$_4$, 5 mM K$^+$-succinate, 0.2 µg/mL antimycin, and 0.05 mg/mL SMP protein for the RET test; and 50 mM K$^+$-HEPES buffer (pH 7.5), 6 mM MgSO$_4$, and 0.033 mg/mL SMP protein for the ETr test. The reduction of NAD$^+$ (RET) or the oxidation of NADH (ETr) is measured at 340 nm after the addition of ATP/NADH to activate the assay.

In the experimental setup, concentrated solutions are prepared and frozen at –20°C, viz. 43 mL 1 M K$^+$-HEPES buffer (pH 7.5), 13 mL 0.4 M MgSO$_4$, and 21.5 mL 0.2 M K$^+$-succinate for the RET test; and 43 mL 1 M K$^+$-HEPES buffer (pH 7.5) and 13 mL 0.4 M MgSO$_4$ for the ETr test. Reaction mixture solutions are added to a 100-mL beaker, adjusted to pH 7.5 with KOH, transferred to a 100-mL volumetric flask and diluted to volume; 15-mL aliquots are dispensed and frozen. On the day of use 129 µL 200 µg/mL antimycin and 0.85 g β-NAD$^+$ (as a powder) are added per thawed 15-mL aliquot for the RET test; no additions are needed for the ETr test. SMP are prepared by thawing SMP concentrate (24 mg/mL protein, obtained frozen in 110 µL aliquots in sealed vials from BioRenewal Technologies Inc., Madison, WI) and adding 100 µL to 1.5 mL RET reaction mixture (enough for 160 microplate wells in the RET test), or 2.3 mL 50 mM K$^+$-HEPES buffer (enough for 240 wells in the ETr test). Toxicant dilutions are included in the DDW or DDW/solvent addition used to bring each solution up to volume. Vehicle solvents (ethanol or dimethyl sulfoxide) are not to exceed 1% of the total volume, i.e., 3 µL. Thus each well contains: 35 µL reaction mixture, 245 µL DDW or toxicant solution, 10 µL SMP, and 10 µL ATP or NADH to activate the assay, at a total volume of 300 µL.

In practice, for each run, 35 µL of the desired assay's reaction mixture is added to each well in the microplate, together with 245 µL DDW containing appropriate concentrations (dilutions) of toxicant or control vehicle (DDW, or DDW + EtOH or DMSO), and 10 µL SMP. Four baseline readings are taken at 340 nm at 2-min intervals, with a 10-sec shake before each reading, on a Dynatech MRX microplate reader (Chantilly, VA) coupled to a computer interface utilizing Dynatech's Biolinx software in the kinetics mode. Just before the fifth reading at 10 minutes, a multi-channel pipettor is used to add 10 µL 50 mM ATP to each well for the RET test, or 10 µL 4.1 mM NADH per well for the ETr test. Further readings are taken every 2 min for a total of 20 readings, with a 10-second shake before each reading. Reaction rates in toxicant-containing wells are compared to those in control wells to determine percent inhibition.

VIII. CORRELATIONS OF SMP TOXICITY DATA ON TOXICITY DATA FROM OTHER TESTS

Table 3.1 shows a list of SMP median effective concentrations (EC$_{50}$ values) measured in our laboratory for 162 chemical species, together with corresponding data for *Vibrio fischeri* in the Microtox®* assay and median lethality concentrations (96-h LC$_{50}$ values) toward fathead minnows (or bluegill sunfish, if fathead minnow values were unavailable) and EC$_{50}$s for 34 compounds measured with frozen/thawed beef heart mitochondria. Comparisons were made among these values by logarithmically transforming the four data sets in Table 3.1 and applying linear regression. A long-standing goal of *in vitro* testing with SMP or Microtox is to demonstrate the capacity to predict toxicity of chemicals to whole aquatic organisms. Acute toxicity testing with fish, particularly the fathead minnow (*Pimephales promelas*), has a long history, and ample data are available. An ideal regression of log-transformed *in vitro* test EC$_{50}$s on fish LC$_{50}$s (both in µg/L) would exhibit a

---

* Microtox is a registered trademark of AZUR Environmental, Carlsbad, CA.

Table 3.1 Median effective concentrations (EC$_{50}$s, μg/L) for submitochondrial particles (SMP), *Vibrio fischeri* (Microtox), and beef-heart mitochondria (BHM), and median lethal concentrations (96-hr LC$_{50}$s, μg/L) for fish (fathead minnow or bluegill) of 164 chemicals, identified by name and Chemical Abstracts Service Registry Number (CAS#). Data are new or from previous publications for SMP,[4,8–10,13] fish,[4,8,9,24–32] *V. fischeri*,[4,8,9,33,34] and frozen/thawed BHM.[35]

Chemical	CAS#	Class	Subclass	SMP	Test	Fish	Microtox	BHM
Acetone	67-64-1	solvent	ketone	26,000,000	RET	8,120,000	17,100,000	10,000,000
Acetylsalicylic acid	50-78-2	pharmaceutical	analgesic	936,000	RET	NA	26,100	NA
Adriamycin	23214-92-8	pharmaceutical	antibiotic	6,300	FEW	NA	NA	NA
Aflatoxin B1	1162-65-8	mycotoxin		10,000	RET	NA	21,250	NA
Alachlor	15972-60-8	herbicide	acetanilide	55,000	ETr	5,000	NA	NA
Aldicarb	116-06-3	insecticide	carbamate	1,120,000	RET	861	NA	NA
p-Aminophenol	123-30-8	phenol	phenol	62,300	RET	24,000	773	NA
Amitryptyline HCl	549-18-8	pharmaceutical	antidepressant	13,100	RET	NA	24,400	NA
Amphetamine sulfate	51-63-8	pharmaceutical	stimulant	2,208,000	RET	NA	NA	NA
Amytal	57-43-2	pharmaceutical	sedative	117,000	ETr	85,400	1,011,000	NA
Antimycin A	642-15-9	mito inhibitor	antibiotic	6	ETr	1	NA	NA
Aroclor-1242	53469-21-9	PCBs	chloroaromatic	590	ETr	300	700	NA
Aroclor-1248	12672-29-6	PCBs	chloroaromatic	890	ETr	690	NA	NA
Arsenic (V)	7440-38-2	metal	metal	35,000	RET	27,000	38,000	NA
Arsenic (III)	1327-53-3	metal	metal	44,500	RET	NA	2,810,000	NA
Atrazine	1912-24-9	herbicide	triazine	132,000	RET	16,000	NA	NA
Atropine sulfate	5908-99-6	pharmaceutical	anticholinergic	371,000	ETr	NA	5,570,000	NA
BHA	25013-16-5	misc. organic	antioxidant	15,800	RET	NA	NA	NA
Barium (II)	10022-31-8	metal		>700,000	RET	NA	>52,000,000	NA
Benzo[a]pyrene	50-32-8	misc. organic	PAH	5,000	ETr	NA	NA	NA
p-Bromoanisole	104-92-7	misc. organic		132,000	ETr	NA	2,580	NA
Bromoxynil	1689-84-5	phenol	herbicide	1,000	RET	600	NA	NA
n-Butanol	71-36-3	n-alkanol		4,260,000	RET	1,900,000	3,715,000	NA
Cadmium	7440-43-9	metal		520	RET	630	41,400	158
Caffeine	58-08-2	pharmaceutical	stimulant	3,500,000	RET	151,000	62,800	NA
Calcium	7440-70-2	metal		870,000	RET	NA	NA	NA
Captan	133-06-2	fungicide	phthalimide	10,000	RET	134	NA	NA
Carbaryl	63-25-2	insecticide	carbamate	62,900	RET	5,290	2,565	NA
Carbon tetrachloride	56-23-5	solvent	chlorinated	114,000	RET	41,400	19,640	NA
Carboxin	5234-68-4	mito inhibitor	fungicide	480	RET	84	NA	NA
Chloramphenicol	56-75-7	pharmaceutical	antibiotic	165,000	RET	NA	1,700,000	NA
Chlordane	57-74-9	insecticide	organochlorine	1,260	ETr	115	600	NA

Table 3.1 (continued) Median effective concentrations (EC_{50}s, μg/L) for submitochondrial particles (SMP), *Vibrio fischeri* (Microtox), and beef-heart mitochondria (BHM), and median lethal concentrations (96-hr LC_{50}s, μg/L) for fish (fathead minnow or bluegill) of 164 chemicals, identified by name and Chemical Abstracts Service Registry Number (CAS#). Data are new or from previous publications for SMP,[4,8–10,13] fish,[4,8,9,33,34] *V. fischeri*,[4,8,9,33,34] and frozen/thawed BHM.[35]

Chemical	CAS#	Class	Subclass	SMP	Test	Fish	Microtox	BHM
Chlorobenzene	108-90-7	solvent	chloroaromatic	85,000	RET	38,000	11,300	3,160
Chloroform	67-66-3	solvent	chlorinated	450,000	RET	58,000	671,000	NA
Chloroquine	54-05-7	pharmaceutical	antimalarial	17,000	ETr	NA	537,000	NA
Chlorpyrifos	2921-88-2	insecticide	organophosphate	3,400	ETr	3	46,200	NA
Chromium (III)	7440-47-3	metal		33,000	RET	33,200	15,300	31,600
Copper (II)	7440-50-8	metal		300	ETr	530	9,300	93
m-Cresol	108-39-4	phenol		115,000	RET	55,900	7,830	69,200
o-Cresol	95-48-7	phenol		110,000	RET	18,200	26,500	50,100
p-Cresol	106-44-5	phenol		144,000	RET	23,800	1,720	69,180
Cyanide	151-50-8	mito inhibitor	inorganic anion	230	ETr	114	1,500	40
Cyhexatin	13121-70-5	acaricide	organometallic	34	RET	7	385	NA
Deethylatrazine		herbicide	triazine	7,000,000	RET	NA	NA	NA
DDE	72-55-9	insecticide	organochlorine	550	Etr	240	NA	794
DDT	50-29-3	insecticide	organochlorine	112	ETr	12	10,400	250
n-Decanol	112-30-1	n-alkanol		9,390	RET	2,400	1,480	NA
Diazepam	439-14-4	pharmaceutical	antianxiety	49,200	RET	NA	NA	NA
Diazinon	333-41-5	insecticide	organophosphate	39,000	ETr	7,460	9,850	NA
Dibutyl phthalate	84-74-2	plasticizer	phthalate	8,500	ETr	2,600	17,000	NA
o-Dichlorobenzene	95-50-1	solvent	chloroaromatic	21,000	RET	9,500	5,990	1,000
Dichloromethane	75-09-2	solvent	chlorinated	33,700	RET	330,000	2,878,000	NA
2,4-D	94-75-7	herbicide	phenoxy acid	562,000	RET	133,000	98,660	NA
Dieldrin	60-57-1	insecticide	organochlorine	150	ETr	7	NA	NA
Diethylhexyl phosphoric acid	298-07-7	plasticizer		60,000	ETr	NA	NA	NA
Diethylhexyl phthalate	117-81-7	plasticizer	phthalate	1,400	ETr	2,000	798,000	NA
Diethylstilbesterol	56-53-1	pharmaceutical	estrogen mimic	832	RET	NA	NA	NA
Digoxin	71-63-6	pharmaceutical	cardiotonic	>1,000,000	RET	NA	>10,000,000	NA
Phenytoin	57-41-0	pharmaceutical	anticonvulsant	20,200	ETr	NA	NA	NA
Dimethylformamide	68-12-2	solvent		14,500,000	ETr	13,000,000	20,100,000	25,000,000
Dimethyl sulfoxide	67-68-5	solvent		37,000,000	RET	37,000,000	98,400,000	31,600,000
Dinitro-o-cresol	534-52-1	defoliant	nitrophenol	440	RET	1,540	6,120	NA
2,4-Dinitrophenol	51-28-5	defoliant	nitrophenol	2,300	RET	6,580	10,600	NA
Dinoseb	88-85-7	defoliant	nitrophenol	60	RET	67	NA	126
Diquat	6385-62-2	herbicide	bipyridilium	1,600	FEW	14,000	NA	NA

Compound	CAS No.	Use	Class		Method			
Disulfoton	298-04-4	insecticide	organophosphate	9,500	ETr	4,300	NA	NA
n-Dodecanol	112-53-8	n-alkanol		3,900	RET	NA	198	NA
Endrin	72-20-8	insecticide	organochlorine	673	ETr	2	NA	NA
EPTC	759-94-4	herbicide	thiocarbamate	200,000	RET	27,000	49,800	NA
Ethyl alcohol	64-17-5	n-alkanol		36,650,000	RET	13,500,000	35,000,000	100,000,000
Ethylene glycol	107-21-1	misc. organic	antifreeze	218,000,000	RET	>100,000	621,000	NA
Fenaminosulf	140-56-7	mito inhibitor	fungicide	90	RET	60	NA	NA
Fentin hydroxide	76-87-9	oxidizer		10	RET	15	NA	NA
Formaldehyde	50-00-0	misc. organic		21,000	ETr	24,100	8,460	NA
Glyphosate	1071-83-6	herbicide		>33,000	RET	97,000	7,730	NA
Guthion	86-50-0	insecticide	organophosphate	18,000	ETr	1,900	NA	7,400
Halothane	151-67-7	solvent	anesthetic	134,000	RET	NA	NA	NA
Heptachlor	76-44-8	insecticide	organochlorine	1,800	ETr	23	NA	NA
n-Heptadecanol	1454-85-9	n-alkanol		>30,200	RET	NA	NA	NA
n-Heptanol	111-70-6	n-alkanol		231,000	RET	37,800	9,890	NA
Hexachlorobenzene	118-74-1	insecticide	organochlorine	>33,000	ETr	22,000	7,900	NA
Hexachlorophene	70-30-4	phenol	chlorophenol	4	RET	NA	7,980	NA
n-Hexadecanol	36653-82-4	n-alkanol		>17,600	RET	NA	NA	NA
n-Hexanol	111-27-3	n-alkanol		588,000	RET	117,000	27,500	NA
Isoniazide	54-85-3	pharmaceutical	antibiotic	18,757,000	RET	NA	10,900,000	NA
Isopropyl alcohol	67-63-0	sec-alkanol		17,400,000	RET	6,550,000	38,500,000	NA
Kelthane	115-32-2	insecticide	organochlorine	575	ETr	510	445	NA
Lead (II)	7439-92-1	metal		2,000	RET	1,390	11,000	80
Lindane	58-89-9	insecticide	organochlorine	26,000	ETr	7,000	5,670	6,310
Lithium (I)	10377-48-7	metal	pharmaceutical	649,000	ETr	NA	2,360,000	NA
Malathion	121-75-5	insecticide	organophosphate	31,900	RET	14,100	117,500	NA
Mancozeb	8018-01-7	fungicide	organometallic	5,400	ETr	1,900	80	NA
Mercury (II)	7439-97-6	metal		130	RET	170	59	126
Methyl alcohol	67-56-1	n-alkanol		41,700,000	ETr	29,400,000	320,000,000	NA
Methoxychlor	72-43-5	insecticide	organochlorine	724	ETr	8	NA	NA
Methyl mercury	593-74-8	misc. organic	organometallic	422	ETr	NA	NA	NA
Methyl isothiocyanate	556-61-6	fumigant		50,000	RET	130	NA	NA
Metolachlor	51218-45-2	herbicide	acetanilide	80,000	RET	10,000	NA	NA
Metribuzin	21087-64-9	herbicide	triazine	165,000	RET	80,000	NA	NA
Mirex	2385-85-5	insecticide	organochlorine	17,000	ETr	>100,000	NA	NA
n-Myristyl alcohol	112-72-1	n-alkanol		19,600	RET	NA	NA	NA
Nabam	142-59-6	fungicide	organometallic	8,000	RET	5,800	102,000	NA

Table 3.1 (continued) Median effective concentrations ($EC_{50}S$, μg/L) for submitochondrial particles (SMP), *Vibrio fischeri* (Microtox), and beef-heart mitochondria (BHM), and median lethal concentrations (96-hr $LC_{50}S$, μg/L) for fish (fathead minnow or bluegill) of 164 chemicals, identified by name and Chemical Abstracts Service Registry Number (CAS#). Data are new or from previous publications for SMP,[4,8–10,13] fish,[4,8,9,24-32] *V. fischeri*,[4,8,9,33,34] and frozen/thawed BHM.[35]

Chemical	CAS#	Class	Subclass	SMP	Test	Fish	Microtox	BHM
Nickel	7440-02-0	metal		2,200	RET	27,000	251,000	9,100
Nicotine	54-11-5	insecticide	alkaloid	1,490,000	ETr	13,900	37,600	NA
4-Nitroquinoline N-oxide	56-57-5	mutagen		1,600	FEW	NA	615	NA
n-Nonanol	143-08-8	n-alkanol		28,820	RET	5,700	NA	NA
n-Octanol	111-87-5	n-alkanol		67,000	RET	12,610	7,320	NA
Orphenadrine HCl	341-69-5	pharmaceutical	muscle relaxant	58,900	RET	NA	370,000	NA
Paracetamol	103-90-2	pharmaceutical	analgesic	752,000	RET	NA	999,000	NA
Paraquat	1910-42-5	herbicide	bipyridilium	2,800	FEW	45,500	1,320,000	NA
Pentachlorophenol	87-86-5	fungicide	phenol	81	RET	240	519	400
n-Pentadecanol	629-76-5	n-alkanol		>20,050	RET	NA	NA	NA
n-Pentanol	71-41-0	n-alkanol		1,410,000	RET	370,000	394,000	NA
Perfluorodecanoic acid	335-76-2	surfactant	ionic	1,400	RET	NA	NA	NA
Perfluorooctanoic acid	335-67-1	surfactant	ionic	10,500	RET	NA	NA	NA
Permethrin	52645-53-1	insecticide	pyrethroid	220	ETr	16	566	NA
Phenobarbital	56-06-6	pharmaceutical	sedative	672,000	ETr	484,000	3,280,000	NA
Phenol	108-95-2	phenol	phenol	133,000	ETr	34,000	35,800	100,000
o-Phenylphenol	90-43-7	fungicide	phenol	7,200	ETr	6,000	2,050	NA
Phenytoin	57-41-0	pharmaceutical	anticonvulsant	19,930	ETr	NA	NA	NA
Potassium chloride	7447-40-7	salt		7,110,000	RET	NA	36,500,000	5,000,000
n-Propanol	71-23-8	n-alkanol		11,800,000	RET	4,630,000	9,310,000	NA
Propoxyphene HCl	1639-60-7	pharmaceutical	analgesic	228,000	RET	NA	62,800	NA
Propranolol	525-66-6	pharmaceutical	adrenergic block	33,100	ETr	NA	184,000	NA
Quinidine sulfate	6591-63-5	pharmaceutical	antimalarial	91,000	RET	NA	837,000	NA
Rotenone	83-79-4	mito inhibitor	insecticide	3	ETr	7	NA	4
Selenium	7782-49-2	metal		229,000	ETr	40,000	33,400	NA
Silver (I)	7440-22-4	metal		140	RET	11	20,600	148
Simazine	122-34-9	herbicide	triazine	80,000	RET	>100,000	237,000	NA
Sodium azide	26628-22-8	mito inhibitor	inorganic anion	2,000	RET	5,460	NA	NA
Sodium chloride	7647-14-5	salt		4,460,000	RET	NA	32,900,000	NA
Sodium dodecyl sulfate	151-21-3	surfactant	anionic	1,900	RET	4,700	1,200	NA
Sodium fluoride	7681-49-4	insecticide		284,000	RET	NA	26,500,000	NA
Sodium oxalate	62-76-0	misc. organic		1,840,000	RET	NA	>10^8	NA
n-Stearyl alcohol	112-92-5	n-alkanol		>15,040	RET	NA	NA	NA

Name	CAS	Class	Type					
Sulfonazo III	68504-35-8	mutagen	diazo dye	320	FEW	NA	NA	NA
Terbufos	13071-79-9	insecticide	organophosphate	7,000	ETr	13	NA	NA
2,3,4,6-Tetrachlorophenol	58-90-2	chlorophenol	phenol	447	RET	580	1,270	1,580
2,3,5,6-Tetrachlorophenol	935-95-5	chlorophenol	phenol	500	RET	400	2,210	NA
Thallium sulfate	7446-18-6	metal	metal	585,000	RET	NA	3,600,000	NA
Thenoyltrifluoroacetone	326-91-0	mito inhibitor		17,400	RET	NA	3,770	NA
Theophylline	58-55-9	pharmaceutical	muscle relaxant	1,550,000	RET	NA	2,490,000	NA
Thioridazine HCl	130-61-0	pharmaceutical	antipsychotic	5,610	RET	NA	692	NA
1,2,3-Trichlorobenzene	87-61-6	chlorobenzene	chloroaromatic	5,300	RET	3,380	3,150	2,960
1,2,4-Trichlorobenzene	120-82-1	chlorobenzene	chloroaromatic	1,820	RET	1,770	3,970	19,900
1,3,5-Trichlorobenzene	108-70-3	chlorobenzene	chloroaromatic	50,000	ETr	99,700	14,100	NA
1,1,1-Trichloroethane	71-55-6	solvent	chloralkane	583,000	RET	42,300	57,000	NA
1,1,2-Trichloroethylene	79-01-6	solvent	chloroalkene	190,000	RET	44,100	117,000	25,100
2,3,5-Trichlorophenol	933-78-8	phenol	chlorophenol	362	RET	800	1,110	NA
2,3,6-Trichlorophenol	933-75-5	phenol	chlorophenol	8,800	RET	5,100	12,700	6,900
2,4,5-Trichlorophenol	95-95-4	phenol	chlorophenol	600	RET	610	1,270	6,310
2,4,6-Trichlorophenol	88-06-2	phenol	chlorophenol	10,000	RET	9,160	7,680	10,000
2,4,5-T	93-76-5	herbicide	phenoxy acid	38,000	RET	10,400	52,200	NA
n-Tridecanol	112-70-9	n-alkanol		4,240	RET	NA	NA	NA
Trifluralin	1582-09-8	herbicide	dinitro aniline	6,900	ETr	130	NA	NA
Triton-X100	9002-93-1	surfactant	nonionic	7,700	RET	9,600	210	NA
n-Undecanol	112-42-5	n-alkanol		5,460	RET	NA	NA	NA
Verapamil HCl	152-11-4	pharmaceutical	vasodilator	25,800	ETr	NA	403,000	NA
Warfarin	81-81-2	rodenticide	anticoagulant	348,000	RET	NA	47,800	NA
Xylenes	1330-20-7	solvent	aromatic	19,600	RET	13,400	16,100	NA
Zinc (II)	7440-66-6	metal		1,700	RET	2,990	33,000	80
Zineb	12122-67-7	fungicide	organometallic	3,200	RET	970	2,090	NA

Table 3.2   Linear regression statistics from comparisons made among logarithmically transformed sets of the bioassay data presented in Table 3.1. SMP = submitochondrial particle tests, Mtx = Microtox *V. fischeri* tests, Fish = fish (fathead minnow or bluegill) tests, BHM = beef heart mitochondria tests.

Independent variable	Fish	Fish	Fish	Fish	Fish	Fish	Fish	SMP	Fish	Fish	SMP
Dependent variable	SMP	SMP	SMP	Mtx	SMP	Mtx	Mtx	Mtx	BHM	SMP	BHM
Data excluded	none	see[a]	see[b]	none	see[c]	see[d]	see[e]	none	none	see[f]	none
No. of chemicals	104	94	85	77	77	67	67	104	34	34	34
R^2	0.763	0.893	0.859	0.603	0.837	0.835	0.680	0.547	0.864	0.939	0.883
Intercept	1.088	0.388	0.407	1.402	0.637	0.181	0.854	1.258	0.150	0.162	0.073
Intercept error	0.759	0.524	0.609	0.915	0.617	0.595	0.850	1.022	0.686	0.461	0.637
Slope	0.836	0.976	0.972	0.747	0.926	1.003	0.865	0.750	0.972	1.014	0.939
Slope error	0.046	0.035	0.043	0.070	0.047	0.055	0.074	0.068	0.068	0.046	0.061

[a] 10 SMP outliers, i.e., log (value) diverges from log (fish value) by more than ±log (25): Aldicarb, chlorpyrifos, terbufos, methyl isothiocyanate, endrin, nicotine, methoxychlor, heptachlor, captan, trifluralin.

[b] 19 neurotoxicants: Aldicarb, carbaryl, chlordane, chlorpyrifos, diazinon, dieldrin, disulfoton, DDE, DDT, endrin, guthion, heptachlor, kelthane, lindane, malathion, methoxychlor, nicotine, permethrin, terbufos.

[c] 27 toxicants for which Microtox data are unavailable.

[d] 10 Microtox outliers, see [a]: Chlorpyrifos, silver, DDT, diethylhexyl phthalate, cadmium, cyhexatin, triton X-100, permethrin, *p*-aminophenol, paraquat.

[e] 10 neurotoxicants: Chlorpyrifos, carbaryl, chlordane, diazinon, DDT, kelthane, lindane, malathion, nicotine, permethrin

[f] 70 toxicants for which bovine heart mitochondria data are unavailable.

coefficient of variance (r^2) of 1.0, a slope of 1.0, and an intercept of zero. Regression of SMP on available fish data (for 104 chemicals) shows a reasonable correlation ($r^2 = 0.76$) and a slope of 0.84 (Table 3.2). An arbitrary value (log 25) was chosen for the minimum difference between any given pair of the log-transformed values to designate outliers. For the SMP-fish regression, 10 chemicals (Table 3.2, footnote[a]) exceed this criterion; for all of these, SMP are less sensitive than fish. An SMP on fish regression excluding these chemicals yields an r^2 of 0.89 and a slope of 0.98. Seven of the excluded chemicals are neurotoxicants, a class for which SMP are not expected to perform well. SMP tests indicated toxic responses to most neurotoxicants tested, but not of the same magnitude as fish. Excluding all 18 neurotoxicants from the data set yields r^2 of 0.86 and a slope of 0.97. Applying the same manipulations to the Microtox data reveals a poorer correlation with the fish test (Table 3.2) and 10 outliers from a more varied set of chemicals (Table 3.2, footnote[d]). Excluding neurotoxicants does not markedly improve this correlation. Applying the SMP test to the 77 chemicals for which data are available for all three tests also shows a better correlation between SMP and fish than for Microtox and fish (Table 3.2).

Polarographic analysis using the Clark oxygen electrode is a standard procedure for measuring the effects of toxicants on respiration rates in freshly isolated mitochondria.[36-39] A group of Italian researchers applied this technique to frozen beef-heart mitochondria, to permit selective determination of toxicants or toxicant mixtures.[35] This study in essence used a type of SMP preparation, because freezing and thawing rupture mitochondria, and for the 37 materials examined a good correlation of LC_{50} values with SMP was found ($r^2 = 0.87$). However, no correlation to fish toxicity was reported. A correlation performed by us using data from this report (see Table 3.2, column fish *v.* BHM) was weaker than one obtained using SMP EC_{50} values for the same substances from Table 3.1 (see Table 3.2, column fish *v.* SMP), suggesting that SMP prepared by sonication and centrifugal fractionation are preferable to unpurified fragments obtained by freeze/thaw cycling. A clever adaptation of the basic protocol in the Italian work involved the use of bovine serum albumin to sorb organic toxins and EDTA to complex divalent metals and thus eliminate toxicity from these sources in separate tests.[35]

IX. APPLICATIONS OF SMP TESTS

Bioassays using SMP have been applied to several environmental problems involving water pollution by toxic substances or mixtures of known or unknown composition, sometimes in comparison with other bioassays. Results of several studies have not been published because of the confidential nature of the investigations in connection with regulatory actions. Applications have been restricted because, until recently, SMP have not been commercially available.

A few of the first published applications investigated the toxicity of industrial effluents,[40] measurements of municipal sewage treatment plant samples to help determine the best technology to reduce the toxicity of the plant effluent,[40,41] and investigations of water quality in the Illinois,[40,42] Minnesota,[43] and Menominee[44] Rivers and some of their tributaries. The method was later applied for selective detection of phenols in water,[44] for monitoring heavy metals in water,[45] to assess the potency and long-term toxicity of leachates from rubber tires in water,[46] as a model to compare the toxicity of tributyltin toward fish and algae,[47] as biosensors for chlorophenols,[48,49] for measuring the toxicity of surfactants and their breakdown products,[50] for examining uncharacterized environmental samples including storm water runoff, landfill leachate, soil and sediment extracts,[51] as well as river water and sediment samples collected in the Temuco and Rapel River Basin, Chile.[52] SMP tests have also been used to demonstrate the efficacy of oxidations with hydrogen peroxide catalyzed by ultraviolet light for detoxifying mixed pesticide wastes.[53]

The ease, rapidity, and low cost of the SMP tests are extolled in all of these applications. These features allow SMP tests to be conducted rapidly on many individuals rather than a few composite samples, as are usually required for whole-organism toxicity tests. Consequently a better real-time picture of changes in toxicity can be obtained in dynamic systems. Their application provided some surprising indicators of practical importance that would have not been readily recognized by other methods. For example, the tests helped select the best tertiary treatment process for a municipal wastewater treatment plant,[41] indicated that nonionic surfactants are a greater environmental hazard than linear alkylbenzene sulfonates,[50] and that antifouling paints were more likely to be toxic to fish than to algae.[47] However, SMP tests were found to have poor sensitivity to ammonia, an important aquatic toxicant in many settings, particularly in monitoring sewage treatment plant outfalls.[40,41]

One limitation in the ETr test, activity stimulation, was recognized through some of these studies.[40] Investigation into this phenomenon revealed that this was not a manifestation of toxicity but appeared to be caused by ion composition. Activity in the ETr test increases with ionic strength, but this effect is also ion-specific. In fresh surface- and ground waters, calcium is probably the principal cause. Moderate levels of salts also appeared to alleviate inhibition of some toxicants, when compared to a control with no added salt.[40] These effects call for caution when testing environmental samples, but may be compensated for by employing controls with similar ion composition, so that stimulation in samples is matched by stimulation in controls.

X. DISCUSSION AND CONCLUSIONS

SMP tests are clearly becoming a popular and efficient method for rapid determination of toxicity in water samples or aqueous extracts and are gaining wider acceptance for practical applications in environmental monitoring. However, studies with a single group of compounds, chlorophenols, highlight the need for use of a uniform biological preparation and more closely standardized test protocols.

Differences in respiration rates between mitochondria from rat embryos and maternal rat livers[38] indicate that toxicity values may vary significantly depending upon the source if whole mitochondria are used. Differences are also observed in toxic responses toward identical chlorophenols depending upon whether whole mitochondria or SMP are used,[9,35,38,39,48] and what protocol is followed in SMP tests.[4,8,9,47,48,50] Apparently, to economize on SMP consumption, RET measurements were made by Italian researchers in one and the same cuvette by increasing the toxicant concentration stepwise in sequence,[47-50] whereas parallel measurements are made simultaneously in separate cuvettes in our protocols.[4,8-12] Sequential toxicant additions led to much lower LC_{50} values (compare BHM data with corresponding SMP values in Table 3.1). The best economy and most uniform results can now probably be achieved most conveniently by using the automated microtiter-plate procedure with commercially produced SMP, which have been fractionated and purified by differential centrifugation. For practical applications, rapid throughput of a series of samples is achievable by a single operator using this modern equipment. However, because an electron flux is generated by SMP, it is conceivable that, in time, a system for continuous monitoring of the toxicity of waste streams or natural bodies of water using SMP integrated with an electrode may be developed.

REFERENCES

1. Tzagoloff, A., *Mitochondria*, Plenum Press, New York, 1982.
2. Tyler, D. D., *The Mitochondrion in Health and Disease*, Verlag Chemie, New York, 1992.
3. Blondin, G. A., Knobeloch, L., and Harkin, J. M., Bioassay of toxic substances in water, *Wis. Acad. Rev.* 32, 31, 1985.
4. Blondin, G. A., Knobeloch, L. M., Read, H. W., and Harkin, J. M., Mammalian mitochondria as *in vitro* monitors of water quality, *Bull. Environ. Contam. Toxicol.*, 38, 467, 1987.
5. Green, D. E., Wharton, D. C., Tzagoloff, A,.., Rieske, J. F., and Brierley, G. P., in *Oxidases and Related Redox Systems*, King, T. E., Mason, H. S., and Morrison, M., Eds., John Wiley & Sons, New York, Vol. II, 1965, 1032.
6. Hansen, M. and Smith, A. L., Studies on the mechanism of oxidative phosphorylation. VII. Preparation of a submitochondrial particle (ETP_H) which is capable of fully coupled oxidative phosphorylation, *Arch. Biochem. Biophys.*, 81, 214, 1964.
7. Blondin, G. A., Vail, W. J., and Green, D. E., The mechanism of mitochondrial swelling. II. Pseudoenergized swelling in the presence of alkali metal salts, *Arch. Biochem. Biophys.*, 129, 158, 1969.
8. Knobeloch, L. M., Blondin, G. A., Read, H. W., and Harkin, J. M., Assessment of chemical toxicity using mammalian mitochondrial electron transport particles, *Arch. Environ. Contam. Toxicol.* 19, 828, 1990.
9. Blondin, G. A., Knobeloch, L. M., Read, H. W., and Harkin, J. M., An *in vitro* submitochondrial bioassay for predicting acute toxicity in fish, in *Aquatic Toxicology and Environmental Fate: Eleventh Volume, ASTM, STP 1007*, Suter, G. W. and Lewis, M. A., Eds., American Society for Testing and Materials, Philadelphia, 1989, 551.
10. Knobeloch, L. M., Blondin, G. A., and Harkin, J. M., Use of submitochondrial particles for prediction of chemical toxicity in man, *Bull. Environ. Contam. Toxicol.*, 44, 661, 1990.
11. Knobeloch, L., Blondin, G., and Harkin, J., A rapid bioassay for toxicity assessment of chemicals: Reverse electron transport assay, *Environ. Toxicol. Water Qual.*, 9, 231, 1994.
12. Knobeloch, L., Blondin, G., and Harkin, J., A rapid bioassay for toxicity assessment of chemicals: Forward electron transport assay, *Environ. Toxicol. Water Qual.*, 9, 79, 1994.
13. Knobeloch, L. M., Blondin, G. A., Lyford, S. B., and Harkin, J. M., A rapid bioassay for chemicals that induce pro-oxidant states, *J. Appl. Toxicol.*, 10, 1, 1990.
14. Cerutti, P. A., Prooxidant states and tumor promotion, *Science*, 227, 375, 1985.
15. Read, H.W., Sub-mitochondrial particle oxidative phosphorylation assay for water quality testing, Wisconsin Alumni Research Foundation, Madison WI, patent application P94164US, 1997.
16. Pullman, M.A. and Racker, E., Spectrophotometric studies of oxidative phosphorylation, *Science*, 123, 1105, 1956.
17. Beechey, R.B., Roberton, A.M., Holloway, C.T., and Knight, I.G., The properties of dicyclohexylcarbodiimide as an inhibitor of oxidative phosphorylation, *Biochem.* 6, 3867, 1967.

18. LeMasters, J.J. and Hackenbrock, C.R., Continuous measurement and rapid kinetics of ATP synthesis in rat liver mitochondria, mitoplasts and inner membrane vesicles determined by firefly-luciferase luminescence, *Eur. J. Biochem.*, 67, 1, 1976.

19. Klöhn, P.C, Massalha, H., and Neumann, H-G., A metabolite of carcinogenic 2-aminoflurene, 2-nitrosoflurene, induces redox cycling in mitochondria, *Biochem. Biophys. Acta* 1229, 363, 1995.

20. Van Der Hoeven, T. A. and Coon, M. J., Preparation and properties of partially purified cytochrome P-450 and reduced nicotinamide adenine dinucleotide phosphate-cytochrome P-450 reductase from rabbit liver microsomes, *J. Biol. Chem.*, 249, 6302, 1974.

21. Otto, S., Marcus, C., Pidgeon, C., and Jefcoate, C., A novel adrenocorticotropin-inducible cytochrome P450 from rat adrenal microsomes catalyzes polycyclic aromatic hydrocarbon metabolism, *Endocrin.*, 129,970, 1991.

22. Anderson, W. M. and Fisher, R. R., Purification and partial characterization of bovine heart mitochondrial pyridine dinucleotide transhydrogenase, *Arch. Biochem. Biophys.*, 187, 180, 1978.

23. Lee, C.-P., and Ernster, L., The energy-linked nicotinamide nucleotide transhydrogenase reaction: Its characteristics and its use as a tool for the study of oxidative phosphorylation, in *Regulation of Metabolic Processes in Mitochondria, Biochim. Biophys. Acta* Library, Vol. 7, Elsevier, Amsterdam, 1966, p. 218.

24. Verscheuren, K., *Handbook of Environmental Data on Organic Chemicals*, Van Nostrand Reinhold, 2nd Ed., New York, 1983.

25. Brooke, L. T., Call, D. J. Geiger, D. L., and Nothcott, C. E., Acute toxicities of organic chemicals to fathead minnows (*Pimephales promelas*), Center for Lake Superior Environmental Studies, University of Wisconsin-Superior, 1984, 414 pp.

26. Geiger, D. L., Nothcott, C. E., Call, D. J., and Brooke, L. T., Acute toxicities of organic chemicals to fathead minnows (*Pimephales promelas*), Center for Lake Superior Environmental Studies, University of Wisconsin-Superior, Vol. II, 1985, 326 pp.

27. Geiger, D. L., Poirier, S. H., Brooke, L. T., and Call, D. J., Acute toxicities of organic chemicals to fathead minnows (*Pimephales promelas*), Center for Lake Superior Environmental Studies, University of Wisconsin-Superior, Vol. III, 1986, 328 pp.

28. Geiger, D. L., Call, D. J., and Brooke, L. T., Acute toxicities of organic chemicals to fathead minnows (*Pimephales promelas*), Center for Lake Superior Environmental Studies, University of Wisconsin-Superior, 1988, Vol. IV, 355 pp.

29. Geiger, D. L., Brooke, L. T., and Call, D. J., Acute toxicities of organic chemicals to fathead minnows (*Pimephales promelas*), Center for Lake Superior Environmental Studies, University of Wisconsin-Superior, 1990, Vol. V, 332 pp.

30. WSSA, *Herbicide Handbook*, Weed Science Society of America, Champaign, IL, 7th Edit., 1994.

31. Mayer, F. L. and Ellersieck, M. R., Manual of acute toxicity, interpretation and data base for 410 chemicals and 66 species of freshwater animals, *Resource Publication 160*, U.S. Department of the Interior, Fish and Wildlife Service, Washington, DC, 1986, 506 pp.

32. Johnson, W. W. and Finley, M. T., Handbook of acute toxicity of chemicals to fish and aquatic invertebrates, *Resource Publication 137*, U.S. Department of the Interior, Fish and Wildlife Service, Washington, DC, 1980, 98 pp.

33. Kaiser, K. L. E. and Palabrica, V. S., *Photobacterium phosphoreum* toxicity data index, *Water Poll. Res. J. Canada,* 26, 361, 1991.

34. Calleja, M.C., Persoone, G., and Geladi, P., Human acute toxicity prediction of the first 50 MEIC chemicals by a battery of ecotoxicological tests and physicochemical properties. Fd. Chem. Toxic., 32, 173, 1994.

35. Bragadin, M. and Dell'Antone, P., A new *in vitro* test based on the response to toxic substances in solution of mitochondria from beef heart. *Arch. Environ. Contam. Toxicol.*, 27, 410, 1994.

36. Asimakis, G. K. and Aprille, J. R., In vitro alteration of the size of the liver mitochondrial pool: correlation with respiratory functions, *Arch. Biochem. Biophys.*, 203, 307, 1980.

37. Bhat, H. K., Asimakis, G. K., and Ansari, G. A. S., Uncoupling of oxidative phosphorylation in rat liver mitochondria by chloroethanols, *Toxicol. Letts.*, 59, 203, 1991.

38. Narasimhan, T. K., Mayura, K., Clement, B. A., Safe, S. H., and Phillips, T. D., Effects of chlorinated phenols on rat embryonic and hepatic mitochondrial phosphorylation, *Environ. Toxicol. Chem.*, 11, 805 (1992).

39. Shannon, R. D., Boardman, G. D., Dietrich, A. M., and Bevan, D. R., Mitochondrial response to chlorophenols as a short-term toxicity assay, *Environ. Toxicol. Chem.* 10, 57, 1991.
40. Read, H. W., Testing of chemicals and environmental water samples for toxicity with mitochondria-based assays, Ph. D. thesis, University of Wisconsin-Madison, 1988.
41. DeBauche, J. W., Kennedy, J., and Guthrie, M., The examination of advanced wastewater treatment technologies for toxicity elimination in the Green Bay Metropolitan Sewerage District, *GBMSD Report*, Green Bay, WI, USA, 1988.
42. Arthur, J.W., Katko, A., Walbridge, C.T., and Read, H.W., Ambient toxicity assessments in the upper Illinois River Basin, *Interim Report*, US-Environmental Protection Agency, Environmental Research Laboratory-Duluth, Duluth, MN, USA, 1989.
43. Arthur, J.W., Thompson, J.A., Walbridge, C.T., and Read, H.W., Ambient toxicity assessments in the Minnesota River Basin, In *Minnesota River Assessment Project Report Vol. III: Biological and Toxicological Assessment*, Minnesota Pollution Control Agency, N. St. Paul, MN, USA, 1994.
44. Bridgetown, M., Argese, E., and Orsega, E. F., A simple *in vitro* method for selective detection of the presence of phenols in water using the mitochondrial membrane from rat liver, *Environ. Technol.* 12, 777, 1991.
45. Bridgetown, M., Argese, E., Nicolli, A., and Bernardi, P., A simple *in vitro* test to monitor trace-metal toxicity in aqueous samples, *Environ. Technol.* 13, 779, 1992.
46. Day, K. E., Holtze, K. E., Metcalfe-Smith, J. L., Bishop, C. T., and Dutka, B. J., Toxicity of leachate from automobile tires to aquatic biota, *Chemosphere,* 27, 665, 1993.
47. Miani, P., Scotto, S., Perin, G., and Argese, E., Sensitivity of *Selenastrum capricornutum, Daphnia magna* and submitochondrial particles to tributyltin, *Environ. Technol.,* 14, 175, 1993.
48. Argese, E., Bettiol, C., Ghelli, A., Todeschini, R., and Miana, P., Submitochondrial particles as toxicity biosensors of chlorophenols, *Environ. Toxicol. Chem.* 14, 363, 1995.
49. Todeschini, R., Bettiol, C., Giurin, G., Gramatica, P., Miana, P., and Argese, E., Modeling and prediction by using WHIM descriptors in QSAR studies — submitochondrial particles (SMP) as toxicity biosensors of chlorophenols, *Chemosphere,* 33, 71, 1996.
50. Argese, E., Marcomini, A., Miana, P., Bettiol, C., and Perin, G., Submitochondrial particle response to linear alkylbenzene sulfonates, nonylphenol polyethoxylates and their biodegradation derivatives, *Environ. Toxicol. Chem.,* 13, 737, 1994.
51. Bettermann, A. D., Lazorchak, J. M., and Dorofi, J. C., Profile of toxic responses to sediments using whole-animal and *in-vitro* submitochondrial particle (SMP) assays, *Environ. Toxicol Chem.* 15, 319, 1996.
52. Dutka, B.J., McInnis, R., Jurkovic, A., Liu, D., and Castillo, G., Water and sediment ecotoxicity studies in Temuco and Rapel River Basin, Chile, *Environ Toxicol. Water. Qual.*, 11, 237, 1996
53. Chen, C.-P., Read, H. W., and Harkin, J. M., Detoxification and decomposition of mixed pesticide wastes by base-catalyzed hydrolysis and hydrogen-peroxide-assisted photo-oxidation, in preparation.

CHAPTER **4**

Bioassays to Measure MFO Inducers in Effluents

Peter V. Hodson, Jim Sherry, and Joanne Parrott

CONTENTS

I. INTRODUCTION

The mixed function oxygenase (MFO) enzymes represent a "super family" of more than 200 genes.[1] They are involved in the metabolism (oxidative, peroxidative, and reductive) of endogenous and "foreign" compounds, including hormones, fatty acids, drugs, plant products, and xenobiotics such as polynuclear and halogenated aromatic hydrocarbons (PAH and HAH). In fish, the most studied of the MFO enzymes are membrane-bound cytochrome proteins that metabolize aromatic xenobiotics to facilitate their excretion. These enzymes are inducible, i.e., their catalytic activity increases with exposure to polyaromatic compounds such as PAHs and polychlorinated dibenzo(p)dioxins (PCDDs). Induction *in vivo* or *in vitro* can be used to indicate the presence and amount of those compounds in environmental samples, and the potential for their toxic effects.

This chapter will review the biochemistry and toxicology of the MFO enzymes, and the use of MFO bioassays to indicate the presence and potency of individual inducing compounds or mixtures of compounds in liquid effluents and natural matrices. Table 4.1 provides a glossary of acronyms.

Table 4.1   A Glossary of Acronyms Used in This Chapter

Ah	arylhydrocarbon protein receptor
AHH	arylhydrocarbon hydroxylase
AOX	adsorbable organic halide
Arnt	Ah-receptor nuclear translocator protein
BaP	Benzo[a]pyrene
βNF	β-naphthoflavone
BKME	bleached kraft mill effluent
CB	chlorobiphenyl
CYP	cytochrome protein
DRE	dioxin-responsive element
EC_{50}	median effective concentration
ED_{50}	median effective dose
ELISA	enzyme-linked immunosorbent assay
EROD	ethoxyresorufin-o-deethylase
GC-MS	gas chromatography-mass spectrometry
HAH	halogenated aromatic hydrocarbon
H411E	rat hepatoma cell line
HPLC	high pressure liquid chromatography
I-TEF	international toxic equivalent factor
K_{ow}	octanol-water partition coefficient
MFO	mixed function oxygenase
PAH	polynuclear aromatic hydrocarbon
PCB	polychlorobiphenyl
PCDD	polychlorinated dibenzo(p)dioxin
cPCR	competitive polymerase chain reaction
PLHC-1	*Poeciliopsis lucida* cell line
PPO	diphenyloxazole
RMT	receptor-mediated toxicity
RT	reverse transcriptase
RTG-2	rainbow trout gonad cell line
RTH-149	rainbow trout hepatoma cell line
SAR	structure activity relationship
SPMD	semipermeable membrane device
TEF	toxic equivalent factor
TEQ	toxic equivalent quantity
TCBP	tetrachlorobiphenyl
2,3,7,8-TCDD	2,3,7,8-tetrachlorodibenzo(p)dioxin

A. Biochemistry and Toxicology of MFO Induction

MFO enzymes are found in virtually all taxa, although their types and relative abundance vary considerably among species.[1] The enzymes were first described as cytochrome P-450 proteins because of characteristic light absorption at 450 nm under reducing conditions.[2] As new proteins were discovered, they were classified broadly according to compounds that caused their induction (e.g., phenobarbital or 3-methyl cholanthrene inducible enzymes) or the substrates they metabolized (e.g., ethoxyresorufin-o-deethylase (EROD)).[3] With the identification of genes coding for individual enzymes, they are named CYP (cytochrome protein) enzymes and classified by number, according to amino acid sequences. Groups sharing less than 40% of identical amino acid sequences with other CYP proteins are termed a "family,"[1] and CYP1A1 is the specific family of interest in fish for studies of polyaromatic xenobiotics. Fish also appear to have CYP2B1 proteins, a form induced by phenobarbital in mammals but not in fish.[4] This review will follow the recommended nomenclature for *CYP* genes, DNA, RNA, and proteins.[1]

MFO induction by specific compounds is primarily transcriptional, i.e., increased activity due to gene activation, although post-transcriptional or post-translational regulation also occurs.[1] The current model of induction describes the binding of an inducing ligand to an arylhydrocarbon or Ah protein receptor in the cytoplasm to form an Ah receptor (AhR) complex. Concurrent release of heat-shock proteins from the receptor and subsequent binding of the complex to an Ah-receptor nuclear translocator (Arnt) protein allows transport of the AhR complex into the nucleus. The AhR-Arnt complex interacts with a dioxin-responsive element (DRE) upstream of the *CYP* gene to activate gene transcription to *CYP* mRNA. This mRNA moves to the endoplasmic reticulum to initiate gene translation and CYP protein synthesis.[5]

The potency of MFO inducers varies with Ah receptor binding affinity and persistence.[5] In most species, 2,3,7,8-tetrachlorodibenzo(p)dioxin (2,3,7,8-TCDD) is the most potent inducer of CYP1A1 enzymes because it has the greatest binding affinity and is very persistent,[5,6] although recent studies show that some pentachloro congeners are most potent in fish.[7] In fish, CYP1A1 activity is inducible by some PAHs, PCDDs, polychlorinated dibenzofurans (PCDFs), coplanar congeners of PCBs, and by some plant flavones and wood extracts.[8,9]

MFO enzymes enable the addition of oxygen across double bonds in aromatic compounds, forming epoxides and dihydrodiols.[10] This Phase I metabolic conversion of xenobiotics increases the reactivity, polarity, and water solubility of the substrate. In Phase II reactions, these metabolites are conjugated to endogenous substrates such as glucuronic acid by glucuronosoyl transferase. Conjugation further increases water solubility and rates of excretion.[10]

Detoxification is not the only result of induction; the increased reactivity of oxygenated metabolites of aromatic compounds can lead to genetic damage. Highly reactive epoxides can bind to DNA, forming adducts that cause mutagenicity and carcinogenicity by interfering with gene replication. Benzo[a]pyrene (BaP) is a well-known procarcinogen activated to a carcinogenic diol epoxide by MFO enzymes. PAH such as BaP generally have short tissue half-lives due to metabolism by MFO enzymes; induction is rapid and transient following a brief exposure to PAH,[5] but toxic effects may be delayed by months or years due to the latency of carcinogenic responses.

In contrast, most chlorinated inducers, such as PCDDs, PCDFs, and PCBs, persist in tissues because they are metabolized slowly by MFO enzymes. Because they also bind strongly to the Ah receptor, MFO induction is rapid and prolonged;[5,11,12] induction is also part of a suite of toxicological symptoms known as receptor-mediated toxicity (RMT) associated with continuous and prolonged gene activation. These symptoms include thymic involution, endothelial cell pathology, edema, hemorrhaging, and reduced growth. The type, number, and severity of lesions are species dependent, but features common to all vertebrates include a high sensitivity of young developing animals. While MFO enzymes occur in all tissues of fish, highest activities are found in liver. Endothelial cells often display higher activity than other cell types, and toxicity associated with induction may start with endothelial cell damage.[13]

The rapid induction of CYP1A1 enzymes by highly toxic compounds, the sensitivity of induction to low-level exposure, and its association with well-known mechanisms of toxicity has stimulated interest in using induction in fish to indicate chemical exposure and possible effects. Among the first experiments was the demonstration of arylhydrocarbon hydroxylase (AHH, or BaP hydroxylase) induction in cunner (*Tautogolabrus adspersus*) exposed to oil.[14] Since then, there have been many reports describing the MFO enzymes of fish, their distribution among tissues and cell types, their relative abundance among fish species, and their response to xenobiotics in the lab or at contaminated sites.[13] Induction *in situ* has been associated with oil spills,[15] industrial waste,[16] pulp mill effluents,[17] PCBs,[18] and dioxins.[19] The presence of inducers in specific effluents or natural substrates was confirmed by caging studies[20] and by laboratory exposures of fish to contaminated sediments and effluents.[9,21]

There are several landmark studies that linked MFO induction to chemical toxicity *in situ*. Downstream of a steel mill, Baumann and co-workers demonstrated a sequence of responses in brown bullhead (*Ictalurus nebulosus*) that were consistent with PAH-induced carcinogenicity.[22] Compared to reference fish, bullhead exposed to PAH-contaminated sediments contained higher concentrations of tissue PAH, a higher activity of liver MFO enzymes, higher concentrations of bile conjugates of PAH, higher concentrations of DNA adducts of BaP, a higher prevalence of neoplastic and preneoplastic lesions in liver and bile ducts, a higher prevalence of tumors and preneoplastic lesions on the skin and barbels, and a greater mortality rate of older individuals.[22,23] Closure of the mill was followed by a reduction in exposure, a reduction in prevalence of most lesions, and more normal rates of mortality.

Similarly, the exposure of English sole (*Parophrys vetulus*) to contaminated marine sediments of Puget Sound was associated with increased tissue concentrations of PCBs and PAH, increased MFO activity and elevated concentrations of tissue metabolites, increased prevalence of BaP DNA adducts, increased prevalence of tumors, and decreased reproductive success.[10,24,25] These responses could be replicated in laboratory exposures of fish to extracts of contaminated sediments.[25]

Most recently, numerous studies have demonstrated MFO induction in fish downstream of bleached kraft pulp mill effluent (BKME) discharges. Induction coincided with reproductive impairment due to changes in concentrations of serum steroid sex hormones during sexual maturation, delayed age to maturity, reduced gonad size, and reduced fecundity.[26-29] Early studies found a correlation of effects with the discharge of chlorinated compounds in BKME, notably adsorbable organic halides (AOX), PCDDs, and PCDFs.[30-32] However, these responses did not disappear following reductions in the discharge of chlorinated compounds,[33] and they were also found in fish near mills that did not use chlorine.[34-36]

The interest in pulp mill effluents and the search for the compounds causing those effects has led to standard bioassay protocols for MFO induction in fish exposed to effluents, extracts of effluents, and single compounds. The logistics of handling large volumes of effluent stimulated the miniaturizing of standard catalytic assays and the development of cell culture methods in place of whole fish. As well, the problems of highly toxic effluent streams and the difficulties of exposing fish to effluents *in situ* at remote sites has led to the adoption of semipermeable membrane devices (SPMDs) as artificial samplers of MFO-inducing compounds. The rest of this chapter describes these recently developed methods.

B. Fish Bioassays for MFO Induction

For bioassays of MFO-inducing compounds, fish may be exposed via injection, feeding, or respiration, but uptake and distribution kinetics will vary considerably according to the route of exposure. Hence, to monitor environmental effects, it is important to replicate the route of exposure of fish *in situ*. For short-term exposures to effluents, the primary exposure will be to waterborne materials taken up via the gills from respiratory water flow. MFO activity increases with time of exposure to effluents, reaching equilibrium within 48 to 96 h;[9,37] log (activity) increases linearly with log (effluent concentration), often to a peak activity.[9,37] Because peak activity occurs at concentrations of effluent or pure compounds that are close to lethal, the decline in activity is likely due to toxicity, either by direct inhibition of the enzyme or by cytotoxicity.[38]

Aqueous bioassays of MFO induction are no different than bioassays of acute lethality. The extent of response will vary according to physical, chemical, and biological characteristics of the test that affect exposure (e.g., the concentration and availability of test compounds), chemical potency (e.g., structure), and the ability of fish to respond (e.g., temperature, dissolved oxygen, pH, and size, age, and sex of fish[9,37,39]). Therefore, methods for bioassays of MFO induction in fish should follow standard protocols for acute lethality tests (e.g., see Sprague[40]). Accordingly, only immature fish or fish of one sex and of a uniform size should be tested under conditions of constant and optimal temperature, dissolved oxygen, pH, and conductivity; constant concentrations of test substances; and low concentrations of excreted waste products of test fish. These conditions are best obtained by constant or intermittent renewal of test solutions at a rate of 1 L of water for every gram of test fish per day. These recommendations were developed for salmonids of about 1 to 2 g each, and volume requirements may vary with species and fish size according to metabolic rate per gram of fish.

Williams[41] applied MFO bioassays on a large scale to compare the potencies of pulp mill effluents, or fractions of effluents, for inducing EROD activity in juvenile 10 to 50 g rainbow trout (*Oncorhynchus mykiss*). EROD activity was measured following the standard method of Pohl and Fouts.[42] He demonstrated that results were quite reproducible, both within his lab and between labs. This bioassay was used by Schnell et al.[43] to identify the sources within a pulp mill of MFO inducing agents, and to demonstrate the removal of inducers by enhanced effluent treatment. Williams' protocol[41] required fish 10 g or larger in order to generate sufficient liver for the assay of EROD activity in 3 mL cuvettes. Hence, about 50 L of effluent per day was needed to expose five 10 g fish at 1 L of solution per g of fish per d. Under these conditions, the availability and cost of test effluents in experimental or monitoring programs become limiting.

Martel et al.[36] also used this trout bioassay to compare potencies of effluents among mills with different bleaching and pulping technologies. Due to the high cost of shipping effluents and a desire to test "environmentally realistic" concentrations, these authors tested one fixed concentration of 10% effluent. There is a risk of false negatives in tests with only one effluent concentration. If the concentration tested falls below the threshold concentration for induction or above the threshold concentration for toxicity to induction, conclusions about the presence or absence of inducers and about the relative potency of samples will be incorrect. False negatives can be avoided by testing multiple concentrations to establish dilution-response curves and threshold concentrations. However, this alternative again requires large numbers of test animals and large volumes of effluent.

Gagné and Blaise[44] tried to circumvent the problem of fish size by pooling livers from 2 g trout. They used the survivors of acute lethality tests to derive additional information about sublethal responses to effluent exposure. However, the pooling of fish livers eliminated statistical comparisons, since the sample size for each treatment was one. Further, the inhibition of induction at near-lethal effluent concentrations increased the risk of false negatives.

To standardize assay conditions, Martel et al.[37] varied bioassay conditions to determine which would most affect MFO induction. They found that activity decreased as loading rates (gram of biomass per liter of test solution) increased, conforming to general practice in lethality bioassays. Paradoxically, when loading rates were held constant, MFO activity increased with fish size. There was no ready explanation of this observation, which might be a function of development and sexual maturation. Alternatively, the diet of larger fish may contain inducing substances, or there may be genetic differences among batches of fish.

Feeding fish during bioassays may also affect the outcome. In one study, endogenous EROD activity of trout declined after 48 h of fasting,[37] suggesting the presence of an inducer in the feed. In a second report, fasting did not change absolute activity of control or β-naphthoflavone (βNF)-exposed fish, but variability of measured activity declined with fasting, perhaps due to reduced levels of particulates associated with feeding and excretion that could sequester waterborne βNF.[9] Other procedures, such as anesthetizing fish before sacrifice, do not affect the outcome of MFO bioassays, and methanol may be used as a carrier for hydrophobic compounds added to water, provided that concentrations do not exceed 5 mL/L (20% of the lethal level of 25.6 mL/L).[9]

C. MFO Enzyme Assays

MFO induction can be measured by assays of mRNA, CYP protein, or catalytic activity;[8] enzyme activity is most commonly used and the least expensive. However, induction may occur without increased activity if the enzyme is inhibited by the inducing compound or by the product of the reaction, or if the enzyme was denatured through poor sample handling. Simple exposure-response bioassays can detect and avoid bias caused by enzyme inhibition, but enzyme-linked immunosorbent assays (ELISA) are a better measure of protein, and hence of induction, when protein is denatured.[45] Another emerging technology is measurement of CYP1A1 mRNA with reverse transcriptase competitive polymerase chain reaction (RT cPCR). In this method, mRNA is copied to its corresponding cDNA which is amplified by PCR to concentrations that can be separated and quantified by gel electrophoresis and ethidium bromide staining.[46]

CYP1A1 catalytic activity has usually been assayed with three substrates: BaP and diphenyl-oxazole (PPO) for arylhydrocarbon hydroxylase (AHH) activity[8] and ethoxyresorufin for EROD activity. Due to the hazards associated with using BaP as a substrate, most recent studies employ the EROD assay. This assay has evolved from a method published by Pohl and Fouts[42] and summarized by Hodson et al.[8] In the majority of cases, liver samples are homogenized in four volumes of ice-cold HEPES buffer and then centrifuged. The 9000 × g supernatant (S-9 fraction), or the microsomes prepared from high-speed centrifugation (100,000 × g) of the S-9 fraction provide the source of the enzyme.

Enzyme activity can be measured by incubating the S-9 fraction or microsomes in buffer containing both substrate and cofactors, the reaction being initiated by addition of reduced nicotineamide adenine dinucleotide phosphate (NADPH).[8] Activity is measured fluorometrically since both the substrates and their products fluoresce, although at different excitation and emission wavelengths. Activity is usually measured kinetically, from the slope of fluorescence vs. time as the reaction product is created, or as an endpoint, from the total fluorescence present after a fixed reaction time.[8] Standard curves of the reaction product (e.g., resorufin for EROD activity) are prepared to calculate a molar concentration, and molar concentration is divided by reaction time to give activity per minute. Protein in the S-9 or microsomal fraction is also measured, since dilutions are approximate when preparing homogenates, and variations in water, fiber, and lipid content of liver can bias activities based on weight. Therefore, the final unit of measurement is specific activity (pmoles of resorufin per mg protein per minute).

These methods typically employ standard 3.0 mL cuvettes, necessitating relatively large volumes of reaction mixtures (1 to 3 mL), liver (100 to 500 mg diluted 1 in 4 with buffer), and expensive substrates and cofactors; about one third of the material cost of an EROD assay is attributed to NADPH. As indicated above, large volumes of liver require large fish (10 to 50 g), larger tank volumes, and larger amounts of effluent or test compounds. All of these requirements translate to higher costs for materials, equipment, and space.

For a monitoring or experimental program involving numerous treatments and samples, these costs become excessive. The advent of microplate fluorometers using fiber-optic technologies for excitation and emission has provided highly sensitive, semiautomated instruments for either kinetic or endpoint assays.[9,47] A number of methods have been developed for these instruments, and the kinetic EROD assay published by Hodson et al.[9] demonstrates the advantages.

In this method, liver is diluted from 1:20 to 1:50 when homogenized, and total reaction volumes are only 110 µL in each well of a 96-well plate. Consequently, only 20 to 30 mg of liver is needed for triplicate determinations of EROD activity because of the high dilution allowed by the instrument's sensitivity. Experimental costs are reduced because only small fish (1 to 3 g) are needed, reagent volumes are five- to 15-fold less, and the instrument can repeatedly scan 96 wells per minute to generate slopes of fluorescence vs. time. Using multiple plates, more than 100 samples can be analyzed in triplicate, including standards, in less than a morning, a task requiring several days using cuvettes. Savings in fish size generate savings in labor and cost associated with smaller volumes of effluent (15 L/d for five 3 g fish) and encourage larger sample sizes for increased statistical power. Storage space for frozen tissues is also reduced and samples spend less time in storage.

In some enzyme assay protocols, protein can also be measured with fluorescing dyes that are added to the same microplate wells and read with the microplate fluorometer.[48] This saves time and reduces error compared to pipetting further aliquots of S-9 fractions for protein assays by cuvette or microplate assays with colorimeters.

The power and utility of these methods has been demonstrated by direct comparisons with cuvette methods[9,47] and by two round-robin comparisons among labs. In the first, 11 labs used individual variants of the cuvette-based fluorometric technique of Pohl and Fouts[42] to measure EROD activity of 11 aliquots of homogeneous liver samples from white sucker (*Catostomus commersoni*) captured at clean reference sites or sites contaminated by pulp mill effluent.[49] While measurements of activity on the same samples varied considerably among labs, the ratios of activities between induced and reference fish (i.e., EROD induction) were quite constant. Without any attempts at standardization, these 11 labs could all discriminate treated from reference fish.

In the second round-robin, greater care was taken to standardize techniques, and 6 of 14 participating labs used microplate methods instead of the cuvette method.[50] With three exceptions, measurements of activity were remarkably constant among labs, and precision and performance of microplate methods were as good or better than those of cuvette methods. Of the three labs that did not perform well, one used a spectrophotometric method, which is relatively insensitive, and the other two had not optimized their techniques.

Enzyme assays based on microplate fluorometers remove some of the barriers to effective monitoring of effluents and make the testing of dilution series practical. However, even these assays may require too much effluent or test substance. Valid bioassays use large volumes of test solution to maintain the concentration of test substances despite uptake into fish; hence most of the substance is not accumulated and is wasted. For testing highly purified materials such as effluent fractions generated by HPLC,[51] the cost of test material becomes prohibitive. Hence, test systems based on *in vitro* cell culture are more practical, providing that *in vitro* tests are validated periodically against *in vivo* fish tests. The next section of this review describes *in vitro* MFO bioassays using cells in culture.

D. Summary

MFO induction in fish by industrial effluents can be measured using standard bioassay protocols. Provided that solutions are changed daily to allow 1 to 2 L of solution per g of test fish per day, four-day exposures will give reliable results with both standard test substances, such as βNF, and effluents. Microplate fluorometers allow the MFO enzyme assays to be miniaturized and semi-automated, which gives considerable savings in analytical time and in the overall cost of bioassays. Most important, these bioassays require fish of only 1 to 3 g, which reduces the need for large volumes of test effluents.

Table 4.2   Performance of *In Vitro* MFO Induction Assays

Cells	Enzyme	Linear Range	EC/ED$_{50}$ (TCDD)	Detection Limit (TCDD)	Reference
Rat 1° hepatocyte	AHH	<10 pM–10 nM		≤10 pM (DMSO)	82
Rat H411E	AHH		376 pM (DMSO) 263 pM (DMSO)	19 pM (DMSO)[1]	59,109-111
		20–380 pM	34 pM (isooct.)	7.6 pM (isooct)[1]	
Rat H411E	EROD	10–80 pM	17 pM	3 pM	61,62
Rat H411E	AHH EROD	20- 260 pM	100 pM	>12.4 pM	60,112,113
Rat H411E	EROD	21–210 pM	47 ± 16 pM	10 pM [2]	64,114
Chicken embryo hepatocytes	AHH EROD		79 pM EROD 24 pM AHH		78
Chicken embryo hepatocytes	EROD	1–100 pM	15 pM	0.9 pM	88,89
Rat H411E	EROD	6–120 pM	23.8 pM		68,76
RTL-W1	EROD	2–25 pM 2–10 pM	15.3 pM	about 2 pM	68,76
PLHC-1	EROD	10–5000 pM	130 pM	about 10 pM	73
HepaG2 101L	Luciferase	<100 pM–1000 pM	100 pM	0.05 ng/g sediment	92
Hepa1c1c7	Luciferase	5 pM–1000 pM	50 pM	50 pM	93
Hepa1c1c7	PAP	100–500 pM	350 pM	100 pM	91
Trout 1° liver	EROD	2.5–100 pM		4 pM	80,115

[1] upper confidence limit of analyte free control.
[2] Mean of blanks = 3 × S.D.

II. CELL CULTURE ASSAYS

In vitro assays that use either cell lines or primary cultures of liver hepatocytes have been developed to assay the presence and activity of MFO inducing compounds. The performance characteristics of some of the assays are summarized in Table 4.2. Cells that have been isolated from a parent tissue and maintained in a viable state in culture medium without being subcultured are termed "primary cultures." "Cell lines" are axenic cultures that have been subcultured repeatedly[52] and may be finite or continuous.[53] Several readily available fish cell lines have a functional MFO system.[54]

As indicated in the previous section, *in vitro* MFO induction bioassays solve several practical problems associated with their whole fish counterparts. They are less costly, logistically simpler, and less labor intensive than whole fish tests. *In vitro* assays allow a higher sample throughput, as multiple tests can be run in a few microwell plates, and most *in vitro* techniques can be semiautomated by means of liquid handling devices and automated plate readers. Many experiments that would be difficult using whole fish bioassays are made possible by *in vitro* bioassays because miniaturized tests radically reduce the amount of sample needed. *In vitro* MFO tests also minimize the use of live fish in screening and monitoring, thus alleviating concerns about animal welfare.

In vitro assays based on fish organs, isolated cells, or cell lines can provide much of the same information as whole fish bioassays, particularly if there is a mechanistic link between *in vitro* responses and known effects in live fish. *In vitro* exposures, however, bear little resemblance to the complexities of natural or environmental exposures. Thus, they may not account for many absorption, metabolic, and pharmacokinetic processes in intact fish, and effects at the organism level.[54] Thus, data from *in vitro* assays often need confirmation by *in vivo* exposures.

A. Early Cell Line Bioassays for MFO Inducers

The first cell line bioassays for inducers of CYP1A1 were developed as analytical tools for detecting small amounts of 2,3,7,8-TCDD. Based on the H411E rat liver hepatoma cell line, those assays could detect as little as 10 pM of 2,3,7,8-TCDD.[55] The H411E assay has been used to elucidate Structure Activity Relationships (SAR) for PCDDs,[56] PCDFs,[57] and PCBs.[58] Generally there is a good correlation between the ability to induce AHH or EROD activity and *in vivo* toxicity. Such SARs are important in risk estimation and in the development of health and environmental guidelines.

These assays have been reversed, i.e., the extent of induction caused by unknown mixtures has been used to estimate the concentration of inducers in the mixture, based on the response to 2,3,7,8-TCDD as a standard. Results are often expressed as the amount of TCDD-equivalents or Toxic Equivalent Quantities (TEQs) present. The expression of potency as TCDD equivalents does not imply that the mixture contained 2,3,7,8-TCDD; rather, the compounds in the mixture are as potent as, or equivalent to, a certain amount of 2,3,7,8-TCDD. A less confusing unit is EROD potency-equivalents (EROD potency-EQ), where TCDD is the standard but not mentioned in the units, and toxicity is not implied, since the constituents causing the response are unknown. Nevertheless, the following review uses the term TCDD-equivalents or TEQs because these were the units reported in the papers cited.

B. Environmental Samples

Casterline et al.[59] used the H411E assay to screen extracts of freshwater fish that were prepared by fractionation on silica gel. Although the fractionation scheme was not optimized for the recovery of planar HAHs, there was good agreement between the assay and the gas chromatography-electron capture detector data. The H411E induction assay was later modified so that both EROD and AHH activity could be measured in suspensions of whole cells, paving the way for microtiter plates and miniaturization. The modified assay was used to screen extracts of 25 fish from the Great Lakes.[60] Bioassay estimates of TEQs in fish from most lakes differed by less than twofold from TEQs estimated from chemical analyses using gas chromatography-mass spectrometry (GC-MS). In contrast, bioassay TEQs for fish from lakes Ontario and Erie were more than fourfold higher than GC-MS derived values, perhaps due to synergistic effects or unidentified inducers.[60]

The H411E bioassay was systematically characterized and evaluated by Tillitt et al.[61] using EROD activity as an endpoint. The assay's performance was validated for a set of bird egg extracts spiked with 2,3,7,8-TCDD; the regression of observed vs. expected potency was 0.949. This version of the H411E assay was used to screen extracts of eggs from colonies of Great Lakes fish-eating birds,[62] and to show that TEQs in chinook salmon (*Oncorhynchus tshawytscha*) were transferred from mother to eggs.[63] The Swedish Dioxin Survey's evaluation of the H411E assay[64] also showed good bioassay-derived TEQs for herring, and osprey samples were closely correlated to TEQs estimated from chemical analyses by GC-MS. However, H411E data for whitefish did not compare as well with GC-MS data, possibly because of interference by unidentified planar compounds.

The H411E bioassay was also used to compare TEQs among white suckers caught near eight pulp mills and five reference sites.[65] Liver extracts of the fish were tested in the H411E bioassay and analyzed chemically by conventional techniques. The bioassay-derived TEQs were highest in fish caught near mills combining the kraft pulping process with chlorine bleaching. The bioassay TEQs were strongly correlated to concentrations of PCDD and PCDF congeners, although extracts from fish downstream of two mills that used little chlorine caused greater induction than predicted by PCDD/F levels, suggesting the presence of other types of inducers.

C. Fish Cell Lines

The use of mammalian cell cultures to predict the potency of chemicals in fish raises concerns about possible interspecies differences in the induction process. Studies using recombinant cell lines have shown interspecies differences in the functional response of the Ah receptor to ligands.[66] Fish cell lines may provide more realistic estimates because the inducible MFO enzymes in fish are limited to CYP1A1, in contrast to multiple inducible forms in mammals. There are also important differences among species in ligand-Ah receptor binding affinities,[67] which may explain species differences in congener toxicity and MFO induction potency. While 2,3,7,8-TCDD may be the most potent congener in mammals, some pentachloro-congeners are more potent in fish.[7,68]

Tritiated and photoaffinity ligands have been used to identify the Ah receptor in the rainbow trout cell lines RTH-149[69] and RTG-2,[70] in a variety of teleost species, and in the topminnow (*Poeciliopsis lucida*) cell line (PLHC-1).[71] 3,3′,4,4′-tetrachlorobiphenyl, a coplanar PCB, could induce EROD activity in a PLHC-1 bioassay, showing an exposure–response relationship from 0.01 μM to 0.1 μM TCBP after which the induced activity decreased;[72] the decrease in activity was not caused by cytotoxicity. The amount of induced protein, measured immunologically, continued to increase up to 10 μM, indicating that inhibition or inactivation of the enzyme could lead to underestimates of Toxic Equivalent Factors (TEFs) or TEQs. Direct quantification of the enzyme protein, or its mRNA, may solve this problem.

Hahn et al.[73] confirmed the utility of the PLHC-1 assay by assessing the ability of 2,3,7,8-TCDD, CB-126, and CB-77 to induce EROD and CYP1A1 protein. As shown in Figure 4.1, EROD activity again declined at lower exposure concentrations than did the protein. The estimated TEFs for CB-126 and CB-77 were about an order of magnitude less when calculated on the basis of the CYP1A1 protein rather than EROD activity.

Finnish researchers used the PLHC-1 cell line to detect MFO inducers in pulp and pulp mill effluents.[74] Pulp bleached without chlorine caused the highest induction of EROD, suggesting that chlorinated compounds may not be the sole inducers.

A nonhepatoma cell line, RTL-W1, that expresses inducible EROD activity has been developed from rainbow trout liver,[75] and was used to derive sets of TEFs for a series of PCDD/PCDF/PCB congeners.[68,76] For the PCDD/Fs most of the RTL-W1-derived TEFs were substantially higher (2 to 8×) than H411E TEFs measured in the same laboratory, although the rank order potencies were the same. The RTL-W1-derived TEFs correlated well with those previously reported for live fish, such as orally dosed rainbow trout.[19] Most important, they corresponded to TEFs based on chronic toxicity of dioxin and furan congeners to larval salmonids. The potency of 1,2,3,7,8-pentachlorodibenzo(p)dioxin for MFO induction in fish cell lines and chronic toxicity to larval salmonids was greater than that of 2,3,7,8-TCDD.

For both live fish and fish cell lines, fish-derived TEFs were higher than internationally agreed factors (I-TEFs) established to assess dioxin risks to humans. For PCBs, H411E TEFs tended to be higher than those from RTL-W1, particularly for the nonortho congeners, suggesting that TEFs for assessing risks to fish would be more appropriately derived from fish-based assays. Thus, the RTL-W1 cell line assay may be more useful than H411E assays for predicting which congeners or samples would induce MFO activity in fish. Although the PLHC-1 assay responds at higher concentrations of inducers than the H411E and RTL-W1 assays, its response signal is at least as strong as that of the RTL-W1 assay.

D. Hepatocyte Assays for MFO Inducers

Many primary cultures of hepatocytes are responsive to MFO-inducing chemicals. Hepatocytes from chicken embryos,[77-79] rainbow trout,[80,81] and rat[82] have been used to test complex mixtures of chemicals and pure compounds. Methods for preparing primary cell cultures from various fish

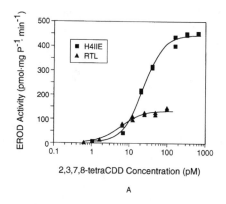

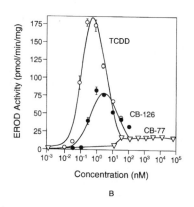 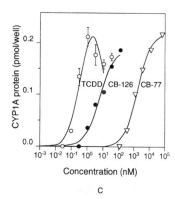

Figure 4.1 A) Assay calibration curves for the induction of EROD activity by 2,3,7,8-TCDD in the H411E and RTL-W1 bioassays. (From Hahn, M. E., Woodward, B. L., Stegeman J. J., and Kennedy, S. W., *Environ. Toxicol. Chem.*, 15, 582, 1996. With permission.) B and C) Induction of kinetically measured EROD activity (B) and immunologically measured CYP1A protein (C) in the PLHC-1 bioassay by 2,3,7,8-TCDD (TCDD) and 3,3′,4,4′5-pentachlorobiphenyl (CB-126), and 3,3′,4,4′-tetrachlorobiphenyl (CB-77). (From Swanson, H. I. and Perdew, G. H., *Toxicol. Lett.*, 58, 85, 1991. With permission.)

tissues have been reviewed elsewhere,[53,83] and cultured hepatocytes can be exposed in microwell plates and assayed using kinetic and immunological techniques. Hepatocyte assays have been used to assess the ability of pulp mill effluents to induce fish liver MFO enzymes[35,81,84,85] and will likely play a role in the identification of inducers in effluents and environmental matrices. Fluorescent-labeled DNA probes have also been used to assess CYP1A1 induction at the transcriptional level in rainbow trout hepatocytes, using flow cytometry as a detection method.[86] Hepatocytes were prepared in the standard manner and exposed in 24-well microplates to wastewaters from a major urban discharge and effluents from a petrochemical plant and pulp and paper mills. The preliminary data suggest the technique has promise as a screen for the presence of CYP1A1 inducers.

E. Calculation Methods

TEFs are usually expressed as the ratio of the ED_{50} or EC_{50} of 2,3,7,8-TCDD to the ED_{50} or EC_{50} of the test chemical or mixture.[61] If a sample's maximum response is lower than that of TCDD, or if maximum activity is not achieved, the estimate of TEQs can be inflated. Hanberg et al.[64] recommend that the TEF ratio be calculated from a point on a sample's dose-response curve at an EROD activity corresponding to the linear portion of the 2,3,7,8-TCDD calibration curve. If the sample and 2,3,7,8-TCDD have the same maximum response, comparisons can be made with the ED_{50} value for 2,3,7,8-TCDD. Tillitt et al.[61] suggest calculating TEFs by a slope ratio method which is unbiased by any failure to attain a maximum response. Sample potency is calculated from the ratio of the slope of its dose-response curve (EROD activity/μL extract) to the slope of the 2,3,7,8-TCDD standard curve (EROD activity/pg of 2,3,7,8-TCDD).[63,87]

F. Miniaturization and Automation

In vitro MFO assays have been made more attractive by simplifying and automating several assay steps. Kennedy et al. modified a chick hepatocyte assay to use with 48-well culture plates that can be read directly in a fluorescence plate reader.[88,89] Assay speed was increased by 100-fold by eliminating two assay steps: harvesting and transfer of treated hepatocytes, and manual measurement of fluorescence. The method's excellent sensitivity (Table 4.1) was attributed to small volumes of culture medium and high sensitivity of chicken embryo hepatocytes. Fluorescamine, a protein probe, has also been incorporated in this assay so that both protein and EROD activity can be measured in the same microplate.[90]

RTL-W1 and PLHC-1 assays have also been modified for 48-well culture plates.[71,76] Special care was taken in the PLHC-1 assay to ensure even distribution of cells in the assay wells. The assay was systematically optimized for cell density, volume of culture medium, composition of the reaction mixture, time course of induction, and the effect of cell passage number on responsiveness. Extracts from individual wells were assayed immunologically using a Western blotting procedure.

G. Recombinant Cell Lines

Several promising bioassays based on recombinant cell lines can detect chemicals that bind to the Ah receptor.[91-93] These cell lines contain a reporter gene downstream of a DRE. When the host cell's Ah receptors are activated by a ligand, the reporter system activates the reporter gene. Since the reporter construct is independent of the CYP1A1 catalytic system, it is not prone to inhibition or inactivation artifacts. Assays based on recombinant cell lines should accurately reflect interspecies functional differences in the Ah receptor because DREs are conserved among species.[67,94,95]

El-Fouly et al. developed a recombinant cell line assay by stably transfecting mouse hepatoma cells with a gene construct that used human placental alkaline phosphatase as the reporter gene.[91] The vector can also be transfected into cell lines derived from a variety of tissues and species, including fish. Anderson et al. used a recombinant cell line based on a human liver cell line promoter gene and a luciferase reporter system.[92] The *CYP1A1* promoter gene, which retained three attached DREs and was positioned upstream of the firefly luciferase gene, was stably transfected into the human hepatoma cell line HepG2 and used to test extracts of aquatic sediments, porewater samples from contaminated marine sediments, and a PAH mixture. Another recombinant cell line was used to detect dioxin-like activity in an extract of pulp mill effluent and in a black liquor sample.[93] Such recombinant cell line assays should be adaptable for use with microplate readers making them more suited to large-scale studies.

H. Future Research

There are subtle differences in the properties of the Ah receptor from various mammalian species;[95] recombinant cell lines show that some compounds, such as 2,2′,5,5′-tetrachlorobiphenyl, act as antagonists for the Ah receptor in a species-specific manner.[66] Interspecies variations in the functionality of the Ah receptor are probably best dealt with by batteries of multispecies tests, or by using cell cultures derived from the species of interest. *In vitro* assays based on cell lines are ideal for use in test batteries. More fundamental research is needed on the biological and environmental significance of interspecies variations in the functionality of the Ah receptor.

If TEFs are calculated from bioassays in which EROD activity is inhibited, estimates of exposure to inducers, and hence estimates of risk, will be biased. Therefore, we need a better understanding of the relationship between the induction of EROD activity and CYP1A1 protein. Sadar et al.[96] have reported that phenobarbital, which is not known to bind to the Ah receptor, can induce CYP1A1 activity and CYP1A1 protein in rainbow trout hepatocytes. The induction was additive when the cultures were exposed to excess levels of 2,3,7,8-TCDD and phenobarbital; the authors suggested the existence of an alternative mechanism for CYP1A1 induction. The ability of some target compounds and environmental samples to induce EROD and CYP1A1 protein should be confirmed in receptor binding assays or in bioassays that are highly specific for the liganded Ah receptor, such as recombinant cell line assays. Sidhu et al. described a promising laser cytometer technique for the observation and measurement of *in situ* CYP1A1 activity in intact single cells[97]. This technique could become a powerful means of unraveling subcellular processes.

I. Summary

In vitro MFO induction assays are cost effective, well suited to large numbers of samples, and allow multiple simultaneous assays without boosting costs. They integrate additive, synergistic, or antagonistic effects that occur in complex mixtures of chemicals and provide an overall estimate of the mixture's potency. In combination with cell culture assays for other responses such as vitellogenin induction, toxicity, membrane disruption, or oxidative stress, *in vitro* techniques will play a key role in elucidating the mechanisms of MFO induction, and help to unravel the links between Ah binding, MFO induction, and other subcellular processes.

III. SPMDs

Semipermeable membrane devices (SPMDs) were developed by Huckins et al.[98] as passive *in situ* samplers containing purified triolein, a substance that constitutes a major fraction of the neutral lipid of fish. When immersed in water, SPMDs absorb water-insoluble chemicals with log $K_{ow} > 1$, size < 10 Å, and a molecular weight of about 600 or less. These characteristics correspond to those of known MFO-inducing compounds, so that SPMDs can be used to collect inducers from effluents or surface waters.

Standard SPMDs are 91 cm long × 2.5 cm wide low-density polyethylene layflat tubes (wall thickness 0.80 μm) containing 1 mL (0.915 g) of high purity (95%) synthetic triolein. Freely dissolved neutral organic chemicals diffuse through pores in the polyethylene membrane and dissolve in the triolein, simulating the diffusion of compounds across a live fish gill membrane.[98] The lipid can be analyzed by traditional chemical techniques to give a list of chemicals absorbed, their concentrations in the SPMDs, and by back-calculation, their concentrations in water or sediment. Because SPMDs accumulate organic compounds in a fashion similar to live fish, tissue concentrations in fish from the site can also be estimated and compared to toxicity data for individual chemicals, to determine the potential for adverse effects. However, the actual biological activity of the complex SPMD mixture remains unknown unless tested by bioassay.

The use of SPMDs as concentrating devices for biological testing is relatively new, although research is expanding the types of tests used and the applications of the technique. The advantage of bioassays of SPMD extracts is that measured potency expresses all potential mixture interactions. Further, if specific classes of chemicals such as MFO inducers are monitored with SPMDs, bioassays of extracts provide a rapid and inexpensive way to identify those samples needing expensive chemical analyses. SPMD extracts have been successfully used in the Microtox® and Mutatox® assays to concentrate toxicants and mutagens from urban stream water, Antarctic sediments, and PAH-spiked sediments.[99-101] Huckins et al.[102] have used SPMDs to concentrate organic chemicals from marine sediments for toxicity tests on *Mysidopsis bahia* and *Ampelisca abdita*. Recently,

SPMD extracts from the Detroit River and Lake Erie have been used for SOS chromotest, *Daphnia magna* lethality tests, and Japanese medaka (*Oryzias latipes*) embryo toxicity tests.[103]

SPMDs have several benefits over water sampling and fish caging for monitoring the presence of specific compounds:

1. SPMDs provide time-integrated samples. Deployments of 6 d or longer provide a representative sample that is less vulnerable than a single grab sample to bias caused by pulses of chemicals.
2. SPMDs are easy to handle and deploy. They can be made to any size, shipped by mail, and deployed from shore, by wading, or from any size boat. To monitor chemicals in effluents or surface waters, SPMDs require much less effort than sampling and extracting water, caging fish *in situ*, or exposing fish to samples shipped to the lab.
3. SPMDs can tolerate extreme conditions. They can be deployed in wastewaters or effluents that would kill bioassay fish due to toxicity or to extremes of temperature, pH, oxygen, or suspended solids.

The disadvantages of SPMDs relate to membrane selectivity and the validity of substituting SPMDs for fish: only freely dissolved neutral organic compounds are sampled. While this selectivity mimics that of a fish membrane, SPMDs lack the active and facilitated transport processes of a living membrane. Charged ions are not taken up due to resistance to passage through the neutral polymer membrane, and SPMDs cannot metabolize labile compounds. While this is an advantage for analytical detection, it must be recognized that compounds accumulated by SPMDs may not be accumulated to the same extent by living organisms, because the organisms may have the ability to metabolize and excrete the chemicals. SPMDs also mimic only the waterborne uptake of chemicals into an organism. If the food chain is the main source of chemical uptake, SPMDs will not predict bioaccumulation.

Analysis of the extracts of SPMDs by MFO assays using *in vitro* cell culture was initiated by Don Tillitt of the Environmental and Contaminants Research Center, Columbia, MO.[99] Measuring MFO induction by SPMD extracts can enhance understanding of the chemical properties of the inducing substance. The maximum EROD activity observed in cells gives clues about the type(s) of inducing chemicals: PCDDs and PCDFs cause high maximum EROD activity while PAH-type inducers cause lower maxima. Reactive clean up of extracts can also identify which classes of chemicals are responsible for MFO induction because H_2SO_4/silica columns destroy labile PAHs (and MFO induction in cell culture) but not PCDDs and PCDFs.

A. Factors Affecting Chemical Accumulation by SPMDs

The accumulation of compounds within SPMDs is complex and related to each compound's octanol-water partition coefficient (K_{ow}) and molecular size. Uptake rates can be expressed as the liters of water sampled by the SPMD per day. For naphthalene the rate of sampling is about 0.3 L/day, and the time to reach 90% of equilibrium concentration is 7 d.[99] The low sampling rate reflects the low K_{ow} of naphthalene: the small molecular size allows easy passage through membrane pores, but limited solubility in triolein means equilibrium is reached rapidly.

The higher the K_{ow}, the more compound will accumulate in triolein, and the longer it will take for concentrations to reach equilibrium with the concentration in water. Phenanthrene, with a higher K_{ow} than naphthalene, requires longer to reach equilibrium (21 d) and the sampling rate is about 4 L/day, while chrysene and pyrene are sampled at about 5 to 6 L/day.[99] For compounds larger than chrysene and pyrene (four aromatic rings), size becomes a limiting factor. The high K_{ow} favors accumulation of the compounds in the triolein, but molecular size impedes rapid transfer through pores in the polyethylene membrane, and accumulation is slowed. SPMDs concentrate large PAHs, such as BaP and benzo[g,h,i]perylene, at rates of about 3.7 and 2.2 L/day, respectively.[99]

Because of the long time to reach equilibrium, over 21 d for PAHs larger than phenanthrene, MFO inducer(s) concentrated by SPMDs in short deployments (6 to 14 d) are probably not in equilibrium with water concentrations, but this should not affect conclusions about relative potencies. Deployments of one to two weeks are sufficient to concentrate inducers. During this time, chemicals diffusing into SPMDs are in the linear portion of the uptake phase, so comparisons of uptake can be made among studies. For example, potencies of SPMDs deployed for 6 to 7 d in pulp mill effluents can be doubled to roughly compare with potencies of SPMDs deployed for 14 d.

Flow and current speed of effluent over SPMDs vary among field deployment sites. The water velocity past the membrane does not influence the concentrations of inducers in the SPMDs, as the rate-limiting step for uptake of compounds into the SPMDs is membrane transfer.[99] In field studies, it is not difficult to ensure that water flow past SPMDs is greater than the highest sampling rates (about 6 L/day) reported for four ring PAHs.[99] Hence, site to site differences in flow should not affect SPMD uptake of inducers from effluents and wastewaters. In laboratory studies of SPMD uptake of inducers from effluents, the volume of solution should exceed the highest sampling rates.

Fouling of the membrane will also affect sampling rate, but effects are not as dramatic as expected after visual examination of fouled membranes. Huckins et al.[99] left SPMDs in the Upper Mississippi for 58 d and removed the fouled membranes to laboratory water to study their uptake properties. Uptake of phenanthrene in the fouled membranes was 35% less than uptake into unfouled membranes. While fouling would slow uptake of compounds from effluents and river waters, this effect would be minimal in deployments of less than 14 d. The effect of slight membrane fouling would be to underestimate potency of the effluents and wastewaters to an unknown degree.

Water temperatures vary among sampling sites, a factor that affects uptake of certain classes of chemicals by SPMDs. Huckins et al.[99] found a 1.3- to twofold increase in rate of uptake of organochlorine pesticides with every 5°C increase in temperature. They proposed that the temperature effect is due to increased molecular diffusion and polymer free volume, which permits pesticides to more easily enter membrane pores. However, the temperature effect was not consistent; for more rigid molecules such as PAHs, the effect is minimal.[99,104] Because compounds that cause MFO induction (e.g., PCDDs, PCDFs, PAHs) are planar and rigid, their accumulation by SPMDs should not be biased by temperature variation.

B. Monitoring with SPMDs

Deployment devices are chosen to suit the location: plastic laundry baskets, metal rotisserie baskets, or fiberglass mesh tubes for calm conditions, and weighted steel tubes for fast-flowing rivers. In the lab, SPMDs can be suspended in barrels of stirred effluent or in glass aquaria with fish. Upon removal, SPMDs are cut from deployment devices, sealed in clean paint tins and frozen until extraction and biological analysis in the laboratory.

During deployment, precautions must be taken to prevent contact of SPMDs with contaminated field equipment. Gloves are used while handling SPMDs and deployment devices, and deployment should be as rapid as possible to reduce exposure to air and contaminants during handling. SPMDs exposed to air and handled for the same amount of time as deployed SPMDs, but then immediately resealed in the can, are recommended as trip blanks.

In the laboratory, external fouling of SPMDs is removed by scrubbing in water prior to methanol and hexane rinses. Membranes are dialyzed for 48 h at 19°C in 1 L hexane, which is subsequently rotary evaporated to about 5 mL and filtered through anhydrous sodium sulfate. The eluent is concentrated to 1 mL under nitrogen, and the compounds of interest separated by size exclusion high pressure liquid chromatography (HPLC). The HPLC fraction collected is rotary evaporated to about 5 mL, solvent exchanged with trimethyl pentane to a volume of 1 mL, and concentrated to 200 μL. The concentrate is dosed to cells in culture, such as PLHC-1 or H411E cells[99] for bioassays of EROD induction to estimate the concentration of MFO inducers in the extract.

PLHC-1 bioassay procedures are a slight modification of the H411E bioassay methods for 96-well microtiter plates described in Tillitt et al.[61] PLHC-1 cells are dosed for 72 h with sample extracts or standards in isooctane or DMSO, and EROD activity is determined fluorimetrically. Testing EROD induction in fish cell lines exposed to SPMD extracts before and after destructive clean up on a silica-acid column will determine whether the primary compound(s) causing induction were PAHs.

The concentration of inducers is expressed per g SPMD, based on the whole weight of the SPMD (5 g; the 4 g polyethylene membrane and the 1 g of triolein both absorb the accumulated compounds). The units are in pg TCDD equivalents per g SPMD, calculated by comparing EROD activity induced by the extract to the EROD activity induced in cells exposed to 2,3,7,8-TCDD.

C. Experimental Applications of SPMDs

SPMDs have been deployed in a variety of pulp mill and refinery effluents and in river waters, both in the field and in the lab.[99,105-107]

In the lab, SPMDs were placed in 200-L barrels of BKME at 15 to 22°C for 6 d, with fresh effluent renewed every 2 d. They accumulated MFO inducers within 4 d.[105,106] Dialysates were run through a reactive cleanup column (H_2SO_4/silica gel) and retested for induction potency in PLHC-1 cell cultures to determine the types of compounds causing induction in the SPMD extracts. SPMD extracts that were passed through a H_2SO_4/silica column did not induce EROD, while precleanup extract contained potent MFO inducers (400 to 500 pg TCDD/g SPMD[105]). This suggests that the inducer(s) was not a dioxin or furan-type of compound, as these chemicals would have passed through the H_2SO_4/silica column. Destruction of the inducer by the reactive column suggests a more labile type of inducer, such as a PAH.

Lab experiments tested SPMD uptake of MFO inducers from solutions of black liquor (0.0032, 0.01, or 0.032%) or βNF (10 μg/L) and compared induction in fish cell lines exposed to SPMD extracts to induction in small rainbow trout exposed to the same solutions. SPMD extracts dosed to fish cells show similar MFO induction as live fish that had been exposed with the SPMDs. SPMDs and small rainbow trout placed for 4 d in solutions of black liquor or βNF detected MFO inducers. Solutions that induced fish EROD tenfold, when concentrated by SPMDs and tested in PLHC-1, contained 27 (black liquor) to 178 (βNF) pg TCDD-EQ/g SPMD.[108]

Both fish and fish cells exposed to SPMD extracts showed dose-related increases in EROD activity with increasing concentrations of black liquor; the live fish appeared more sensitive than PLHC-1 cells.[108] βNF was concentrated by SPMDs, and induced PLHC-1 more than live fish. The differences in sensitivity may be due to variability in fish EROD activities, differences in how βNF and black liquor are processed within fish, the preferential concentration of βNF from solution compared to inducers in black liquor, differing sensitivities of rainbow trout and topminnow to MFO inducers, differences between cell line and whole organism responses, or any combination of these factors. In short, fish and PLHC-1 cells dosed with SPMD extracts appear to respond similarly to solutions of inducers, but much more comparative work is needed to assess the extent of the similarities and causes of any differences.

In the field, SPMD uptake of inducers was tested over 6 d at several sites along the 15 km effluent stream of a bleached kraft mill and at one site on a control stream. SPMDs accumulated the highest concentrations of inducers prior to or in secondary treatment ponds, up to 2600 pg TCDD-EQ/g SPMD.[105] SPMD induction potency decreased as deployments were farther from the source. SPMDs 5 km downstream contained 165 to 268 pg TCDD-EQ/g SPMD, while those 15 km downstream contained 115 to 212 pg TCDD-EQ/g SPMD.

SPMDs can be used to trace the source and production of MFO inducers throughout the pulp mill process. SPMDs immersed for 14 d in effluent streams from several stages of a bleached sulfite mill accumulated more MFO inducers from effluent prior to clarification and secondary treatment (8,895 pg TCDD-EQ/g SPMD). SPMDs from the secondary treatment ponds contained 302 and 743 pg TCDD-EQ/g SPMD, while those early in the pulping process, in the effluent from the hypochlorite bleaching stage, or from the extraction stage, contained no detectable inducers.

In Alberta, SPMDs were deployed for 14 d in pulp mill and refinery effluents and in the Athabasca and Lesser Slave Rivers upstream and downstream of those effluents. Extracts of SPMDs held in effluents from secondary treatment ponds of the four pulp mills contained only 23 to 62 pg TCDD-EQ/g SPMD. In the same study, SPMDs deployed for 14 d in oil sands mining and refining wastewater contained over 16,000 pg TCDD-EQ/g SPMD.[106] SPMDs deployed in river waters in the oil sands area around Fort McMurray, Alberta, concentrated more MFO inducers (up to 728 pg TCDD-EQ/g SPMD) than SPMDs deployed in the Athabasca River upstream of the oil sands area (about 13 pg TCDD-EQ/g SPMD). It is possible that natural erosion or seepage from the oil sands contributes MFO inducers to the Athabasca River.

Villeneuve et al.[107] used SPMDs to concentrate PAHs from Lincoln Creek, a small urban creek in Milwaukee, Wisconsin. Extracts of SPMDs deployed for 2 to 30 d induced EROD in PLHC-1, and the log TCDD-EQ was correlated with total PAH concentrations ($r^2 = 0.80$). SPMD extracts were potent EROD inducers, containing 2200 pg TCDD-EQ/g SPMD after 14-d exposures.[107] Comparisons of extracts between SPMDs exposed during low or high flow conditions showed greatest potency during floods, suggesting that inducers were released from sediments and carried in urban runoff during storm events.

Whether the inducer is similar to a PAH, PCDD, or PCDF can be ascertained by the maximum EROD activity in PLHC-1 or other cell lines. SPMDs deployed in a small urban stream suspected to be contaminated with TCDD, contained compounds that induced H411E EROD activity by more than 50 pmol/mg/min, activities typical of 2,3,7,8-TCDD standards. This suggested the inducer present in the SPMDs was a PCDD or PCDF.[99] Conversely, PLHC-1 exposed to extracts of BKME-exposed SPMDs show induction maxima of 10 to 20 pmol/mg/min, typical of PAHs.[99,105]

D. Conclusions

SPMDs accumulate compounds that induce MFO activity *in vitro*. These samplers are easy-to-use, time-integrative concentrating devices that can withstand conditions unsuitable for fish caging. SPMDs can detect spatial gradients of MFO inducers and can be used to detect where MFO inducers are produced in industrial processes. Reactive columns and maximal MFO responses of the cells also give clues as to the nature of the inducing compound. Preliminary studies show that compounds that induce MFO in fish will also induce MFO in fish cell lines dosed with extracts of SPMDs exposed to solutions of these compounds.

E. Future Applications

SPMDs are a useful tool for the concentration of natural lipophilic organic substances. Their best use is for screening, preliminary assessment, and for studies of the presence and distribution of MFO-inducing compounds, or organic chemicals in general. Several features of SPMD's concentration of MFO inducers are unknown. The use of SPMDs within industries appears useful, but should be examined more closely. It is unknown whether the severe conditions found in many of the in-plant process streams (pH 12, 60°C) may affect uptake of inducers. MFO inducers are relatively lipid-soluble, nonpolar molecules, so their concentration by SPMDs is feasible. If toxicity or mutagenicity were the endpoints of interest, SPMDs may not concentrate charged or less lipophilic mutagens or toxicants. More work is needed to relate induction of MFO in cell lines exposed to SPMD extracts to MFO induction in fish exposed to similar solutions. SPMDs are a relatively inexpensive and simple tool for estimating concentrations of MFO inducers in water. However, they cannot replace fish caging studies or surveys of wild fish, as they fail to integrate ecosystem and habitat effects. SPMDs can provide preliminary data on presence of MFO inducers, and can be used to rank sites or effluents for further study. They are useful for assessment of effluents under extreme conditions where fish could not survive. Their advantages ensure continued and expanded use of SPMDs as samplers. Rather than replacing current bioconcentration and caging tests, these devices will add another approach to sampling and concentrating lipophilic contaminants for biotesting.

IV. SUMMARY

This review has described practical methods for measuring the presence in effluent of compounds that induce the activity of CYP1A1 enzymes. These methods enable the assessment of: bioavailability of inducing compounds (fish MFO induction assay); amounts extractable from liquids, solids, or biota (cell line MFO induction assays); and total amounts present in effluents that would be toxic to fish or difficult to extract by liquid–liquid or liquid–solid methods (SPMDs combined with cell line assays). With this suite of tools, experiments on the sources, fates, transport, forms, and effects of MFO-inducing compounds are much more practical. While these tools have hastened research in these areas, it is clear that new techniques are emerging, particularly with genetically engineered cell lines. The cost, detection limits, and practicality of methods for detecting inducing compounds will continue to improve.

REFERENCES

1. Nelson, D. R., Kamataki, T., Waxman, D. J., Guengerich, F. P., Estabrook, R. W., Feyereisen, R., Gonzalez, F. J., Coon, M. J., Gunsalus, I. C., Gotoh, O., Okuda, K., and Nebert, D. W., The P450 superfamily: update on new sequences, gene mapping, accession numbers, early trivial names of enzymes, and nomenclature, *DNA Cell Biol.* 12, 1, 1993.
2. Addison, R. F., Hepatic mixed function oxidase (MFO) induction in fish as a possible biological monitoring system, in *Contaminant Effects on Fisheries*, Cairns, V. W., Hodson, P. V., and Nriagu, J. O., Eds., Wiley-Interscience, New York NY, 1984, 51.
3. Denison, M. S. and Whitlock, J. P., Xenobiotic-inducible transcription of cytochrome P450 genes, *J. Biol. Chem.* 270, 18175, 1995.
4. Stegeman, J. J., Cytochrome 450 forms in fish, in *Cytochrome P450*, Schenkman, J. B. and Greim, H, Eds., Vol. 105 of Handbook of Experimental Pharmacology, Springer Verlag, New York, 1992, Chap. 18.
5. Okey, A.B., Riddick, D. S., and Harper, P. A., The Ah receptor: Mediator of the toxicity of 2,3,7,8-tetrachlorodibenzo-*p*-dioxin (TCDD) and related compounds, *Tox. Lett.*, 70, 1, 1994.
6. Safe, S., Development, validation and limitations of toxic equivalency factors, *Chemosphere,* 25, 61, 1992.
7. Parrott, J. L., Hodson, P. V., Servos, M. R., Huestis, S. L., and Dixon, D. G., Relative potencies of polychlorinated dibenzo-*p*-dioxins and dibenzofurans for inducing mixed function oxygenase activity in rainbow trout, *Env. Toxicol. Chem.,* 14, 1041, 1995.
8. Hodson, P. V., Kloepper-Sams, P. J., Munkittrick, K. R., Lockhart, W. L., Metner, D. A., Luxon, L., Smith, I. R., Gagnon, M. M., Servos. M., and Payne J. F., Protocols for measuring mixed function oxygenases of fish liver, *Can. Tech. Rep. Fish. Aquat. Sci.* 1829, 49, 1991.
9. Hodson, P. V., Efler, S., Wilson, J. Y., El-Shaarawi, A., Maj, M., and Williams, T. G., Measuring the potency of pulp mill effluents for induction of hepatic mixed function oxygenase activity in fish, *J. Toxicol. Env. Health*, 49, 101, 1996.
10. Collier, T. K., Stein, J. E., Wallace, R. J., and Varanasi, U., Xenobiotic metabolizing enzymes in spawning English sole exposed to organic-solvent extracts of marine sediments from contaminated and reference areas, *Comp. Biochem. Physiol.* 84C, 291, 1986.
11. Hahn, M. E. and Stegeman, J. J., Regulations of cytochrome P450 1A1 in teleosts: sustained induction of CYP 1A1 mRNA, protein, and catalytic activity by 2,3,7,8 - tetrachlorodibenzofuran in the marine fish *Stenotomus chrysops*, *Toxicol. Appl. Pharmacol.*, 127, 187, 1994.
12. Muir, D. C. G., Yarechewski, A. L., Metner, D. A., Lockhart, L., Webster, G. B. R., and Friesen, K. J., Dietary accumulation and sustained hepatic mixed function oxidase enzyme induction by 2,3,4,7,8-pentachlorodibenzofuran in rainbow trout, *Environ. Toxicol. Chem.,* 9, 1463, 1990.
13. Lester, S. M., Braunbeck, T. A., Teh, S. J., Stegeman, J. J., Miller, M. R., and Hinton, D. E., Hepatic cellular distribution of cytochrome P-450 IA1 in rainbow trout (*Oncorhynchus mykiss*): an immuno-histo- and cytochemical study, *Cancer Res.*, 53, 3700, 1992.

14. Payne, J. F., Field evaluation of benzopyrene hydroxylase induction as a monitor for marine petroleum pollution, *Science*, 191, 945, 1976.

15. Payne, J. F., Fancey, L. L., Rahimtula, A. D., and Porter, E. L., Review and perspective on the use of mixed-function oxygenase enzymes in biological monitoring, *Comp. Biochem. Physiol*, 86C, 233, 1987.

16. Holdway, D. A., Brennan, S. E. and Ahokas, J. T., Use of hepatic MFO and blood enzyme biomarkers in sand flathead (*Platycephalus bassensis*) as indicators of pollution in Port Philip Bay, Australia, *Mar. Poll. Bull.*, 28, 683, 1994.

17. Andersson, T., Förlin, L., Härdig, J., and Larsson, A, Physiological disturbances in fish living in the receiving body of water of a kraft mill industry, *Can. J. Fish. Aquat Sci.*, 45, 1525, 1988.

18. Newsted, J. L., Giesy, J. P., Ankley, G.T., Tillitt, D.E., Crawford, R.A., Gooch, J.W., Jones, P.D., and Denison, M.S., Development of toxic equivalency factors for PCB congeners and the assessment of TCDD and PCB mixtures in rainbow trout, *Environ. Toxicol. Chem.*, 14, 861, 1995.

19. Parrott, J.L., Hodson, P.V., and Servos, M.R., Relative potency of polychlorinated dibenzo-p-dioxins and dibenzofurans for inducing mixed-function oxygenase activity in rainbow trout, *Environ. Toxicol. Chem.*, 14, 1041, 1995.

20. Lindström-Seppä, P. and Oikari, A., Biotransformation and other toxicological and physiological responses in rainbow trout (*Salmo gairdneri* Richardson) caged in a lake receiving effluents of pulp and paper industry, *Aquat. Toxicol.*, 16, 187, 1990.

21. Munkittrick, J.R., Blunt, B.R., Leggett, M., Huestis, S., and McCarthy, L.H., Development of a sediment bioassay to determine bioavailability of PAHs to fish, *J. Aquat. Ecosystem Health.* 4, 169, 1995.

22. Baumann, P.C., Harshbarger, J.C., and Hartman, K.J., Relationship between liver tumors and age in brown bullhead populations from two Lake Erie tributaries, *Sci. Total Envir.*, 94, 71, 1990.

23. Fabacher, D.L. and Baumann, P.C., Enlarged livers and hepatic microsomal mixed-function oxidase components in tumor-bearing brown bullheads from a chemically contaminated river, *Environ. Toxicol. Chem.*, 4, 703, 1985.

24. Collier, T. K., Stein, J.E., Sanborn, H.R., Hom, T., Myers, M.S., and Varanasi, U., Field studies of reproductive success and bioindicators of maternal contaminant exposure in English sole (*Parophrys vetulus*), *Sci. Total Envir.*, 116, 169, 1992.

25. Johnson, L., Casillas, E., Sol, S., Collier, T., Stein, J., and Varanasi, U., Contaminant effects on reproductive success in selected benthic fish, *Mar. Environ. Res.*, 35, 165, 1993.

26. McMaster, M. E., Van Der Kraak, G. J., Portt, C. B., Munkittrick, K. R., Sibley, P. K., Smith I. R., and Dixon, D. G., Changes in hepatic mixed-function oxygenase (MFO), plasma steroid levels and age at maturity of a white sucker (*Catostomus commersoni*) population exposed to bleached kraft mill effluent, *Aquat. Tox.*, 21, 199, 1991.

27. Gagnon, M.M., Bussières, D., Dodson J.J., and Hodson. P.V., White sucker (*Catostomus commersoni*) growth, maturation and reproduction in pulp mill contaminated and reference rivers, *Environ. Toxicol. Chem.*, 14, 317, 1995.

28. Munkittrick, K. R., Portt, C., Van Der Kraak, G. J., Smith, I. R., and Rokosh, D. A., Impact of bleached kraft mill effluent on liver MFO activity, serum steroid levels and population characteristics of a Lake Superior white sucker population, *Can. J. Fish. Aquat. Sci.*, 48, 1371, 1991.

29. Munkittrick, K. R., McMaster, M. E., Portt, C. B., Van Der Kraak, G. J., Smith, I. R., and Dixon, D. G., Changes in maturity, plasma sex steroid levels, hepatic MFO activity and the presence of external lesions in lake whitefish exposed to bleached kraft mill effluent, *Can. J. Fish. Aquat. Sci.*, 49, 1560, 1992.

30. Lehtinen, K-J, Kierkegaard, A., Jakobsson, E., and Wandell, A., Physiological effects in fish exposed to effluents from mills with six different bleaching processes, *Ecotox. Environ. Safety*, 19, 33, 1990.

31. Sodergren, A., Adolfsson, M., Bengtsson, B.-E., Jonsson, P., Lagergren, S., Rahm, L., and Wulff, F., Environmental effect of bleached pulp mill effluents discharged into the Baltic Sea, in *Environmental Fate and Effects of Bleached Pulpmill Effluent*, Swedish EPA Report 4031, Södergren, A., Ed., Proceedings of a Swedish EPA conference in Stockholm, Sweden, November 19-21, 1992

32. Hodson, P. V., McWhirter, M., Ralph, K., Gray, B., Thivierge, D., Carey, J., Van Der Kraak, G., Whittle, D. M., and Levesque, M.-C., Effects of bleached kraft mill effluent on fish in the St. Maurice River, Quebec, *Environ. Toxicol. Chem.*, 11, 1635, 1992.

33. Munkittrick, K. R., Van Der Kraak, G. J., McMaster M. E., and Portt, C. B., Relative benefit of secondary treatment and mill shutdown on mitigating impacts of bleached kraft mill effluent (BKME) on MFO activity and serum steroids in fish, *Environ. Toxicol. Chem.*, 11, 1427, 1992.

34. Munkittrick, K. R., Van Der Kraak, G. J., McMaster, M. E., Portt, C., van den Heuvel, M. R., and Servos, M. R., Survey of receiving water environmental impacts associated with discharges from pulp mills. II. Gonad size, liver size, hepatic EROD activity and plasma sex steroid levels in white sucker, *Environ. Toxicol. Chem.*, 13, 1089, 1994.

35. Lindstrom-Seppa, P., Huuskonen, S., Pesonen, M., Muona P., and Hanninen, O., Unbleached pulp mill effluents affect cytochrome-P450 monooxygenase enzyme activities, *Mar. Environ. Res.*, 34, 157, 1992.

36. Martel, P. H., Kovacs., T. G., O'Connor, B. I., and Voss, R. H., A survey of pulp and paper mill effluents for their potential to induce mixed function oxidase enzyme activity in fish, *Water Res.*, 28, 1835, 1994.

37. Martel, P. H., Kovacs, T. G., O'Connor B. I., and Voss, R. H., A laboratory exposure procedure for screening pulp and paper mill effluents for the potential of causing increased mixed function oxidase activity in fish, *Environ. Poll.*, 89, 229, 1995.

38. Hodson, P. V., Maj, M. K., Efler, S., Burnison, B. K., Van Heiningen, A. R. P., Girard, R., and Carey, J.H., MFO induction in fish by spent cooking liquors from kraft pulp mills. *Environ. Toxicol. Chem.*, 16, 908, 1997.

39. Forlin, L. and Haux, C., Sex differences in hepatic cytochrome P-450 monoxygenase activities in rainbow trout during an annual reproductive cycle, *J. Endocrinol.*, 124, 207, 1990.

40. Sprague, J. B. Measurement of pollutant toxicity to fish I. Bioassay methods for acute toxicity, *Water Res.*, 3, 793, 1969.

41. Williams, T. G., A comparative laboratory based assessment of EROD enzyme induction in fish exposed to pulp mill effluents. M.Sc. Thesis. Dept. of Biology, U. of Waterloo, Waterloo, Ontario. 85p p, 1993.

42. Pohl, R. J. and Fouts, J. R., A rapid method for assaying the metabolism of 7-ethoxyresorufin by microsomal subcellular fractions, *Anal. Biochem.*, 107, 150, 1980.

43. Schnell, A., Hodson, P. V., Steel, P., Melcer H., and Carey, J. H., Optimized biological treatment of bleached kraft mill effluents for the enhanced removal of toxic compounds and MFO induction response in fish. Proceedings of the 1993 Environment Conference of the Canadian Pulp and Paper Association, Thunder Bay, Oct. 26-28, 1993. 15 pp.

44. Gagné, F. and Blaise, C., Hepatic metallothionein level and mixed function oxidase activity in fingerling rainbow trout (*Oncorhynchus mykiss*) after acute exposure to pulp and paper mill effluents, *Water Res.*, 27, 1669, 1993.

45. Haasch, M.L., Quardokus, E. M., Sutherland, L. A., Goodrich, M. S., and Lech, J. J., Hepatic CYP1A1 induction in rainbow trout by continuous flowthrough exposure to beta-naphthoflavone, *Fund. Appl. Toxicol.*, 20, 72, 1993.

46. Campbell, P. M., Kruzynski, G. M., Birtwell I. K., and Devlin, R. H., CYP1A1 expression in BKME-exposed juvenile chinook salmon, *Environ. Toxicol. Chem.*, 15, 1119, 1996.

47. Eggans, M. L. and Galgani, F., Ethoxyresorufin-O-deethylase (EROD) activity in flatfish: fast determination with a fluorescence plate-reader, *Mar. Environ. Res.*, 33, 213, 1992.

48. Lorenzen, A., and Kennedy, S. W., A fluorescence-based protein assay for use with a microplate reader, *Anal. Biochem.*, 214, 346, 1993.

49. Munkittrick, K. R., van den Heuvel, M. R., Metner, D. A., Lockhart, W. L., and Stegeman, J. J., Interlaboratory comparison and optimization of hepatic microsomal ethoxyresorufin-o-deethylase activity in white sucker (*Catostomus commersoni*) exposed to bleached kraft pulp mill effluent, *Environ. Toxicol. Chem.*, 12, 1273, 1993.

50. van den Heuvel, M. R., Munkittrick, K. R., Stegeman, J. J., and Dixon, D. G., Second-round interlaboratory comparison of hepatic ethoxyresorufin-o-deethylase activity in white sucker (*Catostomus commersoni*) exposed to bleached kraft pulp mill effluent, *Env. Tox. Chem.*, 14, 1513, 1995.

51. Burnison, B. K., Hodson, P. V., Nuttley, D. J., and Efler, S., A BKME fraction causing induction of a fish mixed function oxygenase (MFO) enzyme, *Environ. Toxicol. Chem.* 15, 1524, 1996.

52. Freshney, R. I., *Culture of Animal Cells, A Manual of Basic Technique*, Wiley-Liss, New York, 1987.

53. Bols, N. C. and Lee, L. E. J., Technology and uses of cell cultures from the tissues and organs of bony fish, *Cytotechnology*, 6, 163, 1991.

54. Isomaa, B., Lilius H., and Rabergh, C., Aquatic toxicology *in vitro* — a brief review, *ATLA*, 22, 243, 1994.

55. Niwa, A., Kumaki K., and Nebert, D. W., Induction of aryl hydrocarbon hydroxylase activity in various cell cultures by 2,3,7,8-tetrachlorodibenzo-p-dioxin, *Molec. Pharmacol.*, 11, 399, 1975

56. Mason, G., Farrell, K., Keys, B., Piskorska-Pliszczynska, J., Safe L., and Safe, S., Polychlorinated dibenzo-p-dioxins: quantitative *in vitro* and *in vivo* structure-activity relationships, *Toxicology*, 41, 21, 1986.

57. Bandiera, S., Sawyer, T., Romkes, M., Zmudzka, B., Safe, L., Mason, G., Keys B., and Safe, S., Polychlorinated dibenzofurans, (PCDFs): effects of structure on binding to the 2,3,7,8-TCDD cytosolic receptor protein, AHH induction and toxicity, *Toxicology*, 32, 131, 1984.

58. Sawyer, T. and Safe, S., PCB isomers and congeners: Induction of aryl hydrocarbon hydroxylase and ethoxyresorufin-o-deethylase activities in rat hepatoma cells, *Toxicol. Lett.*, 13, 87, 1982.

59. Casterline, J. J., Bradlaw, J. A., Puma B. J., and Ku, Y., Screening of fresh water fish extracts for enzyme inducing substances by an aryl hydrocarbon hydroxylase induction bioassay technique, *J. Assoc. Off. Anal. Chem.*, 66, 1136, 1983.

60. Zacharewski, T., Safe, L., Safe, S., Chittim, B., DeVault, D., Wiberg, K., Berqvist P.-A., and Rappe, C., Comparative analysis of polychlorinated dibenzo-p-dioxin and dibenzofuran congeners in Great Lakes fish extracts by gas chromatography-mass spectrometry and *in vitro* enzyme induction activities, *Environ. Sci. Technol.*, 23, 730, 1989.

61. Tillitt, D. E., Giesy, J. P., and Ankley, G. T., Characterization of the H411E rat hepatoma cell bioassay as a tool for assessing toxic potency of planar halogenated hydrocarbons in environmental samples, *Environ. Sci. Technol.*, 25, 87, 1991.

62. Tillitt, D. E., Ankley, G. T., Verbrugge, D. A., Giesy, J. P., Ludwig J. P., and Kubiak, T. J., H411E rat hepatoma cell bioassay-derived 2,3,7,8-tetrachlorodibenzo-p-dioxin equivalents in colonial fish eating waterbird eggs from the Great Lakes, *Arch. Env. Contam. Toxicol.*, 21, 91, 1991a.

63. Ankley, G. T., Tillitt, D. E., Giesy, J. P., Jones P. D., and Verbrugge, D. A., Bioassay-derived 2,3,7,8-tetrachlorodibenzo-p-dioxin equivalents in PCB-containing extracts from the flesh and eggs of Lake Michigan chinook salmon (*Oncorhynchus tshawytscha*) and possible implications for reproduction, *Can. J. Fish. Aquat. Sci*, 48, 1685, 1991.

64. Hanberg, A., Stahlberg, M., Georgellis, A., de Witt, C., and Ahlborg, U. G., Swedish dioxin survey: evaluation of the H411E bioassay for screening environmental samples for dioxin-like enzyme induction, *Pharmacol. Toxicol.*, 69, 442, 1991.

65. Van den Heuvel, M. R., Munkittrick, K. R., Van der Kraak, G. J., McMaster, M. E., Portt, C. B., Servos, M. R., and Dixon, D. G., Survey of receiving-water environmental impacts associated with discharges from pulp mills. 4. Bioassay-derived 2,3,7,8-tetrachlorodibenzo-p-dioxin toxic equivalent concentration in white sucker (*Catostomus commersoni*) in relation to biochemical indicators of impact, *Environ. Toxicol. Chem.*, 13, 1117, 1994.

66. Denison, M. S., Garrison, P. M., Aarts, J. M. M. J. G., Tullis, K., Schalk, J. A. C., Cox M. A., and Brouwer, A., Species-specific differences in Ah receptor ligand binding and transcriptional activation: implications for bioassays for the detection of dioxin-like activity, *Organohalogen Compounds*, 23, 225, 1995.

67. Denison, M. S., Phelps, C. L., DeHoog, J., Kim, H. J., Bank P. A., and Yao, E. F., Species variation in Ah receptor transformation and DNA binding. In M.A. Gallo, R.J. Schuplein and K.A. Van Der Heijen, eds., *Biological Basis of Risk Assessment of Dioxins and Related Compounds*, Cold Spring Harbor Press, Cold Spring Harbor, NY, 1991, 337.

68. Clemons, J. H., Van den Heuvel, M. R., Stegeman, J. J., Dixon D. G., and Bols, N. C., Comparison of toxic equivalent factors for selected dioxin and furan congeners derived using fish and mammalian liver cell lines, *Can. J. Fish. Aquatic. Sci.*, 51, 1577, 1994.

69. Lorenzen, A. and Okey, A. B., Detection and characterization of [^{3}H]2,3,7,8-tetrachlorodibenzo-p-dioxin binding to Ah receptor in a rainbow trout hepatoma cell line, *Toxicol. Appl. Pharmacol.*, 106, 53, 1990.

70. Swanson, H. I. and Perdew, G. H., Detection of the *Ah* receptor in rainbow trout: use of 2-azido-3-[^{125}I]iodo-7,8-dibromodibenzo-p-dioxin in cell culture, *Toxicol. Lett.*, 58, 85, 1991.

71. Hahn, M. E., Poland, A., Glover E., and Stegeman, J. J., The Ah receptor in marine animals — phylogenetic distribution and relationship to cytochrome P-450 1A inducibility, *Mar. Environ. Res.*, 34, 87, 1992.

72. Hahn, M. E., Lamb, T. M., Schultz, M. E., Smolowitz R. M., and Stegeman, J. J., Cytochrome-P4501a induction and inhibition by 3,3',4,4'-tetrachlorobiphenyl in an Ah receptor-containing fish hepatoma cell line (PLHC-1), *Aquat. Toxicol.*, 26, 185, 1993.

73. Hahn, M. E., Woodward, B. L., Stegeman J. J., and Kennedy, S. W., Rapid assessment of induced cytochrome P4501A (CYP1A) protein and catalytic activity in fish hepatoma cells grown in multi-well plates, *Environ. Toxicol. Chem.*, 15, 582, 1996.

74. Huuskonen, S. and Lindstrom-Seppa, P., *in vitro* biomonitoring of pulp and paper industry with fish hepatoma cell line (PLHC-1). Abstract P61, Fifth SETAC-Europe Congress, Copenhagen 25-28 June, 1995, SETAC-Europe, Brussels.

75. Lee, L. E. J., Clemons, J. H., Bechtel, D. G., Caldwell, S. J., Han, K. B., Pasitschniakarts, M., Mosser D. D., and Bols, N. C., Development and characterization of a rainbow trout liver cell line expressing cytochrome P450-dependent monooxygenase activity, *Cell Biol. Toxicol.*, 9, 279, 1993.

76. Clemons, J. H., Lee, L. E. J., Myers, C. R., Dixon D. G., and Bols, N. C., Cytochrome P4501A1 induction by polychlorinated biphenyls (PCBs) in liver cell lines from rat and trout and the derivation of toxic equivalency factors (TEFs), *Can. J. Fish. Aquat. Sci.*, 53, 1177, 1996.

77. Rodman, L. E., Shedlofsky, S. I., Swim A. T., and Robertson, L. W., Effects of polychlorinated biphenyls on cytochrome P450 induction in the chick embryo hepatocyte culture, *Arch. Biochem. Biophys.*, 275, 252, 1989.

78. Yao, C., Panigraphy B., and Safe, S., Utilization of cultured chick embryo hepatocytes as *in vitro* bioassays for polychlorinated biphenyls, (PCBs): quantitative structure activity relationships, *Chemosphere*, 21, 1007, 1990.

79. Kennedy, S. W., Lorenzen, A., James C. A., and Norstrom, R. J., Ethoxyresorufin-o-deethylase, (EROD) and porphyria induction in chicken embryo hepatocyte cultures — a new bioassay of PCB, PCDD, and related chemical contamination in wildlife, *Chemosphere*, 25, 193, 1992.

80. Pesonen, M. and Andersson, T., Characterization and induction of xenobiotic metabolizing enzyme activities in a primary culture of rainbow trout hepatocytes, *Xenobiotica*, 21, 461, 1991.

81. Pesonen, M. and Andersson, T., Toxic effects of bleached and unbleached paper mill effluents in primary cultures of rainbow trout hepatocytes, *Ecotox. Environ. Safety*, 24, 63, 1992.

82. Jansing, R. L. and Shain, W., Aryl hydrocarbon hydroxylase induction in adult rat hepatocytes in primary culture by several chlorinated aromatic hydrocarbons including 2,3,7,8-tetrachlorodibenzo-p-dioxin, *Fundam. Appl. Toxicol.*, 5, 713, 1985.

83. Baksi, S. M. and Frazier, J. M., Isolated fish hepatocytes — model systems for toxicology research, *Aquat. Toxicol.*, 16, 229, 1990.

84. Pesonen, M., Goksoyr A., and Andersson, T., Induction of xenobiotic metabolizing enzyme activities in primary culture of rainbow trout hepatocytes, *Resp. Marine Org. Pollutants*, 28, 113, 1989.

85. Hanninen, O., Lindstrom-Seppa, P., Pesonen, M., Huuskonen S., and Muona, P., Use of biotransformation activity in fish and fish hepatocytes in the monitoring of aquatic pollution caused by pulp industry, in *Bioindicators and Environmental Management*, Jeffrey, D.W. and Madden, B., Eds., Academic Press, London, 1991, 13.

86. Gagné, F. and Blaise, C., Fluorescence *in situ* hybridization *en* suspension using biotin-labeled DNA probes for measuring genetic expression of metallothionein and cytochrome P-450 1A1 (CYP1A1) in rainbow trout hepatocytes exposed to wastewaters, in *Environmental Toxicology Assessment*, Richardson, M., Ed., Taylor & Francis, Hertfordshire, UK, 1996, 41.

87. Tillitt, D. E., Kubiak, T. J., Ankley G. T., and Giesy, J. P., Dioxin-like toxic potency in Forster's tern eggs from Green Bay, Lake Michigan, North-America, *Chemosphere*, 26, 2079, 1993.

88. Kennedy, S., Lorenzen A., and James, C. A., A rapid and sensitive cell culture bioassay for measuring ethoxyresorufin-o-deethylase (EROD) activity in cultured hepatocytes exposed to halogenated aromatic hydrocarbons extracted from wild bird eggs, *Chemosphere*, 27, 367, 1993a.

89. Kennedy, S.W., Lorenzen, A., James, C.A., and Collins, B.T., Ethoxyresorufin-o-deethylase and porphyrin analysis in chicken embryo hepatocyte cultures with a fluorescence multi-well plate reader. *Anal. Biochem.* 211, 102, 1993.

90. Kennedy, S. W., Jones S. P., and Bastien, L. J., Efficient analysis of cytochrome P4501A catalytic activity, porphyrins, and total proteins in chicken embryo hepatocyte cultures with a fluorescence plate reader, *Anal. Biochem.*, 226, 362, 1995.

91. El Fouley, M. H., Richter, C., Giesy J. P., and Denison, M. S., Production of a novel recombinant cell line for use as a bioassay system for detection of 2,3,7,8-tetrachlorodibenzo-p-dioxin-like chemicals, *Environ. Toxicol. Chem.*, 13, 1581, 1994.

92. Anderson, J. W., Rossi, S. S., Tukey, R. H., Vu, T., and Quattrochi, L. C., A biomarker, P450 RGS, for assessing the induction potential of environmental samples, *Environ. Toxicol. Chem.,* 14, 1159, 1995.

93. Zacharewski, T. R., Berhane, K., Gillesby B. E., and Burnison, B. K., Detection of estrogen- and dioxin-like activity in pulp and paper mill black liquor and effluent using *in vitro* recombinant receptor reporter gene assays, *Environ. Sci. Technol.,* 29, 2140, 1995.

94. Bank, P. A., Yao, E. F., Phelps, C. L., Harper P. A., and Denison, M. S., Species-specific binding of transformed Ah receptor to a dioxin responsive transcriptional enhancer, *Eur. J. Pharmacol. Environ. Toxicol.,* 258, 85, 1992.

95. Denison, M. S., Wilkinson C. F., and Okey, A. B., Ah receptor for 2,3,7,8-tetrachlorodibenzo-p-dioxin: Comparative studies in mammalian and non-mammalian species, *Chemosphere,* 15, 1665, 1986.

96. Sadar, M., Ash R., and Andersson, T., Phenobarbital induction of CYP1A1 gene expression in a primary culture of rainbow trout hepatocytes, Abstract, p 354, 10th International Symposium on Microsomes and Drug Oxidations, Toronto, Ontario, July 18-21, 1994.

97. Sidhu, J. S., Kavanagh, T. J., Reilly M. T., and Omiecinski, C. J., Direct determination of functional activity of cytochrome P4501A1 and NADPH DT-diaphorase in hepatoma cell lines using noninvasive scanning laser cytometry, *J. Toxicol. Env. Health.,* 40, 177, 1993.

98. Huckins, J. N., Tubergen M. W., and Manuweera, G. K., Semipermeable membrane devices containing model lipid: A new approach to monitoring the bioavailability of lipophilic contaminants and estimating their bioconcentration potential, *Chemosphere,* 20, 533, 1990.

99. Huckins, J. N., Petty, J. D., Lebo, J. A., Orazio, C. E., Prest, H. F., Tillitt, D. E., Ellis, G. S., Johnson, B. T., and Manuweera, G. K., Semipermeable membrane devices (SPMDs) for the concentration and assessment of bioavailable organic contaminants in aquatic environments," in *Techniques in Aquatic Toxicology,* Ostrander, G., Ed., Lewis Publishers, Boca Raton, 1995.

100. Johnson, B. T., Collection and detection of petroleum products from air, water, and sediment: SPMDs and Microbics toxicity test systems as a screening tool. Presented at 3rd Annual SPMD Workshop and Symposium, Columbia, Missouri, USA, June 13-15, 1995.

101. Johnson, T., Huckins, J., Petty J., and Butorin, A., Rapid collection, detection, and assessment of environmental polycyclic aromatic hydrocarbons (PAHs). Abstract PH 282, Conference Abstracts of the 2nd SETAC World Congress, November 5-9, 1995, Vancouver, B.C., Society of Environmental Toxicology and Chemistry, Pensacola, Florida.

102. Lebo, J. A., Huckins, J. N., Petty, J. D., Orazio, C. E., Gibson V. L., and Ho, K., Investigations into methods of removing from marine sediments that toxicity attributable to organic contaminants. Abstract PW 173, Conference Abstracts of the 2nd SETAC World Congress, November 5-9, 1995, Vancouver, B.C., Society of Environmental Toxicology and Chemistry, Pensacola, Florida.

103. Metcalfe, T. L., White, P., Mackay D., and Metcalfe, C., Development of bioassay techniques with extracts from semi-permeable membrane devices (SPMDs). Abstract PH 293, Conference Abstracts of the 2nd SETAC World Congress, November 5-9, 1995, Vancouver, B.C., Society of Environmental Toxicology and Chemistry, Pensacola, Florida.

104. Huckins, J. N., Petty, J. D., Orazio, C. E., Lebo, J. A., Gibson V. L., Clark, R. C., Rostad, C. E., Gala, W. R., and Kaiser, E. M., Effects of temperature, concentration and biofouling on semipermeable membrane device (SPMD) sampling rates in water. Abstract PH 283, Conference Abstracts of the 2nd SETAC World Congress, November 5-9, 1995, Vancouver, B.C., Society of Environmental Toxicology and Chemistry, Pensacola, Florida.

105. Parrott, J. L., Tillitt, D. E., Hodson, P. V., Bennie, D. T., Huckins, J. N., and Petty, J. D., Semipermeable membrane devices (SPMDs) accumulate inducer(s) of fish mixed function oxygenase (MFO) from pulp mill effluents. Abstract, p 61 of the Conference Abstracts, 2nd International Conference on Environmental Fate and Effects of Bleached Pulp Mill Effluents, Vancouver, British Columbia, Canada, November 6-9, 1994, National Water Research Institute, Environment Canada, Burlington, Ontario.

106. Parrott, J. L., Hodson, P. V., Tillitt, D. E., Bennie D. T., and Comba, M. E., Semipermeable membrane devices (SPMDs) accumulate inducers of fish MFO from pulp mill and oil refinery effluents. Conference Abstracts, 22nd Annual Aquatic Toxicity Workshop, St. Andrews, New Brunswick, Canada, Oct. 1-5, 1995, Department of Fisheries and Oceans, St. Andrews Biological Station, St. Andrews, New Brunswick.

107. Villeneuve, D. L., Crunkilton R. L., and DeVita, W. M., Assessment of toxic potency of complex mixtures of polynuclear aromatic hydrocarbons (PAHs) from Lincoln Creek, Milwaukee, WI. Abstract PH 036, Conference Abstracts of the 2nd SETAC World Congress, November 5-9, 1995, Vancouver, B.C., Society of Environmental Toxicology and Chemistry, Pensacola, Florida.
108. Parrott, J. L. and Hodson, P. V., unpublished data.
109. Bradlaw, J. A., Garthoff, L. H., Hurley N. E., and Firestone, D., Aryl hydrocarbon hydroxylase activity of twenty-three halogenated dibenzo-p-dioxins, *Toxicol. Appl. Pharmacol.*, 37, 119, 1976.
110. Bradlaw, J. A., Garthoff, L. H., Hurley N. E., and Firestone, D., Comparative induction of aryl hydrocarbon hydroxylase activity *in vitro* by analogues of dibenzo-p-dioxin, *Food Cosmet. Toxicol.*, 18, 627, 1980.
111. Bradlaw, J. A. and Casterline, J. L., Jr, Induction of enzyme activity in cell culture: A rapid screen for detection of planar polychlorinated organic compounds, *J. Assoc. Off. Anal. Chem.*, 62, 904, 1979.
112. Safe, S., Mason, G., Farrell, K., Keys, B., Piskorsa-Pilszezynska, J., Madge J. A., and Chittim, B., Validation of *in vitro* bioassays for 2,3,7,8, TCDD-equivalents, *Chemosphere*, 16, 1723, 1987.
113. Zacharewski, T., Harris, M., Safe, S., Thomas, H., and Hutzinger, O., Applications of the *in vitro* aryl hydrocarbon induction assay for determining "2,3,7,8-tetrachlorodibenzo-p-dioxin equivalents": pyrolyzed brominated flame retardants, *Toxicol.*, 51, 177, 1988.
114. Hanberg, A., Waern, F., Asplund, L., Haglund E., and Safe, S., Swedish dioxin survey: determination of 2,3,7,8-TCDD toxic equivalent factors for some polychlorinated biphenyls and naphthalenes using biological tests, *Chemosphere*, 20, 1161, 1990.
115. Pesonen, M., Goksöyr A., and Andersson, T., Expression of P4501A1 in a primary culture of rainbow trout hepatocytes exposed to beta-naphthoflavone or 2,3,7,8-tetrachlorodibenzo-p-dioxin, *Arch. Biochem. Biophys.*, 292, 228, 1992.

CHAPTER 5

Enzyme Inhibition for Examination of Toxic Effects in Aquatic Systems

U. Obst, A. Wessler, and M. Wiegand-Rosinus

CONTENTS

I. INTRODUCTION

Enzymes are known to play a crucial role in the metabolism of all organisms. Therefore, we must turn our attention to enzymatic analysis. The following paper presents two applications of enzymatic analysis considering toxic effects in aquatic systems. On one hand, enzymes can be used as biologic catalysts for the determination of reactive substances, e.g., biological compounds or substances influencing biological reactions, i.e., coenzymes, drugs, pesticides etc., and so take part in the metabolism of living cells. In this respect enzymatic analysis is a division of analytical chemistry[1] and described as *enzyme systems in vitro* using commercially available enzymes. On the other hand, the determination of the catalytic activities of enzymes in various biological matter is useful for diagnostic purposes. Alteration of enzymatic activities determined by an *in vivo test system* — for example, the direct determination of enzyme activities of a microbial population in water samples — can be used to identify toxic effects in the ecosystem. Thus, several authors are convinced that the functioning of an aquatic ecosystem, e.g., seas, oceans, or rivers, cannot be understood without the participation of enzymatic processes.[2,3,4]

In vitro enzyme systems are used as diagnostic aids in medical laboratories to determine blood constituents and activities of catalytic enzymes. In food chemistry, carbohydrates, organic acids, and alcohols are quantified by enzymatic analysis.[1] For these applications the commercially available enzymes are stabilized to guarantee the optimal substrate turnover without being influenced by matrix. The enzyme tests which are introduced in the following pages are based on the specific inhibition of the enzymatic reaction by single substances or specific groups of substances. Those tests are widely used to screen for high efficiency in the field of developing new pesticides and new pharmaceutics. (Unfortunately they are not published.) But they can also be used to detect inhibitors of the enzymes. The three tests described — acetylcholinesterase inhibition test, aldehyde dehydrogenase inhibition test, and urease inhibition test — can be used as screening tests in aquatic systems.

Strong inhibitors of acetylcholinesterase are esters of phosphorous acids and carbamates,[5,6] which are used worldwide as insecticides.[7,8] This specific inhibition can be used to screen environmental samples for phosphorous acetic acids and carbamates.[9] The inhibition of acetaldehyde dehydrogenase in human cells, known as the Antabus-Effect, is performed by calcium cyanamide, an artificial manure, or tetramethylthiuram disulfide, a medication against alcohol abuse.[5] The inhibition of acetaldehyde dehydrogenase in environmental samples indicates fungicides of the ethenbithiocarbamate type. The urease of living cells is very sensitive to heavy metal.[10,11] The sensitivity of the isolated enzyme can be used to detect heavy metals in the environment.

Though the isolated enzymes are very sensitive to special inhibitors, the measured inhibition cannot exclude unspecific inhibition by other substances. Therefore those tests are screening tests demanding further chemical identification of the inhibitor.

The capability to measure *enzymatic activities in vivo* in water samples is mainly based on degradation processes of contaminants by microorganisms. Microorganisms degrade natural and anthropogenic contaminants in water enzymatically for their own anabolism and the production of energy. They catabolize the water contaminants extracellularly by exo-enzymes or they transport them into the lumen of the cell for intracellular metabolism. Thus, the bacterial activity has a strong influence on the concentration and speciation of dissolved organic molecules in water.[2,14] The determination of enzymatic activities *in vivo* is therefore an indicator for microbial degradation of water contaminants and allows the control of natural biological purification processes in water samples.

Recently, the importance of enzymatic processes in aquatic habitats has received increasing recognition, and the methods proposed are now well established.[15-22] Enzyme activities of the microorganisms in a water system are simple to measure with biochemical methods. These are based on the application of chromogenic or fluorogenic model substrates to the native water samples and their enzymatical cleavage. They have become a powerful tool for investigations into the biological processes mentioned above.[18,22,23] Sometimes, however, pollutants inhibit the microbial enzyme activities and thus impair the effectiveness of the enzymatic transformation and degradation of water contaminants.[13,23-25] As a result of our work in Germany, the following chapter describes only a few examples of our experiences and of the application of *in vitro* and *in vivo* enzyme tests.

II. METHODS

A. *In Vitro* Enzyme Activity Tests

The tests are based on measuring the activity of a specific solvated enzyme of defined concentration and purity in a water sample. To detect an inhibitor it is necessary to register the uninhibited enzymatic activity as well as the enzymatic activity inhibited by a known and powerful inhibitor. To detect inhibitors in the aquatic ecosystem in concentrations as low as possible, the test conditions are chosen to optimize the inhibitory effect.

1. Description of the Acetylcholinesterase Inhibition Test

The test is based on the cleavage of the substrate acetylthiocholine iodide into thiocholine and acetic acid by acetylcholinesterase.[26] Thiocholine reacts with Ellman's reagent (5,5'-Dithio-(2-nitro benzoic acid) to 5-thio-2-nitrobenzoic acid, a dye photometrically measurable (wavelength 420 mn).[27] The test protocol described below refers to the German Industry Norm (DIN).[28]

The test has to be carried out in four steps:

Step 1: **Concentration** of insecticide phosphorous acid esters and carbamates in water samples to increase the sensitivity (concentration factor of 14):

Adsorb 20 mL of the sample on a column filled with an adsorbing resin (EXTRELUT*).
Elute with 80 mL of dichloromethane for about 30 minutes.
Concentrate in a rotating evaporator 40°C and dry in a stream of nitrogen
Dissolve the residue in 1 mL of deionized water

Step 2: **Oxidation** of thiophosphorous pesticides increases the toxicity of these substances toward the cholinesterases[29]

Oxidate with N-bromosuccinimide (0.05 mL of a solution of 400 mg/L)
Reduce residual bromosuccine with ascorbic acid (0.05 mL of a solution of 4 g/L) in a phosphate buffer (1.5 mol/L, pH 7.2)

Step 3: **(Inhibition test)**

Incubate the enzyme (0.1 mL of acetylcholinesterase (800 U/L)) with the concentrated and oxidized sample (0.05 mL) for half an hour at 25°C (inhibition mixture).

Step 4: **(Enzymatic reaction)**

2.9 mL of phosphate buffer (0.05 mol/L, pH 7.2) plus 0.1 mL acetylthiocholine iodide (78 mmol/L) and 0.1 mL reagent 5,5'-Dithio-(2-nitrobenzoic acid) (7.8 mmol/L)
Incubate for 10 min at 25°C
Add 0.1 mL of the inhibition mixture (starting the reaction)
Measure the extinction E_0 2 min after starting the reaction
Monitor the extinction after 30, 60, and 90 sec.

If the difference in the extinction of two consecutive measurements is almost constant, the average value is taken to calculate the enzymatic activity. All substances and cuvettes have to be equilibrated to 25°C. During the enzymatic reaction, this temperature must be kept constant.

For a single substance, the inhibitory effect on the enzyme can be expressed as the inhibition constant K_i in $L \times mol^{-1} \times min^{-1}$. The higher the constant, the more toxic the substance is. Because the measured inhibition effect depends not only on the concentration of a single substance but might also depend on the interfering effects of a mixture of substances with different toxicities, the result cannot be given as a sum of concentrations in mg/L. The toxicity of a reference substance — paraoxon — has to be used as a standard. Therefore the inhibitory effect measured in the test is given as "equivalents of paraoxon" in μg/L.[28]

2. Description of the Aldehyde Dehydrogenase Inhibition Test

Aldehyde dehydrogenase oxidizes acetic aldehyde to acetic acid. Because the reduction of NAD to NADH is coupled to this reaction, the enzyme activity can be measured by the increasing extinction of NADH.

* Registered trademark of Merck AG, Darmstadt, Germany.

Table 5.1   Aldehyde Dehydrogenase Inhibition Test: Test Protocols for Cuvettes and Microtiter Plates

		Cuvettes	Microtiter Plates
Control	Deionized water	1900 μL	100 μL
	Phosphate buffer	1000 μL (96 mm)	50 μL (95 mm)
	Enzyme	20 μL (24 U/L)	10 μL (12 U/L)
	Substrate	100 μL (73 μm)	50 μL (26 μm)
Sample	Deionized water	900 μL	—
	Sample	1000 μL	100 μL
	Phosphate buffer	1000 μL (96 mm)	50 μL (95 mm)
	Enzyme	20 μL (24 U/L)	10 μL (12 U/L)
	Substrate	100 μL (73 μm)	50 μL (26 μm)
	Coenzyme	100 μL (1.57 mm)	10 μL (0.75 mm)

Data from Wiegand-Rosinus, M., Untersuchungen zur Hemmbarkeit von Enzymes in vitro durch Pestizide, Doctoral Thesis, Fachbereich Biologie, U. Johannes-Gutenberg-Universitat, Mainz, 1991.

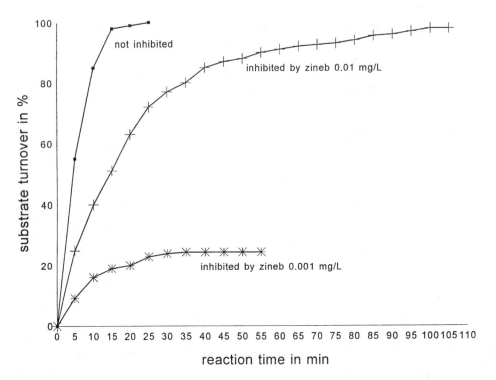

Figure 5.1   Substrate turnover per time. (From Wiegand-Rosinus, M., Untersuchungen zur Hemmbarkeit von Enzymen *in vitro* durch Pestizide, *Doctorial Thesis*, Fachbereich Biologie d. Johannes-Gutenberg-Universität Mainz, 1991.)

The dissolved enzyme is incubated with the water sample for half an hour. The enzyme reaction is started with NAD; the developing NADH is monitored for almost 2 hours. The detailed parameters of the test protocol are presented in Table 5.1.[31] Besides the sample itself, it is necessary to measure a control sample which is the uninhibited enzyme activity, a blank containing enzyme and coenzyme in buffer, and a sample blank containing sample, enzyme, and coenzyme in buffer as well. The substrate is the rate-limiting parameter shown in the kinetic reaction in Figure 5.1.[31] The turnover

Table 5.2   Urease Inhibition Test

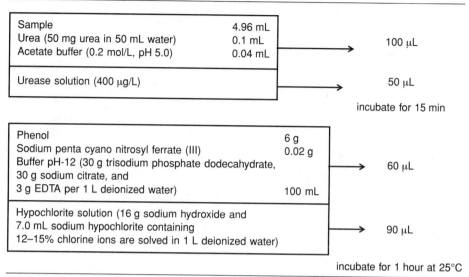

Sample	4.96 mL	
Urea (50 mg urea in 50 mL water)	0.1 mL	→ 100 µL
Acetate buffer (0.2 mol/L, pH 5.0)	0.04 mL	
Urease solution (400 µg/L)		→ 50 µL

incubate for 15 min

Phenol	6 g	
Sodium penta cyano nitrosyl ferrate (III)	0.02 g	
Buffer pH-12 (30 g trisodium phosphate dodecahydrate, 30 g sodium citrate, and		→ 60 µL
3 g EDTA per 1 L deionized water)	100 mL	
Hypochlorite solution (16 g sodium hydroxide and 7.0 mL sodium hypochlorite containing 12–15% chlorine ions are solved in 1 L deionized water)		→ 90 µL

incubate for 1 hour at 25°C

Data from Wittekindt, E., Werner, A., Reinicke, A., Herbert, A., and Hansen, P.-D., *Environ. Technol.,* Vol. 17, 597, 1996.

of the whole substrate should be accomplished within 20 minutes. The shape of the inhibition curve depends on the inhibitor and/or on its concentration. Because the velocity of the turnover can be decreased and/or the turnover itself is incomplete, the evaluation of the results can be done in different ways. One way is to consider only the decreased turnover rate announcing the concentration of inhibitor which causes a certain inhibition (20, 50, 80%) at 20 minutes reaction time; another is to measure the NOEC (no observed effective concentration) of the inhibiting substance.

3. Description of the Urease Inhibition Test [32]

Urease catalyzes the cleavage of urea to ammonia. Ammonia can be detected by phenol-nitroprussite, resulting in a blue color. The detailed description of the test schedule can be found in Table 5.2. The extinction of the blue dye is measured at 630 nm. The degree of inhibition is tested by dilution series to yield dose effect curves. It is necessary to measure the enzyme reaction without inhibition as an activity control and the inhibitory effect of a strong inhibitor of urease, e.g., of mercury, as a positive control.

B. *In Vivo* Enzyme Activity Tests

The applied methods for measuring enzyme activities *in vivo* are based on the addition of chromogenic or fluorogenic substrates to the native water samples. After incubation under standardized test conditions, the color of the enzymatically cleaved dyes is measured spectrometrically, and the results are calculated by the aid of calibration curves. In order to confirm results and to carry out numerous tests with dilutions of the samples, the tests are modified to an application on microtiter plates.[33] Several test protocols are available for measuring the activities of many steps of the microbial metabolism such as phosphatases, α-glucosidases, β-glucosidases, cellulases, peptidases, and so on (Table 5.3). According to the experiences we have gained in Germany in the last few years, in most cases esterase, alanine-aminopeptidase, and β-glucosidase activities were able to be measured. The detailed description of the test procedures is presented in Table 5.4.[22,38]

Table 5.3   Test Protocol for Measuring Esterase, Alanine Aminopeptidase, and β-Glucosidase Activities

	Esterases	Alanine Aminopeptidases	β-Glucosidases
Volume of the samples per assay	200 μL	200 μL	200 μL
Substrate	Fluorescein-diacetate	L-Alanine-4-nitroanilide hydrochloride	4-Nitrophenyl-β-D-glucopyranoside
Concentration of the substrate per assay	mol/assay 2.4×10^{-8}	mol/assay 1.2×10^{-7}	mol/assay 3.32×10^{-8}
Incubation time	24 h	24 h	24 h
Incubation temperature	room temperature	30°C	30°C
Stop-reagent		40 μL Trichloracetic acid (10%)	40 μL Sodium carbonate solution
Wavelength	490 nm	405 nm	405 nm

Data from Wessler, A., Dämgen, K., and Obst, U., AWWA Conference Proceedings of 1995 Water Quality Technology Conference, 905–919, 1995.

C. Detection of Inhibition Effects of Enzymatic Activities *In Vivo*

The harmful influence of pollutants is normally measured by dose-response curves in ecotoxicological studies.[5,39] Analogous to these dose-response curves, a similar test protocol to detect inhibition effects of enzymatic activities has been developed. Inhibition of enzymatic activities is detected using linear dilution of the water being tested and the correlation of the enzyme activities in each dilution step to the concentration of the particular samples. Therefore, the native water sample will be diluted in 12 linear steps with physiological sodium chloride solution. The enzyme activities are measured in each dilution step. The activity is plotted versus the dilution. A linear correlation shows an uninhibited enzyme activity. A decreased slope in higher concentrations of the sample indicates inhibition of enzyme activity (Figure 5.2).[40]

D. Assessment of Inhibited Enzymatic Activities *In Vivo*

Assessment of enzymatic inhibition is based on the dose-response curve. Five classes of inhibition are defined: class 1: no inhibition (straight line of the dilution slope or deviation from linearity between sample portion 81 to 100%); class 2: slight inhibition (deviation from linearity between sample portion 61 to 80%); class 3: moderate inhibition (deviation from linearity between sample portion 41 to 60%); class 4: severe inhibition (deviation from linearity between sample portion 21 to 40%); class 5: more severe inhibition (deviation from linearity between sample portion 0 to 20%) (Figure 5.2). The deviation from the linear relationship of a dose-response curve (graphical analysis) and the corresponding sample concentration determines the inhibition class into which the specific water sample will be grouped. Different locations are sampled several times and grouped into one of the five inhibition classes. The frequency with which the inhibition class will occur at one location is given in percent. In order to assess not only the frequency, but also the intensity of the detected inhibition, the defined inhibition classes are additionally graduated with a factor from zero to four. The factors will be multiplied with the frequency of the inhibition classes. The sum of all values is defined as an index of inhibition. The indices of inhibition range from 0 to 400. If all samples (100%) of one sample station measured several times never show enzymatic inhibition effects, the index of inhibition will be zero. If 50% of samples from a particular site (measured

Table 5.4 Microbial Enzyme Activities — Pathways, Enzymes, and Useful Photometrical and Fluorogenic Substrates

Metabolic Pathway	Enzymes	Photometrical Substrates	Fluorogenic Substrates	References
Nonspecific hydrolases	Esterases Phosphatases	Fluoresceindiacetate Nitrophenylphosphate	Fluoresceindiacetate MUF-phosphate	Obst 1995[22] Obst 1995[22]; Chrost 1991[23]
Carbohydrate metabolism	Amylases Cellulases Glucosidases	Amylopectin azure Carboxymethylcellulose Nitrophenyl-α/β-glucosides	MUF-cellobiopyranoside MUF-α/β-glucoside	Obst 1995[22] Obst 1995[22] Someville and Billen 1984[34]; Holzapfel et al. 1985[35]; Meyer-Reil 1986[21]; Marxen and Witzel 1990[14]
Protein metabolism	Proteases Alaninpeptidases Leucinpeptidases	Hide powder azure Alaninnitroanilidehydrochloride Leucin-p-nitroanilide	Casein resorufin L-Alanine-4-methoxy-β-naphthylamide L-Leucine-4-methyl-7-coumarinyl-amidehydrochloride	Obst 1995[22] Holzapfel et al. 1985[35] Jacobsen and Rai 1988[36] Reinheimer et al. 1989[37]
Electron transport system	Dehydrogenases	Iodonitrotetrazoliumchloride		Heininger and Tippmann 1995[49]

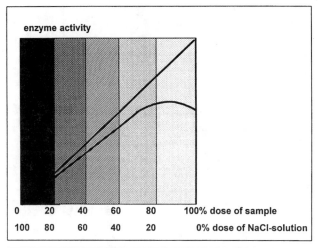

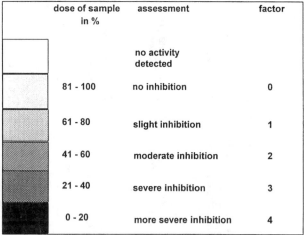

Figure 5.2   Assessment of enzymatic inhibition effects. (From Wessler, A., Dämgen, K., and Obst, U., AWWA Conference Proceedings of 1995 Water Quality Technology Conference, 905–919, 1995. With permission.)

several times) show slight inhibition effects (factor 1) and 50% of the samples show moderate inhibition effects (factor 2), the index of inhibition will be 150 (50 × 1 + 50 × 2 = 150). If all samples (100%) of a distinct sampling site show the severest inhibition effects, the index of inhibition will be 400. The greater the value of the index of inhibition, the worse is the water quality at the specific sampling site.[38]

E. Flow Injection Analysis of Microbial Enzyme Activity[41]

In vivo enzyme activity can also be measured in a flow stream of a bioreactor. The bioreactor consists of a glass column (height 10 cm, diameter 6 cm) filled with inert glass bowls (diameter 2 to 4 mm) populated with the microflora under test. The microorganisms are immobilized by pumping the water of interest through this fixed-bed reactor. The activity of the immobilized microorganisms can be determined by injection of a synthetic substrate flowing through the glass column. The microbial enzymes cleave the substrate dye bond, and the released dye can be detected by a fluorimetric spectrometer. For the measurement of fluorochromes it is necessary to buffer the sample. The buffer is injected into the flow stream behind the bioreactor (Figure 5.3).

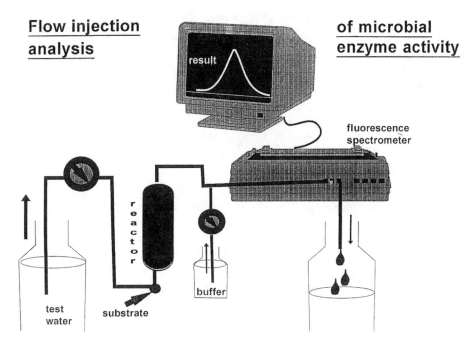

Figure 5.3   Flow diagram of the flow injection analysis of microbial enzyme activity (FAME).

III. RESULTS

A. Examples for Applications of *In Vitro* Enzyme Activity Tests

The sensitivity of the enzymes toward inhibitors has to be examined by testing single substances before screening natural samples.

1. *Acetylcholinesterase Inhibition Test*

Numerous choline esterases exist in the animal kingdom, partly specific to butyl or to acetyl groups as substrates. But besides acetyl or butyl other molecules, especially phosphorous acetic acids and carbamates, can interfere with the active center of the enzymes inhibiting the normal enzyme reaction.

In Table 5.5 the inhibition constants (K_i) of some specific inhibitors of acetylcholine esterase (from bovine erythrocytes) are listed.[28] The higher the K_i value, the more inhibiting the substance is. Esters of phosphorous acids and carbamates are applied as insecticides. Their toxic effects are not only specific to insects — since the 1930s, esters of phosphorous and phosphonic acids have been known to be powerful biocides and were used as chemical warfare agents. Therefore, it is of interest to monitor surface and ground waters which are often used as drinking water sources.

2. *Aldehyde Dehydrogenase Inhibition Test*

The aldehyde dehydrogenase used in this test is isolated from yeast. Acetaldehyde dehydrogenase detoxifies the acetaldehyde produced in cells of all organisms during oxidation of acetic alcohol. The product of the reaction is acetic acid. While acetaldehyde is oxidized, the coenzyme NAD is reduced; NADH can be measured photometrically and fluorimetrically.

Table 5.5   Inhibition Constants (K_i) of Acetylcholinesterase from Bovine Erythrocytes

Substance	K_I (L $\times$ mol^{-1} $\times$ min^{-1})	Substance	K_I (L $\times$ mol^{-1} $\times$ min^{-1})
Phosphoric Acid Esters		**Carbamate Insecticides**	
Azinphosethyloxone	1.0×10^6	Aldicarb	1.3×10^4
Chlorfenvinphos	0.9×10^4	Carbaryl	1.8×10^4
Demeton-s	2.4×10^4	Carbofuran	0.7×10^6
Diazinon	3.4×10^4	Mercaptodimet	5.7×10^4
Dichlorvos	2.3×10^4	Methidathion	0.7×10^5
Paraoxonethyl	0.6×10^6	Propoxur	3.4×10^4
Paraoxonmethyl	2.1×10^6		
Triazophos	0.7×10^6		

Data from German Standard methods for the examination of water, waste, water and sludge; subanimal testing DIN 38415 part 1, Determination of cholinesterase inhibiting organophosphorus and carbamate pesticides (cholinesterase inhibition test) T 1, *VCH Verlaggesellschaft*, Weinheim, 1995.

Table 5.6   Effect of Pesticides on the Activity of Aldehyde Dehydrogenase

Substance	No Observed Effective Concentration in mol/L	
	Cuvette Test	Microtiter Plate Test
Atrazine	1.2×10^{-4}	
Captafol	1.4×10^{-7}	$<2.4 \times 10^{-8}$
Carbendazim	5×10^{-4}	1.6×10^{-5}
Diuron	2.2×10^{-4}	2.2×10^{-6}
Dichlofluamide	3×10^{-8}	$<2.7 \times 10^{-8}$
Ferbam	2.4×10^{-9}	
Maneb	$<3.8 \times 10^{-9}$	$<2.6 \times 10^{-8}$
Metolachlor	1.8×10^{-4}	
Metazachlor		3.2×10^{-6}
Monuron	5×10^{-5}	1.3×10^{-5}
Thiram	2.1×10^{-7}	2.9×10^{-8}
Zineb	$<3.6 \times 10^{-9}$	1.8×10^{-7}
Ziram	1.6×10^{-5}	1.6×10^{-7}

Data from Wiegand-Rosinus, M., Untersuchungen zur Hemmbarkeit von Enzymen *in vitro* durch Pestizide, *Doctorial Thesis*, Fachbereich Biologie d. Johannes-Gutenberg-Universität Mainz, 1991.

As can be seen in Table 5.6 the enzyme is inhibited by dithiocarbamates, especially ethendithiocarbamates in very low concentrations.[42] The NOEC for maneb and zineb — two fungicides — is lower than 1 μg/L. Other organics, e.g., phenol, cannot inhibit this enzyme even at high concentrations.

3. Urease Inhibition Test

The first enzyme to be isolated, crystallized, and characterized is urease from leguminosae.[43] The enzyme catalyzes the cleavage of urea to carbon dioxide and ammonia. The test published in 1985 uses the change in pH for measurement of the enzyme activity.[44] The test described here is based on the photometric detection of the indophenol dye developing from the reaction of ammonia and hypochlorite and phenol. The test can be performed in microtiter plates.[32]

Urease is very sensitive to heavy metals such as mercury and copper. The other metals and some organic substances inhibit the enzyme in higher concentrations (Table 5.7).

Table 5.7   Concentrations of Inorganics and Organics Causing no Effect (NOEC) on the Urease

Substance	NOEC in mg/L	Substance	NOEC in mg/L
Cadmium	45	Acetone	<200
Chromium	11	DMSO	500
Copper	9.5×10^{-2}	Catechol	<0.5
Manganese	slight effect (0.5–100)	Methen methylsulfonic acid	1
Mercury	8×10^{-4}	1-methyl-3-nitro-1-nitrosoguanidin	9
Zinc	15	n-nitroso diethylamine	9
		2-aminoanthracene*	0.05
		4-nitroquinone-n-oxide*	1.5

* Genotoxic organic substances

B. Examples for an Application of Enzyme Activity Tests *In Vivo*

As a result of our work, three examples for an application of enzyme activity tests *in vivo* are given as follows. The first example describes the monitoring of the three tested enzymes in water samples of the River Main (Germany) over a period of two years. In the second example, the application of the new criteria for the assessment of water quality is shown. In the last, the control of toxic influences on specific biochemical processes using a flow injection analysis of microbial enzyme activities is presented.

In Figure 5.4 esterase, alanine aminopeptidase, and β-glucosidase activities of a natural (undiluted) water sample at a particular site in the River Main near Frankfurt (Germany) are shown. Sampling took place in 1993 every fortnight and in 1994 monthly in collaboration with the Hessische Landesanstalt für Umwelt. The activity of the three tested enzymes depends on seasonal changes and weather as expected. The esterase and alanine aminopeptidase activities increase in spring and summer and are lower in autumn and winter. The increasing values of these activities in spring and summer may be due to a much more pronounced production of a variety of primary metabolites and photosynthetic products during the active growth of phytoplankton measurable by the chlorophyll-a content. The increasing concentrations of utilizable substrates support bacterial growth and metabolism. The activity of ectoenzymes is generally inhibited by UDOM (utilizable dissolved organic matter). The synthesis of ectoenzymes in heterotrophic bacteria is repressed. The release of high amounts of polymeric dissolved organic matter (DOM) caused by great biomass turnover at the same time, however, induces the synthesis of ectoenzymes.[23] The results shown in Figure 5.4 indicate the pronounced effect of biomass turnover and of higher concentrations of polymeric organic matter. Obviously the inhibiting UDOM play a minor role. The β-glucosidase activities show a parallel trend to biomass determined as DNA-concentrations in the respective water samples. The activities are high in early spring, decrease in summer, and increase again in autumn. The β-glucosidase activities probably depend on the influence of weather as well. The activities increase significantly during and after rainy days depending on erosion, the whirl up of sediment, and the input of nutrients into the water.

Figure 5.5 shows the frequencies of the classes of inhibition of all tested enzymes from samples collected in the River Main near Frankfurt (Germany). In these samples the influence of the discharge of the industrial wastewater treatment plant of Hoechst AG is observed. The sampling site at Nied, for example, is located upstream of this discharge, whereas the sampling site at Eddersheim is located downstream. The sampling site Bischofsheim is also located downstream at a greater distance. The frequencies of inhibition effects of esterases in samples at the sampling sites near Nied and Bischofsheim are significantly lower than the frequency of inhibition in samples at Eddersheim. Measuring alanine aminopeptidase activities, in samples from Eddersheim the intensity of the detected inhibition effects is much more pronounced than in samples from the other

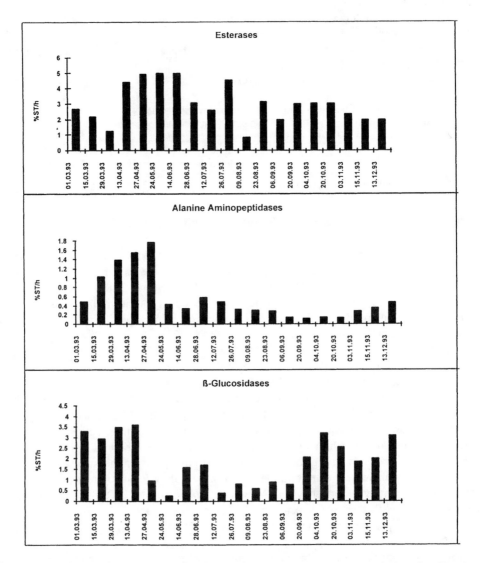

Figure 5.4 Substrate turnover in percent per hour (%ST/h) of the tested enzyme activities of undiluted water samples over a period of two years from the River Main at Frankfurt-Nied (Germany).

two sites. Similar results could also be observed when measuring the β-glucosidase activities. The indices of inhibition at the gauging stations of the River Main (presented in Table 5.8) allow a more pronounced assessment of the water quality. In contrast to the gauging stations near Nied and Bischofsheim the station near Eddersheim shows distinctly higher indices of inhibition for all tested enzymes. An observably higher index of inhibition of the alanine aminopeptidase activities at the gauging station at Bischofsheim than at the station at Nied upstream is also detectable. In 1994 a significant increase of the indices of inhibition of the alanine aminopeptidase activities of all tested gauging stations is observable. The supposition is that an increase in alanine aminopeptidase inhibiting substances is responsible for the detected deterioration.

Assessing the gauging stations of the River Main with the new criteria, it is obvious that downstream of the discharge of the industrial wastewater treatment plant of the Hoechst AG, a deterioration of the water quality and biological self purification is detectable. The frequency and the intensity of the observed inhibition effects as well as the index of inhibition increase downstream

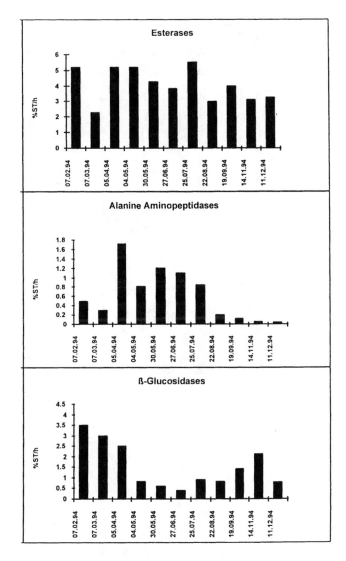

Figure 5.4 (continued)

of this discharge. After a more or less quiet reach of the river, the frequency and intensity of enzymatic inhibition effects decrease. These results indicate that the impairment of the water and the recreational value of the aquatic ecosystem is influenced by the discharge mentioned above. On the one hand, the impairing effect of the pollutants decreases as a result of dilution, but on the other hand, an increase in the microbial transformation and degradation processes is detectable downstream of the industrial discharges, reducing the concentration of pollutants. These degradation processes are measurable by an intensified enzymatic activity (data not shown).

Another application for the *in vivo* activity tests is to characterize the activity of an immobilized microbial population. In order to immobilize microorganisms on glass bowls, water containing the microflora to be tested has to be pumped through the glass column. The inoculation time depends on the density of the natural population. It can be controlled by measuring the enzymatic activity of the immobilized microorganisms. In Figure 5.6 the results of increasing esterase activity during inoculation are shown (velocity of flow: 20 mL/min). Peak height and area depend not only on the inoculation time but also on the speed of flow-through and on the substrate and the enzyme to be

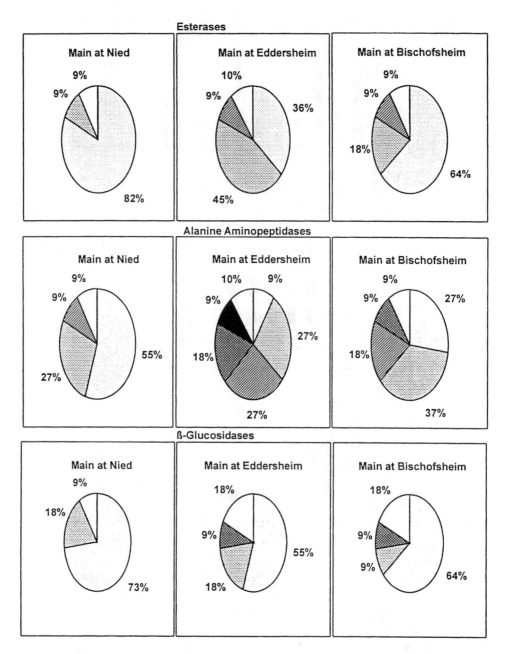

Figure 5.5   Frequencies (graduated in percent) of the detected inhibition classes of all tested enzyme activities of the sampling sites at the River Main in 1994: Sampling in 1994: 11 times a year. (Data from Wessler, A., Dämgen, K., and Obst, U., AWWA Conference Proceedings of 1995 Water Quality Technology Conference, 905–919, 1995.)

tested. After optimizing these parameters for samples of different origin, single substances can be injected and can interfere with the immobilized microorganisms. An inhibition of an enzyme is registered in the turnover curve as a decrease of height and area compared to the graph without inhibition. In Figure 5.7 the inhibiting effect of 1 mg/L penicillin on the microflora can be seen. If the immobilized microorganisms can regenerate, a successive testing of various substances or test waters with the same microflora is possible.

Table 5.8 Indices of Inhibition of All Gauging Stations at the Main River, Germany

Gauging Station	Esterases		Alanine Aminopeptidases		β-Glucosidases	
	1993	1994	1993	1994	1993	1994
Main at Nied	28	27	22	72	33	36
Main at Eddersheim	89	63	72	144	50	45
Main at Bischofsheim	28	36	47	99	39	36

Note: Sampling in 1993: 18 times. Sampling in 1994: 11 times a year.

Data from Wessler, A., Dämgen, K., and Obst, U., AWWA Conference Proceedings of 1995 Water Quality Technology Conference, 905–919, 1995.

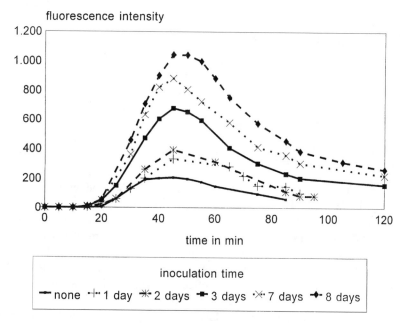

Figure 5.6 FAME. Enzyme activity during inoculating the bioreactor. (Data from Wiegand-Rosinus, M., Obst, U., Herzberger, E., and Haberer, K., *Proceedings DECHEMA Biotechnology Conferences*, 5, 903, 1992.)

IV. DISCUSSION AND PREDICTION OF ENZYME ACTIVITY TESTS

A. Discussion and Prediction of *In Vitro* Enzyme Activity Tests

The inhibition tests using enzymes *in vitro* indicate the effects of samples of unknown composition. In most cases the effect on the enzyme is an inhibition of the enzymatic activity. The substrate turnover can be slowed down; the reaction rate decreases, and/or the amount of product after reaction is smaller than without inhibitor. The inhibitory effect of a sample on an enzyme indicates a potentially toxic component for the organisms that rely on these enzymes. Although the inhibition can be very specific, unspecific reactions cannot be excluded. Therefore, these tests are screening tests indicating toxic effects. As the affinity of the single enzymes toward different inhibitors or different chemical groups of inhibitors is different (e.g., acetylcholinesterase is especially inhibited by phosphorous acid esters and carbamates; aldehyde dehydrogenase especially by ethendithiocarbamates;

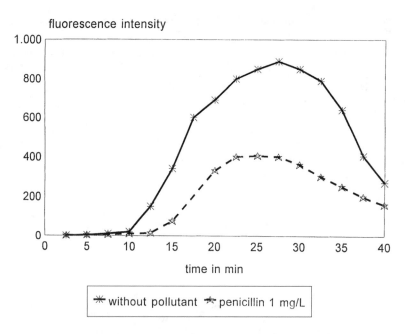

Figure 5.7   FAME: Inhibition of the activity of phosphatases by penicillin. (Data from Herzberger, E., Haberer, K., and Obst, U., *Vom Wasser,* 76, 199, 1991.)

urease by heavy metals), the inhibition of the enzyme is a clue to the chemical or biological character of the toxic water component. The sensitivity of the three tests against their specific inhibitors in water samples differs. The acetylcholinesterase test responds to inhibiting substances in the range of more than 1 µg/L and is therefore not practical within the German and EU drinking water regulation, where the limit concentration for one pesticide is 0.1 µg/L. But the test is helpful in screening surface water and wastewater influents. As the commercially available enzyme preparations are stabilized to be insensitive to matrix effects of samples (inhibition is not desired), the sensitivity toward the inhibitors can become a problem. Sometimes this problem can be solved by enhanced diluting of the enzyme, but it is absolutely necessary to control the sensitivity by a known concentration of a known inhibitor.

B. Discussion and Prediction of *In Vivo* Enzyme Activity Tests

Criteria for evaluation, characterization, and assessment of the quality of surface water are badly needed. Up to now, biological quality mapping of surface water has been based on the saprobic system.[46,47] Chemical analysis of single contaminants and summarizing parameters such as DOC and AOX is a standard to measure the quality of surface water. But mostly the water quality determined using the saprobic system or chemical analysis is much better now than in previous years. In the last few years trace contaminants and microorganics are the main source of pollution. Often, these pollutants are responsible for the inhibition of biological self purification, measurable by the recommended biochemical methods. Therefore, the methods suggested in this chapter, including the criteria for evaluation and assessment, provide an additional aspect of the biological water quality. The methods are applicable to biochemical monitoring of seasonal changes and influences of weather and, therefore, the dependent shift in nutrients and inhibiting substances. They allow the control of natural treatment processes in surface water as well as the control of discharges of wastewater treatment plants into the receiving water or in water after the subsoil run-off.[48] They can be helpful in controlling the purification in wastewater treatment plants. The methods are also applicable to investigations of sediments and suspended matter.[49] In contrast to the methods mentioned above, the flow injection analysis of microbial enzyme activity (FAME) was developed as a semi on-line control of specific biological processes in test water. The FAME allows the routine observation of influences on microorganisms by harmful substances. It can be applied to simulate accidents during drinking water conditioning or wastewater purification processes. But the method was not developed as a very sensitive toxicity assay.

REFERENCES

1. Bergmeyer, H. U., *Methods of Enzymatic Analysis*, Volume I, Verlag Chemie, Weinheim, 1983.
2. Overbeck, J. and Babezien, H.D., Über den Nachweis von freien Enzymen im Gewässer, *Archiv für Hydrobiologie*, 60, 107, 1964.
3. Hoppe, H.-G., Kim, S.-J. and Glocke, K., Microbial decomposition in aquatic environments: combined process of extracellular enzyme activity and substrate uptake, *Appl. Environ. Microbiol.*, 54, 784, 1988.
4. Overbeck, J., Early studies on ecto-and extracellular enzymes in aquatic environments, in *Microbial Enzymes in Aquatic Environments*, Chrost, R.J., Ed., Springer Verlag, Berlin, 1991.
5. Forth, W., Henschler, D., and Rummel, W., *Pharmakologie und Toxikologie*, BI Wissenschaftsverlag, Mannheim, 1990.
6. Watts, P. and Wilkinson, R. G., The interaction of carbamates with acetylcholinesterase, *Biochem. Pharmacol.*, 26, 757, 1977.
7. Schrader, G., *Die Entwicklung neuer insektizide Phosphorsäureester*, Verlag Chemie, Weinheim/Bergstraße, 1963.
8. Büchel, K. H., *Pflanzenschutz und Schädlingbekämpfungsmittel*, Georg Thieme Verlag, Stuttgart, 1977.
9. Herzsprung, P., Methodische Grundlagen des Nachweises und der Bestimmung von insektiziden Phopshorsäureestern und Carbamaten im Wasser mittels Cholinesterasehemmung, *Doctoral Thesis*, Inst.f. Hydrogeol., Hydrochem. u. Umweltanalytik d. TU, München, 1991.
10. Shaw, W. H. R., Raoul D. N., The inhibition of urease by metal ions at pH 8.9, *J. Am. Chem. Soc.*, 83, 3184, 1961.
11. Hughs, R. B., Katz, S.A., and Stubbins, S.E., Inhibition of urease by metal ions, *Enzymologia*, 36, 332, 1969.
12. Reinheimer, G., *Mikrobiologie der Gewässer*, Gustav Fischer Verlag, Jena, 1991.
13. Hoppe, H.-G., Microbial extracellular enzyme activity: A new key parameter in aquatic ecology, in *Microbial Enzymes in Aquatic Environments*, Chrost, R. J., Ed., Springer Verlag, Berlin, 1991.
14. Marxen, J. and Witzel, K.-P., Measurement of exoenzymatic activity in streambed sediment using methylumbelliferyl-substrates, *Arch. Hydrobiol. Beihefte Ergebnisse Limnol.*, 34, 21, 1990.
15. Admiral, W. and Tubbing, G., Extracellular enzyme activity associated with suspended matter in the River Rhine, *Freshwater Biology*, 26, 507, 1991.
16. Bergmeyer, H. U., Enzymatische analyse neuer generation, *Fresenius Z. Anal. Chem.*, 319, 883, 1984.
17. Brodish, K., Teuber, M. and Hegemann, W., Enzymaktivitäten als Kennwerte des belebten Schlammes, *gwf-Wass. Abw.*, 120, 524, 1979.
18. Teuber, M. and Brodish, K., Enzymatic activities of activated sludge, *Eur. J. Appl. Microb.*, 4, 185, 1977.
19. Bucksteeg, W. and Thiele, H., Bestimmung der biochemischen Aktivität von Schlämmen und Abwässern, *gwf-Wass. Abw.*, 105, 46, 1964.
20. Burton, G.A., Sediment microbial activity tests for the detection of toxicant impacts. American Society for Testing and Materials, *Special Technical Publication*, 864, 214, 1984.
21. Meyer-Reil, L.A., Measurements of hydrolytic activity and incorporation of dissolved organic substrates by microorganisms in marine sediments, *Marine Ecology Progress Series*, 31, 143, 1986.
22. Obst, U., *Enzymatische Tests für die Wasseranalytik*, Oldenbourg Verlag, München Wien, 1995.
23. Chrost, R. J., Ed., *Microbial Enzymes in Aquatic Environments*, Springer Verlag, Berlin, 1991.
24. Barnhart, C.L.H. and Vestal, J.R., Effects of environmental toxicants on metabolic activity of natural microbial communities, *Appl. Environ. Microbiol.*, 46, 970, 1983.

25. Burton, G.A. and Lanca, G.R., Sediment microbial activity tests for the detection of toxicant impacts, in *Aquatic Toxicology and Hazard Assessment: 7th Symposium*, Cardwell, R.D., Purdy, R., and Bahner, R.C., Eds., ASTM STP 854, 1985, 214.

26. Weil, L., Enzymatischer Test zur Bestimmung von Organophosphorpestiziden, *Hydrochem. hydrogeol. Mitt.* 1, 111, 1974.

27. Menzel, H., Ohnesorge, F.K., *Hydrochem.hydrogeol.Mitt.* 3, 351, 1978.

28. German standard methods for the examination of water, waste water and sludge; subanimal testing DIN 38415 part 1, Determination of cholinesterase inhibiting organophosphorus and carbamate pesticides (cholinesterase inhibition test) T 1, *VCH Verlaggesellschaft*, Weinheim, 1995.

29. Singh, J. and Lapointe, M. R. J., *AOAC*, 57, 1285, 1981.

30. Wiegand-Rosinus, M., Obst, U., Herzberger, E., and Haberer, K., Miniaturized and half automatic methods to determine enzymatic activities *in vivo* for water research, *Proceedings DECHEMA Biotechnology Conferences*, 5, 903, 1992.

31. Wiegand-Rosinus, M., Untersuchungen zur Hemmbarkeit von Enzymen *in vitro* durch Pestizide, *Doctorial Thesis*, Fachbereich Biologie d. Johannes-Gutenberg-Universität Mainz, 1991.

32. Wittekindt, E., Werner, A., Reinicke, A., Herbert, A., and Hansen, P.-D., A microtiter-plate urease inhibition assay — sensitive, rapid and cost-effective screening for heavy metals in water, *Environ. Technol.*, Vol. 17, 597, 1996.

33. Ashour, Mohammed-Bassern et al., Use of a 96-well microplate reader for measuring routine enzyme activities, *Analytical Biochemistry*, 166, 335, 1987.

34. Someville, M. and Billen, G., A method for determining exoproteolytic activity in natural waters, *Limnology and Oceanography*, 28, 190, 1983.

35. Holzapfel-Pschorn, A., Obst, U., and Haberer, K., Fluoreszenzspektroskopie zum empfindlichen Nachweis von Enzymaktivitäten in der Trinkwasseraufbereitung, *Gewässerschutz, Wasser. Abwasser*, 79, 352, 1985.

36. Jacobsen, T.R. and Rai, H., Determination of aminopeptidases activity in lake water by a short term kinetic assay and its application in two lakes of differing eutrophication, *Archiv. für Hydrobiologie*, 113, 359, 1988.

37. Reinheimer, G., Glocke, K., and Hoppe, H.-G., Vertical distribution of microbiological and hydrographic-chemical parameters in different areas of the Baltic Sea., *Mar. Ecol. Prog. Ser.*, 52, 55, 1989.

38. Wessler, A., Dämgen, K., Hryk, R., and Obst, U., Quality mapping of surface water and assessment of treated and untreated wastewater inputs into the Rhine and Main Rivers based on microbial enzyme activities, *Acta Hydrochim. Hydrobiol.*, 25, 17, 1997.

39. Dekant, W. and Vamvakas, S., *Toxikologie*, Spektrum Akademischer Verlag, Heidelberg, 1994.

40. Wessler, A., Dämgen, K., and Obst, U., Surveys on microbial enzyme activities in river water of different water quality in addition to the biological quality mapping of surface water, *Vom Wasser*, 83, 69, 1994.

41. Herzberger, E., Haberer, K., and Obst, U., Monitoring of the effect of pollutants on a waterwork-microflora by flow analysis of microbial enzymes (FAME), *Vom Wasser*, 76, 199, 1991.

42. Wiegand-Rosinus, M., Obst, U., Haberer, K., and Wild A., Enzymes *in vitro* as indicators for pesticides. An examination, *Environ. Toxicol. Water Qual.*, 7, 313, 1992.

43. Summer, J.B., The isolation and crystallisation of the enzyme urease, *J. Biol. Chem.*, 69, 435 1926.

44. Obst, U., Resch, K., and Feuerstein, Th., A simple biochemical toxicity test to examine water and waste water, *Vom Wasser*, 65, 199, 1985.

45. Wessler, A., Dämgen, K., and Obst, U., Biochemical assessment of the biological self purification in surface water — practical application for the drinking water obtainment, AWWA Conference Proceedings of 1995 Water Quality Technology Conference, 905–919, 1995.

46. Pantle, K. and Buch, H., Die biologische Überwachung der Gewässer und die Darstellung der Ergebnisse, *Gas- und Wasserfach. Wasser/Abwasser*, 96, 609, 1955.

47. Liebmann, H., *Handbuch der Frischwasser- und Abwasserbiologie*, Gustav Fischer Verlag, Jena, 1962.

48. Wiegand-Rosinus, M., Grollius, H.-H., Gerlitzki, E., and Obst, U., Investigation of municipal waste water treatment with enzymatic activity tests *in vivo*, *Vom Wasser*, 84, 143, 1995.

49. Heiniger, P. and Tippmann, P., Determination of enzymatic activities for the characterization of sediments, *Toxicol. Env. Chem.*, 52, 25, 1995.

Tissue Culture Assays

CHAPTER 6

Sponge Cells and Tissue as Biological Monitors of Aquatic Pollution

Werner E.G. Müller and Isabel M. Müller

CONTENTS

I. INTRODUCTION

Sponges (Porifera) are one of the major phyla found in the marine hard-substrate benthos, both with respect to the number of species and their biomass.[1] Despite the ecological dominance of the sponges, especially Demospongiae, due to their diversity of form, structure, and physiological adaptation (for review see reference 2), most species are confined to relatively little or unpolluted areas.

All sponges live in the aquatic environment; adult specimens are sessile filter-feeders able to ingest particles of sizes between 5 and 50 μm through the cells of the mesohyl and the pinacoderm; microparticles (0.3 to 1 μm) are taken up by the cells of the choanocyte chambers. A specimen of 1 kg, e.g., a small *Geodia cydonium*, filters ≈ 24,000 liters every day.[3] Some sponges are found in relatively shallow zones (to 100 m), e.g., Calcarea, whereas others, such as some *Hexactinellida*, prefer to live in deep waters. Demospongiae are found in depths ranging from the coastal areas down to 8600 m. Hexactinellid sponges colonize sand and mud substrates; and rock substrates are preferred by demospongid sponges. Some species even penetrate calcareous material (shells or soft rocks).

Sponges react to environmental stress with the induction of a series of metabolic pathways, e.g., the heat shock system, the polyphosphate turnover, the multixenobiotic resistance transport system, and programmed cell death (apoptosis). For the quantification of the physiological adjustments to fluctuating environments, *in vitro* cultures (regenerating sponge cubes; single cell cultures) both from freshwater and marine sponges have been introduced.

Until recently, sponges as whole organisms, or sponge cell cultures have seldom been used for environmental monitoring in spite of their environmental sensitivity. Previously, cubes cut from sponge specimens have been maintained in special incubators and exposed to a variety of pollutants.[4] Recently we succeeded in establishing short-term cell cultures from the freshwater sponge *Ephydatia muelleri*[5] and the marine sponge *Suberites domuncula*[6] *in vitro* to screen for potential environmental contaminants (see Section III).

II. SPONGES AND THEIR METAZOAN GENES

The genes characterized (cDNAs) to date from the marine sponge *Geodia cydonium* display all of the characteristics of the corresponding genes from higher animals. For example, one adhesion molecule, an S-type lectin, has been isolated with high similarity to other metazoan lectins in its amino acid (aa) sequence at the carbohydrate-binding site.[7] In 1994 two groups identified homeobox-genes from the freshwater sponge *Ephydatia fluviatilis*.[8,9] These coordinate genes are required for the determination of the anteroposterior and dorsoventral axes of the future embryo. A further transcription factor, the serum response factor, was cloned from *G. cydonium*.[10]

Recent evidence indicates that in sponges, certain receptors and linked molecules that are involved in signal transduction, are present that have high homology to those found in higher animals. A receptor protein-tyrosine kinases gene from the siliceous sponge *G. cydonium* has been described.[11-13] The deduced aa sequence shows all characteristic features known from vertebrate sequences. The sponge receptor has — very unexpectedly — at least two Ig-like domains in the extracellular portion of the receptor.

Receptors of type III belong to a family of seven-pass transmembrane receptor proteins. Receptors of types I and II have not yet been identified and cloned in sponges. After binding to the ligand, the type III receptor undergoes a conformational change and transmits the "information" to G-proteins (GTP-binding proteins). The coupling of a given receptor with distinct G-proteins results in a modulation of adenylate cyclase or of phospholipase C activity. Subsequently, either cAMP, Ca^{2+}, or diacylglycerol and inositol-trisphosphate are formed, which act as second messengers. In turn, the secondary effector systems, either cAMP-dependent protein kinase(s) (protein kinase A) or Ca^{2+}- and/or diacylglycerol-dependent protein kinase(s) (protein kinase C; PKC) are activated. The genes for these kinases have also been sequenced from *G. cydonium*[14] and show the distinct traits or building blocks of higher invertebrate and vertebrate PKCs.

A fourth gene, a stress protein, the heat shock protein of 70 kDa (hsp70) was isolated[15] from *G. cydonium,* which might be applicable as a molecular probe for quantitative risk assessment.

As summarized in Figure 6.1 these genes have to be considered as ultimate prerequisites in "Evolved Protozoa" to transform them into metazoan organization. Based on these new molecular biological data obtained from the sponge *G. cydonium,* it can be 1) concluded that all Metazoa are of monophyletic origin[16] and 2) assumed that some of the biomarkers found in sponges might be applied as diagnostic tools to screen for environmental loadings. Their responses to pollution might be extrapolated to other animals.

III. *IN VITRO* CULTURES

Sponge cells and tissue cubes are characterized by a pronounced reaggregation[17] or regeneration potency.[18] These properties have been used to establish *in vitro* cultures from sponges.

A. Regenerating Sponge Cubes

The technique of culturing sponge cubes was described in 1977 and 1981.[4,19] Specimens of *G. cydonium* were collected at depths of 20 to 30 m near Rovinj (Croatia) and were kept in tanks at 14°C until use. Subsequently, cubes with 1 cm edges were cut and incubated in filtered, oxygenated sea water at 14°C.

A special incubator was designed for these cubes, because sponges cannot be oxygenated by direct aeration without their channel system being affected by air bubbles (Figure 6.2). The incubator is hooked to a rack, filled with 50 mL of sea water, and submersed with its aeration compartment into a cold bath of approximately 5°C. It is agitated by 15-sec pulses from an air compressor at 30-sec intervals. The air is pressed through an inlet into the lower part of the incubator, the aeration compartment, where the air reaches the dialysis tubing mounted at the conical, upper end of the aeration compartment. A slotted tube hangs in the center of the dialysis tubing; at the lower base of the aeration compartment the tubing is tied together with the slotted tube. After an air pulse, the dialysis tubing in the aeration compartment is compressed around the slotted tube with the consequence that 42 mL of freshly oxygenated sea water is present within the dialysis tubing through the small circular opening at the conical upper end of the aeration compartment into the incubation cylinder. Seven grams of sponge cubes are placed within this incubation cylinder. The oxygen concentration is adjusted to 6.4 mL O_2/L. Under these conditions, cubes of *G. cydonium* can be kept for at least five weeks. The temperature in the incubation cylinder is adjusted to 14°C.

B. Single Cell Culture

1. Marine Sponges

Cells from the marine sponge *Suberites domuncula* Olivi (Porifera, Demospongiae, Hadromerida) are especially suitable for short-term cell cultures. Animals collected in the northern Adriatic Sea near Rovinj (Croatia) were kept in aquariums at Mainz (Germany) for more than six months at 16°C.

The sponge cells were obtained by mechanical dissociation. The sponge was cut into cubes, incubated for 30 min at 20°C in sea water containing penicillin, streptomycin, and nystatin and then pressed through a nylon net. The single cells obtained were suspended in sea water and centrifuged again. Finally they were resuspended in sea water (plus penicillin and streptomycin). The cells were cultured at 16°C in a temperature-constant incubator. One to two days later the cells were used for the experiments. The single cell suspension comprises over 50% choanocytes and approximately 20% spherulous cells.[6]

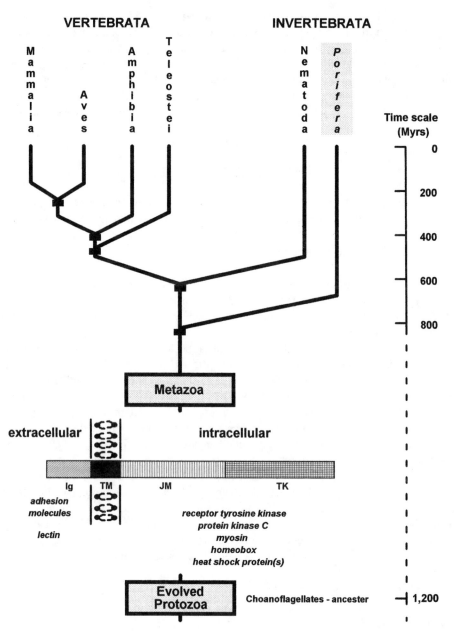

Figure 6.1   The phylogenetic diagram. Based on (1) aa sequence data obtained from sponges (Porifera; we used *Geodia cydonium*), the simplest multicellular organisms, and (2) the vertebrate genes, including those from Mammalia (rat), Aves (chicken), Amphibia (*Xenopus*), and Teleostei (the bone fish *Conger*), as well as (3) from the invertebrate Nematoda (*Caenorhabditis*) a common ancestor of all metazoa has to be deduced. The tree was constructed by sequence comparison of one adhesion molecule, a galactose-binding lectin from *G. cydonium,* with those from the other animals. With respect to sponges, the time of divergence has been calculated on the basis of the lectin as well as the receptor tyrosine kinase, carrying the immunoglobulin domain. All these animals have the protein kinase C, the nonmuscle myosin type II; all sequences have features typical of Metazoa. The divergence time has been calculated and is given in millions of years [Myrs]. It is outlined that during transition from the evolved protozoan species to the first Metazoa — a process which might have taken place around 1.2 Myrs ago[44] — adhesion molecules (lectins), cell surface receptor (e.g., a receptor tyrosine kinase as shown here with its extracellular domain (Ig: Immunoglobulin domain), the transmembrane region (TM), the juxtamembrane region (JM; intracellular domain I) and the catalytic region (intracellular tyrosine kinase — TK) domain], and signal transduction molecules (protein kinase C) have developed.

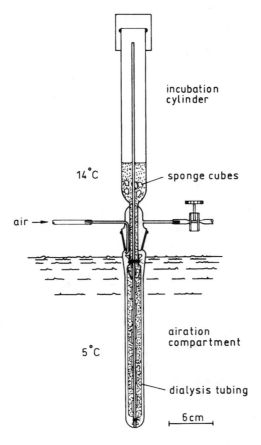

incubation
cylinder

14°C — sponge cubes

air →

5°C

airation
compartment

dialysis tubing

6 cm

Figure 6.2 Incubator for sponge cubes. Details are given in the text.

2. Freshwater Sponges

Short-term cultures from freshwater sponges have been established as follows.[5] Gemmules of *Ephydatia muelleri* (Porifera, Demospongiae, Spongillidae) collected in the River Wied near Neustadt (Germany), were kept at 4°C in tap water until use. After a short treatment with NaOCl and ethanol, approximately 100 gemmules were incubated per plastic Petri dish containing 10 mL deionized and autoclaved water. Under these conditions, after 48 h, at 20°C 90% of the gemmules started to hatch. The hatched gemmules were disrupted by squeezing through a 10-mL syringe. The cells were then immediately suspended in the culture medium consisting of sterile "M" medium,[20] supplemented with HEPES (pH 7.5), and BME-amino acids, BME-nonessential amino acids, and BME-vitamins, as well as 0.1% chicken embryo extract. To avoid bacterial contamination, the antibiotics penicillin and streptomycin were added to the medium. The cultures of single cells could be kept for up to three months; after this period the number of living cells decreased gradually. The mitotic index determined in the aggregates at day 21 was found to be 0.5 to 1%.

IV. ASSESSMENT OF AQUATIC POLLUTION USING SPONGES AS MODEL ORGANISMS

The effects of a series of pollutants on biomarkers have been studied *in vitro* using both regenerating sponge cubes and primary cells.

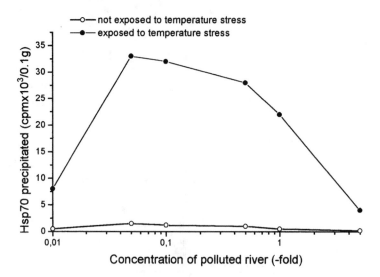

Figure 6.3   Metabolic labeling of hsp70 in tissue samples of *E. fluviatilis* in relation to different incubation temperatures and/or different levels of concentrate from the Schwarzbach River. Levels of hsp70 in sponge samples after treatment with different amounts of river concentrate. Sponges were incubated at 18°C during the complete incubation period [○] or were exposed first to temperature stress [33°C; 2h] and second to different concentrations of river extract at 18°C [●]. The results are means of five experiments each; the SD was less than 20%.

A. Regenerating Sponge Cubes

1. Heat Shock Proteins

Studies on the function of heat shock proteins (hsps) in vertebrates, indicate that hsps may protect cells against environmental stressors.[21]

We have determined — for the first time — in sponges the response to temperature stress by inducing the production of the stress protein hsp70.[22] For this study the freshwater sponge *Ephydatia fluviatilis* was used. These animals respond differently with regard to expression of hsp70 if they are treated with environmental xenobiotics (compounds not produced by the species used) alone or if a short temperature stress precedes the exposure to environmental xenobiotics.

Expression of hsp70 in sponge tissue was assayed by metabolic labeling at different incubation temperatures. Lysates were prepared and hsp70 was immunoprecipitated with anti-hsp70 antibodies. The data showed that a low level of hsp70 protein could be detected up to a temperature of 25°C. At 30°C or 35°C, the amount of immunoreacting material increased by 4.5-fold and 15-fold, respectively. At 40°C, the amount of labeled material was significantly reduced, but it was still higher than that measured at temperatures below 30°C.[22]

The interactive effects of 1) a nonionic organic extract from a highly polluted river (River Schwarzbach, Hesse, Germany), concentrated by adsorption of water samples to Amberlite XAD7,[22] and 2) heat stress on hsp expression were determined. At normal temperature (18°C) the extract caused a low twofold increase of hsp70 expression within the range of 0.04-fold to 0.4-fold ambient Schwarzbach levels. However, if the cultures were pretreated by heat (28°C) for 2 h and then incubated with different concentrations of the nonionic organic extract (extract levels at 0.04-fold to onefold ambient levels) a > tenfold increase of hsp70 synthesis was recorded (Figure 6.3). Proliferation studies revealed that temperature stress alone resulted in a moderate decrease of sponge cell proliferation (25%); increasing river concentrates (two- to fourfold) resulted in a 53.6% to 99.4% decrease in cell proliferation, respectively. However, temperature stress together with treatment by

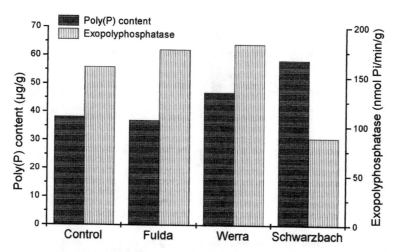

Figure 6.4 Changes of poly(P) content and of exopolyphosphatase activity in *Ephydatia muelleri* after exposure to differentially polluted waters from rivers. Gemmules (4 days after germination) were incubated for 5 days in water from an unpolluted site (Fulda), a moderately polluted river (Werra), and a heavily polluted site (Schwarzbach); as "Control" the sponges have been incubated in phosphate-buffered freshwater.

nonionic organic pollutants has improved cell proliferation by an order of magnitude as compared with temperature stress alone. These data imply that a sublethal treatment of sponges with heat results in a higher tolerance of the animals to chemical stressors.

2. Polyphosphates

Inorganic polyphosphates [poly(P)] are long linear polymers of orthophosphates (reviewed in reference 23), which have been assumed to function as: 1) storage molecules for energy-rich phosphate, 2) chelators for divalent cations, 3) counterions for basic amino acids in vacuoles, or 4) donors of phosphate for certain sugar kinases. In addition, changes in the size of poly(P) chains may be associated with the response of cells to 5) alkaline stress or 6) osmotic stress.[24,25]

Several enzymes involved in the synthesis and degradation of poly(P) have been identified, including a poly(P) kinase, which catalyzes the transfer of phosphate from ATP to poly(P),[26] and a number of polyphosphatases (both exo- and endo-enzymes), which hydrolyze poly(P).[27] Recently, two exopolyphosphatases from the marine sponge *Tethya lyncurium* were purified.[28]

Significant amounts of inorganic poly(P) ($\approx$55 μg of polyphosphate/g of wet weight) were found in the freshwater sponge *Ephydatia muelleri*, particularly in the gemmules (260 μg/g).[29] An increase in polyphosphate content and a decrease in exopolyphosphatase activity was observed during tissue regression when hatched sponges were exposed to polluted waters from rivers. For these studies, samples of river water from a nearly unpolluted site (River Fulda near Wahnhausen, Germany), a moderately polluted site (River Werra near Heldra), and from a heavily polluted site (River Schwarzbach near Trebur) were used. The water samples were prepared as described.[22] Small sponge samples obtained from gemmules 5 days after hatching were incubated in water from differently polluted rivers for 4 days, and then poly(P) content and exopolyphosphatase activity were determined. Controls were cultivated in phosphate-buffered freshwater-HEPES. As summarized in Figure 6.4, exposure of *E. muelleri* to water from the rivers Fulda and Werra, containing only a very low or moderate pollutional load, resulted in only small or nonsignificant changes in poly(P) content. However, a marked increase in poly(P) content (79%) was observed when *E. muelleri* was exposed to highly polluted water from the River Schwarzbach. The increase in poly(P) content of *E. muelleri* after exposure to polluted water (River Schwarzbach) was accompanied by a decrease in exopolyphosphatase activity

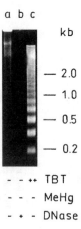

Figure 6.5 Effects of tributyltin (TBT) on apoptotic DNA fragmentation in sponge tissue from *G. cydonium*. Sponge cubes were incubated with 0 (lane a) or 3 μ*M* TBT (lane c). In the control experiments, DNA preparations were treated with DNAse I (lane b). These experiments have been performed in the absence of methyl mercury (MeHg).

(Figure 6.4). Therefore, we suppose that under incubation conditions with polluted water the exopolyphosphatase activity is inhibited in the sponge.

From these results we conclude that changes in poly(P) metabolism in sponge tissue, which can be measured by simple methods, may be a sensitive indicator of polluted environments.

3. Detergents

The regenerating cube system from *G. cydonium* was used to assess the effect of detergent pollution.[4] Na-dodecylsulfate as well as the commercial laundry detergents Faks® and Radion® were found to reduce incorporation of [³H]dThd, [³H]Urd, and [³H]Phe into DNA, RNA, and protein at concentrations as low as 0.1 ppm (10^{-8} g/L).

4. Benzo[a]pyrene

The model pollutant benzo[a]pyrene (BaP) was studied[19] using the sponge *Tethya lyncurim* (cubes). At concentrations of 1 to 6 ppm both the polyamine concentrations and the activity of the key enzyme in polyamine metabolism, the ornithine decarboxylase, increased during a one-day incubation. BaP associated not only with protein but also with DNA and RNA during the course of incubation.[30] The possible repair mechanisms occurring during the incubation with BaP were studied.[31]

5. Organic Derivatives of Heavy Metals

The effect of conservative organic derivatives of heavy metals, such as tin (e.g., tributyltin; TBT) or mercury (methyl mercury; MeHg) have been tested in the *G. cydonium* cube system. It was found that the water pollutants TBT (>1 μ*M*) and MeHg (>3 μ*M*) induce apoptosis in tissue of the marine sponge *G. cydonium* (Figure 6.5). Apoptosis is a special form of cell death in which the cells actively participate in the process of dying. At concentrations of >5 μ*M*, MeHg causes alkaline labile sites in DNA. At the lower dose of 0.3 μ*M*, MeHg abolishes the TBT-induced apoptosis in sponge tissue. Incubation of sponge tissue with 3 μ*M* of TBT induces not only apoptosis but also an expression of heat shock protein-70 (Figure 6.6).[32]

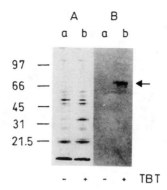

Figure 6.6 Expression of hsp70 protein in *G. cydonium* tissue after incubation with TBT. Sponge cubes were pretreated for 2 h with 0 or 3 µM TBT and then incubated in the absence of the compound for 10 h at 16°C. The proteins were extracted and the resulting protein fractions were size-separated by polyacrylamide gel electrophoresis. The proteins from sponges treated with 0 (lanes a) or 3 µM TBT (lanes b) were either stained by Coomassie Brilliant Blue (*A*) or were analyzed for the presence of hsp70 by Western blotting technique using antibodies raised against the protein hsp70 (*B*). The M_r of the protein species crossreacting with human anti-hsp70 antibody is marked with an arrow.

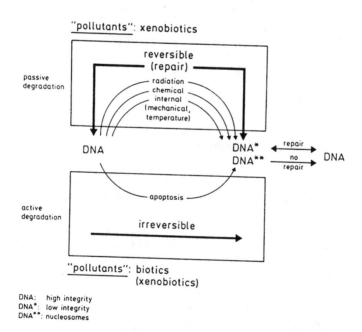

Figure 6.7 DNA can be disintegrated from a state of high integrity via a state of low integrity down to nucleosomes (1) by passive degradation or (2) by active degradation. Passive degradation of DNA, caused by radiation, chemicals, or internal factors (xenobiotics) can be repaired by the cellular DNA repair systems, while active degradation by apoptosis (biotics/xenobiotics) is an irreversible event.

This study demonstrates that in addition to the passive form of DNA degradation (necrosis), the active degradation process (apoptosis) can also be activated by genotoxic agents, such as TBT. Whereas the latter effect is irreversible, passive degradation of DNA can in some cases be repaired. A survey of the two modes of DNA fragmentation (necrosis toward apoptosis) is given in Figure 6.7.

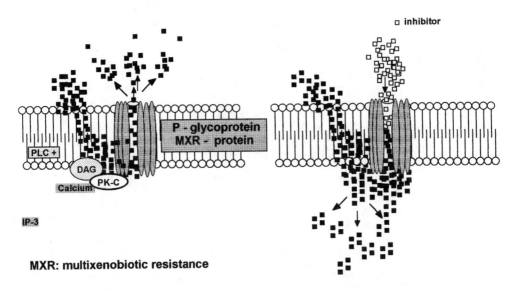

MXR: multixenobiotic resistance

Figure 6.8   The P-glycoprotein pump is located in the plasma membranes and is presumably regulated by phosphorylation/dephosphorylation reactions mediated by protein kinase C (PKC). In this model it is outlined that cytostatic drugs or xenobiotics [■] activate plasma membrane-associated phospholipase C (PL-C), resulting in a local formation of diacylglycerol (DAG) and inositol-trisphosphate (IP-3) and an ultimate translocation of PKC (only shown in the left scheme). The latter enzyme might activate the P-glycoprotein and extrudes the compound — some xenobiotics (*left*). Inhibitors of the P-glycoprotein [□] compete for the transport system and block extrusion of the compound (*right*). We identified a P-glycoprotein-like pump in marine sponges and termed it multixenobiotic resistance protein (MXR protein).

B. Single Cell Culture from Marine Sponges

1. Multixenobiotic Resistance (MXR) Pump

Chemotherapy has been widely and successfully used in the treatment of a variety of tumors in humans. However, one significant problem is the acquisition of a simultaneous resistance to a variety of structurally unrelated cytotoxic drugs in human malignancies reviewed in reference 33. The drug resistance phenotype is often connected to the emergence of the multixenobiotic transporter (a molecular pump which extrudes xenobiotics out of the cell), also termed P-glycoprotein, in the plasma membrane,[34] reviewed in reference 35. The P-glycoprotein catalyzes an ATP-dependent extrusion of a series of anticancer drugs (e.g., *Vinca* alkaloids, anthracyclines, epipodophyllotoxins), and of other cytotoxic agents reviewed in reference 33; Figure 6.8.

We searched for a P-glycoprotein-related extrusion pump among aquatic invertebrates, which might lower the environmental load emanating from toxic agents. The pump was identified in the marine sponge *G. cydonium* as a 125 kDa protein species[36] and as a 135 kDa molecule in the clam *Corbicula fluminea*.[37] In these animals the P-glycoprotein-like pump extrudes xenobiotics from their cells; this pump was termed multixenobiotic resistance protein (MXR protein).[37]

Recently, we observed that the sponge *Suberites domuncula* also has a plasma membrane extrusion pump, an MXR protein.[6] Cells from this sponge were used to demonstrate first, the existence of the pump, and second, their use for the establishment of a potential test system. As indicator dye calcein-AM was used, it is taken up by sponge cells and extruded in the same, verapamil-sensitive manner (like in mammalian cell systems, which express the P-glycoprotein) from the cells (Figure 6.9).[6]

Until now, no methods have been introduced to identify compounds in the marine environment that act as toxic secondary metabolites or as harmful xenobiotics to the multixenobiotic transporter.

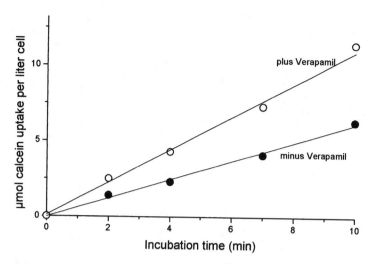

Figure 6.9 Uptake of calcein-AM into *S. domuncula* cells. A suspension of 1 × 10⁶ cells/mL of sea water was incubated in the presence of 0.25 μ*M* calcein-AM in 96-well microtiter plates and immediately subjected to analysis using a fluorescence microtiter reader at excitation and emission wavelengths of 485 nm and 538 nm. Where indicated, the cell suspension was preincubated with 20 μ*M* of verapamil for 10 min before the loading period. Incubation was performed in the absence (●------●) or presence of 20 μ*M* of verapamil (○------○).

No such cellular assay system was available until now that could be incorporated into the test batteries within the framework of environmental monitoring of aquatic environments. It is reasonable to assume that many compounds are present in the sea that inactivate the MXR protein in sponges and hence increase the toxicity of those compounds that are otherwise extruded via the MXR protein.

Recent data from our group support this idea. Inhibition of the MXR pump by xenobiotics, e.g., by verapamil[37] or XAD-7 extracts from River Rhein,[38] was shown to enhance the toxicity of low levels of carcinogens, e.g., 2-acetylaminofluorene. A scheme outlining the "cooperative" action of the two xenobiotics is shown in Figure 6.10.

V. POTENTIAL MOLECULAR PROBES AS BIOMARKERS

A. Heat Shock Proteins

In general, five families of stress proteins are found in eukaryotes; four of them are grouped according to their molecular weights as hsp90, hsp70, hsp58-60, and hsp20-30, whereas the fifth hsp is termed ubiquitin.[39] We have cloned two genes coding for proteins of the hsp family: hsp70 and ubiquitin from the sponge *G. cydonium*.

1. Heat Shock Protein-70 [hsp70]

The cDNA from the sponge *G. cydonium* has been cloned and used to determine the expression of hsp70 as a reaction of the organism to temperature stress. As shown in Figure 6.11, exposure of *G. cydonium* to heat stress resulted in the synthesis of mRNA coding for hsp70 (left panel: lane b); the controls did not contain this mRNA species (lane a). To establish if the mRNA is also transcribed, the hsp70 protein was determined immunochemically (left panel: lanes c and d). This protein was detectable only in samples from stressed animals. The complete sequence of the hsp70 is published elsewhere.[15]

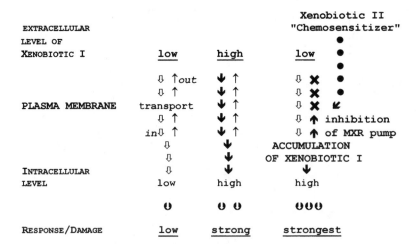

Figure 6.10   Scheme of the function of a xenobiotic acting as a "chemosensitizer" of the MXR pump. A classical xenobiotic (here termed xenobiotic I) is taken up by the cell through the plasma membrane; the compound is again exported to some extent. If low levels of xenobiotic I are present in the extracellular space, the intracellular concentrations remain low; ultimately the response or damage of the cells is also low. In turn, higher extracellular concentrations result in a higher intracellular level of xenobiotic I and a more severe effect to the cell metabolism. If the cells are exposed to low levels of a xenobiotic (here xenobiotic I) in the presence of an inhibitor of the MXR pump, the so-called "chemosensitizer" (here xenobiotic II), a severe effect will occur due to the accumulation of xenobiotic I.

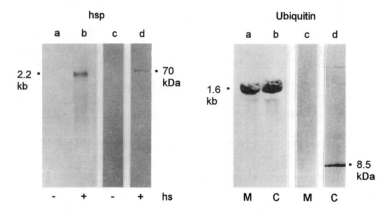

Figure 6.11   Expression of hsp70 and ubiquitin in cells from the sponge *G. cydonium:* Left panel: Heat shock protein-70 (hsp70): The sponges collected at an environmental temperature of 14°C remained either untreated (hs: −) or were exposed to 24°C for 30 min and subsequently incubated at 14°C for an additional 3 hr (hs: +). The mRNA, coding for hsp70 (2.2 kb), was detected by Northern blotting (lanes a and b) and the hsp70 protein by Western blotting (70 kDa) (lanes c and d). Right panel: Ubiquitin: Identification of ubiquitin in extracts from the sponge *G. cydonium.* Extracts from the medulla region (M; lanes a and c) and from the cortex region (C; lanes b and d) were prepared. For the analysis of the ubiquitin protein (lanes c and d) the resulting protein fractions were size-separated by polyacrylamide gel electrophoresis. Then the proteins were transferred to nitrocellulose sheets and incubated first with anti-ubiquitin antibodies (rabbit) and second, with the alkaline phosphatase conjugated anti-rabbit IgG-BCIP/NBT detection system. 50 μg of protein each was analyzed. The expression of mRNA, coding for ubiquitin, (lanes a and b) was determined by Northern blotting using the sponge ubiquitin cDNA as probe. The size of the ubiquitin mRNA (1.6 kb) as well as the apparent molecular mass of the ubiquitin protein (8.5 kD) are given.

2. *Ubiquitin*

We demonstrated that the ubiquitin pathway exists in sponges.[40] The expression of this protein is regulated by cell adhesion molecules during reaggregation of cells. As in other organisms, the level of ubiquitin in sponges varies in accordance with physiological and pathophysiological conditions.[41] Therefore, it was interesting to determine if the ubiquitin protein is expressed differentially within the organism.

It was found that detectable amounts of ubiquitin protein are present only in the cortex of the animal (the surface of the sponge exposed to the environment) and not in the medulla region (the internal zone of the sponge) (Figure 6.11; right panel: lanes c and d), whereas the level of ubiquitin mRNA was almost identical in both compartments (right panel: lanes a and b). This finding might reflect a rapid turnover of ubiquitin in the medulla region. To the best of our knowledge the expression of ubiquitin is regulated in other systems primarily at the transcriptional level;[42] in the *G. cydonium* system, translation efficiency of ubiquitin mRNA appears to be under tight control.

B. Members of the Signal Transduction Pathway

1. *Protein Kinase C*

Regenerating cubes from *G. cydonium* were used to determine the effect of pollution in the marine environment on selected parameters of the intracellular signal transduction pathway.[43] The activity of protein kinase C (an initial component of one of the signal transduction pathways) and DNA polymerase-α, (one endpoint of the chain) was determined. It was found that at low to moderate pollution, an activation of protein kinase C enzyme occurs, whereas at sites of heavy pollution this change was not observed. In contrast, DNA polymerase-α activity decreased gradually with increasing loads of pollutants. We have sequenced the sponge protein kinase C gene.[14]

2. Receptor Tyrosine Kinase

The receptor tyrosine kinase,[13] another gene coding for a further inducible and adaptive protein, was cloned from *G. cydonium*.

If the expression rates of protein kinase C and/or receptor tyrosine kinase change in sponge cells with respect to different environmental loads, then this system can be used for environmental monitoring.

VI. CONCLUSIONS

It was not established until recently on the basis of molecular DNA sequencing data that sponges (Porifera) evolved from the same ancestor as other Metazoa. They have the basic structural elements and the signal transduction pathways present in other multicellular organisms. Sponges are primarily filter-feeders and lack effective structural defense systems. Hence, these animals must have developed powerful metabolic strategies that enable them to endure unfavorable environmental conditions. The outline of these pathways and their subsequent application in microscale assays for monitoring aquatic pollution was the subject of this study.

Furthermore, the molecular probes have been isolated and characterized from a marine sponge, *Geodia cydonium*. Heat shock protein hsp70, ubiquitin, protein kinase C, and receptor tyrosine kinase may have future applications as biomarkers for stress response reactions.

ACKNOWLEDGMENTS

We are grateful to Dr. Renato Batel, Dr. Vera Gamulin, and Dr. Branko Kurelec (Institute Ruder Boskovic; Zagreb and Rovinj, Croatia) for their advice and for the results obtained in a collaborative work in the field of environmental contamination. Moreover, we would like to thank our co-workers for their contributions, especially H. Schäcke, M. Kruse, C. Koziol, and C. Wagner. This work is supported by grants from the Bundesministerium für Bildung und Wissenschaft (project STRESS-TOX) and from the Commission of the European Communities (93/Avi 001).

REFERENCES

1. Sarà, M. and Vacelet, J., Ecologie des Démosponges, *Traité de zoologie: Spongiaires. Tome III (1).* Ed. Grassé, P. P., Masson, Paris, 1973, 462.
2. Bergquist, P. R., *Sponges*, Hutchinson, London, 1978.
3. Vogel, S., Current-induced flow through living sponges in nature, *Proc. Natl. Acad. Sci. USA,* 74, 2069, 1977.
4. Zahn, R. K., Zahn, G., Müller, W. E. G., Müller, I., Beyer, R., Müller-Berger, U., Kurelec, B., Rijavec, M., and Britvic, S., Consequences of detergent pollution of the sea: Effects on regenerating sponge cubes of *Geodia*, *Sci. Total Environm.*, 8, 109, 1977.
5. Imsiecke, G., Steffen, R., Custodio, M., Borojevic, R., and Müller, W. E. G., Formation of spicules by sclerocytes from the freshwater sponge *Ephydatia muelleri* in short-term cultures *in vitro*, *In Vitro Cell. Dev. Biol.*, 31, 528, 1995.
6. Müller, W. E. G., Steffen, R., Rinkevich, B., and Kurelec, B., Multixenobiotic resistance mechanism in the marine sponge *Suberites domuncula*: Its potential application in assessment of environmental pollution, *Marine Biol.*, 125, 165, 1996.
7. Pfeifer, K., Haasemann, M., Gamulin, V., Bretting, H., Fahrenholz, F., and Müller, W. E. G., S-type lectins occur also in invertebrates: high conservation of the carbohydrate recognition domain in the lectin genes from the marine sponge *Geodia cydonium*, *Glycobiol.*, 3, 179, 1993.
8. Coutinho, C., Vissers, S., and Van de Vyver, G., Evidence of homeobox genes in the freshwater sponge *Ephydatia fluviatilis*, *Sponges in Time and Space,* Eds. v. Soest, R. W. M., v. Kempen, T. M. G., and Braekman, J. C., Balkema Press, Rotterdam, 1994, 47.
9. Seimiya, M., Ishiguro, H., Miura, K., Watanabe, Y., and Kurosawa, Y., Homeobox-containing genes in the most primitive metazoa, the sponges, *Eur. J. Biochem.*, 221, 219, 1994.
10. Scheffer, U., Krasko, A., Pancer, Z. and Müller, W.E.G., Isolation and characterization of the cDNA clone encoding the serum response factor homolog in the marine sponge *Geodia cydonium:* High conservation of this transcription factor within Metazoa, *Biol. J. Linnean Soc.,* in press.
11. Schäcke, H., Rinkevich, B., Gamulin, V., Müller, I. M., and Müller, W. E. G., Immunoglobulin-like domain is present in the extracellular part of the receptor tyrosine kinase from the marine sponge *Geodia cydonium*, *J. Molec. Recognition*, 7, 272, 1994.
12. Schäcke, H., Müller W. E. G., Gamulin, V., and Rinkevich, B., The Ig superfamily includes members from the lowest invertebrates to the highest vertebrates, *Immunology Today*, 15, 497, 1994.
13. Schäcke, H., Schröder, H. C., Gamulin, V., Rinkevich, B., Müller, I. M., and Müller, W. E. G., Molecular cloning of a tyrosine kinase gene from the marine sponge *Geodia cydonium*: a new member belonging to the receptor tyrosine kinase class II family, *Mol. Memb. Biol.*, 11, 101, 1994.
14. Kruse, M., Gamulin, V., Cetkovic, H., Pancer, Z., Müller I.M., and Müller, W.E.G., Molecular evolution of the metazoan protein kinase C multigene family, *J. Molec. Evol.*, 43, 374, 1997.
15. Koziol, C., Kruse, M., Wagner-Hülsmann, C., Pancer, Z., Batel, R., and Müller, W. E. G., Cloning of the heat-inducible stress protein (hsp70) from the marine sponge *Geodia cydonium, Can. Techn. Rep. Fish. & Aquatic Science,* 2093, 104, 1996.
16. Müller, W. E. G., Molecular phylogeny of metazoa [animals]: Monophyletic origin, *Naturwiss.*, 82, 321, 1995.
17. Müller, W. E. G. and Zahn, R. K., Purification and characterization of a species-specific aggregation factor in sponges, *Exptl. Cell Res.*, 80, 95, 1973.

18. Müller, W. E. G., Zahn, R. K., Kurelec, B., Müller, I., Vaith P., and Uhlenbruck, G., Aggregation of sponges cells. Isolation and characterization of an anti-aggregation receptor from the cell surface, *Eur. J. Biochem.*, 97, 585, 1979.

19. Zahn, R. K., Zahn, G., Müller, W. E. G., Kurelec, B., Rijavec, M., Batel, R., and Given, R., Assessing consequences of marine pollution by hydrocarbons using sponges as model organisms, *Sci. Tot. Environm.*, 20, 147, 1981.

20. Rasmont, R., Une technique de culture des éponges d'eau douce en milieu controlé, A*nn. Soc. Roy. Zool. Belg.*, 91, 147, 1961.

21. Lindquist, S. and Craig. E. A., The heat shock proteins, *Annu. Rev. Genet.*, 22, 631, 1988.

22. Müller, W. E. G., Koziol, C., Kurelec, B., Dapper, J., Batel, R., and Rinkevich, B., Combinatory effects of temperature stress and nonionic organic pollutants on stress protein (hsp70) gene expression in the fresh water sponge *Ephydatia fluviatilis, Arch. Environm. Contam. Toxicol.*, 14,1203, 1995.

23. Wood, H. G. and Clark, J. E., Biological aspects of inorganic polyphosphates, *Ann. Rev. Biochem.*, 57, 235, 1988.

24. Bental, M., Pick, U., Avron, M., and Degani, H., Metabolic studies with NMR spectroscopy of the alga *Dunaliella salina* trapped within agarose beads, *Eur. J. Biochem.*, 188, 111, 1990.

25. Leitao, J. M., Lorenz, B., Bachinski, N., Wilhelm, C., Müller, W. E. G., and Schröder, H. C., Osmotic stress-induced synthesis and degradation of inorganic polyphosphates in the alga *Phaeodactylum tricornutum, Marine Ecol. Prog. Ser.*, 121, 279 1995.

26. Ahn, K. and Kornberg, A., Polyphosphate kinase from *Escherichia coli*. Purification and demonstration of a phosphoenzyme intermediate, *J. Biol. Chem.*, 265, 11734, 1990.

27. Lorenz, B., Müller, W.E.G., Kulaev, I. S., and Schröder, H. C., Purification and characterization of an exopolyphosphatase activity from *Saccharomyces cerevisiae, J. Biol. Chem.*, 269, 22198, 1994.

28. Lorenz, B., Batel, R., Bachinski, N., Müller, W. E. G., and Schröder, H. C., Purification of two exopolyphosphatases from the marine sponge *Tethya lyncurium, Biochim. Biophys. Acta*, 1245, 17, 1995.

29. Imsiecke, G., Münkner, J., Lorenz, B., Bachinski, N., Müller, W. E. G., and Schröder, H. C., Inorganic polyphosphates in the developing freshwater sponge *Ephydatia muelleri*: Effect of stress by polluted waters. *Environ. Toxicol. Chem.*, 15, 1329, 1996.

30. Zahn, R. K., Kurelec, B., Zahn-Daimler, G., Müller, W. E. G., Rijavec, M., Batel, R., Given, R., Pondeljak, V., and Beyer, R., The effect of benzo(a)pyrene on sponges as model organisms in marine pollution, *Chem.-Biol. Interaction*, 39, 205, 1982.

31. Zahn, R. K., Zahn, G., Müller, W. E. G., Michaelis, M. L., Kurelec, B., Rijavec, M., Batel R., and Bihari, N., DNA damage by PAH and repair in a marine sponge, *Sci. Total Environm.*, 26, 137, 1983.

32. Batel, R., Bihari, N., Rinkevich, B., Dapper, J., Schäcke, H., Schröder H. C., and Müller, W. E. G., Modulation of organotin-induced apoptosis by the water pollutant methyl mercury in a human lymphoblastoid tumor cell line and a marine sponge, *Marine Ecol.*, 93, 245, 1993.

33. Gottesman, M. M. and Pastan, I., Biochemistry of multidrug resistance mediated by the multidrug transporter, *Annu. Rev. Biochem.*, 62, 385, 1993.

34. Juliano, R. L. and Ling, V., A surface glycoprotein modulating drug permeability in chinese hamster ovary cell mutants, *Biochim. Biophys. Acta,* 455, 152, 1976.

35. Juranka, P. F., Zastawny, R. L., and Ling, L., P-glycoprotein: multidrug-resistance and a superfamily of membrane-associated transport proteins, *FASEB J.*, 3, 2583, 1989.

36. Kurelec, B., Krca, S., Pivcevic, B., Ugarkovic, D., Bachmann, M., Imsiecke, G., and Müller, W. E. G., Expression of P-glycoprotein gene in marine sponges. Identification and characterization of the 125-kDa drug-binding glycoprotein, *Carcinogenesis,* 13, 69, 1992.

37. Waldmann, P., Pivcevic, B., Müller, W. E. G., Zahn, R. K., and Kurelec, B., Increased genotoxicity of aminoanthracene by modulators of multixenobiotic resistance mechanism: Studies with the fresh water clam *Corbicula fluminea, Mutation Res.,* 342, 113, 1995.

38. Kurelec, B., Pivcevic, B., and Müller, W. E. G., Determination of pollutants with xenobiotic-resistance inhibiting properties, *Marine Environ. Res.,* 39, 261,1995.

39. Schlesinger, M. J., Ashburner, M., and Tissieres, A., *Heat Shock. From Bacteria to Man*, Cold Spring Harbor, N.Y., Cold Spring Harbor Laboratory, 1982, 440.

40. Pfeifer, K., Frank, W., Schröder, H. C., Gamulin, V., Rinkevich, B., Batel, R., Müller I. M., and Müller, W. E. G., Cloning of the polyubiquitin cDNA from the marine sponge *Geodia cydonium* and its preferential expression during reaggregation of cells, *J. Cell Sci.*, 106, 545, 1993.

41. Mayer, R.J. Arnold, J., Laszlo, L., Landon, M., and Lowe, J., Ubiquitin in health and disease. *Biochim. Biophys. Acta*, 1089, 141, 1991.
42. Schwartz, L. M., Myer, A., Kosz, L., Engelstein, M., and Maier, C., Activation of polyubiquitin gene expression during developmentally programmed cell death, *Neuron*, 5, 411, 1990.
43. Ugarkovic, D., Kurelec, B., Krca, S., Batel, R., Robitzki, A., Müller W. E. G., and Schröder, H. C., Alterations in *ras*-gene expression and intracellular distribution of protein kinase C in the sponge *Geodia cydonium* in response to marine pollution, *Marine Biol.*, 107, 191, 1990.
44. Knoll, A. H., Proterozoic and Early Cambrian protists: Evidence for accelerating evolutionary tempo, *Proc. Natl. Acad. Sci. USA*, 91, 6743, 1994.

CHAPTER 7

The Use of Fish Cells in the Toxicological Evaluation of Environmental Contaminants

Francine Denizeau

CONTENTS

I. INTRODUCTION

Over the last decade, there has been tremendous progress made in the development of *in vitro* cell systems for toxicological evaluation. Although a lot of effort has been devoted to applications in medical sciences, there has been increasing interest in environmental toxicology. *In vitro* cell systems offer many advantages including low cost, rapidity, and availability of a wide range of cell types, organisms, or endpoints. Such systems provide a unique opportunity for studying human cells. In addition, they are compatible with complex environmental mixtures and have a significant potential to reduce the number of animals in evaluation protocols. They are also particularly suitable for investigation of mechanisms of toxicity. However, a major challenge is the validation of cellular systems for their acceptance into regulatory toxicology.[1-5]

In aquatic toxicology, fish cells have been used mainly to study the mechanisms of action of environmental toxicants. Numerous biochemical processes have been investigated such as xenobiotic metabolism, induction of DNA damage and repair, membrane transport, ionic homeostasis perturbation, oxidative stress, and expression of metallothioneins and stress proteins. The action of xenobiotics on specialized cell functions, in particular those of immune cells, as well as the response to estrogenic compounds, has received some attention. Work has also been performed to validate test systems for environmental monitoring. The purpose of this review is to present the advances that have been made in these areas of research with emphasis on the progress made over the last five years, as reviews have appeared on the subject a few years ago.[6-8]

II. FISH CELLS IN THE STUDY OF ACTION MECHANISMS OF XENOBIOTICS

A. Biochemical Systems and Responses

1. Regulation of Xenobiotic-Metabolizing Enzymes

Regulation via the Ah Receptor

A number of studies on fish cells have focused on the mechanisms that regulate the expression of xenobiotic-metabolizing enzymes. Regulation via activation of the aryl hydrocarbon (Ah) receptor by environmental contaminants such as polycyclic aromatic hydrocarbons (PAHs) and halogenated aromatic hydrocarbons (HAHs) has received particular attention. A significant step in this area was the demonstration of the presence of the Ah receptor in fish liver cells by Lorenzen and Okey.[9] These authors reported the detection and characterization of the Ah receptor in the rainbow trout hepatoma cell line RTH-149. Using sucrose density gradient analysis, they found that the receptor sediments at about 9 S. The receptor binds several ligands known to be agonists for the Ah receptor in mammalian systems, namely 2,3,7,8-tetrachlorodibenzo-p-dioxin (TCDD), 3-methylcholanthrene (MC), and benzo[a]pyrene (BaP). The RTH-149 cells show a low concentration of high-affinity receptor binding sites. The translocation of the Ah receptor-ligand complex to the nucleus as well as the induction of aryl hydrocarbon hydroxylase (AHH) activity, which is an Ah receptor-dependent response, also takes place in these hepatoma cells.[9]

Subsequent studies confirmed the response of fish cells to compounds that act through the Ah receptor. The induction of 7-ethoxyresorufin O-deethylase (EROD), a cytochrome P-450-dependent activity which is associated with the isoform 1A1 (CYP1A1) and which corresponds to AHH, has been shown in primary cultures of rainbow trout hepatocytes following exposure to β-naphthoflavone (BNF) and TCDD.[10] In parallel with induced EROD activity, the amount of the protein CYP1A1, as determined by ELISA using anti-cod P-450 1A1, is increased, whereas an elevated level of CYP1A1 mRNA is observable before that of the protein.[11] Similar results on EROD induction by BP have been presented by Masfaraud et al.[12] Furthermore, in trout hepatocytes, the induced EROD activity can be potentiated by glucocorticoids as previously observed in mammalian cells.[13]

Acetaminophen is another agent that is capable of modulating cytochrome P-450 1A1 in trout liver cells. Indeed, addition of acetaminophen to liver cell cultures prepared from fish treated with the potent CYP1A1 inducer BNF help maintain elevated levels of P-450 1A1. The effects can be attributed to stabilization of P-450 1A1 mRNA, although the mechanisms involved are not fully understood.[14]

Effects of Temperature Acclimation

The activity of xenobiotic-metabolizing enzymes in fish is influenced by environmental variables such as temperature, and isolated cells have been used to study this phenomenon. The metabolism of BP and the rate of 7-ethoxycoumarin deethylation have been measured in liver cells obtained from cold-acclimated or warm-acclimated rainbow trout. The results show that the ability of rainbow trout to oxidize xenobiotics by cytochrome P-450-dependent metabolism remains constant during acclimation from warm to cold temperatures, whereas the ability to conjugate substances with glucuronic acid decreases during the same period.[15] Similarly, temperature effects can be seen in a model using the gulf toadfish *Opsanus beta* and cultured hepatocytes from this species. The toadfish were acclimated at 18 and 28°C, and hepatocytes prepared from these animals were incubated at the previous temperatures. Under these conditions, temperature adaptations are seen in cytochrome P-450 content, AHH, epoxide hydrolase, and UDP-glucuronyltransferase (UDPG) activity but not in glutathione-S-transferase (GST). Because incomplete temperature acclimation is observed in cultured cells as compared to what occurs in whole organisms, it appears that systemic factors which are absent *in vitro* may be involved *in vivo*.[16] The fundamentals of temperature-dependent responses in fish cells have been reviewed by Bols et al.[17]

2. Biotransformation of Xenobiotics

Biotransformation pathways of BP, a representative PAH, have been studied in fish cells and compared to those observed *in vivo*. Metabolites of BP have been detected in BP-exposed English sole (*Parophrys vetulus*) hepatocyte cultures.[18] The metabolites are found in the form of glucuronide and glutathione conjugates and to a lesser extent as the unconjugated BP-9,10-dihydrodiol. After hydrolysis, the glucuronide conjugates yield BP-7,8-dihydrodiol, 1-hydroxy BP, and 3-hydroxy BP. These metabolites are similar to those present in the bile and liver of English sole exposed to BP, indicating that isolated hepatocytes from this fish can be used to study the mechanisms of xenobiotic metabolism.

In addition to BP, pristane and chloramphenicol have been studied in trout hepatocytes to demonstrate their biotransformation. Pristane, a widespread branched alkane found in the aquatic environment, has been shown to be metabolized *in vitro* to pristanol, pristanediol, and pristanic acid. It is worth mentioning that isolated hepatocytes give a metabolic profile that better correlates with the *in vivo* situation than liver microsomes.[19] The biotransformation pathways of chloramphenicol, a well-known toxicant, have been investigated in rat hepatocytes in addition to trout hepatocytes. HPLC analysis shows that the major metabolite is the glucuronide conjugate in both species, while some differences appear between the rat and the trout in the first phase of metabolism (phase 1) prior to conjugation. The data obtained with the previous compounds suggest that cell systems would be suitable for predicting metabolic pathways in the whole animal.[20]

3. Induction of DNA Damage and Repair

The genotoxic response of cultured fish cells to both physical and chemical agents has been described, although lately, a limited number of reports have been published on this topic. DNA adduct formation following exposure to the carcinogens BP and 7,12-dimethylbenz[*a*]anthracene (DMBA) has been studied in cell lines derived from bluegill fry (BF-2), rainbow trout (RTG-2), and brown bullhead (BB).[21] The major BP-DNA adduct in the BB and BF-2 cells is that formed by reaction of (+)*anti*-PB-7,8-diol-9,10 epoxide [(+)*anti*-BPDE] with deoxyguanosine. In RTG-2 cells, the previous adduct is also found, but to a lesser extent, while other unidentified adducts are present. DMBA-3,4-diol-1,2-epoxide adducts are detected in BB cells but not in significant amounts in RTG-2 and BF-2 cell DNA. The ability to form glucuronide conjugates with BP-7,8-diol is inversely correlated with BP-DNA binding activity through BPDE, suggesting that conjugation protects against genotoxicity. In primary cultures of rainbow trout hepatocytes, BP has also been shown to produce DNA adducts, as detected by the ^{32}P-postlabeling assay. In this case, the DNA-adduct pattern is very similar to that found in the liver after treatment *in vivo*.[12] The previous results illustrate that fish cells are good models for studying the mechanisms of action of genotoxic carcinogens.

Fish cells have also been exposed to DNA-damaging radiation, and resistance mechanisms have been described. Kurihara et al.[22] have irradiated mudminnow (ULF-23) and goldfish (CAF-31) cells with low doses of X-rays and looked at chromosomal aberrations and micronucleus formation. It was found that cells exposed to the radiation become less sensitive to subsequent exposures to high doses of X-rays. In addition, the adaptive response is greater when the first dose is given during the G1 phase as compared to the S phase of the cell cycle. This response involves poly (ADP-ribose) polymerase and DNA polymerase α, as indicated by the effects of inhibitors of these enzymes, suggesting that a DNA excision repair system is involved. These data confirm that fish cells possess radio-adaptive response (RAR) mechanisms similar to those previously described for mammalian cells.[22] Moreover, fish cells exhibit DNA photorepair ability. This was concluded from goldfish cells (erythrophoroma line GEM 218) exposed to UV radiation. The induction of pyrimidine dimers, a major cytotoxicological lesion due to UV, was followed and the results indicate that fish cells have an efficient photoreactivation system at wavelength >304 nm that reverses cytotoxicity and dimer formation after exposure to irradiation at a shorter wavelength (l > 290 nm). Such a response has not been detected in normal human fibroblasts used for comparison with fish cells in this study.[23]

4. S-phase DNA Synthesis

Peroxisome proliferators are a class of nongenotoxic carcinogens that have been well studied in rodents. A number of such agents show mitogenic properties in rat hepatocytes and this property has been related to their carcinogenic potential. In fish, very little is known regarding the action of peroxisome proliferators. With the aim of filling this knowledge gap, Baldwin et al.[24] examined the ability of a series of structurally diverse peroxisome proliferators to induce S-phase synthesis in primary cultures of rainbow trout and medaka hepatocytes. *In vivo/in vitro* comparisons were carried out with lead nitrate, one of the compounds tested, which is a potent rodent liver mitogen. It was found that none of the peroxisome proliferators induces S-phase DNA synthesis in the fish cultures. These results support the view that the fish species studied differ from rodents in their response to peroxisome proliferators.

5. Function of Cell Membrane Transport Systems

In fish cells, the transport of xenobiotics as well as the effects of xenobiotics on the transport of endogenous agents have been looked at. The uptake of cadmium (Cd) by respiratory epithelial gill cells from rainbow trout has been described. Cells were prepared using trypsinization, grown in Petri dishes, and uptake experiments were performed on 13- to 15-day-old cultures. Interestingly, attention was paid to the influence of Cd speciation on the transport of Cd. The results suggest that Cd is mainly taken up by the cells as Cd^{2+}, although Cd can also be transported through the membrane in the form of $CdCl_2$. In addition, as lanthanum fails to affect Cd uptake, it appears that the latter would not occur via lanthanum-sensitive calcium channels. However in this system, the availability of Cd is increased by xanthates, indicating the formation of hydrophobic metal complexes that penetrate the lipid bilayer of the plasma membrane.[25]

Membrane ion transport systems have been shown to be affected by aquatic contaminants. Using rainbow trout hepatocytes, Råbergh et al.[26] investigated the action on cell membrane systems of two common resin acids, dehydroabietic acid (DHAA) and isopimaric acid (IPA). Resin acids are toxic compounds found as contaminants in the effluents of pulp and paper mill industries, and DHAA and IPA are among the most abundant. The effects on potassium efflux, potassium influx through the K^+/Na^+ pump, and on the uptake of two bile acids, cholic acid (CHA) and taurocholic acid (TCHA), were examined. It was found that both resin acids inhibit K^+/Na^+-pump activity and the uptake of bile acids, although the latter occurred at lower concentrations of resin acids, suggesting that it is not merely a consequence of K^+/Na^+ pump blockade. These results provide information that can be useful in elucidating the mechanisms of action of resin acids that cause liver dysfunction and jaundice in fish *in vivo*. Inhibition of uptake of bilirubin by resin acids in a way similar to that observed with bile acids could be an important factor in the development of jaundice in fish exposed to these contaminants.

6. Perturbation of Ionic Homeostasis

In recent years, there has been a lot of discussion about the role of Ca^{2+} homeostasis alterations in the process of cell death caused by toxic agents in mammalian cells.[27,28] Investigations of the perturbations of ionic homeostasis and ion metabolism by xenobiotics in fish cells are very few. A detailed study has been carried out with tri-n-butyltin (TBT). TBT is among the most effective bactericides and fungicides known and, as a result of its use in antifouling paints to combat the proliferation of aquatic organisms which encrust on ship hulls, it is found as a highly toxic contaminant in aquatic ecosystems. The effect of TBT on intracellular free Ca^{2+} levels ($[Ca^{2+}]$indice) was examined in isolated rainbow trout hepatocytes using flow cytometry. This powerful methodological approach allows live and dead cells to be discriminated and changes *in situ* to be detected in individual cells in heterogeneous populations. TBT induces a sustained elevation of $[Ca^{2+}]$indice in trout hepatocytes before loss of cell viability is detectable. At the same time, a severe depletion of cellular thiols is seen.[29] In addition to perturbations in $[Ca^{2+}]$indice, TBT causes a significant reduction in intracellular pH which appears to be caused by the combination of intracellular Ca^{2+} mobilization and by direct action of the compound.[30] It can be hypothesized that lowering the intracellular pH protects the cells against the action of degradative enzymes which have an optimum pH around 7. In another series of experiments, Virkii and Nikinmaa[31] demonstrated that TBT affects the adrenergic response in erythrocytes of rainbow trout. More specifically, TBT inhibits the isoproterenol-induced efflux of protons, uptake of Na^+ and Cl^-, and cell swelling, indicating an interference with the Na^+/H^+ exchanger. The previous results stress the importance of considering several parameters pertaining to ionic homeostasis because they are interrelated and can reciprocally influence each other.

7. Expression of Ionic Mimicry

Regarding the interactions of environmental contaminants with Ca^{2+} metabolism in fish cells, an interesting phenomenon has been described with Cd. It is recognized that Cd^{2+} resembles Ca^{2+} in at least two ways: these ions have the same charge and approximately the same ionic radius. Cd^{2+} therefore has the potential to replace Ca^{2+} at various sites in the cell. One such site is calmodulin, a ubiquitous calcium-binding protein that regulates many calcium-dependent processes. It has been shown that Cd^{2+} can interact with calmodulin. In the rainbow trout gonadal cells RTG-2, calmodulin accounts for about 0.7% of the soluble proteins. Calcium-dependent phosphorylation of numerous cytosolic proteins of RTG-2 cells can be observed *in vitro*. It is quite striking to see that Cd is as effective as calcium in stimulating calmodulin-dependent protein phosphorylation in cytosolic fractions of RTG-2 cells.[32] These results point to a possible action of Cd as a signal transduction disruptor in fish cells via interference with calcium-mediated signals. The implications of such an action at the physiological level could be very important.

8. Induction of Oxidative Stress

Oxidative stress is one of the serious consequences of intoxication of cells by numerous xenobiotics including environmental agents. The herbicide paraquat belongs to this group of agents. Data from mammalian systems have shown that it is an inducer of redox-cycling reactions and of oxidative stress. Bluegill sunfish BF-2 cells have been used to study oxidative stress in fish cells exposed to paraquat and H_2O_2, and a comparison with mammalian cells was made. It was found that the cells are sensitive to both agents and that chelation of Fe^{2+} reduces their cytotoxicity, presumably by preventing hydroxyl radical formation in the Fenton reaction.[33] The cytotoxic response in the case of H_2O_2 is antagonized by quin-2, AM, a calcium chelating agent, suggesting that calcium homeostasis is involved in the cytotoxicity of this oxidative stress-inducing agent in fish cells. It is also interesting to note that the mammalian cells examined in this study are approximately ten times more sensitive than the fish cells. It can be concluded that, although oxidative stress responses are shared by mammalian and lower vertebrate cells, fish systems appear sufficiently different to deserve attention on their own.

9. Expression of Metallothioneins and Stress Proteins

Metallothioneins

Proteins such as metallothioneins (MT), and to a lesser extent stress proteins, which are induced in response to various stressors, have received special attention, in particular in view of the fact that they could serve as biomarkers in environmental monitoring. MT induction and metal homeostasis in rainbow trout hepatocytes have been described.[34,35] Trout hepatocytes exposed to $HgCl_2$ significantly accumulate the metal which mainly binds to low-molecular-weight components present in the cells. Although MT synthesis is induced, it does not appear to play a significant role in the sequestration of Hg. In parallel, elevated levels of Cu and Zn are seen which could be related to MT induction.[34] In a series of experiments similar to those performed for Hg, it was observed that Cd induces the synthesis of MT in trout hepatocytes. After uptake by the cells, in contrast to Hg, Cd is mainly associated with cytosolic high- and middle-molecular-weight components, the latter including MT. As similarly detected for Hg, an increase in intracellular Cu and Zn is measured as a result of exposure to Cd and, presumably, MT induction.[35] Olsson et al.[36] also reported the induction of MT in primary cultures of trout hepatocytes which occurs in the presence of Zn, cortisol, and corticosterone. In cultures treated with Zn or cortisol, there is a high temporal correlation between MT mRNA and MT protein levels. In contrast to the hepatocytes, the cell line RTH-149 did not respond to the steroid hormones, indicating that primary cultures could provide a better model than immortal cell lines for investigating the inducibility of MT.

In the same line of investigation, George and his colleagues[37] demonstrated the induction of MT mRNA and MT levels by Cd in a fibroblast cell line derived from a marine flatfish, the turbot *Scophthalmus maximus*. Cu, Hg, and Zn also produce the induction of MT, whereas Pb is inactive. Furthermore, the effects of hormones on MT synthesis were verified using the same turbot cell line and RTG-2 cells. The data obtained show that only the turbot cells respond to glucocorticoid and progesterone treatments with modest induction of MT (1.5- to twofold increases), whereas neither cell line responds to estradiol. From the results of their study, the authors point to an important fact pertaining to the applicability of MT as a biomarker in environmental monitoring. Indeed, they indicate that levels less than threefold higher than reference sites in MT levels must be disregarded for heavy-metal pollution monitoring.[38]

The regulation of trout MT isoforms has been studied at the genetic level in fish cell culture systems. By applying a novel technique of primer extension and DNA sequencing, it was established that there is a higher response of MTa to $ZnCl_2$ and $CdCl_2$ compared to MTb in the trout gonadal cell line RTG-2.[39] Furthermore, the contribution of the two metal-responsive elements (MREs) of the rainbow trout MT-B gene has been evaluated using gene fusion and transfection of the RTH-149 cell line. It was found that both elements cooperate to elicit a significant response to Zn.[40]

Stress Proteins

In addition to MT metabolism, the response to metals involving the induction of stress proteins such as hsp70 has been modeled in fish cells. Recently, Ryan and Hightower[41] took advantage of two fish cell culture systems, the PLHC-1 cells, derived from a hepatocellular carcinoma in the topminnow *Poeciliopsis lucida*, and primary cultures of renal tubule cells obtained from the winter flounder *Pleuronectes americanus*. In PLHC-1 cells, increases in stress proteins are detectable following Cd (six proteins) and Cu exposure (four proteins). In flounder kidney cells, similar changes are observed except that there is a smaller number of proteins that can be detected (three proteins). In both cell systems, two proteins, hsp70 and a 27-kDa protein, are identifiable following exposure to either Cd or Cu. It remains to be seen whether potential molecular markers expressed *in vitro* will behave similarly *in vivo* and thus be useful as environmental biomarkers.[41]

B. Morphological Changes in the Cytotoxic Response

Ultrastructural analysis has been put forward as a tool for investigating the action of toxic substances on fish cells. Mayer et al.,[42] using the R1 cell line established from liver tissue of rainbow trout, showed ultrastructural alterations induced by urea, copper sulfate, and sodium nitrite. A segmented nucleus is the most evident change seen after exposure to high levels of urea, whereas cytoplasmic changes are observed with copper sulfate. Sodium nitrite leads to both nuclear and cytoplasmic changes. From their data, Mayer et al.[42] proposed that electron microscopical analysis be used as a complement in *in vitro* screening procedures. Similarly, an attempt was made to evaluate the substance-specificity of hepatocellular reactions *in vitro*. For this purpose, Zahn and Braunbeck[43] exposed rainbow trout hepatocytes to dinitro-*o*-cresol and 2,4-dichlorophenol. They found that both specific and unspecific changes occur, the latter involving heterochromatin, RER cisternae, lysosomes, glycogenosomes, myelinated bodies, and large vacuoles. These authors pointed out that unspecific changes could serve as rapid biomarkers of contamination by chemicals, and specific changes as markers for the diagnosis of toxicants.[43] However, discrepancies between cytological alteration *in vivo* and *in vitro* in rainbow trout hepatocytes have been observed following treatment with 4-chloroaniline.[44] These discrepancies have been attributed to differences in metabolic pathway activation or modulation of responses by systemic factors *in vivo*.

In another study, the morphological alterations caused by acetaminophen in liver cells of rainbow trout were investigated in an effort to better characterize the hepatotoxicity of this compound. The rearrangement of mitochondria is a striking change induced by acetaminophen in these cells. An interesting feature is the absence of detectable DNA fragmentation in contrast to what is seen in mouse liver cells. It therefore appears that trout cells provide a system in which acetaminophen cytotoxicity can be studied without the effects that are secondary to DNA fragmentation.[45]

C. Action of Xenobiotics on Specialized Cell Functions

1. Immune Cell Function

In vitro approaches have been applied to study the action of xenobiotics on the cells of the immune system. So far as aquatic pollutants are concerned, the cytotoxicity and functional impairment of blood and head kidney leukocytes and phagocytes of rainbow trout exposed to mercury *in vitro* have been reported.[46] Cell viability, mixed leukocyte reaction, leukocyte mitogenic response, phagocytic response, and oxidative burst of phagocytes were monitored over a wide range of mercuric chloride and methylmercury concentrations. The effects on functional activities were

Table 7.1   Fish Cells Used in Toxicity Testing

Name	Type	Origin
R1	FCL[1]	Rainbow trout liver
BB		Posterior trunk tissue from brown bullhead
RTG-2	FCL[1]	Rainbow trout gonads
Hepatocytes	PC[2]	Rainbow trout liver
GFS	FCL[1]	Goldfish scale
BG/F	ECL[3]	Fin tissue from bluegill sunfish
Gill, fin, and gonad tissue cells	PC[2]	Brown bullhead
FHM	ECL[3]	Tissue posterior to the anus from fathead minnow
CCB	FCL[1]	Carp brain
OLF-136	FCL[1]	Mekada fin
PLHC-1	HCL[4]	Topminnow

[1] FCL = fibroblastic cell line
[2] PC = primary culture
[3] ECL = epithelioid cell line
[4] HCL = hepatoma cell line

almost exclusively seen at cytotoxic concentrations. It was concluded that the mercurials cause a nonspecific impairment of immune cellular functions. Fish leukocytes have also been used to investigate the signal transduction pathways involved in mitogen stimulation. The system involved stimulation of channel catfish peripheral blood leukocytes with the model compound 12-O-tetradecanoylphorbol 13-acetate (TPA).[47] The results obtained suggest that the phosphatidylinositol bisphosphate pathway is involved in transmembrane signaling, as is the case in mammalian lymphocytes. Invaluable information on the mechanisms of action of xenobiotics that affect immune functions could be gained by employing similar models for environmental contaminants.

2. Estrogenic Response

The response of rainbow trout hepatocytes to estrogens has been studied *in vitro*. When exposed to estradiol-17 β (E_2), these cells produce vitellogenin, a large protein that is sequestered by the oocytes of oviparous vertebrates and stored as food supply for the future embryo. Pelissero et al.[48] have tested a number of hormones and phytoestrogens in the trout hepatocyte culture system. They found that the cells respond to E_2 by expressing vitellogenin in a concentration-dependent manner, whereas the phytoestrogens are also active but much less potent than E_2. The observation that tamoxifen inhibits the induction of vitellogenin supports the conclusion that this function is estrogen-dependent. The authors propose that trout hepatocyte cultures could be used as the basis of a test to detect the estrogenicity of chemicals. This system is all the more interesting since liver cells could perform biotransformations that may be necessary for the estrogenicity of certain compounds to be expressed.

III. FISH CELLS IN TOXICITY TESTING

A. Validation of Established Systems

The principles for the validation of *in vitro* toxicology methods have been discussed previously.[1,49] With fish cells, the emphasis has been put on *in vivo/in vitro* comparisons. About a dozen cell types have been used, and established cell lines have been mostly involved (Table 7.1). Studies have investigated series of chemically-related compounds and series of various reference compounds. In addition, tests on environmental samples have been performed. Tables 7.2 and 7.3 give an overview of the studies carried out with series of chemicals. It can be seen that neutral red

Table 7.2 Validation of Fish Cell Assays with Series of Chemically-Related Compounds

Test Chemicals	Cell Type	In Vitro Assay	In Vivo Data Used for Comparison	Correlation In Vivo/In Vitro	Reference
Inorganic and organic lead compounds	BG/F	NR[1]	Published LC$_{50}$ for P. platessa	Good	Babich and Borenfreund[63]
Phenol + 14 chlorophenols	GFS	NR[1]	Published LC$_{50}$ for goldfish, guppy, killifish, and killifish embryo	Good (r > 0.9)	Saito et al.[64]
45 pesticides	GFS	NR[1]	Published LC$_{50}$ for carp	Good	Saito et al.[65]
Polycyclic aromatic hydrocarbons	BB RTG-2	NR[1]	—	—	Martin-Alguacil et al.[66]
Chlorinated pesticides	Primary cultures of cells from gill, fin, and gonad tissue of brown bullhead.				
19 lipophilic solvents	FHM	EC$_{50}$ = concentration required for 50% reduction in protein content	Published LC$_{50}$ for golden orfe	Poor	Dierickx[50]
13 anilines 10 aldehydes	GFS	NR[1]	Published LC$_{50}$ for guppies	Good	Saito et al.[67]
Phenol + 14 chlorophenols	CCB OLF-136	NR[1]	—	—	Saito and Shigeoka[68]
13 metal salts	R1	NR[1]	Published LC$_{50}$ for golden ide, bluegill sunfish, and rainbow trout	Good for cationic metals (0.64 < r < 0.92) No correlation for metal complexes	Segner et al.[51]
14 chlorophenols	GFS	MTT[2] LDH[3]	Published LC$_{50}$ for goldfish, guppy, and medaka	Good (0.89 < r < 0.97)	Saito et al.[69]
21 organotin compounds	PLHC-1	NR[1] MTT[2] BrdU[4] CV[5]	Published LC$_{50}$ LC$_{50}$ for red killifish	Good (0.80 < r < 0.86)	Brüschweiler et al.[70]

[1] NR = Neutral red; [2] MTT = tetrazolium salt reduction; [3] LDH = lactate dehydrogenase; [4] BrdU = bromodeoxyridine; [5] CV = crystal violet

uptake is by far the most widely used assay. In the vast majority of cases, good *in vivo/in vitro* correlations are obtained when comparisons of IC$_{50}$s (concentrations that inhibit the response by 50% as determined *in vitro*) with LC$_{50}$s (found in the literature) are made. However, it is worth mentioning that in most cases, interspecies rather that intraspecies comparisons are performed. This indicates that the *in vivo* database lacks information and therefore is not completely suitable for the validation process, a handicap already encountered in mammalian toxicology. Despite the generally good *in vivo/in vitro* correlations, a number of exceptions are reported, however, in particular for lipophilic solvents[50] and metal complexes,[51] the cause of which is not understood.

Table 7.3 Validation of Fish Cell Assays with Series of Various Reference Chemicals

Number of Compounds	Cell System	Assay	In Vivo Data Used for Comparison	In Vivo/In Vitro Correlation	Reference
15	R 1	NR[1] CV[2] MTT[3]	Published LC_{50} for golden ide	Good	Lenz et al.[71]
29	FHM	NR[1]	Published LC_{50} for golden orfe	Good (r = 0.93)	Dierickx and Van De Vyver[72]
50	FHM	NR[1]	Published LC_{50} for golden orfe	Good (r = 0.89)	Brandão et al.[73]
109	GFS	NR[1]	Published LC_{50} for fathead minnow and guppy	Good (r = 0.95)	Saito et al.[74]
30	R1	CV[2]	Published LC_{50} for golden ide	Good (r = 0.84, n = 21)	Segner and Lenz[75]
25	FHM	Protein content	Published LC_{50} for golden orfe	Good (r = 0.90)	Dierickx[76]
50	Primary cultures of rainbow trout hepatocytes	[86]Rb leakage	Immobility in *Daphnia magna*	Poor (r = 0.71)	Lilius et al.[77]
10	RTG-2	NR[1] MTT[3]	Published LC_{50} for fathead minnow, *Brachydanio rerio,* and *Phoxinus neogaeus*	Inconclusive	Lange et al.[78]
30	Freshly isolated gill epithelial cells and rainbow trout hepatocytes	Viability with Calcein-NM	Published LC_{50} for rainbow, fathead minnow, guppy, and bluegill sunfish	Good (r^2 = 0.87 for epithelial cells)	Lilius et al.[79]

[1] NR = Neutral red
[2] CV = Crystal violet
[3] MTT = Tratrazalium salt reduction

These exceptions ought to be especially studied in order to elucidate why such discrepancies are observed. This knowledge should further our assessment of *in vitro* test system limitations and provide unique opportunities to use these systems in an optimum way.

The attempts that have been made to test the response of fish cells to environmental samples have also been considered successful. Recent studies in this area are presented in Table 7.4. Rusche and Kohlpoth[52] found a very good correspondence between *in vitro* results and those obtained in parallel with fish tests using golden ide, supporting the view evolving from data on single chemicals. In the other studies, although there was no direct *in vivo* evaluation, the authors draw conclusions about the usefulness of the models for detection of toxic potential of complex samples. In addition to the search for *in vivo/in vitro* relationships, there has been work aimed at comparing fish cells to mammalian cells in *in vitro* toxicity evaluation. The investigation of Devlin and Mottet,[53] based on the use of RTG-2 and Chinese hamster ovary (CHO) cells exposed to methylmercury, provides a good example of such an approach. It underlines the fact that *in vitro* systems may facilitate interspecies comparisons.

Recently, Gagné and Blaise[54] developed a particularly interesting strategy for the validation of the rainbow trout hepatocyte model for testing industrial wastewater. They applied artificial network modeling to analyze *in vitro* cytotoxicity data to predict whole-fish toxicity. They found that artificial neural network modeling of *in vitro* data allows the investigator to classify toxic effluents in a way that is predictive of what is observed *in vivo* with trout in the 96-h bioassay. This approach could be quite useful for screening large numbers of wastewater samples and to proceed to *in vivo* evaluation only when it is deemed necessary.[54]

Table 7.4   Bioassay Systems Based on Fish Cells to Evaluate the Toxicity of Environmental Samples

Samples Tested	Cell Type	Assay	Remarks by Authors	Evaluation of Model(s)	Reference
Paper mill effluents	Primary cultures of rainbow trout hepatocytes	LDH[1] Glutathione Cytochrome P-450 activity		Primary hepatocytes provide convenient method for screening cytochrome P-450-inducing activity and study of cell injury mechanisms	Pesonen and Andersson[80]
362 wastewater samples	R 1	CV[2]	86.8% correspondence with fish test	Alternative method to solve conflict between German Wastewater Act and German Animal Welfare Act.	Rusche and Kohlpoth[52]
Sediment extracts in two fractions (two groups of chemicals)	BB	NR[3] UDS[4]	Synergistic interaction between the two fractions	Good potential for model system	Ali et al.[81]
Extracts of organic compounds from fish-canning factory effluents	RTG-2	Cell viability Intracellular ATP	Six toxic fractions out of 210 from HPLC fractionation of toxic concentrates	Allow risk assessment of fish canning effluents	Vega et al.[82]
Sediment extracts in the vicinity of a creosote-treated wharf	Primary cultures of rainbow trout hepatocytes	Cell viability, nick translation, and alkaline precipitation (genotoxicity)	Positive correlation between polycyclic aromatic hydrocarbons in samples and genotoxicity	Pertinent model to detect (geno)toxic compounds in contaminated sediments	Gagné et al.[83]
Seepage waters from garbage dumps	Primary cultures of rainbow trout hepatocytes	LDH[1] EM[5], NR[3], CV[2]		Multitiered approach proposed with permanent cell lines and hepatocytes	Zahn et al.[84]

[1] LDH = lactate dehydrogenase; [2]CV = crystal violet; [3]MR = neutral; [4]UDS = unscheduled DNA synthesis; [5]EM = electron microscopy

B. Development of Novel Systems

Although many well-established fish cell lines and primary culture systems have been used in toxicity testing, there is ongoing development of systems proposing novel ways to exploit existing models or dealing with the use of new cell types. For example, rainbow trout hepatocytes have been applied to the study of the cytotoxicity of binary mixtures of metals in an effort to detect possible antagonistic or synergistic interactions. The data based on the release of lactate dehydrogenase (LDH) reveal increased sensitivity to Cd when Cu is present; this modulation cannot be attributed to changes in uptake of metals by the cells.[55]

In a different facet of the investigation, the RTG-2 cell line has been shown to provide a good bioindicator for the toxicity of fractions of complex environmental mixtures. Indeed, in combination with concentration/fractionation procedures (HPLC), these cells allow the identification of the toxic fractions in an effluent. This approach has many advantages: it requires only small quantities of material, whereas *in vivo* tests require amounts that may be difficult and costly to obtain. Further-more, it gives ecotoxicological information without the need to identify all the chemicals in the sample.[56,57] Biodetection such as that performed with RTG-2 cells should be used more in the future.

A number of researchers have also devoted efforts to characterizing new fish cell systems. Roche et al.[58] worked on erythrocytes of a marine fish (*Dicentrarchus labrax*) and investigated the culture conditions and the metabolic status of the cells in order to propose a new *in vitro* toxicity test. For their part, Collodi et al.[59] developed methods for the culture of cells from blastula-stage diploid and haploid zebrafish embryos and from various tissues of adult fish. In addition, this group has shown that gene transfection as performed in mammalian cells can also be applied to zebrafish cell cultures, thereby providing a way to modify the genotype and phenotype of the cells. In another study, Chen et al.[60] demonstrated the culture of grass carp lip and embryo cells on Cytodex 3 and GT-2 microcarriers in a bioreactor. Such a system allows the propagation of virus for vaccine production. Recently, Lee et al.[61] developed and characterized a new rainbow trout liver cell line (RTL-W1) expressing cytochrome P-450-dependent activity that can be useful for assessing the toxic potential of environmental contaminants.

A particularly interesting cell culture system has been developed by Williams et al.[62] to examine the possible role of exposure to aromatic hydrocarbons (AH) in the development of eye lens cataract in some fish species. They have established a procedure for the *in vitro* growth and differentiation of eye lens epithelial cells from the marine teleost spot *Leiostomus xanthurus*. This system allows the direct effects of toxic pollutants on these cells to be studied, as opposed to the result of reactions at the level of the whole organism. Exposure of the cultured lens cells to the BP metabolite benzo[α]pyrene-7,8-dihydrodiol leads to interference with the synthesis of DNA, RNA, and pro-teins. The impact of these biochemical changes on the physiology of the lens epithelium remains to be investigated further.

IV. CONCLUSIONS

A survey of the literature on fish cells in toxicological evaluation highlights important features. First, the diversity of mechanisms and biochemical endpoints being considered is expanding steadily, showing dynamic development that is expected to continue in the years to come. Second, validation has taken a significant step forward and the conclusions reached so far are positive. A consensus is indeed emerging that cytotoxicity assays could provide reliable information in envi-ronmental surveillance programs. Third, an area that is clearly deserving is that of toxic mechanisms that interfere with specialized cell functions. More efforts should be devoted to improving our knowledge and investigation methods in this area. Overall, fish cell models have now been firmly established as invaluable tools in toxicological evaluation, and the place that they have taken can only grow in importance in the future.

REFERENCES

1. Frazier, J. M. Ed., *In vitro Toxicity Testing: Applications to Safety Evaluation,* Marcel Dekker, New York, 1992.
2. Fielder, R. J., Acceptance of *in vitro* studies by regulatory authorities, *Toxicol. In Vitro,* 8, 911, 1994.
3. Fentem, J. H. and Balls, M., Why, when and how *in vitro* tests should be accepted into regulatory toxicology, *Toxic. In Vitro,* 8, 923, 1994.
4. Williams, P. D., Use of *in vitro* techniques in toxicology: When, why, how? *J. Am. Col. Toxicol.,* 13, 302, 1994.
5. Combes, R. D., Regulatory genotoxicity testing: A critical appraisal, *ATLA,* 23, 352, 1995.

6. Baksi, S. M. and Frazier, J. M., Review: Isolated fish hepatocytes — model systems for toxicology research, *Aquatic. Toxicol.,* 16, 229, 1990.
7. Babich, H. and Borenfreund, E., Cytotoxicity and genotoxicity assays with cultured fish cells: A review, *Toxic. In Vitro,* 5, 91, 1991.
8. Isomaa, B., Lilius, H., and Råbergh, C., Aquatic toxicology *in vitro*: A brief review, *ATLA,* 22, 243, 1994.
9. Lorenzen, A. and Okey, A. B., Detection and characterization of [^{3}H]2,3,7,8-tetrachlorodibenzo-p-dioxin binding to Ah receptor in a rainbow trout hepatoma cell line, *Toxicol. Appl. Pharmacol.,* 106, 53, 1990.
10. Pesonen, M. and Andersson, T., Characterization and induction of xenobiotic metabolizing enzyme activities in a primary culture of rainbow trout hepatocytes, *Xenobiotica,* 21, 461, 1991.
11. Pesonen, M., Goksøyr, A., and Andersson, T. Expression of P4501A1 in a primary culture of rainbow trout hepatocytes exposed to β-naphthoflavone or 2,3,7,8-tetrachlorodibenzo-p-dioxin, *Arch. Biochem. Biophys.,* 292, 228, 1992.
12. Masfaraud, J.- F., Devaux, A., Pfohl-Leszkowicz, A., Malaveille, C., and Monod, G., DNA adduct formation and 7-ethoxyresorufin-o-deethylase induction in primary culture of rainbow trout hepatocytes exposed to benzo[a]pyrene, *Toxic. In Vitro,* 6, 523, 1994.
13. Devaux, A., Pesonen, M., Monod, G., and Andersson, T., Glucocorticoid-mediated potentiation of P450 induction in primary culture of rainbow trout hepatocytes, *Bioch. Pharmacol.,* 43(4), 898, 1992.
14. Miller, M. R., Saito, N., Blair, J. B., and Hinton, D. E., Acetaminophen toxicity in cultured trout liver cells II. Maintenance of cytochrome P450 1A1, *Exp. Mol. Pathol.,* 58, 127, 1993.
15. Andersson, T. and Koivusaari, U., Oxidative and conjugative metabolism of xenobiotics in isolated liver cells from thermally acclimated rainbow trout, *Aquatic Toxicol.,* 8, 85, 1986.
16. Kennedy, C. J., Gill, K. A. and Walsh, P. J., Temperature acclimation of xenobiotic metabolizing enzymes in cultured hepatocytes and whole liver of the gulf toadfish, *Opsanus beta, Can. J. Fish. Aquatic Sci.,* 48, 1212, 1991.
17. Bols, N. C., Mosser, D. D. and Steels, G. B., Mini Review: Temperature studies and recent advances with fish cells *in vitro, Comp. Biochem. Physiol.,* 103A, 1, 1992.
18. Nishimoto, M., Yanagida, G. K., Stein, J. E., Baird, W. M., and Varanasi, U., The metabolism of benzo(a)pyrene by English sole (*Parophrys vetulus*): Comparison between isolated hepatocytes *in vitro* and liver *in vivo, Xenobiotica,* 22, 949, 1992.
19. Cravedi, J. P., Perdu-Durand, E., Baradat, M., and Tulliez, J., Hydroxylation of pristane by isolated hepatocytes of rainbow trout: A comparison with *in vivo* metabolism and biotransformation by liver microsomes, *Mar. Environ. Res.,* 28, 15, 1989.
20. Cravedi, J. P. and Baradat, M., Comparative metabolic profiling of chloramphenicol by isolated hepatocytes from rat and trout (*Oncorhynchus mykiss*), *Comp. Biochem. Physiol.,* 100C, 649, 1991.
21. Smolarek, T. A., Morgan, S. L., Moynihan, C. G., Lee, H., Harvey, R. G., and Baird, W. M., Metabolism and DNA adduct formation of benzo[a]pyrene and 7,12-dimethylbenz[a]anthracene in fish cell lines in culture, *Carcinogenesis,* 8, 1501, 1987.
22. Kurihara, Y., Rienkjkarn, M., and Etoh, H., Cytogenetic adaptive response of cultured fish cells to low doses of X-rays, *J. Radiat. Res.,* 33, 267, 1992.
23. Ahmed, F. E., Setlow, R. B., Grist, E., and Setlow, N., DNA damage, photorepair, and survival in fish and human cells exposed to UV radiation, *Environ. Mol. Mut.,* 22, 18, 1993.
24. Baldwin, L. A., Kostecki, P. T., and Calabrese, E. J., Oxidative stress in fish cells: *in vitro* studies, *Ecotoxicol. Environ. Safety,* 25, 193, 1993.
25. Block, M. and Pärt, P., Uptake of ^{109}Cd by cultured gill epithelial cells from rainbow trout (*Oncorhynchus mykiss*), *Aquatic Toxicol.,* 23, 137, 1992.
26. Råbergh, C. M. I., Isomaa, B., and Eriksson, J. E., The resin acids dehydroabietic acid and isopimaric acid inhibit bile acid uptake and perturb potassium transport in isolated hepatocytes from rainbow trout, *Aquatic Toxicol.,* 23, 169, 1992.
27. Orrenius, S. and Nicotera, P., The calcium ion and cell death, *J. Neural. Transm.,* Supp. 43, 1, 1994.
28. Harman, A. W. and Maxwell, M. J., An evaluation of the role of calcium in cell injury, *Annu. Rev. Pharmacol. Toxicol.,* 35, 129, 1995.

29. Reader, S., Marion, M., and Denizeau, F., Flow cytometric analysis of the effects of tri-*n*-butyltin chloride on cytosolic free calcium and thiol levels in isolated rainbow trout hepatocytes, *Toxicology,* 80, 117, 1993.

30. Reader, S., Steen, H. B., and Denizeau, F., Intracellular calcium and pH alterations induced by tri-n-butyltin chloride in isolated rainbow trout hepatocytes: A flow cytometric analysis, *Arch. Biochem. Biophys.,* 312(2), 407, 1994.

31. Virkii, L. and Nikinmaa, M., Tributyltin inhibition of adrenergically activated sodium/proton exchange in erythrocytes of rainbow trout (*Oncorhynchus mykiss*), *Aquatic Toxicol.,* 25, 139, 1993.

32. Behra, R. and Gall, R., Calcium/calmodulin-dependent phosphorylation and the effect of cadmium in cultured fish cells, *Comp. Biochem. Physiol.,* 100C, 191, 1991.

33. Babich, H., Palace, M. R., and Stern, A. Oxidative stress in fish cells: *in vitro* studies, *Arch. Environ. Contam. Toxicol.,* 24, 173, 1993.

34. Gagné, F., Marion, M., and Denizeau, F., Metal homeostasis and metallothionein induction in rainbow trout hepatocytes exposed to cadmium, *Fund. Appl. Toxicol.,* 14, 429, 1990.

35. Gagné, F., Marion, M., and Denizeau, F., Metallothionein induction and metal homeostasis in rainbow trout hepatocytes exposed to mercury, *Toxicol. Lett.,* 51, 99, 1990.

36. Olsson, P.- E., Hyllner, S.J., Zafarullah, M., Andersson, T., and Gedamu, L., Differences in metal-lothionein gene expression in primary cultures of rainbow hepatocytes and the RTH-149 cell line, *Biochim. Biophys. Acta.,* 1049, 78, 1990.

37. George, S., Burgess, D., Leaver, M., and Frerichs, N., Metallothionein induction in cultured fibroblasts and liver of a marine flatfish, the turbot, *Scophthalmus maximus, Fish Physiol. Biochem.,* 10, 43, 1992.

38. Burgess, D., Frerichs, N., and George, S., Control of metallothionein expression by hormones and stressors in cultured fish cell lines, *Marine Environ. Res.,* 35, 24, 1993.

39. Zafarullah, M., Olsson, P.-E., and Gedamu, L., Differential regulation of metallothionein genes in rainbow trout fibroblasts, RTG-2, *Biochim. Biophys. Acta,* 1049, 318, 1990.

40. Samson, S. L. A. and Gedamu, L., Metal-responsive elements of the rainbow trout metallothionein-B gene function for basal and metal-induced activity, *J. Biol. Chem.,* 270, 6864, 1995.

41. Ryan, J. A. and Hightower, L. E., Evaluation of heavy-metal ion toxicity in fish cells using a combined stress protein and cytotoxicity assay, *Environ. Toxicol. Chem.,* 13, 1231, 1994.

42. Mayer, D., Ahne, W., and Storch, V., Cytotoxicity of chemicals to fibroblastic fish cell cultures (R1 cells) investigated by electron microscopy, *Z. Angew Zool.,* 75, 147, 1988.

43. Zahn, T. and Braunbeck, T., Isolated fish hepatocytes as a tool in aquatic toxicology: sublethal effects of dinitro-o-cresol and 2,4-dichlorophenol, *Sci. Total Environ.,* Supplement, 721, 1993.

44. Braunbeck, T., Cytological alterations in isolated hepatocytes from rainbow trout (*Oncorhynchus mykiss*) exposed *in vitro* to 4-chloroaniline, *Aquatic Toxicol.,* 25, 83, 1993.

45. Miller, M. R., Wentz, E., Blair, J. B., Pack, D., and Hinton, D. E., Acetaminophen toxicity in cultured trout liver cells I. Morphological alterations and effects on cytochrome P450 1A1, *Exp. Mol. Pathol.,* 58, 114, 1993b.

46. Voccia, I., Krzystyniak, K., Dunier, M., Flipo, D., and Fournier, M., *In vitro* mercury-related cytotox-icity and functional impairment of the immune cells of rainbow trout (*Oncorhynchus mykiss*), *Aquatic Toxicol.,* 29, 37, 1994.

47. Lin, G. L., Ellsaesser, C. F., Clem, L. W., and Miller, N. W., Phorbol ester/calcium ionophore activate fish leukocytes and induce long-term cultures, *Dev. Compar. Immunol.,* 16, 153, 1992.

48. Pelissero, C., Flouriot, G., Foucher, J. L., Bennetau, B., Dunoguès, J., Le Gac, F., and Sumpter, J. P., Vitellogenin synthesis in cultured hepatocytes; an *in vitro* test for the estrogenic potency of chemicals, *J. Steroid Biochem. Molec. Biol.,* 44, 263, 1993.

49. Walum, E., Clemedson, C., and Ekwall, B. Principles for the validation of *in vitro* toxicology test methods, *Toxic. In Vitro,* 8, 807, 1994.

50. Dierickx, P. J., Comparison between fish lethality data and the *in vitro* cytotoxicity of lipophilic solvents to cultured fish cells in a two-compartment model, *Chemosphere,* 27, 1511, 1993.

51. Segner, H., Lenz, D., Hanke, W., and Schuurmann, G. Cytotoxicity of metals toward rainbow trout R1 cell line, *Environ. Toxicol. Water Qual.: Internat. J.,* 9, 273, 1994.

52. Rusche, B. and Kohlpoth, M., The R1-cytotoxicity test as a replacement for the fish test stipulated in the German Waste Water Act, in: T. Braunbeck, W. Hanke and H. Segner, Eds. *Fish — Ecotoxicol. Ecophysiol.,* VCH, Weinheim, 1993, 81.

53. Devlin, E. A. and Mottet, N. K., Comparative methyl mercury toxicity on a mammalian and a teleostean cell line, *In vitro Toxicol.*, 5, 83, 1992.

54. Gagné, F. and Blaise, C., Validation of the rainbow trout hepatocyte model for ecotoxicity testing of industrial wastewater, Society of Toxicology of Canada, Twenty-ninth Annual Symposium, Montreal, Canada, Dec. 5-6, 1995.

55. Denizeau, F. and Marion, M., Toxicity of cadmium, copper, and mercury to isolated trout hepatocytes, *Can. J. Fish. Aqua. Sci.*, 47, 1038, 1990.

56. Tarazona, J. V., Castano, A., and Gallego, B., Detection of organic toxic pollutants in water and waste-water by liquid chromatography and *in vitro* cytotoxicity tests, *Anal. Chim. Acta.*, 234, 193, 1990.

57. Castaño, A., Vega, M., Blazquez, T., and Tarazona, J. V., Biological alternatives to chemical identification for the ecotoxicological assessment of industrial effluents: The RTG-2 *in vitro* cytotoxicity test, *Environ. Toxicol. Chem.*, 13, 1607, 1994.

58. Roche, H., Cransac, H., and Peres, G., Intérêt et modalités pratiques d'utilisation des érythrocytes d'un poisson marin (*Dicentrarchus labrax*) en écotoxicologie, *Ichtyophysiol. Acta.*, 14, 23, 1991.

59. Collodi, P., Kamei, Y., Ernst, T., Miranda, C., Buhler, D. R., and Barnes, D. W., Culture of cells from zebrafish (*Brachydanio rerio*) embryo and adult tissues, *Cell Biol. Toxicol.*, 8, 43, 1992.

60. Chen, Z., Chen, Y., and Shi, Y., Microcarrier culture of fish cells and viruses in cell culture bioreactor, *Can. J. Microbiol.*, 38, 222, 1992.

61. Lee, L. E. J., Clemons, J. H., Bechtel, D. G., Caldwell, S. J., Han, K.-B., Pasitschniak-Arts, M., Mosser, D. D., and Bols, N. C., Development and characterization of a rainbow trout liver cell line expressing cytochrome P450-dependent monooxygenase activity, *Cell Biol. Toxicol.*, 9, 279, 1993.

62. Williams, C. D., Faisal, M., and Huggett, R. J., Polynuclear aromatic hydrocarbons and fish lens cataract: Effects of benzo[a]pyrene-7,8-dihydrodiol on the macromolecular synthesis of cultured eye cells, *Mar. Environ. Res.*, 34, 333, 1992.

63. Babich, H. and Borenfreund, E., *In vitro* cytotoxicities of inorganic lead and di- and trialkyl lead compounds to fish cells, *Bull. Environ. Contam. Toxicol.*, 44, 456, 1990.

64. Saito, H., Sudo, M., Shigeoka, T., and Yamauchi, F., *In vitro* cytotoxicity of chlorophenols to goldfish GF-scale (GFS) cells and quantitative structure-activity relationships, *Environ. Toxicol. Chem.*, 10, 235, 1991.

65. Saito, H., Iwami, S., and Shigeoka, T., *In vitro* cytotoxicity of 45 pesticides to goldfish GF-scale (GFS) cells, *Chemosphere,* 23, 525, 1991.

66. Martin-Alguacil, N., Babich, H., Rosenberg, D. W., and Borenfreund, E., *In vitro* response of the brown bullhead catfish cell line, BB, to aquatic pollutants, *Arch. Environ. Contam. Toxicol.*, 20, 113, 1991.

67. Saito, H., Koyasu, J., and Shigeoka, T., Cytotoxicity of anilines and aldehydes to goldfish GFS cells and relationships with 1-octanol/water partition coefficients, *Chemosphere,* 27, 1553, 1993.

68. Saito, H. and Shigeoka, T., Comparative cytotoxicity of chlorophenols to cultured fish cells, *Environ. Toxicol. Chem.*, 13, 1649, 1994.

69. Saito, H., Koyasu, J. Shigeoka, T., and Tomita, I., Cytotoxicity of chlorophenols to goldfish GFS cells with the MTT and LDH assays, *Toxic. In Vitro,* 1107, 1994.

70. Brüschweiler, B. J., Wurgler, F. E. and Fent, K. Cytotoxicity *in vitro* of organotin compounds to fish hepatoma cells PLHC-1 (*Poeciliopsis lucida*), *Aquatic Toxicol.*, 32, 143, 1995.

71. Lenz, D., Segner, H., and Hanke, W., Comparison of different endpoint methods for acute cytotoxicity tests with the R1 cell line. *Ecotoxicol. Ecophysiol.* T. Briaunbeck, W. Itanke and H. Segner, Eds., VCH, Weinheim, 1992, 93.

72. Diereckx, P. J. and Van De Vyver, I. E., Correlation of the neutral red uptake inhibition assay of cultured fathead minnow fish cells with fish lethality tests, *Bull. Environ. Contam. Toxicol.*, 46, 649, 1991.

73. Brandan, J. C., Bohets, H. H. L., Van De Vyver, I. E. and Dierickx, P. J., Correlation between the *in vitro* cytotoxicity to cultured fathead minnow fish cells and fish lethality data for 50 chemicals, *Chemosphere,* 25, 553, 1992.

74. Saito, H., Koyasu, J., Yoshida, K., Shigeoka, T., and Koike, S., Cytotoxicity of 109 chemicals to goldfish GFS cells and relationships with 1-octanol/water partition coefficients, *Chemosphere,* 26(5), 1015, 1993b.

75. Segner, H. and Lenz, D., Cytotoxicity assays with the rainbow trout R1 cell line, *Toxic. In Vitro*, 7, 537, 1993.

76. Dierickx, P. J., Correlation between the reduction of protein content in cultured FHM fish cells and fish lethality data, *Toxic In Vitro,* 7, 527, 1993.
77. Lilius, H., Isomaa, B., and Holmstrom, T., A comparison of the toxicity of 50 reference chemicals to freshly isolated rainbow trout hepatocytes and *Daphnia magna, Aquatic Toxicol.,* 30, 47, 1994.
78. Lange, M., Gebauer, W., Markl, J., and Nagel, R., Comparison of testing acute toxicity on embryos of zebrafish, *Brachydanio rerio,* and RTG-2 cytotoxicity as possible alternatives to the acute fish test, *Chemosphere,* 30, 2087, 1995.
79. Lilius, H., Sandbacka, M., and Isomaa, B., The use of freshly isolated gill epithelial cells in toxicity testing, *Toxic. In Vitro,* 9, 299, 1995.
80. Pesonen, M. and Andersson, T., Toxic effects of bleached and unbleached paper mill effluents in primary cultures of rainbow trout hepatocytes, *Ecotoxicol. Environ. Safety,* 24, 63, 1992.
81. Ali, F., Lazar, R., Haffner, D., and Adeli, K., Development of a rapid and simple genotoxicity assay using a brown bullhead fish cell-line: application to toxicological surveys of sediments in the Huron-Erie corridor, *J. Great Lakes Res.,* 19, 342, 1993.
82. Vega, M. M., Castaño, A., Blazquez, T., and Tarazona, J. V., Assessing organic toxic pollutants in fish-canning factory effluents using cultured fish cells, *Ecotoxicol.,* 3, 79, 1994.
83. Gagné, F., Trottier, S., Blaise, C., Sproull, J., and Ernst, B., Genotoxicity of sediment extracts obtained in the vicinity of a creosote-treated wharf to rainbow trout hepatocytes, *Toxicol. Lett.,* 78, 175, 1995.
84. Zahn, T., Hauck, C., Holzschuh, J., and Braunbeck, T., Acute and sublethal toxicity of seepage waters from garbage dumps to permanent cell lines and primary cultures of hepatocytes from rainbow trout (*Oncorhynchus mykiss*): A novel approach to environmental risk assessment for chemicals and chemical mixtures, *Zbl. Hyg.,* 196, 455, 1995.

CHAPTER 8

In Vitro Assays for Detection of Nongenotoxic Carcinogens

M. J. Durand, V. Cruciani, S. Alexandre, J. C. Hoflack, H. Bessi,
C. Blaise, and P. Vasseur

CONTENTS

I. INTRODUCTION

Cancer is not specific to humans and may also affect wildlife. A number of isolated as well as epizootic neoplasms have been reported since the mid-20th century in aquatic populations. The first case of liver cancer in wild fish populations, where pollution was suspected as a possible etiological factor, was noticed in white suckers in Maryland.[1] Later studies confirmed that liver tumors appeared to be associated with pollution more than any other tumor type in flatfish,[2-5] though skin neoplasms besides epidermal lesions have also been described.[6-8] Polycyclic aromatic hydrocarbons could be identified as partly responsible for the increase in neoplasm prevalence in fish living in contaminated areas.[9] However, correlations between exposure to known genotoxic pollutants and prevalence of cancer in aquatic species was poor. It was thus suggested that a number of unidentified chemical factors acting as carcinogens, cocarcinogens, or tumor promoters could be involved in the disease.[10] These chemical substances can be anthropogenic contaminants, but may also have a natural origin. Some strong tumor promoters, such as okadaic acid produced by sponges, can be found among other aquatic toxins of fungi, blue green algae, and dinoflagellates. Little attention has been paid up to now to nongenotoxic carcinogens, compared to the genotoxic ones. Yet, their relation to cancer may be important and certainly deserves to be taken into account.

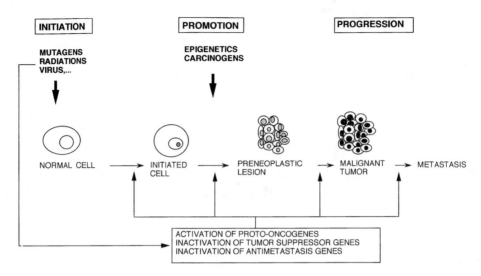

Figure 8.1 Multistage process of carcinogenesis according to Harris. (Adapted from Harris, C. C., *Cancer Res.*, 51, 5023s, 1991.)

A few models based on the use of mammalian cells have been developed to screen nongenotoxic carcinogens. These mammalian models still need to be standardized and are not yet currently applied in toxicity screening. However, it should be emphasized that *in vitro* screening for aquatic toxicity could benefit from the experience gained in working with mammalian cells. This could then be useful for developing cellular models derived from aquatic species.

In the following chapter, we shall focus on mechanisms of carcinogenesis from which arose *in vitro* mammalian assays to detect carcinogens, especially the nongenotoxic ones. As an introduction to these methods, it seems logical to recall the basic concepts of carcinogenesis and the problems posed by nongenotoxic carcinogens.

II. GENOTOXIC AND NONGENOTOXIC CARCINOGENS

At present, the assessment of the biological risk due to environmental chemical carcinogenesis is based largely on short-term genotoxicity assays using prokaryotic and eukaryotic cells. While this approach is useful, it can only detect genotoxic chemicals. Since nearly 40% of carcinogens are nongenotoxic (epigenetic carcinogens) according to the U.S. National Toxicology Program, they will escape the *in vitro* screening currently applied. For such chemicals, carcinogenicity can only be detected by means of long-term animal experiments or epidemiological studies. Unfortunately, the use of animal experiments is time consuming and costly, and thus cannot be applied to all chemicals and to environmental contaminants. There is an urgent need, therefore, to develop and to validate *in vitro* assays for the detection of nongenotoxic carcinogens in order to complete the battery of short-term bioassays for carcinogenicity.

Carcinogenesis is a multistage process resulting from complex interactions of genetic and epigenetic factors (Figure 8.1). At the moment, there is a consensus to classify carcinogens into two groups:

1. Genotoxic carcinogens including mutagens, which may react with DNA and irreversibly alter the genome. They are capable of initiating cells and are named initiators. Initiators involve chemicals, radiations (UV, X-rays, etc.), as well as viruses causing genetic damage. The initiated cells have a decreased responsiveness to inter- and intracellular signals that maintain normal tissue homeostasis.

Initiated cells may be less responsive to negative growth factors, to inducers of terminal cell differentiation and/or programmed cell death.[11] DNA damage able to alter protooncogenes and/or tumor suppressor gene expression enhances the probability for initiated cells to convert to malignancy. Progression to malignancy can also be increased when the genes involved in the control of cell cycle and mitosis are mutated.

2. Nongenotoxic carcinogens, which may alter the proliferation and the differentiation patterns in cells.[12] They may perturb DNA repair systems leading to a fixation of DNA lesions,[13] thus increasing the survival of initiated cells. Cellular targets which lead to a disturbance of these processes are multiple and involve mechanisms such as:

- Interference with transduction signals which control normal cell growth and differentiation
- Activation of protein kinase C and inhibition of protein phosphatases
- Activation of the metabolism of polyamines (activating ornithine decarboxylase, the rate-limiting enzyme of the synthesis of polyamines which are directly linked to cell proliferation)[14]
- Increase in the production of free radicals
- Structural and functional alterations of cytoplasmic membranes and disturbances of intercellular communications
- Inhibition of mechanisms involved in apoptosis[15]
- Modification of the DNA methylation profile[16]

Cyclin-dependent kinases which play an important role in eukaryotic cell cycle regulation, for example by interaction with transcription factors, by phosphorylation of histone H1, or by phosphorylation of retinoblastoma susceptibility gene product (pRb), may also contribute to the development of neoplastic disorders.[17,18] These crucial events should be considered in the development of *in vitro* assays for carcinogenicity.

The variety of the mechanisms involved in the control of cell cycle and mitosis explains the difficulty in screening nongenotoxic carcinogens. Several endpoints would have to be investigated to detect these properties, unless a criterion integrating the effects of these various disturbances can be used. Morphological transformation seems to fulfill this last characteristic. Indeed, malignant cells exhibit progressive phenotypic changes during tumor progression: loss of contact inhibition, increase in the ratio of nucleus/cytoplasm volume, change in the morphology of cells and colonies due to cytoskeleton alterations. These changes either precede or occur concomitantly with genomic disturbances. Indeed, cancer cells display chromosomal instability, abnormal number (aneuploidy), and chromosomal structural changes such as amplification and translocation of genes.[11] Thus, these changes in cell morphology can be criteria for carcinogenicity.

In the following sections, we shall focus on three relevant and promising assays for detection of carcinogens. The Syrian hamster embryo (SHE) cell transformation system and measurements of intercellular communication through gap junctions will be especially described. The former has recently been proposed as an Organization for Economic Cooperation and Development (OECD) guideline for the testing of chemicals. The gap junctional intercellular communication (GJIC) assays are also considered within the objectives of OECD. Poly(ADP ribosyl)ation involved in DNA repair will also be considered due to its importance in maintaining the integrity of the genome and its relevance as a criterion for nongenotoxic carcinogenicity, though this field of research has not been investigated nearly as broadly as the former two.

III. *IN VITRO* CELL TRANSFORMATION ASSAY

Various *in vitro* cell transformation systems have been developed using established cell lines such as BALB/3T3,[19] C3H10T1/2,[20] or primary or secondary SHE cells.[21,22] The endpoint of such systems is the formation of either transformed foci for cell lines (BALB/3T3, C3H10T1/2) or morphologically transformed colonies for SHE cells. Morphological transformation (MT) is a recognized step in the multistage process of the conversion of a nontumorigenic to a neoplastic

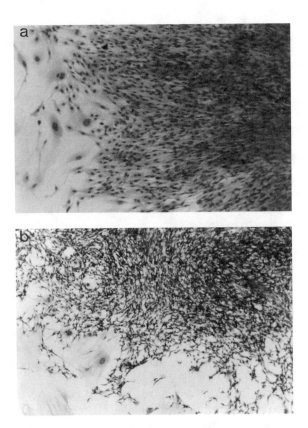

Figure 8.2   Colonies of Syrian hamster embryo cells after 7 days of growth at cloning density in Dulbecco's modified Eagle's medium supplemented with 15% fetal calf serum. Normal colony of cells (×80) (a). Morphologically transformed colony of SHE cells (×60) (b).

state. MT phenotype is characterized by loss of contact inhibition, pattern of growth in three dimensions within the colony resulting in random (criss cross) orientation. Cells in MT colonies are more basophilic and display an increased nuclear to cytoplasmic ratio (Figure 8.2).

Though established cells lines can be grown easily *in vitro*, they have some disadvantages due to their subtetraploidy, their limited metabolizing capacities, their high level of spontaneous transformation frequency and the long time needed (three to eight weeks) for the transformation to be expressed.[19,20]

In contrast, primary SHE cells are normal diploid cells with metabolizing capacities and low spontaneous transformation frequency. Morphological transformation of SHE cells closely mimics the first steps of the carcinogenic process. Several studies showed a good correlation between the multistage neoplastic transformation of SHE cells and *in vivo* carcinogenesis.[23-27] Cells derived from transformed colonies were shown to induce tumor formation when inoculated in an isogenic animal.[21]

Moreover the SHE cell morphological transformation assay may detect nongenotoxic as well as genotoxic carcinogens. This argues in favor of the use of this assay for the screening of carcinogens. The synergistic effects between genotoxic and nongenotoxic chemicals could be shown in the SHE cell assay by using a sequential treatment consisting of an initiation-promotion procedure (Figure 8.3).

In order to assess the transforming capacity of a chemical, SHE cells are seeded at clonal density on a monolayer of feeder cells and treated for one to seven days with the tested compound. The latter can be applied alone or associated with other carcinogens, initiators, or tumor promoters, in order to detect interacting effects. Coexposure or sequential treatment is carried out with

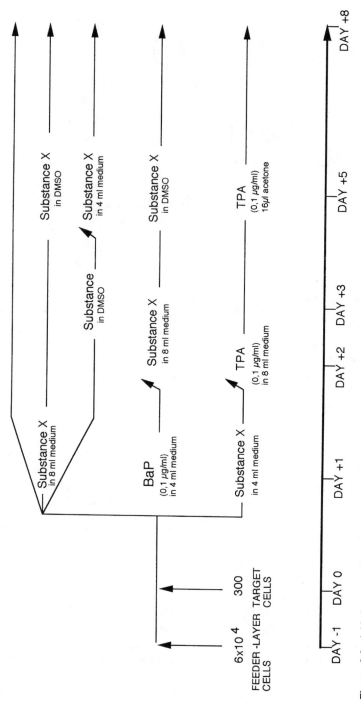

Figure 8.3 Initiation-promotion protocol used for the detection of epigenetic carcinogens. ✔ indicates the removal of the medium which is replaced by new medium supplemented with the substance to test or TPA according to the protocol used.

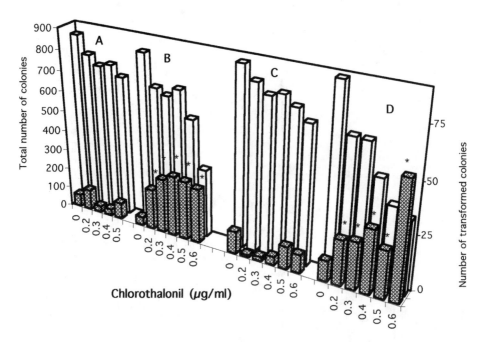

Figure 8.4 Effects of chlorothalonil on cloning efficiency and morphological transformation of primary SHE cells. Cells were exposed for 24 h to: (A) 0.1 μg/mL BaP followed by single treatment of chlorothalonil at day 2; (B) 0.1 μg/mL BaP followed by two applications of chlorothalonil at day 2 and day 5; (C) 0.2 μg/mL BaP followed by single treatment of chlorothalonil at day 2; (D) 0.2 μg/mL BaP followed by two applications of chlorothalonil at day 2 and day 5. * = values significantly different from their respective controls (P<0.05, test Mann-Whitney).

(1) benzo(a)pyrene (BaP) used at low concentrations (0.4 μM) as an initiator to detect a promoting effect of the tested substance, or (2) 12-*O*-tetradecanoylphorbol-13-acetate (TPA, 0.16 μM) to evaluate an initiating capacity.

Cells are incubated for a 7-d period to allow the development of colonies and phenotypic modifications. Cells are then fixed and stained with Giemsa. Microscopic examination is performed to identify transformed colonies. The number of morphologically transformed colonies relative to normal colonies in chemically treated groups is compared to controls.[24,26,28-30]

We showed that chlorothalonil, a fungicide classified as a nongenotoxic carcinogen, promoted MT of SHE cells previously initiated by BaP which caused no effect on its own at the low concentration used (Figure 8.4).

Chlordane, a cyclodiene pesticide reported to be carcinogenic in rodents but devoid of genotoxic properties, could induce MT of SHE cells at noncytotoxic concentrations. Chlordane potentiated TPA effects and enhanced MT frequency in SHE cells. These results underlined that interactions between nongenotoxic substances may also have dramatic consequences on cell transformation.[30]

In considering all the available data,[28-32] it is clear that the SHE cell transformation assay appears to be a relevant model for carcinogenicity. The initiation–promotion protocol increases the capacity of the model to detect carcinogenic properties. Nevertheless, this assay suffers some drawbacks:

- The low transformation frequency induced by chemical carcinogens requires the examination of a large number of colonies
- The plating efficiency and the transformation frequency are dependent on growth factors provided by the fetal calf serum added to the culture medium. These variability factors can be lowered by pretesting the serum and the batches of embryo cells in order to optimize the above parameters.

The development of synthetic serum allowing growth of cells at clonal density will certainly help to solve this problem in the future

- The phenotypic changes are detected by microscopic examination of colonies. Thus, criteria for transformation need to be clearly defined in order to limit subjectivity in scoring the morphologically transformed colonies. Biomarkers of MT phenotype in addition to morphological criteria could be useful to identify cell transformation. At the moment, such biochemical indicators are not available and research is needed in this area.

Even though this transformation cell assay still requires some improvements experimentally, it constitutes a good alternative to animal experiments for the detection of nongenotoxic carcinogens. Another application of SHE cells could be envisaged owing to the fact that SHE cells are composed of a mixture of fibroblastic and epithelial cells and since most cancers are epithelioma (carcinoma). It would therefore be interesting to recover epithelial cells by selective growth media and to use these in testing. At the moment, morphological cell transformation assays use mammalian cells. Yet it would certainly be of interest to adapt this test system to cells derived from aquatic vertebrates or invertebrates in the future.

IV. ROLE OF GAP JUNCTIONAL INTERCELLULAR COMMUNICATION FOR THE DETECTION OF NONGENOTOXIC CARCINOGENS

Gap junctional intercellular communication (GJIC) is one of the major means by which multicellular organisms mediate the direct intercellular exchange of cellular signal factors (small molecules and ions) from the interior of one cell to neighboring cells. GJIC plays a crucial role in the maintenance of cell homeostasis, and aberrant GJIC is often involved in carcinogenesis. Gap junctions are formed by the contact of two hemichannels (connexons) penetrating the plasma membranes of adjacent cells (Figure 8.5).[33] Hemichannels are formed by six protein subunits named connexins (Cx).[34]

It has been shown that many tumor-promoting agents alter the regulation of GJIC. The inhibition of GJIC is an epigenetic event which may be related to cell transformation and proliferation of cancer cells.[13,35] There are several levels of regulation of GJIC. Permeability decrease or inhibition may result from changes in the expression of Cx during transcription and translation. Alteration of post-translational processing and connexon assembly and insertion into the plasma membrane can also be involved in the deregulation of GJIC. Most transformed and tumorigenic cells show aberrant GJIC, but the transfection of connexin genes into these tumorigenic cells may restore the normal phenotype. This emphasizes the relevance of GJIC in cancer promotion. All these topics have been extensively reviewed.[35-38]

Budunova[34] reported that 60% of a set of 80 substances with known carcinogenic activity inhibit GJIC. Antipromoting and anticarcinogenic agents such as retinoic acid, glucocorticoids, quercetin, etc., may prevent the inhibition of GJIC induced in vitro by the tumor promoter TPA.[34] These results strongly support the hypothesis that alteration of GJIC is a critical event in carcinogenesis.

Several methods were developed to study GJIC. One approach, called metabolic cooperation, is based on the transfer of small molecules between cells. The first experimental assay used to study metabolic cooperation is the coculture of 6-thioguanine (6-TG) resistant and 6-TG sensitive V79 cells.[13] 6-TG sensitive cells are able to phosphoribosylate this synthetic purine. The phosphoribosylated purine is a toxic metabolite which kills 6-TG resistant cells after its transfer through operational gap junctions. The number of resistant cells recovered after exposure of the coculture to the test substance is a quantitative indicator of its capacity to inhibit GJIC. Another approach to study GJIC consists of dye transfer techniques. These methods register the spread of fluorescent dye (lucifer yellow) microinjected into one of the cells (Figure 8.6). These techniques are easy to perform and give rapid results, which make them powerful tools in screening.

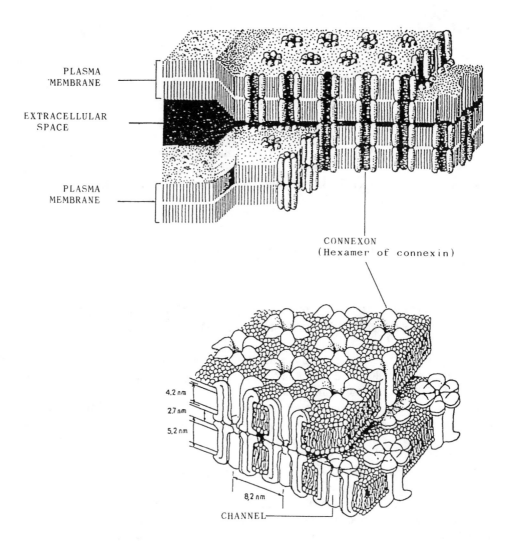

PLASMA MEMBRANE

EXTRACELLULAR SPACE

PLASMA MEMBRANE

CONNEXON
(Hexamer of connexin)

4.2 nm
2.7 nm
5.2 nm

8.2 nm

CHANNEL

Figure 8.5 Molecular organization of a gap junction.

We showed that chlordane which induced the morphological transformation of SHE cells also inhibited intercellular communication. Inhibition of GJIC by chlordane was revealed by a dye transfer technique in SHE cells and with the metabolic cooperation assay with V79. These results emphasized the epigenetic character of this pesticide which disturbs homeostasis by several mechanisms.[30]

It was initially postulated that inhibition of GJIC could be a common property of nongenotoxic carcinogens. Some of them, however, do not comply with this postulate. As an example, chlorothalonil was not found to inhibit intercellular communication either in V79 cells or in SHE cells, while it promoted MT in SHE cells.[29] Thus, caution is required in interpreting negative results, which are not a guarantee of safety. Indeed, GJIC is a complex process controlled by many cellular and environmental factors. In the case of responses demonstrating an inhibition of intercellular communication, GJIC remains a useful criterion which may relate to carcinogenic potential.

The above examples illustrate the wide spectrum of activities of epigenetic carcinogens and emphasize the need for developing and validating several types of assays for the evaluation of nongenotoxic carcinogens. Another area worth investigating is the complex DNA repair system. This system can be the target of chemicals which will indirectly increase genomic alterations.

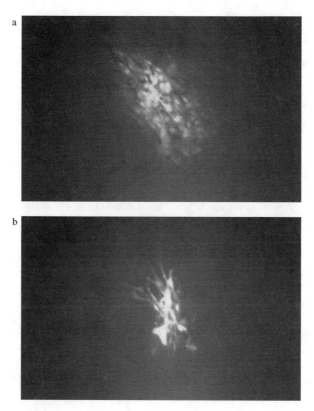

Figure 8.6 GJIC measured by dye transfer of lucifer yellow after microinjection of single SHE cell. (a) Control cells. (b) Inhibition of GJIC by TPA (1 hour exposure 0.1 µg/mL).

V. IMPACT OF EPIGENETIC CARCINOGENS ON POLY(ADP-RIBOSE) POLYMERASE

Poly(ADP-ribose) polymerase (PARP; EC 2.4.2.30.) is a zinc-binding nuclear enzyme that catalyzes the synthesis of the polymer poly(ADP-ribose) (pADPr) from its substrate, nicotinamide adenine dinucleotide (NAD+).[39] The intracellular synthesis of poly(ADP-ribose), a process called poly(ADP-ribosyl)ation, is induced by strand breaks in DNA.[40] The polymer pADPr will bind to nuclear proteins and histones, leading to chromatin decondensation which facilitates the activity of DNA repair enzymes.[41]

Thus, PARP through poly(ADP ribosyl)ation of nuclear proteins plays an essential role in DNA metabolism (replication and repair) in eukaryotic cells. Cell cycle, cell transformation, cell differentiation, spermatogenesis, and embryogenesis are also controlled by PARP.[42-45] An inhibition of poly(ADP-ribose) synthesis may be deleterious to DNA metabolism and integrity, and to cell growth and differentiation. Thus, poly(ADP ribosyl)ation pattern in cells could be used as an early and general endpoint in genetic toxicology.

Specific inhibitors of PARP are not genotoxic but can disturb the cell cycle in S and G2 phases.[45] 3-aminobenzamide (3-AB), an inhibitor of poly(ADP-ribose) synthesis, was shown to induce sister chromatid exchanges,[46] and also to potentiate the genotoxic effects of clastogens.[47] At present, there is a consensus suggesting that PARP inhibitors could induce variations in carcinogenicity. The mechanisms by which poly(ADP-ribosyl)ation interferes in this process remain to be elucidated. Considering the importance of PARP cellular functions, poly(ADP-ribosyl)ation deserves to be more thoroughly studied.

Because epigenetic carcinogens may change PARP expression and synthesis of pADPr,[48] the relationships between poly(ADP-ribosyl)ation and cell transformation is another area worthy of investigation. These effects could be monitored in cells by measuring poly(ADP-ribose) levels according to the method of Shah et al.[49]

VI. CONCLUSIONS

One of the major challenges of our times is to prevent the environmental consequences of pollution for human and environmental safety. Efficient prevention requires that the regulation of new chemicals be improved in order to fill the present gap concerning carcinogenicity. The variety of mechanisms involved in the control of cell cycle and mitosis explains the difficulty in screening nongenotoxic carcinogens. This contrasts with genotoxic substances easily detected by mutations, and genomic and chromosomal alterations.

The detection of genotoxic carcinogens by *in vitro* short-term assays is quite efficient, but at this time no test has been validated for nongenotoxic carcinogens. Cell morphological transformation assays and GJIC measurements are relevant methods which deserve to be considered in a battery of *in vitro* short-term assays for carcinogenicity. It is imperative that these methods become standardized in the near future.

Environmental problems result in a great part from the interactions between a multitude of contaminants in the ecosystems. These interactions cannot be underestimated in the context of a multistage and multifactorial carcinogenic process. They may involve genetic as well as epigenetic factors, acting directly or not on DNA metabolism and on cellular homeostasis. Up to now, scientists have mainly focused on the study of DNA damaging agents, i.e., initiators. It now appears essential that nongenotoxic carcinogens and the combined effects of initiators and tumor promoters be investigated more extensively.

At present, *in vitro* genotoxicity assays for chemicals use mainly mammalian cells. A logical next step would be to develop new techniques from selected aquatic (in)vertebrate cells. The promotion of such microscale techniques should be encouraged so as to offer more relevant diagnostic tools to ensure better protection for the aquatic environment.

REFERENCES

1. Dawe, C. J., Stanton, M. F., and Schwartz, F. J., Hepatic neoplasms in native bottom feeding fish of Deep Creek Lake, Maryland, *Cancer Res.*, 24, 1194, 1964.
2. Pierce, K. V., McCain, B. B., and Wellings, S. R., Pathology of hepatomas and other liver abnormalities in English sole (*Parophrys vetulus*) from the Duwamish river estuary, Seattle, Washington, *J. Natl. Cancer Inst.*, 60, 1445, 1978.
3. Murchelano, R. A. and Wolke, R. E., Epizootic carcinoma in the winter flounder *Pseudopleuronectes americanus*, *Science*, 228, 587, 1985.
4. Bucke, D., Watermann, B., and Feist, S., Histological variations of hepato-splenic organs from the North Sea dab, *Limanda limanda* L., *J. Fish. Dis.*, 7, 255, 1984.
5. Malins, D. C., Krahn, M. M., Brown, D. W., Rhodes, L. D., Myers, M. S., McCain, B. B., and Chan, S. L., Toxic chemicals in marine sediment and biota from Mukiteo, Washington: relationships with hepatic neoplasms and other hepatic lesions in English sole (*Parophrys vetulus*), *J. Natl. Cancer Inst.*, 74, 487, 1985.
6. Black J.J., Field and laboratory studies of environmental carcinogenesis in Niagara river fish, *J. Great Lakes Res.*, 9, 326, 1983.
7. Dethlefsen V., Observations on fish diseases in the German Bight and their possible relation to pollution, *Rapp P-v Réun. Cons. Perm. Int. Explor. Mer.*, 179, 110, 1980.
8. Vethaak A.D., Fish disease and marine pollution, *Thesis Universiteit van Amsterdam*, 1993.

9. Myers, M. S., Landahl, J. T., Krahn, M. M., and McCain, B. B., Relationships between hepatic neoplasms and related lesions and exposure to toxic chemicals in marine fish from the U.S. West Coast, *Environ. Health Perspect.*, 90, 7, 1991.

10. Vasseur, P., Godet, F., Bessi, H., and Lambolez, L., Indices for carcinogenicity in aquatic ecosystems: significance and development, in *Evaluating and Monitoring the Health of Large-Scale Ecosystems*, Rapport, D. J., Gaudet, C. L., and Calow, P., Eds., Springer-Verlag, Berlin Heidelberg, 1995, 179.

11. Harris, C. C., Chemical and physical carcinogenesis: advances and perspectives for the 1990s, *Cancer Res.*, 51, 5023, 1991.

12. Weinstein, I. B., The origins of human cancer: molecular mechanisms of carcinogenesis and their implication for cancer prevention and treatment, *Cancer Res.*, 48, 4135, 1988.

13. Trosko, J. E., Jone, C., and Chang, C. C., The use of *in vitro* assays to study and to detect tumor promoters, in *Models, Mechanism and Etiology of Tumor Promotion*, IARC Sc. Publ., 56, Lyon, 1984.

14. Raunio, H. and Pelkonen, O., Effect of polycyclic aromatic compounds and phorbol esters on ornithine decarboxylase and aryl hydrocarbon hydroxylase activities in mouse liver, *Cancer Res.*, 43, 782, 1983.

15. Bayly, A. C., Roberts, R. A., and Dive, C., Suppression of liver cell apoptosis *in vitro* by the non-genotoxic hepatocarcinogen and peroxisome proliferator nafenopin, *J. Cell Biol.*, 125, 197, 1994.

16. Counts, J. L. and Goodman, J. I., Hypomethylation of DNA: a nongenotoxic mechanism involved in tumor promotion, *Toxicol. Lett.*, 82/83, 663, 1995.

17. Wolowiec, D. and French, M., Kinases dépendantes des cyclines: rôle biologique et implications dans la pathologie humaine, *Médecine/Sciences*, 12, 165, 1996.

18. Tyson, J. J., Novak, B., Odell, G. M., Chen, K., and Thron, C. D., Chemical kinetic theory: understanding cell-cycle regulation, *Trends Biochem. Sci.*, 21, 89, 1996.

19. Kakunaga, T., A quantitative system for assay of malignant transformation by chemical carcinogens using a clone derived from Balb/3T3, *Int. J. Cancer*, 12, 463, 1973.

20. Reznikoff, C. A., Bertram, J. S., Brankow, D., and Heidelberger, C., Quantitative and qualitative studies of chemical transformation of cloned C3H mouse embryo cells sensitive to post confluence inhibition of cell division, *Cancer Res.*, 33, 3239, 1973.

21. Berwald, Y. and Sachs, L., *In vitro* cell transformation with chemical carcinogens, *Nature*, 200, 1182, 1963.

22. Di Paolo, J. A., Quantitative *in vitro* transformation of Syrian golden hamster embryo cells with the use of frozen stored cells, *J. Natl. Cancer Inst.*, 64, 1485, 1980.

23. Barrett, J. C., Hesterberg, T. W., and Thomassen, D. G., Use of cell transformation systems for carcinogenicity testing and mechanistic studies of carcinogenicity, *Pharmacol. Rev.*, 36, 535, 1984.

24. Pienta, R. J., Poiley, J. A., and Lebherz, W. B., Morphological transformation of early passage golden Syrian hamster embryo cells derived from cryopreserved primary cultures as reliable *in vitro* bioassay for identifying diverse carcinogens, *Int. J. Cancer*, 19, 642, 1977.

25. Chouroulinkov, I. and Lasne, C., Two stage (initiation-promotion) carcinogenesis *in vivo* and *in vitro*, *Bulletin du cancer*, 65, 255, 1978.

26. Pienta, R. J., Lebherz, W. B., and Schuman, R. F., The use of cryopreserved Syrian hamster embryo cells in a transformation test for detecting chemical carcinogens, in *Short term tests for chemical carcinogens*, Stick, H. F., and San, R. H. C., Eds., Springer-Verlag, New York, 1981, 323.

27. LeBoeuf, R. A., Kerckaert, G. A., Poiley, J. A., and Raineri, R., An interlaboratory comparison of enhanced morphological transformation of Syrian hamster embryo cells cultured under conditions of reduced bicarbonate concentration and pH, *Mutat. Res.*, 222, 205, 1989.

28. Elias, Z., Poirot, O., Pezerat, H., Suquet, H., Schneider, O., Daniere, M. C., Terzetti, F., Baruthio, F., Fournier, M., and Cavelier, C., Cytotoxic and neoplastic transforming effects of industrial hexavalent chromium pigments in Syrian hamster embryo cells, *Carcinogenesis*, 10, 2043, 1989.

29. Bessi, H., Rast, C., Nguyen-Ba, G., and Vasseur, P., Chlorothalonil promotes morphological transformation in hamster embryo cells but does not inhibit GAP junctional intercellular communication either in SHE cells or in the V79 cell line, *Cancer J.*, 7, 248, 1994.

30. Bessi, H., Rast, C., Rether, B., Nguyen-Ba, G., and Vasseur, P., Synergistic effects of chlordane and TPA in multistage morphological transformation of SHE cells, *Carcinogenesis*, 16, 237, 1995.

31. Rivedal, E., Mikalsen, S. O., Roseng, L. E., Sanner, T., and Eide, I., Effects of hydrocarbons on transformation and intercellular communication in Syrian Hamster Embryo Cells, *Pharmacol. & Toxicol.*, 71, 57, 1992.

32. Lasne, C., Gentil, A., and Chouroulinkov, I., Two stage malignant transformation of Syrian hamster cells, *Nature*, 247, 490, 1974.

33. Maillet, M., *Biologie Cellulaire*, 7th Edition, Masson, Paris, 1995.

34. Budunova, I. V., Alteration of gap junctional intercellular communication during carcinogenesis, *Cancer J.*, 7, 228, 1994.

35. Yamasaki, H., Gap junctional intercellular communication and carcinogenesis, *Carcinogenesis*, 11, 1051, 1990.

36. Yamasaki, H., Non-genotoxic mechanisms of carcinogenesis: studies of cell transformation and gap junctional intercellular communication, *Toxicol Lett.*, 77, 55, 1995.

37. Mesnil, M. and Yamasaki, H., Cell-cell communication and growth control of normal and cancer cells: evidence and hypothesis, *Mol. Carcinogen.*, 7, 14, 1993.

38. Trosko, J. E. and Chang, C. C., Nongenotoxic mechanisms in carcinogenesis: role of inhibited intercellular communication, *Banbury Rep.*, 31, 139, 1988.

39. Chambon, P., Weill, J. D., Doly, J., Strosser, M. T., and Mandel, P., On the formation of a novel adenylic compound by enzymatic extracts of liver nuclei, *Biochem. Biophys. Res. Comm.*, 25, 638, 1966.

40. Benjamin, R. C. and Gill, D. M., Poly(ADP-ribose) synthesis *in vitro* programmed by damaged DNA, *J. Biol. Chem.*, 255, 10502, 1980.

41. Ding, R. and Smulson, M., Depletion of nuclear poly(ADP-ribose) polymerase by antisense RNA expression: influences on genomic stability, chromatin organization and carcinogen cytotoxicity, *Cancer Res.*, 54, 4627, 1994.

42. Sugimura, T. and Miwa, M., Poly(ADP-ribose) and cancer research, *Carcinogenesis*, 12, 1503, 1983.

43. Jacobson, E. L., Meadows, R., and Measel, J., Cell cycle perturbations following DNA damage in presence of ADP-ribosylation inhibitors, *Carcinogenesis*, 6, 711, 1985.

44. Cleaver, J. E. and Morgan, W. F., Poly(ADP-ribose)polymerase: a perplexing participant in cellular responses to DNA breakage, *Mutat. Res.*, 257, 1, 1991.

45. De Murcia, G., ADP-ribosylation reactions: Mechanisms and biological significance, Part I, *Biochimie*, 77, 5, 1995.

46. Morgan, W. F. and Cleaver, J. E., 3-Aminobenzamide synergistically increases sister-chromatid exchanges in cells exposed to methyl methanesulfonate but not to ultraviolet light, *Mutat. Res.*, 104, 361, 1982.

47. Schwartz, J. L., Morgan., W. F., and Weichselbaum, R. R., Different efficiencies of interaction between 3-aminobenzamide and various monofunctional alkylating agents in the induction of sister chromatid exchanges, *Carcinogenesis*, 6, 699, 1985.

48. Singh, N., Poirier, G., and Cerutti, P., Tumor promoter phorbol-12-myristate-13-acetate induces poly(ADP-ribosyl)ation in fibroblasts, *EMBO J.*, 4, 1491, 1985.

49. Shah, G. M., Poirier, D., Duchaine, C., Brochu, G., Desnoyers, S., Lagueux, J., Verreault, A., Hoflack, J. C., Kirkland, J. B., and Poirier, G. G., Methods for biochemical study of poly(ADP-ribose) metabolism *in vitro* and *in vivo*, *Anal. Biochem.*, 227, 1, 1995.

Bacteria

CHAPTER 9

Microbial Enzyme Assays for the Detection of Heavy Metal Toxicity

Gabriel Bitton and Jean Louis Morel

CONTENTS

I. INTRODUCTION

The main sources of heavy metal pollution are geological weathering and industrial activity (e.g., mining, electroplating, battery manufacturing, and paints and dyes).[1] There are concerns over the persistence of heavy metals in the environment and their adverse effects on humans and the biota in general, resulting in the inclusion of some toxic metals in the U.S. Environmental Protection Agency's list of 129 priority pollutants. The main problems caused by discharges of toxic pollutants to receiving waters are toxicity to aquatic organisms and restrictions on human use of these waters. Metals may be present in a wide variety of physicochemical forms in aquatic environments, and their speciation has a profound effect on their bioavailability and toxicity to aquatic organisms. Several physical, chemical, and biological factors influence the bioavailability and thus, the toxicity of metals in the environment. These factors include pH, suspended particles, redox potential (Eh), water hardness, and organic and inorganic compounds.[2] The free ionic forms of metals are generally the most toxic to the biota, while the precipitated, sorbed, and complexed forms appear to be nontoxic, or at least are considerably less toxic than free metal ions.[3] The topic of toxicity testing

0-8493-2626/5/98/$0.00+$.50

Table 9.1   Short-term Toxicity Assays Based on Enzyme Activity or Biosynthesis

Enzyme	Endpoint Measured	Comments
Dehydrogenases	Measure reduction of oxidoreduction dyes such as INT or TTC	Widely tested in water, wastewater, soils, sediments
ATPase	Measure phosphate concentration using ATP as a substrate	*In vivo* and *in vitro* tests have been used
Esterases	Nonfluorescent substrates degraded to fluorescent products	Acetylcholinesterase sensitive to organophosphates and carbamates
Phosphatases	Measure organic portion of substrate (e.g., phenol) or inorganic phosphate	Little sensitivity to heavy metals
Urease	Measure ammonia production from urea	Studied mostly in soils and more recently in water
Luciferase	Measure light production using ATP as a substrate	Used in ATP-TOX bioassay in conjunction with inhibition of ATP levels in a bacterial culture
β-galactosidase	Measure hydrolysis of *o*-nitrophenyl-D-galactoside or other substrates	Toxicant effect on both enzyme activity and biosynthesis was tested
α-glucosidase	*p*-nitrophenyl-α-D-glucoside	Toxicant effect on enzyme biosynthesis has been tested
Tryptophanase	Add Ehrlich's reagent and measure absorbance at 568 nm	Toxicant effect on enzyme biosynthesis has been tested

Adapted from Bitton, G. and Koopman, B., *Rev. Environ. Contam. Toxicol.*, 125, 1,1992.

for heavy metals, using fish, invertebrates, plants, microorganisms, and enzymes, has been extensively reviewed by Kong et al.[3]

In wastewater treatment plants, heavy metal sources include industrial discharges and urban storm water runoffs. These toxicants affect both aerobic (e.g., nitrification) and anaerobic (e.g., methanogenesis) treatment processes. While several metals are quite toxic to methanogenesis (although trace levels of Ni or Co may be stimulatory to methanogenic bacteria), they may cause deflocculation in activated sludge, leading to sludge separation problems.[4-6]

A wide range of toxicity bioassays has been developed for assessing the impact of toxic chemicals on natural and man-made ecosystems. These bioassays utilize fish, zooplankton, algae, bacteria, and fungi as test organisms. Several enzyme and microbial assays have been developed and proposed to respond to the need for carrying out rapid and inexpensive toxicity assessment of environmental samples.[7,8] These assays, when packaged as test kits, are convenient for the screening of large numbers of samples in the laboratory as well as under field conditions.[9,10] Most of the proposed kits measure the general toxicity of a given sample. Thus, our research focused on the development of toxicity assays for the determination of heavy metal contamination.

II. USE OF ENZYMES IN TOXICITY TESTING

Dehydrogenases and hydrolases (e.g., ATPase, phosphatase, esterase, urease, luciferase, β-galactosidase, α-glucosidase) are the main classes of enzymes which have been considered for toxicity testing in aquatic environments, wastewater effluents, sludges, soils, and sediments.[11] Table 9.1 gives a compilation of the main enzymes used in toxicity assessment of environmental samples. Enzymatic toxicity assays essentially fall into two categories: 1) assays based on induction of enzyme biosynthesis and 2) assays based on enzyme activity.[11-13]

It was suggested that the inhibition of *de novo enzyme biosynthesis* can serve as an endpoint in toxicity testing[14,15]. Toxi-Chromotest™ is a toxicity test kit based on the inhibition of *de novo* synthesis of β-galactosidase.[15] However, this test was found to be less sensitive than *Daphnia pulex* and *Ceriodaphnia dubia* toxicity tests (based on immobilization of the daphnids) or Microtox® bioassay (based on inhibition of bioluminescence).[16-18] The inhibition of biosynthesis of other inducible enzymes has also been explored as a basis of chemical contamination testing. A test based on inhibition of tryptophanase biosynthesis induction in *E. coli* was found to be less sensitive than the one based on inhibition of α-glucosidase biosynthesis induction in *Bacillus licheniformis*.[19] The latter test was used to determine the toxicity of sediment elutriates from hazardous waste sites in Florida. For the ten most inhibitory sediment elutriates, the α-glucosidase induction assay consistently gave lower EC_{50}s (EC_{50} = toxicant concentration which causes 50% inhibition of the enzyme) than Microtox.[20]

Toxicity tests based on the inhibition of *enzyme activity* have also been suggested for assessing chemical toxicity in aquatic environments. They include phosphatases,[21] esterases (e.g., acetylcholinesterase),[22,23] ATPase,[24] peroxidase,[25] urease,[26-27] luciferase,[13,28] β-galactosidase,[29-30] protease, amylase, and α-glucosidase.[13]

III. USE OF ENZYMES FOR THE SPECIFIC DETERMINATION OF HEAVY METAL TOXICITY

The specific determination of the presence of heavy metals in environmental samples is possible using microbial or enzyme biosensors. Some of the biosensors detect the presence of specific organic toxicants such as phenol and related compounds[31] or organophosphorous pesticides and carbaryl.[32] In some other biosensors the test bacteria respond to specific metals via induction of specific genes such as the *lux* gene which has been fused with the *mer* operon.[33-36] However, there are some problems associated with microbial biosensors, namely, the lack of response at high metal concentration and, sometimes, the lack of selectivity. (See Kong et al.[3] for further discussion.) Moreover, these biosensor assays are not toxicity assays *per se* and merely indicate the presence of a given metal.

We have tested the inhibition of the activity of hydrolytic enzymes (α-glucosidase, urease, phosphatase, peroxidase) as an indication of heavy metal toxicity (Jung and Bitton, unpublished data). α-glucosidase responded somewhat to all of the metals tested (Cd, Cr, Cu, Ni, Hg, Zn) except Pb. Alkaline phosphatase responded to some heavy metals (Cr, Hg, and Zn), but was insensitive to others tested (e.g., Cu, Cd). Tyler's study[21] showed that, indeed, this enzyme was hardly inhibited by heavy metals. It was also found that peroxidase was quite insensitive to heavy metal toxicity, confirming the results of Guibault et al.[25]

As a result of our preliminary screening of several enzymes for sensitivity and specificity to heavy metal toxicity, we have retained two hydrolases for further testing, urease and β-galactosidase.

A. Urease

It was previously reported that some of the widely used pesticides (malathion, captan, diazinon, carbaryl) as well as other organic toxicants (dithiocarbamates, sodium *p*-chloromercuribenzoate, hydroxamates, catechol, hydroquinone, *p*-benzoquinone) do not significantly affect urease activity,[26,37,38] while metal ions inhibit urease by reacting with the thiol groups of the enzyme.[39,40] We developed a Urease Toxicity Assay (UTA) that was found to be quite sensitive to heavy metals but was relatively insensitive to organic toxicants.[27] UTA was most sensitive to Hg^{2+} and Cu^{2+} (EC_{50}s = 0.008 and 0.013 mg/L, respectively) and least sensitive to lead (EC_{50} = 2.5 mg/L) (Table 9.2). The toxicity of the metals tested followed the following sequence: Hg > Cu > Cd > Zn > Ni > Cr > Pb. This sequence is similar to the one reported by Zhylyak et al.[41] who used a urease-based conductimetric biosensor. A comparison of UTA EC_{50}s to those obtained with other toxicity tests revealed that this bioassay was comparable in sensitivity to acute *Daphnia magna* bioassay and was more sensitive than other microbial or enzymatic toxicity bioassays.[27] However, since the end product of urease action on urea is ammonia, UTA use is questionable in some environmental samples, especially wastewater, due to the interfering effect of ammonia. We have attempted to remove this interference by immobilizing the enzyme on glass beads, but this practice helped protect

Table 9.2   Sensitivity of Free Urease to Selected Heavy Metals and Organic Toxicants

Toxicant	EC_{50} (mg/L)
2,4-Dichlorophenol	>500
2,4-D	>150
Hydrothol	>200
Lindane	>3,000
Methanol	>100,000
Parathion	>500
Pentachlorophenol	>200
Phenol	>3,000
SDS	>500
2,4,6-Trichlorophenol	>500
Cd	0.12 ± 0.04*
Cr(III)	0.82 ± 0.11
Cu	0.013 ± 0.006
Pb	2.5 ± 1.2
Hg	0.008 ± 0.002
Ni	0.51 ± 0.25
Zn	0.18 ± 0.074

* mean ± standard deviation

Adapted from Jung, K., Bitton, G., and Koopman, B., *Water Res.*, 29, 1929, 1995.

the enzyme from heavy metal insult and, therefore, would not be suitable for toxicity testing. The protective effect of immobilization is confirmed by the findings of other investigators.[41,42] Thus, UTA[27] and urease conductimetric biosensors[41] may not be suitable candidate assays for assessing heavy metal toxicity in some environmental samples.

B. β-galactosidase

Early in our investigations, we observed that the activity of β-galactosidase was insensitive to organic toxicants while being relatively sensitive to heavy metals.[29] It was thought that this phenomenon may be used as a basis for a toxicity assay which would be specific for heavy metal toxicity in environmental samples, thus avoiding sample fractionation to identify metal toxicity. We have thus developed two new kits, MetPAD™ and MetPLATE™, for the specific determination of heavy metal toxicity. We will now briefly review our findings concerning the development of these test kits and their application to the determination of metal toxicity in aquatic and solid environmental samples.

IV. DESCRIPTION OF THE METPAD AND METPLATE ASSAY KITS

MetPAD and MetPLATE toxicity test kits are based on the specific inhibition of the activity of β-galactosidase in an *E. coli* strain by heavy metals. The MetPAD kit comprises the bacterial reagent (a mutant strain of *E. coli*), diluent, buffer, and assay pads containing a dried β-galactosidase substrate (chlorophenol red-β-D-galactopyranoside). The MetPLATE kit contains the same ingredients as MetPAD except that the assay pads are replaced by a freeze-dried enzyme substrate and a 96-well microplate. Figure 9.1 shows the flow chart of the MetPAD and MetPLATE toxicity bioassays. One MetPLATE kit is sufficient for running three complete assays for EC_{50} determination or for screening for the heavy metal toxicity of up to 20 samples.

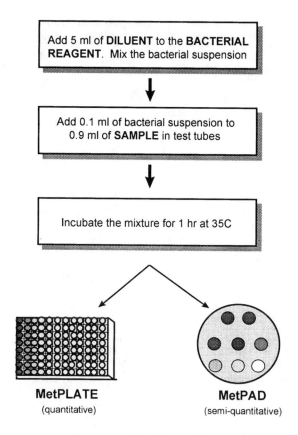

Figure 9.1 Flowchart of MetPAD/MetPLATE toxicity tests.

A. MetPAD: a Semiquantitative Test for Heavy Metal Toxicity

We have tested the response of MetPAD to several heavy metals and organic toxicants. While the kit responds to all heavy metals tested, it was relatively insensitive to the organic toxicants examined (e.g., phenol, pentachlorophenol, sodium dodecyl sulfate). For example, the kit did not respond to 3000 mg/L of phenol or sodium dodecyl sulfate.[43] We have concluded that the MetPAD kit can be used for the direct and specific detection of heavy metals in environmental samples without the need for treatment of the samples with EDTA or ion exchange resins.[44-46]

Metal analysis and MetPAD assays were run concurrently to assess the heavy metal content and toxicity of wastewater effluents (Table 9.3).[43,47] Five out of eight wastewater effluents were found to be toxic. Furthermore, except for sample #3, the toxicity generally increased with the heavy metal content of the samples. MetPAD showed no toxicity when some of the samples were passed through a cation exchange resin to remove heavy metals,[48] confirming that this kit is indeed specific for heavy metal toxicity. MetPAD, when run concurrently with Microtox, was useful in pinpointing sediment elutriates with heavy metal toxicity. Table 9.3 shows the results given by MetPAD as regards the toxicity of sediment elutriates from hazardous waste sites in Florida.[47] All toxic elutriates contained heavy metals as shown by chemical analysis, while nontoxic samples contained only trace amounts of heavy metals. Since MetPAD does not give an EC_{50} for a sample (however, there is the possibility of determining an MIC = minimum inhibitory concentration), there was a need to develop a more quantitative assay for heavy metal toxicity.

**Table 9.3   Use of MetPAD™ for Detecting Metal Toxicity
in Wastewater Effluents and Sediment Elutriates**

Sample Code	MetPAD™ Results*	Metal (metal content in mg/L)
Wastewater		
WW1	Nontoxic	Cu (0.08)
WW2	Nontoxic	Zn (0.1)
WW3	Nontoxic	Cu (1.1)
WW4	+++++	Cr (6.6)
WW5	++++	Zn (1.5)
WW6	+++++	Zn (23.1); Cr (2.8); Ni (1.9)
WW7	+++++	Zn (6.7); Ni (1.6)
Sediment Elutriates		
SE1	Nontoxic	Zn (<0.1)
SE2	+++	Cu (0.99); Zn (3.9)
SE3	+++++	Cu (5.12); Zn (4.1)
SE4	+++++	Cu (3.2); Zn (3.4)
SE5	+++++	Cu (1.9); Zn (5.5); Pb (2.9)
SE6	Nontoxic	Pb (<0.1)
SE7	++++	Cu (2.8); Zn (1.1)

* Degree of toxicity for MetPAD™: +++++ = 100% toxic (no purple color development); + = slightly toxic; nontoxic = purple spots of similar intensity as the control (moderately hard water)

Adapted from Bitton, G., Koopman, B., and Agami, O., *Water Environ. Res.*, 64, 834, 1992.

B. MetPLATE: a Quantitative Test for Heavy Metal Toxicity

MetPLATE (see Figure 9.1 for methodology), using a 96-well microplate format, was developed to respond to the need for a quantitative assay for heavy metal toxicity.[49] As regards its sensitivity to heavy metals (Table 9.4), MetPLATE compares favorably with Microtox, *Daphnia magna*, and fish toxicity bioassays. When compared to another microbial toxicity test, MetPLATE was similar to Microtox concerning its response to mercury, but was more sensitive to the other heavy metals tested, in particular cadmium and nickel. Ten industrial effluents or process waters were tested for heavy metal toxicity by MetPLATE (Table 9.5). Toxicity was observed in six samples and was confirmed by chemical analysis, which showed the presence of heavy metals. An exception was the effluent from newspaper printing, which contained 1.9 mg/L Cu but was not toxic. This may be due to the fact that copper was not in a bioavailable form.

It is generally recognized that much lower concentrations of reactants are needed in fluorogenic enzyme assays.[50] The use of fluorogenic substrates helps enhance the sensitivity of methods for determining microbial activities in water samples, as compared to chromogenic substrates.[51,52] This is why we have investigated the use of a fluorogenic enzyme substrate in an attempt to increase the sensitivity of the enzyme assay. Therefore, methyl umbelliferyl-β-D-galactopyranoside (MUGA) was used in lieu of the chromogenic substrate used in MetPLATE. Table 9.6 shows that, except for lead, the fluorogenic version of MetPLATE, hereafter called FluoroMetPLATE™, was quite sensitive to the heavy metals tested but was also insensitive to organic compounds. Four of the seven heavy metals tested (Cd, Cu, Hg, Zn) had EC_{50}s of less than 0.1 mg/L. Except for lead, the EC_{50}s for metals were comparable to those obtained with a 48-h *Ceriodaphnia dubia* bioassay. Monitoring of industrial samples showed that FluoroMetPLATE gave results similar to the daphnid toxicity assay for 22 out of 29 samples. It was also demonstrated that the toxicity of most of the remaining samples was due to organic toxicants, confirming the specificity of FluoroMetPLATE for heavy metals.[53]

Table 9.4 Sensitivity of MetPLATE™ to Heavy Metals in Comparison to Microtox™, *Daphnia,* and Fish Bioassays

Metal	EC$_{50}$ (mg/L)			
	MetPLATE	15 min-Microtox*	48-hr* *Daphnia magna*	96-hr* Rainbow trout
Cd	0.029 ± 0.001**	19–220	0.041–1.9	0.15–2.5
Cr(III)	6.9 ± 0.31	13	0.10–1.8	11
Cu	0.22 ± 0.042	0.076–3.8	0.020–0.093	0.25
Pb	10 ± 0.3	1.7–30	3.6	8.0
Hg	0.038 ± 0.001	0.029–0.05	0.0052–0.21	0.033–0.2
Ni	0.97 ± 0.020	23	7.6	36
Zn	0.11 ± 0.001	0.27–29	0.54–5.1	0.55–2.2

* Data drawn from the literature (original references in Bitton[9]).
**Mean ± 1.0 S. D.

Adapted from Bitton, G., *Wastewater Microbiology,* Wiley-Liss, New York, 1994.

Table 9.5 Toxicity and Heavy Metal Content of Selected Industrial Wastewaters and Process Waters, using MetPLATE™

Industry	MetPLATE, % inhibition	Metals (mg/L)
Battery recycling	100	Zn (6.1), Cd (0.1) Cu (0.01), Hg (0.03)
Brewery	Nontoxic	Zn (0.06)
Newspaper printing	Nontoxic	Cu (1.86)
Organic chemicals	Nontoxic	Zn (0.56), Cu (0.06)
Pulp and paper	16.7 ± 7.5*	Zn (0.28)
Pulp and paper	11.3 ± 8.0	NT**
Stationery products	Nontoxic	NT
Wire and cable	62.1 ± 5.0	Zn (0.16), Cu (0.01)
Wood preserving	13.9 ± 9.3	NT
Wood preserving	85.4 ± 0.7	Zn (0.95), Cu (176) Hg (0.01)

* Mean ± 1.0 S. D.
**NT = not tested for metals

From Bitton, G., Jung, K., and Koopman, B., *Arch. Environ. Contam. Toxicol.,* 27, 25, 1994. With permission.

Table 9.6 Comparison of the Sensitivity of FluoroMetPLATE™ to 48-Hour Acute *Ceriodaphnia* Bioassay

Toxicant	FluoroMetPLATE EC$_{50}$ (mg/L)	48-hr *Ceriodaphnia dubia* bioassay EC$_{50}$ (mg/L)
Cadmium(II)	0.0029 ± 0.0003	0.054 ± 0.0026*
Copper(II)	0.0124 ± 0.0007	0.011 ± 0.0010*
Lead(II)	1.8675 ± 0.1998	0.118 ± 0.0045*
Mercury(II)	0.0037 ± 0.0004	0.013 ± 0.0005
Zinc(II)	0.0521 ± 0.0037	0.060 ± 0.0122*
SDS	>2500	10 ± 2.9**
Phenol	>1250	14 ± 7.1**
Pentachlorophenol	>500	0.33 ± 0.058**
2,4,6-Trichlorophenol	>625	4.0 ± 0.53**

* Data from Jung, K., Bitton, G., and Koopman, B., *J. Environ. Toxicol. Chem.,* 15, 711-714.
**Data from Koopman, B., Bitton, G., Delfino, J.J., Mazidji, C., Voiland, G., and Neita, N., Toxicity testing in wastewater systems, Contract No. WM-222, Report to the Florida Depart. Environ. Regulation, Tallahassee, FL, 1989.

V. CONCLUSIONS AND CONTINUING WORK

The following conclusions can be drawn:

- A common feature of currently available microscale bioassays is their use in *general toxicity testing* (i.e., they respond to both metallic and organic toxicants). There is a need to develop microbial and enzymatic tests for the direct measurement of specific categories of toxicants (e.g., heavy metals).
- Certain enzymes (e.g., urease, β-galactosidase) are quite sensitive to heavy metal toxicity and do not respond well to organic toxicants. Thus, bioassay kits based on the inhibition of β-galactosidase activity (e.g., MetPAD/MetPLATE/FluoroMetPLATE) are designed for the specific determination of metal toxicity in environmental samples. These kits should be part of a battery of tests for assessing the toxicity of environmental samples.
- These kits offer a simpler and faster approach for detecting heavy metal toxicity than the proposed effluent fractionation schemes which are part of the Toxicity Reduction Evaluation proposed by the U.S. EPA.[44,45] Furthermore, some have observed inherent problems in the fractionation procedures.[54]
- These kits can be used to screen and pinpoint the source of heavy metals in wastewater collection systems and evaluate pretreatment options for metal removal.
- Continuing work includes the use of MetPLATE for the rapid identification of the physicochemical characteristics contributing to the detoxification of heavy metals in aquatic environments and engineered systems. A solid-phase methodology has been developed to test heavy metal toxicity associated with solids such as soils, sludges, and sediments.[55] The solid-phase assay has been used to assess the success of soil treatment options for heavy metal remediation.[56] Preliminary work is showing that these enzymatic toxicity assays can assess the bioavailability of metals in soils and their uptake by plants.[57] More work is needed to extend this research to other soils and plants.

ACKNOWLEDGMENTS

This research was supported in part by grant number BCS-9117267 from the National Science Foundation, USA, and by contract number 93099 from the Ministere de l'Environnement, Direction de la Recherche et des Affaires Economiques et Internationales (DREAI), France.

REFERENCES

1. Forstner, U. and Prosi, F., Heavy metal pollution in freshwater ecosystems, in *Biological Aspects of Freshwater Pollution*, Ravera, O., Ed., Pergamon Press, Oxford, 1979, 129.
2. Babich, H. and Stotzky, G., Environmental factors that affect the utility of microbial assays for the toxicity and mutagenicity of chemical pollutants, in *Toxicity Testing Using Microorganisms*, Dutka, B.J. and Bitton, G., Eds., Vol. 2, CRC Press, Boca Raton, FL., 1986, 9.
3. Kong, I-C., Bitton, G., Koopman, B., and Jung, K., Heavy metal toxicity testing in environmental samples, *Rev. Environ. Contam. Toxicol.*, 142, 119, 1995.
4. Barth, E.F., The effects and removal of heavy metals in biological treatment: Discussion, in: *Heavy Metals in the Aquatic Environment*, Krenkel, P.A., Ed., Pergamon Press, Oxford, 1975.
5. Henney, R.C., Fralish, M.C., and Lacina, W.V., Shock load of chromium (VI), *J. Water Pollut. Control Fed.*, 52, 2755, 1980.
6. Neufield, R.D., Heavy metal induced deflocculation of activated sludge, *J. Water Poll. Control Fed.*, 48, 1940, 1976.
7. Bitton, G. and Dutka, B.J., Eds., *Toxicity Testing Using Microorganisms*, Vol. 1. CRC Press, Boca Raton, FL, 1986.
8. Dutka, B. J. and Bitton, G., Eds., *Toxicity Testing Using Microorganisms*, Vol. 2. CRC Press, Boca Raton, FL, 1986.

9. Bitton, G., *Wastewater Microbiology*, Wiley-Liss, New York, 1994.

10. Bitton, G. and Koopman, B., Bacterial and enzymatic bioassays for toxicity testing in the environment, *Rev. Environ. Contam. Toxicol.*, 125, 1,1992.

11. Bitton, G. and Koopman, B., Biochemical tests for toxicity screening, in *Toxicity Testing Using Microorganisms*, Vol.1, Bitton, G. and Dutka, B.J., Eds., CRC Press, Boca Raton, FL, 1986, 27.

12. Christensen, G.M., Olson, D., and Riedel, B., Chemical effects on the activity of eight enzymes: a review and a discussion relevant to environmental monitoring, *Environ. Res.*, 29, 247, 1982.

13. Obst, U., Holzapfel-Pschorn, A., and Wiegand-Rosinus, M., Application of enzyme assays for toxicological water testing, *Toxicol. Assess.*, 3, 81, 1988.

14. Cenci, G., Morozzi, G., and Caldini, G., Injury by heavy metals in *Escherichia coli, Bull. Environ. Contam. Toxicol.*, 34,188, 1985.

15. Reinhartz, A., Lampert, I., Herzberg, M., and Fish, F., A new short-term, sensitive bacterial assay kit for the detection of toxicants, *Toxicity Assess.*, 2,193, 1987.

16. Koopman, B., Bitton, G., Dutton, R.J., and Logue, C.L., Toxicity testing in wastewater systems: Application of a short-term assay based on induction of the *lac* operon in *E. coli, Water Sci. Tech.*, 20(11/12), 137, 1988.

17. Koopman, B., Bitton, G., Delfino, J.J., Mazidji, C., Voiland, G., and Neita, N., Toxicity testing in wastewater systems, Contract No. WM-222, Report to the Florida Depart. Environ. Regulation, Tallahassee, FL, 1989.

18. Logue, C.L, Koopman, B., Brown, G.K., and Bitton, G., Toxicity screening in a large, municipal wastewater system, *J. Water Poll. Control Fed.*, 61, 632, 1989.

19. Dutton, R.J, Bitton, G., Koopman, B., and Agami, O., Effect of environmental toxicants on enzyme biosynthesis: A comparison of β-galactosidase, α-glucosidase and tryptophanase, *Arch. Environ. Contam. Toxicol.*, 19, 395, 1990.

20. Campbell, M., Bitton, G., and Koopman, B., Toxicity testing of sediment elutriates based on inhibition of α-glucosidase biosynthesis in *Bacillus licheniformis, Arch. Environ. Contam. Toxicol.*, 24, 469, 1993.

21. Tyler, G., Heavy metal pollution, phosphatase activity, and mineralization of organic phosphorus in forest soils, *Soil Biol. Biochem.*, 8, 327, 1976.

22. Guibault, G.G, and Kramer, D.N., Fluorimetric determination of lipase, acylase, α- and g-chymotrypsin and inhibitors of these enzymes, *Anal. Chem.*, 36, 409, 1964.

23. Holland, H.T, Coppage, D.L., and Butler, P.A., Use of fish brain acetylcholinesterase to monitor pollution by organophosphorus pesticides, *Bull. Environ. Contam. Toxicol.,* 2, 156, 1967.

24. Reidel, B. and Christensen, G., Effect of selected water toxicants and other chemicals upon adenosine triphosphatase activity *in vitro, Bull. Environ. Contam. Toxicol.*, 23, 365, 1979.

25. Guibault, G.G., Brignac, P., and Zimmer, M., Homovanillic acid as a fluorometric substrate for oxidative enzymes, *Anal. Chem.*, 40, 190, 1966.

26. Douglas, L.A. and Bremner, J.M., A rapid method of evaluating different compounds as inhibitors of urease activity in soils, *Soil. Biol. Biochem.*, 3, 309, 1971.

27. Jung, K., Bitton, G., and Koopman, B., Assessment of urease inhibition assays for measuring toxicity of environmental samples, *Water Res.*, 29, 1929, 1995.

28. Xu, H. and Dutka, B.J., ATP-TOX system: a new rapid sensitive bacterial toxicity screening system based on the determination of ATP, *Tox. Assess.*, 2, 149, 1987.

29. Dutton R.J., Bitton, G., and Koopman, B., Enzyme biosynthesis versus enzyme activity as a basis for microbial toxicity testing, *Tox. Assess.*, 3, 245, 1988.

30. Katayama-Hirayama, K., Inhibition of the activities of β-galactosidase and dehydrogenases of activated sludge by heavy metals, *Wat. Res.*, 20, 491, 1986.

31. Rusgas, T., Emneus, J., Gorton, L., and Marko-Varga, G., The development of a peroxidase biosensor for monitoring phenol and related aromatic compounds, *Anal. Chim. Acta*, 311, 245, 1995.

32. Marty, J.L., Mionetto, N., Lacorte, S., and Barcelo, D., Validation of an enzymatic biosensor with various liquid chromatographic techniques for determining organophosphorus pesticides and carbaryl in freeze-dried waters, *Anal. Chim. Acta*, 311, 265, 1995.

33. Corbisier, P., Diels, L., van der Lelie, D., and Mergeay, M., Bioluminescent biosensors for the detection of heavy metals or xenobiotic compounds, in *Proc. Sixth International Symposium Toxicity Assessment and On-line Monitoring*, Berlin University of Technology, May 10-14, 1993, 58.

34. Paton, G.I., Campbell, C.D., Glover, L.A., and Killham, K., Assessment of bioavailability of heavy metals using *lux* modified constructs of *Pseudomonas fluorescens, Lett. Appl. Microbiol.*, 20, 52, 1995.

35. Selifonova, O., Burlage, R., and Barkay, T., Bioluminescent sensors for the detection of bioavailable Hg (II) in the environment, *Appl. Environ. Microbiol.*, 59, 3083, 1993.

36. Tescione L. and Belfort, G., Construction and evaluation of a metal ion detector, *Biotechnol. Bioeng.*, 42, 945, 1993.

37. Olson, D.L. and Christensen, G.M., Effect of selected environmental pollutants and other chemicals on the activity of urease (*in vitro*), *Bull. Environ. Contam. Toxicol.*, 28, 439, 1982.

38. Summer, J.B., Urease, in *Metabolic Inhibitors*, Vol. III, Hochster, R.M. and Quastel, J.H., Eds., Academic Press, New York, 1963, 882.

39. Andrews, A. and Reithel, F.J., The thiol groups of jack bean urease, *Arch. Biochem. Biophys.*, 141, 538, 1970.

40. Shaw, W.H.R. and Raval, D.N., The inhibition of urease by metal ions at pH 8.9, *J. Amer. Chem. Soc.*, 83, 3184, 1961.

41. Zhylyak, G.A., Dzyadevich, S.V., Korpan, Y.I., Soldatkin, A.P., and El'skaya, A.V., Application of urease conductimetric biosensor for heavy metal ion determination, *Sensors & Actuators B*, 24-25, 145, 1995.

42. Kallury, K.M.R., Lee, W.E., and Thompson, M., Enhanced stability of urease immobilized onto phospholipid covalently bound to silica, tungsten, fluoropolymer surfaces, *Anal. Chem.*, 65, 2459, 1993.

43. Bitton, G., Koopman, B., and Agami, O., MetPAD™: a bioassay for rapid assessment of heavy metal toxicity in wastewater, *Water Environ. Res.*, 64, 834, 1992.

44. US EPA, *Methods for aquatic toxicity identification evaluations: Phase I Toxicity Characterization Procedures,* US Environmental Protection Agency, Duluth, MN, Report EPA-600/3-88-034, 1988.

45. US EPA, *Methods for aquatic toxicity identification evaluations: Phase II Toxicity Identification Procedures.* US Environmental Protection Agency, Duluth, MN, Report EPA-600/3-88-035, 1989.

46. Wong, S.L, Wainwright, J.F., and Pimenta, J., Quantification of total and metal toxicity in waste using algal bioassays, *Aquatic Toxicol.*, 31, 57, 1995.

47. Bitton, G., Campbell, M., and Koopman, B., MetPAD™: A bioassay kit for the specific determination of heavy metal toxicity in sediments from hazardous waste sites, *Environ. Toxicol. Water Qual.*, 7, 323, 1992.

48. Mazidji, C.N., Koopman, B., and Bitton, G., Chelating resin versus ion-exchange resin for heavy metal removal in toxicity fractionation, *Water Sci. Technol.*, 26, 189, 1992.

49. Bitton, G., Jung, K., and Koopman, B., Evaluation of a microplate assay specific for heavy metal toxicity, *Arch. Environ. Contam. Toxicol.*, 27, 25, 1994.

50. Guibault, G.G., Fluorescence in enzymology, in *Practical Fluorescence,* Guibault, G.G., Ed., Marcel Dekker, New York, 1990, 683.

51. Holzapfel-Pschorn, A., Obst, U., and Haberer, K., Sensitive methods for the determination of microbial activities in water samples using fluorogenic substrates, *Fresenius Z. Anal. Chem.*, 327, 521, 1987.

52. Obst, U., Test instructions for measuring the microbial metabolic activity in water samples, *Fresenius Z. Anal. Chem.*, 321, 166, 1985.

53. Jung, K., Bitton, G., and Koopman, B., Selective assay for heavy metal toxicity using a fluorogenic substrate, *Environ. Toxicol. Chem.*, 15, 711-714, 1996.

54. Schubauer-Berigan, M. K., Amato, J. R., Ankley, G. T., Baker, S. E., Burkhard, L. P., Dierkes, J.R., Jenson, J. J., Lukasewycz, M. T., and Norberg-King, T. J., The behavior and identification of toxic metals in complex mixtures: Examples from effluent and sediment pore water toxicity identification evaluations, *Arch. Environ. Contam. Toxicol.*, 24, 298, 1993.

55. Bitton, G., Garland, E., Kong, I.C., Morel, J.L., and Koopman, B., A direct solid phase assay for heavy metal toxicity. I. Methodology. *J. Soil Contam.*, 5, 385, 1996.

56. Boularbah, A., Morel, J.L., Bitton, G., and Mench, M., A direct solid phase assay for heavy metal toxicity. II. Assessment of heavy metal immobilization in soils and bioavailability to plants. *J. Soil Contam.*, 5, 395, 1996.

57. Hosy, C., Morel, J.L., and Bitton, G., Microbiological test to assess the bioavailability of metals to plants, *in Third Intern. Conference on the Biogeochemistry of Trace Elements*, May 15-19, 1995, Versailles, France.

CHAPTER 10

ATP-TOX System — A Review

H. Howard Xu

CONTENTS

I. INTRODUCTION

Adenosine triphosphate (ATP) is the most important high-energy phosphate compound in living organisms because it is the primary energy carrier in living cells. Organisms use both chemical energy (heterotrophs and chemoautotrophs) as well as light energy (phototrophs) to synthesize ATP. The free energy of the high-energy phosphate bonds on ATP is most commonly used to drive biosynthetic reactions and other aspects of cell function through carefully regulated processes.

Bacterial cells contain a fairly constant amount of ATP[1] which is rapidly destroyed after cell death; thus it is an ideal parameter of live biomass. ATP has been used as an indicator of cell biomass in: pure bacterial cultures,[2] freshwater,[3] marine environments,[4] soils,[5,6,7] and sewage sludge.[8] In sewage treatment, ATP was used as biomass indicator to monitor changes in operating conditions as well as in measuring live biomass in effluents.[9] In freely suspended and immobilized cell bioreactors, ATP was used to characterize biomass viability.[10] ATP was also used by Brezonik and Patterson[11] in testing the toxicity in activated sludge.

During the early 1980s, microbial toxicity testing systems had just been increasingly employed as alternatives to more traditional bioassay procedures to assess toxicants in water, effluents, and sediment extracts.[12,13] These microbial systems had the advantages of being rapid, sensitive, and economical. It was at this juncture that the ATP-TOX System was developed.[14,15] This test measures the effects of toxicants on the growth of bacterial cells as well as on the activity of firefly luciferase. It employs a luminescence/luminometer system that combines an ATP-releasing reagent and firefly luciferin/luciferase (Turner Designs, Mountain View, CA), without the usual step of heat extraction of ATP.

II. PROCEDURES

The ATP-TOX System can be performed using any one or combinations of a few bacterial species, such as *Escherichia coli, Salmonella typhimurium,* and *Pseudomonas fluorescens. E. coli* strain K12 PQ37 and *S. typhimurium* strain TA-98 were chosen because of their partial loss of the lipopolysaccharide barrier which renders them more permeable to toxicants of a hydrophobic nature.[16,17] When rapidly growing bacterial cells are exposed to low concentrations of toxic agents, growth is usually adversely affected. After a number of life cycles, the inhibitory effect is exhibited by the difference in biomass between the affected culture and the control culture (negative control). The relative bacterial biomass can be conveniently determined by measuring light emission using a luminometer following the reaction of luciferin with ATP in the presence of Mg^{++} and luciferase.[18] However, some toxicants not only inhibit cell growth, but also affect the luciferase activity during the determination of ATP concentrations. Thus, the observed light output reduction caused by a specific toxic solution compared to the control is the net result of both cell growth inhibition and enzyme activity inhibition (termed "total inhibition"). A brief protocol of the ATP-TOX System is described below, since detailed experimental procedures have been described previously.[14,15]

To determine the total inhibition, an overnight culture is diluted 1:100 with fresh medium. One milliliter of sample solution to be tested is combined with 1 mL of diluted bacterial culture in a 12 × 100 mm sterile glass tube. Distilled water is used as a negative control (no toxicant). The mixtures are incubated with shaking at an appropriate temperature for 5 hours.

At the end of the incubation period, 50 μL of culture is taken from each tube and placed into a 1.6 mL polypropylene tube specifically made for Model TD-20e Luminometer (Turner Designs). Also added to the tube are 50 μL of ATP-releasing reagent and 50 μL ATP HEPES buffer. The mixed tube is placed into the Luminometer, and 50 μL of luciferin/luciferase solution is immediately injected into the tube. The light output is recorded. During every batch of tests, a standard control (50 μL of standard ATP solution) and a blank (50 μL of sterile distilled water) are included.

To estimate the luciferase inhibition, the toxic solution (50 μL) at a specific concentration as well as distilled water (50 μL, negative control) are separately combined with 50 μL of standard ATP and 50 μL HEPES buffer. Light output for each mixture is recorded following the addition of the luciferin/luciferase solution.

The percent inhibition of toxicants on the total ATP-TOX System or on the luciferase activity is calculated based on the following equation:

$$I(\%) = \frac{C - U}{C - B} \times 100 \tag{10.1}$$

where I = inhibition percentage; C = reading for negative control; B = reading for blank; and U = reading for unknown samples or chemicals.

Once the total inhibition (I_t) of the ATP-TOX System and the enzyme activity inhibition (I_e) by the toxicant at a specific concentration are known, its inhibitory effect on bacterial growth (I_g) is calculated by the following equation:

$$I_t = I_g + I_e \tag{10.2}$$

E. coli strain K-12 PQ37 appeared to be the most sensitive species of the three bacterial species tested in the ATP-TOX System using nutrient broth as the growth medium (Figure 10.1). The sensitivity of *S. typhimurium* strain TA-98 is slightly lower than that of the *E. coli* strain. The higher sensitivity of the two enteric bacteria to toxicants compared to *P. fluorescens* may be due to their distinct environmental niches. An environmental isolate, *P. fluorescens,* might have been exposed to higher concentrations of toxicants during its course of evolution than have *E. coli* and *S. typhimurium.*

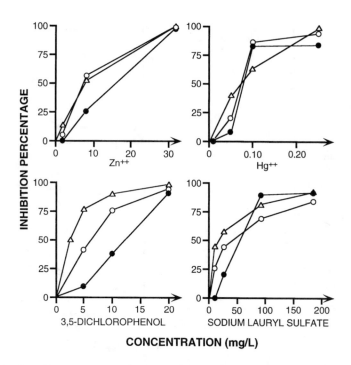

Figure 10.1   Evaluation of bacterial sensitivity in the ATP-TOX System (total inhibition): *Pseudomonas fluore-scens*, ●—●; *Salmonella typhimurium* TA-98, ○—○; *Escherichia coli* K-12 PQ37, Δ—Δ (based on 5-h exposure period, in nutrient broth). (From Xu, H. and Dutka, B. J., *Toxicity Assessment*, 2, 149, 1987. © John Wiley & Sons, Inc. With permission.)

For some toxicants, growth inhibition was their main mode of action, but for others, such as sodium lauryl sulfate, the total inhibition was primarily the result of inhibition on luciferase activity (Figure 10.2). Combining both inhibition of growth and inhibition of enzyme activity in the ATP-TOX System provides a convenient assay that detects the toxicity of two categories of chemicals: one that mainly exerts its action through effect on growth, the other (such as sodium lauryl sulfate) which has no adverse effect on growth but is very potent in disrupting active sites and/or tertiary structures of enzymes.

The ATP-TOX Systems were used to determine toxicity of 12 toxic chemicals. Table 10.1 summarizes the IC_{50} values of these chemicals in the ATP-TOX Systems using *P. fluorescens* and *E. coli* K-12 PQ37 as tester strains. In addition, the sensitivity of the ATP-TOX Systems was compared to that of the Microtox test and the *Spirillum volutans* test. It is apparent that the ATP-TOX Systems were more sensitive than the *S. volutans* test. The sensitivity of the ATP-TOX Systems (combining results of two tester strains) was comparable to and often greater than that of the Microtox test. The Microtox test still has a few advantages over the ATP-TOX Systems. It is a much faster test (requiring 45 min as compared to about 6 h for the ATP-TOX System), has fewer manipulation procedures, and a huge worldwide database. Nevertheless, the ATP-TOX System appears to complement the Microtox data very well. The ATP-TOX System determines the longer-term effects of toxicants on growing cells (over several generations) compared to the Microtox test, while being equally sensitive and more economical to perform.

The ATP-TOX Systems were first applied to examine the toxicity of four sediment extracts and were compared to the results obtained by the Microtox test (Figure 10.3). While all three toxicity screening procedures showed different sensitivity patterns, the Microtox test and ATP-TOX (*E. coli* K12 PQ37) System both appeared to be more sensitive than the ATP-TOX (*P. fluorescens*) System. Extracts from samples 1 and 2 were more toxic in the Microtox test, while extracts from samples

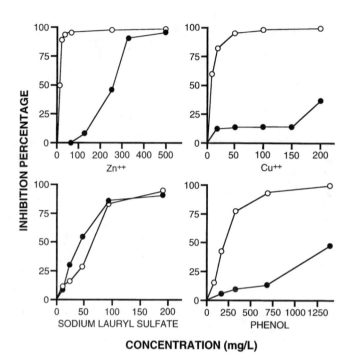

Figure 10.2   Inhibition of luciferase activity by chemicals (●—●) as compared to the total inhibition on the ATP-TOX System (*P. fluorescens*, 5-h exposure, ○—○). *P. fluorescens* was grown in minimal medium. (From Xu, H. and Dutka, B. J., *Toxicity Assessment*, 2, 149, 1987. © John Wiley & Sons, Inc. With permission.)

Table 10.1   Sensitivity of Three Toxicity Screening Procedures to Various Chemicals

	Concentration in mg/L to Give Typical Endpoint			
		ATP-TOX IC$_{50}$* (5 h)		
Chemicals	Microtox EC$_{50}$ (15 m)	*P. fluorescens*	*E. coli* K12 PQ37	*S. volutans* (120 m)[b]
Hg^{++}	0.05	0.02[a]	0.07	0.20
Zn^{++}	5.50	4.00[a]	6.80	11.60
Cu^{++}	2.30[a]	8.10	11.90	10.00
Ni^{++}	22.50	4.33[a]	10.40	20.00
Cd^{++}	18.80	2.35	1.36[a]	—
α-Naphthol	6.25[a]	6.30	42.50	10.00
Sodium lauryl sulfate	1.80[a]	65.80	10.00	43.00
3,5-Dichlorophenol	3.95	7.81	2.49[a]	5.00
Phenol	31.00[a]	218.00	448.00	300.00
p-Nitrophenol	9.40	5.00[a]	35.00	—
Cetyltrimethyl ammonium chloride	1.50	0.59[a]	—	1.45
2,4-Dichlorophenoxyacetic acid	31.25	20.30	12.50[a]	95.00

* IC$_{50}$ is defined as the concentration of toxicant causing 50% inhibition (total inhibition) as compared to the negative control. *P. fluorescens* was grown in minimal medium, while *E. coli* strain K12 PQ37 was grown in nutrient broth.
a Most sensitive
b Data for *S. volutans* from Dutka and Kwan.[26]

From Xu, H. and Dutka, B. J., *Toxicity Assessment*, 2, 149, 1987. © John Wiley & Sons, Inc. With permission.

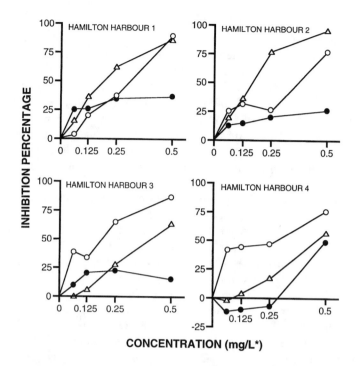

Figure 10.3   Toxicity of four Hamilton Harbor sediment extracts as measured by Microtox test (Δ—Δ) and ATP-TOX Systems (total inhibition; *E. coli* K-12 PQ37, o—o; *P. fluorescens*, •—•). *Sample concentration corresponds to the wet weight of sediment per unit volume from which chemicals were extracted. (From Xu, H. and Dutka, B. J., *Toxicity Assessment*, 2, 149, 1987. © John Wiley & Sons, Inc. With permission.)

3 and 4 were more toxic in the ATP-TOX (*E. coli* K12 PQ37) System. Again, the results indicate that the Microtox test and ATP TOX Systems were complementary to each other.

III. APPLICATIONS

The ATP-TOX System has been used primarily by Dutka and colleagues since its establishment.[19-22] Shortly after the development of this test, it was incorporated into a battery of microbiological, biochemical, and toxicity tests applied to identify highly polluted aquatic environments.[19,20] When samples from the Detroit and Niagara Rivers and Lake Erie inshore were examined, none of the 40 water samples produced observable inhibitory effects in the Microtox test. Surprisingly, all but five of these samples showed toxic effects in the ATP-TOX System.[19] Similarly, the ATP-TOX System was more sensitive to sediment samples than the Microtox test.[19] Only one sediment sample was found toxic by both the Microtox test and the ATP-TOX System, suggesting the complementarity of these two tests. These results indicated that the ATP-TOX System is a sensitive and easily performed screening test, and it is at least as sensitive as the Microtox test and adds an additional dimension, life cycle effects, to rapid toxicity screening test batteries.

Water and sediment samples from the Saint John River Basin (New Brunswick, Canada) and inshore marine waters influenced by the Saint John River were screened for polluted areas using microbial and toxicity tests including the ATP-TOX System.[20] The ATP-TOX System was the most sensitive test compared to three other toxicity tests (the Microtox test, the algal-ATP test, and the *Spirillum volutans* test). Moreover, only three of 38 sediment samples exhibited positive toxic effects in all four toxicity tests. There were nine sites positive for both the Microtox test and the

ATP-TOX System, 12 sites by the algal-ATP test and the ATP-TOX System, and 8 sites by the *S. volutans* test and the ATP-TOX System. These results reinforced the concept of using the "battery of screening tests" approach to evaluate and prioritize polluted environments.

The ATP-TOX System was included in a battery of microbial and toxicant screening tests in an attempt to analyze the degree of degradation of the Fraser River estuary in western Canada.[21] Of the toxicant screening tests used, both the *S. volutans* test and algal-ATP test were negative in all 40 water samples tested. Seven of 40 water samples were positive in the Microtox test. Twenty-seven water samples exhibited toxic effects in the ATP-TOX System, although none of the inhibition was over 50% and many were only slightly above the background. The *Daphnia magna* test was the most sensitive for detecting the presence of toxicants.[21] There appeared to be a relationship between ATP-TOX System values greater than 20% inhibition and the finding of a toxic response with the *Daphnia magna* test. None of the sediment samples was positive in the Microtox test and *S. volutans* test. Most of these sediment samples tested negative in the ATP-TOX System; eight of 40 samples did produce an inhibitory effect of between 10 and 28%.

Data from The Yamaska River study also indicated that the ATP-TOX System and *Daphnia magna* test were the most sensitive or responsive tests to the river contaminants.[22] The ATP-TOX System was positive in every water sample with inhibition ranging from 5 to 100%. The ATP-TOX System and the *Daphnia magna* test were the only toxicant screening tests not affected by seasonal variability in the samples from the Yamaska River. When tested against organically extracted sediment samples, both the Microtox test and the ATP-TOX System gave positive results in all samples, and the Microtox test appeared to be responsive to a group of chemicals different from those triggering the response of the ATP-TOX System, further confirming the complementarity of these two sensitive tests.[22]

Toxicity testing studies based on determination of ATP have also been carried out using a number of other microorganisms. The toxic effects of 16 chemicals were examined based on light emission by two cell-free bioluminescent systems (firefly and *Vibrio fischeri*) and by whole cells of *Photobacterium phosphoreum*.[23] An ATP-based toxicity testing procedure similar to the ATP-TOX System was used by Seyfried and Desjardins to assess toxic effects of metal ions on bacteria isolated from two geographically different marsh treatment systems.[24] Blaise et al. developed an algal toxicant screening test based on the inhibition of ATP production by the green alga *Selenestrum capricornutum*[25] (also see Chapter 18 by Blaise et al.).

IV. CONCLUSIONS

The ATP-TOX System is a sensitive, fast, and simple toxicity screening test. It has been proven to be at least as sensitive as the Microtox test. It adds an additional dimension (life cycle effects) to rapid toxicity tests and properly complements the results of other toxicity screening tests. Thus it has become an integral part of a battery of microbiological, biochemical, and toxicity screening tests to evaluate the severity of environmental pollution/degradation.

However, there is need for improvement. The application of microtitration techniques to the ATP-TOX System may further increase replicate numbers of each exposure, thus increasing the reproducibility. This may also reduce the minimal time required to finish a batch test. This is very significant when large numbers of environmental samples are to be screened in a short period of time.

ACKNOWLEDGMENTS

The ATP-TOX System was developed while I was a visiting scholar at the laboratory of B. J. Dutka, whose help is greatly appreciated. The financial support of a scholarship from the Chinese Academy of Sciences is acknowledged. I thank M. Zianni for critical reading of the manuscript.

REFERENCES

1. Knowles, E. J., Microbial metabolic regulation by adenine nucleotide pools, in *Microbial Energetics*, Haddock, B. A. and Hamilton, W. A., Eds., Cambridge University Press, Cambridge, 1977, 241.
2. Ng, L.-K., Taylor, D. E., and Stiles, M. E., Estimation of *Campylobacter* spp. in broth culture by bioluminescence assay of ATP, *Applied and Environmental Microbiology*, 49, 730, 1985.
3. Cavari, B., ATP in Lake Kinneret: indicator of microbial biomass or of phosphorus deficiency, *Limnology & Oceanography*, 21, 231, 1976.
4. Holm-Hansen, O. and Booth, C. R., The measurement of adenosine triphosphate in the ocean and its ecological significance, *Limnology & Oceanography*, 11, 510, 1966.
5. Eiland, F., An improved method for determination of adenosine triphosphate (ATP) in soil, *Soil Biology and Biochemistry*, 11, 31, 1979.
6. Bååth, E., Arnebrant, K., and Nordgren, A. Microbial biomass and ATP in smelter-polluted forest humus, *Bulletin of Environmental Contamination and Toxicology*, 47, 278, 1991.
7. Post, R. D. and Beeby, A. N., Microbial biomass in suburban roadside soils: estimates based on extracted microbial C and ATP, *Soil Biology and Biochemistry*, 25, 199, 1993.
8. Patterson, J. W., Brezonik, P. L., and Putnam, H. D., Measurement and significance of adenosine triphosphate in activated sludge, *Environmental Science and Technology*, 4, 569, 1970.
9. Levin, G. V., Schrot, J. R., and Hess, W. C., Methodology and application of adenosine triphosphate determination in wastewater treatment, *Environmental Science and Technology*, 9, 961, 1975.
10. Gikas, P. and Livingston, A. G., Use of ATP to characterize biomass viability in freely suspended and immobilized cell bioreactors, *Biotechnology and Bioengineering*, 42, 1337, 1993.
11. Brezonik, P. L. and Patterson, J. W., Activated sludge ATP: effects of environmental stress, *J. San. Eng. Div. Proc.* ASCE, 97, 813, 1971.
12. Liu, D. and Dutka, B. J., Toxicity screening procedures using bacterial systems, Marcel Dekker, Inc., New York, 1984.
13. Bitton, G. and Dutka, B. J., *Toxicity Testing Using Microorganisms*, Vol. 1, CRC Press, Boca Raton, 1986.
14. Xu, H. and Dutka, B. J., ATP-TOX System — a new, rapid, sensitive bacterial toxicity screening system based on the determination of ATP, *Toxicity Assessment*, 2, 149, 1987.
15. Xu, H. and Dutka, B. J., ATP-TOX System — a bacterial toxicity screening procedure, *Toxicity Assessment*, 2, 357, 1987.
16. Quillardet, P., Huisman, O., D'ari, R., and Hofnung, M., SOS chromotest, a direct assay of induction of an SOS function in *Escherichia coli* K-12 to measure genotoxicity, *Proceedings of National Academy of Sciences (U.S.A.)*, 79, 5971, 1982.
17. Ames, B. N., Lee, F. D., and Durston, W. E., An improved bacterial test system for the detection and classification of mutagens and carcinogens, *Proceedings of National Academy of Sciences (U.S. A.)*, 70, 782, 1973.
18. Holm-Hansen, O., Determination of total microbial biomass by measurement of adenosine triphosphate, in *Estuarine Microbial Ecology*, Sterenson, L. H. and Colwell, R. R., Eds., University of South Carolina Press, Columbia, 1973.
19. Dutka, B. J., Jones, K., Xu, H., Kwan, K. K., and McInnis, R., Priority site selection for degraded areas in the aquatic environment, *Water Pollution Research Journal of Canada*, 22, 326, 1987.
20. Dutka, B. J., Jones, K., Kwan, K. K., Bailey, H., and McInnis, R., Use of microbial and toxicant screening tests for priority site selection of degraded areas in water bodies, *Water Research*, 22, 503, 1988.
21. Dutka, B. J., Tuominen, T., Churchland, L., and Kwan, K. K., Fraser river sediments and waters evaluated by the battery of screening tests technique, *Hydrobiologia*, 188/189, 301, 1989.
22. Dutka, B. J., Kwan, K. K., Rao, S. S., Jurkovic, A., and Liu, D., River evaluation using ecotoxicological and microbiological procedures, *Environmental Monitoring and Assessment*, 16, 287, 1991.
23. Surowitz, K. G., Burke, B. E., and Pfister, R. M., Comparison of cell-free and whole cell luminescence assays in toxicity testing, *Toxicity Assessment*, 2, 17, 1987.
24. Seyfried, P. L. and Desjardins, R. M., Use of the ATP bioassay to assess toxic effects in marsh treatment systems, *Toxicity Assessment*, 2, 29, 1987.
25. Blaise, C., Legault, R., Bermingham, N., Van Coillie, R., and Vasseur, P., Microtest mesurant l'inhibition de la croissance des algues (CI50) par le dosage de l'ATP, *Sc. et Tech. de l'Eau*, 17, 245, 1984.

CHAPTER 11

Luminescent Bacterial Biosensors for the Rapid Detection of Toxicants

Scott F. Briscoe, Jie Cai, and Michael S. DuBow

CONTENTS

I. INTRODUCTION

For many years now, an array of physicochemical systems has been developed to assay for specific toxic agents in contaminated environmental samples.[1-4] These tests typically offer a high degree of specificity, sensitivity, and selectivity. More recently, a series of *in vitro* biochemical assays has been added to this array.[5-8] These assays are generally less technically demanding and less costly to perform. Biological assays, based on the effects of the toxicant on a wide variety of organisms (from bacteria to the common water flea *Daphnia magna* to rainbow trout), have also been developed.[9-15] These types of assay systems have proven to be effective in the determination of total toxicant concentration in contaminated environments. They have also demonstrated the differential effects pollutants can have throughout the various stages in the life cycles of certain organisms. However, there is much current concern regarding another parameter of such contaminated areas: the bioavailable toxicant concentration. It is this concentration of the toxicant — that which is available to living organisms — that has the potential to affect them, making knowledge of this parameter important for meaningful risk and environmental quality analyses to be performed. Moreover, there is also a need for information regarding the genetic and physiological parameters which can influence the sensitivity or resistance of an organism to a particular toxicant.

Biological assays provide important data regarding the bioavailability of toxic agents to living organisms. However, observable effects caused by any particular agent under a given set of conditions can be obscured by other toxic agents which may also be present in the environmental sample. In addition, the fundamental mechanisms of toxicity (e.g., chemical transformations, cellular targets, etc.) of a particular compound are frequently difficult to assess from these assays, especially given the low concentrations typically found in the environment and the long observation periods that are often required before the effects of such compounds can be seen. In order to determine these parameters with the high degree of specificity and sensitivity offered by chemical assays, and the ease of use and low cost of *in vitro* biochemical tests, research efforts have focused on the development of cellular- and molecular-based biosensors. These biosensors are constructed such that the changes an organism senses in its environment are transduced to an easily assayed signal. Moreover, due to the "engineering" required for their construction, much information about the genetic and biochemical effects of particular compounds (or formulations) on living cells/organisms can be obtained from their use.

II. BACTERIA AS BIOSENSORS

The main advantages of using bacteria for this purpose are that they can be produced rapidly, abundantly, and inexpensively. The maintenance of a genetically homogeneous population of these organisms is straightforward, which is vital for their use as biosensors, because mutations can alter their response. Verification of homogeneity can be accomplished both phenotypically and genotypically, and *in vivo* or *in vitro* reconstruction of the bacterial clones can be performed if necessary. Bacteria also offer ease of transport and storage via lyophilization, with subsequent rehydration requiring only a few minutes.

Like all cells, bacteria have evolved mechanisms that enable them to withstand a wide range of environmental conditions, including increased concentrations of toxic substances. In many cases, these mechanisms are conserved throughout evolution, and involve the post-exposure reorientation of cellular physiology via changes in gene expression.[16-18] Frequently, these changes in gene expression are proportionate to the level of toxicant the organism senses in its environment. Thus, measurement of the expression of such genes will not only indicate if the specific toxicant is present, but may also demonstrate the concentration of the toxic agent that is available to organisms living in the particular environment from which the sample was taken. In addition, the identification of these genes can provide important clues to the action of a particular chemical at the molecular level. In turn, this information can be used to extrapolate these molecular effects to organisms from the plant and animal kingdoms.

III. REPORTER GENES

The expression of such toxicant-inducible (or repressible) genes can be measured, even if nothing is known about the gene itself (in terms of its DNA sequence or the protein product it encodes), via the use of reporter genes and gene fusion technology. This method can even be used to search for new genes that are not yet known. Reporter genes are known DNA sequences whose protein products are easily assayed. The reporter gene, minus its own regulatory sequences, is inserted within the coding sequence of the gene whose expression is to be monitored. As such, expression of the reporter gene is now under the regulatory control of the gene into which it has been inserted. Since the reporter gene product is easily assayed, it thus serves as a "reporter" of the regulation of the toxicant-responsive gene.

Two types of reporter genes exist. One is deprived of its transcriptional regulatory sequences but retains the signals for its correct translation into a protein. This type is the more commonly used. The other type is devoid of both its transcriptional and translational regulatory sequences, and its function requires that it be inserted in the correct translational reading frame of the protein whose expression is to be measured.

A. *lux*AB Reporter Genes

One of the newest of these reporter genes is *lux*AB. Naturally occurring in a number of marine bacteria, these genes encode the enzyme luciferase, which catalyzes the oxidation of a reduced flavin mononucleotide (FMNH$_2$) and a long-chain fatty aldehyde by molecular oxygen, according to the following bioluminescent reaction:

$$FMNH_2 + RCHO + O_2 \rightarrow FMN + RCOOH + H_2O + light^{[19]}$$

Since cells expressing bacterial luciferase usually contain sufficient intracellular FMNH$_2$ and O$_2$ for bioluminescence (although these parameters also reflect the physiological status of the cell), all that is subsequently required to measure luciferase activity is the addition of a long-chain aldehyde substrate such as decanal, either in solution or as a vapor. The amount of light thus produced is proportional to the amount of luciferase expressed from the toxicant-responsive gene.[20-22]

1. Measurement of luxAB Expression

Photon emission from this luciferase-catalyzed reaction results in a glow at 490 nm, making it easy to detect and measure. Qualitative measurements can be made using high-speed instant photographic film, X-ray film, or even by simple visual observation. If quantified data are desired over a wider dynamic range, luminometers can be used to count the photons produced by the reaction. These machines come in a wide variety of formats. Some hold a single sample in a small disposable cuvette, while others can handle multiple samples using microtiter plates. Injection systems are also available, allowing addition of the aldehyde substrate (or other reagents) while the sample is in the photon-detection chamber. Computer systems may also be added to such luminometers, allowing data tabulation and analysis to be carried out both quickly and easily. Luminometers also currently exist that are sufficiently small and portable to be utilized on site in the field, alleviating the need for sample transport back to the laboratory.

2. Advantages of the luxAB Reporter Gene System

A wide variety of other reporter genes also exist and have been used in a number of different applications. Each system has specific analytical methods to detect the protein expressed by these genes, including colorimetry, fluorescence, radiometry, and immunoassays (Table 11.1). The key advantages of the bacterial *lux*AB reporter gene systems are the speed, simplicity, and high degree of sensitivity with which the gene product can be measured. The activity of the enzyme can encompass changes in gene expression of over six orders of magnitude,[19] allowing low-level or even cryptic responses to be discovered and measured.[39] The luminescence assays are relatively inexpensive to perform, and allow for real-time, noninvasive measurement of the expression of specific toxicant-responsive genes. Thus, the capacity to create "luminescent bacterial biosensors" for the detection and measurement of environmental toxicants is proving to be both feasible and practical.

Table 11.1 Reporter Genes and Detection Methods for the Products They Encode

Reporter Gene	Detection Method	Reference
Acid phosphatase (TRAP)	Colorimetry	23
Aequorin (*phot*)	Bioluminescence	24
Alcohol dehydrogenase (ADH)	Colorimetry	25
Alkaline phosphatase (ALP)	Colorimetry	26
Bacterial luciferase (*lux*AB)	Bioluminescence	19
Catechol 2,3 dioxygenase (*xyl*E)	Colorimetry	27
Chloramphenicol acetyltransferase (CAT)	Radiometry	28
	Fluorescence	28
	Immunoassay	29
Firefly luciferase (*luc*)	Bioluminescence	30
β-galactosidase (*lacZ*)	Colorimetry	31,32
	Fluorescence	33,34
β-glucuronidase (*uid*A)	Colorimetry	35
	Fluorescence	35
Green fluorescent protein (GFP)	Fluorescence	36
Ice nucleation (*ina*Z)	Freezing assay	37
Vargula hilgendorfii luciferase (*luc*)	Bioluminescence	38

Table 11.2 Bacterial Biosensors Created Using Reporter Genes

A. Insertion of Reporter Genes Into Previously Known Toxicant-Responsive Genes

Toxicant	Responsive Gene	Reporter Gene	Reference
Naphthalene	*nah*G	*lux*AB	40
Salicylate	*nah*G	*lux*AB	40
Mercury	*mer*	*lux*AB	41
DNA damaging agents	*sul*A	*lacZ*	42

B. Insertion of Reporter Genes Into Previously Unknown Toxicant-Responsive Genes

Toxicant	Responsive Gene	Reporter Gene	Reference
Metal cations	*fliC*	*lux*AB	44
Nickel	*celF*	*lux*AB	39
Arsenic	*ars*	*lacZ*	45
Selenium	?	*lux*AB	46
Tributyltin	?	*lux*AB	47
Dimethyl sulfoxide	?	*lux*AB	47

IV. CREATION OF BIOSENSORS

For some toxic agents, genes are already known to exist which have their transcription levels altered in response to the particular toxicant. Thus, the creation of biosensors for these toxicants can be accomplished by inserting reporter genes into the coding sequence of the known responsive gene(s) (Table 11.2A). However, for other toxic agents, genes have yet to be identified which have their expression altered upon cellular exposure to the specific compound. We have searched for such toxicant-responsive genes in *Escherichia coli* using a library of approximately 3000 *E. coli* clones,[43] each containing a single copy of the *lux*AB reporter genes from the marine bacterium *Vibrio harveyi* located at a random site within the chromosome (Figure 11.1). By monitoring changes in light emission from these clones in the presence (as opposed to the absence) of specific added toxicants, clones (and ultimately genes) have been identified which respond to these specific toxic agents (Table 11.2B). These luminescent biosensors can thus be added to the "battery of microbiotests," an approach which is gaining increased use for toxicity assessment.[48]

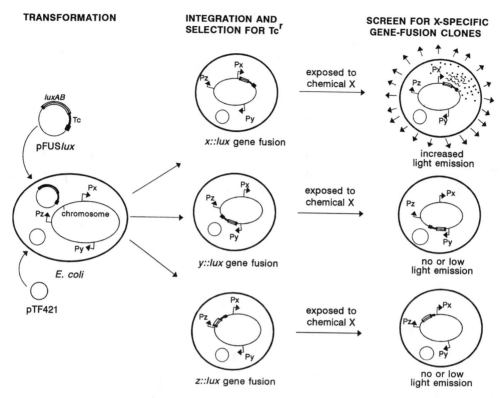

Figure 11.1 Creation of an *Escherichia coli* gene-fusion library to search for toxicant-responsive genes. *E. coli* DH1 cells were transformed with plasmid pFUS*lux*, containing the *luxAB* genes of *Vibrio harveyi* along with a tetracycline resistance gene (*Tc*), all within a truncated Tn*5* transposon. Plasmid pTF421 was used to inhibit replication of pFUS*lux*, thus ensuring that only one copy of *luxAB* integrated into the bacterial chromosome via Tn*5*-mediated transposition.[43] A library of approximately 3000 gene-fusion clones was thus created. Three such clones are schematically depicted, with *luxAB* inserted within three different genes: *x*, *y*, and *z*. Assuming that gene *x* shows an increase in transcription in response to chemical X, then screening the library by growing each clone in the presence of chemical X will result in increased light emission from the *x::lux* gene-fusion clone (due to the increase in transcription of *luxAB* reporter genes within gene *x*), but not from the other gene-fusion clones.

V. USE OF BIOSENSORS

A. Measurement of Toxicants in Environmental Samples

The clones developed in our laboratory,[39,44-47] plus others,[40,41] are now available for use as highly sensitive luminescent bacterial biosensors of the specific toxicant to which each responds. For example, an aqueous sample can first be tested with a cytotoxicity-measuring "control" biosensor (constructed by inserting the *lux*AB genes into a universal stress gene of *E. coli*, such as *usp*A[49]) to determine the level of cytotoxic pollutants. Transcription of this gene is augmented whenever the cell growth rate falls below the maximal rate supported by the growth medium, regardless of the condition inhibiting growth.[50] If this test indicates that a toxicity problem does exist, then specific biosensors can be used to determine which toxicant(s) are present in the aqueous sample and at what concentrations they are available to living organisms (Figure 11.2). To control for other types of stress-induced changes in cellular physiology that might be brought about by addition of the environmental sample and ultimately alter luciferase expression, a second control biosensor

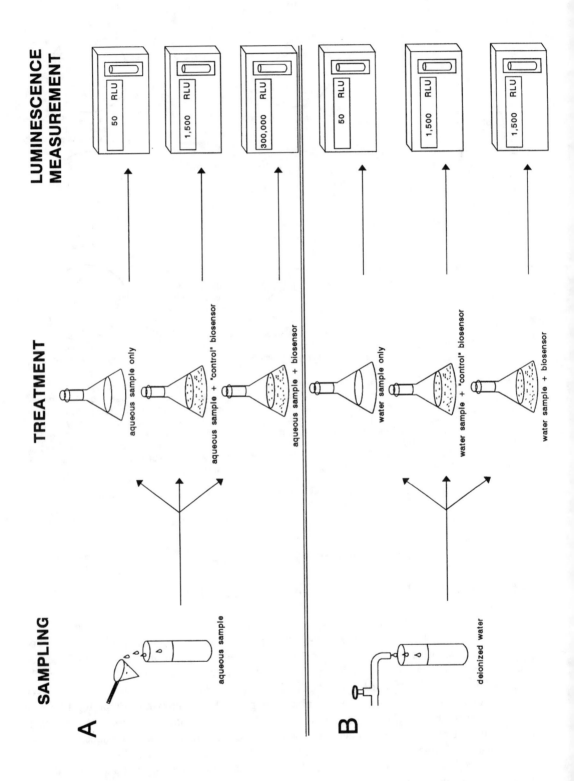

can be used, with the *lux*AB genes inserted within a constitutively expressed *E. coli* gene, such as *xap*R.[51] Comparison of light emission from this clone before and after addition of the environmental sample being tested would allow for the detection and compensation of any such generalized stress-induced effects. In this manner, even in the absence of detectable cytotoxicity, these biosensor clones can be used to determine if elevated levels of bioavailable toxic agents are present, and to investigate their effects on cellular physiology, thus allowing the early detection of potential risks.

B. Elucidation of Molecular Mechanisms of Toxicant Action

Not only can these clones be used as biosensors to detect specific toxicants in polluted environmental samples, but they can also be utilized to obtain information regarding the precise biochemical mechanisms underlying the effects of these toxicants on organisms that are exposed to them. Each of these toxicant-responsive genes can be cloned and sequenced for the purpose of identification. Further characterization of these genes, their upstream regulatory elements, and the proteins they encode can then be performed. These clones may also allow for the determination of the speciation of metal contaminants in environmental samples, and for the identification of active and inactive forms of the contaminant.[52] Organic components of organometallic contaminants can also be distinguished. For example, tributyltin can be detected among other organotins using our tributyltin-responsive clones.[47] These clones can also be used to determine the existence of potentiators or inhibitors of cellular responses to the chemical, as well as the structure/activity coefficients of related compounds.

Results from such studies will provide a better understanding of the effects of each of these agents at the molecular level, giving insight into how each may affect organisms that are exposed to them. In the case of xenobiotic agents, including many herbicides and pesticides, this new-found knowledge of exactly how such compounds affect the molecular events of living cells may allow these chemicals to be altered such that they still have the desired effect on their targets, but less detrimental effects on nontarget organisms. These assays will also be important in reducing the need for animals to be used in toxicity assessment tests.

Thus, these clones have great potential not only as highly sensitive sensors of biologically available amounts of specific toxicants in field tests, but also as highly informative research tools within the laboratory, to answer many of the as yet unsolved molecular puzzles involving such specific toxic agents and their interactions with living organisms.

ACKNOWLEDGMENTS

Work in our laboratory is supported by a grant (96035R2) from the Center for the Alternatives to Animal Testing (USA).

FIGURE 11.2 Use of luminescent bacterial biosensors to detect toxicants in aqueous samples. A) An aqueous sample is divided into three aliquots. To the first aliquot nothing is added. To the second aliquot a "control" biosensor is added. To the third aliquot, a biosensor is added for the toxicant of interest. Luminometry readings are taken. The first reading (sample alone) gives the background level of luminescence from the aqueous sample. The second reading (sample + control biosensor) shows that the increase in luminescence seen in the third reading (sample + toxicant-specific biosensor) is specific for the toxicant-responsive biosensor and not a nonspecific effect of the aqueous sample on the luciferase bioluminescent reaction. B) The same procedure is carried out as in A, using deionized water in place of the aqueous sample. This test indicates the basal level of light emission from each of the biosensors used, above the background level of luminescence from the sample itself. Thus, these luminescent readings indicate contamination of the sample with the specific toxicant assayed. (RLU = relative light units).

REFERENCES

1. Ford, J. H., McDaniel, C. A., White, F. C., Vest, R. E., and Roberts, R. E., Sampling and analysis of pesticides in the environment, *Journal of Chromatographic Science*, 13, 291, 1975.
2. Aldridge, W. N. and Street, B. W., Spectrophotometric and fluorometric determination of tri- and di-organotin and organolead compounds using dithizone and 3-hydroxyflavone, *Analyst*, 106, 60, 1981.
3. Andreae, M. O., Determination of the chemical species of some of the 'hydride elements' (arsenic, antimony, tin, and germanium) in seawater: methodology and results, in *Trace Metals in Seawater*, Wong, C. S., Boyle, E., Bruland, K. W., Burton, J. D., and Goldberg, E. D., Eds., Plenum Press, New York, 1983, Chap. 1.
4. Wylie, P. L. and Oguchi, R., Pesticide analysis by gas chromatography with a novel atomic emission detector, *Journal of Chromatography*, 517, 131, 1990.
5. Olson, D. L. and Christensen, G. M., Effect of selected environmental pollutants and other chemicals on the activity of urease (*in vitro*), *Bulletin of Environmental Contamination and Toxicology*, 28, 439, 1982.
6. Durand, P., Nicaud, J. M., and Mallevialle, J., Detection of organophosphorous pesticides with an immobilized cholinesterase electrode, *Journal of Analytical Toxicology*, 8, 112, 1984.
7. Durand, P. and Thomas, D., Use of immobilized enzyme coupled with an electrochemical sensor for the detection of organophosphate and carbamate pesticides, *Journal of Environmental Pathology, Toxicology, and Oncology*, 5, 51, 1984.
8. Skladal, P. and Mascini, M., Sensitive detection of pesticides using amperometric sensors based on cobalt phthalocyanine-modified composite electrodes and immobilized cholinesterases, *Biosensors and Bioelectronics*, 7, 335, 1992.
9. Nacci, E., Jackim, E., and Walsh, R., Comparative evaluation of three rapid marine toxicity tests: sea urchin early embryo growth test, sea urchin sperm cell toxicity test and Microtox, *Environmental Toxicology and Chemistry*, 5, 521, 1986.
10. Khangarot, B. S. and Ray, P. K., Correlation between heavy metal acute toxicity values in *Daphnia magna* and fish, *Bulletin of Environmental Contamination and Toxicology*, 38, 722, 1987.
11. Deneer, J. W., van Leeuwen, C. J., Seinen, W., Maas-Diepeveen, J. L., and Hermens, J. L. M., A QSAR study of the toxicity of nitrobenzene derivatives toward *Daphnia magna*, *Chlorella pyrenoidosa* and *Photobacterium phosphoreum*, *Aquatic Toxicology*, 15, 83, 1989.
12. Giesy, J. P., Rosiu, C. J., Graney, R. L., and Henry, M. G., Benthic invertebrate bioassays with toxic sediment and pore water, *Environmental Toxicology and Chemistry*, 9, 233, 1990.
13. Dutka, B. J. and Kwan, K. K., Comparison of three microbial toxicity screening tests with the Microtox test, *Bulletin of Environmental Contamination and Toxicology*, 27, 753, 1981.
14. Ribo, J. M. and Kaiser, K. L. E., Effects of selected chemicals on photoluminescent bacteria and their correlations with acute and sublethal effects on other organisms, *Chemosphere*, 12, 1421, 1983.
15. Maas-Diepeveen, J. L. and van Leeuwen, C. J., Toxicity of methylene-bisthiocyanate (MTB) to several freshwater organisms, *Bulletin of Environmental Contamination and Toxicology*, 40, 517, 1988.
16. Little, J. W. and Mount, D. W., The SOS regulatory system of *Escherichia coli*, *Cell*, 29, 11, 1982.
17. Gottesman, S., Bacterial regulation: global regulatory networks, *Annual Review of Genetics*, 18, 415, 1984.
18. Demple, B., Regulation of bacterial oxidative stress genes, *Annual Review of Genetics*, 25, 315, 1991.
19. Meighen, E. A., Molecular biology of bacterial luminescence, *Microbiological Reviews*, 55, 123, 1991.
20. Stewart, G. S. A. B. and Williams, P., *lux* genes and the applications of bacterial bioluminescence, *Journal of General Microbiology*, 138, 1289, 1992.
21. Wolk, C. P., Cai, Y., and Panoff, J. M., Use of a transposon with luciferase as a reporter to identify environmentally responsive genes in cyanobacterium, *Proceedings of the National Academy of Sciences of the United States of America*, 88, 5355, 1991.
22. Kragelund, L., Christoffersen, B., Nybroe, O., and De Bruijn, F. J., Isolation of *lux* reporter gene fusions in *Pseudomonas fluorescens* DF57 inducible by nitrogen or phosphorus starvation, *FEMS Microbiology Ecology*, 17, 95, 1995.
23. Reddy, S. V., Takahashi, S., Haipek, C., Chirgwin, J. M., and Roodman, G. D., Tartrate-resistant acid phosphatase gene expression as a facile reporter gene for screening transfection efficiency in mammalian cell cultures, *Biotechniques*, 15, 444, 1993.

24. Tanahashi, H., Ito, T., Inouye, S., Tsuji, F. I., and Sakaki, Y., Photoprotein aequorin: use as a reporter enzyme in studying gene expression in mammalian cells, *Gene*, 96, 249, 1990.

25. Lin, W. C. and Culp, L. A., Selectable plasmid vectors with alternative and ultrasensitive histochemical marker genes, *Biotechniques*, 11, 344, 1991.

26. Yoon, K., Thiede, M. A., and Rodan, G. A., Alkaline phosphatase as a reporter enzyme, *Gene*, 66, 11, 1988.

27. Curcic, R., Dhandayuthapani, S., and Deretic, V., Gene expression in mycobacteria: transcriptional fusions based on *xyl*E and analysis of the promoter region of the response regulator *mtr*A from *Mycobacterium tuberculosis*, *Molecular Microbiology*, 13, 1057, 1994.

28. Hruby, D. E. and Wilson, E. M., Use of fluorescent chloramphenicol derivative as a substrate for chloramphenicol acetyltransferase assays, *Methods in Enzymology*, 216, 369, 1992.

29. Porsch, P., Merkelbach, S., Gehlen, J., and Fladung, M., The nonradioactive chloramphenicol acetyl-transferase-enzyme-linked immunosorbent assay test is suited for promoter activity studies in plant protoplasts, *Analytical Biochemistry*, 211, 113, 1993.

30. de Wet, J. R., Wood, K. V., Helinski, D. R., and DeLuca, M., Cloning firefly luciferase, *Methods in Enzymology*, 133, 3, 1986.

31. Hidalgo, C., Reyes, J., and Goldschmidt, R., Induction and general properties of beta-galactosidase and beta-galactoside permease in *Pseudomonas* BAL-31, *Journal of Bacteriology*, 129, 821, 1977.

32. Lin, W. C., Pretlow, T. P., Pretlow, T. G., and Culp, L. A., Development of micrometastases: earliest events detected with bacterial *lacZ* gene-tagged tumor cells, *Journal of the National Cancer Institute*, 82, 1497, 1990.

33. Jongkind, J. F., Verkerk, A., and Sernetz, M., Detection of acid-beta-galactosidase activity in viable human fibroblasts by flow cytometry, *Cytometry*, 7, 463, 1986.

34. Lo, J., Mukerji, K., Awasthi, Y. C., Hanada, E., Suzuki, K., and Srivastava, S. K., Purification and properties of sphingolipid beta-galactosidases from human placenta, *Journal of Biological Chemistry*, 254, 6710, 1979.

35. Jefferson, R. A., Kavanagh, T. A., and Bevan, M. W., GUS fusions: β-glucuronidase as a sensitive and versatile gene fusion marker in higher plants, *EMBO Journal*, 6, 3901, 1987.

36. Kremer, L., Baulard, A., Estaquier, J., Poulain-Godefroy, O., and Locht, C., Green fluorescent protein as a new expression marker in mycobacteria, *Molecular Microbiology*, 17, 913, 1995.

37. Lindgren, P. B., Frederick, R., Govindarajan, A. G., Panopoulos, N. J., Staskawicz, B. J., and Lindow, S. E., An ice nucleation reporter gene system: identification of inducible pathogenicity genes in *Pseudomonas syringae* pv. phaseolicola, *EMBO Journal*, 8, 1291, 1989.

38. Thompson, E. M., Nagata, S., and Tsuji, F. I., Cloning and expression of cDNA for the luciferase from the marine ostracod *Vargula hilgendorfii*, *Proceedings of the National Academy of Sciences of the United States of America*, 86, 6567, 1989.

39. Guzzo, A. and DuBow. M. S., A *lux*AB transcriptional fusion to the cryptic *cel*F gene of *E. coli* displays increased luminescence in the presence of nickel, *Molecular and General Genetics*, 242, 455, 1994.

40. Heitzer, A., Malachowsky, K., Thonnard, J. E., Bienkowski, P. R., White, D. C., and Sayler, G. S., Optical biosensor for environmental on-line monitoring of naphthalene and salicylate bioavailability with an immobilized bioluminescent catabolic reporter bacterium, *Applied and Environmental Microbiology*, 60, 1487, 1994.

41. Selifonova, O., Burlage, R., and Barkay, T., Bioluminescent sensors for detection of bioavailable Hg(II) in the environment, *Applied and Environmental Microbiology*, 59, 3083, 1993.

42. Quillardet, P., Huisman, O., D'Ari, R., and Hofnung, M., SOS chromotest, a direct assay of induction of an SOS function in *Escherichia coli* K-12 to measure genotoxicity, *Proceedings of the National Academy of Sciences of the United States of America*, 79, 5971, 1982.

43. Guzzo, A. and DuBow, M. S., Construction of stable, single-copy luciferase gene fusions in *Escherichia coli*, *Archives of Microbiology*, 156, 444, 1991.

44. Guzzo, A., Diorio, C., and DuBow, M. S., Transcription of the *Escherichia coli fli*C gene is regulated by metal ions, *Applied Environmental Microbiology*, 57, 2255, 1991.

45. Diorio, C., Cai, J., Marmor, J., Shinder, R., and DuBow, M. S., An *Escherichia coli* chromosomal *ars* operon homolog is functional in arsenic detoxification and is conserved in Gram-negative bacteria, *Journal of Bacteriology*, 177, 2050, 1995.

46. Guzzo, A. and DuBow, M. S., Identification and characterization of genetically programmed responses to toxic metal exposure in *Escherichia coli*, *FEMS Microbiology Reviews*, 14, 369, 1994.
47. Briscoe, S. F., Diorio, C., and DuBow, M. S., Luminescent biosensors for the detection of tributyltin and dimethyl sulfoxide and the elucidation of their mechanisms of toxicity, in *Environmental Biotechnology: Principles and Practice*, Moo-Young, M., Anderson, W. A., and Chakrabarty, A. M., Eds., Kluwer Academic Publishers, Dordrecht, 1996, 645.
48. Blaise, C., Microbiotests in aquatic ecotoxicology: characteristics, utility, and prospects, *Environmental Toxicology and Water Quality*, 6, 145, 1991.
49. Nystrom, T. and Neidhardt, F. C., Cloning, mapping and nucleotide sequencing of a gene encoding a universal stress protein in *Escherichia coli*, *Molecular Microbiology*, 6, 3187, 1992.
50. Nystrom, T. and Neidhardt, F. C., Expression and role of the universal stress protein, UspA, of *Escherichia coli* during growth arrest, *Molecular Microbiology*, 11, 537, 1994.
51. Seeger, C., Poulsen, C., and Dandanell, G., Identification and characterization of genes (*xap*A, *xap*B, and *xap*R) involved in xanthosine catabolism in *Escherichia coli*, *Journal of Bacteriology*, 177, 5506, 1995.
52. Sanz-Medel, A., Beyond total element analysis of biological systems with atomic spectrometric techniques, *Analyst*, 120, 799, 1995.

CHAPTER 12

Stress-Responsive Luminous Bacteria for Toxicity and Genotoxicity Monitoring

Shimshon Belkin

CONTENTS

I. INTRODUCTION

A. Genetically Engineered Microorganisms as Biosensors

With the increasing need for real-time toxicity testing and the resulting advances in microscale assays, bacteria have gained increasing importance as test organisms. The obvious factors which combine to make bacteria highly useful tools for toxicity bioassays have been often pointed out in the past:[1] large population sizes, low growth and maintenance costs, high reproducibility, and fast responses. Bacteria possess, however, an additional characteristic that conveys another unique advantage: being amenable to sophisticated genetic manipulations, they can be "programmed" to respond in a specific manner to different classes of target compounds. This may allow at least a partial combination, in a single organism, of the two traditional approaches to environmental monitoring: analytical identification of the pollutant and a bioassay for the determination of its effects. Thus, an observed response should both indicate the existence of a toxic compound as well as provide some information on its nature.

Several components have to be included in the design of such a microbial biosensor; foremost of these are the sensing mechanism, the reporting element, and the host cell. Each of the first two may be coded for by a genetic element from a different organism, but both should be chosen and/or designed so that they are properly linked and expressed in the third. In the biosensors described in this chapter *Escherichia coli* has been chosen as the host, for reasons based mostly on convenience of manipulation and the wealth of available genetic, biochemical, and physiological information.

Recent advances in microbial genetics allow us to select both sensing and reporting elements from a battery of readily available options. The choice of the sensing element determines the specificity of the response, and the selection of the reporter dictates the signal detection methodology. We have chosen to use bioluminescence as our reporting system, employing *lux* genes of *Vibrio fischeri*. As opposed to other genetic reporting tools, bioluminescence is unique in that it allows real-time monitoring of the signal. This methodology has recently been utilized in several cases to construct microbial tools capable of sensitively reporting on the presence of naphthalene,[2] hydrophobic pollutants,[3] mercury,[4] or other heavy metals.[5-7] In our case, however, we have used sensors of a more general type, able to detect a wide range of chemical pollutants; rather than selecting highly specific systems as in the examples above, we have chosen *E. coli*-derived promoters of global regulatory stress circuits. Both of these elements, the stress promoters and the bioluminescence genes, were combined in a single plasmid, with which *E. coli* could be transformed into the desired biosensor.[8-13]

B. Bioluminescence as a Reporter System

Luminous bacteria are naturally occurring light-emitting microorganisms, mostly of marine origin. The luminous systems of a few of these species have been extensively investigated on the biochemical as well as on the molecular levels.[14,15] The *lux* system of *V. fischeri*, probably the best-studied microbial bioluminescence apparatus, is composed of seven genes. *LuxA* and *B* code for the protein actually involved in the light emission process — the luciferase; the *luxC,D*, and *E* gene products are responsible for the synthesis of an aldehyde, the luciferase substrate; *luxR* and *luxI* are part of the regulatory system controlling transcription of the *luxCDABE* operon. Since light is easy to monitor and quantify, *lux* genes are gaining increasing usage as real-time reporter systems for many cellular functions.[16,17] As a result, plasmids containing *lux* genes in different configurations are available. The two most common types are those containing either *luxAB* or *luxCDABE*. For visualization of the bioluminescent expression of the former, but not of the latter plasmid, an aldehyde has to be added to the experimental setup. For this reason, in the construction of the strains described in this chapter we have used a "self-sufficient" plasmid, pUCD615,[18] containing the entire five-genes operon. Since these genes were placed under the control of a different promoter, there was no need in the final construct for the regulatory elements, *luxR* and *luxI*.

C. Global Regulatory Circuits and Their Significance

Bacteria have developed a variety of systems to regulate gene expression in response to changes in environmental conditions. While some of these are very specific in nature, others exhibit a broader response: some stimuli simultaneously activate many previously silent operons, leading to *de novo* synthesis of many proteins. Though in many cases the roles of some of these proteins remain unknown, it is clear that their synthesis is controlled by a single regulatory protein, and that as a group they are meant to combat different aspects of a single type of cellular damage. Such regulatory systems are known as global circuits, or global stress responses. A few examples of global circuits of potential environmental significance are the "heat shock" response to temperature elevation and many chemical insults,[19] the *oxyR* and *soxRS* systems dealing with oxidative damage,[20-22] the *rpoS* regulon activated by starvation/stationary phase conditions,[23] and the SOS system for combating DNA damage.[24]

The strategy adopted for the construction of the *E. coli* strains described in this chapter was to select, as the sensing element, promoters under the control of different global regulatory circuits. Once the environmental insult is "felt," and the global defense system is activated, protein synthesis should be initiated at all the coregulated promoters. Among these would be the selected promoter, fused to the *lux* reporter; its activation will lead to the synthesis of functional Lux proteins and hence to light generation. The applicability of the approach to the monitoring and quantifying of microbial stress responses, including those caused by environmentally significant chemicals, has been demonstrated.[8-13] This chapter is dedicated to its implementation in microscale ecotoxicity testing.

II. EXPERIMENTAL

A. Design and Construction of the Desired Microbial Sensor

Details concerning the actual design and construction of the desired *lux* fusions have been reported[8-13] or will be published elsewhere, and are not within the scope of this chapter. The general approach was based upon the following stages:

- Select a sequenced *E. coli* gene, under the control of the desired regulatory circuit.
- Obtain the sequence, design and synthesize suitable PCR primers, and amplify the desired segment from *E. coli* DNA.
- Ligate the PCR product into the multiple cloning site of plasmid pUCD615, upstream of the *lux* genes.
- Transform *E. coli*, selecting for kanamycin resistance and for basal luminescence as indicators of successful transformation (in most cases, some baseline luminescence could be detected even in the uninduced organism).
- Check for luminescence induction under specific and general stress conditions.

B. Assay Procedures

The procedure described below has been developed to suit luminescence measurement with a microplate luminometer, an instrument which allows real-time monitoring of luminescence in all 96 wells of an opaque microtiter plate. Only minor modifications would be needed, however, to adapt the systems to other available light-measuring devices. All assays described below were carried out at 26°C, a compromise between the optimal temperature for the host (37°C) and that for the luminescence apparatus (<20°C). This was also the temperature at which, prior to the assay, the bacteria were grown overnight with shaking in LB broth.[25] The culture was then diluted to approximately 10^7 cells/mL and regrown under the same conditions for two to three generations.

All samples tested were at pH 7.0; when necessary, wastewater samples were clarified by filtration (0.22 μm pore size). A twofold dilution series in LB was prepared in opaque white microtiter plates (Dynatech, VA, USA), to a final volume of 50 μL per well. LB broth served as a control. To each of the wells, 50 μL of the early exponential cell suspension were added, and the plates were incubated in a temperature-controlled (26°C) microtiter plate luminometer. Two such instruments were used in the course of this study — a Dynatech (VA, USA) ML3000 and an Anthos

Table 12.1   *Escherichia coli* Luminescent Biosensors

Expected Stimulus	Regulatory Gene	Circuit	Promoter Used	Strain	Reference
Protein damage	*rpo*H	Heat shock	*grp*E	TV1061	8
DNA damage	*lex*A, *rec*A	SOS	*rec*A	DPD2794	28
Oxidative stress					
peroxides	*oxy*R		*kat*G	DPD2511	29
superoxide	*sox*RS		*mic*F	DPD2515	29
Membrane damage	*fad*R	Fatty acid synthesis	*fab*A	DPD2540	29
Any	unknown	"universal"	*usp*A	DE135	12

(Salzburg, Austria) Lucy 1. Both instruments are linear for over six decades of measurement and allow continuous real-time monitoring of the emitted light.

Luminescence values are presented as arbitrary relative light units (RLU). In most cases, the results reported here were obtained with the Dynatech instrument; since different RLUs were produced by each of the luminometers, the use of the Anthos type is reported when relevant. In some of the cases, responses are reported as the ratio of the luminescence of the induced sample to that of the uninduced control (response ratio). Results of independent experiments rarely varied by more than 15%.

Metabolic activation by a rat liver microsomal fraction was conducted using freeze-dried preparations obtained from Molecular Toxicology, Inc. (Annapolis, MD, USA), according to standard activation procedures.[26,27]

C. Tester Strains

Based on the principles discussed above, a large number of plasmids was constructed, combining different *E. coli* stress promoters with the *V. fischeri lux* genes.[8-13] Six of those are referred to in this communication, and are described in Table 12.1. Further modifications to the system, not listed in this table, are presently being investigated.[29-31] These include variations in the host bacterium (*E. coli* strains exhibiting enhanced sensitivities, or representatives from other genera), luminescence genes from sources other than *V. fischeri*, and integration of the fusion into the bacterial chromosome, rather than its residence on a plasmid.

III. DATA MANIPULATION

A. The EC$_{200}$ Concept and Calculation of Sample Toxicity

All the constructs listed in Table 12.1 respond to their designated environmental challenges by increased light production. However, both baseline and induced luminescence levels varied greatly among the different strains, depending upon the regulatory circuit, the promoter used, and the stimulus applied. It is thus advisable to follow the kinetics of luminescence development for at least two hours after exposure to the toxicant and, for toxicity calculation, choose a luminescence value at a time point suitable for each strain (see below). The availability of microtiter plate luminometers allows the collection of such data in real time for all 96 wells, thereby facilitating the simultaneous screening of many combinations of samples, concentrations, and tester strains.

The example in Figure 12.1 depicts the luminescent response of strain DPD2511 to a series of concentrations of the organic peroxide cumene hydroperoxide. The data presented reflect the actual

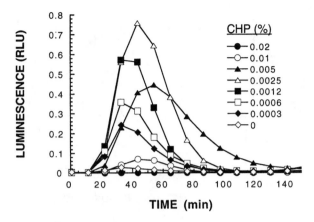

Figure 12.1 Induction of luminescence in strain DPD2511 by cumene hydroperoxide.

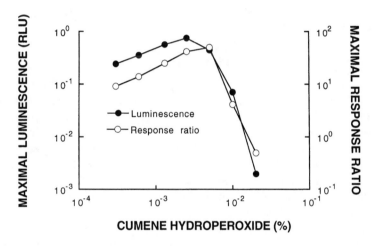

Figure 12.2 Peak responses of strain DPD2511 to cumene hydroperoxide.

luminescence measured; the same data may also be presented as the ratio to the uninduced control (not shown).

In order to quantify the inductive effect of any compound on a studied construct, a dose-dependency has to be established. For this purpose, the luminescence at a single time point may be plotted as a function of sample concentration. Since, however, the lag periods preceding light development may also vary between sample concentrations, it is often more convenient to use, as a basis for further calculations, the maximal luminescence observed during the assay period. For the response of strain DPD2511 to cumene hydroperoxide (Figure 12.1), that would be the peak observed at 30 to 60 minutes; for other constructs, the peak may occur at later times, or there may be no decline in luminescence for several hours.

The peak response values from Figure 12.1 are plotted in Figure 12.2 as a function of sample concentration, in two modes: measured maximal luminescence values reached and calculated maximal response ratios.There are several possible ways to handle these data so that a specific value can be assigned to each compound/tester strain combination. This value should denote the toxicity of the tested sample as well as the sensitivity to this compound of the test organism. In an analogy to the widely used EC_{50}[32-35] notation, we have chosen to define an EC_{200} value. In a similar manner in which EC_{50} denotes the sample concentration causing a 50% *decrease* in the measured activity, EC_{200} depicts the sample concentration causing a twofold *increase* in luminescence. An

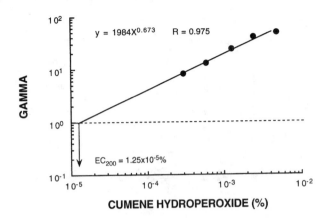

Figure 12.3   Calculation of EC_{200} for cumene hydroperoxide, with DPD2511 as the tester strain.

EC_{200} determination can be conveniently carried out using a gamma (Γ) plot as for EC_{50}; in this case, however, gamma is calculated as:

$$\Gamma = \left(I_s - I_o\right)/I_o \qquad (12.1)$$

I_s is the maximal luminescence obtained for the given sample concentration s, I_o is the luminescence of the control at the same time point, and s = EC_{200} when $\Gamma = 1$. Both luminescence and response data can be used for calculation with similar results; in the latter case, I_s is the response ratio for concentration s, and I_o is equal to 1. Similar to the EC_{50} concept, lower EC_{200} values signify higher toxicities.

In Figure 12.3, data from Figures 12.1 and 12.2 representing the effect of cumene hydroperoxide on strain DPD2511, are used as an example for EC_{200} determination. The general form of the best fit line to the gamma values can be given as:

$$Y = aX^b \qquad (12.2)$$

From this equation, EC_{200} values can be calculated by solving for X for Y = 1; alternatively, the value can be extracted graphically from the intercept of the best-fit curve with the $\Gamma = 1$ line. In the example in Figure 12.3, an EC_{200} of 0.13 mg/L is obtained. This value is over tenfold lower than the reported Microtox™* EC_{50} for this compound.[35]

B. Data Selection Criteria

Several criteria were arbitrarily set in order to ensure the reproducibility and validity of the calculated EC_{200} values: (1) at least four sample concentrations showing an enhancement in luminescence are required for calculation; (2) at least two of those should have a gamma value higher than one; and (3) the R value for the best-fit curve should be at least 0.97. At the present communication, only data answering to these criteria are presented. It should be emphasized, however, that different data selection criteria may be employed, modifying the stringency requirements set upon the calculations. For instance, the desire to avoid false positives would lead to stricter criteria, whereas if false negatives are to be avoided looser restrictions could be applied.

* Registered trademark of AZUR Environmental, Carlsbad, CA.

IV. EXAMPLES

A. "Designer" Strains: Specific Responses to Specific Stimuli

Figure 12.4 presents light development in four strains, each in response to induction by a compound of the type expected to activate the selected regulatory circuit in that construct (see Table 12.1). The response kinetics varied, but in each case a very marked increase in luminescence above the background level was observed: Strain DPD2511, the peroxide sensor, responded to hydrogen peroxide; DPD2515, the superoxide sensor, was greatly induced by the presence of paraquat, a redox-cycling herbicide which generates oxygen radicals; TV1061 luminesced in response to ethanol, the "classic" chemical inducer of the heat shock response; DPD2794, the SOS construct, reacted strongly to mitomycin C, a genotoxicant. In addition, Figure 12.5 summarizes the response ratios of five constructs to two oxidants, cumene hydroperoxide and paraquat, as a function of inducer concentration. Three strains responded to the peroxide: DPD2511, which was "designed" to sense such stress, TV1061, since the heat-shock apparatus is induced by peroxides,[19] and DPD2794 due to the suspected genotoxic effect of the inducer. The pattern was markedly different when the same five strains were challenged with paraquat. A minimal response was exhibited by all strains, but that of DPD2515, designed to respond to superoxides (intracellularly generated by paraquat), was nearly a thousand-fold greater than all the rest. Clearly, in some cases there is a broader specificity of the response, so that some induction occurs even without an intended activation of the control circuit controlling the *lux* genes. Nevertheless, the response of the strains designed to be induced by a specific type of compounds is significantly higher than in the others.

B. Phenols, Halomethanes, and Oxidants

Table 12.2 presents EC_{200} values obtained using two bacterial strains for three groups of compounds of environmental significance: phenols, halomethanes, and oxidants. The two constructs used were the "peroxide sensor" DPD2511 and the "heat-shock sensor" TV1061.[8] Also exhibited are the Microtox EC_{50} values for the same compounds. As has already been published for TV1061,[8,9,12] it can be seen from the data in Table 12.2 that it sensitively responds to all of the compounds tested. This may be explained by the universality of the heat-shock response, activated upon exposure to many environmental insults. Strain TV1061 may thus be considered a general toxicity sensor, fulfilling a function similar to that of the Microtox assay, but with a generally higher sensitivity.

A different response pattern is exhibited by strain DPD2511. It expectedly responded to the three oxidants tested, while remaining indifferent to the presence of halomethanes (except for a slight response to methylene bromide). As seen for strain TV1061, the response of strain DPD2511 was elicited at lower concentrations than that detected by the Microtox assay. It may also be observed that while the response to cumene hydroperoxide of both TV1061 and DPD2511 was of a similar magnitude (Figure 12.5), the sensitivity of the latter was much higher (as evidenced by a 30-fold lower EC_{200}). This serves as an example of a general trend often observed with the use of the different constructs: the specific responses are mostly much more sensitive than the general ones. Curiously, a very clear and sensitive luminescence induction in strain DPD2511 was also caused by the various phenols tested. This phenomenon is presently under investigation; preliminary indications point to the involvement of these compounds in the exertion of oxidative stress on *E. coli*.

C. Genotoxicants

Several of the promoters selected to drive luminescence in the presented panel are inducible by potential DNA damage (see Table 12.1). As such, they may serve as convenient biosensors for the presence of potential genotoxicants. Since the response may be observed and quantified within a few hours, the approach may prove to be useful for the rapid screening of numerous samples.

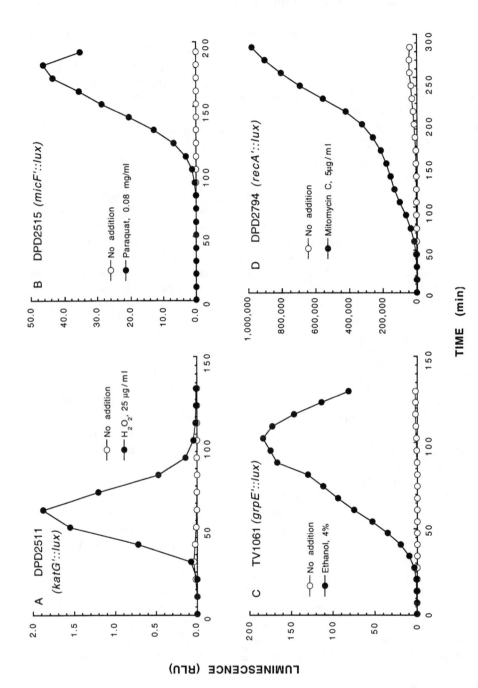

Figure 12.4   Responses of four different tester strains to their "designated" inducers. Data for panel D were obtained using Anthos Lucy 1.

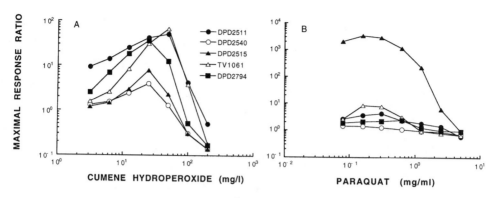

Figure 12.5 Responses of five different strains (see Table 12.1 for details) to cumene hydroperoxide (A) and paraquat (B).

Table 12.2 Calculated EC$_{200}$ Values for Various Phenols, Halomethanes, and Oxidants

Compound	EPA PP[a]	DPD2511 (*katG'::lux*) EC$_{200}$	TV1061 (*grpE'::lux*) EC$_{200}$	Microtox EC$_{50}$[c]
Phenol	+	11.5	8.2	21–36
4–Nitrophenol	+	5.5	2.6	10–14
2–Nitrophenol	+	2.9	11	35
2–Chlorophenol	+	11	9	21–34
4–Bromophenol	−	0.04	0.1	0.5
Methylene chloride	+	−	170	1000–2900
Methylene bromide	−	±[b]	185	NA
Bromodichloromethane	+	−	14.7	NA
Bromoform	+	−	1.0	NA
Chloroform	+	−	30	433
Methyl viologen	−	20	±	780–1300
Hydrogen peroxide	−	0.1	1	—
Cumene hydroperoxide	−	0.13	3.8	1.5

[a] US Environmental Protection Agency Priority Pollutant.
[b] A positive yet weak reaction.
[c] Microtox data from Kaiser, K. L. E. and Palabrica, V. S., *Water Poll. Res. J. Canada,* 26, 361, 1991. NA = data not available.

More costly and elaborate tests may then be applied to positively responding samples, suspected of genotoxicity. One such example is presented in Figure 12.4D, which depicts the effect of mitomycin C, a known SOS-inducing agent, on strain DPD2794 (*recA'::lux*). An additional example is presented in Figure 12.6, in which the effects on the same strain of two compounds, 1- and 2-nitropropane, are described. The results reflect the response ratio for each of the compounds, with or without metabolic (S9) activation, calculated three hours after exposure. This short period sufficed to yield results which were in excellent agreement with published *Salmonella* mutagenicity test data:[36] 1-nitropropane does not show any activity, while 2-nitropropane was positive both with and without metabolic activation, but more so in the activated case.

V. INDUSTRIAL WASTEWATER TESTING USING A PANEL APPROACH

As described in the introduction, one of the appealing aspects of the system is the potential ability not only to discern the existence of toxicity, but also to provide some indication as to its

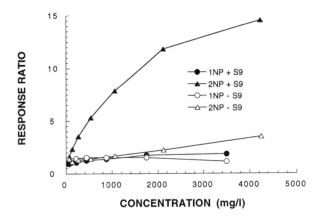

Figure 12.6 Response of the genotoxicant sensor strain DPD2794 to 1- and 2-nitropropane, with or without S9 metabolic activation. Data obtained using Anthos Lucy 1.

Table 12.3   Toxicity of a Chemical Industrial Wastewater Before and After Treatment

| | | Influent | | |
Strain	Stress Sensed	Induction[a] (EC$_{200}$)	Toxicity[b] (EC$_{50}$)	Effluent
DPD2511 (*katG'::lux*)	Oxidative (peroxides)		0.45	ND[c]
DPD2515 (*micF'::lux*)	Oxidative (oxygen radicals)	2	+[d]	ND
DPD2794 (*recA'::lux*)	DNA damage	1.8	+	ND
TV1061 (*grpE'::lux*)	"Heat shock"		1.5	ND
DE135 (*uspA'::lux*)	General stress		1.4	ND
Microtox	General toxicity		1.8	ND

Note: All values are in percent (v/v).

[a] EC$_{200}$ calculated as described in the text.
[b] EC$_{50}$ calculated from the inhibition of baseline luminescence.
[c] No toxicity detected.
[d] Toxicity evident, but EC$_{50}$ not calculable due to masking by luminescence induction.

nature. Such information can be obtained by the parallel use of several indicator strains. In order to test this possibility, chemical industrial wastewater streams were tested using a panel of selected strains. As an example, representative results for the influent and effluent of a biological wastewater treatment plant of a large chemical factory are presented in Table 12.3. In addition to the EC$_{200}$ parameter calculated as described above, a conventional EC$_{50}$ value was also determined, based upon the decrease in baseline luminescence of the same biosensor caused by the sample. Also presented are the Microtox data obtained for the same samples.

All five tester strains were positive for the influents and negative for the effluents, indicating that whatever toxic compounds are present in the raw wastewaters were effectively removed or neutralized during treatment. All five tests indicated that the raw sample was toxic, causing an inhibition of luminescence. For two strains, the toxicity EC$_{50}$ values were very close to that of the Microtox test — 1.5% and 1.4% for the heat shock and universal stress sensors, respectively, compared to 1.8% for Microtox; in another, the peroxide sensor, the value was lower (0.45%). In two strains, however, while a toxic response could still be observed, it was at least partially masked by a clear and specific inductive response. In both cases, the oxygen radical (DPD2515) and the DNA damage (DPD2794) sensors, very similar EC_{200} values were calculated: 2% and 1.8%, respectively. It is tempting to hypothesize that the same compounds which were responsible for the creation of the oxidative hazard are the ones which induced the SOS system; this, however, was not proven. What was nevertheless very clear was that the use of the five-membered panel allowed the determination of the potential oxidative and genotoxic nature of the sample, which could have remained hidden if a single "general toxicity" indicator strain was used.

VI. SUMMARY

The relative ease with which bacteria can be genetically manipulated to quantitatively respond to certain stimuli opens a new horizon for the monitoring of ecotoxicity. By the fusion of sensor and reporter elements, bacteria can be "programmed" to emit a danger signal upon sensing either general or specific environmental stress factors. In this chapter, one approach to the exploitation of this potential was described: the fusion of *Vibrio fischeri* bioluminescence (*lux*) genes to promoters of *Escherichia coli* global regulatory stress-circuits. The resulting constructs, upon exposure to environmental hazards, emit light which is easy to monitor and quantify. The response may be of varying specificity, depending upon the nature of the promoter, but in all cases is very rapid: normally a very clear pattern emerges after one or two hours. There are several advantages to the described approach, beyond those shared with other microbial-based microscale assays. The most important of these are data availability in real time, the ability to discern toxic effects of specific types (such as oxidative agents), and the potential for a very rapid and low-cost determination of genotoxicity potential.

The strains described in this chapter are but a small and preliminary example of the potential inherent in the ability to design the bacterial response according to environmental needs. Better, more sensitive, or just different bioassays can be designed according to different requirements, in which each of the biosensor elements may be modified or replaced. Such work is presently being conducted.

ACKNOWLEDGMENTS

A large part of the work summarized here was carried out in the laboratory of, and in cooperation with, R. A. LaRossa from DuPont, Central Research and Development, Wilmington, DE, USA. His contribution, and that of his co-workers, T. K. Van Dyk, D. R. Smulski, and A. C. Vollmer (Swarthmore College, PA, USA) were in many ways more significant than that of the author and are gratefully acknowledged.

REFERENCES

1. Bitton G. and Dutka, B. J., Introduction and review of microbial and biochemical toxicity screening procedures, in *Toxicity Testing Using Microorganisms*, Bitton, G. and Dutka, B. J., Eds., CRC Press, Boca Raton, FL, 1986, 1.
2. King, J. M. H., DiGrazia, P. M., Applegate, B., Burlage, R., Sanseverino, J., Dunbar, P., Larimer, F., and Sayler, G. S., Rapid, sensitive bioluminescent reporter technology for naphthalene exposure and biodegradation, *Science*, 249, 778, 1990.

3. Selifonova, O. V. and Eaton, R. W., Use of an *ipb-lux* fusion to study regulation of the isopropylbenzene catabolism operon of *Pseudomonas putida* RE204 and detect hydrophobic pollutants in the environment, *Appl. Environ. Microbiol.,* 58, 4068, 1995.

4. Selifonova, O., Burlage, R., and Barkay, T., Bioluminescent sensors for detection of bioavailable Hg (II) in the environment, *Appl. Environ. Microbiol.,* 59, 3083, 1993.

5. Corbisier, P., Ji, G., Nuyts, G., Mergeay, M., and Silver, S. *LuxAB* gene fusions with the arsenic and cadmium resistance operons of *Staphylococcus aureus* plasmid pI258, *FEMS Microbiol. Lett.,* 110, 231, 1993.

6. Guzzo, A. and DuBow, M. S., A *luxAB* transcriptional fusion to the cryptic *celF* gene of *E. coli* displays increased luminescence in the presence of nickel, *Mol. Gen. Genet.,* 242, 455, 1994.

7. Guzzo, J., Guzzo, A., and DuBow, M. S., Characterization of the effects of aluminium on luciferase biosensors for the detection of ecotoxicity, *Toxicol. Lett.,* 64-65, 687, 1992.

8. Van Dyk, T. K., Majarian, W. R., Konstantinov, K. B., Young, R. M., Dhurjati, P. S., and LaRossa, R. A., Rapid and sensitive pollutant detection by induction of heat shock gene-bioluminescence gene fusions, *Appl. Environ. Microbiol.,* 60, 1414, 1994.

9. Van Dyk, T. K., Belkin, S., Vollmer, A. C., Smulski, D. R., Reed, T. R., and LaRossa, R. A., Fusions of *Vibrio fischeri lux* genes to *Escherichia coli* stress promoters: detection of environmental stress, in *Bioluminescence and Chemiluminescence: Fundamentals and Applied Aspects,* Campbell, A. K., Kricka, L. J., and Stanley, P. E., Eds., John Wiley & Sons, Chichester, 1994, 147.

10. Belkin, S., Vollmer, A. C., Van Dyk, T. K., Smulski, D. R., Reed, T. R., and LaRossa, R. A., Oxidative and DNA damaging agents induce luminescence in *E. coli* harboring *lux* fusions to stress promoters, in *Bioluminescence and Chemiluminescence: Fundamentals and Applied Aspects,* Campbell, A. K., Kricka, L. J., and Stanley, P. E., Eds., John Wiley & Sons, Chichester, 1994, 509.

11. Van Dyk, T. K., Reed, T. R., Vollmer, A. C., and LaRossa, R. A., Synergistic induction of the heat shock response in *Escherichia coli* by simultaneous treatment with chemical inducers., *J. Bacteriol.,* 177, 6001, 1995.

12. Van Dyk, T. K., Smulski, D. R., Reed, T. R., Belkin, S., Vollmer, A. C., and LaRossa, R. A., Responses to toxicants of an *Escherichia coli* strain carrying a *uspA'::lux* genetic fusion and an *E. coli* strain carrying a *grpE'::lux* fusion are similar, *Appl. Environ. Microbiol.,* 61, 4124, 1995.

13. Belkin, S., Van Dyk, T. K., Vollmer, A. C., Smulski, D. R., and LaRossa, R. A., Monitoring sub-toxic environmental hazards by stress-responsive luminous bacteria, *Environ. Toxicol. Water Qual.,* 11, 179, 1996.

14. Meighen E.M. and Dunlap, P.V., Physiological, biochemical and genetic control of bacterial bioluminescence, *Adv. Microb. Physiol.,* 34, 1, 1993.

15. Meighen, E. A., Genetics of bacterial bioluminescence, *Annu. Rev. Genet.,* 28, 117, 1994.

16. Stewart G.S.A.B. and Williams, P., *lux* genes and the applications of bacterial bioluminescence, *J. Gen. Microbiol.,* 138, 1289, 1992.

17. Chatterjee, J. and Meighen, E. A., Biotechnological applications of bacterial bioluminescence (*lux*) genes, *Photochem. Photobiol.,* 62, 641, 1995.

18. Rogowsky, P.M., Close, T.J., Chimera, J. A., Shaw, J. J., and Kado, C.I., Regulation of the *vir* genes of *Agrobacterium tumefaciens* plasmid pTiC58, *J. Bacteriol.,* 169, 5101, 1987.

19. Yura, T., Nagai, H. and Mori, H., Regulation of the heat-shock response in bacteria, *Annu. Rev. Microbiol.,* 47, 321, 1993.

20. Storz, G., Tartaglia, L. A., Farr, S. B., and Ames, B. N., Bacterial defenses against oxidative stress, *Trends Genet.,* 6, 363, 1990.

21. Demple, B., Regulation of bacterial oxidative stress genes, *Ann. Rev. Genet.,* 25, 315, 1991

22. Storz, G. and Altuvia, S., OxyR regulon, *Methods Enzymol.,* 234, 217, 1994.

23. Loewen, P. C. and Hengge-Aronis, R., The role of the sigma factor σ^s (KatF) in bacterial global regulation, *Annu. Rev. Microbiol.,* 48, 53, 1994.

24. Walker, G. C., The SOS response of *Escherichia coli,* in *Escherichia coli and Salmonella typhimurium: Cellular and Molecular Biology,* Neidhardt, F. C., Ingraham, J. L., Low, K. B., Magasanik, B., Schaechter, M., and Umbarger, H. E., Eds., ASM, Washington, DC, 1987, 1346.

25. Miller, J.H., *Experiments in Molecular Genetics,* Cold Spring Harbor Laboratory Press, Cold Spring Harbor, NY, 1972.

26. Ames, B. N., McCann, J., and Yamasaki, E., Methods for detecting carcinogens and mutagens with the *Salmonella*/mammalian-microsome mutagenicity test, *Mutat. Res.*, 31, 347, 1975.
27. Maron, D. M. and Ames, B. N., Revised methods for the *Salmonella* mutagenicity test, *Mutat. Res.*, 113, 173, 1983.
28. Vollmer, A. C., Belkin, S., Smulski, D. R., Van Dyk, T. K., and LaRossa, R. A., unpublished, 1996.
29. Belkin, S., Smulski, D. R., Vollmer, A. C., Van Dyk, T. K., and LaRossa, R. A., unpublished, 1996.
30. Van Dyk, T. K. et al., unpublished, 1996.
31. Elsmore, D. A. et al., unpublished, 1996.
32. Bulich, A.,A., Bioluminescence assays, in *Toxicity Testing Using Microorganisms*, Bitton, G. and Dutka, B. J., Eds., CRC Press, Boca Raton, FL, 1986, 57.
33. Qureshi, A. A. and Bulich, A. A., Microtox toxicity test systems — where they stand today, *Microscale Testing in Aquatic Toxicology: Advances, Techniques, and Practice,* Wells, Peter G., Lee, Kenneth, and Blaise, Christian, Eds., CRC Press, Boca Raton, FL, 1998, Chap. 13.
34. Ribo, J. M. and Kaiser, K. L. E., *Photobacterium phosphoreum* toxicity bioassay, I. Test procedures and applications, *Tox. Assess.*, 2, 305, 1987.
35. Kaiser, K. L. E. and Palabrica, V. S., *Photobacterium phosphoreum* toxicity data index, *Water Poll. Res. J. Canada*, 26, 361, 1991.
36. Haworth, S., Lawlor, T., Mortelmans, K., Speck, W., and Zeiger, E., *Salmonella* mutagenicity test results for 250 chemicals, *Environ. Mutagen. Suppl.*, 1, 3, 1983.

CHAPTER 13

Microtox* Toxicity Test Systems — Where They Stand Today

Ansar A. Qureshi, Anthony A. Bulich, and Don L. Isenberg

CONTENTS

I. BACKGROUND

One of the first microbiotests to be commercialized (Bulich 1979) was Beckman Instruments' Microtox system.** It resulted from an industry plea for an acute aquatic toxicity bioassay which would be better than the 96-h fish test as a work-a-day method for measuring and controlling general aquatic toxicity. Beckman's answer to this seemed outrageous to many, at first; what does an instrument have to do with bioassays? The world now knows. Based on bacterial luminescence as it reflects the overall health of the organisms and as measured by a photometer, the Microtox test system gave fast, cost-effective answers to the question — "How toxic is this sample?" Because of the simplicity, speed, economy, convenience, reproducibility, and other virtues of the test, the Microtox test is now one of the most thoroughly characterized and validated aquatic bioassays in

---

* Microtox® is a registered trademark of AZUR Environmental, Carlsbad, California.
** The rights to the Microtox patents, trademarks, and technology were purchased from Beckman in 1985 by Microbics Corporation, which in 1996 changed its name to AZUR Environmental.

the world today, engendering more than 500 publications which testify to its place in the arena of toxicity testing.

This chapter provides an overview of the development and evolution of the Microtox test and summarizes some of the important findings from the Microtox literature as well as regulatory and standards achievements. The Microtox test, as originally presented, still flourishes and will be the focus of this chapter. However, new procedures, new tests, and new platforms are continually being developed and will also be discussed.

II. GENERAL DESCRIPTION OF THE MICROTOX TEST

The Microtox test is based on measuring changes in the light emitted by a nonpathogenic naturally luminescent marine bacterium (*Vibrio fischeri*,* NRRL B-11117) upon exposure to a toxic substance or sample containing toxic materials. The Microtox test is a short-term acute toxicity bioassay that combines the advantages of a biological test with the speed and ease of use of a laboratory instrument.

The usual expression of sample toxicity measured by the Microtox test is EC_{50}: the effective concentration of a sample that causes a 50% decrease in the light output of the test organisms under controlled experimental conditions (normally after 15 min at 15°C). According to Environment Canada's standard test method (1992), the Microtox results should be reported in terms of an inhibitory concentration (IC) which measures the quantitative reduction in light production by test bacteria relative to control. For practical purposes EC_{50} and IC_{50} are considered synonymous, but in most of the Microtox literature the investigators have used the term EC_{50}. Additionally, the Microtox test results, particularly those of screening tests, can simply be expressed qualitatively as: toxic or nontoxic, positive or negative, and presence or absence of toxicity.

In the Microtox test, like other toxicity bioassays, the relationship between the exposure time and bacterial response (in terms of light production) is dependent on the nature of the specific compound or sample being tested. For example, some samples and chemicals need a longer time to interact with the test organisms, while others react immediately to provide a response within seconds of exposure. Therefore, based on the calculated effect, the Microtox test results can be described as EC_{80}, EC_{50}, EC_{20}, and EC_{10} representing 80, 50, 20, and 10% effects, respectively. Similarly EC_5 and EC_1 could be used to represent the lowest observed effect concentration (LOEC). Furthermore, based on the exposure time, the Microtox test results may also be expressed as EC_{50} (5), EC_{50} (15), EC_{20} (5), and EC_{20} (15) to indicate 50 and 20% effects after 5 and 15 min exposure of a sample to the test organism.

The Microtox test can be extended to 30 or 60 minutes to allow the determination of effects of slightly longer exposure to a specific chemical or particular sample. In practice, however, for many chemicals and samples no significant differences have been observed between the 5 and 30 min EC_{50} values. Alternatively, Qureshi et al. (1984), while studying the toxicity of various metallic ions, observed that EC_{50} values decreased substantially up to 15 min with little change occurring between 15 and 30 min exposure. *Based on their results, Qureshi* et al. *(1984) recommended the 15 min EC_{50} as the standard for toxicity testing and assessment of all chemicals, effluents, and natural samples.* Although in the literature both the EC_{50} (5) and EC_{50} (15) data have been provided by various investigators, the use of the EC_{50} (15) endpoint has been widely accepted and now applied as a standard (norm) for expressing the results of Microtox testing. Also with respect to Microtox data interpretation, it should be noted that (similar to other toxicity bioassays) high values (e.g., EC_{50} of 70%) would indicate lower toxicity, and inversely, low values (e.g., EC_{50} of 12%) are indicative of high toxicity.

* Originally identified as *Photobacterium phosphoreum*, it has recently been classified as a strain of *Vibrio fischeri*. Deposit with the Northern Regional Research Laboratory (NRRL), Peoria, IL, makes the strain freely available, making classification somewhat academic.

The Microtox test is a simple, rapid, reproducible, sensitive, practical, and cost-effective bio-assay. It has been extensively used worldwide for over 18 years for toxicity screening of chemicals and effluents, water quality and sediment contamination surveys, and environmental risk assessment. More specifically the Microtox test has been effectively used in the toxicity monitoring and assessment of complex industrial effluents, domestic wastewaters, sewage and sludges, lake and river waters, agricultural and storm water runoffs, leachates, and aqueous extracts of contaminated soils and sediments, groundwater, drilling muds and sump fluids, diverse industrial inorganic and organic chemicals, herbicides, pesticides, mycotoxins, landfill leachates and mixtures of contami-nants and chemicals (Bulich 1979, 1982; Bulich et al., 1981; Curtis et al., 1982; Qureshi et al., 1982; Yates and Porter, 1982; Plotkin and Ram, 1984; Liu and Dutka, 1984; Strosher, 1984; Bitton and Dutka, 1986; Ribo and Kaiser, 1987; Blaise et al., 1988; Kaiser and Ribo, 1988; Kaiser et al., 1988; E.V.S. Consultant, 1989; Mazidji et al., 1990; Blaise, 1991; Hankenson and Schaeffer, 1991; Kaiser and Palabrica, 1991; Richardson, 1993; Bengtsson and Triet, 1994; Galli et al., 1994; Gaggi et al., 1995; Cook and Wells, 1996; Gosh et al., 1996; Newman and McClosky, 1996; McClosky et al., 1996).

The Microtox test has also been used extensively to produce toxicity data for the prediction assessment, and relationships of chemicals (through computer modeling) based on quantitative structure–activity relationships (QSAR) (Kaiser and Ribo, 1985; Kaiser et al., 1984, 1987; Kaiser, 1993; Zhao et al., 1993; Shultz and Cronin, 1997). Also, it is a useful predictor of the outcomes of other bioassays, chemical testing, and process changes. The study of potential interactions of combinations of toxic substances generally present in industrial effluents would not be feasible with the fish acute lethality test, but could be conducted using the Microtox assay (Michaud et al., 1990). In fact, Qureshi et al. (1984) examined toxicity patterns of binary metal mixtures and found that various combinations of metallic ions exhibited a variety of synergistic, additive, and antago-nistic responses with the Microtox test.

III. STANDARDIZATION OF MICROTOX METHODS

The Microtox test is performed by reconstituting freeze-dried reagent (containing about 10^8 bacteria/vial) and determining the initial light emission (before the addition of test sample) of homogenized and stabilized luminescent bacterial suspensions. Appropriate aliquots of osmotically adjusted sample dilutions are then added to bacterial suspensions, and light output measurements are made at specific intervals (mostly 5 and 15 min, or perhaps after 30 min for slow-acting toxicants). The light readings are corrected according to change in the dilution control (blank) to allow for natural time-dependent drifts in light output and small dilution effects. The EC_{50} and other desirable endpoints are calculated by log-linear plotting of sample concentration (dose) versus percent light decrease (response) or more precisely by log-log plotting of gamma versus concen-tration (Johnson et al., 1974). In practice, the gamma (which is the corrected ratio of the amount of light lost to the amount remaining) and corresponding EC_{50} values are calculated using various computer programs and data reduction systems.

The Microtox test procedures were described by Bulich and co-workers (1979, 1980, 1981, 1982) and were detailed in the original Operating Manual (#015-55879). Numerous investigators used these test procedures for toxicity assessment of diverse chemicals, complex effluents, and a wide variety of environmental samples. As a result, within a short period of time, extensive Microtox data became available in the published literature, and some in unpublished work. Unfortunately, considerable variation and discrepancy were observed in the EC_{50} values of various toxicants reported by different investigators. Qureshi et al. (1984) and Greene et al. (1985) initially defined the assessment of factors which could introduce variability in the early Microtox data. These factors include: (1) use of different compounds/formulations of specific toxicants; (2) different endpoints and exposure time (i.e., 5, 10, 15, 20, or 30 min); (3) methods used in measuring/determining

Table 13.1   Major Sources of Standardized Methods for the Microtox Bioassay

1. Microtox Assay Procedure, Part 3, Section 2. *Microbiological Methods Manual.* AECV90-M2. A.A. Qureshi (Ed.) 1990. Alberta Environmental Centre, Vegreville, AB.
2. Luminescent Bacteria — Microtox DIN38412-L34, 1991. Deutsche Institute für Normung (DIN) — German Institute for Standardization (Hansen, 1993).
3. Microtox Manual, Vol. 1 to 5. 1992. Microbics Corporation, Carlsbad, CA.
4. Environment Canada (EC) 1992. *Biological Test Method: Toxicity Test Using Luminescent Bacteria* (*Photobacterium phosphoreum*), EPS1/RM/24.
5. Western Canada Microtox User Committee (WCMUC) *Standard Procedure for Microtox Analysis.* AECV 94-G1. I.D. Gaudet (Ed.) 1994. Alberta Environmental Centre, Vegreville, AB.
6. APHA, AWWA, WEF 1995 *Standard Methods for the Examination of Water and Wastewater,* Part 8050 Bacterial Luminescence.
7. ASTM 1995 *Standard Test Method for Assessing the Microbial Detoxification of Chemically Contaminated Water and Soil Using a Toxicity Test with a Luminescent Marine Bacterium.* D-5660-95.
8. Microtox Bioassay, Test Requirements and Specifications, Appendix 4. Guide 50 Drilling Waste Management. 1996. Alberta Energy and Utilities Board (AEUB), Calgary, AB.

toxicant concentrations; (4) operator inaccuracies in diluting and pipetting samples; (5) sample pH and solubility; (6) age of bacterial reagent; (7) salt concentration in osmotic adjustment; (8) test temperature; (9) improper storage and handling of bacterial and other reagents; and (10) dissimilar data reduction and analysis procedures.

Subsequently, various investigators (Qureshi et al., 1984; Vasseur et al., 1984a; Yates and Porter, 1984; Greene et al., 1985; Ribo and Kaiser, 1987; Ghosh and Doctor, 1992; Carlson-Ekvall and Morrison, 1995) critically studied and evaluated the test methods, procedures, conditions, and various physical and chemical factors influencing the Microtox data and results. Based on the findings and recommendations of these investigators, several initiatives were undertaken by various agencies and groups to develop standardized Microtox test methods and procedures. After extensive literature reviews, evaluation of various test procedures and variables, and several interlaboratory comparisons, the Microtox test methods have been standardized in both the U.S. and Canada, as well as internationally. As a result, the Microtox test has been accepted and adopted for toxicity testing in many countries. The standard Microtox procedures are available in the open literature and also from organizations that have developed the standardized methodologies. Some of the major sources of standardized methods for conducting Microtox toxicity testing are summarized in Table 13.1.

It should be emphasized that, while the standardized Microtox methods from various sources have many common elements and procedures, they differ somewhat from each other in scope, content, experimental setup, protocol types, and test results applications. Nevertheless, the use of standardized Microtox methods and procedures has improved test precision.

As a positive outcome of the test methods standardization, there is now a general agreement on the types of Microtox protocols that can be used for toxicity assessment of chemicals, wastes, and environmental samples. The selection of the appropriate procedure depends, among other factors, on the type and physical characteristics of the sample, relative toxicity of the sample, information needed according to testing objectives, and intended application of the results. Some of the standardized Microtox protocols are: 1) screening test (2, 32, and 90% concentrations); 2) basic test (45% concentration); 3) basic test (82% concentration); 4) 100% test (90% concentration); 5) inhibition test (90% concentration); and 6) comparison test (82% concentration).

Specific details for all of these Microtox protocols may be found in the Microtox Manual (Vols. 2 and 3 of AZUR Environmental) as well as the listed sources and documents of standardized methods summarized in Table 13.1.

The precision, sensitivity and general robustness of the Microtox system has led to the incorporation of variations of the basic Microtox test into several written standards and governmental regulations. The initial work of Casseri et al. (1983) was the basis for the first request for a consensus standard for the Microtox test, which eventually resulted in ASTM Standard D-5660 (ASTM, 1995). Over the years, many standards and regulations have been finalized or are now in process, which are based upon the Microtox acute test method. These are summarized in Table 13.2.

Table 13.2   Microtox Status — Regulations and Standards

Organization	Application	Status
Deutsches Institut für Normung — Germany	Effluent testing	Standard 9/91
Sverige Natur Verkar — Sweden	Effluent testing	Issued 4/90
Netherlands Normalization Institut — Netherlands	Effluent testing	Final 6/93
Inter-governmental Aquatic Toxicity Group — Canada	Effluent testing	Final 1/92
Energy Resources Conservation Board — Canada*	Drilling waste testing	Guide 6/93
International Standards Organization-International	Effluent testing	In Process TC 147/SC
Environment Agency — UK	Effluent testing	In Process
L'Association Francaise de Normalisation — France	Effluent testing	Standard 8/91
National Government Laboratory and Research Institute — Italy	Effluent testing	In Process
Environmental Protection Agency — Spain	Soil leachates	Standard 1991
Environmental Protection Agency — Mexico	Wastewater	Standard 10/96 Sedesol
United States Public Health Service	Wastewater	Issued 10/95 Standard Methods #8050
American Society for Testing and Materials — US	Wastewater	Issued 2/96 D-5660
American Society for Testing and Materials — US	Sediments	In Process E47.01
American Society for Testing and Materials — US	Sediments	In Process E47.03
United States Environmental Protection Agency	Effluent testing	In Process

* Guide 50 Drilling Waste Management (10/96), Alberta Energy and Utilities Board — Canada.

IV. ADVANTAGES OF THE MICROTOX TEST

Of the many different microbiotests, the Microtox test has been most widely used worldwide for toxicity screening and assessment of chemicals, wastes, leachates, effluents, and diverse environmental samples. Many researchers and investigators favor the use of the Microtox test as an effective alternative to fish and other aquatic bioassays for a variety of reasons. Indeed there are many attractive features of the Microtox test which have contributed to its present widespread utility. Major advantages of the Microtox test are listed in Table 13.3.

V. PRECISION OF THE MICROTOX TEST

The precision of the Microtox test has been addressed in many studies. Results of such studies, involving both the inter- and intralaboratory trials, demonstrated excellent reproducibility (between-lab precision) and repeatability (within-lab precision) of the Microtox test. There seems to be a broad consensus from numerous studies reporting the Microtox test coefficients of variation (CV) between 10 and 20%.

The precision of the Microtox assay was evaluated in an exhaustive study conducted by the Canadian Petroleum Association (Strosher 1984) which involved the testing of 29 waste drilling fluids by three laboratories. The average CV for EC_{50} (5) and EC_{50} (15) values were 11 and 13%, respectively, with a maximum of 31% for both endpoints. As a comparison, in the same study, the results of fish bioassays from the three laboratories showed a maximum CV of 98% with a mean of 30%.

Another study by Casseri et al. (1983) found the Microtox data to be very reproducible (with CVs between 5 and 10%) not only for duplicated tests, but also for testing conducted at different times on split samples of effluents. Also, Vasseur et al. (1984a) reported CVs ranging from 3 to 20% (with an average of 12%) based on the Microtox testing of 39 effluents. Curtis et al. (1982) examined 68 chemicals for toxicity assessment using the Microtox test. Based on duplicate tests of seven of these chemicals, they found that overall replicates deviated from the EC_{50} (5) values by only 10%. In a similar study involving pure compounds, De Zwart and Slooff (1983) reported a CV of 10% for their Microtox reproducibility data. Vasseur et al. (1984b) reported the average CV of 27.6% for all EC_{50} values obtained from triplicate testing of 55 different water samples.

Table 13.3   Major Advantages of the Microtox Test

- Easy to use and convenient
- Simple, fast, and practical
- Highly sensitive, reliable, and reproducible
- Precise and accurate data/information
- High degree of standardization
- Excellent quality control
- Suitable for interlab round robins
- Economical (low cost/test, labor saving)
- Time/Cost-effective
- Fast turn-around time; results in two hours or less
- Instant test capability (lab testing or field studies)
- Ease of test reagent availability and storage
- Bacterial reagent availability, consistency, and stability (long term use)
- Excellent correlation with common acute toxicity tests
- Applicable to testing of solid and liquid samples
- Applicable to testing of highly turbid and colored samples
- Statistical advantage in using large number (10^5) of test organisms
- Allows screening of large number of samples in relatively short time
- Requires small sample volumes (2.0 mL)
- Requires little lab space (<10 sq. ft.)
- Does not require elaborate lab facilities
- Specialized expertise/skilled manpower not essential
- Suitable for compliance monitoring/new product testing
- Ecologically and environmentally relevant
- Applicable and practical for regulatory testing
- Suitable for toxicity identification evaluation (TIE) and toxicity reduction evaluation (TRE) studies
- One test for many applications

Walker (1988) reviewed the Microtox data for phenol from eight different laboratories and found that the $EC_{50}(5)$ values ranged from 22.0 to 40.2 mg/L with a mean of 26.2 ± 6.3 mg/L. The CV of 21.8% was highly acceptable considering the large number of tests and diversity of laboratories involved.

One of the most comprehensive Microtox interlaboratory studies was conducted by Qureshi et al. (1987, 1990) to evaluate data reproducibility and variability. The study involved 18 laboratories in four round robins, during which five blind samples of the same toxicants were tested using the same bacterial reagent belonging to two different lots. Based on the pooled data of four round robins, 87 (96.7%) of the 90 test EC_{50} (15) values (18 labs × 5 samples) were within X ± 2 S.D. limits, indicating excellent data reproducibility between laboratories. The CV for the pooled data set ranged from 14.29 to 18.57, while the overall CV (regardless of sample) was 17.8%. These results further demonstrated excellent precision of the Microtox data produced by a very large number (18) of diverse laboratories.

With respect to the Microtox test repeatability (within-laboratory precision), many of the laboratories involved in Microtox testing now use zinc sulfate (as the reference toxicant) to produce internal quality control (QC) data and also to evaluate variability among different batches of the bacterial reagent. The EC_{50} (15) data, obtained from replicate analyses, are then used to develop and establish QC charts for monitoring laboratory precision and performance of the Microtox test. Such charts include upper and lower warning limits (UWL, LWL) and upper and lower control limits (UCL, LCL) based respectively on 2± and 3± standard deviations of long-term means.

VI. COMPARISON OF THE MICROTOX TEST SYSTEMS WITH OTHER BIOASSAYS

The Microtox test has been used and compared with other toxicity bioassays in numerous studies during the last several years. In fact, most of the available data on correlations between

Table 13.4   Summary of Microtox Correlation Coefficients with Three Common Acute Toxicity Bioassays

Bioassays	Correlation Coefficient (r)	References
Fathead minnows	0.41, 0.80, 0.80, 0.85, 0.85, 0.85, 0.86, 0.90, 0.91, 1.00	Chang et al., 1981; Lebsack et al., 1981; Curtis et al., 1982; Indorato et al., 1984; Kaiser and Esterby, 1991; Isenberg, 1993.
Rainbow trout	0.74, 0.81, 0.84, 0.85, 0.89	Lebsack et al., 1981; Ribo and Kaiser, 1983; Strosher et al., 1984; Kaiser and Esterby, 1991; Isenberg, 1993.
Daphnids	0.80, 0.85, 0.85, 0.85, 0.85 0.85, 0.86, 0.87	Ribo and Kaiser, 1983; Kaiser and Esterby, 1991; Isenberg, 1993.

microbiotests and other aquatic toxicity bioassays deal with the Microtox test. Most of these comparative investigations, from laboratories around the world, have evaluated and quantitated the relative sensitivity and correlations of the Microtox test toward pure chemicals (both organic and inorganic), industrial and municipal effluents, and diverse and complex environmental samples.

The Microtox comparative studies have involved the use of over 50 different test organisms, species, and systems, but have focused mainly on the three most common acute lethality bioassays, i.e., rainbow trout, fathead minnow, and daphnids. To date, the sensitivity of the Microtox test has been quantified for over 1300 individual compounds (Kaiser and Palabrica, 1991).

While conducting comparative studies, most of the investigators elected to compare results of various bioassays (involving diverse test organisms) using correlation coefficients (r). In some cases they reported results in terms of percent agreement using various assessment and classification systems (e.g., the log units ranking system used by Bulich, 1982). Various other authors summarized LC_{50} and EC_{50} data but did not quantitate the correlations and instead made general comments about data comparability.

A description of all studies which have compared the Microtox test with at least one other acute toxicity bioassay is exhaustive and certainly outside the scope of this chapter. For additional detailed information and to appreciate the quality and breadth of Microtox comparative data, interested readers are directed to consult many excellent reviews and articles available in the literature. Several relevant publications and specific papers include: Bulich et al., 1981; Chang et al., 1981; Dutka and Kwan, 1981, 1982, 1984; Lebsack et al., 1981; Curtis et al., 1982; Qureshi et al., 1982; Casseri et al., 1983; DeZwart and Sloof, 1983; Dutka et al., 1983; McFeters et al., 1983; Ribo and Kaiser, 1983; Sloof et al., 1983; Liu and Dutka, 1984; Strosher et al., 1984; Vasseur et al., 1984a; Coleman and Qureshi, 1985; Greene et al., 1985; Blaise et al., 1988; Elnabarawy et al., 1988; E.V.S. Consultant, 1989; Blaise, 1991; Kaiser and Esterby, 1991; Kaiser and Palabrica, 1991; Munkittrick et al., 1991; Fort, 1992; Richardson, 1993; Kwan and Dutka, 1995; Toussaint et al., 1995; and Vismara et al., 1996.

Results of comparative studies and correlations between various toxicity tests must be interpreted carefully because a perfect correlation is neither desirable nor necessarily the optimum case. Instead, a clear evidence of colinearity, short of complete equivalency, is important. Such a result would demonstrate that any given two tests have certain general similarities (i.e., they both measure and indicate toxicity), but that each also has its own particular characteristics (i.e., high sensitivity to certain groups of compounds and low sensitivity to other types of samples). In the context of correlations, Table 13.4 summarizes the correlation coefficients (r) of Microtox results with three other common acute toxicity bioassays as obtained in two dozen independent investigations (Isenberg, 1993). The r values ranged from 0.41 to 1.00, 0.80 to 0.87, and 0.74 to 0.89 for fathead minnows, daphnids, and rainbow trout bioassays, respectively. The average correlation of 85% (r = 0.85) appears to be as good as for any interspecies correlation and demonstrates the same or increased sensitivity range for the Microtox test as obtained with other traditional toxicity bioassays.

While correlation coefficients give an indication of the degree of similarity of two parallel sets of results, they provide no information on relative sensitivity of two toxicity tests. The Microtox test has been found to be too sensitive by some investigators, not sensitive enough by others, but sufficiently sensitive by most investigators. The level of sensitivity can become a legitimate concern in a threshold detection test (e.g., rainbow trout), where a simple yes or no answer is often sufficient. Microtox, however, is a quantitative test in which each use determines baseline toxicity limits to serve the needs and objectives of specific applications. Regarding the sensitivity issue, as pointed out by Isenberg (1993), "too sensitive" had less to do with the Microtox technology than with politics.

In general, the data sets of bioassay comparative studies invariably show Microtox as being more sensitive to certain groups of chemicals and less or equally sensitive to others. It appears that the Microtox reagent demonstrates increased sensitivity to some organic chemicals because they can cross cell walls of bacteria easily and rapidly, while with other compounds (like metals) somewhat decreased sensitivity is observed because of the presence of salt in the test medium (Hinwood and McCormick, 1987). Irrespective of the type of toxicant, however, the results of most of the comparative studies indicated that the relative (average) sensitivity of the Microtox test is well within the same order of magnitude as the sensitivity of other toxicity bioassays. In this regard, it should be emphasized that Microtox EC_{50} values were used in most of these sensitivity comparisons. The degree of Microtox sensitivity can easily be increased by selecting another endpoint. For example, if chosen, the EC_{20} values will be 2.5 times more sensitive than the EC_{50} data. Also, such a selection of more sensitive endpoints is possible with the Microtox assay, because it is a quantitative and functional test which involves measuring the integrated response of approximately one million individual bacterial cells during each experimental unit exposure (Ross, 1993).

In an excellent review, Munkittrick et al. (1991) compared the relative sensitivity of Microtox to daphnids, rainbow trout, and fathead minnow acute lethality bioassays for the toxicity assessment of various chemical compounds, complex effluents, sediments, and other environmental samples. The study results suggested that despite considerable differences and variability in the relative sensitivity of the Microtox assay and the other three acute lethality bioassays, the Microtox test appears to be the best available choice for rapid screening and assessment (presence or absence) of toxicity of diverse environmental samples, pure compounds, and complex effluents.

VII. USES AND APPLICATIONS OF THE MICROTOX TEST SYSTEM

Since the development of the Microtox test in 1979 (Bulich, 1979), microbiologists, ecologists, biologists, ecotoxicologists, regulators, and other investigators worldwide have used it extensively for toxicity assessment of chemicals, wastewaters, industrial effluents, and a wide variety of environmental samples. The Microtox test system has also been used in water quality monitoring, soil extract testing, contaminated sediment, or site surveys and environmental impact and risk assessment studies (Matthews et al., 1987; Symons et al., 1988; U.S. EPA, 1989; Loehr, 1989; Giesy et al., 1989; Dombroski et al., 1996).

In industrial applications, the Microtox test has been proved to be a useful predictor of the outcome of other bioassays, chemical testing, and process changes. Microtox could also be used effectively as a screening test for the monitoring and testing of large numbers of samples in a very cost-effective manner, or as an early warning system (EWS) to detect the presence of toxic materials in the aquatic environment before they can cause adverse effects (Coleman and Qureshi, 1985). The EWS applications may include the detection of abruptly increased concentrations of toxicants in effluents, process change impacts, waste spills, and for the monitoring of wastewater and

Table 13.5   Applications and Uses of the Microtox Test

- Rapid toxicity screening of complex effluents and receiving waters
- Influent monitoring and biomass protection in water and wastewater treatment plants
- Toxicity testing of sewage, sludges, and contaminated soils and sediments
- Toxicity monitoring and evaluation of agricultural, storm water, and combined sewer runoffs
- Toxicity assessment of groundwater, surface waters, and drinking waters
- Toxicity screening of inorganics and organic chemicals individually and based on quantitative structure-activity relationships (QSAR)
- Prediction of other bioassay results and process changes
- Determining the toxicity of potentially hazardous wastes
- Screening and testing of diverse environmental samples
- Toxicity screening of septage and landfill leachates
- Monitoring efficiency and effectiveness of drinking water treatment operations and systems
- Detection and prediction of ocular irritancy in substances and products designed for industrial, pharmaceutical, and cosmetic uses
- Environmental effects monitoring (EEM)
- Regulatory decision-making and compliance monitoring
- Toxicity screening of drilling muds and additives and sump waste fluids
- Detection and control of toxic trade effluents and spillages
- Identification and prioritization of effluents and discharges for toxicity-based control
- Comparison and correlations with other traditional and nontraditional toxicity bioassays
- Ecotoxicological monitoring and testing
- Toxicity identification evaluation (TIE) and toxicity reduction evaluation (TRE) studies and investigations
- Toxicity assessment of fossil fuel process waters and phenolic constituents
- Toxicity testing of biocides, pesticides, herbicides, and bactericides
- Industrial waste streams and process control monitoring
- Quality control monitoring of raw materials and new chemicals and formulations
- Determining toxicity of plasticizers, stabilizers, antioxidants, and surfactants
- Toxicity screening of mycotoxins and biological toxins
- Provision of marine toxicity data for offshore chemical control schemes
- Toxicity testing of wastewater and effluents from plastics, resins, wood, pulp and paper, textile, and leather industrial sectors
- Determining and assessing synergistic, additive, and antagonistic effects of metal mixtures and other toxicants present in aquatic environments
- Toxicity detection of wastewaters associated with oil refinery and petroleum industry
- Biological effects monitoring of air and pollutants in livestock operations
- Bioreactivity and safety evaluation and assessment of biomaterials
- Research and developmental studies in diverse areas
- Evaluation and assessment of genotoxicity and chronic toxicity

industrial plant influents and source waters for drinking water supplies. The Microtox technology has also been applied to nonenvironmental uses such as an *in vitro* alternative to animal testing (Bulich et al., 1989).

The multiplicity of uses and applications have resulted in the production of voluminous literature on the Microtox test comprising over 500 publications. For details and specific information on particular applications, readers are directed to many of the references cited in Section II. Some of the major uses and applications of the Microtox test are listed in Table 13.5. The variety of samples that have been analyzed by the Microtox systems is not only unique, but cannot be matched by any other toxicity bioassay.

VIII. NEW DEVELOPMENTS BASED ON LUMINESCENT BACTERIA

The use of light as a biological endpoint offers many advantages, as demonstrated by the success of the Microtox acute bioassay. Over the years many additional Microtox bioassays and specialized systems have been developed to take advantage of this unique attribute of luminescent bacteria. This section will briefly describe some of these new and exciting developments.

Methods have been provided for adapting the Microtox system to direct contact measurement of toxicity of soils, sediments, and sludges. The procedures avoid the necessity of using solvent extraction or the use of pore water, although these practices are not incompatible with the Microtox test and are still used for specific situations and applications. The Microtox Solid-Phase Tests (SPT) provide direct contact between test organisms and sample particles, increasing the probability for the measurement of the responses to particle bound and marginally soluble toxicants. Although care must be taken in interpreting results of the Microtox Solid-Phase test because of possible nonspecific interferences, a procedure is provided for correcting results to minimize these potential nonspecific factors. This procedure uses a presumed nontoxic sample of soil or sediment, with characteristics similar to the unknown sample, for the correction. In fact such a correction is only necessary for samples having marginal toxicity. Additionally the Microtox SPT has been found to be useful for toxicity assessment of contaminated sediments (Kwan and Dutka, 1995; Day et al., 1995; Cook and Wells, 1996).

Another major development was the Mutatox®* genotoxicity test system. This bioassay is designed to detect the effect of DNA-damaging agents by measuring the light output of a specially selected dark variant of *Photobacterium leiognathi* (Ulitzur, 1986). When these organisms are grown in appropriate sample concentrations, they begin to produce light in proportion to the presence of genotoxicity after an incubation period of 16 to 24 hours. The test may be performed with and without S-9 activation. The Mutatox test has been used and compared with other assays for genotoxicity screening (Kwan et al., 1990; Legault et al., 1994; Jarvis et al., 1996).

With currently increasing concern over chronic toxicity, the Microtox Chronic Toxicity Test (Bulich et al., 1996) offers sensitivity comparable to the popular *Ceriodaphnia* chronic test over a wide range of toxicants, with better precision and faster results. The Microtox chronic test is a 22-h growth inhibition/light induction assay, with light output once again being the endpoint. Software is provided for calculating the lowest observed effect concentration (LOEC) and the highest no observed effect concentration (NOEC).

In addition to these new tests the Microtox family of instrumental platforms is increasing. The original photometer included in the Microtox system introduced in 1979, the Model 2055, was a benchtop, research-oriented instrument. In 1988, the Microtox Model 500 was introduced as a more dedicated and simpler-to-operate instrument with more throughput capacity. The methods and protocols mentioned above were designed to be compatible and work with both instruments.

The first departure from benchtop-oriented systems resulted from a joint development between the Microbics Corporation and Compagnie Générale des Eaux (a major French water company). The system has been in use since 1990 for continuous, on-line, unattended monitoring of surface waters for the purpose of protecting drinking water sources. The systems are monitored and controlled by remote access. More recently a joint development between AZUR Environmental, Seimens Environmental plc, and Yorkshire Water plc (a major U.K. water company) was announced that would extend the same on-line capability to monitoring influents and effluents to and from water and sewage treatment plants, as well as in-plant process control. This system, the Microtox®-OS On-line System, is currently undergoing beta site testing and is expected to be commercially available in late 1997. The Microtox-OS is designed to sample and test at 15-min intervals and will operate unattended for 14 days.

Finally, a major new capability is the development and introduction of the DeltaTox™** PS1 Test System. It combines a new, field-portable instrument, having an unusually wide photometric dynamic range, with a specially selected strain of *Photobacterium leiognathi*. The focus of this system is onsite toxicity screening and nonregulatory applications for monitoring the quality of influents and wastewaters. It can be easily and conveniently used in remote field environments at ambient (sample) temperatures (10 to 35°C). Its dynamic range and dual function capability make

---

* Mutatox is a registered trademark of AZUR Environmental.
** DeltaTox is a trademark of AZUR Environmental.

it suitable, not only for toxicity screening, but also for biomass estimation with conventional ATP reagents and methods.

IX. CONCLUSIONS

The Microtox system for acute, aquatic toxicity testing, first introduced in 1979, pioneered commercial availability of microbiotests, and thereby created a new, exciting, and viable market. Its virtues of simplicity, fast results, economy, precision, and flexibility of adapting to specific applications have established a new testing standard. Most of all, the Microtox acute test is generally standardized worldwide and has the largest pure compound database of any aquatic toxicity bioassay. The embodied technology has spawned additional capabilities which extend the utility to other important test systems for measuring the responses of bioreactive substances which routinely invade our precarious ecosystems. It is also evident that the Microtox and other biolumi-nescence-based assays will continue to be important as an integral component of multispecies and multitrophic level toxicity tests for environmental monitoring, regulations, and ecotoxicological investigations.

ACKNOWLEDGMENTS

Selected portions and content of this work were originally included in a paper prepared, as part of the Government of Canada Contract No. K2608-5-0016 for Environment Canada, by Ansar Qureshi (AQ) while he was with Norwest Labs, Edmonton, Alberta, Canada. The support of these two organizations is gratefully acknowledged by AQ.

REFERENCES

APHA. 1995. *Standard Methods for the Examination of Water and Wastewater*, Part 8050 Bacterial Biolumi-nescence. A.D. Eaton, L.D. Clesceri and A.E. Greenberg (Eds.) 19th ed., Washington, DC

ASTM. 1995. Standard Method for Assessing the Microbial Detoxification of Chemically Contaminated Water and Soil Using a Toxicity Test With a Luminescent Marine Bacterium. D5660-95. American Society for Testing and Materials, Philadelphia, PA.

Bengtsson, B-E. and T. Triet. 1994. Tapioca starch wastewater characterized by Microtox and duckweed tests. *Ambio* 23(8):471-477.

Bitton, G. and B.J. Dutka. (Eds.). 1986. *Toxicity Testing Using Microorganisms*, Vol 1. CRC Press, Boca Raton, FL, 163 pp.

Blaise, C. 1991. Microbiotests in aquatic ecotoxicology: characteristics, utility and prospects. *Environ. Toxicol. Water Quality.* 6:145-155.

Blaise, C., G. Sergy, P. Wells, N. Bermingham and R. Van Coillie. 1988. Biological testing -Development, application, and trends in Canadian environmental protection laboratories. *Tox. Assess.* 3:385-406.

Bulich, A.A. 1979. Use of luminescent bacteria for determining toxicity in aquatic environments. In *Aquatic Toxicology: Second Conference*, L.L. Marking and R.A. Kimerle (Eds.), ASTM STP 667. American Society for Testing and Materials, Philadelphia, PA. pp. 98-106.

Bulich, A.A. 1982. A practical and reliable method for monitoring the toxicity of aquatic samples. *Process Biochem.* March/April: 45-47.

Bulich, A.A. and D.L. Isenberg. 1980. Use of the luminescent bacterial system for the rapid assessment of aquatic toxicology. *Adv. Instrum.* 35:35-40.

Bulich, A.A., M.M. Greene and D.L. Isenberg. 1981. Reliability of the bacterial luminescence assay for determination of the toxicity of pure compounds and complex effluents. In *Aquatic Toxicology and Hazard Assessment: Fourth Conference*, D.R. Branson and K.L. Dickson (Eds.), ASTM STP 737, American Society for Testing and Materials, Philadelphia, PA., pp. 338-347.

Bulich, A.A., K. Tung and G. Scheibner. 1989. The luminescent bacteria toxicity test: its potential as an *in vitro* alternative. *J. Biolum. Chemilum.* 5:71-77.

Bulich, A.A., H. Huynh, and S. Ulitzur. 1996. The use of luminescent bacteria for measuring chronic toxicity. In *Techniques in Aquatic Toxicology*, G. K. Ostrander (Ed.), CRC Press, Boca Raton, FL, pp. 3-12.

Carson-Ekvall, C. E. A. and G. M. Morrison. 1995. Contact toxicity of metals in sewage sludges; evaluation of alternatives to sodium chloride in the Microtox® assay. *Environ. Tox. Chem.* 14(1):17-22.

Casseri, N.A., W.C. Ying and S.A. Sojka. 1983. Use of a rapid bioassay for the assessment of industrial wastewater treatment effectiveness. In *Proceedings of the 38th Purdue Industrial Wastewater Conference.* J.M. Bell (Ed.), Butterworth Publishers, Woburn, MA. pp. 867-878.

Chang, J.C., P.B. Taylor and F.R. Leach. 1981. Use of the Microtox assay system for environmental samples. *Bull. Environ. Contam. Toxicol.* 26:150-156.

Coleman, R.N., and A.A. Qureshi. 1985. Microtox and *Spirillum volutans* tests for assessing toxicity of environmental samples. *Bull. Environ. Contam. Toxicol.* 35:443-451.

Cook, N. H. and P. G. Wells. 1996. Toxicity of Halifax harbour sediments: an evaluation of the Microtox® solid-phase test. *Water Qual. Res. J. Canada.* 31:673-708.

Curtis, C., A. Lima, S.J. Lozano and G.D. Veith. 1982. Evaluation of a bacterial bioluminescence bioassay as a method for predicting acute toxicity of organic chemicals to fish. In *Aquatic Toxicology and Hazard Assessment: Fifth Conference*, J.G. Pearson, R.B. Foster and W.E. Bishop (Eds.), ASTM STP 766, American Society for Testing and Materials, Philadelphia, PA. pp 170-178

Day, K. E., B. J. Dutka, K. K. Kwan, N. Batista, T. B. Reynoldson and J. L. Metcalfe-Smith. 1995. Correlations between solid phase microbial screening assays, whole sediment toxicity tests with macroinvertebrates and *in-situ* benthic community structure. *J. Great Lakes Res.* 21(2):192-206.

De Zwart, D. and W. Slooff. 1983. The Microtox as an alternative assay in the acute toxicity assessment of water pollutants. *Aquat. Toxicol.* 4:129-138.

Dombroski, E. C., I. D. Gaudet, L. Z. Florence and A. A. Qureshi. 1996. A comparison of techniques used to extract solid samples prior to acute toxicity analysis using the Microtox® test. *Environ. Tox. and Water Qual.: An Int'l. J.* 11:121-128.

Dutka, B.J. 1988. Proposed ranking scheme and battery of tests for evaluating hazards in Canadian waters and sediments. N.W.R.I. Contribution No. 88-80, Unpublished Manuscript.

Dutka, B.J. and K.K. Kwan. 1981. Comparison of three microbial toxicity screening tests with the Microtox test. *Bull. Environ. Contam. Toxicol.* 27:753-757.

Dutka, B.J. and K.K. Kwan. 1982. Application of four bacterial screening procedures to assess changes in the toxicity of chemicals in mixtures. *Environ. Pollut.* 29A:125-134.

Dutka, B.J. and K.K. Kwan. 1984. Studies on a synthetic activated sludge toxicity screening procedure with comparison to three microbial toxicity tests. In *Toxicity Screening Procedures Using Bacterial Systems*, D. Liu and B.J. Dutka, (Eds.) Marcel Dekker, New York, pp. 125-138.

Dutka, B.J., N. Nyholm and J. Peterson. 1983. Comparison of several microbiological toxicity screening tests. *Water Res.* 17:1363-1368.

EPA, U.S., Environmental Research Laboratory, Corvallis, Oregon. 1989. *Ecological assessment of hazardous waste sites.* EPA-600/3-89/-13.

E.V.S. Consultant. 1989. An evaluation of the sensitivity of microassays relative to trout and daphnid acute lethality tests. Unpublished Report of the Environmental Protection Directorate, Environment Canada, 55 pp.

Elnabarawy, M.T., R.R. Robideau and S.A. Beach. 1988. Comparison of three rapid toxicity test procedures: Microtox, polytox, and activated sludge respiration inhibition. *Tox. Assess.* 3:361-370.

Energy Resources Conservation Board (ERCB). 1993. *Drilling Waste Management Guide G-50* (Guide 50 Drilling Waste Management, October 1996). ERCB (AEUB) Calgary, AB.

Environment Canada (EC). 1990C. Biological Test Method: Acute Lethality Test Using *Daphnia spp.* Conservation and Protection, Ottawa, ON. EPS I/RM/11, 57 pp.

Environment Canada (EC). 1992. Biological Test Method: Toxicity Test Using Luminescent Bacteria (*Photobacterium phosphoreum*). Conservation and Protection, Ottawa, ON. EPS 1/RM/24, 61pp.

Fort, F. L. 1992. Correlation of Microtox EC_{50} with Mouse LD50. *In vitro Toxicol.* 5(2):73-82.

Gaggi, C., G. Sbrilli, A. M. Hasab El Naby, M. Bucci, M. Duccini and E. Bacci. 1995. Toxicity and hazard ranking of triazine herbicides using Microtox®, two green algae species and a marine crustacean. *Environ. Toxicol. Chem.* 14(2):203-208.

Galli, R., C. D. Munz and R. Scholtz. 1994. Evaluation and application of aquatic toxicity tests: use of the Microtox test for the prediction of toxicity based upon concentrations of contaminants in soil. *Hydrobiologia* 273:179.

Gaudet, I.D. 1994. *WCMUC Standard Procedure for Microtox Analysis.* AECV94-G1. Alberta Environmental Centre, Vegreville, AB, 63 pp.

Ghosh, S.K. and P.B. Doctor. 1992. Toxicity screening of phenol using Microtox. *Environ. Toxicol. Water Qual.* 7:157-163.

Ghosh, S. K., P. B. Doctor and P. K. Kulkarni. 1996. Toxicity of zinc in three microbial test systems. *Environ. Toxicol. and Water Qual.* 9:13-19.

Giesy, J.P. and R.A. Hoke. 1989. Freshwater sediment toxicity bioassessment: rationale for species selection and test design. *J. Great Lakes Res.* 15(4):539-569.

Greene, J.C., W.E. Miller, M.K. Debacon, M.A. Long and C.L. Bartels. 1985. A comparison of three microbial assay procedures for measuring toxicity of chemical residues. *Arch. Environ. Contam. Toxicol.* 14:659-667.

Hansen, P.D. 1993. Regulatory significance of toxicological monitoring by summarizing effect parameters. In M. Richardson (Ed.) *Ecotoxicology Monitoring.* VCH Publishers, New York, pp. 273-286.

Hinwood, A.L., and M.L McCormick. 1987. The effect of ionic solutes on EC_{50} values measured using the Microtox test. *Tox. Assess.* 2:449-461.

Indorato, A.M., K.B. Snyder and P.B. Usinowicz. 1984. Toxicity screening using Microtox analyzer. In *Toxicity Procedures Using Bacterial Systems,* D. Liu and B.J. Dutka (Eds.). Marcel Dekker, New York, pp. 37-54.

Isenberg, D.L. 1993. The Microtox® Toxicity Test — A Developer's Commentary. In *Ecotoxicology Monitoring,* M. Richardson (Ed.). VCH Publishers, New York, pp. 3-15.

Jarvis, A. S., M. E. Honeycutt, V. A. McFarland, A. A. Bulich and H. C. Bonds. 1996. A comparison of the Ames assay and Mutatox® in assessing the mutagenic potential of contaminated dredged sediment. *Ecotoxicol. and Environ. Safety* 33:193-200.

Johnson, F.H., H. Eyring and B.J. Stover. 1974. *The Theory of Rate Processes in Biology and Medicine,* John Wiley & Sons, N.Y., 703 p.

Johnson, I., R. Butler, R. Milne and C.J. Redshaw. 1993. The role of Microtox in the monitoring and control of effluents. In *Ecotoxicology Monitoring,* M. Richardson (Ed.). VCH Publishers, New York, pp. 309-317.

Hankenson, K. and D. J. Schaffer. 1991. Microtox assay of trinitrotoluene, diaminonitrotoluene, and dinitromethylanaline mixtures. *Bull. Environ. Contam. Toxicol.* 46:550-553.

Kaiser, K.L.E. 1993. Qualitative and quantitative relationships of Microtox data with toxicity data for other aquatic species. In *Ecotoxicology Monitoring,* M. Richardson (Ed.). VCH Publishers, New York, pp. 197-211.

Kaiser, K.L.E. and J.M. Ribo. 1985. QSAR of chlorinated aromatic compounds. In *QSAR in Toxicology and Xenobiochemistry.* M. Tichy (Ed.), Elsevier, Amsterdam, pp. 27-38.

Kaiser, K.L.E. and J.M. Ribo. 1988. *Photobacterium phosphoreum* toxicity bioassay, II. Toxicity data compilation. *Tox. Assess.* 3:195-237.

Kaiser, K.L.E. and S.R. Esterby. 1991. Regression and clear cluster analysis of the acute toxicity of 267 chemicals to six species of biota and the octanol/water partition coefficient. *Sci. Total Environ.* 109/110, 499-514.

Kaiser, K.L.E. and V.S. Palabrica. 1991. *Photobacterium phosphoreum* toxicity data index. *Water Pollut. Res. J. Canada,* 26:361-431.

Kaiser, K.L.E., P.V. Hodson and D.G. Dixon. 1984. QSAR studies on chlorophenols, chlorobenzenes and para-substituted phenols. In *QSAR in Environmental Toxicology.* K.L.E. Kaiser (Ed.), D. Reidel Publ. Co. Dordrecht, pp. 189-206.

Kaiser, K.L.E., K.R. Lum and V.S. Palabrica. 1988. A review of field applications of the Microtox test in Great Lakes and St. Lawrence River Waters. *Water Poll. Res. J. Canada.* 23:270-278.

Kaiser, K.L.E., V.S. Palabrica and J.M. Ribo. 1987. QSAR of acute toxicity of mono-substituted benzene derivatives to *Photobacterium phosphoreum.* In *QSAR in Environmental Toxicology, II.* K.L.E. Kaiser (Ed.), D. Reidel Publ. Co. Dordrecht, pp. 153-168.

Kross, B.C. and K. Cherryholmes. 1993. Toxicity screening of sanitary landfill leachates: a comparative evaluation with Microtox analyses, chemical and other toxicity screening methods. In *Ecotoxicology Monitoring,* M. Richardson (Ed.). VCH Publishers, New York, pp. 225-249.

Kwan, K. K. and B. J. Dutka. 1995. Comparative assessment of two toxicity bioassays: The direct sediment toxicity testing procedure (DSTTP) and the Microtox® Solid Phase Test (SPT). *Bull. Environ. Contam. Toxicol.* 55:338-346.

Kwan, K. K., B. J. Dutka, S. S. Rao and D. Liu. 1990. Mutatox test: A new test for monitoring environmental genotoxic agents. *Environ. Pollut.* 65:323-332.

Lebsack, M.E., A.D. Anderson, G.M. DeGrueve and H.L. Bergman. 1981. Comparison of bacterial luminescence and fish bioassay results for fossil-fuel process waters and phenolic constituents. In *Aquatic Toxicology and Hazard Assessment: Fourth Conference*, D.R. Branson and K.L. Dickson (Eds.), ASTM STP 737, American Society for Testing and Materials, Philadelphia, PA. pp. 348-356.

Legault, R. C., C. Blaise, D. Rokosh and R. Chong-Kit. 1994. Comparative assessment of the SOS Chromotest and the Mutatox test with the *Salmonella* plate incorporation (Ames Test) and fluctuation tests for screening genotoxic agents. *Environ. Toxicol. and Water Qual.*, 9(1):45-57.

Liu, D. and B.J. Dutka (Eds.) 1984. *Toxicity Screening Procedures Using Bacterial Systems.* Marcel Dekker, New York, 476 pp.

Loehr, R.C. 1989. *Treatability potential for EPA listed hazardous wastes in soil.* U.S. EPA-600/S2-89/011 (Sept.).

Matthews, J.E. and L. Hastings. 1987. Evaluation of a toxicity test procedure for screening treatability potential of waste in soil. *Tox. Assess.* 2:265-281.

Mazidji, C. N., B. Koopman, G. Bitton, G. Voiland and C. Logue. 1990. Use of Microtox and *Ceriodaphnia* bioassays in wastewater fractionation. *Tox. Assess. Int. J.* 5:265-277.

McFeters, G.A., P.J. Bond, S.B. Olson and Y.T. Tchan. 1983. A comparison of microbial bioassays for the detection of aquatic toxicants. *Water Res.* 17:1757-1762.

McClosky, J. T., M. C. Newmann and S. B. Clark. 1996. Predicting the relative toxicity of metal ions using ion characteristics: Microtox® bioluminescence assay. *Environ. Toxicol. Chem.* 15(10):1730-1737.

Microtox Microbics Manual Vo. 1 to 5. 1992. Microbics Corporation (Azur Environmental), Carlsbad, CA.

Microtox Bioassay, Test Requirements and Specifications, Appendix 4 Toxicity protocols, in Guide 50 Drilling Waste Management. 1996. Alberta Energy and Utilities Board (AEUB), Calgary, AB.

Munkittrick, K.R., E.A. Power and G.A. Sergy. 1991. The relative sensitivity of Microtox®, daphnid, rainbow trout and fathead minnow acute lethality tests. *Env. Toxicol. Water Quality.* 6:35-62.

Newman, M.C. And J.T. McCloskey. 1996. Predicting relative toxicity and interactions of divalent metal ions: Microtox® bioluminescence assay. *Environ. Toxicol. Chem.* 15(3):275-281.

OECD. 1984. *Organization for Economic Cooperation and Development Guidelines for Testing of Chemicals: Earthworm Acute Toxicity Tests.* OECD Guidelines No. 207, Paris, France, 15 pp.

Plotkin, S and N.M. Ram. 1984. Multiple bioassays to assess the toxicity of a sanitary landfill leachate. *Arch. Environ. Contam. Toxicol.* 13:197-206.

Qureshi, A.A. (Ed.) 1990. Microtox Assay Procedure, Part 3, Section 2. In *Microbiological Methods Manual.* AECV90-M2. Alberta Environmental Centre, Vegreville, AB. 483 pp.

Qureshi, A.A., K.W. Flood, S.R. Thompson, S.M. Janhurst, C.S. Inniss and D. A Rokosh. 1982. Comparison of a luminescent bacterial test with other bioassays for determining toxicity of pure compounds and complex effluents. In *Aquatic Toxicology and Hazard Assessment: Fifth Conference*, J.G. Pearson, R.B. Foster and W.E. Bishop (Eds.), ASTM STP 766, American Society for Testing and Materials, Philadelphia, PA, pp. 179-195.

Qureshi, A.A., R.N. Coleman and J.H. Paran. 1984. Evaluation and refinement of the Microtox test for use in toxicity screening. In *Toxicity Procedures Using Bacterial Systems,* D. Liu and B. J. Dutka (Eds.), Marcel Dekker, New York, pp. 1-22.

Qureshi, A.A., A.K. Sharma and J.H. Paran. 1987. Microtox quality control collaborative study: A unique and enlightening experience. (Abstr. p.11). Presented at the Third International Symposium on Toxicity Testing Using Microbial Systems, Valencia, Spain, May 1987.

Ribo, J.M. and K.L.E. Kaiser. 1983. Effects of selected chemicals to photoluminescent bacteria and their correlations with acute and sublethal effects on other organisms. *Chemosphere*, 12:1421-1442.

Ribo, J.M. and K.L.E. Kaiser. 1987. *Photobacterium phosphoreum* toxicity bioassay, I. Test procedures and applications. *Tox. Assess.* 2:305-323.

Richardson, M. (Ed.) 1993. *Ecotoxicology Monitoring.* VCH Publishers, New York, 384 pp.

Ross, P. 1993. The use of bacterial luminescence systems in aquatic toxicity testing. In *Ecotoxicology Monitoring,* M. Richardson (Ed.), VCH Publishers, New York, pp. 185-195.

Schultz, T. W. and M. T. D. Cronin. 1997. Quantitative structure-activity relationships for weak acid respiratory uncouplers to *Vibrio fischeri*. *Environ. Toxicol. Chem.* 16:357-360.

Sloof, W., J.H. Canton and J.L. Hermens. 1983. Comparison of the susceptibility of 22 freshwater species to 15 chemical compounds. I. (Sub) acute toxicity tests. *Aquatic Toxicol.* 4:113-128.

Strosher, M.T. 1984. *A comparison of biological testing methods in association with chemical analyses to evaluate toxicity of waste drilling fluids in Alberta, Volume 1*, Canadian Petroleum Association, Calgary, AB. pp. 1-32.

Toussaint, M. W., T. R. Shedd, W. H. Van Der Schalie and G. R. Leather. 1995. A comparison of standard acute toxicity tests. *Environ. Tox. Chem.* 14(5):907-915.

Ulitzur, S. 1986. Bioluminescence test for genotoxic agents. In *Bioluminescence and Chemiluminescence, Methods in Enzymology*, M. A. Deluca and W. D. McElroy, Ed., Academic Press, 133:264-274.

Vasseur, P., J.F. Ferard, C. Rast and G. Larbaigt. 1984a. Luminescent marine bacteria in acute toxicity testing. In *Ecotoxicological Testing for the Marine Environment, Vol. 2*, G. Persoone, E. Jaspers and C. Claus (Eds.), State Univ. Ghent and Inst. Mar. Scient. Res., Bredene, Belgium. 2:381-396.

Vasseur, P., J.F. Ferard, C. Rast and G. Larbaigt. 1984b. Luminescent marine bacteria in ecotoxicity screening tests of complex effluents. In *Toxicity Procedures Using Bacterial Systems*, D. Liu and B.J. Dutka (Eds.), Marcel Dekker, New York, pp. 23-35.

Vismara, C., C. Rossetti, E. Bolzacchini, M. Orlandi, A. Luperini and G. Bernardini. 1996. Toxicity evaluation of 4-chloro-2-methylphenoxyacetic acid by Microtox® and comparison with FETAX. *Bull. Environ. Contam. Toxicol.* 56:85-89.

Walker, J.D. 1988. Relative sensitivity of algae, bacteria, invertebrates, and fish to phenol: Analysis of 234 tests conducted for more than 149 species. *Tox. Assess.* 3: 415-447.

Yates, I.E. and J.K. Porter. 1982. Bacterial bioluminescence as a bioassay for mycotoxins. *Appl. Environ. Microbiol.* 44:1072-1075.

Yates, I.E. and J.K. Porter. 1984. Temperature and pH affect the toxicological potential of mycotoxins in the bacterial bioluminescence assay. In *Toxicity Procedures Using Bacterial Systems*, D. Liu and B.J. Dutka (Eds.), Marcel Dekker, New York. pp. 77-88.

Young, W., R. Butler and I. Johnson. 1992. *Review of the Microtox toxicity test*. National Rivers Authority Interim Report 049/31W, 82 pp.

Zhao, Y. L., L. Wang, H. Gao and Z. Zhang. 1993. Quantitative structure-activity relationships: relationship between toxicity of organic chemicals to fish and to *Photobacterium phosphoreum*. *Chemosphere* 26(11):1997-1979.

CHAPTER 14

Microtox® Toxicity Test System — New Developments and Applications

B. Thomas Johnson

CONTENTS

I. INTRODUCTION

The present challenge in environmental toxicity is to apply objective criteria to determine which tests can be used most effectively, to agree on critical aspects of test protocols and data interpretation, and to define, as clearly as possible, how results will be interpreted in order to help identify potential areas of concern.[1,2] In the last two decades microscale bacterial *in vitro* toxicity assays have emerged as important ecotoxicological screening tools to monitor the hazards of chemical contaminants in the biosphere.[3] The Microtox Toxicity Test System which uses strains of the bioluminescent bacteria,

Table 14.1   Microtox® Toxicity Test System

System Attributes	Microtox Basic Test	Microtox Solid-Phase Toxicity Test	Microtox Chronic Toxicity Test	Mutatox Genotoxicity Test
Toxicity Test	Acute	Acute	Chronic	Genotoxic
Test Organism: *Photobacterium/Vibrio*	Selected wild-type strains	Selected wild-type strains	Selected wild-type strains	Selected dark mutant strains
Sample Type	Liquid[a]	Solid[b]	Liquid[a]	Liquid[a]
Test Medium	Buffer	Buffer	Growth nutrients	Growth nutrients[c]
Growth Phase	Stationary	Stationary	Log	Log
Experimental Design	Dose-response	Dose-response	Dose-response	Dose-response
Test Duration	≤30 min	≤30 min	<24 h	<24 h
Test Endpoint	Reduced light emission	Reduced light emission	Reduced light emission	Increased light emission
Toxicological Designation[d]	EC_{50}	EC_{50}	LOEC	Genotoxic
Computer Software	Yes	Yes	Developmental	Yes
Scientific Development	In common use	Validation — experimental phase	Introductory phase	Validation — experimental phase
Sensitivity	Broad spectrum	Experimental[e]	Experimental	High selectivity[f]
Database	Broad	Limited	Experimental	Expanding

[a] Elutriate, porewater, compatible organic solvent extracts.
[b] Soil or sediment samples.
[c] Rat hepatic S9 fractions added for metabolic activation phase.
[d] EC_{50} = effective concentration with 50% loss of light; LOEC = Lowest Observable Effect Concentration; Genotoxic = two or more positive responses/dilution series.
[e] Clay and turbidity questions.
[f] High sensitivity to polyaromatic hydrocarbons.

Photobacterium/Vibrio, has played an important role in this revolution and offers a simple, cost-effective, and sensitive alternative to traditional and costly whole-animal tests with fish and invertebrates (also see Chapter 13).

The Microtox Toxicity Test System (Table 14.1) includes four toxicity tests: the Microtox Acute Toxicity Test, the Microtox Solid-Phase Toxicity Test, the Microtox Chronic Toxicity Test, and the Mutatox Genotoxicity Test. An extensive and expanding bibliography is available on computer disk from the manufacturer, AZUR Environmental in Carlsbad, California. The tests are primarily used as Tier I screening assays to detect the presence of toxic substances in the biosphere — water, soil, sediment, and air. The bioassays differ in bioluminescent strain selection (wild-type vs. dark mutant), sample presentation (liquid vs. solid phase), growth cycle (stationary vs. log phase), duration (minutes vs. hours), bioluminescent emissions (decrease vs. increase), and toxicological endpoints (lethality vs. genotoxicity). It must be emphasized that this system is a series of toxicity tests, not surrogate toxicity tests for fish or invertebrates (i.e., a test to simulate an animal response); no ecological relevance should be applied or implied by comparing toxicity data from animals and bacteria, nor should any be concluded.[4,5] The number of test organisms, cell wall composition, lipid content, metabolism, life cycles, and bioavailability potential make such comparison spurious. The Microtox Toxicity Test System is designed to detect biotoxicity in the laboratory and in the field.

Environmental risk assessments are traditionally made with toxicity bioassays. The assay essentially consists of four components: the organism, the sample, the exposure, and the analysis — all predicated on a dose-response experimental design. This chapter will present an overview of the Microtox Toxicity Test System, highlights of the four individual tests, and topics such as assay validation, sediment toxicity, collection of water-, sediment-, and airborne samples, contaminant bioavailability, tandem uses of toxicological bioassays, metabolic activation of toxicants, and photoenhanced chemicals. Discussion will emphasize new approaches and applications of the Microtox System and focus on methods for the collection and detection of environmental biotoxins used at the Environmental and Contaminants Research Center (ECRC), Columbia, MO.

IN VITRO TOXICITY TEST: EXPERIMENTAL DESIGN

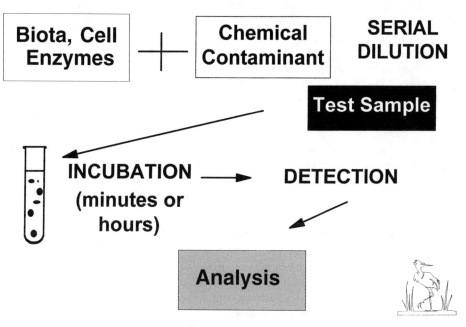

Figure 14.1 Experimental design for Microtox Toxicity Test System.

II. MICROTOX TOXICITY TEST SYSTEM

The Microtox Toxicity Test System monitors the toxicological effects of chemical contaminants on strains of bioluminescent bacteria by measuring their light emissions in response to toxicological stress. The protocols are simple, well defined, and technician friendly. The assay's test organisms (*Photobacterium/Vibrio*) are stored freeze-dried, eliminating the tedium and cost of continuous culture; the bacteria are clonal cultures ensuring quality control of the tester strain and diminishing genetic differences. Media and glassware are standardized, and the quantity is minimal, dramatically reducing the disposal cost of toxic waste materials. No aseptic technique is required. The endpoints of all tests are based on a dose-response experimental design, typically one control and four dilutions (Figure 14.1). As the toxicant's concentration increases, bacterial light emissions decrease or increase in a dose-dependent manner. Light emissions are measured with a temperature-controlled luminometer simplifying data recovery and instrumentation. Data analysis is supported by computer software and easily quantified; statistical compilations and summary are typically presented in a standard format (i.e., EC_{50} or genotoxicity designation). Most important, all toxicity bioassays are available on demand and require no preculturing of the test organisms. A portable temperature-controlled luminometer and laptop computer with software make field analyses feasible and simple. Typically, tests are completed and data available in <24 h and in some instances <30 min. This system permits a rapid response time to the toxicologist's needs — chemical spills, detection of unstable or transitory toxicants, or other emergencies that require immediate evaluations either in the laboratory or in the field.

A. Microtox Acute Toxicity Test (Basic Test)

The basic test determines the acute toxicity of samples from surface water, ground water, wastewaters, leachates, and organic or aqueous sediment extracts by measuring the changes of light produced by bioluminescent bacteria. Liquid phase (aqueous or organic) samples are prepared in a standard four-tube 1:2 dilution series with control. The carrier solvent should not exceed 5% of the sample's volume; osmotic correction with NaCl may be necessary with freshwater samples. Glowing luminescent bacteria in stationary growth phase are added to the test substance and incubated in a Microbics temperature-controlled Model 500 Analyzer (luminometer). Readings are taken with the luminometer typically after five and 15 minutes. The amount of light remaining in the sample is used to determine the sample's relative toxicity, which may be compared to standard reference toxicants. The standard dose-response curve method is used to determine a 50% loss of light in the test bacteria, i.e., the effective concentration (EC_{50}). Some samples with low toxicity may require an expanded dilution series (eight to 10 dilutions with two controls). The luminometer and supporting computer software with a standard log-linear model are used to calculate EC_{50} values. All EC_{50} values are expressed as weight/mL, typically at ECRC reported as the mean of three pseudoreplicates performed on separate days; pseudoreplicates are a statistical measurement of the test's precision. The toxicological endpoint of the basic test is defined with an EC_{50} value and 95% confidence intervals. *The lower the EC_{50} value, the greater the toxicity of the sample.*

B. Microtox Solid-Phase Toxicity Test (Solid-Phase Test)

The solid-phase test, similar to the basic test in experimental design, exposes bioluminescent bacteria directly to sediment-bound chemical contaminants in an aqueous suspension of a test sample. The sediment sample is first centrifuged to remove the pore water and the remaining residual homogenized. A 300-mg sample is placed with solid-phase NaCl diluent in a solid phase tube (SPT), stirred with a vortex mixer, and used to prepare a three-control, 12-tube 1:2 dilution series. Glowing luminescent bacteria in stationary growth phase are now introduced into each SPT tube directly exposing the bacteria to the test sample in the water–soil matrix. This sample is briefly blended with a mixer and incubated for 20 min at 15°C in a temperature-controlled water bath. A special filter column is inserted into the SPT tube to facilitate the separation of solid and liquid materials. The supernatant containing treated bioluminescent bacteria is transferred into standard cuvettes that are placed in a temperature-controlled luminometer for a 5-min stabilization period. Next, the light emissions are read. The standard dose-response curve method is used to determine a 50% loss of light in the test bacteria. The luminometer and supporting computer software with a standard log-linear model are used to calculate EC_{50} values. The toxicological endpoint of the solid-phase test is typically defined with an EC_{50} value expressed as sediment dry weight/mL, in parts per million, or as percent of sample. (Note the 25-min total exposure period is only used for the solid-phase test.)

C. Microtox Chronic Toxicity Test (Chronic Test)

The chronic test is designed to measure adverse biological effects that result when an organism is exposed over a long period to toxicant stress. Test samples are introduced in liquid phase (aqueous or organic) from surface water, ground water, wastewaters, leachates, and organic or aqueous extracts. A standard multiple-tube 1:2 dilution series with parallel controls is incubated with growth medium at 27°C for 22 h. Glowing luminescent bacteria in a growth cycle (i.e., stationary through log phase) are exposed to a test substance at sublethal concentrations, and the amount of light remaining after incubation is measured. Light emissions of bacteria in the control series are compared with light production of bacteria in the test series. The test endpoint is calculated on the lowest observable effect concentration (LOEC), defined as that sample's concentration which after a 22-h exposure period displays an observable light production that is at least two standard deviations lower than the observable light production from the controls. The chronic test is a recent release[6] and has not been used at ECRC.

D. Mutatox® Genotoxicity Test

Mutatox determines the genotoxicity of samples in liquid phase (aqueous or organic) from surface water, ground water, wastewaters, leachates, and organic or aqueous sediment extracts by measuring the changes of light produced by bioluminescent bacteria. The test can be used as a screening tool to detect the presence of genotoxins (DNA-damaging substances) in complex mixtures (i.e., waters, sediments, wastewaters, or leachates) and in new products and their metabolites, or as an analytical tool to evaluate chemicals for structural alerts of DNA-reactivity. Nonglowing luminescent bacteria (dark mutant strain) in a growth cycle (i.e., stationary through log phase) are exposed to a test substance at sublethal concentrations, and the amount of light produced after an exposure of 16 to 24 h is measured with a modified Model 500 Analyzer (similar to the unit used for Microtox tests but with extended light sensitivity). The relative genotoxicity of the sample is determined by comparing the stress-induced increases in light emissions with controls.

Because most prokaryotic cells, such as the dark mutant strain of *Photobacterium/Vibrio,* fail to duplicate vertebrate biotransformation of progenotoxins into potential DNA-damaging agents, a mammalian metabolic activation system is incorporated.[7] Test samples are serially diluted in a mixture of bacteria–1%S9–nutrient–buffer over a 100-fold dose range, preincubated in a water bath at 37°C for 15 min for the activation phase of the test, and then grown at 27°C overnight.[7] Dimethyl sulfoxide (DMSO) is routinely used as carrier solvent with Mutatox;[8] it has previously been shown to be compatible with the test system. Benzo(a)pyrene and DMSO are used as the standard controls. A positive genotoxic response is defined as a light value of 100 or more and at least three times the light intensity of the bacteria (reagent) control blank. The dose-response number is defined as the number of positive responses recorded at different concentrations in a single dilution series. The sample is designated "genotoxic" when replicate dilution series contain two or more positive responses at two or more different concentrations, "suspect" when replicate series each contain one positive response, and "negative" when neither series contains a positive response. Mutatox defines each test sample as genotoxic, suspect, or negative only after three pseudoreplicate dilution series are performed on different days. The toxicological endpoint of Mutatox is qualitative, providing a yes–no assessment of DNA-damaging substances.

III. APPLICATIONS

A. Sediment Toxicity

Most water- and sediment-borne xenobiotic contaminants are lipophilic and commonly found in the environment only in nanogram or microgram/kg concentrations, generally far below the sensitivity range of most bioassays. Concentration of the contaminant, whether naturally by biota or by passive absorptive devices, or by direct organic extraction, is essential to the development of meaningful risk assessments. Organic extractions of surficial sediment samples from Pensacola Bay in Florida[9] were prepared using strong lipophilic solvents such as hexane, methylene chloride, etc. Extracts were expressed as units of sediment (wet weight) equivalent per unit of solvent (mg eq/mL). DMSO was used as the universal carrier solvent; high lipophilic solubility, high vapor pressure, low freezing point, and low cytotoxicity makes DMSO an attractive carrier solvent for *in vitro* bioassays. These organic extracts were evaluated with the basic test and Mutatox to determine the toxicity of the sediments (Table 14.2).

These questions were considered: What are the sensitivity and selectivity of basic test and Mutatox as broad tier-one screening tools to detect acutely toxic and genotoxic substances in sediments? What is a "toxic" sediment? Does the use of the basic test and Mutatox in tandem assist the researcher or resource manager in "red flagging" or prioritizing specific areas for more comprehensive toxicological profiles? Does tandem testing broaden the risk assessment profile?

Table 14.2   Sediment Toxicity Site Profile of Pensacola Bay in Florida with Microtox Basic Test and Mutatox

Sample[a]		EC$_{50}$[b]		Toxicity Indices		Genotoxicity[c]
Region	Site	Mean	SD	Control[d]	Phenol[e]	Designation
A	Bayou Grande	0.37*	0.35	133.8	14.1	Positive/Suspect
B	Bayou Chico	0.48*	0.21	103.1	10.8	Positive
C	Bayou Texar	0.68*	0.47	72.8	7.6	Positive/Negative
D	Warrington	7.29*	1.92	6.8	0.7	Negative
E	Bayou Channel	4.72*	1.65	10.5	1.1	Suspect/Negative
F	Inner Harbor	1.97*	1.92	25.2	2.6	Positive
G	Harbor Channel	10.48*	0.19	4.7	0.5	Negative
H	Lower Bay	10.36*	1.60	4.8	0.5	Suspect/Negative
I	Central Bay	1.84*	0.25	26.9	2.8	Positive
J	East Bay	1.11	0.24	44.6	4.7	Negative
K	East Bay Extension	2.49*	1.23	19.9	2.1	Negative
L	Blackwater Bay	3.25*	4.14	15.3	1.6	Suspect/Negative
M	Escambia Bay	4.72*	1.02	10.5	1.1	Suspect/Negative
N	I-10	1.46*	1.15	33.9	3.6	Suspect
O	River Delta	6.75*	3.04	7.3	0.8	Positive/Suspect/Negative
P	Floridatown	3.39*	1.56	14.6	1.5	Positive/Suspect
Control	ECRC Sediment	49.5	3.95	1.0	—	Negative

[a] Dichloromethane extraction in DMSO carrier solvent.
[b] 5 min EC$_{50}$ = mg equivalent sediment wet weight organic extract/mL; S.D. = standard deviations;(*) = significant difference from control (p>0.05); n = 3.
[c] Rat hepatic S9 activation.
[d] Control Index = control EC$_{50}$/sample EC$_{50}$.
[e] Phenol index = spiked control EC$_{50}$/sample EC$_{50}$.

1. Validation: Sensitivity and Selectivity

Validation studies with single compounds and complex mixtures of more than 50 EPA priority organic pollutants[10] and petroleum products were used to determine the sensitivity and selectivity of the basic test and Mutatox (Tables 14.3, 14.4, and 14.5). DMSO was again used as the universal carrier solvent for all model lipophilic compounds; complex mixtures were prepared on a weight per weight ratio. A number of specific toxicological patterns were observed with the basic test. The 5-min EC$_{50}$ values of insecticides, petroleum products, and polychlorinated biphenyls (PCBs) all tended to group around the mean EC$_{50}$ value of one µg/mL (1.2 ± 0.8). Herbicides, however, were at least fivefold less toxic than insecticides. Interestingly, the toxicity and the ring structure of polyaromatic hydrocarbons (PAHs) tended to correlate (r = 0.85, p < 0.0001). For example, the EC$_{50}$ value of 0.50 µg/mL indicated that the two-ringed PAH fluorene was about 20-fold more toxic than the five-ringed compound benzo(a)pyrene with an EC$_{50}$ value of 10.7; on the other hand, pyrene with four-rings unsubstituted was not detected at 500 µg/mL (Table 14.4). The mean EC$_{50}$ value was 1.6 µg/mL for 12 mixtures containing various combinations of PCBs, insecticides, and PAHs (weight/weight ratio i.e., 1:1). The basic test indicated that the toxicity values of single and complex mixtures were similar (Table 14.5). For example, the EC$_{50}$ value (µg/mL) of kepone was 1.41, aldrin 0.88, lindane 1.56, DDT 1.25, and PCB 1254 1.01, whereas the observed EC$_{50}$ value of the complex mixture was 1.6 (Table 14.10). Four- and five-ringed PAHs, such as pyrene and benzo(a)pyrene, tended to be more acutely toxic in complex mixtures containing pesticides and PCBs; others showed no detectable interactions (Table 14.5). One would suspect some type of induced activation of the PAHs in the complex mixtures, but there was no evidence to support this idea.

Mutatox, more selective than the basic test, detected genotoxins in PAH and petroleum products, but none in either pesticides or PCBs at test concentrations ≤10 µg/mL. Genotoxins were found in complex mixtures of PAHs, PCBs, and pesticides only when PAHs were present. The sensitivity of the activated Mutatox to pesticides, PCBs, PAHs, and petroleum products was in general similar

Table 14.3 Toxicological Evaluation of Polychlorinated Biphenyls (PCBs), Pesticides, and Petroleum Products with Microtox Basic Test and Mutatox

Compound	Microtox EC_{50}[a]	95% CI[a]	Mutatox Genotoxicity[b]
Insecticides: Organochlorine			
Aldrin	0.88	0.75–1.05	Negative
Chlorodane (T)	1.3	1.1–1.5	Negative
DDD	0.82	0.73–0.92	Negative
DDE	0.97	0.8–1.2	Negative
DDT	1.25	1.04–1.52	Negative
Dieldrin	1.3	1.1–1.7	Negative
Endrin	>50		Negative
Heptachlor	0.95	0.69–1.31	Negative
Hexachlorbenzene	0.63	0.54–0.73	Negative
Kepone	1.41	1.08–1.83	Negative
Lindane	1.56	1.22–1.99	Negative
Methoxychlor	0.86	0.78–0.94	Negative
Mirex	1.2	1.2–1.28	Negative
Pentachlorophenol	0.83	0.77–0.90	Negative
Toxaphene	4.9	2.6–9.5	Negative
Insecticides: Organophosphate			
Dyfonate	2.1	2.0–2.1	Negative
Malathion	0.85	0.64–1.1	Negative
Parathion	0.72	0.68–1.1	Negative
Phorate	3.1	2.6–3.7	Negative
Insecticides: Carbamate			
Carbaryl	0.57	0.52–0.62	Suspect
Carbofuran	0.91	0.74–1.11	Negative
Insecticides: Pyrethroid			
Permethrin	1.56	1.38–1.75	Suspect
Herbicides: Triazine			
Atrazine	3.8	2.9–4.7	Negative
Simazine	4.4	3.3–5.8	Negative
Herbicides: Trifluralin			
Treflan	3.7	2.2–6.3	Negative
Herbicides: Others			
Dacthal	1.3	1.0–1.6	Negative
Industrial: PCBs			
PCB 1242	1.2	0.89–1.6	Negative
PCB 1248	0.55	0.51–0.59	Negative
PCB 1254	1.01	0.87–1.2	Negative
PCB 1254-1260	0.75	0.59–0.95	Negative
Industrial:Others			
Dihexyl phthalate	82	42.2–159.4	Negative
Nonylphenol	0.44	0.29–0.65	Negative
Phenol	15.1	14.2–16.3	Negative
Petroleum Products			
Fuel oil #2	0.06	0.04–0.10	Positive
Jet fuel JP4	0.12	0.10–0.13	Positive
Recycled motor oil	1.0	0.82–1.2	Positive
Gasoline	0.16	0.12–0.21	Positive
Crude oil	0.4	0.25–0.64	Positive

[a] 5 min EC_{50} = µg/mL; CI = confidence interval; n = 3; DMSO carrier solvent.
[b] 1% rat S9 activation.

to that reported for the Ames test.[11] Figure 14.2 shows a single data set with raw data and histogram of 2-aminoanthracene (2-AA), a known genotoxin frequently used as a standard; note the maximum detected concentration was 5 µg/cuvette, the lowest detected concentration was 0.07 µg/cuvette,

Table 14.4   Toxicological Evaluation of Polyaromatic Hydrocarbons (PAHs) with Microtox Basic Test and Mutatox

| Compound | Microtox | | Mutatox |
	EC_{50}[a]	CI[a]	Genotoxicity[b]
Acenaphthylene	0.34	0.25–0.47	Positive
Phenanthrene	0.48	0.33–0.68	Positive
Fluorene	0.50	0.35–0.70	Positive
Anthracene	0.64	0.53–0.78	Positive
Benzo(a)anthracene	0.73	0.65–0.81	Positive
Acenaphthene	0.75	0.69–0.81	Positive
2-Aminoanthracene	0.75	0.49–1.2	Positive
Fluoroanthrene	0.83	0.63–1.08	Positive
Naphthalene	0.90	0.85–0.99	Positive
Chrysene	0.92	0.85–0.99	Positive
2-Aminonaphthalene	1.3	1.1–1.52	Positive
2-Acetamidofluorene	2.3	1.26–4.09	Positive
2-Aminofluorene	4.1	2.5–6.4	Positive
Benzo(a)pyrene	10.7	6.4–18.2	Positive
3-Methylcholanthracene	19.9	18.3–21.5	Positive
7,12-Dimethyl benzanthracene	33.1	14.6–74.7	Positive
Pyrene	>500		Positive
DMSO (Control)	ND[c]		Negative

[a] 5 min EC_{50} = µg/mL, CI = 95% confidence interval, n = 3, DMSO carrier solvent.
[b] 1% rat S9 activation.
[c] ND = not detected.

Table 14.5   Toxicological Evaluation of Complex Mixtures Containing Polychlorinated Biphenyls (PCBs), Polyaromatic Hydrocarbons, and Pesticides with Microtox Basic Test

Complex Mixture	EC_{50}[a]	CI[a]
PCBs:1242+1248+1254+1260	0.9	0.85–0.95
DDT+DDE+DDD	1.5	1.3–1.7
Kepone+Aldrin+Lindane+DDT+PCB1254	1.6	1.4–1.7
Phenanthrene+Chrysene+ Anthracene+Benzo(a)pyrene	0.6	0.56–0.59
Aminoanthracene+Benzo(a)pyrene+Aminofluorene+3-methylcholine	3	2.1–4.4
Aminoanthracene+Benzo(a)pyrene+Aldrin+DDT	1.8	1.6–2.0
Aldrin+DDT+Heptachlor+Endrin	1.6	1.1–2.2
Atrazine+DDT+Aldrin+PCB1254+Pyrene	1.7	1.4–2.1
DDT+Benzo(a)pyrene+PCB1254+1260+Atrazine	2.2	1.6–2.9
Carbofuran+Carbaryl+Atrazine+Treflan	1.7	1.4–2.1
Carbofuran+Carbaryl+Atrazine+Permethrin	1.2	0.94–1.5
Carbofuran+DDT+Atrazine+ Permethrin	1.6	1.5–1.6

[a] 5 min EC_{50} = µg/mL, CI = 95% confidence interval, complex mixture = weight/weight, DMSO carrier solvent.

and the dose-response number was seven. As a result, the sample was designated genotoxic, therefore confirming with Mutatox that 2-AA was a genotoxin.

2. Mutatox and Ames Test Comparison

Mutatox was compared with the Salmonella Mutagenicity Test[12] (usually referred to as the Ames test), the benchmark assay to detect DNA-damaging agents in the environment (Table 14.6). Mutatox and the Ames test are both *in vitro* genotoxicity assays that use bacterial tester strains and metabolic activation to detect genotoxins. A battery of tests with EPA priority PAHs showed that

[Raw Data Set: Light Emission Values Recorded]
2-Aminoanthracene

Ten-tube dilution series (microgram per cuvette)

Concentration	10	5	2.5	1.2	0.6	0.3	0.15	0.07	0.03	0.01
2-AA	0	1060	930	410	380	280	170	140	80	10
Control	0	2	2	3	3	3	5	6	6	10

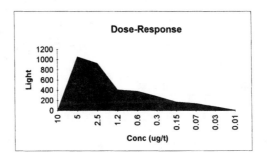

Summary:

Maximum detected concentration = 5 micrograms per cuvette

Lowest detected concentration = 0.07 microgram per cuvette

Dose-response number (DRN) = 7

Conclusion: 2-Aminoanthracene is genotoxic.

Figure 14.2　Genotoxicity of 2-aminoanthracene (2-AA) determined with Mutatox.

Mutatox compared favorably with a similar spectra of sensitivity.[11] Mutatox, however, was available on demand and had a short incubation period of 24 h, instead of 72 h. In addition, Mutatox tended to be user friendly, requiring minimal technical and microbiological expertise. Axenic conditions were not necessary for the preparation of activation enzymes which allowed the use of nonsterile, nonmammalian systems. Unlike the Ames test, Mutatox was an easier assay to perform because it used prepackaged test materials that included freeze-dried test organisms, glassware, and computer software to handle data.

3. Sediment Data Analysis: Toxicity Index

One could define a toxicant as any agent capable of producing a deleterious response in a biological system, e.g., cellular dysfunction or lethality. This is not, however, a useful working definition for sediment toxicity because virtually every chemical or xenobiotic has the potential to produce injury or death if it is present in a sufficient amount. Paracelsus (1493–1541) phrased this well when he noted, "What is there that is not poison? All things are poison and nothing [is] without poison. Solely the dose determines that a thing is not a poison."[21] It must be remembered that an EC_{50} value can be derived for any sediment. The problem is to determine the degree to which a poison is toxic. Two concepts were explored to determine when the poison is harmful to the specific resource of interest: the control comparison and Toxicity Index.

The ECRC sediment control is an agricultural soil sample from Florissant, Missouri, with an EC_{50} value of 49.7 mg eq/mL and no detectable evidence of chemical contamination (i.e., no detectable PCBs, PAHs, or organochlorine pesticides). Any sediment sample with an EC_{50} value that was significantly ($p \leq 0.05$) lower than that of this control was designated acutely toxic. To separate endogenous normal background sediment toxicity and present a toxicological reference point, the control was spiked with 100 mg/L of phenol to make an acute toxicity reference sediment.

Table 14.6   Comparison of the Mutatox Genotoxicity Test with the Ames *Salmonella* Mutagenicity Test

Test Organism	Mutatox *Photobacterium/Vibrio*	Ames *Salmonella*
Bacterial Requirements	One isolate	Usually one to four isolates
Test Endpoint	Light emission	Colony formation
Exogenous Activation[a]	Optional	Optional
Test Duration	16–24 h	48–72 h
Test Temperature	25 ± 2°C	37°C
Sensitivity[b]	≤1.0 µg per tube	≤1.0 µg per plate
Sterility Requirements	Optional	Essential
Procedure[c]	Simple	Complex
Instrumentation	Luminometer or scintillation counter	Manual plate counter or electronic particle counter
Cost[c]	Low	High
Scientific Development	Validation phase	In common use[d]

[a] Activation system with rat hepatic S9 fractions.
[b] Polyaromatic hydrocarbon comparison.
[c] Labor and materials.
[d] Extensive literature and validation.

This spiked control's basic test-derived 5-min EC_{50} value of 5.2 mg eq wet weight sediment/mL provided the basis for the Toxicity Reference Index[9] and was given the number one. Test samples were compared to this index for sediment analysis. For example, the Bayou Grande site in Florida[9] had a toxicity reference index number of 14.1 (phenol-spiked control's EC_{50} value/test sample's EC_{50} value = toxicity reference index number; 5.2/0.37 = 14.1), indicating that this sample was about 14-fold more acutely toxic than the phenol-spiked control (Table 14.2).

4. Tandem Toxicity Testing

Gregus and Klaassen[13] have noted that measures of acute lethality such as the EC_{50} value may not accurately reflect the spectrum of toxicity, or hazard, associated with exposure to a chemical contaminant. For example, PAHs may have carcinogenic or mutagenic effects at doses that produce no evidence of acute toxicity. Table 14.12 shows the dosage of PAHs needed to produce death in 50% of bacteria (EC_{50}). For example, acenaphthene, acenaphthylene, and fluorene produced low 5-min EC_{50} values of <1 µg/mL and were considered acutely toxic. On the other hand, benzo(a)pyrene (BaP) and pyrene that produced 5-min EC_{50} values greater than 10 µg/mL (far exceeding their water solubility) would be considered essentially harmless aquatic xenobiotics. However, all five PAHs tested as genotoxic and are potentially carcinogenic agents.[11] Pyrene and BaP confirmed the pattern described by Gregus and Klaassen.[13]

Tandem use of acute toxicity and genotoxicity tests significantly broadened the scope of environmental risk assessment. The Pensacola Bay data[9] sets produced by the basic test and Mutatox and analyzed with Spearman Rank Correlation showed a significant correlation of acute toxicity and genotoxicity ($p \leq 0.001$) (Table 14.2). Bayou Grande, Bayou Chico, and Bayou Texar in Pensacola Bay with toxicity reference index numbers >1 were also designated genotoxic and identified as areas of concern warranting further toxicological characterizations. The basic test and Mutatox as sister toxicity assays were sensitive, specific, and predictable in providing a broad toxicity assessment of sediments.

B. Solid-Phase Test: Artificial Sediment

Because the solid-phase test needs a reference sediment to produce baseline data, the use of artificial sediment was explored. A formulated substrate developed by Kemble et al.[14] and based

Table 14.7   Toxicological Evaluation of Sediment Samples from Waukegan Harbor in Michigan with Microtox Solid-Phase Toxicity Test

Sample Number	EC_{50}[a]	95% CI[b]	Toxicity Reference Index[c]	% Clay
1	0.5	0.23–0.79	2.2	38.3
2	0.27	0.25–0.29	4.1	38.2
3	1.1	1.0–1.1	1.0	27.3
4	1.1	0.9–1.3	1.0	28.1
5	3.4	2.6–4.4	0.32	32.5
6	5.4	2.9–9.8	0.20	36.3
7	1.1	0.9–1.4	1.0	NR
8	0.34	0.23–0.51	2.7	28.2
9	2.3	1.5–3.4	0.48	33.5
10	0.37	0.34–0.40	2.9	NR
11	0.14	0.10–0.18	7.9	29.3
11R	0.4	0.58–0.84	2.75	35.4
12	0.11	0.06–0.19	10.0	58
13	0.68	0.47–0.97	1.62	37.1
14	0.25	0.16–0.41	4.4	49.8
15	0.25	0.23–0.27	4.4	26.5
16	0.1	0.09–0.11	11.0	29.9
17	0.44	0.32–0.59	2.5	39.2
18	0.53	0.37–0.76	2.1	52.4
19	0.37	0.25–0.55	2.9	NR
Artificial sediment[d]	7.3	2.4–12.6	0.15	17.1
ECRC sediment[e]	4.4	3.6–5.3	0.25	NR
Phenol spiked artificial sediment	1.1	0.9–1.3	1.0	NR

[a] EC_{50} = mg wet weight sediment/mL; n = 3; incubation time = 25 min; NR = not recorded.
[b] CI = confidence interval
[c] Toxicity Reference Index = EC_{50} value of the phenol spiked artificial sediment/the EC_{50} value of the test sample.
[d] Developed after Kemble et al.[14]
[e] ECRC Sediment = Standard Environmental and Contaminants Research Center control sediment.[9]

on alpha-cellulose as an organic carbon source was used as artificial sediment to prepare negative and positive controls for the study of Waukegan Harbor, Michigan, sediments. A baseline toxicity reference sediment was prepared by phenol-spiking the artificial sediment as described in Section III.A.3. Table 14.7 shows the EC_{50} values of the Waukegan samples as well as the EC_{50} values for the artificial sediment, ECRC sediment control, and the phenol-spiked artificial sediment; these values were used to formulate the toxicity reference index numbers. The EC_{50} value of the phenol-spiked artificial sediment was divided by the EC_{50} value of each Waukegan sediment sample to determine the toxicity reference index numbers of the Waukegan samples. Note, for example, that Waukegan sediment sample number 12 was tenfold more toxic than the phenol-spiked artificial sediment; sample number five was about fourfold less toxic. No correlation was found between clay content and EC_{50} values in these studies (Table 14.7).

C. Assessment of Photoactivated Environmental Pollutants

A study was designed to learn if the increased ultraviolet (UV) radiation of the Earth's surfaces, resulting from depletion of atmospheric ozone, may transform anthropogenic materials into hazardous substances that threaten terrestrial and aquatic biota.[15] This investigation simulated stratospheric ozone depletion to assess the UV-altered toxicity of environmental pollutants. Aqueous samples of single and complex mixtures of insecticides, herbicides, and PAHs in DMSO were

Table 14.8   Acute Toxicity of Ultraviolet Light Irradiated Polyaromatic Hydrocarbons as Determined with Microtox Basic Test

Rings	Compound	EC_{50}[a] 0 h	1 h	24 h
2	Acenaphthene	0.6	1.4	6.3
2	Acenaphthylene	0.4	0.7	4.0
2	Fluorene	0.8	1.8	2.8
2	Naphthalene	0.8	0.96	1.2
3	Anthracene	1	0.81	1
3	Fluoranthene	0.8	0.56	3.5
3	Phenanthrene	0.4	0.69	0.8
4	Chrysene	1.2	0.78	0.8
4	Benz(a)anthracene	1.4	1.4	1
4	Benzo(b)fluoranthene	9.2	2.0	1.3
4	Pyrene	>500	0.93	1.5
5	Benzo(a)pyrene	13	1.5	0.5
5	Benzo(ghi)perylene	12.3	1.1	1.3
5	Dibenzo(a,h)anthracene	11.2	1	1.3
5	Indeno(1,2,3-c,d)pyrene	11.7	0.9	1
5	3-Methylcholanthrene	22.9	0.42	0.41

[a] 5 min EC_{50} = µg/mL; 0 h = control; n = 3; carrier solvent = DMSO.

Table 14.9   Acute Toxicity of Ultraviolet Light Irradiated Agricultural Contaminants as Determined with Microtox Basic Test

Compound	0 h EC_{50}[a]	24 h EC_{50}	CI[b]
Insecticides			
Carbaryl	0.7	1.3	0.8–1.9
Carbofuran	1	6.4	5.4–7.5
Malathion	0.88	1.6	1.5–1.7
Parathion	0.80	1.7	1.0–3.1
Permethrin	1.4	11.9	6.4–22.2
DDT	1.5	8	2.1–30.2
Pentachlorophenol	0.85	0.6	0.48–0.75
Herbicides			
Atrazine	4.5	26.3	8.3–93.0
Simazine	4.3	20.2	6.5–62.3
Dacthal	1.3	8.1	4.5–14.7
Treflan	4.5	11.7	1.7–83.4

[a] 5 min EC_{50} = µg/mL; 0 h = control; n = 3; DMSO = carrier solvent.
[b] CI = 95% confidence interval.

irradiated with UV for periods of one and 24 h. Acute toxicity values were determined with the basic test immediately after UV exposure. (A caveat: because UV radiation is a strong bactericide, the Microtox bacteria cannot be introduced with the test samples nor exposed to UV irradiation.)

The basic test was performed at 0 h, 1 h, and 24 h to determine the acute toxicity values (5-min EC_{50}) of the photoenhanced contaminants (Tables 14.8, 14.9, 14.10). Enhanced acute toxicity occurred in PAHs after one hour of irradiation with no appreciable increase after an additional 23-h exposure. Seven of the 14 PAHs showed significantly enhanced toxicity with a two- to 500-fold increase (Table 14.8). Four- and five-ringed PAHs were most affected; for example, the EC_{50} value for pyrene (four rings) changed from >500 to 1 µg/mL. Lower molecular weight PAHs showed

Table 14.10 Acute Toxicity of Ultraviolet Light Irradiated Complex Mixtures of Polyaromatic Hydrocarbons, Polychlorinated Biphenyl (PCBs), and Pesticides as Determined with Microtox Basic Test

Sample	0 h		24 h	
	EC_{50}[a]	CI[b]	EC_{50}	CI
PCBs: 1242, 1248, 1254, 1260	0.9	0.85–0.95	4	2.9–5.7
DDT, DDE, DDD	1.5	1.3–1.7	4.4	3.5–5.6
Kepone, Aldrin, Lindane, DDT, 1254	1.6	1.4–1.7	11.6	9.4–114.4
Phenanthrene, Chrysene, Anthracene, Benzo(a)pyrene	0.6	0.56–0.59	1.3	1.0–1.7
Aminoanthracene, Aminofluorene, Benzo(a)pyrene, 3-Methylcholanthrene	3	2.1–4.4	6.4	3.4–12.0
Aminoanthracene, DDT, Aldrin, Benzo(a)pyrene,	1.8	1.6–2.0	5	3.1–7.7
Aldrin, Dieldrin, Heptachlor, Endrin	1.6	1.1–2.2	12.3	8.6–17.7
Atrazine, DDT, Aldrin, 1254, Pyrene	1.7	1.4–2.1	5.9	3.4–10.4
DDT, Benzo(a)pyrene, 1254, 1260, Atrazine	2.2	1.6–2.9	7	4.1–11.9
Carbofuran, Carbaryl, Treflan, Atrazine	1.7	1.4–2.1	3.8	2.4–5.8
Carbaryl, DDT, Atrazine, Permethrin	1.2	0.94–1.5	3.6	3.0–4.3
Carbofuran, Carbaryl, Permethrin, Atrazine	1.6	1.6–1.8	2.9	2.6–3.3

[a] 5 min EC_{50} = µg/mL; 0 h = control; n = 3; DMSO carrier solvent.
[b] CI = 95% confidence interval.

either a significant decrease or no change in acute toxicity after UV irradiation. Interestingly, UV exposure did not enhance the toxicity of agricultural contaminants (Tables 14.9 and 14.10), but tended to decrease toxicity by two- to tenfold.

D. Polyphylogenetic Metabolic Activation

Most environmental genotoxins must be metabolically activated to an electrophilic form to become an active DNA-damaging agent in biota.[16] The feasibility of using different organisms including birds, fish, invertebrates, algae, and bacteria instead of the rat as exogenous metabolic activation systems for Mutatox was evaluated with five classes of known genotoxins — PAHs, mycotoxins, nitrosoamines, phosphamides, and azo-aromatic dyes. All five classes of genotoxins were detected within acceptable sensitivity (≤1.0 µg/cuvette). Table 14.11 showed that polyphylogenetic activation systems were compatible with Mutatox.[17] For example, the sensitivity of birds, fish, and algae was similar to that of rats. The incorporation of a mammalian (rodent) hepatic metabolic activation system — the postmitochondrial supernatant fraction (commonly abbreviated as the S9 fraction) — significantly improved Mutatox's sensitivity to a broader spectrum of known environmental genotoxins.[8]

E. Bioavailability: Passive Absorbers

Unlike traditional assays with fish and invertebrates, the Microtox tests frequently require microgram amounts of a toxicant to elicit measurable responses. This limitation significantly affects the detection of many chemical contaminants directly from water. Since most lipophilic chemical contaminants are present in water only at trace concentrations (i.e., low µg or ng/L), direct use of *in vitro* assays may not be feasible unless samples are extracted with organic solvents (i.e., hexane, methylene chloride, etc.) to concentrate the lipophilic contaminants and increase the assay's sensitivity range. This approach tends to focus only on the sample's potential toxicity, which may be over estimated by the extraction of sediment-bound contaminants with strong scouring types of solvents, and does not address the all-important toxicological factor — bioavailability.

The tendency of aquatic organisms to accumulate and concentrate lipophilic chemical contaminants from the environment is well-known.[18] To mimic this bioconcentration process Huckins

Table 14.11 Detection of Genotoxins with Mutatox Genotoxicity Test Using Polyphylogenetic Activation Systems

Organism	Genotoxins[a]				
	BaP	AFT	DMN	CP	AZT
Mammals[b]					
Rat	Yes	Yes	Yes	Yes	Yes
Birds[b]					
Mallard duck	Yes	Yes	Yes	Yes	Yes
Fish[b]					
Bullhead catfish	Yes	Yes	Yes	Yes	Yes
Channel catfish	Yes	Yes	Yes	Yes	Yes
Invertebrates[c]					
Chiromonas sp.	ND	ND	Yes	Yes	Yes
Daphnia magna	ND	ND	Yes	Yes	Yes
Algae[c]					
Selenastrum capricornutum	Yes	Yes	Yes	Yes	Yes
Bacteria[c]					
Aeromonas liquefaciens	ND[d]	ND	Yes	Yes	Yes
Bacillus cereus	ND	ND	Yes	Yes	Yes
Bacillus substilis	ND	ND	Yes	Yes	Yes
Pseudomonas fluorescens	ND	ND	Yes	Yes	Yes

[a] Genotoxins: BaP = Benzo(a)pyrene; AFT = aflatoxin; DMN = dime-thylnitrosoamine; CP = cyclophosphamide; AZT = o-aminoazotolu-ene; sensitivity = ≤ 1 μg/cuvette; n = 3; DMSO = carrier solvent.
[b] Hepatic S9 fractions.
[c] Cell-free homogenate.
[d] ND = Not detected at 10 μg/cuvette.

et al.[19] designed and patented a passive absorber — the semipermeable polymeric membrane device (SPMD). This device, a low-density polyethylene lay-flat tube, contains a lipid to passively sample *in situ* bioavailable organic chemical contaminants (Figures 14.3 and 14.4). Several studies used SPMDs with the basic test and Mutatox to monitor the bioavailability and acute toxicity of environmental chemical contaminants in water, sediment, and air.[20-22]

In validation experiments, SPMDs were exposed to 15 EPA priority PAHs in a flow-through diluter system for 21 days. SPMDs readily sequestered PAHs with concentration factors ranging from 400× to more than 20,000× and EC_{50} values ranging from 0.4 to >500 μg/mL. Note the correlation between the SPMD concentration factor and the acute toxicity value (r = 0.84, p < 0.0001). For example, acenaphthylene with a concentration factor of 400× had an EC_{50} value of 0.38 μg/mL while benzo(a)pyrene with a concentration factor of 20,090× had an EC_{50} value of 13 μg/mL. Significantly, all 15 PAHs at ≤ 1 μg/cuvette were detected as strong genotoxins. Table 14.12 shows the SPMD concentration factors and the toxicity data derived from this study.[20]

In a crude oil spill that was simulated in sediment/water microcosms, SPMDs were used to monitor the changes that occurred in toxicity each week over eight weeks. Instead of the slower dialysis method, analytes were directly extracted by cutting the SPMDs into small fragments and dissolving the enclosed lipid into acetone-DMSO; analyses were performed with the basic test (Figure 14.4). Figure 14.5 shows that the spill's toxicity decreased over time as the crude oil dissipated.[21] Similarly, the bioavailability of airborne toxicants was assessed by combining SPMDs with the basic test and Mutatox for detection of acute toxicity and genotoxicity.[22] SPMDs collected both primary and secondary tobacco smoke from microcosm (Figure 14.6), from clothing, and from room air. Table 14.13 shows the toxicity and movement of tobacco smoke as an airborne environmental contaminant. SPMDs significantly enhanced the collection of chemical contaminants for bioavailability assessment.

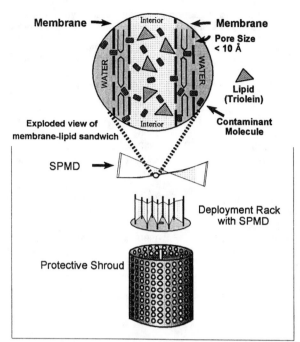

The lipid containing semipermeable membrane device (SPMD) and a typical deployment apparatus.

Figure 14.3   Semipermeable membrane device (SPMD) with protective shroud. (Diagram courtesy of Randal Clark and Carl Orazio, Environmental and Contaminants Research Center.)

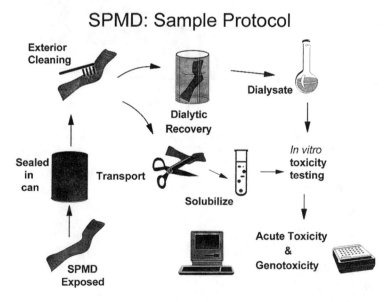

Figure 14.4   Semipermeable membrane device (SPMD) sample protocol with Microtox Basic Test and Mutatox. (Diagram courtesy of Randal Clark, Environmental and Contaminants Research Center.)

Table 14.12 Collection and Detection of Polyaromatic Hydrocarbons from Water with SPMDs, Microtox Basic Test, and Mutatox

Ring	Compound	SPMD/Water Conc. Factor[a]	Microtox EC_{50}[b]	Mutatox[c]
2	Acenaphthene	3700	0.6	Genotoxic
2	Acenaphthylene	400	0.4	Genotoxic
2	Fluorene	6500	0.8	Genotoxic
2	Naphthalene	8100	0.8	Genotoxic
3	Anthracene	10000	1	Genotoxic
3	Fluoranthene	9800	0.8	Genotoxic
3	Phenanthrene	2300	0.4	Genotoxic
4	Chrysene	13900	1.2	Genotoxic
4	Benz(a)anthracene	14800	1.4	Genotoxic
4	Benzo(b)fluoranthene	17400	9.2	Genotoxic
4	Pyrene	24100	>500	Genotoxic
5	Benzo(a)pyrene	20900	13	Genotoxic
5	Benzo(ghi)perylene	20300	12.3	Genotoxic
5	Dibenzo(a,h)anthracene	17600	11.2	Genotoxic
5	Indeno(1,2,3-c,d)pyrene	18400	11.7	Genotoxic

[a] Concentration factor.
[b] 5 min EC_{50} = μg/mL; n = 3; DMSO = carrier solvent.
[c] n = 3; rat S9 activation.

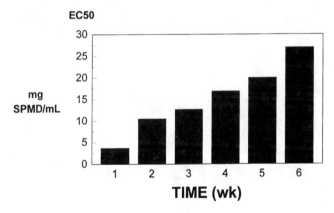

Risk Assessment: Acute Toxicity

Figure 14.5 Acute toxicity of a crude oil spill in a soil-water microcosm, measured with semipermeable membrane device and Microtox Basic Test.

IV. SUMMARY

The use of the Microtox Toxicity Test System is essentially a function of the researcher's creativity. New applications, specific sampling approaches, detection capabilities, and data interpretation were discussed in the protocols and applications of the system.

These highlights emerged:

1. The Microtox Toxicity Test System contains four bioassays that are laboratory-ready to perform on demand; results are available in hours and frequently in minutes.
2. The basic test provides a quantitative assessment, that is acute toxicity values of chemical contaminants. Mutatox provides a qualitative yes–no assessment, that is the genotoxicity designation of

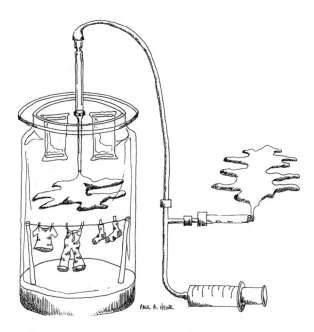

Figure 14.6 Detection of environment tobacco smoke in microcosm with semipermeable membrane device and Microtox Basic Test and Mutatox. (Drawing courtesy of Paul Heine, Environmental and Contaminants Research Center.)

Table 14.13 Collection and Detection of Environmental Tobacco Smoke with SPMDs, Microtox Basic Test, and Mutatox

Environmental Tobacco Smoke	Microtox EC$_{50}$[a]	Mutatox[b]
Primary Exposure		
Brand A (USA)	13.7	Genotoxic
Brand B (USA)	2.4	Genotoxic
Brand C (Russia)	6.0	Genotoxic
Brand D (Denmark)	4.1	Genotoxic
Brand E (USA)	2.7	Genotoxic
Secondary Exposure		
Bar	13.1	Genotoxic
Cotton T-Shirt	7.5	Genotoxic
Cotton Socks	7.7	Genotoxic

[a] 5 min EC$_{50}$ = mg SPMD/mL; n = 3; DMSO = carrier solvent.
[b] n = 3; rat S9 activation.

chemical contaminants. The tandem use of these tests delineates areas of concern and broadens toxicity risk assessments.

3. The toxicity reference index produces a baseline for determining the relevancy of a toxicant.
4. A new discipline of aerotoxicology was initiated to assess airborne chemical contaminants from both primary and secondary sources.
5. The use of SPMDs to collect and concentrate a substance and the Microtox system to monitor its movement, determine its bioavailability, and assess its toxicity gives the researcher a valuable toxicological tool.

In vitro microscale toxicity tests can play an important role in environmental toxicology. Cost and sample numbers, not whether one is enamored with bacteria or cellular enzymes, will determine which assay is used for environmental risk assessment. Such *in vitro* tests will address the complex toxicological challenges of the next century.

REFERENCES

1. Lyman, W. J., Establishing sediment criteria for chemicals — industrial perspectives, in Dickson, K. I., Maki, A. W., and Brungs, W. A., Eds., *Fate and Effects of Sediment-Bound Chemicals in Aquatic Systems*, Pergamon Press, New York, 1984, 378.

2. Lee, G. I., and Jones, R. A., Water quality significance of contaminants associated with sediments: an overview, in Dickson, K. I., Maki, A. W., and Brungs, W. A., Eds., *Fate and Effects of Sediment-Bound Chemicals in Aquatic Systems*, Pergamon Press, New York, 1984, 3.

3. Burton, G. A., Jr., Assessing toxicity of freshwater sediments, *Environ. Toxicol. Chem.*, 10, 1585, 1991.

4. Giesy, J. P. and Hoke, R. A., Freshwater sediment toxicity bioassessment: rationale for species selection and test design, *J. Great Lakes Res.*, 15, 539, 1989.

5. Burton, G. A. Jr., Nelson, M. K., and Ingersoll, C. G., Freshwater benthic toxicity tests, in *Sediment Toxicity Assessment,* Burton, G. A. Jr., Ed., Lewis Publishers, Inc., Chelsea, MI, 1992, 213.

6. Bulich, A. A., Huynh, H., and Ulitzur, S., The use of luminescent bacteria for measuring chronic toxicity, in *Techniques in Aquatic Toxicology,* Ostrander, G. K., Ed., CRC Lewis Publishers, New York, 1996, 3.

7. Johnson, B. T., Activated Mutatox® Assay for detection of genotoxic substances, *Environ. Toxicol. Water Qual.*, 8, 103, 1993.

8. Johnson, B. T., An evaluation of a genotoxicity assay with liver S9 for activation and luminescent bacteria for detection, *Environ. Toxicol. Chem.*, 11, 473, 1992.

9. Johnson, B. T. and Long, E. R., Rapid toxicity assessment of sediments from large estuarine ecosystems: a new tandem *in vitro* testing approach, *Environ. Toxicol. Chem.*, in press.

10. Richards, D. J. and Shieh, W. K., Biological fate of organic priority pollutants in the aquatic environment, *Water. Res.*, 20, 1077, 1986.

11. Zeiger, E., Anderson, B., Haworth, S., Lawlor, T., and Mortelmans, K., *Salmonella* mutagenicity tests: V. Results from the testing of 311 chemicals, *Environ. Mol. Mutagen.*, 19, Supplement 21, 2, 1992.

12. Ames, B. N., Durston, W. E., Yamasaki, E., and Lee, F. D., Carcinogens are mutagens: a single test system combining liver homogenates for activations and bacteria for detection, *Proc. Natl. Acad. Sci. (USA)*, 70, 2281, 1973.

13. Gregus, A. and Klaassen, C. D., Principles of toxicology, in *Casarett and Doull's Toxicology: the Basic Science of Poisons,* Klaassen, C. D., Ed., McGraw-Hill Companies, New York, 1996, 13.

14. Kemble, N. E., Dwyer, F. J., and Ingersoll, C. G., Development of a formulated sediment for use in whole-sediment toxicity tests with freshwater organisms, *Environ. Toxicol. Chem.*, in press.

15. Johnson, B. T., Butorin, A. N., and Little, E. E., Rapid toxicity assessment of photoactivated environmental pollutants with bioluminescent bacteria, *95th Annual Meeting, American Society for Microbiology*, Washington, DC (Abstr.), 1995.

16. Johnson, B. T., unpublished data.

17. Miller, E. C. and Miller, J. A., The metabolism of chemical carcinogenesis to reactive electrophiles and their possible mechanisms of action in carcinogenesis, in *Chemical Carcinogens*, Searle, C. E., Ed., American Chemical Society, Washington, DC, 1976, 737.

18. Spacie, A. and Hamelink, J. L., Bioaccumulation, in *Fundamentals of Aquatic Toxicology*, Rand, G. M. and Petrocelli, S. R., Eds., Hemisphere Publishing Corporation, New York, 1985, 495.

19. Huckins, J. N., Petty, J. D., Lebo, J. A., Orazio, C. E., Prest, H. F., Tillitt, D. E., Ellis, G. S., Johnson, B. T., and Manuweera, G. K., Semipermeable membrane devices (SPMDs) for the concentration and assessment of bioavailable organic contaminants in aquatic environments, in *Techniques in Aquatic Toxicology*, Ostrander, G. K., Ed., CRC Lewis Publishers, New York, 1996, 625.

20. Johnson, B. T., Huckins, J. N., Petty, J. D., and Butorin, A. N., Rapid collection, detection, and assessment of environmental polycyclic aromatic hydrocarbons (PAHs), *Second World Congress, Society of Toxicology and Chemistry*, Vancouver, British Columbia, Canada. (Abstr.), 1995.

21. Johnson, B. T., Huckins, J. N., and Petty, J. D., Rapid toxicological assessment of environmental airborne chemical contaminants: a new monitoring approach, *Sixth Annual Meeting, Society of Toxicology and Chemistry-Europe*, Taormina, Sicily, Italy. (Abstr.), 1996.

22. Johnson, B. T., Petty, J. D., and Huckins, J. N., Toxicological assessment of environmental tobacco smoke (ets) in buildings: a new monitoring approach, *96th General Meeting of the American Society for Microbiology*, New Orleans, LA. (Abstr.), 1996.

CHAPTER 15

Measurement of Microbial Exoenzyme Activity in Sediments for Environmental Impact Assessment

K. Lee and K.L. Tay

CONTENTS

I. INTRODUCTION

The environmental impact of contaminants in aquatic ecosystems cannot be assessed accurately with a single species biotest, as it cannot represent the range of sensitivity of all biota within an ecosystem.[1] Even for the microbes alone, patterns of sensitivity to various classes of toxicants differ between and within species of test organisms.[2-3] To expand the application and ecological relevance of biotests, it is now recommended that future efforts be based on a test battery approach that includes at least two to three genera and two to four species,[4] multitrophic level biological responses,[5] and the effects on indigenous biota from the area of concern.[6]

A. Importance of Bacteria

Since sediments are a sink for contaminants, the benthic environment has become the site of choice for monitoring environmental impacts in the aquatic environment. However, while chemical analyses of contaminants within sediments will provide information on the degree and nature of contamination, they provide little or no direct evidence in regard to biological availability or effects.

Biotests with microorganisms are now considered an essential component in environmental impact assessments. Bacteria provide a major source of particulate organic carbon to higher trophic levels and mediate vital processes, including organic matter degradation and nutrient regeneration.[7-10] Any adverse effect on the composition or activity of the microbial community will be detrimental to the aquatic ecosystem as a whole.

Bacterial isolates have been used routinely in sublethal bioassays since they grow rapidly, are inexpensive, easy to maintain in culture, have enzymatic processes common to those of higher organisms, and respond rapidly to alterations to their surroundings.[4,11-14] Unfortunately, due to the absence of appropriate microbiological techniques in the past, *in situ* methods of assessing the effects of toxic chemicals on the activity and community structure of microorganisms in the aquatic ecosystem are still limited in number.

B. Importance of Enzymes

To fully assess the potential impact of physical disturbances and contaminant additions on the environment, effects must be examined at the ecosystem level, not just the population and community level. In this regard, there is now a need to develop biotests with endpoints based on the identification of changes in the primary metabolic processes which will, if disturbed, cause significant detrimental effects throughout the ecosystem.

The toxicity of most pollutants is attributed to enzyme inhibition.[15-19] A number of techniques has been reported for the measurement of specific enzyme activities in cultures and natural populations of bacteria, including that of the hydrolases and oxidoreductases which are of major importance in aquatic sediments.[18,20-24] Toxicant effects are frequently monitored with inducible dehydrogenase enzymes which are responsible for feeding electrons into the electron transport system to maintain vital cellular anabolic and catabolic processes.[25,26] Bioassays based on enzyme activity are generally more sensitive than population level (LD_{50}) tests,[12] faster and more reliable than measurements of whole organism activity.[25] A correlation between the results of *in vitro* enzyme inhibition tests and bioassays with higher organisms has been reported.[16]

Differences in sensitivity to contaminants has been observed among various enzyme assays evaluated. For example, Armant et al.[27] reported that glucose-6-phosphate dehydrogenase was much less sensitive to toxicant additions than either phosphatase or amylase. Griffiths et al.[28] reported a depression in phosphatase, cellulase, and chitobiase activities and stimulation of laminarinase, amylase, and alginase activities in marine sediments in response to crude oil. Dehydrogenase activity and growth rates have been shown to be relatively sensitive to chlorophenols while β-galactosidase was not.[29] Obst et al.[18] reported a range of EC_{50} values in response to Hg and Cd additions to water samples from the Rhine River for a number of *in vivo* enzymes including phosphatases, β-glucosidase, amylases, proteases, alanine-aminopeptidase, and esterases. The observed inconsistency in sensitivity patterns between the various assays is undoubtedly due to the physicochemical variation between enzymes and differences in the modes of action by toxicants. Therefore, the influence of contaminants on the microbiota is better estimated with a spectrum of enzyme tests than with a single one.

1. Exoenzymes

Organic matter in aquatic environments consists mostly of large, high-molecular-weight compounds, such as proteins, polysaccharides, nucleic acids, and humic materials, that cannot be incorporated directly by bacteria.[30,31] The penetration of organic molecules across the cytoplasmic membrane of bacteria is an active process occurring with the intervention of specific enzymes called permeases. However, since only small organic molecules (monomers or small polymers) with a maximum molecular weight of approximately 600 daltons can be incorporated in this manner,[32] bacteria produce surface-bound (ectoenzymes) and extracellular enzymes (exoenzymes) to degrade polymeric materials prior to transport into the cell.[33,34] In terms of ecological significance, extracellular hydrolysis of compounds to monomers or oligomers is the first and often the rate-limiting step in organic matter degradation.[35-41]

II. TOXICITY TEST DEVELOPMENT

To assess the potential damage of contaminants at ocean disposal sites, Lee et al.[6] proposed that bioassay tests should be conducted with fresh samples of water and/or sediment from the specific sites in question. Use of a microbiotest based on metabolic processes associated with indigenous microorganisms has high ecological relevance because the approach evaluates the response of the whole microbial community — a true multispecies test. It was proposed that quantification of contaminant-induced changes in exoenzyme activity within the samples could be used to: 1) monitor conditions at specific sites of interest to ascertain conditions before and after industrial operations; 2) prescreen wastes for toxicity and determine allowable limits for oceanic disposal; 3) determine the size of benthic impact zones; and 4) monitor the recovery of sites following impact.

A. Toxicity Test Criteria

The following criteria were outlined for the development of a new biotest based on the effects of contaminants on microbial exoenzyme activity[6]:

1. Based on the activity of a primary metabolic process
2. Measures contaminant impact on indigenous biota under natural conditions
3. Rapid, inexpensive, and simple to interpret
4. Short incubation periods with no adjustment of pH in the test system
5. Highly enzyme specific to avoid interference from natural chemical reactions
6. Reliable and reproducible results
7. Sensitivity to a wide range of contaminants and correlation with existing bioassays and/or chemical contamination
8. Feasible for use under field conditions
9. Suitable for use in current toxicity/impact assessment programs

B. Assay Development and Sensitivity Testing

The activity of specific enzymes, including that of exoenzymes in natural waters, has been measured with the use of nonfluorescent substrates conjugated with fluorophores such as methylumbelliferone (MUF) or methylcoumarinylamide (MCA).[37,42,43] Bacterial enzymes acting on MUF compounds in natural water are described as extracellular[37] and are at least partially cell bound, since some enzymatic activity is retained during the filtration (0.22 μm pore-size) of cell suspensions. Excellent nanomolar sensitivity is achieved with these substrates due to the release of highly

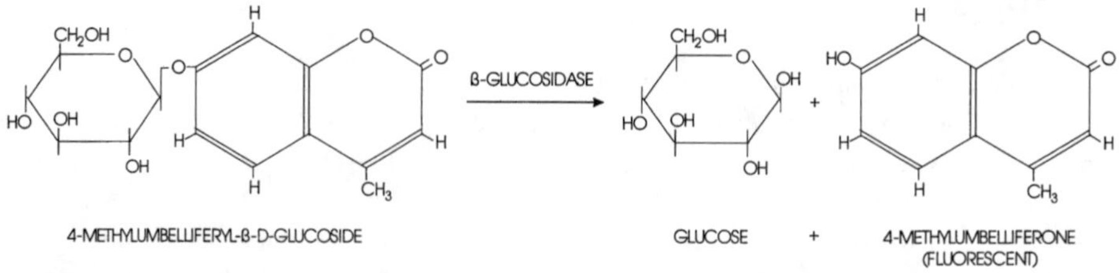

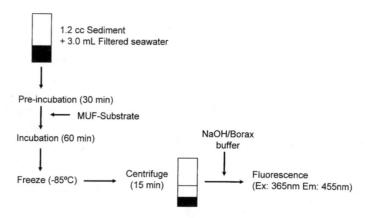

Figure 15.1   Schematic representation of the chemical reaction on which the exoenzyme assay is based. The nonfluorescent substrate, MUF-β-D-glucoside, is split by the exoenzyme β-D-glucosidase. Glucose is liberated from the compound, and the equimolar concentration of free MUF is measured by fluorescence.

Figure 15.2   Test procedure for the determination of exoenzyme activity in sediment slurries.

fluorescent MUF or MCA following enzymatic hydrolysis (Figure 15.1). This allows the measurement of enzymatic activity with short incubation times, at ambient temperatures, and at substrate concentrations similar to those found in nature.[43] There is no need to account for a chemical breakdown of substrates during assays because MUF substrates are highly enzyme specific.[37] The activity of bacterial exoenzymes, including phosphatase,[44,45] glucosidase,[37,46] protease,[47,48] and chitinase,[49] has been measured fluorometrically using commercially available fluorescent substrate analogs.

The hydrolysis of these fluorogenic analogs can be used to determine the kinetic properties of various enzyme systems, their spatial and temporal distribution, and their response to key environmental variables.[48,50] Fluorogenic measurements of glucosidase activity have been shown to correlate well with cellular uptake of organic substrates including glucose, thymidine, and amino acids.[6,46,47]

Lee et al.[6] proposed the development of an environmental assessment protocol using both MUF- and MCA-based fluorogenic substrates to assess the impact of toxic contaminants in sediments. MUF-β-glucoside, MUF-phosphate, MCA-leucine, and MCA-alanine were identified as potential substrates for detailed study since they possessed high activity rates within marine sediments. Optimal substrate concentrations (50 μM), incubation times (40 and 90 minutes), methods of sample fixation (low temperature freezing with dry ice, for example), sample volume size, test slurry density, preincubation times, and measuring routines were determined[51] (Figure 15.2). Dose-response curves were generated with various metals classified as Schedule I prohibited and regulated substances under the Ocean Dumping Regulations of the Canadian Environmental Protection Act (CEPA),[52] Part VI, to illustrate the sensitivity of the assay procedure (Figure 15.3). Variations in the values between the substrates under evaluation were most likely due to the differences in

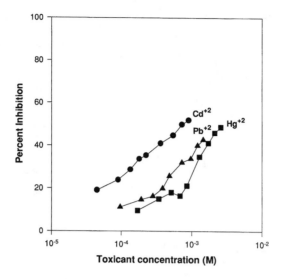

Figure 15.3 Exoenzyme (protease) activity dose response curves for Hg, Cd, and Pb.

sensitivity to toxicants of the various individual enzyme systems. For example, Dutton et al.[53] reported that β-galactosidase was insensitive to organic toxicants while being relatively sensitive to heavy metals. Montuelle et al.[54] reported that the β-glucosidase of a microbial community in river sediments differed in the presence of Pb and Cd. Pb was clearly more toxic (IC_{20} = 182 ppm) than Cd (IC_{20} = not detected). The use of multiple substrates in a test battery approach is thus recommended to give a more accurate estimate of toxicity.

III. ENVIRONMENTAL IMPACT ASSESSMENTS

Preliminary feasibility studies on the potential use of microbial exoenzyme activity measurements for environmental impact and dumpsite recovery has been supported by the Ocean Dumping Research Fund of Environment Canada.[6,55,56] Laboratory studies proved the sensitivity of the assay procedures to a wide range of toxic metals and hydrocarbons (the water soluble fraction of crude oils and condensates). Field studies showed that long-term impacts and recovery at dredge spoil disposal sites could be assessed by comparing microbial exoenzyme activity in sediments impacted by sediment disposal operations with those found in adjacent, nonimpacted areas.

A. Monitoring Biological Effects: The Heron Island Ocean Disposal Site

Harbors and waterways throughout the world are dredged for navigation on a regular basis. Sediments from these sites, usually contaminated by industrial activities, may pose a risk to the benthic biota and at ocean disposal sites following dredge spoils disposal. The effects of pollutants in an ecosystem depend on the conditions experienced. For example, differences in parameters such as organic carbon content can greatly affect pollutant toxicity.[57] Environmental impact assessment procedures carried out in unenclosed natural systems are thus the most realistic, since few extrapolations are needed. Although variability is generally higher in the results of field studies, subtle effects and interactions not present or easily detected in simplified systems can often be detected. Lee et al.[6] suggested that, even if predisposal microbial activity is unknown, comparison of microbial community activity in the impacted area with activity in unimpacted but similar adjacent areas can yield meaningful results.[6] This method of comparing microbial exoenzyme activity in an impacted dumpsite with activity in surrounding areas was subsequently used to assess

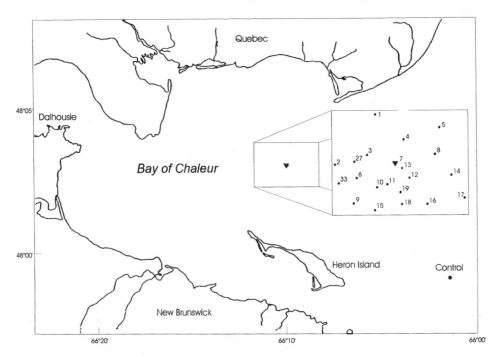

Figure 15.4   Location of sampling stations at the Heron Island ocean disposal site (Bay of Chaleur, New Brunswick, Canada).

the extent of habitat recovery at an ocean dumpsite in the Bay of Chaleur, New Brunswick, Canada, 10 years after the cessation of disposal activities.

Dalhousie, New Brunswick, Canada, is a major industrial port at the mouth of the Restigouche River on the Bay of Chaleur. The sediments in the harbor of this port are contaminated by industrial wastes associated with the operations of a pulp mill and an ore loading facility. Due to high siltation rates in the Restigouche River, the harbor requires dredging on a regular basis. Beginning in 1972, anoxic sediments contaminated by wood pulp and fibers and trace metals including Cd, Pb, and Zn were dredged from the harbor for disposal at an open-water dumpsite near Heron Island, 15 km from Dalhousie, in the Bay of Chaleur (Figure 15.4). Elevated metal concentrations in the dumpsite sediments (e.g., Cd at 1 to 2 mg kg^{-1}) forced the closure of the open-water disposal site in 1978.[58] Chemical analysis showed that a large proportion of the total metal concentration present was bound in organic complexes (mostly wood and bark fibers) and that approximately 40% of the total metal was released in a weak acid leach and deemed to be readily bioavailable. Macrobenthos were found to be relatively rare at the dumpsite, and those organisms present were characteristic of highly perturbed environments.[59] Geochemical and biological surveys conducted in 1980[58] and 1981[60-62] showed a reduction in metal concentrations at the dumpsite center, especially in the surficial sediment, since a thin layer of relatively clean sediment had been deposited over the old dredge spoils. However, benthic community diversity and structure were reported to be significantly different at the dumpsite, relative to surrounding areas.

As part of a large-scale geochemical and biological survey, ten years after the closure of the Heron Island dumpsite, levels of bacterial enzyme activity in sediments in the dumpsite area were assessed to determine the degree of habitat recovery at the dumpsite. Microbial exoenzyme activity (cellulase, phosphatase, and protease) was measured in slurries of freshly collected sediment. The data obtained from the dumpsite survey was plotted as topographical figures and three-dimensional surface figures (e.g., Figure 15.5). Protease activity at the dumpsite center was significantly lower

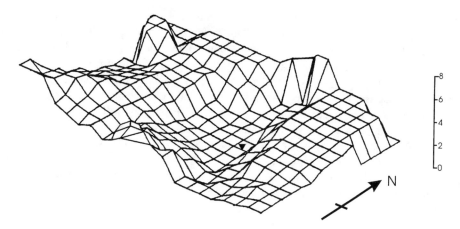

Figure 15.5 Three-dimensional representation of protease activity (μM/g/h) at the ocean disposal site area (Dumpsite center ▼).

($p < 0.05$) than at three control stations, as was observed with phosphatase. Analysis along a NW–SE transect that passed through the dumpsite zone showed a significant positive correlation between protease activity and distance from the dumpsite center. These results, and that of increased cellulase activity at the center of the dumpsite zone, relative to surrounding areas, support the theory of a persistent effect caused by the disposal of dredge spoils.[55] The enhancement of cellulase activity was most likely due to high concentrations of wood pulp and fibers in the dumpsite sediments.

The results of this preliminary study supported the premise that bacterial activity measurements can be used to monitor conditions at specific dump sites, to ascertain conditions before and after waste disposal events, to prescreen wastes to determine suitability for ocean disposal, and to assess the potential impact of disposal activities on large-scale processes.

B. Screening for Toxicity: The Saint John Harbour Black Point Ocean Disposal Site

A comprehensive marine sediment test battery approach has recently been used to evaluate the disposal impact of dredge spoils at the Black Point ocean disposal site, Saint John, New Brunswick.[63] The Bacterial Exoenzyme Test was evaluated against three sediment bioassay tests recommended by Environment Canada for dredge spoil assessment under the Canadian Environmental Protection Act, Part VI. These three bioassays are the Microtox® Porewater and Solid-Phase Tests,[64,65] the Echinoid Fertilization Test,[66] and the Amphipod Survival Test.[67] A total of 26 stations was sampled at the site to obtain baseline exoenzyme activity levels. Varying levels of exoenzyme activity inhibition were observed at these stations. The endpoints of the exoenzyme test were correlated with those of the Microtox Solid-Phase Test ($r = 0.79$, $p = 0.0002$, $n = 24$) and the two amphipod survival tests ($r = 0.85$, $p = 0.005$, $n = 12$ for *Amphiporeia virginiana* and $r = 0.68$, $p = 0.02$, $n = 12$ for *Rhepoxynius abronius*). The relation was slightly less but significant between the endpoints of exoenzyme and sea urchin tests ($r = 0.49$, $p = 0.02$, $n = 24$). Correlation between the endpoints of the exoenzyme test and the sediment heavy metal concentrations were also quite significant (Pb: $r = -0.48$, $p = 0.02$, $n = 24$; Zn: $r = -0.63$, $p = 0.003$; As: $r = -0.58$, $p = 0.005$; and Cu: $r = -0.50$, $p = 0.02$). Since the exoenzyme test is based on the microbial response of the indigenous bacterial community, it has the potential to be more site-relevant than current regulatory bioassays based on the response of single nonindigenous species isolates.

The Bacterial Exoenzyme Test was based on the modification of the method for sediment slurries (final substrate concentration of 50 μ*M*; incubation: 60 minutes) as described by Lee et al.[6] The determination of toxicity was based on the assumption that contaminants in the "test sediment" will have an effect (inhibition or stimulation) on the exoenzyme activity of a clean "uncontaminated" sediment. Theoretical activity values corrected for differences from dilution were determined by plotting a line between the measured values for the "uncontaminated reference sediment" (0%) and 100% "contaminated experimental sediment." Any deviation of greater than 20% from the theoretical value following the addition of "contaminated test sediment" at the 25% level was deemed a positive toxic response. On the basis of this operational endpoint, in comparison to the other regulatory tests used in the toxicity test battery, the Bacterial Exoenzyme Test was identified as the most sensitive procedure in 14 out of 19 samples. The tests were ranked in order of decreasing sensitivity as: Bacterial Exoenzyme Test = Microtox Solid-Phase Test > Echinoid Fertilization Test > Amphipod Survival Test > Microtox Pore Water Test.[68,69] The sensitivities of the bioassay tests and their environmental relevance to regulatory decision-making processes related to ocean disposal of dredged materials was demonstrated by this exercise.

IV. STANDARD PROTOCOL FOR THE DETERMINATION OF IC_X

The acute toxicity of compounds is traditionally defined in bioassay tests by the dosage which causes 50% mortality of test organisms (LD_{50}). However, contaminants in the environment are often present at concentrations, not acutely lethal, but that cause chronic or sublethal effects, which cannot be quantified by the LD_{50} parameter.[26] Parameters such as respiration, enzyme inhibition, and reproductive success are used to assess the sublethal effects of pollutants Chapman and Long[70] defined by the effective or inhibitory concentration which causes a known X% reduction in the rate of the process studied (i.e., EC_X or IC_X).

A. Test Method Description

The bioassay test protocol to determine IC_X values is based on the determination of exoenzyme activity rates in a sediment slurry dilution series prepared by mixing "test" and "uncontaminated reference" sediments in various proportions. Inhibitory effects (IC_X values) attributed to toxicity associated with the "test sediments" is determined by quantifying the difference in exoenzyme activity rates from theoretical values attributed to dilution alone (calculated from the measured values of the "uncontaminated reference" and "test" sediments).

The procedure can be broken down into five major components (Figure 15.6; Appendix A):

1. Sediment sample preparation (dilution series)
2. Substrate addition and incubation
3. Termination of exoenzyme activity and sample storage
4. Fluorescence analysis and dry weight determination
5. Calculations and endpoint determination

B. Test Validation: Baie Des Anglais

An inter-test validation exercise was completed in 1995 with an independent third-party private-sector laboratory (Beak Consultants Limited, Canada) to validate the standard operating procedure (SOP) described. The Bacterial Exoenzyme Test was compared to "regulatory" biotest procedures that included the Echinoid Fertilization Test,[66] the Toxi-ChromoPad® Test,[71,72] and the Microtox Solid Phase Test.[64]

The validation exercise was conducted with a series of sediment samples predominantly contaminated by polycyclic aromatic hydrocarbons, polychlorinated biphenyls and polychlorinated dibenzofurans.[73,74] These surficial (0 to 10 cm depth) sediment samples were collected along a

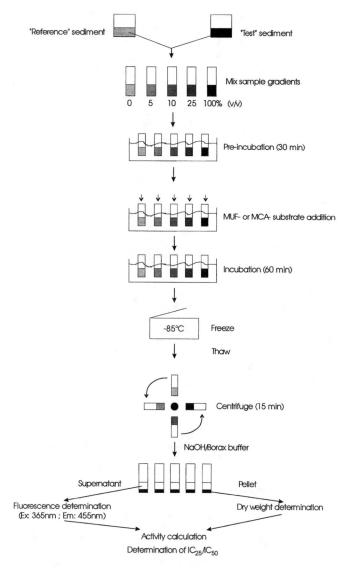

Figure 15.6 Procedure for the determination of IC_X values in marine sediments.

transect from the highly industrialized (aluminum refinery/pulp and paper mill) town of Baie Comeau located on the shore of Baie des Anglais, Québec, Canada (Figure 15.7). Chemical analysis of the sediment samples (SED-P-1, SED-P-2, SED-P-3) showed a toxicity gradient with distance from the shore (Table 15.1).

The order of toxicity ranking for the Bacterial Exoenzyme Test was similar to that obtained with the other "regulatory" tests (Table 15.1). The biotest results were also in agreement with the chemical data for total PAH concentrations.

V. CONCLUSIONS

Open-water disposal of solid or liquid wastes has become a cost-effective method of waste disposal worldwide, since it is cheaper than land filling or confined aquatic disposal.[75] Regulators

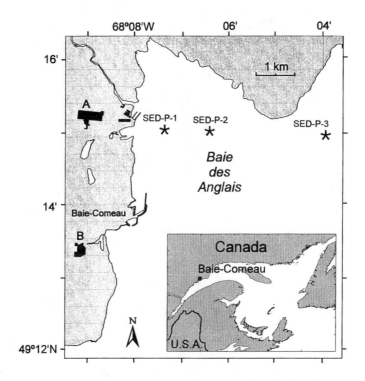

Figure 15.7 Sampling locations in the Baie des Anglais, Québec, Canada (A = aluminum smelter; B = pulp and paper mill).

Table 15.1 **Biotest Results and Corresponding Chemical Analysis Obtained for Contaminated Surface Sediments Collected from Baie des Anglais, Québec, Canada**

	Bacterial Exoenzyme Test (MCA-LEU)	Echinoid Fertilization Test	Toxi-ChromoPad Test		Microtox		Total PAH
	IC_{25} (%)	IC_{25} (%)	EC_{100} (%)	NOEC[1] (%)	IC_{50} (mg/L)[2]	IC_{50} (μL/L)[3]	(ng/g)
SED-P-1	5	22.1	25	3.1	2,878	150	41,101
SED-P-2	15	33.8	25	3.1	8,482	561	3,171
SED-P-3	25	69.5	ND	6.3	86,565	1,419	418

[1] No observable effect concentration.
[2] Microtox® Solid-phase test.
[3] Microtox® 100% test on solvent extract.
ND = not detected.

must now have an improved means of obtaining information that will clearly identify those sites where actual (biological) harm to the receiving environment has occurred or is occurring due to contamination. Bulk chemical analysis measures contamination and cannot be used by itself to determine pollution. As a result, there exists a great need for simple and cost-effective tests to evaluate the toxicity of waste materials to be disposed of at sea.[16,76]

There is now considerable concern that many bioassays using specific strains of indicator species overestimate "real world" impacts by using inappropriate organisms that do not actually live in the affected areas, or by having endpoints that are not indicative of *in situ* effects. Furthermore, laboratory test organisms may be insensitive to the pollutant of concern, as the ease of availability for many laboratory culture organisms is correlated to their inherent stress resistance. To address these problems and the current emphasis on improving the simplicity, cost-effectiveness, and reliability of microbiotests to enable the large-scale screening of xenobiotics (singularly or in mixtures) in the environment,[77] we have supported the development of a sediment bioassay test with an endpoint based on the inhibition of microbial exoenzyme activity by contaminants. To fully assess the impact of any polluting agent on the biological activity and chemistry of an impacted area, effects on microbial activity must be taken into account.[78] Preliminary studies have shown that the activity of microbial exoenzymes is sensitive to a range of toxic metals and organic contaminants (both organic and inorganic) in complex liquid/sediment samples. The biotest procedures described have high ecological relevance, as any inhibition on the exoenzyme activity of bacteria will have a negative effect on the mineralization of organic matter within the ecosystem.

The bacterial exoenzyme assay procedures described can be used in bioassays and as tools for monitoring the long-term effects of contaminants on marine environmental quality. The use of MUF/MCA substrates provides a rapid, inexpensive, and sensitive bioassay procedure that has many advantages over conventional sediment tests. MUF/MCA enzyme substrate analogs are available for a wide range of exoenzyme systems. Thus, a specific substrate can be chosen for many basic processes, such as nutrient cycling, carbohydrate or protein degradation. The high sensitivity of this method enables the use of short incubation periods (matter of hours) with low substrate concentrations. Traditional bioassay tests using bacteria with endpoints based on the measurement of growth rates, oxygen consumption, or colony formation[12] require lengthy incubations during which a single concentration of toxicant is exposed to increasing biomass.[79] These long periods of incubation increase the risk of selecting resistant forms and, thus, effects of a toxicant on these forms, not the whole population, are measured.[5]

We suggest that biochemical measurements of bacterial exoenzyme activity in natural sediments can be used to monitor conditions at ocean disposal and/or industrial sites such as offshore drilling platforms to ascertain conditions both before and after operations; prescreen wastes for toxicity and to determine allowable limits for oceanic disposal; determine the size of benthic impact zones; and monitor the recovery of sites following impact. Furthermore, since the proposed test procedures provide a rapid means of determining the status of the benthic environment (results obtained in a matter of hours), it can be used as an early warning system that allows remedial action to be taken before permanent damage to living resources occurs.

Simultaneous implementation of the Bacterial Exoenzyme Test with other procedures (based on other metabolic pathways) would enhance the reliability of our predictive capability. No single test will be suitable in all situations,[17,80] and it is advisable to employ a battery of tests in disposal impact assessment.[77] The choice of tests to be used depends on time, data requirements, biological and cost considerations.[81] Use of a wide spectrum of tests increases reliability and ease of interpretation of results in situations where sediment heterogeneity or differences in exposure routes may cause ambiguities. For routine monitoring of dump sites or effluents, microbial bioassays may be used for primary screening of wastes to indicate if further tests with higher levels of diversity are required. The Sediment Quality Triad approach, based on synoptic measurements of sediment contamination by chemical analysis, sediment toxicity through bioassays, and benthic infaunal community structure through taxonomic analysis of macrofauna, has been proposed as the ideal means of assessing marine pollution.[82,83] The procedure developed can be applied within existing monitoring programs to identify areas of concern where immediate regulatory action may be required. The assay can become part of a set of standard techniques used to regulate ocean disposal operations and to monitor recovery rates at current previously impacted sites.

Although much of the development for the Bacterial Exoenzyme Test has been completed, there is a continued effort to improve the analytical method[84] and to determine its operational limitations within the aquatic environment for the identification of environmental impacts caused by toxic contaminants.

ACKNOWLEDGMENTS

Funds for the development of the microbial exoenzyme biotest have been provided by the Ocean Dumping Research Fund (Environment Canada), the St. Lawrence Action Plan II, and the Panel of Energy Research and Development (PERD) Canada. Experimental results described in this chapter were attained with the aid of C. Bélanger, S.E. Cobanli, K. Doe, B. Desrosiers, C.N. Ewing, J. Gauthier, R. Larocque, and A.J. MacDonald.

REFERENCES

1. Cairns, J. Jr., Are single species toxicity tests alone adequate for estimating environmental hazard?, *Environ. Monitoring and Assess.*, 4, 259, 1984.
2. Bringmann, G. and Kuhn, R., Comparisons of the toxicity thresholds of water pollutants to bacteria, algae and protozoa in the cell multiplication inhibition test, *Water Res.*, 14, 231, 1980.
3. Qureshi, A. A., Flood, K. W., Thompson, S. R., Janhurst, S. M., Inniss, C. S., and Rokosh, D. A., Comparison of a luminescent bacterial test with other bioassays for determining toxicity of pure compounds and complex effluents, *Aquatic Toxicology and Hazard Assessment*: 5th Conf. ASTM STP 766, 1982, 179.
4. Dutka, B. J. and Kwan, K. K., Comparison of three microbial toxicity screening tests with the Microtox test., *Bull. Environ. Contam. Toxicol.*, 27, 753, 1981.
5. Barnhart, C. L. H. and Vestal, J. R., Effects of environmental toxicants on metabolic activity of natural microbial communities, *Appl. Environ. Microb.*, 46, 970, 1983.
6. Lee, K., Tay, K. L., Ewing, C. N., and Levy, E. M., Toxicity and environmental impact assessment tests based on the activity of indigenous bacteria, *Ocean Dumping Report #4*, Environment Canada, 1990a, 138.
7. Degobbis, D., Homme-Maslowska, E., Orio, A. A., Donazzolo, R., and Pavoni, B., The role of alkaline phosphatase in the sediments of Venice Lagoon on nutrient regeneration, *Estuar., Coast. Shelf Sci.*, 22, 425, 1986.
8. Atlas, R. M. and Bartha, R., *Microbial Ecology: Fundamentals and Applications*, The Benjamin/Cummings Publishing Company, Inc., Menlo Park, California, USA, 1987, 533.
9. Azam, F., Fenchel, T., Gray J. G., Meyer-Reil, L. A., and Thingstad, F., The ecological role of water-column microbes in the sea, *Mar. Ecol. Prog. Ser.*, 10, 257, 1983.
10. Boetius, A. and Lochte, K., Regulation of microbial enzymatic degradation of organic matter in deep-sea sediments, *Mar. Ecol. Prog. Ser.*, 104, 299, 1994.
11. Bitton, G., Bacterial and biochemical tests for assessing chemical toxicity in the aquatic environment: a review, *CRC Crit. Rev. Environ. Control*, 13, 51, 1983.
12. Liu, D., A rapid biochemical test for measuring chemical toxicity, *Bull. Environ. Contam. Toxicol.*, 26, 145, 1981.
13. White, P. A. and Côté, C., Investigating the sources and fate of genotoxic substances in aquatic ecosystems with the SOS Chromotest, in *Microscale Testing in Aquatic Toxicology: Advances, Techniques and Practice*, Wells, P. G., Lee, K., and Blaise, C., Eds., CRC Press, Boca Raton, Florida, 1998, 607.
14. Xu, H. H., ATP-TOX System — a review, in *Microscale Testing in Aquatic Toxicology: Advances, Techniques and Practice.*, Wells, P. G., Lee, K., and Blaise, C., Eds., CRC Press, Boca Raton, Florida, 1998, 153.
15. Riedel, B., and Christensen, G., Effect of selected water toxicants and other chemicals upon adenosine triphosphatase activity *in vitro*, *Bull. Environ. Contam. Toxicol.*, 23, 365, 1979.
16. Buikema, A. L., Rutherford, C. L., and Cairns, J. Jr., Screening sediments for potential toxicity by *in vitro* enzyme inhibition, in *Contaminants and Sediments*, Baker, R. A., Ed., Ann Arbor Sci., Ann Arbor, Michigan, 1980, 463.
17. Vives-Rego, J., Vague, D., and Martinez, J., Effect of heavy metals and surfactants on glucose metabolism, thymidine incorporation and exoproteolytic activity in sea water, *Water Res.*, 20, 1411, 1986.

18. Obst, U., Holzapfel-Pschorn, A., and Wiegand-Rosinus, M., Application of enzyme assays for toxicological water testing, *Toxicol. Assess.*, 3, 81, 1988.
19. Snell, T. W., Mitchell, J. L., and Burbank, S. E., Rapid toxicity assessment with microalgae using *in vivo* esterase inhibition, in *Techniques in Aquatic Toxicology*, Ostrander, G. K. Ed., CRC Press, Boca Raton, Florida, 1996, 13.
20. Duddridge, J. E. and Wainwright, M., Enzyme activity and kinetics in substrate amended river sediments, *Water Res.*, 16, 329, 1982.
21. Burton, G. A., Jr. and Lanza, G. R., Sediment microbial activity tests for the detection of toxicant impacts, in *Aquatic Toxicology and Hazard Assessment: 7th Symposium*, Cardwell, R. D., Purdy, R., and Bahner, R. C., Eds., ASTM STP 854, 1985, 214.
22. Bitton, G. and Koopman, B., Biochemical tests for toxicity screening, in *Toxicity Testing Using Microorganisms*, 1, Bitton, G. and Dutka, B. J., Eds., CRC Press, Boca Raton, Florida, 1986, 27.
23. Kanazawa, S. and Filip, Z., Distribution of microorganisms, total biomass and enzyme activities in different particles of brown soil, *Microb. Ecol.*, 12, 205, 1986.
24. Chrost, R. J., Microbial ectoenzymes in aquatic environments, in *Aquatic Microbial Ecology: Biochemical and Molecular Approaches*, Overbeck, J. and Chrost, R. J., Eds., Springer-Verlag, New York, 1990, 47.
25. Wieser, W. and Zech, M., Dehydrogenases as tools in the study of marine sediments, *Mar. Biol.*, 36, 113, 1976.
26. Liu, D. and Thomson, K., Toxicity assessment of chlorobenzenes using bacteria, *Bull. Environ. Contam. Toxicol.*, 31, 105, 1983.
27. Armant, D. R., Buikema Jr., A. L., Rutherford, C. L., and Cairns, J. Jr., Evaluation of *in vitro* enzyme inhibition for screening petroleum effluents, *Bull. Environ. Contam. Toxicol.*, 24, 244, 1980.
28. Griffiths, R. P., Caldwell, B. A., Broich, W. A., and Morita, R. Y., Long-term effects of crude oil on microbial processes in subarctic marine sediments: Studies on sediments amended with organic nutrients, *Mar. Pollut. Bull.*, 13, 273, 1982.
29. Cenci, G., Caldini, G., and Morozzi, G., Chlorinated phenol toxicity by bacterial and biochemical tests, *Bull. Environ. Contam. Toxicol.*, 38, 868, 1987.
30. Allen, H. L., Dissolved organic matter in lake water: characteristics of molecular weight size-fractions and ecological implications, *Oikos*, 27, 64, 1976.
31. Munster, U., Investigations about structure, distribution and dynamics of different organic substrates in the DOM of Lake Plubsee, *Arch. Hydrobiol. Suppl.*, 70, 429, 1985.
32. Weiss, M. S., Abele, U., Weckesser, I., Welte, W., Schiltz, E., and Schulz, G. E., Molecular architecture and electrostatic properties of a bacterial porin, *Science*, 254, 1627, 1991.
33. Priest, F. G., *Extracellular Enzymes*, American Society for Microbiology, Washington, DC, 1984, 79.
34. Chrost, R. J., Environmental control of the synthesis and activity of microbial ectoenzymes, in *Microbial Enzymes in Aquatic Environments*, Chrost, R. J., Ed., Springer-Verlag, New York, 1991, 29.
35. Rogers, H. J., The dissimilation of high molecular weight organic substances, in *The Bacteria.*, Gunsalus, I. C. and Stanier, R. Y., Eds., Academic Press, New York, 1961, 261.
36. Billen, G., Modelling the processes of organic matter degradation and nutrients recycling in sedimentary systems, in *Sediment Microbiology*, Nedwell, D. B. and Brown, C. M., Eds., Academic Press, London, 1982, 15.
37. Hoppe, H.-G., Significance of exoenzymatic activities in the ecology of brackish water: measurements by means of methylumbelliferyl substrates, *Mar. Ecol. Prog. Ser.*, 11, 299, 1983.
38. King, G. M., Characterization of B-glucosidase activity in intertidal marine sediments, *Appl. Environ. Microbiol.*, 51, 373, 1986.
39. Meyer-Reil, L.-A., Ecological aspects of enzymatic activity in marine sediments, in *Microbial Enzymes in Aquatic Environments*, Chrost, R. J., Ed., Springer-Verlag, Berlin, 1991, 84.
40. Smith, D.C., Simon, M., Alldredge, A. L., and Azam, F., Intense hydrolytic enzyme activity on marine aggregates and implications for rapid particle dissolution, *Nature*, 359, 139, 1992.
41. Hoppe, H.-G., Ducklow, H., and Karrasch, B., Evidence for dependency of bacterial growth on enzymatic hydrolysis of particulate organic matter in the mesopelagic ocean, *Mar. Ecol. Prog. Ser.*, 93, 277, 1993.
42. Kanaoka, Y., Takahashi, T., and Nakayama, H., A new fluorogenic substrate for aminopeptidase, *Chem. Pharm. Bull.*, 25, 262, 1977.

43. Meyer-Reil, L.-A., Measurement of hydrolytic activity and incorporation of dissolved organic substrates by microorganisms in marine sediments, *Mar. Ecol. Prog. Ser.*, 31, 143, 1986.

44. Perry, M. J., Alkaline phosphatase activity in subtropical Central North Pacific waters using a sensitive fluorometric method, *Mar. Biol.*, 15, 113, 1972.

45. Petterson, K. and Jansson, M., Determination of phosphatase activity in lakewater — a study in methods, *Verh. Internat. Verein. Limnol.*, 20, 1226, 1978.

46. Somville, M., Measurement and study of substrate specificity of exoglucosidase activity in eutrophic water, *Appl. Environ. Microbiol.*, 48, 1181, 1984.

47. Somville, M. and Billen, G., A method for determining exoproteolytic activity in natural waters, *Limnol. Oceanogr.*, 28, 190, 1983.

48. Hoppe, H.-G., Schramm, W., and Bacolod, P., Spatial and temporal distribution of pelagic microorganisms and their proteolytic activity over a partly destroyed coral reef, *Mar. Ecol. Prog. Ser.*, 44, 95, 1988.

49. O'Brien, M. and Colwell, R. R., A rapid test for chitinase activity that uses 4-methylumbelliferyl-N-acetyl-B-D-glucosaminide, *Appl. Environ. Microbiol.*, 53, 1718, 1987.

50. Tubbing, D. M. J. and Admiraal, W., Sensitivity of bacterioplankton in the Rhine River to various toxicants measured by thymidine incorporation and activity of exoenzymes, *Environ. Toxicol. Chem.*, 10(9),1161, 1991.

51. Lee, K., Larocque, R., and Tay, K. L., Development of environmental impact assessment tests based on the exoenzyme activity of indigenous bacteria, *Proceedings of the 21st Annual Aquatic Toxicity Workshop, October 2-5, 1994, Sarnia, Ontario, Can. Tech. Rep. Fish. Aquat. Sci.*, 2050, 173, 1995.

52. Ocean Dumping Control Act, *Ocean Dumping Control Act Regulations*, House of Commons. Canada Gazette, Part II, 109, 1975, 2786.

53. Dutton, R. J., Bitton, G., and Koopman, B., Enzyme biosynthesis versus enzyme activity as a basis for microbial toxicity testing, *Tox. Assess.*, 3, 245, 1988.

54. Montuelle, B., Latour, X., Volat, B., and Gounot, A. M., Toxicity of heavy metals to bacteria in sediments, *Bull. Environ. Contam. Toxicol.*, 53, 753, 1994.

55. Lee, K., Tay, K. L., Levy, E. M., Ewing, C. N., and Cobanli, S. E., Microbial exoenzyme activity at the Heron Island ocean dumpsite (Bay of Chaleur): 10 Years after the disposal of dredged sediment from Dalhousie, New Brunswick, *Ocean Dumping Report #5*, Environment Canada, 1990, 82.

56. Lee, K., Tay, K. L., Levy, E. M., Ewing, C. N., and Cobanli, S. E., Application of microbial exoenzyme activity measurements to assess the impact of dredge spoils disposal in Pictou Harbour, Nova Scotia and the Miramichi River, New Brunswick, *Ocean Dumping Report #6*, Environment Canada. 1990, 95.

57. Cunningham, V. L., Morgan, M. S., and Hannah, R. E., Effect of natural water source on the toxicity of chemicals to aquatic microorganisms, in *Aquatic Toxicology and Environmental Fate*, 9, Poston, T. M. and Purdy, R., Eds., ASTM STP 921, 1986, 436.

58. MacKnight, S. D., The Dalhousie ocean dumpsite, 1978 and 1980: A geochemical view of history, in *3rd Internat. Oceans Dumping Symp.*, Woods Hole, Mass. Abstracts., 1981, 87.

59. MacKnight, S. D. and Schafer, C. T., Geochemistry and benthic ecology at the Dalhousie, New Brunswick ocean dump site, in *Proc. 9th World Dredging Conf.* (WODCON IX), Vancouver, B.C, 1980, 761.

60. Packman, G. A., Tay, K. L., and Berman, C., Environmental monitoring at the Heron Island dump site in the Bay of Chaleur near Dalhousie, *N.B. Environment Canada Atlantic Region Surveillance Report*, EPS-5-AR-84-6, 1984, 91.

61. MacKnight, S. D., Geochemistry of the Dalhousie ocean dumpsite, in *Wastes in the Ocean: Nearshore Waste Disposal*, 6, Ketchum, B., Capuzzo, J., Burt, W., Duedall, I., Park, P., and Kester, D., Eds., 1985, 281.

62. Tay, K. L., An overview of dredged material ocean dump site monitoring and assessment programs and dump site designation conducted by Environment Canada under the Ocean Dumping Control Act in the Atlantic Region, in *Proc. 7th Internat. Ocean Disposal Symp.*, Wolfville, Nova Scotia, Canada, September 21-25, 1987, 292.

63. Tay, K. L., Doe, K. G., MacDonald, A. J., and Lee, K., Monitoring of Black Point Ocean Disposal Site, Saint John Harbour, New Brunswick, 1992-1994. *Ocean Dumping Report #9*, Environment Canada, 1997, 132.

64. Microbics Corporation, *Solid-Phase Test Protocols*, Microbics Corporation, Carlsbad, CA, 1991, 49.

65. Environment Canada, *Biological Test Method: Toxicity Test Using Luminescent Bacteria (Photobacterium phosphoreum)*, Report EPS 1/RM/24, 1992a.

66. Environment Canada, *Biological Test Method: Fertilization Assay Using Echinoids (Sea Urchins and Sand Dollars)*, Report EPS 1/RM/27, 1992b.

67. Environment Canada, *Biological Test Method: Acute Lethality Test Using Marine and Estuarine Amphipods*, Report EPS 1/RM/27., 1992c.

68. Tay, K. L., Doe, K. G., MacDonald, A. J., Wade, S. J., Vaughan, J. D. A., Huybers, A. L., Lee, K., and Larocque, R., Application of sediment bioassays for the environmental assessment of ocean dump sites, in Proceedings of the 20th Annual Aquatic Toxicity Workshop, October 17-21, 1993, Quebec City, Quebec. *Can. Tech. Rep. Fish. Aquat. Sci.*, 1989, van Coillie, R., Roy, Y., Bois, Y., Campbell, P. G. C., Lundahl, P., Martel, L., Michaud, M., Riebel, P., and Thellen, C., Eds., 1994, 48.

69. Tay, K. L., Doe, K. G., MacDonald, A. J., Wade, S. J., Vaughan, J. D. A., Huybers, A. L., Wohlgeschaffen, G. D., Lee, K., and Larocque, R., Environmental assessment of a dredged material dump site using a multitrophic level sediment bioassay and bioaccumulation test, Proceedings of the 21st Annual Aquatic Toxicity Workshop, October 2-5, 1994, Sarnia, Ontario. *Can. Tech. Rep. Fish. Aquat. Sci., 2050,* Westlake, G. F., Parrott, J. L., and Niimi, A. J., Eds., 1995, 50.

70. Chapman, P. M. and Long, E. R., The use of bioassays as part of a comprehensive approach to marine pollution assessment, *Mar. Pollut. Bull.*, 14, 81, 1983.

71. Kwan, K. K., Direct sediment toxicity testing procedure using the Sediment-Chromotest kit, *Environ. Toxicol. Wat. Qual.*, 9, 193, 1995.

72. Kwan K. K, and Dutka, B. J., Evaluation of the Toxi-Chromotest direct sediment toxicity-testing procedure and Microtox solid-phase testing procedure, *Bull. Environ. Contam. Toxicol.,* 49, 656, 1992.

73. Lacroix, A., Nagler, J., Lee, K., Lebeuf, M., Fournier, M., and Cyr, D., Immunotoxicity in plaice exposed to marine sediments in Baie des Anglais on the St. Lawrence Estuary, in *Abstracts of the 1996 Annual Meeting of the Society of Environmental Toxicology and Chemistry (SETAC)*, November 17-21, 1996, Washington, DC, 1996, 78.

74. Lebeuf, M., Moore, S., and Brochu, C., The influence of PCBs and PCDFs in the sediments of Baie des Anglais on the Lower St. Lawrence Estuary, *Organohalogen Compd.*, 28, 243, 1996.

75. Iskandar, I. K., Cragin, J. H., Parker, L. V., and Jenkins, T. F., Impact of dredging on water quality at Kewaunee Harbour, Wisconsin, US Army Corps of Engineers, *CRREL* Rep. 84-21, 1984.

76. Busch, A. W., Bioassay technique for relative toxicity in water pollution control, *J. Water Pollut. Contr. Fed.*, 54, 1152, 1982.

77. Blaise, C., Microbiotests in aquatic ecotoxicology: characteristics, utility, and prospects, *Environ. Toxicol. Wat. Qual.*, 6, 145,1991.

78. Griffiths, R. P., The importance of measuring microbial enzymatic functions while assessing and predicting long-term anthropogenic perturbations, *Mar. Pollut. Bull.*, 14, 162, 1983.

79. Trevors J. T., Mayfield, C. I., and Innis, W. E., A rapid toxicity test using *Pseudomonas fluorescens, Bull. Environ. Contam. Toxicol.*, 26, 433, 1981.

80. Coleman, R. N. and Qureshi, A. A., Microtox and *Spirillum volutans* tests for assessing toxicity of environmental samples, *Bull. Environ. Contam. Toxicol.*, 35, 443, 1995.

81. Williams, L. G., Chapman, P. M., and Ginn, T. C., A comparative evaluation of marine sediment toxicity using bacterial luminescence, oyster embryo and amphipod sediment bioassays, *Mar. Environ. Res.*, 19, 225, 1986.

82. Long, E. R. and Chapman, P. M., A sediment quality triad: measures of sediment contamination, toxicity and infaunal community composition in Puget Sound., *Mar. Pollut. Bull.*, 16, 405, 1985.

83. Chapman, P. M., Dexter, R. N., and Long, E. R., Synoptic measures of sediment contamination, toxicity and infaunal community composition (the Sediment Quality Triad) in San Francisco Bay, *Mar. Ecol. Prog. Ser.*, 37, 75, 1987.

84. Bélanger, C., Desrosiers, B., and Lee, K., Microbial extracellular enzyme activity in marine sediments: extreme pH to terminate reaction and sample storage, *Aquat. Microb. Ecol.*, 13, 187, 1997.

APPENDIX A: BACTERIAL EXOENZYME TEST PROTOCOL

The following method is for the use of the substrates leucine-4-methyl-7-coumarinylamide (MCA-Leu) and 4-methylumbelliferone β-D-glucoside (MUF-Glu):

1. **Stock solution preparation**
   a. 0.22 μm filtered sea water (e.g., Millipore Type GS).
   b. 5.2 mM solution of MCA-Leu (L-Leucine-4-methyl-7-coumarinylamide hydrochloride) or MUF-Glu (4-methylumbelliferone β-D-glucoside) prepared in 25 mL deionized water containing 10% methanol. 1 mL volumes of working stock solutions are frozen in cryovials (–80°C) for storage until use.
   c. 0.05 mM MUF or MCA standard solutions in 100% methanol. Upon preparation, immediately transfer to vials and freeze at –80°C.
   d. NaOH/Borax buffer solution prepared by mixing 1:1 a 0.2 M NaOH solution with a 0.05 M solution of sodium tetraborate (borax) and adjusting the pH to 10.8 using hydrochloric acid.

2. **Sediment sample preparation (dilution series)**
   Note:If the test is not initiated within 6 hours of sample collection, samples destined for analysis should be frozen (≤20°C) immediately after collection.
   a. Label preweighed sterile polypropylene 15 mL centrifuge tubes and lids with sample name and test concentration in triplicate.
   b. The test is conducted in triplicate. Prepare at least 100 mL of each test concentration in polypropylene beakers by mixing the test sample with a "clean" reference sample in the appropriate proportions (i.e., to prepare the 25% test concentration, homogenize 25 mL of test sediment sample and 75 mL of the "clean" reference sample with a stainless steel spatula or a plastic rod).
   c. A 1.2 mL aliquot of each sediment dilution is transferred to six labeled and preweighed test tubes, as follows: three tubes with the test sediment; two tubes with standard and sediment; and, one "blank" tube with sediment and sea water only.
   Repeat this for each of the five gradient concentrations (0, 5, 10, 25, 100%) and the blank in the dilution series.
   d. Make a sediment slurry by adding 3 mL of filtered sea water.
   e. Vortex or mix well (3 to 4 seconds) prior to preincubation (30 minutes) at the *in situ* temperature.

3. **Substrate addition and incubation**
   a. For each gradient concentration, add 40 μL of working MUF-Glu (or MCA-Leu) stock into three centrifuge tubes containing the sediment samples. Add 100 μL of the standard MUF or MCA to two of the tubes containing the sediment samples.
   b. To facilitate the calculation of a correction factor to account for the adsorption of MUF or MCA to sediments, prepare two extra tubes with standard solution and sea water only, by adding 100 μL of MUF/MCA standard to 4.2 mL of 0.22 μm-filtered sea water.
   c. Homogenize all test tubes using a vortex mixer (mix for 3 to 4 seconds) and return to the *in situ* temperature water bath for a 1-hour incubation period.

4. **Termination of exoenzyme activity and sample storage**
   a. After 1 hour of incubation from the time of substrate addition, enzyme activity is terminated by transferring the test tubes into crushed dry ice, a liquid nitrogen bath, or a low temperature (–85°C) freezer.
   b. Tubes are stored at low temperature (–85°C) until the fluorescence analysis.

5. **Fluorescence analysis and dry weight determination**
   a. The test tubes are thawed in batches (maximum 20 tubes per batch). Remove the test tubes containing sediment from the freezer and thaw in a water bath at 5°C. It takes approximately 10 minutes for the samples to thaw.
   b. Centrifuge the tubes for 15 minutes at 2000 g at 2°C. Sample tubes are kept in an ice bath until fluorescence analysis to inhibit exoenzyme activity.
   c. Transfer 3.0 mL of the supernatant from the test tubes held in the low temperature water bath to a 1 cm path-length cuvette (glass or quartz). Add 0.25 mL of the sodium/borax buffer and mix.

    d. Conduct fluorescence analysis (excitation: 365 nm; emission: 455 nm).

    e. Determine the dry weight of residual sediments within each of the tubes after drying at 80°C temperature for 48 hours (or until constant dry weight is obtained).

6. **Calculations and endpoint determination**

    a. Calculation of corrected sediment dry weights:

$$P_{dry\ sed} = PS_{t+s} - P_t$$

where P_{t+s} is the dry sediment and tube weight, and P_t is the tube weight.

    b. Determination of fluorescence *per se*diment dry weight:

$$Fl_w = \frac{Fl_{read}}{P_{dry\ sed}}$$

where Fl_{read} is the fluorescence reading of the sample, and $P_{dry\ sed}$ is the corrected sediment dry weight (as determined above).

    c. Adsorption factor calculation:

$$A = \frac{Fl_{ss}}{Fl_{ss+s}}$$

where Fl_{ss} is the fluorescence reading of the standard (MUF/MCA) in sea water, and Fl_{ss+s} is the fluorescence reading of the sediment sample containing the standard (MUF/MCA).

    d. Fluorescence corrected for adsorption:

$$Fl_{corr} = Fl_w \times A$$

where A is the adsorption factor, and Fl_w is the fluorescence *per se*diment dry weight.

    e. Quantity of MUF/MCA produced (nmol/g dry sediment/h):

$$\frac{Qty.\ MUF/MCA}{g \times h} = \frac{5.0 \times Fl_{corr}}{t \times Fl_{ss}}$$

where 5.0 is a constant (in nmol) and represents the quantity of standard (i.e., 100 μL) of known molarity (i.e., 0.05 mM) added to standard (MUF/MCA) in a sea water sample, Fl_{corr} is the fluorescence per dry weight after corrections for coloration and adsorption (as above), t is the incubation time (h), and Fl_{ss} is the fluorescence reading of the standard (MUF/MCA) in sea water.

    f. Endpoint determination:

    A theoretical "no effect" line representing changes in exoenzyme activity associated with the dilution of the "clean reference" sediment by the "test sediment" is established (activity in the "clean reference" sediment represents 100% activity). Relative values obtained with the dilution series samples (e.g., 5, 10, and 25% of test sediment) are plotted. Values located above the theoretical line represent stimulation of activity by the contaminant while values below the line are indicative of inhibition. IC_x values (based on the inhibition of enzyme activity within the "clean reference" sediment fraction) are determined by graphical analysis of the data, as illustrated in Figure 15.8.

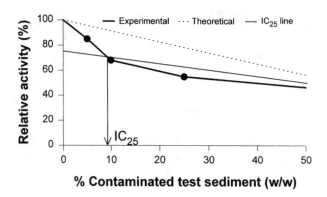

Figure 15.8 Graphical representation illustrating the principal of data analysis for the determination of IC_{25} values.

CHAPTER 16

Small-Scale *In Vitro* Genotoxicity Tests for Bacteria and Invertebrates

P.-D. Hansen and A. Herbert

CONTENTS

0-8493-2626/5/98/$0.00+$.50
© 1998 by CRC Press LLC

I. INTRODUCTION

Monitoring the effects of environmental toxic and genotoxic substances has gained increased importance in recent times. Assessment of environmental samples for the presence of genotoxins (mutagens and carcinogens) has, therefore, become a valuable approach for evaluating the genotoxic potentials of environmental pollutants. Effects at the organismic level, such as DNA-damage and mutagenesis, would eventually result in contaminant genotoxic effects at macromolecular levels. Since environmental genotoxicity assessment approaches provide information on such effects they could be potential environmental monitoring tools.

Table 16.1 provides information on the consequences of DNA damage at the organizational levels of biological systems.

Extended exposure of organisms to environmental genotoxins would result in several physiological disorders such as reproductive impairments and other related abnormalities. Therefore, the response measurements to reproductive toxicity are essential for assessing the effects of anthropogenic sources.

Several biomonitoring approaches currently exist. For the organizational level of bacteria the *umu*C-assay, and for invertebrates (mussels) the DNA-unwinding assay are available as monitoring tools. Both test systems exist as microscale assays (microplate technique). The development of a genotoxicity sensor using the principle of the *umu* test system is currently in progress.

The German Institute for Standardisation (DIN) has recently standardized the *umu*C-assay as an official protocol for monitoring environmental genotoxicity (DIN 38415, part 3). For this test system, the genetically engineered bacterium *Salmonella typhimurium* TA1535/pSK1002 serves as the test organism. The assay will be standardized under the International Standardization Organisation (ISO CD13829).

The test bacteria are exposed under controlled conditions to different concentrations of the samples to be tested. The *umu* assay is based on the capability of genotoxic agents to activate the *umu*C operon in the *Salmonella* strain in response to genotoxic lesions in the DNA.[2-4] Due to its capability to respond to different types of genotoxic lesions, only one single strain is necessary to detect different kinds of genotoxic substances. The activation of the *umu*C operon is thus a measure for the genotoxic effects of the material tested. Since the *umu*C operon is fused with the *lac*Z-gene for the β-galactosidase, the activation of the *umu*C operon can be easily assessed by the determination of β-galactosidase activity.

Table 16.1   Consequences of DNA Damage of Different Organizational Levels of Biological Systems

Level of Biological Organization	Effects
DNA	Mutations
Cell	Cell death
	Disordered proliferation and differentiation
	Neoplastic transformation
Tissue/Organ	Functional defects
	Malformations
	Tumors
Organism	Reduced viability
	Reduced fertility
Population	Reduced population size
	Extinction
Ecosystem	Reduced species diversity

From Hansen, P.-D., *Environmental Toxicology Assessment*, Richardson, M., Ed., Taylor & Francis, London, 1995.

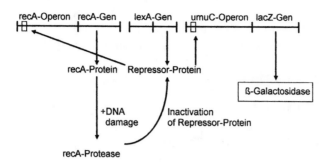

Figure 16.1   Principle of the *umu*C assay.

Microscale *in vitro* genotoxicity tests have to meet some criteria for routine use in environmental regulatory practices. The assay has to generate significant and reproducible results. The assays should be economical and should be done with the unaltered water samples (Figure 16.1).

II. THE *UMU* ASSAY PROTOCOL

A. Principle

The *umu* assay is based on the detection of genotoxicity in a test sample which increases the expression of the SOS-repair system associated with the *umu*C gene, compared to control.

B. Range of Application

The procedure can be used to determine the genotoxic responses in water and wastewater and in other aqueous samples.

C. Interferences

Undissolved substances can provide false test results and/or can hinder reproducibility. In colored and/or turbid effluents, light loss due to absorption can occur in the photometric measurements. In this case, an uninoculated control should be run as a blank. If a sample contains high levels of cytotoxic materials, these may impede cell division and can lead to cell death.

D. Test Organism

Salmonella typhimurium is a Gram-negative, facultative anaerobic bacterium from the Enterobacteriaceae family. *Salmonella typhimurium* TA 1535 is the original test strain. The test organism is *Salmonella typhimurium* TA1535/pSK1002, which carries the plasmid pSK1002 along with the *umu*C-*lac*Z gene, and a marker gene for ampicillin resistance.

E. Stock Culture Preparation and Preservation

Salmonella typhimurium TA1535/pSK1002 is preserved in 150 μL culture medium (TGA) with 10% dimethyl sulfoxide (DMSO) in 2 mL ampules at a temperature not above −80°C. In the preparation of an overnight culture of the bacteria, one ampule at a time is used.

F. Reagents

All chemicals used are of analytical grade. All solutions are prepared in sterile distilled water.

Hydrochloric acid (HCl) = 37 g/L
Sodium hydroxide solution (NaOH) = 40 g/L
Dimethyl sulfoxide (DMSO), $C_2H_6SO_4$

1. Culture Medium (TGA-Medium: Tryptone, Glucose, Ampicillin)

Combine 10 g tryptone, 5 g sodium chloride (NaCl), and 11.9 g HEPES, dissolved in water, pH 7.0 ± 0.2, fill to 980 mL and autoclave 20 min at 121°C. Dissolve 2 g D(+)-glucose anhydrous in 20 mL distilled water and autoclave separately. After autoclaving, mix both batches 1:1 and add 50 mg ampicillin to 1000 mL cooled TGA medium under sterile conditions. If the solution is not to be used on the same day, it can be stored in portions at –20°C for up to 4 weeks.

2. Concentrated (10× TGA) Culture Medium

Can be stored at 4°C for 14 days.

3. For Incubation without S9

Dissolve 10 g tryptone, 5 g sodium chloride (NaCl), and 11.9 g HEPES in 80 mL water. Adjust the pH to 7.0 ± 0.2. Dissolve 2 g D(+)-glucose (water free) in 20 mL water. Autoclave both solutions separately for 20 min at 121°C, cool, and mix the solutions under aseptic conditions. Then add 50 mg ampicillin to 100 mL under sterile conditions.

4. For Incubation with S9

Dissolve 10 g tryptone, 5 g sodium chloride (NaCl), 2.46 g potassium chloride (KCl), 1.63 g magnesium chloride hexahydrate ($MgCl_2 \cdot 6\ H_2O$), and 11.9 g HEPES in 80 mL water. Adjust the pH to 7.0 ± 0.2. Dissolve 2 g D(+)-glucose (water free) in 20 mL distilled water.

Autoclave each solution separately for 20 min at 121°C, cool, mix under microbiologically sterile conditions, and add 50 mg ampicillin (made up in 5 mL highly purified water and filter sterilized) to 100 mL of the mixture under sterile conditions.

5. B-Buffer (Cell Lysis and Reaction Buffer)

Dissolve 20.18 g disodium hydrogen phosphate dihydrate ($Na_2HPO_4 \cdot 2\ H_2O$), 5.5 g sodium dihydrogen phosphate monohydrate ($NaH_2PO_4 \cdot H_2O$), 0.75 g potassium chloride (KCl) and 0.25 g magnesium sulfate heptahydrate ($MgSO_4 \cdot 7\ H_2O$) in water. Adjust the pH to 7.0 ± 0.2. Then add 1.0 g sodium dodecylsulfate (SDS), and make up to 1 L. Before use, add 0.27 mL 2-mercaptoethanol to 100 mL B-buffer and mix.

6. Phosphate Buffer pH (7.0 ± 0.2)

Dissolve 1.086 g disodium hydrogen phosphate dihydrate ($Na_2HPO_4 \cdot 2\ H_2O$) and 0.538 g sodium dihydrogen phosphate monohydrate ($NaH_2PO_4 \cdot H_2O$), in 100 mL sterile distilled water. Adjust the pH value to 7.0 ± 0.2. Autoclave the solution at 121°C.

7. Stop Reagent

Dissolve 105.99 g sodium carbonate (Na_2CO_3), in water and dilute to 1000 mL.

8. o-nitrophenol-β-ᴅ-galactopyranoside (ONPG) Solution

Dissolve 45 mg ONPG in 10 mL phosphate buffer (see Step II.F.6). This solution should be prepared 2 h before use. The solution should be stored in the dark.

9. S9 Fraction

The S9 fraction is induced by Arochlor 1254 and a minimum protein content of 34 mg protein/mL. On the day of the test, thaw it slowly and keep it on ice. Shake briefly before adding to the respective preculture in the TGA. Store the S9-fraction at least at –80°C.

10. Cofactor Solution

The solution should be prepared on the day of the test and should be kept on ice during the test. Dissolve 148 mg NADP (sodium salt) and 76 mg glucose-6-phosphate (disodium salt) in 5 mL 10× TGA.

11. Reference Substances in Dimethyl Sulfoxide (DMSO)

Dissolve 5 mg 4-nitroquinoline-1-oxide (4-NQO) in 5 mL DMSO and dilute to 500 µg 4-NQO/L with a DMSO-/water mixture (3: 7) for stock solution (×10).

Dissolve 5 mg aminoanthracene (2-AA) dissolved in 5 mL DMSO and adjust at 2 mg 2-AA/L for stock solution (×10) with a DMSO-/water mixture (3+7).

G. Apparatus

Storage bottles (250 and 500 mL)
1, 10, and 25 mL pipettes
Multichannel pipettes (8 channels), volume 5 to 50 µL, 50 to 200 µL, 50 to 300 µL
Reservoirs for charging the multichannel pipettes (volume 17 and 34 mL)
Measuring cylinders, 500 mL
Culture vessels, Erlenmeyer flasks, 100 mL
Preservation ampules 2 mL (cryotubes)
Thermometer (measuring in the range 25 to 30°C (±1°C)
Temperature-controlled water bath for maintenance of bacteria suspension
Incubator
Microplate incubator for 28°C and 37°C
Rotator with diagonal mountings for the microplates
Autoclave
Centrifuge
96-well microplates (380 µL capacity) with flat transparent bottoms
Photometer for the microplates
Microplate sorting instrument (with plotting unit if possible)

H. Sample Preparation and Preservation

Water and wastewater samples should be tested as soon as possible after collecting. If immediate testing is not possible, the sample must be stored at 4°C. Deep freezing preservation shall be applied in the preservation of samples only when samples are not processed within 48 h of collection.

Note: In exceptional cases, centrifugation and filtration of samples can be done. This will probably lead to the elimination or reduction of genotoxic substances in the samples. Solvent extraction and concentration should be avoided.

The pH of the sample shall be 7.0 ± 0.2 before incubation. Adjust the pH value of the effluent to 7.0 ± 0.2 by adding hydrochloric acid or sodium hydroxide solution. Select the concentrations of acid or alkali so that the added volumes are as small as possible.

I. Procedure

1. Preparation of the Overnight Preculture

Prepare the inoculum one day before the test under sterile conditions.

Fill the sterilized Erlenmeyer flasks with 20 mL TGA culture medium. Close with air porous sterile stoppers
Thaw the frozen stock culture
Fill ampule with 1 mL TGA medium
Centrifuge the test bacteria in the ampule (10 min, 2,000 g)
Decant the culture supernatant
Resuspend the test bacteria in 1 mL TGA medium
Inoculate the TGA medium in an Erlenmeyer flask with 0.5 mL of the test bacteria
Incubate overnight (no longer than 12 h; use a timer) by shaking in a water bath at 37 ± 1°C. This should give a turbidity of ≥400 FNU (*F*ormazine *N*ephelometric *U*nits)

2. Preparation of the Inoculum

Dilute the overnight culture 1 in 10 with fresh TGA medium.
Continue incubation for approximately 1.5 h in a shaking water bath at a temperature of 37°C. Then measure the turbidity (optical density 600 nm; 1 cm cuvette) and standardize with TGA medium at FNU (340 to 350).

The test organisms are now in the exponential growth phase and are ready for the test. The test has to be carried out immediately.

3. Preparation of the Test Culture Without Addition of S9 Mixture

During the incubation of the preculture, dilute the samples and prepare the test plates. Carry out the test in triplicate for each concentration.

Table 16.2 provides a grid of the pipetting scheme.

a. Charge the microplates with double distilled water or with water of equivalent purity (180 µL each well). Leave out the first three sets of wells (A-F 1-3).
b. Add 360 µL of the test material to the first three sets of wells (three replicates determination) of the same microplate (A-F 1-3).
c. At this point, prepare 1 in 2 dilutions of the test material (see Table 16.3).
d. Discard 180 µL from the last three sets of wells (A-F 10-12).
e. Mark the negative control (180 µL water) on the microplate (G 1-12).
f. Fill the wells for blank reading with 180 µL water (H 7-12).
g. Place 27 µL of the stock solution of the reference substances as positive control (H1-H3). Final concentrations in the wells: 50 ng 4-NQO/mL or 200 ng 2-AA/mL.
h. Place 27 µL of a 30% water/DMSO solution as negative control in wells H4-H6.
i. Fill positive and negative controls (H1- H6) with 153 µL water to 180 µL.
j. Add 20 µL 10× concentrated culture medium *to all wells* (see Step II.F.2).

Table 16.2 Grid of Pipetting Scheme (see Section 3 of Procedure)

	1	2	3	4	5	6	7	8	9	10	11	12
A	S1 1:1.5	S1 1:1.5	S1 1:1.5	S1 1:3	S1 1:3	S1 1:3	S1 1:6	S1 1:6	S1 1:6	S1 1:12	S1 1:12	S1 1:12
B	S2 1:1.5	S2 1:1.5	S2 1:1.5	S2 1:3	S2 1:3	S2 1:3	S2 1:6	S2 1:6	S2 1:6	S2 1:12	S2 1:12	S2 1:12
C	S3 1:1.5	S3 1:1.5	S3 1:1.5	S3 1:3	S3 1:3	S3 1:3	S3 1:6	S3 1:6	S3 1:6	S3 1:12	S3 1:12	S3 1:12
D	S4 1:1.5	S4 1:1.5	S4 1:1.5	S4 1:3	S4 1:3	S4 1:3	S4 1:6	S4 1:6	S4 1:6	S4 1:12	S4 1:12	S4 1:12
E	S5 1:1.5	S5 1:1.5	S5 1:1.5	S5 1:3	S5 1:3	S5 1:3	S5 1:6	S5 1:6	S5 1:6	S5 1:12	S5 1:12	S5 1:12
F	S6 1:1.5	S6 1:1.5	S6 1:1.5	S6 1:3	S6 1:3	S6 1:3	S6 1:6	S6 1:6	S6 1:6	S6 1:12	S6 1:12	S6 1:12
G	NC-S	NC-S	NC-S	NC-S	NC-S	NC-S	NC-S	NC-S	NC-S	NC-S	NC-S	NC-S
H	RB	RB	RB	NC-RB	NC-RB	NC-RB	BL	BL	BL	BL	BL	BL

S1 = Sample 1 and the dilution 1:1.5; 1:3; 1:6; 1:12; S6 = Sample 6 and the dilution 1:1.5; 1:3; 1:6; 1:12
NC-S = Negative Control of the Sample, RB = Reference Batch, NC-RB = Negative Control of the Reference Batch, BL = Blank

Table 16.3   Composition of the Test and Control Batches per Parts of the Procedure

Test Batches

Dilution	Dilution Level D	Volume Effluent µL	Volume Dilution Water µL	Volume 10× Culture Medium (TGA) µL	Volume Inoculum µL
1 in 1,5	1	180	—	20	70
1 in 3	3	90	90	20	70
1 in 6	6	45	135	20	70
1 in 12	12	22.5	157.5	20	70
1 in 24	24	11.25	168.75	20	70

Control Batches

	Volume Effluent Sample µL	Volume Reference Sample µL	Volume Dilution Water µL	Volume 10× Culture Medium (TGA) µL	Volume Inoculum µL	Volume, Simple TGA µL
Blank Reading Batch without Bacteria	—	—	180	20	—	70
Negative Controls	—	—	180	20	70	—
Reference Batch	—	180	—	20	70	—

   k. Pipette 70 µL preculture: bacterial culture to be used to adapt the test bacteria to the test conditions (FNU 340 to 350) and to grow the inoculum for the test batch. Mix by repeatedly using the pipette (from lower to higher concentration).
   l. Pipette 70 µL inoculum to the positive and negative controls (NQO and DMSO) and mix.
   m. Place 70 µL TGA (see Step II.F.1) in the blank batch and mix.

4. Batch of the Test Culture with Addition of S9

Instead of 4-NQO in the reference batch (see Step g above) use a 2-AA solution in 30% DMSO. The preparation of the dilution rows takes place as described above (see Steps a–i) up to the addition of 10× TGA as described.

Add to each well 20 µL cofactor solution.
Slowly thaw S9
Add 450 µL S9 (shake briefly before adding) to 15 mL of the inoculum (FNU 340 to 350) and mix the suspension.
Pipette 70 µL of this suspension with S9 both to the test batches and to the control batches and mix by repeatedly filling and emptying the pipette.
Add 45 µL S9 (shake briefly before adding) to 1.5 mL of the TGA to prepare the blanks + S9 and place 70 µL of this mixture into the blank reading batch.

5. Incubation

Incubate the microplate (A) once it has been preincubated for 2 h at 37 ± 1°C. Transfer the microplate to a vibrating rotor which is positioned at a 45° angle or use a shaker to prevent sedimentation of the bacteria while ensuring no cross contamination.
Toward the end of the first incubation phase (2h), charge a fresh microplate (B) with TGA medium (270 µL each well) and place in an incubator at 37 ± 1°C.
Pipette 30 µL test bacteria suspension from microplate A into the corresponding well of microplate B, i.e., 1 in 10 dilution of the batches.
Incubate microplate B in the same way as microplate A.

J. Measurements

For measurement of the optical density, measure the bacterial growth (A 600) with a photometer used for microplates.

1. Determination of the Induction of the umuC Gene

- Place 120 μL B-buffer in the wells of a new microplate C.
- Preheat the plate at 28 ± 1°C in a microplate incubator for 10 min.
- Place 30 μL test bacteria suspension from each well of microplate B (from the lower to the higher concentration) in the holes of the microplate filled with the B-buffer. Add 30 μL ONPG solution and mix very well.
- Place the microplate immediately into the incubator and incubate 30 min at 28 ± 1°C.
- After 30 min add 120 μL Na_2CO_3 and mix. The reaction is hereby stopped.
- Remove bubbles by means of a cold air stream.
- Measure the final absorption immediately at 420 nm with a photometer for microplates.
- Decontaminate microplates A and B by autoclaving.

K. Calculation and Expression of Results

Prepare a table for showing the individual concentrations, respectively, for the dilution levels, the measured turbidity and absorption values, the induction and β-galactosidase units calculated, and the growth factors.

1. Determination of the Growth Factors (G)

$$\text{Growth factor } G = \frac{A_{600}T - A_{600}B}{A_{600}N - A_{600}B}$$

where $A_{600}T$ is the absorption (mean value) of the test batch at 600 nm; $A_{600}N$ is the absorption (mean value) of the negative control at 600 nm; and $A_{600}B$ is the absorption (mean value) of the blank reading value at 600 nm.

2. Determination of the Induction Rate (IR)

$$\text{Induction rate (IR)} = \frac{1}{G} \times \frac{A_{420}T - A_{420}B}{A_{420}N - A_{420}B}$$

where $A_{420}T$ is the extinction (mean value) of the test batch at 420 nm; $A_{420}N$ is the extinction (mean value) of the negative control at 420 nm; and $A_{420}B$ is the extinction (mean value) of the blank reading value at 420 nm.

The smallest value of D (dilution level) at which an induction rate < 1.5 is measured under the given test conditions, counts as the result (see Table 16.4). If different induction rates are measured after incubation with or without S9, then the higher of the two values counts as the result.

Calculate the activity of the β-galactosidase in relative units (UT):

$$UT = \frac{A_{420}T - A_{420}B}{A_{600}T - A_{600}B}$$

The units of the negative controls (UN) are to be calculated analogous to the above. The values of the reference batches are to be related to negative controls containing solvents.

3. Validity Criteria

The test will be considered valid if the reference batches reach an induction rate of at least 2 under the test conditions. The results cannot be evaluated when a growth factor of <0.5 is obtained. Minimum growth of the control plate B: 140-280 FNU.

L. Test Report

The test protocol should refer to this procedure and contain the following details:

a. Identification of the test material
b. Preparation of the sample:
   pH value, centrifugation, homogenization, dilution substances used
c. Test organism (type, strain)
d. Test conditions. Date of test
e. Statement of the results. The smallest value of D (dilution level) at which an induction rate <1.5 is measured, tabular compilation of the concentration of the substances to be measured, mean values of the induction rates

$$\text{Growth factors, A600 and A420 values } \frac{A420}{A600} - \text{values, standard error.}$$

(see Table 16.4)

Note: In dilution step 1 (original sample) the sample is already diluted by the medium and the reagents 1 in 1.5. By definition D (original sample) = 1.5.

The value of D, by which an induction rate <1.5 is measured under the given test conditions, counts as the result (decisive D value).

If by the addition of S9, differing induction rates are measured, then the higher of the two values counts as the result (decisive D value).

Table 16.4 Final Protocol for the *umu*-Test

Sample/Dilution Step	S-9	Growth Factor W	IR	Induction Rate	D
1 (original sample)	–S9:				
3 (1 orig. + 2 water)	–S9:				
6 (1 orig. + 5 water)	–S9:				
12 (1 orig.+ 11 water)	–S9:				
1 (original sample)	+S9:				
3 (1 orig. + 2 water)	+S9:				
6 (1 orig. + 5 water)	+S9:				
12 (1 orig.+ 11 water)	+S9:				
Positive controls					
4-NQO (50 ng 4-NQO/mL)	–S9:				
2-AA (200 ng 2-AA/mL)	+S9:				

Note: In dilution step 1 (original sample) the sample is already diluted by the medium and the reagents 1: 1.5. By definition D (original sample) = 1

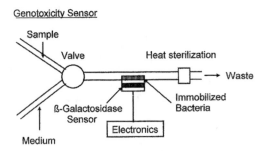

Figure 16.2 Principal components of the *umu*C assay Genotoxicity Sensor.

III. THE *umu*C-ASSAY FOR THE DETECTION OF GENOTOXIC EFFECTS OF EFFLUENTS AND SURFACE WATERS

The *umu*C-assay is a helpful tool for the management of genotoxicity of effluents and surface waters due to the presence of genotoxins. Some studies have been done to validate the *umu* assay.[5,6] There are also some comparative studies with the *umu* genotoxicity assay and the Ames mutagenicity assay using pulp mill effluents.[7] The study shows clearly that the *umu* assay and the Ames mutagenicity assay provided similar results. The Ames mutagenicity assay[8] is now miniaturized. A new development is now extended using the microplate assay by a genotoxicity sensor with the *umu* bacteria and the enzymatic amplification on a so-called sandwich membrane (see Figure 16.2), and can be used on site at the river, at the end of the effluent pipe, or in harbors as an early warning system. A photometric detection similar to the microtiter plate test is used. The flow scheme of the genotoxicity sensor is presented in Figure 16.2.

IV. THE DNA-UNWINDING ASSAY FOR INVERTEBRATES

The problem with the *umu* assay is that it is a prokaryotic test system not applicable to eukaryotic cells. The DNA-unwinding assay however is a eukaryotic test system. The DNA-unwinding assay is known to be a promising tool for detecting the DNA damage due to environmental genotoxins in aquatic animals. Extensive work is currently under way to develop a basis for uniform application of this bioassay for environmental samples. Recent work on the application of this protocol has clearly indicated its usefulness as a genotoxic monitoring tool.[9,10]

DNA-damaging activity in mussel hemocytes (*Mytilus edulis*) was observed after 96-h static exposure to genotoxic chemicals.[11] Recent investigations indicate that increased DNA damage in *Mytilus edulis* was observed in a contaminated area in a harbor,[11] compared to an unpolluted reference station. The DNA-unwinding assay is useful for assessing the impact of environmental contaminants in stressed areas and coastal zones for effective management.

Figure 16.3 shows the unwinding of the DNA double helix starts at natural unwinding points which are strand ends and strand nicks caused by enzymes, i.e., topoisomerases. Additional unwinding points can be found in exogenously damaged DNA. These can be apurinic/apyrimidinic (AP) sites, strand breaks, and repair patches. They enhance the average unwinding rate under controlled conditions. After a defined time, alkaline lysates of tissue homogenates or cells are neutralized, and the fractions of single-stranded and double-stranded DNA are separated by ion exchange chromatography using hydroxyapatite.

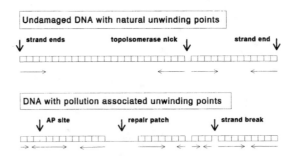

Figure 16.3 Principle of the DNA-unwinding assay.

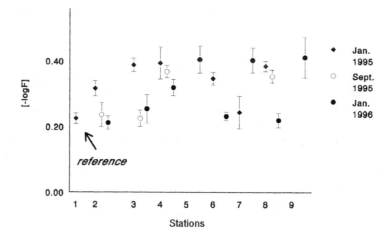

Figure 16.4 DNA-Fragmentation [–log F] of *Dreissena polymorpha* exposed along the waterstretch of the River Elbe. Sampling stations 1-9: 1 = Controls, 2 = Lake Gatow, 3 = Blankenese (City of Hamburg), 4 = Schnackenburg (former border of the DDR), 5 = City of Dessau, 6 = City of Magdeburg, 7 = Zehren (Saxonia), 8 = Schmilka (border to Tschechien), 9 = Obristvi (Tschechien). N = 20 mussels exposed in a flow-through system on site for 2 months (Jan. 1995, Sept. 1995 and Jan. 1996) at the sampling stations 1-9.

The DNA concentration of each fraction is quantified by DNA-specific fluorescence stain in a microplate fluorometer. The degree of unwinding is calculated as F (= double-stranded remaining fraction) and the negative logarithm (–log F) which is derived is proportional to the average number of unwinding points. Increased –log F values in samples over control values indicate genotoxic impact. A practical example for monitoring a river is shown in Figure 16.4.

V. PROTOCOL OF THE MINIATURIZED DNA-UNWINDING ASSAY

A. Required Chemicals

Homogenization buffer: 0.114 mol/L Tris HCl, 0.077 mol/L NaCl, pH 9.00 at 0°C (pH 8.54 at 23°C); store at 0°C or sterilize by autoclaving (to prevent bacterial growth)

Lysis buffer: 0.15 mol/L NaOH, 0.05 mol/L Na_2HPO_4, 1 mol/L NaCl; use 1 mol/L NaOH-Titrisol (Merck)

Neutralizing solution: 0.18 mol/L HCl, 5 mg/L phenol red; use 1 mol/L HCl-Titrisol

SDS solution: 0.5% SDS

Hydroxyapatite: (BioRad DNA-Grade Bio-Gel HTP); wash one volume hydroxyapatite (HAP) twice with ten volumes 0.01 mol/L phosphate buffer (see below), then resuspend in two volumes of the same buffer. Boil the hydroxyapatite suspension for 5 min before use (in a boiling water bath)

Formamide: (purification: 1 L formamide + 10 g Norit A charcoal + 50 g mixed-bed ion exchanger, 2 h stirring at 4°C, filter through a double paper filter; the pH value of a 1:2 dilution with water should be below 7.5 and conductivity below 40 μS)

Potassium phosphate stock buffer: 1 mol/L (0.5 mol/L KH_2PO_4 plus 0.5 mol/L K_2HPO_4, do *not* adjust pH)

0.01 mol/L Phosphate buffer: dilute 1 mol/L phosphate stock buffer

Wash buffer: 0.01 mol/L potassium phosphate buffer, 20% formamide

Elution buffer 1: 0.125 mol/L potassium phosphate buffer, 20% formamide

Elution buffer II: 0.5 mol/L potassium phosphate buffer, 20% formamide; use 1 mol/L phosphate stock buffer for all buffer preparations

Bisbenzimide buffer: 0.05 mol/L potassium phosphate buffer, 2.2 mol/L NaCl, adjust pH 7.4 with 5 mol/L KOH

Bisbenzimide stock solution: 250 μmol/L bisbenzimide in distilled water, store cold (4°C) and dark, stable for at least 3 months

Bisbenzimide dilution: Bisbenzimide stock solution, diluted 1:100 with bisbenzimide buffer

B. Equipment

Tip sonicator

Desktop centrifuge for 1.5 mL reagent tubes

1.5 mL reagent tubes (3 per sample)

2.0 mL reagent tubes (1 per sample)

Reagent tube mixer or shaker

Eppendorf Multipette 4780 (50 μL, 250 μL, 500 μL)

Microplate fluorometer or conventional spectro-/filterfluorometer with wavelength set at 360/450 nm

Adjustable microliter pipettes 10 to 200 μL, 200 to 1000 μL

Glassware (beakers, etc.)

C. Assay Protocol

1. *Preparation and Lysis of Samples*

All steps are carried out with samples placed on ice! (Temperature control is strictly required!)

Transfer tissue samples immediately into 5 volumes homogenization buffer and homogenize (preferentially with a Dounce homogenizer, but others may also be used). Allow debris to settle for 15 sec.

Alternatively for invertebrate hemolymph: Place 1.2 mL hemolymph in 1.5 mL reagent tubes for 5 min on ice and allow hemocytes to settle; discard 0.6 mL of the upper supernatant and resuspend cells in the remaining; carry on immediately.

Transfer 0.2 mL of the homogenized sample or resuspended hemocytes into a 2 mL reagent tube and leave it on ice. (Usually 2 parallels!)

To determine the reagent blank later on, do the same with 0.2 mL homogenization buffer and process it in the same way as the samples. (Usually 2 parallels!)

Add 0.5 mL ice-cold lysis buffer, and leave samples for 30 min on ice; avoid mechanical agitation and exposure to light.

IMPORTANT: Add lysis buffer and later neutralizing solution by forceful injection with the pipette *along* the tube wall and *never* directly into the sample to minimize bubbles and other DNA shearing forces.

To stop alkaline DNA unwinding, add 0.5 mL ice-cold neutralizing solution and 0.5 mL SDS solution (stored at room temperature).

To minimize renaturation of partially unwound DNA immediately (if samples are placed on ice within 20 min after neutralization) sonicate samples for 10 seconds (method of choice) or shear them 4 times through a 27G syringe needle.

After this step, samples can be stored or transported frozen at –20°C or below.

2. Hydroxyapatite Batch Elution to Separate Single- and Double-Stranded DNA

Preparation of reagent tubes: per sample and blank control, prepare and label three 1.5 mL reagent tubes; tubes are used for the hydroxyapatite "batch" elution, single-stranded (ss)-DNA fraction, and double-stranded (ds)-DNA fraction.

Boil the hydroxyapatite suspension for 5 min (in a boiling water bath).

Cool the hydroxyapatite suspension and transfer 0.2 mL of it into a 1.5 mL reagent tube (one per sample and blank).

IMPORTANT: Assure in all following steps, that the hydroxyapatite is completely suspended by mixing.

Add 1.0 mL of the lysed samples and blank controls to the hydroxyapatite and resuspend by repeated mixing (at least 3 times, with 3 min intervals; at low DNA concentrations more mixing, e.g., 6 times, and time, e.g., 15 to 20 min, increases binding efficiency). Centrifuge at 600 to $1200 \times g$ for 2 to 3 seconds; remove the supernatant by turning the reagent tube quickly upside-down (alternatively with a pipette or — for convenience — with a pipette tip connected to a vacuum pump) and discard it. If the DNA content of the samples is expected to be very low, the step may be repeated with the remaining lysate.

Add 1.25 mL wash buffer (preferably with a multistep pipette) and resuspend the hydroxyapatite by mixing twice with an interval of 3 min. Centrifuge and discard the supernatant as described above. Repeat this washing step a second time.

Add 1 mL elution buffer I and resuspend the hydroxyapatite by mixing three times with an interval of 3 min. Centrifuge as above, but save a part of the supernatant with a multistep pipette and dispense four 100 μL aliquots into wells of a 96-well microtiter plate for the determination of the ss-DNA. (Alternatively, if a microtiter plate fluorometer is not available, save 0.4 mL of the supernatant and transfer it to a 1.5 mL reagent tube.) Discard remaining supernatant.

Repeat last step using elution buffer II and keep aliquots of the supernatant for the determination of the ds-DNA.

3. Determination of DNA Concentrations in the ss- and ds-DNA Supernatants

Aliquots of the supernatants are heated in a boiling water bath for 5 min to transfer all DNA to single-stranded conformation to get comparable fluorescence yields; then cool rapidly to room temperature (use 20°C water bath)

Prepare the total required volume of bisbenzimide dilution: dilute bisbenzimide stock solution 1/100 v/v with bisbenzimide buffer.

Add 0.2 mL bisbenzimide dilution to each sample well and determine fluorescence at an excitation wavelength of 360 nm and emission of 450 nm. (Alternatively, if a microtiter plate fluorometer is not available, add 0.8 mL bisbenzimide dilution to each sample tube and determine fluorescence.)

Calculate DNA-dependent fluorescence by subtracting control blanks from sample fluorometer readings; distinguish between blanks obtained from the first and second elution step.

Calculate the fraction F of double-stranded(ds)-DNA by the formula

$$F = \frac{ds}{(ss + ds)} \qquad (16.1)$$

where ss is the relative fluorescence of the single stranded(ss)-DNA fraction and ds that of the double-stranded(ds)-DNA fraction.

The negative logarithm of F (–log F) is proportional to the relative number of DNA unwinding points, and thus to the degree of fragmentation. The relative average number of exposure-related DNA lesions per unwinding unit (i.e., the average distance between unwinding points) can be calculated as follows:

$$n = \frac{-\log F_x}{-\log F_0} \qquad (16.2)$$

where the index x indicates values from polluted and 0 from unpolluted organisms.

4. Possible Causes of Troubles

Tissue or hemolymph samples have been prepared too slowly and insufficiently cooled: Perform initial steps steadily and completely on ice to minimize activities of endogenous nucleases. In homogenates the high pH (9.0) of the homogenization buffer minimizes nuclease activity, however, in some cases the additional use of 20 m$Moles$/l Na_2EDTA may also be considered.

pH changes of the lysis buffer or NaOH solutions after long storage: Avoid CO_2 binding by regular preparation of new solutions and air-tight storage of aliquots. *Note: Reaction conditions of the assay are defined by the given concentrations of solution constituents and not by adjustment to any pH value. The quality of the lysis buffer may, however, be controlled by pH changes from initial values after storage.*

Use of hydroxyapatite of insufficient quality: Only "spherical" hydroxyapatite is suitable, but still not all commercially available preparations will separate ss- and ds-DNA properly. Suitable preparations will sediment slowly after suspension in ten volumes 0.01 mol/L potassium phosphate buffer (about 5 to 10 min); crystal spikes around the crystallization core can be visualized by light or scanning electron microscopy which cannot be seen in preparations of insufficient quality. Avoid mechanical destruction of the crystals, e.g., by use of magnetic stirrers.

Binding of CO_2 to suspended hydroxyapatite: Always boil hydroxyapatite solutions before use!

Insufficient amount of DNA binding to the hydroxyapatite: When low concentrations of DNA in lysates occur, the given time for loading of the samples and the lysate volume of 1 mL may be insufficient. **Alternative:** Extend time given for HAP–DNA binding to 15 or 20 min with additional vortexing. Load the remaining lysate volume in a second step in the same way.

Insufficient quality of formamide and formamide-containing solutions: Be aware of the instability of formamide; prepare formamide-containing solutions fresh at least every four weeks; check quality criteria (see above) before preparation of solutions; clean formamide may be stored in aliquots at –70°C.

VI. THE DNA-UNWINDING ASSAY FOR THE DETECTION OF GENOTOXIC EFFECTS IN SURFACE WATERS AND MUSSELS

There have been several genotoxic studies performed over the years along the water stretch of the River Elbe. A significant increase of genotoxicity in the former DDR (Stations 4-8, see Figure 16.4) and in Tschechien (station 9, see Figure 16.4) was observed.[12] These genotoxicity data are collected along with chemistry data. Information concerning the long-term genotoxicity data of certain river sites is valuable. Genotoxic substances are dangerous to ecosystems and human health and as such should be limited or controlled from their entry into the environment. Future environmental monitoring approaches should therefore include mandatory measurements of DNA damage in mussels and *umu* assays in other environmental samples to identify hot spots.

VII. CONCLUSIONS

1. The *umu*-assay and the DNA-unwinding assays are helpful genotoxicity assessment tools for the management of surface water and effluent quality due to the presence of genotoxic substances.
2. While the *umu*-assay is a prokaryotic test system the DNA-unwinding assay is a eukaryotic test system. The *umu*-assay is currently included in the International Standardization Organisation (ISO).
3. Both test systems are now adapted to microscale assays (microplate technique).

REFERENCES

1. Hansen, P.-D., The potential and limitations of new technical approaches to ecotoxicology monitoring, in *Environmental Toxicology Assessment,* Richardson, M., Ed., Taylor & Francis, London, 1995.
2. Nakamura, S.I., Oda, Y., Shimada, T., Oki, I., and Sugimoto, K., SOS-inducing activity of chemical carcinogens and mutagens in *Salmonella typhimurium* TA1535/pSK1002: examination with 152 chemicals, *Mutation Res.,* 192, 239, 1987.
3. Oda, Y., Nakamura, S.I., Oki, I., Kato, T., and Shinagawa, H., Evaluation of the new system (umu-test) for the detection of environmental mutagens and carcinogens, *Mutation Res.,* 147, 219, 1985.
4. Reifferscheid, G., Heil, J., Oda, Y., and Zahn, R.K., A microplate version of the SOS/umu-test for rapid detection of genotoxins and genotoxic potentials of environmental samples, *Mutation Res.,* 253, 215, 1991.
5. Whong, W.-Z., Wen, Y.-F., Steward, J., and Ong, T., Validation of the SOS/Umu test with mutagenic complex mixtures, *Mutation Res.,* 175, 139, 1986.
6. ISO, International Standardization Organisation, Water Quality — Determination of the Genotoxicity of Water and Waste Water Using the umu-Test. CD13829, 1997.
7. Rao, S.S., Burnison, B.K., Efler, S., Wittekindt, E., Hansen, P.-D., and Roskosh, D.A., Assessment of genotoxic potential of pulp mill effluent and an effluent fraction using AMES-mutagenicity and umuC-genotoxicity assays, *Environ. Toxicol. Water Qual. — An Internat. Jour.,* 10, 301, 1995.
8. Côté, C., Blaise, C., Delisle, C., Meighen, E.A., and Hansen, P.-D., A miniaturized Ames test employing bioluminescent strains of *Salmonella typhimurium, Mutation Res.,* 345, 137, 1995.
9. Herbert, A. and Hansen, P.-D., Erfassung des erbgutverändernden Potentials von Gewässern durch Messung von DNS-Schäden mittels alkalischer Denaturierungsverfahren. Schriftenreihe Wasser-, Boden- und Lufthygiene, Bd. 89, Gustav Fischer Verlag, Stuttgart, 1992.
10. Rao, S.S., Neheli, T.A., Carey, J.H., Herbert, A., and Hansen, P.-D., DNA alkaline unwinding assay for monitoring the impact of environmental genotoxins, *Environ. Toxicol. Water Qual. — An Internat. Jour.,* 11, 351, 1996.
11. Hansen, P.-D., Assessment of ecosystem health: development of tools and approaches, in, *Evaluating and Monitoring the Health of Large-Scale Ecosystems,* Rapport, D., Gaudet, C., and Calow, P., Eds., Series I: *Global Environmental Change,* Vol. 28. Springer-Verlag, Berlin, 1995, 195.
12. Hansen, P.-D., Wittekindt, E, Bonse, S., Saftic, S., and Herbert, A., Die Ermittlung erbgutverändernder Belastungen entlang der Elbe mit dem DNA-Aufwindungstest. 7. Magedeburger Gewässerschutzseminar 23.-26.10.1996, Budweis, Tschechien.

Algae and Microalgae

CHAPTER 17

Recent Advances in Toxicity Test Methods Using Kelp Gametophytes

B.S. Anderson, J.W. Hunt, and W. Piekarski

CONTENTS

I. INTRODUCTION

Toxicity test methods for marine and estuarine environmental monitoring have evolved to include procedures for assessing sublethal effects of toxicants on a variety of species. Included in this "multispecies" approach to toxicity testing are bioassay procedures developed for marine micro- and macroalgae. As these methods are implemented in regulatory programs, it has become evident that protocols for algae are among the most sensitive available for environmental monitoring (see Thursby et al.[1] for a review). While microscale toxicity test protocols have been developed for a variety of alga species, this chapter emphasizes procedures using kelp gametophytes (order Laminariales), particularly *Macrocystis pyrifera*.

The State of California has developed and implemented a macroalgal toxicity test protocol for effluent National Pollution Discharge Elimination System (NPDES) monitoring purposes using the giant kelp *Macrocystis pyrifera*, an important primary producer in near-shore marine communities. *Macrocystis pyrifera* is an appropriate toxicity test species because it is amenable to microscale laboratory culture and relatively sensitive to toxicants. Kelp forests provide a significant ecological and economic resource to California; kelps are among the most productive species in the world, often dominating the physical structure of many coastal marine ecosystems.[2] There is concern over the effects of pollution on *Macrocystis* and its associated marine communities because of the proximity of kelp forests to near-shore pollution sources.[3] *Macrocystis* has been cultured extensively in laboratory studies and its life history has been well described.[4-6] Until recently, the use of *Macrocystis* early life stages in toxicity studies had been limited,[1,7-11] although other Laminarian algae have been used in toxicity testing.[8,12-15]

Two toxicity test protocols have been developed using *Macrocystis* early life-stages: a short-term 48-hour test and a longer-term 16- to 20-day test.[9,16] The 48-h test has two endpoints: germination of the settled zoospores, and initial growth of the "germ-tube" of the developing gametophyte. The longer-term test assesses sporophyte production (i.e., sexual reproduction). After initial development of these protocols, it was recognized that the short-term 48-h test was more appropriate for routine effluent monitoring because of logistical constraints imposed by the longer-term reproductive test. Subsequent research assessed temporal, intrapopulation, and interlaboratory variability of this test using effluents and reference toxicants, primarily zinc and copper.[9,16,17] This protocol is one of several routinely used by the approximately 90 municipal and industrial dischargers releasing an estimated 11 billion gallons of waste per day into California marine waters.[18] More recently, this test has been included in the U.S. Environmental Protection Agency's chronic toxicity methods manual for use in west coast NPDES monitoring.[19]

This paper discusses adaptations of the *Macrocystis* protocol for microscale toxicity test applications and some recent experiments which assess the ecological relevance of the 48-h toxicity test endpoints. In addition, recent development of alternative protocols and endpoints using gametophytes of *Macrocystis* and other kelps, and progress on application of TIE procedures for the 48-h *Macrocystis* protocol are discussed.

II. METHODS

A. Toxicity Test Procedures Using *Macrocystis* Gametophytes

Like all kelps, the life history of *Macrocystis* consists of an alternation between a macroscopic diploid stage (sporophyte) and a microscopic haploid gametophyte stage (Figure 17.1). Biflagellate zoospores are released from specialized reproductive blades (sporophylls) at the base of the adult plant. Amsler and Neushul[20] demonstrated that *Macrocystis* spores are behaviorally active; they are positively chemotactic to nutrients required for growth and negatively chemotactic to compounds which may inhibit reproduction. After release, spores settle onto substrate where they develop into gametophytes. After an initial development period, the dioecious gametophytes produce either eggs (ova) or sperm (antherozooids). Fertilized eggs develop into the sporophyte stage completing the life cycle.

Both short-term and long-term protocols are initiated with motile zoospores. Zoospores are obtained from sporophylls typically collected by skindivers. Sporophylls may be collected up to a day before tests are initiated and are stored damp in a chiller prior to testing.

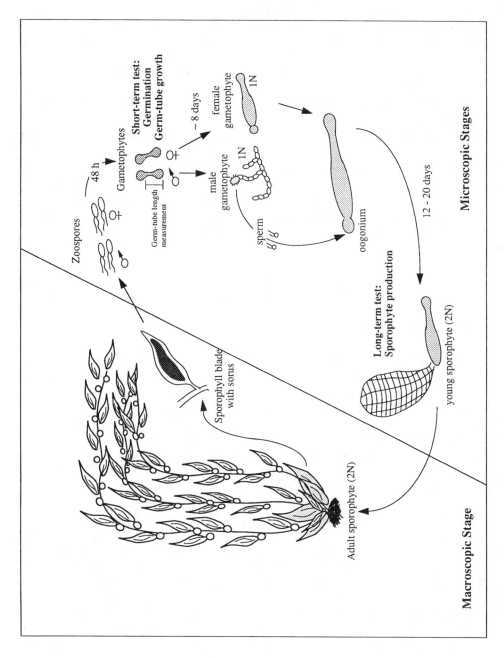

Figure 17.1 Life history of giant kelp *Macrocystis pyrifera* and microscopic life stages used in short-term and long-term toxicity tests.

Spore release is induced by immersing the sporophylls in sea water. A specific density of spores is pipetted into test containers and exposed under cool-white fluorescent lighting (50 $\mu E/m^2/sec$) for 48 h in the short-term protocol. This test has two endpoints: percent germination, measured by counting 100 spores and scoring them as either germinated or not germinated, and germ-tube growth, measured as length from the edge of the spore case to the tip of the germ-tube. Growth is measured with an ocular micrometer on 10 randomly selected spores from each test container. In the longer-term protocol, gametophytes are induced to develop reproductive structures using a combination of full spectra "daylight" illumination and appropriate nutrients for gametogenesis. The endpoint assessed is successful reproduction, measured as sporophyte production (Figure 17.1). This process takes approximately 12 to 20 days in laboratory cultures. Detailed methods for the short-term and longer-term protocols are described in Anderson et al.,[16] and in state and federal methods manuals.[18,19]

Statistical methods for assessing toxicity in the experiments described here follow those described in Chapman et al.[19] In brief, treatments were analyzed using ANOVA followed by Dunnett's test to determine no observed effect concentrations (NOECs). Concentrations causing 50% inhibition of germination and sporophyte production (EC_{50}s) were calculated using the Spearman–Karber method.[21] Concentrations causing 50% reduction in germ-tube length were calculated using the EPA linear interpolation, or ICp method.[22] This is a nonparametric, nonmodel specific, monotonic smoothing method. The EPA program uses bootstrapping procedures to generate a point estimate, a standard error, and confidence intervals.

III. RECENT ADVANCES

A. Adaptations for Microscale Testing

Initial test protocols developed for *Macrocystis* gametophytes used <200 mL exposure volumes. Smaller volume exposures are preferable, particularly when test material is difficult to obtain, as is the case with sediment interstitial water or extracts for toxicity identification evaluations (TIEs). Microscale studies (e.g., ≤30 mL) have been conducted with kelp gametophytes,[10,11,23,24] but investigations of possible artifacts associated with small volume procedures are rare. Small volume containers have larger surface-to-volume ratios which may increase the possibility of adsorption of toxicants onto container walls and subsequent reduction of toxicant bioavailability.

1. Microscale Comparisons Using Copper

Anderson et al.[25] compared results of copper exposures conducted in standard 200 mL polyethylene plastic test containers and 5 mL Labtek® chamber slides consisting of a borosilicate glass slide covered by a polystyrene chamber. All experiments were static exposures conducted with spores from the same spore release. Results showed no difference in copper dose-response between the kelp tests conducted in 200 mL and 5 mL (Figure 17.2). The copper NOECs for germination were 56 µg/L using both test volumes, while the NOECs for germ-tube growth were <10 µg/L using both test volumes. The median effect concentrations (EC_{50}s) for germination were 73.3 and 71.2 µg/L for the 200 mL and 5 mL containers, respectively. The 95% confidence intervals around the germination EC_{50}s for both test volumes overlapped, indicating minimal difference in response. Because a 50% reduction relative to the control was not achieved in the 5 mL test volume, 40% effect concentrations (IC_{40}s) were calculated for germ-tube growth data for both test volumes. The IC_{40}s were 72.9 and 135.5 µg/L for the 200 mL and 5 mL containers, respectively. The 95% confidence intervals around the germ-tube length IC_{40}s for both test volumes overlapped (Figure 17.2).

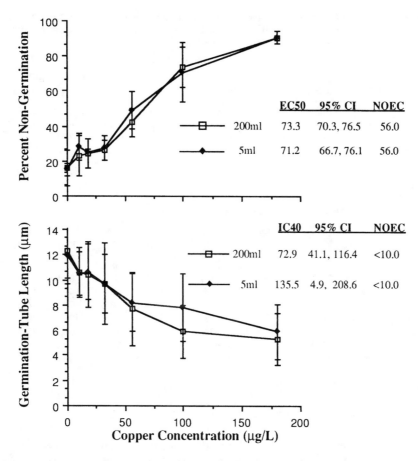

Figure 17.2   Effects of copper on germination and germ-tube growth of kelp gametophytes exposed in 200 mL polyethylene plastic and 5 mL Labtech™ polystyrene plastic test containers (mean ± S.D.; n = 5).

2. Microscale Comparisons Using Azide

We have also recently assessed intralaboratory variability of small and larger volume exposures using sodium azide (NaN_3), a cytochrome oxidase inhibitor. A series of 48-h azide exposures were conducted in 200 mL borosilicate glass test containers using spores collected on different sampling dates. These were compared to experiments conducted in 15 mL polystyrene Petri dishes. Azide concentrations are reported as nominal values. Although azide concentrations were not measured directly, correct dilutions of stock azide concentrations were confirmed using UV spectroscopy (data not included). Statistical comparisons of the EC_{50}s and IC_{50}s generated using the two test volumes consisted of separate variance t-tests using an unbalanced design (n = 4 for the 15 mL tests, vs. n = 8 for the 200 mL tests). Comparisons were made using the SYSTAT statistical software (Version 5 for Windows, SYSTAT, Inc., Evanston, IL).

The results indicate similar ranges in response to azide, although the mean median effect concentrations (EC_{50}s and IC_{50}s) for germination and germ-tube growth were lower using the 15 mL test containers (Table 17.1). There were no statistically significant differences between the germination EC_{50}s (p = 0.063), or between the germ-tube growth IC_{50}s (p = 0.084) from the 15 mL and 200 mL test containers.

These results indicate no obvious artifacts associated with small volume exposures using copper and azide; there was no loss of test sensitivity using smaller volumes. It should be noted that similar

Table 17.1   Results of 48 h *Macrocystis* Germination and Germ-Tube Growth Tests Conducted in 15 mL Polystyrene Petri Dishes and 200 mL Borosilicate Glass Containers Using Sodium Azide as a Reference Toxicant

	15 mL Test Volume							
	Germination				Germ-Tube Length			
Test Date	NOEC	LOEC	EC$_{50}$	95% CI	NOEC	LOEC	IC$_{50}$	95% CI
07/02/93	18.0	32.0	87.9	84.8, 91.2	10.0	18.0	82.9	63.2, 108.6
07/05/95	18.0	56.0	45.1	43.0, 47.3	3.2	5.6	48.6	43.0, 65.9
09/19/95	18.0	56.0	61.1	58.2, 63.8	18.0	56.0	75.5	40.5, 95.0
11/08/95	18.0	56.0	51.5	49.3, 53.8	10.0	18.0	96.3	58.0, 214.4
Mean			61.4				75.8	
CV (%)			30.7				26.5	

	200 mL Test Volume							
	Germination				Germ-Tube Length			
Test Date	NOEC	LOEC	EC$_{50}$	95% CI	NOEC	LOEC	IC$_{50}$	95% CI
02/11/92	18.0	32.0	69.6	65.7, 73.8	18.0	32.0	133.7	81.3, 175.3
02/18/92	32.0	56.0	87.5	83.3, 91.6	18.0	32.0	96.5	80.7, 164.7
06/29/92	32.0	56.0	137.4	133.3, 141.7	32.0	56.0	142.2	101.7, 180.2
07/07/92	10.0	18.0	89.3	85.8, 92.9	10.0	18.0	92.5	74.5, 141.1
07/15/92	18.0	32.0	125.3	117.7, 133.3	18.0	32.0	138.9	108.7, 168.1
07/16/92	10.0	18.0	60.6	57.8, 63.5	5.6	10.0	68.4	48.7, 83.2
07/22/92	18.0	32.0	88.2	83.4, 92.7	10.0	18.0	80.6	70.4, 93.3
10/09/92	5.6	10.0	65.4	62.9, 67.9	5.6	10.0	80.0	68.6, 90.8
Mean			90.4				104.1	
CV (%)			30.7				28.4	

Note: Azide concentrations are reported as mg/L.

comparisons should be made with other compounds, particularly hydrophobic organic compounds and complex effluents, to determine whether fewer water soluble compounds may occur in small volume containers.

B. Comparison of Short-Term and Long-Term Effects

Reproductive endpoints with macroalgae are more sensitive indicators of toxicity than vegetative endpoints.[1,8,9,14-16] However, because of the requirement for longer exposure times and the greater potential for contamination by competing alga species, protocols assessing *Macrocystis* reproduction have been found to be more difficult to conduct on a routine basis.[25] Test success rates were considerably higher for the 48-h protocol, and this protocol was recommended for routine use, recognizing that there was a need for some compromise between sensitivity and the requirement for high test success rates for NPDES monitoring programs.[25]

1. Short-Term and Long-Term Exposure-Recovery Experiments

To expand on previous research investigating the relationship between short-term and longer-term endpoints, a series of experiments was conducted comparing the relative effects of sodium azide on *Macrocystis* spore germination, gametophyte growth, and sporophyte production. Long-term tests measuring sporophyte production were conducted as either continuous exposures or as 48- and 96-h pulse exposures followed by recovery in uncontaminated sea water. These experiments were designed to determine whether vegetative effects observed in short-term exposures persist when cultures are allowed to recover in toxicant-free media, and whether duration of initial exposure affects recovery.

Two separate experiments were conducted using sodium azide. In the first experiment, three separate treatments were compared concurrently: 1) a short-term 48-h exposure, 2) a 48-h exposure followed by a 17-day recovery period (48-h-ER), and 3) a 19-d continuous exposure (CE). The second experiment also involved concurrent comparisons of three separate treatments: 1) a short-term 48-h exposure, 2) a short-term 96-h exposure followed by 20-d recovery period (96-h-ER), and 3) a 24-d continuous exposure (CE).

In Experiment 1, germination and germ-tube growth were assessed as described above after a 48-h exposure. For the 48-h exposure-recovery (48-h-ER) treatment, slides containing gametophyte cultures were transferred from azide-spiked media to containers with control sea water containing nutrients necessary for gametogenesis (PES[26] plus germanium dioxide to control diatoms[27]). In addition, the lighting was changed from cool-white fluorescent to full-spectrum Vitalights® $(50 \ \mu Em^{-2}sec^{-1})$ to induce gametogenesis. Nutrients and full spectrum lighting were also provided to the continuous exposure (CE) treatments after the initial 48 h, and test solutions for both ER and CE treatments were renewed every 96 h thereafter. Gametophyte cultures in the ER and CE treatments were allowed to develop for an additional 17 d, when sporophyte numbers were counted.

Procedures in Experiment 2 were identical except for the following modifications: for the 96-h-ER treatments, slides were exposed to azide-free media after an initial 96-h exposure period, instead of after 48 h. In addition, rather than transferring the slides to new containers for recovery, the azide-spiked media was siphoned out of the original ER test containers and replaced with clean sea water.

All treatments were analyzed using ANOVA followed by Dunnett's test to determine NOECs. Concentrations causing 50% inhibition of germination and sporophyte production (EC_{50}s) were calculated using the Spearman–Karber method.[21] Concentrations causing 50% reduction in germ-tube length were calculated using the EPA ICp program.[22]

As in previous experiments with copper and pentachlorophenate,[9,16] the results using sodium azide indicate that the longer-term sporophyte production test is more sensitive than the 48-h germination and growth test when cultures are continuously exposed. Sporophyte production was significantly inhibited (p = .0001) at the lowest azide concentration (Figure 17.3a, NOEC <3.2 mg/L); the NOECs for germination and germ-tube growth were 18 mg/L (Figure 17.3b and 17.3c). Sporophyte production in the 48-h-ER treatments indicate that gametophytes cultured in azide concentrations which significantly inhibit germination and growth are able to recover and undergo reproduction when transferred after 48 h to azide-free media. Although there was greater than 50% reduction in sporophyte numbers at the two highest concentrations (100 and 180 mg/L), these treatments did not differ significantly from the control because of high variability in the data set.

Although some recovery was indicated, it is apparent that except in the lowest azide concentration, gametogenesis continued to be suppressed in cultures exposed to azide for 96 h and then transferred to azide-free media (Figure 17.4a). The EC_{50} for sporophyte production in the 96-h-ER test was 9.2 mg/L, an order of magnitude lower than the EC_{50} in the 48-h-ER test (128 mg/L; Table 17.2). The EC_{50}s for the CE tests were comparable in both experiments, approximately 1.0 mg/L. As before, germination and germ-tube growth were less sensitive indicators of azide toxicity than sporophyte production (Figures 17.4b and 17.4c). Sporophyte numbers were considerably lower in the second experiment; the reason for this disparity is not clear. It should be noted that although sporophyte numbers were lower in the second experiment there was considerably less between-replicate variability, and therefore greater power to detect statistical differences.

Comparing the relative sensitivity of short-term and longer-term endpoints allows for an estimation of the biological significance of endpoints measured in the 48-h *Macrocystis* germination and growth protocol. These results demonstrate that azide concentrations which inhibit spore germination and gametophyte growth also inhibit kelp sporophyte production, an ecologically important endpoint. These results are consistent with previous experiments indicating kelp spore germination and initial gametophyte growth may be used as short-term indicators of more chronic effects, although it is clear that vegetative endpoints measured after 48-h exposure are considerably

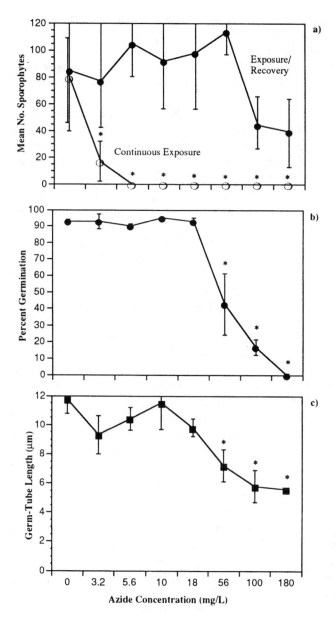

Figure 17.3 Effects of sodium azide on sporophyte production following (a) continuous exposure to azide and sporophyte production after a 48-h azide exposure followed by a 17-day recovery period (exposure recovery) and (b,c) on kelp gametophyte germination and germ-tube growth in 48-h exposures. * Indicates significant difference from control value. See text for details (mean ± S.D.; n = 5).

less sensitive than the reproductive endpoint measured after longer exposures. These data do not allow us to determine whether the increased sensitivity in the protocol assessing sporophyte production is due to increased exposure time or to a more sensitive endpoint (i.e., reproduction); the mechanism of azide toxicity to the various life stages and endpoints used in these experiments is beyond the scope of this study. However, it is possible that the mechanism of azide inhibition of kelp spore germination and germ-tube elongation differs from mechanisms inhibiting kelp reproductive processes. Azide is a cytochrome oxidase inhibitor and therefore blocks ATP production.[28] Pillai et al.[23] have shown that germ-tube elongation is a process involving actin microfilaments.

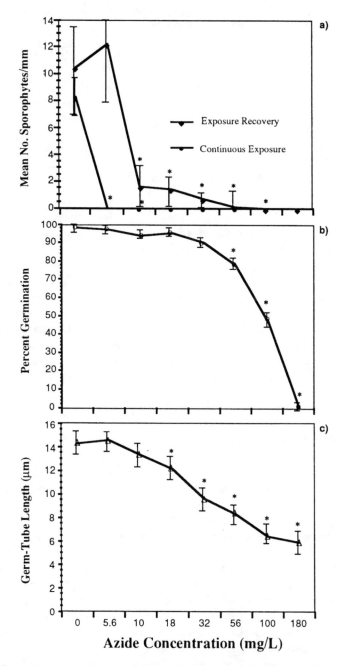

Figure 17.4   Effects of sodium azide on sporophyte production following (a) continuous exposure to azide and sporophyte production after a 96-h azide exposure followed by a 20 day recovery period (Exposure Recovery) and (b,c) on kelp gametophyte germination and germ-tube growth in 48-h exposures. * Indicates significant difference from control value. See text for details (mean ± S.D.; n = 5).

Although the precise mechanism is unknown, azide may inhibit actin function through disruption of actin/myosin interactions requiring ATP.[29] Kelp reproduction involves a series of complex biochemical processes requiring ATP, although it is unclear which specific process is affected by azide. There is evidence that many brown algal species rely on a pheromone system to attract sperm to eggs; this may also be disrupted by azide.

Table 17.2   Results of 48-Hour, Continuous Exposure (CE) and 48- and 96-h Exposure Recovery (ER) Experiments with *Macrocystis pyrifera*

Endpoint	NOEC	EC$_{50}$	95% CI	IC$_{50}$	(95% CI)
Germination (Experiment 1)	18.0	51.53	(49.3–53.8)		
Germ-tube length (Experiment 1)	18.0			96.3	(62.8–208)
Sporophyte production CE (Experiment 1)	<3.2	1.12	(0.9–1.3)		
Sporophyte production 48-h-ER (Experiment 1)	>180.0	127.7	(28.0–580.8)		
Germination (Experiment 2)	18.0	89.9	(NR)		
Germ-tube length (Experiment 2)	10.0			80.1	(NR)
Sporophyte production CE (Experiment 2)	<5.6	1.1	(NR)		
Sporophyte production 96-h-ER (Experiment 2)	5.6	9.2	(8.6–9.8)		

NR = Estimates not reliable due to lack of intermediate response

Although preliminary, these data suggest that both the concentration and the initial exposure duration determines whether kelp gametophytes exposed to sodium azide are competent to recover and successfully reproduce when transferred to uncontaminated media. Cultures exposed for 48 h were able to recover, while those exposed for 96 h were not. The initial exposure concentration apparently affected the ability of the cultures to recover. For example, there was a reduction in sporophyte production in 48-h-ER cultures exposed to the highest azide concentrations (100 and 180 mg/L). Studies with additional toxicants are necessary to determine whether these results are consistent for compounds having different modes of action. In addition, it should be noted that kelp gametophytes are capable of producing sporophytes after prolonged vegetative periods,[24] sometimes on the order of several months. It is possible that gametophytes exposed to greater azide concentrations, or those exposed for longer periods, might become reproductive when provided longer recovery periods.

C. Alternative Protocols Using Kelp Gametophytes

1. Sexual Reproduction Test with Laminaria saccharina

An alternative kelp reproduction protocol is being developed by the U.S. EPA using an east coast clonal isolate of the kelp *Laminaria saccharina*.[24] *Laminaria* has an identical life history to that of *Macrocystis* (Figure 17.1). The protocol under development assesses sporophyte production by exposing mature kelp gametophytes in 15 mL Petri dishes. To minimize problems associated with long-term cultures of kelp gametophytes, this procedure uses clonal cultures of vegetative *Laminaria* gametophytes which have been held under iron-free nutrient conditions to inhibit gametogenesis. The gametophytes are therefore mature but not producing eggs or sperm. Prior to initiation of the assay, separate cultures of male and female gametophytes are exposed to nutrient-enriched media for 4 to 6 d to promote gametogenesis. After the gametophytes have produced reproductive structures, slides with female gametophytes are placed in Petri dishes containing male gametophytes, and these are exposed to toxicant for 48 h. After toxicant exposure, the exposure media is replaced with uncontaminated media containing "recovery" nutrients, and sporophytes are allowed to grow for an additional 2 to 4 d, after which they are counted as described above for *Macrocystis*.[24] The 6-d exposure-recovery period is designed to be short enough to reduce chances for contamination associated with longer-term kelp cultures.[1]

This protocol was evaluated by the State of Washington in a round-robin study designed to compare several toxicity testing protocols for use in monitoring pulp mill effluents.[24,30] This study compared four protocols: a bivalve development test, an echinoid fertilization test, a larval fish growth and survival protocol, and the kelp reproduction test. The results showed that the *Laminaria* test was less sensitive than the bivalve development and echinoid fertilization tests (to pulp mill effluent), but more sensitive than the fish protocol. The authors concluded that the kelp reproduction protocol was promising, but because of lower test success rates, and higher interlaboratory variability, this protocol should not be recommended for routine use in pulp mill effluent monitoring until the procedure underwent further evaluation.[30]

2. Protocols with Alternative Kelp Species

In addition to the protocols described above, recent use of other laminarian species in toxicity testing includes studies using several species indigenous to California and Australia. James et al.[8] compared the relative sensitivity of seven laminarian species to hydrazine, a corrosion inhibitor used to treat power plant boilers. Assessing growth of 6-day-old vegetative gametophytes in Petri dish exposures, these authors ranked the order of sensitivity for these species to hydrazine as follows (from most to least resistant): *Pterygophora californica* > *Eisenia arborea* > *Laminaria farlowii* > *Macrocystis pyrifera* > *Laminaria ephemera* > *Laminaria dentigera*. In addition, the relative sensitivity of several life stages of *Macrocystis pyrifera* to hydrazine was compared. Of the stages tested, sporophyte production after 15-d exposure was by far the most sensitive (LOEC = 0.025 mg/L hydrazine). Gametophyte growth was affected at 0.13 mg/L after 1-d exposure, and spore motility (1-h exposure), antherozooid motility, gametophyte survival, and sporophyte survival (all = 1-d exposure) were all affected at 0.25 mg/L hydrazine. These authors concluded that due to the minimal exposure time necessary to inhibit spore motility (1 hour), this endpoint was potentially quite sensitive.

Burridge et al.[31] and Gunthorpe et al.[32] have reported on the recent development of procedures using brown macroalgae species from the order Fucales indigenous to Australia (*Hormosira baksii, Phyllospora comosa*). Burridge et al.[33] have also recently adapted a gametophyte germination procedure for use with *M. angustifolia*, a congener of *Macrocystis pyrifera*. Preliminary studies with *M. angustifolia* indicate comparable sensitivity to the two fucalean species to primary and secondary sewage effluent and to 2,4 dichlorophenoxyacetic acid (2,4-D).

3. Nuclear Migration in Macrocystis Gametophytes

Garman et al.[10,11] have reported recent modifications of the germination and growth protocol using *Macrocystis pyrifera* gametophytes to include a nuclear migration endpoint. Pillai et al.[23] have shown that kelp zoospores settle and adhere to the substrate within 3 h in laboratory culture; at this time they absorb their flagella and initiate germination. Germination tube growth occurs over a 15- to 20-h period. Nuclear migration is initiated at approximately 20 h and is completed at approximately 42 h. This process includes replication and division of the parent spore's nuclear material to form two nuclei and the translocation of one of the daughter nuclei along the germination tube into the first gametophytic cell. Together, the three endpoints: germination, germ-tube growth, and nuclear migration are termed "nuclear migration." Different toxicants may inhibit nuclear dynamics by affecting actin microfilaments necessary for germ-tube elongation or microtubules necessary for nuclear migration.[10,11] Translocation of the nuclear material to the end of the germination tube is assessed by staining the nucleus with Hoechst 33342, a vital DNA fluorochrome, and migration of stained nuclear material is measured with a conventional bright-field microscope adapted for epifluorescence.

Garman et al.[10] assessed the relative sensitivity of the three "nuclear migration" endpoints in microscale (5 to 10 mL) experiments assessing the toxicity of arsenic (As), copper (Cu), and the water-soluble fraction of produced water (PW), a complex effluent associated with oil production. In addition, exposure recovery experiments were conducted with As and PW to determine whether any of the three endpoints recovered after transfer of cultures to toxicant-free media. Nuclear migration demonstrated a consistently greater magnitude of inhibition by these toxicants, particularly relative to germination. Results of the exposure-recovery experiments were variable and showed no recovery of any of the three endpoints 24 h after transferring cultures to sea water following a 42-h exposure to arsenic (see figure numbers in Garman et al.[10]). Exposure-Recovery experiments with PW showed recovery of nuclear migration after transfer to sea water but no recovery of germ-tube growth; germination was not significantly inhibited by PW. By incorporating measures of germination, germ-tube growth, and nuclear migration in a suite of measures which are affected by specific toxicants, these authors concluded that more precise resolution of the toxic mechanism of contaminants is achieved.

D. Development of TIE Procedures Using Kelp

In using kelp gametophytes for toxicity assessments, it is important that these protocols are amenable to toxicity identification evaluation procedures (TIEs). Although EPA has developed methods for the red alga *Champia parvula*,[34] no formal methods are currently available for kelp protocols. Griffin et al.[35] have used TIE procedures developed by EPA to demonstrate that toxicity of complex effluents could be successfully characterized using the *Macrocystis* 48-h germination and growth protocol. Kelp gametophytes were tolerant of Phase I TIE manipulations (e.g., pH adjustment, aeration, filtration, solid phase extraction, and the addition of various chelators). Kelp gametophytes were tolerant of un-ionized ammonia (NOEC > 4.0 ppm; D. Griffin, MEC Analytical Systems, Tiburon CA, personal communication), an important attribute because it simplifies the TIE process, particularly with environmental samples containing complex mixtures of ammonia and anthropogenic contaminants (e.g., sewage effluents, sediment interstitial waters).

Existing EPA Phase I TIE procedures are being evaluated for a variety of protocols developed for species indigenous to the west coast of the United States. Funded by the San Francisco Bay Area Dischargers Association (BADA), this research will also evaluate the applicability of these methods to the *Macrocystis* 48-h protocol. In addition to guidance on general procedures, this work will provide effect ranges for the various reagent chemicals used in Phase I manipulations and will include results of experiments confirming the efficacy of the procedures using two chemicals, zinc sulfate and sodium dodecyl sulfate (personal communication, R. Berger, EBMUD, Oakland, CA).

IV. FUTURE RESEARCH NEEDS

Toxicity test protocols using kelp early life stages are significant components of state and national pollution monitoring programs in the U.S. and contribute to databases necessary for establishing water-quality objectives for individual chemicals. Kelps are among the most productive plants in the world, and kelp forests and their associated communities are key components of coastal ecosystems.[2,4,36,37] Much is known about the biology and ecology of kelps, particularly the sporophyte generation. The role of biotic and abiotic factors on gametophyte growth and reproduction have also been described.[20,38-44] Although it is clear that the microscopic stages of kelps are sensitive to xenobiotics, the relationship between chemical contaminants and ecological effects are poorly understood. In addition to developing protocols for assessing the effects of contaminants on new species and endpoints, it is important that future research investigate relationships between effects on kelp early life stages as they relate to kelp ecosystem structure.

ACKNOWLEDGMENTS

The manuscript was improved through comments by T. Dean and G. Thursby. Laboratory assistance was provided by M. Englund. This document was prepared through agreement with the California State Water Resources Control Board (CSWRCB Contract No. 2-041-250-2); M. Reeve is the project manager. Primary funding for this study has been provided by the CSWRCB. This does not signify that the contents reflect the views and policies of the U.S. EPA or the CSWRCB, nor does mention of trade names or commercial products constitute endorsement or recommendation for use.

REFERENCES

1. Thursby, G. B., Anderson, B. S., Walsh, G. E., and Steele, R. L., A review of the current status of marine algal toxicity testing in the United States, in *Environmental Toxicology and Risk Assessment,* ASTM STP 1179, Landis, W.G., Hughes, J.S., and Lewis, M.A., Eds., American Society for Testing and Materials, Philadelphia, PA, 1993, 362.

2. Foster, M. S. and Schiel, D. R., The biology of giant kelp forests in California: a community profile. U.S. Fish and Wildlife Service Biological Report 85 (7.2), Washington, DC, 1985.

3. Devinney, J. S. and Volse, L. A., Effects of sediments on the development of *Macrocystis* gametophytes, *Mar. Biol.* (Berl.), 48, 343, 1978.

4. North, W. J., The biology of giant kelp beds (*Macrocystis*) in California. *Nova Hedwigia,* 32, 1, 1971.

5. North, W., Gerard, V. A., and Kuwabara, J.S., Farming *Macrocystis* at coastal and oceanic sites, in *Synthetic and Degradative Processes in Marine Macrophytes,* Srivastava, L., Ed., Walter de Gruyter, Berlin, 1982, 247.

6. Luning, K., Critical levels of light and temperature regulating the gametogenesis of three *Laminaria* species (Phaeophyceae), *J. Phycol.,* 16, 1, 1980.

7. Smith, B. M. and Harrison, F. L., Sensitivity of *Macrocystis* gametophytes to copper. Final report to US Nuclear Regulatory Commission, NRC FIN No. A0119, 1, 1978.

8. James, D. F., Manley, S. L., Carter, M. C., and North, W. J., Effects of PCB's and hydrazine on life processes in microscopic stages of selected brown seaweeds, *Hydrobiologia,* 151/152, 411, 1987.

9. Anderson, B. S. and Hunt, J. W., Bioassay methods for evaluating the toxicity of heavy metals, biocides, and sewage effluent using microscopic stages of giant kelp *Macrocystis pyrifera* (Agardh): a preliminary report, *Mar. Environ. Res.,* 26, 113, 1988.

10. Garman, G. D., Pillai, M. C., and Cherr, G. N., Inhibition of cellular events during early algal gametophyte development: effects of select metals and an aqueous petroleum waste, *Aquat. Tox.,* 28, 127, 1994.

11. Garman, G. D., Pillai, M. C., Goff, L. J., and Cherr, G. N., Nuclear events during early development in gametophytes of *Macrocystis pyrifera,* and the temporal effects of a marine contaminant, *Mar. Biol.,* 121, 355, 1994.

12. Pybus, C., Effects of anionic detergent on the growth of *Laminaria, Mar. Poll. Bull.,* 4, 73, 1973.

13. Hopkin, R. and Kain, J. M., The effects of some pollutants on survival, growth and respiration of *Laminaria hyperborea, Estuar. Coast. Mar. Sci.,* 7, 531, 1978.

14. Thompson, R. S. and Burrows, E. M., The toxicity of copper, zinc, and mercury to the brown macroalga *Laminaria saccharina,* in *Ecotoxicological Testing For The Marine Environment,* Persoone, G., Jaspers, E., and Claus, C., Eds., State Univ. Ghent and Inst. Mar. Scient. Res., Brendene, Belgium. 2, 588 pp, 1984.

15. Chung, I. K. and Brinkhuis, B. H., Copper effects in early life stages of kelp, *Laminaria saccharina, Mar. Poll. Bull.,* 17, 213, 1986.

16. Anderson, B. S., Hunt, J. W., Turpen, S. L., Coulon, A. R., and Martin, M., Copper toxicity to microscopic stages of giant kelp *Macrocystis pyrifera*: interpopulation comparisons and temporal variability, *Mar. Ecol. Prog. Ser.,* 68, 147, 1990.

17. Hunt, J. H., Anderson, B. S., Turpen, S. L., and Barber, H. R., Marine Bioassay Project Sixth report: Interlaboratory comparisons and protocol development with four marine species, Final Report No. 91-21-WQ. California State Water Resources Control, Sacramento, CA, 204 pp, 1991.

18. California State Water Resources Control Board (CSWRCB), Procedures manual for toxicity test protocols developed by the Marine Bioassay Project, Report No. 96-WQ-XX, CSWRCB, Sacramento, CA., 1996.

19. United States Environmental Protection Agency. Short-term methods for estimating the chronic toxicity of effluents and receiving waters to west coast marine and estuarine organisms (first edition), Chapman, G.A., Denton, D.L, and Lazorchak, J.M., Eds., EPA/600/R-95-136. Office of Research and Development, Cincinnati, OH, 657 pp, 1995.

20. Amsler, C. D. and Neushul, M., Chemotactic effects of nutrients on spores of the kelps *Macrocystis pyrifera* and *Pterygophora californica, Mar. Biol.* (Berl.), 102, 557, 1989.

21. Hamilton, M. A., Russo, R. C., and Thurston, R. V., Trimmed Spearman-Karber method for estimating lethal concentrations in toxicity bioassays, *Environ. Sci. Technol.,* 11, 714, 1978.

22. Norberg-King, T., A linear interpolation method for sublethal toxicity: The inhibition concentration (ICp) approach, National Effluent Toxicity Assessment Center Technical Report 03-93, U.S. EPA Environmental Research Laboratory, Duluth, MN, 1993.

23. Pillai, M. C., Baldwin, J., and Cherr, G. N., Early development in an algal gametophyte: regulation of germination and nuclear events by cytoskeletal elements, *Protoplasma,* 170, 34, 1992.

24. Pastorok, R. A., Anderson, J. W., Butcher, M. K., and Sexton, J. E., West coast marine species chronic protocol variability study, Final report by PTI Environmental Services, Bellevue, WA for the Washington Department of Ecology, Olympia, WA, PTI Contract No. C333-03-01. 46 pp, 1994.

25. Anderson, B. S., Hunt, J. W., Turpen, S. L., Coulon, A. R., Martin, M., and Palmer, F. H., Marine Bioassay Project Fifth Report: Protocol development and interlaboratory tests with complex effluents, Final Report No. 90-13-WQ, California State Water Resources Control Board, Sacramento, CA, 183pp, 1990.

26. Provasoli, L., Media and prospects for the cultivation of marine algae, in *Cultures and Collections of Algae,* Watanabe, A. and Hattori, A., Eds., Proceedings of the U.S.-Japan Conference, Hakone, *Japanese Society of Plant Physiology,* 63, 1968.

27. Marham, J. W. and Hagmeier, E., Observations on the effects of germanium dioxide on the growth of macroalgae and diatoms, *Phycologia,* 21(2), 125, 1982.

28. Kidder, G. W. III and Awayda, M. S., Effects of azide on gastric mucosa, *Biochem. Biophys. Acta.,* 973, 59, 1989.

29. Lehninger, A. L., *Biochemistry,* (Second edition), Worth, New York. 1104 pp, 1975.

30. Biomonitoring Science Advisory Board (BSAB), West coast marine species chronic protocol variability study, Washington Department of Ecology. Olympia, WA. 11 pp, 1994.

31. Burridge, T. R., Lavery, T., and Lam, P. K. S., Effects of tributyltin and formaldehyde on the germination and growth of *Phyllospora comosa* (Labillardiere) C. Agardh (Phaeophyta:Fucales), *Bull. Environ. Contam. Toxicol.,* 55, 525, 1995.

32. Gunthorpe, L., Nottage, M, Palmer, D., and Wu, R., The development of a fertilization inhibition assay using gametes of the brown alga *Hormosira banksii, Aust. J. Ecotoxicol.,* 1, 25, 1995.

33. Burridge, T. R., Portelli, T., and Ashton, P., In review. The effect of sewage effluents on germination of three marine brown algal macrophytes, *Austr. J. Mar. Fresh. Res.*

34. Burgess, R. M., Ho, K. T., and Morrison, G. E. Marine toxicity identification evaluation (TIE) guidance document: Phase I (DRAFT), U.S. Environmental Protection Agency, Environmental Research Laboratory, Narragansett, RI, 67 pp, 1993.

35. Griffin, D. M., Kline, K. F., and Targart, L. S., Development of Phase I TIE methods for the giant kelp *Macrocystis pyrifera* and red abalone *Haliotis rufescens,* Abstract of paper presented at the Society of Environmental Toxicology and Chemistry annual meeting, Houston TX, 1993.

36. Mann, K. H., *The Ecology of Coastal Waters,* University of California Press, Berkeley, 322 pp, 1982.

37. Schiel, D. R. and Foster, M. F., The structure of subtidal algal stands in temperate waters, *Oceanogr. Mar. Biol. Annu. Rev.,* 24, 265, 1986.

38. Vadas, R., Ecological implications of culture studies on *Nereocystis luetkeana, J. Phycol.,* 8, 196, 1972.

39. Luning, K. and Neushul, M., Light and temperature demands for growth and reproduction of laminarian gametophytes in southern and central California, *Mar. Biol.* (Berl.), 45, 297, 1978.

40. Deysher, L. E. and Dean, T. A., Critical irradiance levels and the interactive effects of quantum irradiance and quantum dose on gametophytes in the giant kelp, Macrocystis pyrifera, *J. Phycol.,* 20, 520, 1984.

41. Deysher, L. E. and Dean, T. A., *In situ* recruitment of sporophytes of the giant kelp *Macrocystis pyrifera*: effects of physical factors, *J. Exp. Mar. Biol. Ecol.,* 103, 41, 1986.

42. Hsiao, S. I. C. and Druehl, L. D., Environmental control of gametogenesis in *Laminaria saccharina.* II. Correlations of nitrate and phosphate concentrations with gametogenesis and selected metabolites, *Can. J. Bot.,* 51, 829, 1973.

43. Amsler, C. D. and Neushul, M., Photosynthetic physiology and chemical composition of spores of the kelps *Macrocystis pyrifera, Nereocystis luetkeana, Laminaria farlowii*, and *Pterygophora californica* (Phaeophyceae), *J. Phycol.,* 27, 26, 1991.

44. Reed, D. C., Neushul, M., and Ebeling, A. W., Role of settlement density on gametophyte growth and reproduction in the kelps *Pterygophora californica* and *Macrocystis pyrifera* (Phaeophyceae), *J. Phycol.,* 27, 361, 1991.

CHAPTER **18**

Microplate Toxicity Tests with Microalgae: A Review

Christian Blaise, Jean-François Férard, and Paule Vasseur

CONTENTS

I. INTRODUCTION

Biological testing to address ecotoxicity concerns for phototrophic systems has rapidly evolved over the past two decades.[1,2] In particular, phytotoxicity tests carried out with microplates have been around for nearly as long, as verified by our recent search for publications within both the primary and "gray" literature. Recorded in this time-frame have been more than 50 articles, reports, and theses describing various methods, applications, data comparisons, etc., performed with miniaturized protocols involving microalgae (Figure 18.1). These publications confirm the international interest expressed in this specific area of small-scale bioassay testing, as ten countries (Canada, Czech Republic, France, Germany, Japan, New Zealand, Norway, South Africa, Sweden, USA), based on (co)author affiliations, are represented in the process.

In order to circumscribe the present state of knowledge and to set the course for future research activities with microplate technology, we considered it an important and useful task to summarize the significant information reported in this field thus far. After recalling some basic arguments favoring the growing desire to employ simplified microscale tests with algae, the present text recapitulates the major types of microplate procedures described in the literature based on their chronological appearance. For convenience, these procedures have been split into two main categories: 1) tests conducted with unialgal species (by far the most numerous and diverse) and 2) tests undertaken with batteries of unialgal species.

Since microplate-based phytotoxicity tests evolved from earlier (and more laborious) "yardstick" tests performed with flasks, several studies have been devoted to demonstrating comparability of toxicity responses between what might now be called a standard microplate phytotoxicity test and reference flask tests. A section of this review therefore deals with this particular aspect.

Reported microplate testing protocols display marked versatility as far as experimental variables are concerned. Without wishing to appraise their relative value at this time, it appeared relevant to look at their similarities and differences. This section also highlights the diversity of microplate assays and hence the interesting options that an operator has at his disposal to customize tests based on self-imagination and desired objectives. Because of this same diversity in microplate-based protocols, a subsequent section examines toxicity sensitivity responses for chemicals commonly assessed by different procedures.

From a philosophical point of view, it can be said that the perfect bioassay is not of this world, and this certainly holds true for microplate-based assays. We can nevertheless hope for bioassays that will be "as least imperfect as possible" as new knowledge, materials, and equipment, for example, improve their undertaking. We have thus recalled the major experimental factors of importance linked to microplate assays and taken a look at how problems have been resolved, minimized, circumvented, or (at least) discussed by research groups. Based on the above, the final section of this review chapter offers a summary of our major findings and proposes some additional avenues of research for microplate-based phytotoxicity assays.

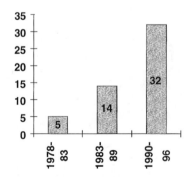

Figure 18.1 Numbers of microplate-based phytotoxicity publications as a function of time (1978 to 1996).

C O L U M N

		1	2	3	4	5	6	7	8	9	10	11	12
	A		WATER ONLY										
	B		T1	T2	T3	T4	T5	T6	T7	T8	T9	T10	
	C		T1	T2	T3	T4	T5	T6	T7	T8	T9	T10	
R O W	D		C	C	C	C	C	C	C	C	C	C	
	E		T1	T2	T3	T4	T5	T6	T7	T8	T9	T10	
	F		T1	T2	T3	T4	T5	T6	T7	T8	T9	T10	
	G		T1	T2	T3	T4	T5	T6	T7	T8	T9	T10	
	H		WATER ONLY										

Figure 18.2 A typical microplate experimental disposition. Peripheral wells are filled with 200 μL of dilution water. There are five replicates per concentration with T1 and T10 being the highest and lowest concentrations, respectively. There are 10 centrally-located control (C) wells (From St-Laurent, D., Blaise, C., MacQuarrie, P., Scroggins, R., and Trottier, B. *Environmental Toxicology and Water Quality*, 7, 35, 1992. © 1992 John Wiley & Sons. With permission.)

II. WHY UNDERTAKE MICROPLATE TOXICITY TESTS?

As producers of oxygen and as a vital food source for other trophic levels, the well-being of microalgae is essential to ensure the stability of aquatic ecosystems. Recognition of the necessity to protect this basic trophic level from man-made insults owing to chemical pollution thus promoted the development of a series of flask phytotoxicity tests from the late 1970s onward.[3-6] While flask-based tests are scientifically sound, they are also labor-intensive and not very cost-effective. Hence, sparked by a genuine need to find more practical alternatives for phytotoxicity testing, 96-well microtitration plates (or microplates) — for which varied applications began in the field of clinical microbiology[7] — were discovered and employed by environmental research groups for the conduct of algal bioassays. Microplate-based tests confer many advantages[8] and, *per se*, these have clearly stimulated the contemporary development of numerous small-scale biological tests undertaken with a variety of microorganisms.[9,10] While experimental configurations for algal microplate assays can be varied, they are usually tailored to meet specific testing objectives (e.g., acute or chronic testing with one or more species) or to offset potential interferences (e.g., toxicant volatility). As an example, Figure 18.2 displays a typical experimental disposition which was recently used to evaluate the phytotoxicity of herbicides.[11]

III. TESTS CONDUCTED WITH UNIALGAL CULTURES

A. Liquid Media Procedures

1. Freshwater Methods

Publications of microplate-based algal assays involving liquid media procedures conducted with monospecific cultures began to appear in the late 1970s. According to our literature search, a team from the University of Bergen (Norway) may well have been the first to employ a miniaturized procedure to report "toxic effect threshold concentrations" for *Chlamydomonas reinhardti* exposed for 48 h to a series of oil dispersants.[12] In this 1978 pioneering paper, the authors were quick to point out the noticeable advantages (simplicity, cost-efficiency, rapid screening potential) linked to the use of microplates as experimental vessels. Next, Blaise et al.[13] described an 8-day growth inhibition algal microassay technique undertaken with the chlorophyte *Selenastrum capricornutum*. Inspired by the Heldal et al. (1978) procedure,[12] a Swedish group studied the effects of various sterilization techniques of wastewaters on the toxicity responses of two chlorophytes, *S. capricornutum* and *Monoraphidium pusillum*, by determining visual EC_{100}s (lowest concentration without detectable growth) after 14 days of growth in microtitration plates.[14] The following year, another chronic exposure miniaturized assay (4- and 8-day tests) carried out with *S. capricornutum* was reported, based on the measurement of intracellular ATP as the endpoint reflecting growth inhibition.[15]

Because chronic exposure tests of a too-long duration can be affected by several factors (e.g., pH change, complexation of chemicals by algal exudates, algal adaptation, algal/bacterial degradation, and/or photodegradation of tested xenobiotics) which can alter the original toxicity of environmental samples,[16] the algal growth inhibition microtest reported above was eventually shortened to a 4-day exposure assay.[8] A technical method detailing the test procedure was also published.[17] A 4-day algal ATP microplate assay technique, again performed with *S. capricornutum*, was next described by Kwan,[18] who reported on the interferences that colored pulp and paper effluents can have in the generation of the luciferin–luciferase light reaction and that appropriate controls must be incorporated into the test procedure.

In 1990, South African researchers, inspired by previous phytotoxicity methodologies,[5,8] developed their own two-step microplate technique with *Selenastrum capricornutum* in which 72-h IC_{50} tests were first performed in 24-well plates followed by transfer of post-exposure sample mixture microvolumes into a 96-well microplate for photometric assessment of algal biomass.[19] This procedure was later recommended as part of a test battery for conducting toxicity appraisals of drinking and environmental waters in South Africa.[20] A 96-well 72-h exposure microplate toxicity test using photometric measurement of algal biomass to record growth of *S. capricornutum* and *Scenedesmus subspicatus* was also developed and optimized in Germany.[21]

This same procedure was recently standardized and proposed as a suitable test for the ecotoxicological examination of chemicals, domestic sewage, and industrial wastes.[22] Starting from the hypothesis that microalgae can be subjected to high chemical concentrations for short periods of time (i.e., a few hours), Hickey et al.[23] developed a 4-h acute exposure microassay to assess toxic impact on *S. capricornutum* under such conditions. In this study, intracellular ATP reduction and inhibition of post-exposure cell recovery (cell growth in nutritive medium for 96 h following the 4-h exposure) were the selected toxicity endpoints. Prior to recommending an appropriate procedure for determining chronic algal growth inhibition to *S. capricornutum* after exposure to toxicants based on chlorophyll *in vivo* fluorescence, Caux et al.[24] conducted experiments to investigate various physical (nontoxic) factors likely to affect the fluorescence response in the microplate assay. During the same year, the government of Canada published a recommended microplate procedure for determining the chronic toxicity of liquid samples toward the green alga *S. capricornutum*.[25]

More recently, a review described practical phytotoxicity applications performed at the Centre Saint-Laurent laboratories (Environment Canada, Montréal, Québec Region) during the past decade which involved various microplate procedures.[26] At the Environmental Sciences Centre (ESC) of the University of Metz, *S. capricornutum* was successfully cryopreserved with the help of polyvinylpyrrolidone, a high-molecular-weight polymer, thereby alleviating laboratory costs associated with culture maintenance for undertaking algal microplate assays.[27] During the same period, a parallel ECS group reported a new semistatic 72-h growth inhibition test, where daily renewals of toxicants were enabled by employing microplates with special membrane-bottomed wells.[28] A new experimental protocol, referred to as "Cryoalgotox," is presently under development to couple this semistatic assay with the cryopreserved algal reagent.[27] In Canada, the very popular

S. capricornutum was used once more to develop a chronic exposure lethality assay (96-h LC_{50}) with the help of flow cytometry. With this instrumental technology, the algicidal status of cells was determined from the intensity of green fluorescence (emission resulting from fluorescein after algal esterase cleavage of fluorescein diacetate) and red fluorescence (emission resulting from chlorophyll) as discriminatory parameters.[29] Finally, two very simple acute exposure microplate phytotoxicity assays were developed. The first is based on motility inhibition of the chlorophyte *Chlamydomonas variabilis*.[30] Attractive features of this 5-h exposure assay are its high sensitivity toward metals as well as its initiation and completion within a normal work day. The second reports rapid microfluorometric measurements of *in vivo* esterase inhibition in *S. capricornutum* and *Synechoccus leopoliensis* after a brief 1-h exposure to selected metals and organics.[31]

2. Marine Methods

Unlike their freshwater counterparts, marine microplate phytotoxicity tests conducted with unialgal species are scarce. In fact, only one type of experimental protocol has been reported so far by two research groups. The first group, located at IFREMER in Nantes (France) reported a sublethal assay used to measure the metabolic status of cells exposed for 5 h to toxicants (herbicides, insecticides, metals) or surface sea water samples.[32-34] More specifically, esterase activity is appraised after post-exposure algae have been allowed to incubate for an additional hour in the presence of fluorescein diacetate (FDA). Nonfluorescent (nonpolar) intracellular FDA is cleaved to fluorescent (polar) fluorescein (F) by nonspecific esterases and F fluorescence intensity is then measured microfluorometrically. A computer program enabling 5-h IC_{50} determinations, after capture of fluorescence data from an automated plate reader, has been detailed.[33] Three unicellular marine algae (*Tetraselmis suecica*, *Skeletonema costatum*, *Prorocentrum lima*) have thus far been used to generate toxicity data, and further studies with additional species are ongoing to assess their relative sensitivities to contaminants.[34] Based on a similar endpoint, a second USA-based group recently reported an even more rapid 1-h exposure esterase inhibition test for the prasinophyte *T. suecica* and the diatom *Cyclotella sp.*[31]

3. Special Applications

Commented on hereafter are reports on 1) algal microplate applications which deal with special methods not directly linked to toxicity assessment[35-37] — but which are nevertheless worthy of mention in this review — and on 2) the utilization of algal microplate methods as practical aids (adjuncts) to assess other methodologies[38] or contaminants.[39-48]

Hassett et al.[35] were the first to describe the development of an ingenious 3-h exposure microplate technique to assess the metal (radionuclides) uptake potential of different species of green and blue-green algae. Several variables likely to influence accumulation (e.g., algal species, culture age, metals, pH) were investigated, and metal concentration factors generated with this microassay closely matched those reported in the literature by more traditional methods. Microalgal bioaccumulation of metals in solution was also assessed by an entirely biological procedure with *C. variabilis* and *S. capricornutum*.[36] In this study, the former alga (an indicator of accumulation) was exposed to metal solutions for 24 h, after which accumulated metals were unbound from organic cell constituents by a photooxidation step. A microplate toxicity assay was then carried out on the irradiated solution with *S. capricornutum*, whose toxic expression reflected mineral sorption by *C. variabilis*. This combined bi-algal accumulation/toxicity assay yielded toxicity responses which closely paralleled metal accumulation as indicated by chemical analysis of *C. variabilis* tissue.

A relatively new area of bioanalytical applications with microalgae in the field of environmental research is that involving interaction studies on the metabolic potential of vegetal systems to modify the genotoxic activity of chemicals. In this area, Rodrigue[37] developed a miniaturized bioassay to measure the genotoxic (de)activation of chemicals owing to algal metabolism. The experimental protocol, performed in special microplates with membrane-bottomed wells, requires a 4-h contact of test chemical with a specific biomass of *S. capricornutum*. After exposure, the interaction filtrate, without the cells, is recovered with a microplate filtering apparatus. Comparison of the pre-exposure (T = 0 h) with the post-exposure (T = 4 h) genotoxicity status of the chemical — as measured with the SOS Chromotest (another microplate-based assay which acts as the genotoxicity indicator in this methodology) — will then determine whether it has been (de)activated by the algae.

Microplate algal procedures have also proven useful to various research groups. Baarschers et al.,[39] employing a somewhat modified Blaise et al.[8] microplate technique, determined the phyto-toxic potential of trichlopyr and related herbicides for *Chlorella vulgaris* and *Chlorella pyrenoidosa*, along with a battery of other microbial assays. The herbicide hexazinone, assessed with a conventional flask method and the Blaise et al.[8] microplate procedure, yielded 18-d and 4-d IC_{50}s of 22.5 and 24 μg/L, respectively, for *S. capricornutum*.[40] The scaled-down version of the flask assay as well as the abbreviated test duration, with respect to the microplate procedure, were pointed out to be useful features by this author. In work studying the potential of algae to modify the genotoxic activity of chemicals and wastewaters, Harwood et al.[38] had to ensure that selected exposure concentrations would be devoid of growth inhibition effects. To obtain this information, they used a 96-h growth inhibition algal microtest[8] to quickly assess the toxicity of their liquid samples, prior to undertaking the algal/sample interaction experiments,. A microplate technique performed with *S. capricornutum* was again employed, within the framework of an integrated bioassay approach, to appraise the potential impact of zirconium ($ZrCl_4$), an efficient phosphorus precipitator considered for use as an anti-eutrophication agent to control agricultural discharges.[41] During a toxicological study on the environmental hazards posed by chromated copper arsenate, pentachlorophenol, and preserved wood, the same microassay technique was applied once more.[42] The ecotoxic potential of tebuthiuron (a substituted urea-class herbicide useful in forestry management to control growth of a variety of herbaceous and woody plants) on nontarget aquatic species was scrutinized with a series of bioassays, one of which was an algal microtest.[43] The PEEP index (Potential Ecotoxic Effects Probe), an environmental management tool recently developed to assess and compare the toxic potential of industrial effluents, incorporates an algal microtest, along with other complementary microbiotests, in a specially designed formula to calculate log_{10} values on toxic loading contributions to aquatic receiving systems.[44] A cost-effective battery of (geno)toxicity microassays, including a phytotoxicity component, was recently assembled to estimate the environmental risk posed by leachates of various solid industrial wastes destined for landfill disposal.[45] Again, toxicity (72-h IC_{50}s) of the herbicide metolachlor to *S. capricornutum* and to *Anabaena cylindrica* was reported[46] using Environment Canada's microplate technique,[25] while the same technique[25] and that of St-Laurent and Blaise[29] were respectively employed to generate 4-day IC_{50}s and 4-day LC_{50}s for the herbicides MCPA, butylate, atrazine, and cyanazine.[47] Finally, a comprehensive ecotoxicological assessment of Lake Kojima (Japan) made use of a microbioassay battery to generate toxicity data for extracts of surface waters, their associated suspended solids, and sediments.[48] In this study, 5-day IC_{50}s for *S. capricornutum* were reported by applying slightly-modified Canadian methodologies.[8,25]

B. Solid Media Procedures

Microplate phytotoxicity testing procedures conducted on solid media have been reported by a research group from the Institute of Botany, Dukelska, Czech Republic.[49-52] After a comprehensive study of experimental factors influencing the growth rate of colonies of *Chlorella kessleri* on agar medium, a microplate technique was proposed to assess eutrophication (algal growth potential converted to dry weight) and toxicity (colonial growth rate) potential of environmental samples.[49] In this article, Lukavsky[49] describes a device for inoculating several thousands of cells (from dense growth in a Petri dish) directly onto the surface of individual microplate wells, each containing agar, nutritive medium and varying toxicant concentrations. Microalgal cells inoculated in this fashion are periodically monitored for growth microscopically and will eventually give rise to giant colonies. Inhibition of colonial growth rate, the selected toxicity endpoint, is therefore based on the speed with which colonies will expand owing to the presence of nutrient medium in the wells. Two to three days of exposure appear to be sufficient to record measurable toxic effects with this procedure. The inhibitory effects on colonial growth rate of *C. kessleri* are given for chromium, copper, and zinc. In a second article, this author details a simple and inexpensive cultivation unit, capable of holding six microplates, for undertaking auxinic or toxic investigations of pollutants by gauging the diametric growth of giant colonies of *C. kessleri*.[50]

More recently, a microplate liquid assay procedure is proposed by Lukavsky once more for eutrophication and toxicity studies, based on nephelometric recordings of *C. kessleri* growth with a plate reader taken over a period of several weeks.[51] In this article, it is suggested that this miniaturized test has value for screening the phytotoxicity potential of surface and groundwaters, but no data are presented to substantiate this claim. Technical details, for both the microplate solid-phase assay[49] and the microplate liquid-phase assay[51] have been summarized.[52] In this last article, concentration-response curves determined from microplate solid phase testing are illustrated for copper, zinc, chromium, and 3,5 dichlorophenol. Concentration-response curves demonstrating strikingly similar toxicity effects on four chlorophytes (*C. kessleri, S. capricornutum, Scenedesmus quadricauda,* and *subspicatus*), determined from microplate liquid phase testing, are also presented for chromium.

To conclude, we have listed the algal species reportedly employed in the conduct of microplate toxicity tests with single species of microalgae (Table 18.1). Of the 16 recorded species, 10 are chlorophytes of freshwater origin. Clearly, *S. capricornutum* tops the list in popularity, no doubt owing to the fact that monospecific microplate testing with microalgae has been used in Canada since the early 1980s and that most publications referenced in Table 18.1 originate from this country. If we also consider those species identified for conducting microplate tests with batteries of unialgal cultures, as will be discussed in the following section, one could surmise that microalgae appear generally well-suited as biological reagents for undertaking microplate-based assays. In a way, this is also corroborated by the absence of literature reports indicating certain species to be unsuitable for such microassays.

Table 18.1 Reported Microalgae Employed in Single Species Microplate Toxicity Testing, as of 1996

Test Species	References
Freshwater	
Anabaena cylindrica	31,46
Chlamydomonas variabilis	30
Chlamydomonas reinhardti	12
Chlorella vulgaris	39
Chlorella sp.	19,20
Chlorella pyrenoidosa	39
Chlorella kessleri	49,50,51,52,59
Monorhaphidium pusillum	14
Scenedesmus quadricauda	19,52
Scenedesmus subspicatus	21,22,52
Selenastrum capricornutum	8,11,14,15,17,18,19,20,21,23,24,25,27,28,29, 36,38,40,41,43,44,46,47,48,52,57,58,60
Synechoccus leopoliensis	31
Marine	
Cyclotella sp.	31
Prorocentrum lima	34
Skeletonema costatum	32,34
Tetraselmis suesica	31,32,33,34

IV. TESTS CONDUCTED WITH BATTERIES OF UNIALGAL CULTURES

The concept and practice of applying microplate-based tests with batteries of microalgae were introduced by a research group from the Swedish University of Göteborg.[53,54] In a major study conducted with 13 different species of microalgae (composed of eight chlorophytes, three xanthophytes, and two cyanophytes) and 19 diverse chemicals, it was clearly demonstrated that a significant amount of phytotoxicity information could be gathered in a cost-effective way thanks to a microplate methodology.[53] Briefly, these authors describe a technique whereby the toxicity of one chemical can be assessed in two microplates with three microalgae. In their experimental disposition, provision is made for four replicates per concentration with a geometric concentration series covering 4.2 orders of magnitude (14 sample concentrations in two plates). Post-exposure growth inhibition measurements are then simply made by visually recording the lowest concentration of test compound with no detectable growth (EC_{100}). One important finding of this work showed that species-dependent variation in algal sensitivity reached over three orders of magnitude with the chemicals tested. Based on their results, these authors concluded that the use of representative algal test batteries should be emphasized over single algal test systems for routine screening, as the information generated by the former is likely more environmentally relevant and useful for decision-making with respect to chemical hazard posed on phototrophic systems.

In a more recent report entitled *The Algal Microtest Battery,* Blanck and Björnsäter[54] detail the technique for undertaking the unialgal test battery and list all material and instrumental requirements. Changes with respect to the Blanck et al. technique,[53] very likely because of methodological improvements after experience of use, include a new microplate disposition (two algae per microplate with provision for appropriate controls that may be required, such as solvent controls), a shortened exposure period (5 days instead of 14 days), and the possibility of determining growth inhibition IC_{50}s based on microfluorometric readings of algal biomass. These authors further state that their method can be employed to study chemical impact on both freshwater and marine algae, and list suggested taxa for setting up appropriate batteries depending on the aquatic environment of concern.

A current publication resulting from a joint cooperation of the University of Göteborg and the Swedish EPA describes the use of an algal test battery (ATB) to assess the toxicity of industrial effluents along with a suite of other short-term bioassays.[55] Here, authors clearly demonstrate the sensitivity of microbial tests and, in particular, of the ATB, in its capacity to detect wastewater toxic potential. Again, it is concluded that applying an ATB over singular unialgal species testing markedly augments ecotoxicological information in terms of sensitivity. Lastly, inspired by the ATB work performed at the U. of Göteborg, another contemporary study published by the Saskatchewan Research Council (Canada) reports the development of a five-species ATB specifically designed to assess phytotoxicity of liquid mining effluents and metal toxicants.[56] Notable features of this microplate-based method include a more modest starting test inoculum for algae (10^4 to 5×10^5/mL) coupled with a shorter exposure time (48 to 72 h), thanks to the use of a modern highly-sensitive microfluorometer for biomass measurements.

V. COMPARATIVE STUDIES BETWEEN MICROPLATE AND FLASK ASSAY PROCEDURES

As genuine adjuncts or alternatives to the earlier, albeit more laborious, flask tests, many felt it important to demonstrate that microplate phytotoxicity assays did indeed reproduce the results of their glassware counterparts. Displaying agreement between the two methodologies would indicate microassay reliability and obviously enhance microtesting acceptance by the scientific community. In this perspective, several studies have been undertaken to establish interprocedural compatibility. An initial investigation undertaken with *S. capricornutum* showed congruity in 8-d IC_{50} results generated with flask and microplate tests for seven industrial effluents.[13] With the same alga, intermethod concordance for 25 other effluents (pulp and paper and chemical production plants) was again borne out for 4-d and 8-d IC_{50} endpoint tests.[8,57] Repeatability and reproducibility of results, as well as agreement in 3- and 5-d IC_{50}s generated with both types of procedures, was further established with *S. capricornutum* following tests carried out with metals (Cu^{2+}, Zn^{2+}, Cd^{2+}, Sn^{2+}) and phenol.[58] In Germany, a 72-h IC_{50} exposure microplate method measuring growth of *Scenedesmus subspicatus* with a photometric assessment endpoint was successfully compared to the corresponding flask method for $K_2Cr_2O_7$ and $KClO_3$.[21] Metals (Cu^{2+}, Zn^{2+}, Cr^{6+}), phenol, and nine herbicides were appraised using both a microplate and flask assay by St-Laurent et al.[11] Barring the herbicide diquat, whose toxicity toward *S. capricornutum* was shown to be almost seven times greater in the microplate assay than in the flask assay, excellent correspondence existed in all other 4-d IC_{50} data produced. Recently, biprocedural 4-d IC_{50}s reported with *Chlorella kessleri* for three metals (Cd^{2+}, Cu^{2+}, Zn^{2+}) and three herbicides were again shown to agree.[59]

The comparative investigations conducted with varied liquid samples (metals, herbicides, phenol, effluents) therefore essentially point out that microplate and flask methodologies are compatible insofar as generation of phytotoxicity data with microalgae is concerned. Each of these studies also presents evidence that microplate tests are reliable because of their repeatability and reproducibility. Only one microplate assay has thus far been subjected to an intercalibration exercise to assess its performance when undertaken by operators of different laboratories.[60] This particular round-robin study, which also examined the influence of specific test variables (algal cultivation technique, presence or absence of Na_2EDTA in nutrient medium, passive/active gas exchange in microplate wells during exposure), again confirmed microtesting reliability and markedly contributed to protocol optimization of this particular procedure.

VI. FEATURES OF MICROPLATE TESTING PROTOCOLS

The basic features of microplate liquid testing protocols are summarized in Table 18.2. Chronic exposure tests clearly outrank acute exposure tests in numbers. Algal inocula requirements for the former range from 10^3 cells/mL to 6×10^5 cells/mL, while those reported for the latter vary from 10^5 to 10^6 cells/mL. Intracellular ATP, esterase activity, and motility have sparingly been employed to report assessment endpoints, but growth inhibition (determined by cell counts, ATP, absorbance, fluorescence, and visual observation) has distinctly been the most popular choice. In terms of measurement endpoints, IC_{50}s have been most frequently determined, but IC_{20}s, IC_{100}s, and NOECs/LOECs have also been worked out. As a supplement to Table 18.2 information, it is pertinent to note that microplates used were invariably made of polystyrene with U- or flat-bottomed wells, except for Gilbert et al.[34] and Radetski et al.[28] whose plate wells were membrane-bottomed. Lighting during exposure was continuous (except for Baarschers et al.[39]: 16/8-h photoperiod; Gilbert et al.[34]: unspecified; Snell et al.[31]: exposure in darkness) and essentially varied from 30 to 95 $mE.m^{-2}.s^{-1}$. Exposure microplates were never stirred barring two exceptions. Daily resuspension of algae was performed with a plate shaker prior to undertaking *in vivo* fluorescence growth measurements for the protocol described by Blanck and Björnsäter.[54] Microplate well contents were mixed during exposure (400 rpm) in the case of the algal test battery protocol developed by the Saskatchewan Research Council.[56]

VII. SENSITIVITY RESPONSES OF MICROPLATE METHODS

Evidently, the toxicity response of any laboratory-based bioassay will be influenced by factors which are procedure- (light conditions, temperature, inoculum, exposure time, medium, endpoint, etc.), laboratory- (air pressure and content, equipment, operator performance, etc.), test sample- (metals, organics, effluents, etc.), and test alga-specific. Hence, the value of undertaking method

Table 18.2   Basic Features of Microplate Liquid Testing Protocols

Test Type	Cell Inoculum	Assessment Endpoint	Measurement Endpoint	References (and Test Samples)[a]
Acute exposure				
1 h	$1\text{--}10 \times 10^5$/mL	esterase inhibition	NOEC/LOEC	31 (M,O)
4 h	10^6/mL	ATP energy loss; cell growth recovery	IC_{50}	23 (M,O)
5 h	unspecified	esterase inhibition	IC_{50}	33 (O); 34 (M,O,S)
5 h	5×10^5/mL	motility inhibition	IC_{20}; NOEC/LOEC	30 (M)
Chronic exposure				
2 d	unspecified	growth inhibition	toxic threshold effect concentration	12 (OD)
2–3 d	$10^4\text{--}5 \times 10^5$/mL	fluorescence as biomass indicator	IC_{50}	56 (E,M)
3 d	$1\text{--}2 \times 10^4$/mL	growth inhibition (cell counts)	IC_{50}	25 (aqueous samples), 27 (M,H), 28 (M,H,L),[b] 45 (L), 46 (H)
3 d	$1\text{--}6 \times 10^5$/mL	absorbance (450 nm) as indicator of biomass inhibition	IC_{20}, IC_{50}	21 (M), 22 (E), 19 (M,O)
4 d	2×10^4/mL	ATP as indicator of biomass inhibition	% loss of relative light units in relation to control	18 (E)
4 d	2×10^4/mL	ATP as indicator of biomass inhibition	IC_{50}	15 (E)
4 d	1×10^4/mL	absorbance (750 nm) as indicator of biomass inhibition	IC_{50}	59 (M,H)
4 d	$1\text{--}2 \times 10^4$/mL	cell death (flow cytometry method)	LC_{50}	29 (M), 47 (H)
4 d	$1\text{-}2 \times 10^4$/mL	growth inhibition (cell counts)	IC_{50}	8 (M,E), 11 (M,H), 17 (aqueous samples), 41(M), 43 (H), 47 (H), 60 (M,O)
5 d	1×10^4/mL	absorbance (420 nm) as indicator of biomass inhibition	IC_{50}	48 (M,O,OE)
5 d	biomass corresponding to 10 ng chlorophyll a/mL	visual observation (v.o.) of total growth inhibition at end of test or daily microfluorimetric measurement of in vivo fluorescence	IC_{100} (v.o.); IC_{50} (fluorescence)	54 (M,O,E)[c]
7 d	unspecified	absorbance (690 nm) as indicator of biomass inhibition	IC_{50}	55 (E)
8 d	10^3/mL	ATP as indicator of biomass inhibition	IC_{50}	8 (M,E), 15 (E)
		growth inhibition (cell counts)	IC_{50}	8 (M,E), 13 (E),
10–14 d	10^4/mL	growth inhibition (cell counts)	IC_{50}	39 (H)
14 d	biomass corresponding to 10 ng chlorophyll a/mL	visual observation (v.o.) of total growth inhibition at end of test	IC_{100}	14 (E), 53 (M,O)
21–30 d	10^4/mL	absorbance (750 nm) as indicator of biomass inhibition	unspecified	51,52 (M,O,G,S)

[a] E (effluents), G (groundwater), H (herbicides), L (solid waste leachates), M (metals), O (organics), S (surface waters), OD (oil dispersants), OE (organic extracts).

[b] Semistatic microplate assay with daily renewal of medium and test substance.

[c] Unialgal battery test method.

Table 18.3 Comparison of the Relative Sensitivity Responses of Microplate Methods Conducted with Single Species for Specific Chemicals

Chemical	Acute Exposure Tests: 4–5 h IC_{50}s	Chronic Exposure Tests: 3, 4, or 5 d IC_{50}s
Cu^{2+}	0.009[a] (30); 0.14, 0.42 (23); 0.02, 0.3, 0.4, 1.0 (31); 8.5, 35 (34); 0.02, 0.3, 0.4,1.0 (31)[b]	0.012, 0.019 (58); 0.022, 0.029 (27); 0.048 (8, 28); 0.054 (59); 0.066 (11); 0.072 (19); 0.078 (59); 0.8[c] (52); 1.0[c] (49)
Cd^{2+}	0.005 (30); 0.32, 0.66 (23); 0.05, 1.0, 1.0, 2.0 (31)[b]	0.017 (58); 0.021 (28); 0,030 (58); 0.032 (27); 0.040 (59); 0.043 (27); 0.050 (8); 0.056 (60); 0.080 (48); 0.080 (59); 0.20 (19)
Zn^{2+}	0.034 (30); 0.46, 8.0 (23); 20.4, 21 (34)	0.025 (58, 19); 0.048 (8); 0.053 (11); 0.054, 0.067 (59); 8.0[c] (49); 20[c] (52)
Cr^{6+}	0.023[a] (30); 0.1, 3.0 (23)	0.04, 0.04, 0.05, 0.21, 0.42 (56)[d]; 0.066 (28); 0.074 (27); 0.13 (11); 0.14 (27); 0.17 (23); 0.33 (48); 3.0[c] (49)
Phenol		27, 47 (58); 63.1 (11); 69.7 (60); 107 (19)
Atrazine		0.026(47); 0.093 (27); 0.12 (28); 0.16 (27)
2, 4-d		19.0[e] (39); 24.2 (11); 56.0[e] (39)
Hexazinone		0.024 (40); 0.028 (11)
Picloram		22.7 (11); >160[e] (39)
Tebuthiuron	>100 (Hickey et al., 1991)	0.080 (43); 0.10 (23)
Metolachlor		0.037, 0.052, 0.056 (46); 0.051 (11)

Note: Endpoint values were generated with various algae and assessment endpoints. All endpoint values are expressed in mg/L. References are parenthesized.

[a] 5 h IC_{20} values.
[b] LOEC values determined with four different algal species after 1-h exposure tests.
[c] Unspecified exposure time for this protocol. Values are toxicity thresholds (≈LOECs) visually determined from concentration-response curves illustrated in Lukavsky, 1983 (49) and Lukavsky, 1994 (52).
[d] IC_{50} values from an algal battery of five different species.
[e] 10–14-d IC_{50} values.

assessments to indicate which are most or least sensitive can be questioned, as such exercises run the risk of falling into the apples and oranges comparison syndrome. We have nevertheless rounded up toxicity endpoint values obtained from microplate methods conducted on similar chemicals to see whether any overall trends might emerge in terms of methods sensitivity (Table 18.3). Unsurprisingly, chronic exposure procedures appeared generally more sensitive than their acute exposure counterparts, as seen with the four metals and the herbicide tebuthiuron for which acute and chronic exposure test results can be compared. A notable exception, however, lies with the 5-h *C. variabilis* flagellar immobilization test reported by Kusui and Blaise,[30] whose toxicity response was either equally (Zn^{2+}) or more sensitive (Cu^{2+}, Cd^{2+}, Cr^{6+}) toward metals in relation to the chronic tests. Chronic exposure test result variation with metals, for which comparative values are more numerous, is reflected by sensitivity factors ranging from 12 (Cd^{2+}) to 800 (Zn^{2+}) between the lowest and highest IC_{50} response generated for each metal. Although toxicity data are scarce for microplate solid media procedures, their sensitivity appears to be markedly lower than that of liquid media procedures, as can be seen with Cu^{2+},[49,52] Zn^{2+},[49,52] and Cr^{6+}.[49] The paucity of comparative microplate methods data for the organic compounds (phenol and herbicides) listed in Table 18.3 does not allow for much to be said other than to signal the respective concentration ranges in which they exhibit toxicity under chronic testing conditions.

VIII. EXPERIMENTAL FACTORS OF CONCERN FOR MICROPLATE ASSAYS

As with all bioassays, microplate-based tests will yield reliable results if proper care is given to their undertaking. Several caveats, influencing factors, potential problems, and/or interferences, linked to these microtests, have been reported in the literature. They are discussed below.

A. Well Evaporation

Unacceptable evaporation will occur in microplate wells under conditions of chronic exposure testing (>1 d) unless care is taken to deal with this factor. This is essentially owing to the fact that microplate lids are not air-tight. Evaporation effects are adequately counteracted, however, by placing microplates with their lids inside a transparent plastic lining during the exposure period (e.g., Blaise et al.[13]; Blanck and Björnsäter[54]). With microplates tucked inside a heat-sealed poly-ethylene liner, well evaporation (<10%) was recorded as being no greater than that of a 125 mL Erlenmeyer conventionally employed in flask phytotoxicity assays after an incubation period of eight days.[13] During the conduct of 10- to 14-d chronic tests, Baarschers et al.[39] also reported a means of minimizing well evaporation by keeping microplates in glass chromatography tanks equipped to prevent drying. A recent technique calls for sealing the lid-microplate interface with parafilm, provided that the microplate lid is "aerated" with small holes at strategic locations to permit gas exchange during exposure.[56] Maintaining high humidity levels (50 to 100%) in closed incubators is yet another means of counteracting evaporation problems.[56,61]

B. Edge Effects

A phenomenon commonly associated with 96-well microtitration plates is the so-called "edge effect," whereby the evaporation rate of circumferential wells tends to be greater than that of centrally-located wells. Several studies have warned about edge effects,[11,24,25,49,51] and recent exper-imental dispositions tend to exclude peripheral wells from testing because they increase variability among replicates. In microplates placed inside a heat-sealed polyethylene liner, Caux et al.[24] experimentally demonstrated that external wells lose 40% more water than internal wells during a 4-d incubation period. Hence, for chronic tests of one day or more, experimental designs are usually built around the 60 internal wells (see Figure 18.2), as several recent publications indicate.[11,25,28,51,52] While peripheral wells are not directly employed for testing, they are nevertheless always filled with water to increase humidity inside the microplate and, in this sense, they contribute positively to the overall experimental protocol. In contrast, the Saskatchewan Research Council 48- to 72-h exposure microplate growth inhibition test (see Section VIII.A above) does employ peripheral wells in its experimental configuration, as its parafilmed microplate with strategically-located lid hole configuration is reported to eliminate edge effects.[56] It is important to emphasize once again that well evaporation and edge effects are factors that have bearing solely upon chronic exposure assays. Obviously, acute exposure microplate assays of several hours will not be affected in this way.

C. Sample to Experimental Vessel Volume Ratio

U.S. EPA flask phytotoxicity testing is classically performed in 125 mL Erlenmeyers, filled with 25 mL of total experimental solution containing sample, nutrient spike, and algal inoculum.[5] This "sample volume to experimental vessel volume ratio" (S/EV) of 0.20 was recommended to avoid CO_2 limitation, as it provides sufficient "head space" to ensure adequate gas exchanges within each flask. Most microplate liquid media protocols exceed this ratio, as they generally call for a 220 µL sample volume to be placed in a 300 µL capacity well (S/EV = 0.73). The influence of different S/EVs (80 µL/300 µL = 0.27; 120 µL/300 µL = 0.40; 150 µL/300 µL = 0.50; 220 µL/300 µL = 0.73) on algal control growth in microplate wells was studied by Blaise et al.[13] who found no significant differences in 8-d cell counts for all S/EV conditions with those of flask tests undertaken with S/EVs of 0.20 (25 mL/125 mL) and 0.32 (40 mL/125 mL). These authors therefore favored the use of 220 µL/300 µL volume ratios (S/EV = 0.73), because they allowed for a greater amount of algal inoculum to be placed in each well. *Selenastrum capricornutum* growth curves were later determined with flask (S/EV = 0.32) and microplate (S/EV = 0.73) procedures and demonstrated similar maximum standing crops after eight days of growth.[57] Based on these results, it appears that an $S/EV = 0.73$ does not limit CO_2 in microplate wells, as confirmed by control algal growth which is commensurate to that found in flask assays. Blaise et al.[13] have suggested that lack of negative influence of an $S/EV = 0.73$ on gas limitation is likely due to the fact that the microplate lid is not air-tight. Hence, the experimental volume (EV) is theoretically that contained within the sealed plastic bag covering the microplate and cover, in which case the true S/EV ratio is well below the 0.20 value recommended by Miller et al.[5]

D. Chemical Adherence to Microplate Wells

Some criticism has been levied against microplate toxicity assays because of the polystyrene make-up of microplates and because 96-well microplate wells markedly increase the area to volume (a/v) ratio. Indeed, the plastic composition of microplates might, on its own, theoretically cause more toxicant adhesion to occur on wells than would occur on glass flasks, and this potential problem could again be exacerbated by a more important a/v well ratio which tends to favor greater adsorption of test substances.[62] Conscious that chemical adsorption to wells might influence their concentration series, Blanck et al.[53] compared endpoint values (ECs) determined by successive dilutions on plates and from premade solutions for 19 compounds. No adsorption effect was found for any of the chemicals, as confirmed by the similarity of EC values obtained by both techniques. The recent development of microplate semistatic phytotoxicity procedures (daily renewal of nutrient and toxicant concentrations) has the advantage of minimizing the potential for chemical adsorption on wells.[27,28] Based on the adequate agreement reported for flask and microplate comparative studies undertaken thus far, however (Section V), there is no direct evidence suggesting that chemical adsorption is greater in microplate wells than it is in glass flasks. Clearly, adsorption affinity for glass or plastic will likely be chemical-specific and neither media will be the most convenient at all times.

E. Volatile Substances

Microplate wells covered with standard (loose-fitting) lids are potentially subject to cross-contamination effects owing to volatile toxicants. Aware of this problem, Blanck and Björnsäter[54] stated that their algal microtest battery procedure might prove unsuitable for testing such substances. Following their round robin exercise with *S. capricornutum*, Thellen et al.[60] reported the marked influence of phenol volatility on algal growth of control wells in a microplate. They nevertheless demonstrated that optimization of microplate experimental disposition, coupled with repeating an assay with lower starting test concentrations of a volatile substance, could reduce or eliminate volatile effects. A simple way to check for potential volatile contamination effects from adjacent test wells is to "sandwich" control wells between a gradient of test concentration wells (see Figure 18.2). If notable volatile effects occur, skewness in algal cell counts will likely be observed, as shown by Thellen et al.[60] Moreover, these same authors concluded that microplate assays can be advantageous by actually providing information on the presence of volatile toxicants. Recently, St-Laurent et al.[63] reported on the use of an adhesive polyester film which hermetically seals each microplate well from the other. Tests with phenol revealed no cross-contamination, as confirmed by algal growth homogeneity obtained in sealed "sandwiched" control wells (as in Figure 18.2). The influence of this adhesive polyester film on final microplate algal growth yield and toxicity response, however, remains to be more fully investigated.

F. Colored Test Samples

It has been argued that chronic algal toxicity tests, whether these be flask- or microplate-based procedures, cannot discriminate between cell growth inhibition owing to the intrinsic chemical toxicity of a sample and that possibly due to the filtering (absorbance) of light rays beneficial for photosynthesis. As pointed out by Kwan[18] in this review, pulp and paper effluents, among other industrial types, can have varying colors (chalky to dark brown) and may interfere in this fashion. Colored samples whose toxicity endpoint is eventually determined in a dilution range devoid of color are obviously not of concern. The problem lies rather with those samples whose toxic effects manifest themselves in a colored (low dilution) concentration zone. It could certainly be held that sample color characteristics are a genuine part of toxicity and that the need to differentiate algal growth inhibition on the basis of chemical toxicity and color interference is unnecessary. Other reasons (toxicity reduction evaluation of an effluent plant, court action, etc.) may nevertheless motivate such an enquiry. Acute exposure tests of several hours, where growth inhibition does not come into play, may offer an interesting means of separating toxic from color effects. As noted in this review, short-term procedures based on exposure-recovery,[23] metabolic activity inhibition,[31,34] and motility inhibition[30] may provide support in this respect. Another alternative may be in the use of a solid medium procedure, such as that reported by Lukavsky,[49] where biomass assessment is apparently not influenced by colored samples.

G. Gas Exchange in Microplate Wells

Flask phytotoxicity tests call for agitation of experimental vessels in order to favor adequate CO_2 exchanges among cells.[5] Rotatory movement of flasks also creates turbulence which enables algal cells to be homogeneously suspended in their confined environment.[62] This then ensures optimal contact between medium and toxicant. In contrast, microplates are seldom stirred during exposure (see Section VI). Thellen et al.[60] investigated the effects of passive (undisturbed micro-plate) and active (repetitive multichannel pipetting of well contents, by 3 to 4 quick volume uptakes and releases, on days 1, 2, and 3 of a 96-h exposure period) gas exchange while undertaking chronic exposure tests with Cd^{2+} and phenol. Their results indicated that significantly more control growth was obtained when active gas exchange was applied during incubation, suggesting that CO_2 limitation may be a factor in the latter stages of a 96-h incubation period under a passive regime as growth and pH increase. Toxicity results (96-h IC_{50}s), however, were unaffected by the gas exchange variable for both chemicals tested. These authors therefore deemed, for the sake of test simplicity, that active gas exchange should not be an absolute requirement in conducting microassays. Radetski et al.[28] also showed that control growth was significantly higher in their newly developed semistatic phytotoxicity microassay as opposed to the static microplate assay for 72-h and 96-h exposure tests. Here, daily renewal of medium and toxicant solutions is commensurate with shaking, in a certain sense, as it guarantees a fresh input of CO_2 for algal cells. A semistatic procedure, although slightly more laborious to perform, may equally contribute greater test sensitivity over a static procedure, as it minimizes potential factors (e.g., toxicant adsorption, complexation, biodegradation, photodegradation) capable of altering the initial toxicity of a test sample.[28]

H. pH

Photosynthetic activities of actively dividing cells cause an increase in pH of test solutions (owing to consumption of CO_2) during static algal bioassays of chronic exposures.[5] Such pH shifts can alter the toxicity of test substances. Vasseur et al.,[64] for example, demonstrated speciation effects on metallic cations with increasing pH resulting in reduced toxicity responses. As part of conditions essential for test validity, standardized procedures require that the initial pH of test solutions should be fixed between 7.0 and 8.5 and that final pH should not have varied by more than 1.5 units in controls.[25,65,66] The Environment Canada protocol[25] also suggests adjusting pH for samples outside the 6 to 9 range to 6.5 or 8.5, as the case may be, but specifies that toxicity testing may be performed concurrently on both adjusted and nonadjusted samples. Despite these precautions, pH shifts in test solutions in microplate wells might be concentration-dependent and could still alter toxicity. While this particular event is unpredictable and uncontrollable, it is nevertheless imperative that microplate techniques comply with recognized procedural standards. Blaise[57] reported pH compliance for controls in 96-h microplate tests performed with an initial inoculum of 2×10^4 cells/mL. Since more recent testing procedures[25,28,56,66] employ even less inoculum (10^4 cells/mL) and shorter exposure periods (72 h), final algal biomass is reduced and consequent changes in pH due to CO_2 depletion should be minimal. As pointed out by Hörnström,[62] an additional benefit of keeping initial cell inoculum low in algal toxicity procedures is that it will tend to increase test sensitivity. Daily renewal of test solutions, as shown by employing a semistatic procedure,[28] is another indirect means of ensuring pH stability in microplate wells. Finally, Arensberg et al.,[61] who developed a minitest performed in 20-mL scintillation glass vials according to ISO standards, have discussed the pH control problem in algal tests and present evidence which suggests that reducing test exposure from 72 h to 48 h would minimize pH shifts (<0.5 units) without altering phytotoxic response. This reduced exposure time could certainly prove beneficial to microplate techniques as well and merits investigation.

I. Miscellaneous Considerations

Are all microplate types and brands adequate for phytotoxicity testing? This question has received little attention thus far, as researchers appear to have made their choice either randomly, based on availability of a certain product, or because of previous use by other groups. With few exceptions, round or flat-bottomed 96-well polystyrene microplates have been the experimental substrates of choice. One study did report that three brands of such round-bottomed microplates (Nunc®, Cooke®, Linbro®) were capable of yielding similar biomasses of *S. capricornutum* after 4 and 8 d of growth in a complete algal growth medium.[57] A recent study recommended the use of plates with round-bottomed wells over flat-bottomed wells when using fluorometry as the biomass endpoint, as the former give lower blank readings.[56] Again, recent studies caution against the use of tissue culture-treated microplates, purported to be toxic to algae and hence unusable.[56,61] Clearly, there is room for further studies in this area.

Phytotoxic assessment of hydrophobic molecules may require the use of carriers for adequate dissolution during exposure. Blanck and Björnsäter[54] mention utilizing minimal concentrations of solvents and the inclusion of appropriate controls to ensure their innocuity toward test algae. St-Laurent et al.[11] employed solvent concentrations corresponding to a 0.5 NOEC value for methanol and acetone in their evaluation of nine herbicides. Preliminary checks were even made with two herbicides, capable of being appraised with or without an organic carrier at 0.5 NOEC concentration of acetone, to certify that herbicide/solvent interactions had no impact on toxicity results.

Reference toxicants are valuable for demonstrating methodological reliability of test results and should be used as part of a quality control strategy. Noted shifts in toxicity endpoint data might then indicate procedural and/or biological reagent problems. NaCl, $ZnSO_4$ and phenol,[25] as well as $K_2Cr_2O_7$,[52,56] have thus far been proposed as reference toxicants for algal microplate assays.

Finally, when running algal test batteries, it is worth mentioning that experimental conditions (e.g., light, temperature) will never be optimal for all algae composing a particular battery, as pointed out by Blanck and Björnsäter.[54] The impact that such experimental stresses might have on the toxicity responses of algae exposed to toxicants under suboptimal conditions are unknown at present.

IX. CONCLUSIONS AND RESEARCH PERSPECTIVES

The compendium of microplate-based phytotoxicity bioassays scrutinized in this review conclusively points out that they have been very valuable for screening, ranking, and prioritizing the potential hazards of chemicals and varied environmental matrices. There is no question that they will continue to provide such services in the years to come, along with other small-scale tests conducted at different trophic levels, spurred on by, among other factors, the pressing need to augment the cost-efficiency and diagnostic capacity of environmental assessment schemes confronted with an escalating demand for sample bioanalysis.

But, where to from here with algal microplate assays, on the basis of what we have seen so far? Clearly, tests conducted with unialgal cultures have enjoyed the greatest popularity so far and there is still much room to maneuver in this area in terms of future research. While comparative studies have confirmed concordance of toxicity results between flask and microplate procedures, it would be erroneous to think that microplate protocols cannot stand further optimization. Investigations to alleviate or eliminate some of the (nontoxic) factors linked to microplate assays, as discussed previously, are certainly justified. For example, appropriate endpoints and techniques for resolving the colored sample issue comprise one area of interest. While it has been shown that volatility problems can be curtailed by special experimental dispositions or adhesive films to seal off individual wells, these may not be ultimate solutions to the problem. New microplates have been recently commercialized, for instance, whose designs have been re-engineered to provide improved individual well-sealing capabilities that merit evaluation in this respect.

Assessment endpoint calibrations also warrant studies as microplate phytotoxicity data individually derived from cell counts, photometry, or fluorometry, for example, have seldom really been the object of rigorous comparisons. Besides generating valuable information on the relative merits of various endpoints, such initiatives would serve to develop much needed automation potential for algal microplate assays at the post-exposure side of things, since photometric, fluorometric, and luminometric plate readers abound. Thus far, only three microfluorometric[33,54,56] and three microphotometric[20,22,55] techniques have been proposed for toxicity endpoint evaluation with fixed inocula of specific algae. There is also a need to further explore new endpoints capable of producing measurable and relevant toxicity responses after acute exposures of only a few hours and to attempt to correlate these with the results of chronic tests. Some acute exposure tests, attractive from the point of view that they do not suffer from potential problems or interferences identified for chronic exposure tests and because they can be initiated and completed in a work day, may hold promise as alternatives to chronic testing with microalgae. Breakthroughs in this area may come from appraising fresh physiological, biochemical, enzymatic, and immunological types of parameters, coupled, possibly, with new advances in instrumental technology. Furthermore, such endeavors would be directly useful for the eventual development of one or more commercialized microplate phytotoxkits for which a definite market exists, if we are to judge by the present success of commercialized bioassay kits (e.g., Microtox®, SOS Chromotest, microinvertebrate cyst-based Toxkits). The recent development of a cryopreserved algal reagent[27] is also a positive step, as it may lead to maintenance-free preparation of algal inocula for microplate assays. This fledgling field requires more study, as does associated research aimed at developing lyophilized or immobilized algal reagents for microassay (and other) uses.

Interspersed with the above work is the necessity to harmonize at least some microplate assay procedures with those flask assays presently recognized by international standards organizations.[65-67] While microplate/flask comparisons have proven conclusive, only one microplate toxicity intercalibration study has been undertaken.[60] Furthermore, acceptance of an algal microassay by a standards organization will only come from successful comparison with its officially recognized testing procedure and after demonstrating that the microassay mimics testing protocol in every aspect. Who is ready to take on this challenge? Another useful venture could involve the development of a microplate liquid medium eutrophication assay procedure, since, quite strangely, none have yet been reported. This neglect may be due to the fact that attention was very much drawn to toxicity assessment at the time (late '70s–early '80s) when microplates were being considered for environmental applications with microalgae. Linked to this development to some extent is that of an appropriate microtest procedure and endpoint(s) to deal with samples that contain both toxic and auxinic substances. Can chronic tests properly address this issue if the action of rapid-acting toxicants are later masked by auxinic effects? For laboratories engaged in the evaluation of environmental samples of diverse origins, this is clearly a point of interest. Developing solid phase phytotoxicity tests to assess contaminated sediments is an additional area for research to follow, perhaps, in the footsteps of recent bacterial assays which have been successful in doing so (e.g., *Escherichia coli* Direct solid phase Toxi-Chromotest and *Vibrio fischeri* Microtox solid phase test). Lukavsky[52] has also suggested that solid medium (agar-based) tests might be useful for developing microassays to detect mutagenic effects based on colonial morphological and color changes.

Of major interest as well, because of their great span in assessing general algal sensitivity, is the use of algal test batteries.[53-56] It is logical to surmise that future efforts will be directed toward improving the predictive values of such batteries. Here as well, novel endpoints and procedural optimization may prove beneficial, as mentioned for unialgal tests. Much more could be said about the way in which future research will be directed with microplate algal techniques. Clearly, this field can only be limited by creativity and imagination. It is expected that much effervescence will continue in this general area, as more and more laboratories become involved with phytotoxicological preoccupations. As this new research unfolds, it will also be important to keep in mind the ways in which phytotoxicity data generated by microtests can best serve the interests of aquatic environment management programs.

ACKNOWLEDGMENTS

This review was undertaken as part of an official project under the Canada/France bilateral agreement for cooperation in science and technology. The first author is grateful to the management of the Centre Saint-Laurent, Environment Canada, for supporting this research initiative and to the Green Plan International Partnership program of Environment Canada for providing travel funds which enabled this work to be accomplished at the University of Metz with his French colleagues. We also wish to thank John Wiley & Sons, Inc., New York, for permission to reproduce Figure 1 (our Figure 18.2) initially published in the St-Laurent et al. (1992) article referenced herein.

REFERENCES

1. Thursby, G. B., A review of the current status of marine algal toxicity testing in the United States, in *Environmental Toxicology and Risk Assessment*, ASTM STP 1179, Landis, W. G., Hughes, J. S., and Lewis, M. A., Eds., American Society for Testing and Materials, Philadelphia, 1993, 362.
2. Lewis, M. A., Use of freshwater plants for phytotoxicity testing: a review, *Environmental Pollution*, 87, 319, 1995.
3. Keighan, E., Caractérisation du niveau d'enrichissement et de la toxicité des eaux du bassin du fleuve St-Laurent, Comité d'étude sur le Fleuve St-Laurent, Environnement Canada, Service de la Protection de l'Environnement, Région du Québec, Rapport technique No. 6, 153 pages, 1977.
4. Chiaudani, G. and Vighi, M., The use of *Selenastrum capricornutum* batch cultures in toxicity studies, *Verhandlungen Internationale Vereinigung Limnologie*, 21, 316, 1978.
5. Miller, W. E., Greene, J. C., and Shiroyama, T, The *Selenastrum capricornutum* Printz algal assay bottle test: Experimental design, application, and data interpretation protocol, US Environmental Protection Agency Report No. EPA-600/9-78-018, Corvallis, OR, 126 pages, 1978.
6. Joubert, G., A bioassay application for quantitative toxicity measurements using the green alga *Selenastrum capricornutum*, *Water Research*, 14, 1759, 1983.
7. Conrath, T.B., Handbook of microtiter procedures, Dynatech Corporation, Cambridge, MA, 475 pages, 1972.
8. Blaise, C., Legault, R., Bermingham, N., van Coillie, R., and Vasseur, P., A simple microplate algal assay technique for aquatic toxicity assessment, *Toxicity Assessment*, 1, 261, 1986.

9. Blaise, C., Microbiotests in aquatic ecotoxicology: characteristics, utility and prospects, *Toxicity Assessment*, 6, 145, 1991.

10. Persoone, G., Blaise, C., Snell, T., Janssen, C., and van Steertegem, M., Cyst-based toxicity tests: II. Report on an international intercalibration exercise with three cost-effective Toxkits, *Zeitschrift für Angewandte Zoologie*, 79, 17, 1992.

11. St-Laurent, D., Blaise, C., MacQuarrie, P., Scroggins, R., and Trottier, B. Comparative assessment of herbicide toxicity to *Selenastrun capricornutum* with microplate and flask bioassay procedures, *Environmental Toxicology and Water Quality*, 7, 35, 1992.

12. Heldal, M., Norland, S., Lien, T., and Knutsen, G., Acute toxicity of several oil dispersants toward the green algae *Chlamydomonas* and *Dunaliella*, *Chemosphere*, 7, 247, 1978.

13. Blaise, C., Legault, R., and Bermingham, N., A simple microassay technique for measuring algal growth inhibition ($EC_{50}s$) in aquatic toxicity studies, *Canadian Technical Report of Fisheries and Aquatic Sciences,* 1163, 1, 1982.

14. Blanck, H., Gustafsson, K., and Adolfsson-Erici, M., Effects of various sterilization methods on toxicity and chemical composition of industrial wastewater samples, *Water Research*, 17, 965, 1983.

15. Blaise, C., Legault, R., Bermingham, N., van Coillie, R. and Vasseur, P., Microtest mesurant l'inhibition de la croissance des algues (CI50) par le dosage de l'ATP, *Sciences et Techniques de l'Eau*, 17, 245, 1984.

16. Couture, P., van Coillie, R., Campbell, P. G. C., and Thellen, C. 1981. Le phytoplancton, un réactif biologique sensible pour détecter rapidement la présence de toxiques, *Institut National de la Santé et de la Recherche Médicale*, 106, 255, 1981.

17. Blaise, C., Micromethod for acute toxicity assessment using the green alga *Selenastrum capricornutum*, *Toxicity Assessment*, 1, 377, 1986.

18. Kwan, K.K., Testing of colored samples for toxicity by the algal-ATP bioassay microplate technique, *Environmental Pollution*, 60, 47, 1989.

19. Slabbert, L. and Hilner, Development of an algal toxicity test for water quality testing, Research Project No. 67027441, CSIR, Division of Water Technology, Pretoria, South Africa, 32 pages, 1990.

20. Slabbert, L., Guidelines for toxicity bioassaying of drinking and environmental waters in South Africa, Contract document for the Water Research Commission, CSIR, Division of Water Technology, Pretoria, South Africa, 38 pages, 1994.

21. Höhne, L., Development of a simple algal test as an express method under article 7a of the Water Economy Act and the Sewage Charges Act (Federal Protection Agency of the Federal Republic of Germany, Umweltbundesampt), Report No. UBA-FB 102 05 151, 45 pp., 1991.

22. Hansen, P.D. and Höhne, L., Standard protocol for testing effluents by the microplate algal assay, Wabolu Technical Report, Institut für Wasser, Boden and Lufthygiene des Bundesgesundheitsamtes, Berlin, Germany, 29 pages, 1995.

23. Hickey, C., Blaise, C., and Costan, G., Microtesting appraisal of ATP and cell recovery endpoints after acute exposure of *Selenastrum capricornutum* to selected chemicals, *Toxicity Assessment*, 6, 383, 1991.

24. Caux, P. Y., Blaise, C., Leblanc, P., and Taché, M., A phytoassay procedure using fluorescence induction. *Environmental Toxicology and Chemistry*, 11, 549, 1992.

25. Environment Canada, Biological test method: Growth inhibition test using the freshwater alga *Selenastrum capricornutum*, Report EPS 1/RM/25, Conservation and Protection, Environment Canada, Ottawa, Canada, pp. 42, 1992.

26. Blaise, C., Practical laboratory applications with micro-algae for hazard assessment of aquatic contaminants, in *Ecotoxicology Monitoring*, Richardson M., Ed., VCH Publishers, New York, 1993, 83.

27. Benhra, A., Validation de l'utilisation d'algues cryopréservées en écotoxicologie, Doctorate thesis in toxicology, University of Metz (France), 141 pages, 1994.

28. Radetski, C. M., Férard, J. F., and Blaise, C., A semi-static microplate-based phytotoxicity test, *Environmental Toxicology and Chemistry*, 14, 299, 1995.

29. St-Laurent, D. and Blaise, C. Validation of a microplate-based lethality test with the help of flow cytometry, in *Environmental Toxicology Assessment*, Richardson, M., Ed., Taylor & Francis, 1995, chap. 11.

30. Kusui, T. and Blaise, C., Acute exposure phytotoxicity assay based on motility inhibition of *Chlamydomonas variabilis*, in *Environmental Toxicology Assessment*, Richardson, M., Ed., Taylor & Francis, London, 1995, chap. 10.

31. Snell, T., Mitchell, J. L., and Burbank, S. E., Rapid toxicity assessment with microalgae using *in vivo* esterase inhibition, in *Techniques in Aquatic Toxicology*, Ostrander, G. K., Ed., CRC Press/Lewis Publishers, 1996, chap. 2.

32. Galgani, F., Gilbert, F., and Cadiou, Y., Test écotoxicologique sur microalgues et microplaques: mesure d'activités enzymatiques sur cellules *in vivo, Océanis*, 17, 223, 1991.

33. Galgani, F., Cadiou, Y., and Gilbert, F., Simultaneous and iterative weighted regression analysis of toxicity tests using a microplate reader, *Ecotoxicology and Environmental Safety*, 23, 237, 1992.

34. Gilbert, F., Galgani, F., and Cadiou, Y., Rapid assessment of metabolic activity in marine microalgae: application in ecotoxicological tests and evaluation of water quality, *Marine Biology*, 112, 199, 1992.

35. Hassett, J. M., Jennett, C., and Smith J.E., Microplate technique for determining accumulation of metals by algae, *Applied Environmental Microbiology*, 41, 1097, 1981.

36. Bisson, S., Blaise, C. and Bermingham, N., Assessment of the inorganic bioaccumulation potential of aqueous samples with two algal bioassays, in *Aquatic Toxicology and Water Quality Management*, Nriagu, J. and Lakshminarayana, L., Eds., Wiley Series Volume 22, 1989, 205.

37. Rodrigue, D., Mise au point d'un protocole d'interaction miniaturisé mettant en évidence l'activation génotoxique de produits chimiques par l'algue *Selenastrum capricornutum*, M.Sc. thesis, U. du Québec (INRS-EAU), Québec City, Canada, 129 pages, 1991.

38. Harwood, M., Blaise, C., and Couture, P., Algal interactions with the genotoxic activity of selected chemicals and complex liquid samples, *Aquatic Toxicology*, 14, 263, 1989.

39. Baarschers, W. H., Donnelly, J. G., and Heitland, H. S., Microbial toxicity of trichlopyr and related herbicides, *Toxicity Assessment*, 3, 127, 1988.

40. Williamson, D.A., Hexazinone residues in surface and groundwater at two sites within Agassiz provincial forest, Manitoba, Canada, *Water Pollution Research Journal of Canada*, 23, 434, 1988.

41. Couture, P., Blaise, C., Cluis, D., and Bastien, C., Zirconium toxicity assessment using bacteria, algae and fish assays, *Water, Air and Soil Pollution*, 47, 87, 1989.

42. Warner, J., Environmental and toxicological effects of chromated copper arsenate, pentachlorophenol and preserved wood, M.Sc. thesis, University of Guelph, Ontario, Canada, 87 pages, 1990.

43. Blaise, C. and Harwood, M., Contribution à l'évaluation écotoxicologique du Tébuthiuron — un herbicide de la classe des urées substituées, *Revue des Sciences de l'Eau*, 4, 121, 1991.

44. Costan, G., Bermingham, N., Blaise, C., and Férard, J. F., Potential ecotoxic effects probe (PEEP): a novel index to assess and compare the toxic potential of industrial effluents, *Environmental Toxicology and Water Quality*, 8, 115, 1993.

45. Lambolez, L., Vasseur, P., Férard, J. F., and Gisbert, T., The environmental risks of industrial waste disposal: an experimental approach including acute and chronic toxicity studies, *Ecotoxicology and Environmental Safety*, 28, 317, 1994.

46. Day, K. and Hodge, V., The toxicity of the herbicide metolachlor, some transformation products and a commercial safener to an alga (*Selenastrum capricornutum*), a cyanophyte (*Anabaena cylindrica*) and a macrophyte (*Lemna gibba*), *Water Quality Research Journal of Canada*, 31, 197, 1996.

47. Caux, P. Y., Ménard, L., and Kent, R. A., A comparative study of the effects of MCPA, butylate, atrazine and cyanazine on *Selenastrum capricornutum*, *Environmental Pollution*, 92, 219, 1996.

48. Okamura, H., Rong, L., Aoyama, I., and Liu, D., Ecotoxicity assessment of the aquatic environment around Lake Kojima–Japan, *Environmental Toxicology and Water Quality*, 1996 (in press).

49. Lukavsky, J., The evaluation of algal growth potential by cultivation on solid media, *Water Research*, 17, 549, 1983.

50. Lukavsky, J., A simple cultivation unit for the evaluation of algal growth potential and toxicity of water, *Water Research*, 19, 269, 1985.

51. Lukavsky, J., The evaluation of algal growth potential (AGP) and toxicity of water by miniaturized growth bioassay, *Water Research*, 26, 1409, 1992.

52. Lukavsky, J., Miniaturized algal growth bioassay, in *Monitoring of Ecological Change in Wetlands of Middle Europe*, Aubrecht, G., Dick, G. and Prentice, C., Eds., Proceedings of an International Workshop, Linz, Austria, 1993. Stapfia 31, Linz, Austria, and IWRB Publication No. 30, Slimbridge, UK, 1994, 151.

53. Blanck, H., Wallin, G., and Wängberg, S. A., Species dependent variation in algal sensitivity to chemical compounds, *Ecotoxicology and Environmental Safety*, 8, 339, 1984.

54. Blanck, H. and Björnsäter, B., The algal microtest battery: a manual for routine tests of growth inhibition, Swedish National Chemicals Inspectorate, Solna, Sweden, Kemi Report No. 3/89, pp. 27, 1989.

55. Wängberg, S., Bergström, B., Blanck, H., and Svanberg, O., The relative sensitivity patterns of short-term toxicity tests applied to industrial wastewaters, *Environmental Toxicology and Water Quality*, 10, 81, 1995.

56. SRC (Saskatchewan Research Council), Development of aquatic plant bioassays for rapid screening and interpretive risk assessments of metal mining wastewaters, SRC Publication No. E-2100-2-C-95, Saskatchewan, Canada, 1995, 191.

57. Blaise, C., Développement de bioessais sublétaux pour les évaluations écotoxicologiques des effluents, Doctorate thesis in toxicology, University of Metz (France), 172 pages, 1984.

58. Cossu, C., Toxicité de micropollutants minéraux et organiques sur l'algue *Selenastrum capricornutum*, Advanced studies diploma report, Environmental Sciences Centre, U. of Metz, France, 31 pp., 1991.

59. Marsalek, B., The comparison of microplate, bottle and fluorescence algal assay procedures, Internal Report of the Institute of Botany, Brno, Czech Republic, 1994.

60. Thellen, C., Blaise, C., Roy, Y., and Hickey, C., Round robin testing with the *Selenastrum capricornutum* microplate toxicity assay, *Hydrobiologia*, 188/189, 259, 1989.

61. Arensberg, P., Hemmingsen, V. H., and Nyholm, N., A miniscale algal toxicity test, *Chemosphere*, 30, 2103, 1995.

62. Hörnström, E., Toxicity test with algae — A discussion on the batch method, *Ecotoxicology and Environmental Safety*, 20, 343, 1990.

63. St-Laurent, D., Costan, G., and Blaise, C., Resolution of variability problems caused by volatile substances in undertaking microplate-based algal toxicity assays, Internal report, Centre Saint-Laurent, Environment Canada, Montréal, Canada, 1994.

64. Vasseur, P., Pandard, P., and Burnel, D., Influence of some experimental factors on metal toxicity to *Selenastrum capricornutum*, *Toxicity Assessment*, 3, 331, 1988.

65. OECD (Organization for Economic Cooperation and Development), Alga growth inhibition test, Test Guideline No. 201, OECD Guidelines for Testing of Chemicals, Paris, 1984.

66. ISO (International Standardization Organisation), Water quality — algal growth inhibition test, No. 8692. ISO, Paris, 1987.

67. EEC (European Economic Community), Algal inhibition test, European Economic Community Directive, Official Journal of the European Communities, No. L133, pages 89-94, 1988.

CHAPTER 19

Ecophysiological Considerations in Microalgal Toxicity Tests

C. Nalewajko and M.M. Olaveson

CONTENTS

I. INTRODUCTION

A. Rationale

Freshwater microalgae are used more frequently in laboratory toxicity tests than any other type of aquatic plant.[1] During the 1960s, increasing concern about the eutrophication of freshwater lakes in North America and Europe provided the impetus for the development of bioassays for the estimation

0-8493-2626/5/98/$0.00+$.50

of algal growth potential.[2] The species selected for these tests were chosen for their reproducible growth under laboratory conditions and their reliable responses to nutrient enrichment. The most commonly used species is the green alga, *Selenastrum capricornutum* Printz (= *Raphidocelis subcapitata* Korsikov), but other commonly used species include other green algae (e.g., *Chlorella vulgaris, Scenedesmus quadricauda, Scenedesmus subspicatus*), cyanobacteria (e.g., *Anabaena flos-aquae, Microcystis aeruginosa*), diatoms (e.g., *Cyclotella* spp., *Navicula pelliculosa, Nitzschia* spp., *Synedra* spp., and chrysophytes (e.g., *Synura petersonii*).[1,3-6] Eventually, the Algal Assay Bottle Test using *Selenastrum* was adapted for widespread use in toxicity tests and became the standard procedure for assessing the potential toxicity of a variety of substances to the microalgal community.[3,5] As a result, most algal bioassays in recent years have relied almost exclusively on *Selenastrum*, leading to its promotion as the algal representative in multispecies testing protocols designed by government regulatory agencies.[3,7,8] But increasing concern has been expressed by a number of researchers and environmental regulators about the trend toward reliance on this one species, *Selenastrum capricornutum*, to represent the responses of microalgae in general.[1,5,9-11] This concern is justified by increasing evidence of significant differences in sensitivities to a variety of contaminants[1,5,12] among algal genera and even within individual microalgal species. This has led investigators to question the assumption that all microalgal species respond in the same way to all contaminants and to recognize that the use of one algal species under highly artificial conditions could lead to misleading conclusions about the impact of contaminants in natural ecosystems. Such concerns are especially important when considering the use of *Selenastrum*, which is a "laboratory weed" but which rarely attains any importance in the natural aquatic ecosystems that are most likely to be impacted by environmental contaminants.

B. Objectives

Potential detrimental impacts of environmental contaminants on the photosynthetic microalgal community, the primary producers at the base of the aquatic food web, must be reliably assessed in order to avoid disastrous consequences at higher levels in the food chain. Therefore, a critical assessment of the use of microalgal bioassays, especially those relying on one or a few species, is in order. Much of the current literature is unclear whether the reported differences in sensitivity among various algal taxa, and indeed within individual species, are real or artifacts caused by differences in media composition and other environmental (e.g., abiotic) factors. In addition, the extent to which the physiological state of the test organism during the bioassays influences the results is rarely considered. Various bioassay endpoints, including yield, growth rate, photosynthetic rate, etc., are used to evaluate bioassay results; the method used can influence the overall outcome of the tests if the physiological state of the test organism and the test conditions are not considered.

While it is highly desirable to design simple laboratory-based biological tests which can be widely used in environmental impact assessments, it is nevertheless imperative that we understand the meaning of the results and are aware of the limitations in applying these results to the real world. It is our contention that an ecophysiological approach is required in designing and conducting laboratory microalgal tests of potential environmental contaminants. Such an approach to microalgal toxicity testing emphasizes the relationship between the physical/chemical characteristics of the environment and the physiological dynamics of the photosynthetic algal cells. The objective of this chapter is to demonstrate the importance of choosing the appropriate test organism(s) of known physiological state and the appropriate physical and chemical test conditions in order to obtain meaningful toxicity assessments.

II. CHOICE OF TEST SPECIES IN MICROALGAL TOXICITY TESTS

The most important step in conducting microalgal toxicity tests is to select an appropriate test organism. Although this sounds simple, this step is often taken with little consideration of the overall goals of the toxicity tests. Frequently, *Selenastrum* is chosen simply because of its wide-spread use in routine toxicity testing.[5,13] As a result, there is relatively little information available on other microalgal species, many of which are equally, if not more, appropriate for many toxicity tests.[1] Recently, attempts have been made to select a group, or battery, of microalgal species which represent a wide range of sensitivities to a broad spectrum of toxic substances. These multispecies tests,[5,10] while promising, do not always represent a realistic or natural combination of species.

Ideally, then, the selection of a test species should involve consideration of: 1) the taxonomy of the test organism; 2) its ecological preferences; 3) its physiological requirements; and 4) its availability from culture collections or other reliable sources. The choice of a microalgal species which can be expected to occur in the natural ecosystems most likely to be affected by the toxic substance will provide more ecologically relevant assessments of potential toxicity. An appropriate matching of the physiological traits of the test organism with the physical/chemical characteristics of the test substances likely to target these traits will give a more sensitive measure of potential toxicity. Finally the availability of the test organism from culture suppliers is crucial to reliable toxicity testing.

A. Taxonomy of Freshwater Microalgae

Freshwater algae are classified into 6 to 11 Divisions (depending on authority) on the basis of biochemical characteristics including types of pigments (chlorophylls, carotenoids, phycobilins), plastid organization, predominant reserve storage products, cell wall composition, and morphological characteristics (e.g., presence and type of flagella).[14,15] The majority of freshwater microalgae belong to one of the following 8 Classes: Cyanophyceae (= Cyanobacteria), Chrysophyceae, Bacillariophyceae, Dinophyceae, Cryptophyceae, Euglenophyceae, and Chlorophyceae. The presence and type of cell wall, as well as the characteristics of cell membranes and membranes around cell organelles, are likely to influence pollutant effects on algae.[16] For instance, the cell wall, which controls xenobiotic adsorption and entry into the cell, is siliceous in Bacillariophyceae (diatoms), calcareous in some Chlorophyceae, cellulose-based in other Chlorophyceae and in Dinophyceae (dinoflagellates), proteinaceous in Euglenophyceae, and entirely absent in some taxa (e.g., some species of Chrysophyceae and Chlorophyceae). Chloroplasts as well as other organelles are absent in the prokaryotic Cyanobacteria (blue-green algae); the plastids in the other eukaryotic classes differ in pigment composition, as well as in the number and type of membranes that envelop them. This is likely to influence the penetration of pollutants into plastids, and their effects on photosynthesis and energy production.[17] Differences in surface charge among species or strains can influence metal adsorption and uptake.[18] The presence of extensive mucilaginous fibrils around the cells of some taxa[19] as well as the variable chemical composition of such materials is likely to influence the cell's capacity for passive adsorption and the exclusion of metals by increasing the apoplastic space. This is particularly important in species in which exclusion is the only metal resistance strategy used.[16] The presence of flagella, spines, and other cellular extensions can be important in increasing the surface area exposed to contaminants.

B. Ecological Considerations from Natural Ecosystems

At any given time, a diverse assemblage of algae and cyanobacteria is present in freshwater, and the species composition exhibits seasonal changes that conform to a recognizable pattern year after year in lakes that are sufficiently deep to exhibit temperature stratification.[20,21] Although typically more than two hundred algal species are present on an annual basis in a lake, only about 20 to 40 may be present at any one time with the 6 to 8 most abundant species accounting for 90% of the total phytoplankton biomass.[22,23] According to Lewis,[5] about 500 genera of algae occur in freshwaters of the U.S. Clearly, real-world relevance of laboratory tests in detection of the impact of toxic substances on growth would be enhanced if species chosen for toxicity tests were those that dominate the receiving waters.

Reynolds[20,21] pointed out that the species which become dominant in a given lake are few in number, and the same species usually occur in a given lake year after year. Using a semiquantitative approach, 12 different phytoplankton assemblages were defined and distinguished by dominant and codominant species or groups, and presented a generalized seasonal successional pathway of these assemblages in oligotrophic, mesotrophic, eutrophic, and hypereutrophic lakes (Figure 19.1). If indeed species-specific responses differ sufficiently to require a variety of species to be used in bioassays, this model could provide a basis for test species selection once the trophic state of the receiving waters has been identified.

Twenty taxa that figure in Reynold's assemblages as dominants include the diatoms: *Asterionella, Fragilaria, Melosira granulata, M. italica, Stephanodiscus,* and *Tabellaria;* the green algae: *Closterium; Coelastrum, Eudorina, Pediastrum, Sphaerocystis, Staurastrum,* and *Volvox;* the cyanobacteria: *Anabaena, Aphanizomenon, Microcystis,* and *Oscillatoria;* the chrysophytes: *Dinobryon* and *Uroglena;* and the dinoflagellate, *Ceratium.* Eighteen additional algal taxa appear as co-dominants and may also warrant attention.

C. Physiology of Microalgae

The fact that algal Divisions and Classes are delineated primarily on a biochemical basis (e.g., cell wall, pigments, storage products)[14,15] leads one to expect less physiological and metabolic variability within, than among, Divisions. Yet there is remarkable diversity of metabolic and physiological types among the Classes of any given Division, as well as among genera of a given family. For example, genera in the Chrysophyceae differ in their ability to utilize sugars, organic acids, amino acids, and alcohols for heterotrophic growth (in the dark), and some members of this Class are obligate phototrophs, which are incapable of utilizing these organic compounds for growth in the absence of light. In fact, such differences in trophic ability have been reported among species within the same genus (e.g., in *Chlorella* species),[24] and differences in pH and salt tolerances have been used to distinguish species of this genus as well.[25] Indeed, it has been well documented that physiological races and ecotypes within a single species respond differently to environmental stresses depending on the physiological and genetic history of the organisms.[26]

In the present context, such differences could significantly affect the outcome of toxicity tests. In situations where the toxic substance specifically altered photosynthetic responses, algal strains capable of facultative heterotrophy would not be affected as severely as obligate phototrophs. Cyanobacterial species which have limited heterotrophic ability, attributed to an incomplete Krebs cycle, would be much more sensitive to substances which target photosynthesis. Dinoflagellates also have rather limited heterotrophic abilities,[24] but their demonstrated phagotrophic capability (allowing the ingestion of organic particles)[15] can complicate bioassay results with these organisms if the culture is not axenic (bacteria-free).

D. Availability of Cultures of Freshwater Microalgae

Service culture collections of algae exist in almost all developed countries as well as in some developing countries; a listing can be found in Takishima et al.[27] Examples in North America include the American Type Culture Collection (ATCC), the University of Texas Culture Collection of Algae (UTEX), and the University of Toronto Culture Collection (UTCC).[27] The number of freshwater species known to science is approximately 11,000 of which about 40% have been successfully cultured.[15] It is likely that most of these are "laboratory weed" species that are relatively undemanding and easily cultured.

However, culture collection acquisitions continue to increase as demand increases and as information becomes available on unusual nutritional requirements of species that were once notoriously difficult to culture. For example, the discovery of a selenium requirement as an essential micronutrient for *Peridinium cinctum*[28] and for *Chrysochromulina breviturrita*[29] has enabled

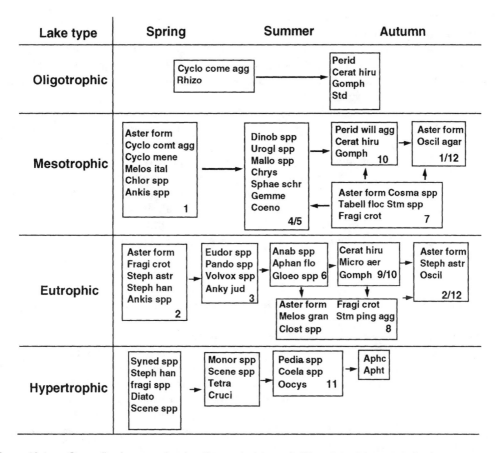

Figure 19.1 Generalized successional pathways in lakes of different trophic status: the boxes represent assemblages of species (as listed below), one or more of which may be abundant; bold numbers refer to assemblage-labels of Reynolds.[19] (From Reynolds, C.S., *Holarctic Ecology*, 3, 141, 1980. With permission.) Abbreviations of species: Ankist = *Ankistrodesmus* spp.; Anky = *Ankyra judayi*; Anab = *Anabaena* spp.; Aphan flo = *Aphanizomenon flos-aquae*; Aphc = *Aphanocapsa*; Apht = *Aphanothece*; Aster form = *Asterionella formosa*; Cerat hir = *Ceratium hirundinella*; Chlor = *Chlorella* spp.; Chrys = *Chrysochromulina*; Clost spp includes *Closterium aciculare, C. acutum, C. tortum*; Coela = *Coelastrum* spp.; Coeno = *Coenococcus*; Cosma = *Cosmarium* spp. (*C. abbreviatum, C. contractum, C. depressum*); Cruci = *Crucigenia* spp.; Cyclo come agg = *Cyclotella comensis* aggregate; Cyclo comt agg = *Cyclotella comta* aggregate; Cyclo mene = *Cylcotella meneghiniana*; Diato = *Diatoma*; Dinob = *Dinobryon* (chiefly *D. divergens*; Eudor spp = *Eudorina* spp.; Fragi spp = *Fragilaria* spp.; *Fragilaria crotonensis* (Fragi crot) is not distinguished; Gemme = *Gemellicystis* cf. *neglecta*; Gloeo = *Gloeotrichia*; Gomph = *Gomphosphaeria* spp. including forms ascribable to *Coelosphaerium*; Mallo = *Mallomonas* (e.g., *M. caudata*); Melos gran = *Melosira granulata*; Melos ital = *Melosira italica*; Micro aer = *Microcystis aeruginosa*; Monor spp = *Monoraphidium* spp.; Oocys spp = *Oocystis* spp. (e.g., *O. borgei*); Oscil = *Oscillatoria* spp., usually ascribable to *O. agardhii* (Oscil agar); Pando spp = *Pandorina* spp. (e.g., *P. morum*); Pedia spp = *Pediastrum* spp. (*P. boryanum, P. duplex, P.tetras*); Perid = *Peridinium* spp.; the *P. willei* aggregate (Perid will agg) is distinguished; Rhizo = *Rhizosolenia*; Scene = *Scenedesmus* spp.; Sphae schr = *Sphaerocystis schroeteri*; Std = *Staurodesmus*; Steph astr = *Stephanodiscus astraea* (= *S. rotula*); Steph han = *Stephanodiscus hantzschii*; Stm spp = *Staurastrum* spp.; the *S. pingue* aggregate (Stm ping agg) is distinguished; Syned spp = *Synedra* spp.; Tabell floc = *Tabellaria flocculosa*; Tetra = *Tetrastrum*; Urogl spp = *Uroglena* spp.; Volvox = *Volvox* spp.

researchers to isolate and successfully culture these organisms, once believed too difficult to grow under laboratory conditions. Vitamin additions are now routinely made in diatom cultures following the documentation that auxotrophy (need for exogenous vitamins) is widespread in this group. In spite of this, prolonged (>8 to 12 months) culturing of pennate diatoms is still uncertain, as the morphology becomes abnormal over time, and eventually the cultures cease to grow. In practice, the use of common diatoms such as *Asterionella formosa, Cyclotella comta, C. meneghiniana,* and *Stephanodiscus astraea* for routine bioassays would not be possible, as no culture collection, to our knowledge, can guarantee continued availability of these species in axenic culture. However, *Asterionella ralfsii* var. *americana,* an indicator species for acid-impacted waters, has been successfully isolated into axenic culture[30] using modified Fraquil medium and is still normal in appearance at the UTCC after five years.

It is desirable to use axenic (bacteria free) algal cultures in bioassays, for a number of reasons, but nonaxenic cultures in mineral media can be used without problems if only few bacteria are present. However, large bacterial populations can confound test results in media with organic compounds. Bacterial contaminants of algal cultures are generally heterotrophs which require organic material for growth, hence population density measurements based on turbidity (e.g., absorbance or optical density [O.D.]) would overestimate the algal population because bacteria would be included as well. Competition for nutrients between algae and bacteria in heavily contaminated cultures could also confound test results if it causes nutrient-limitation of algal growth. Also, the greater ability of bacteria than algae to degrade organic pollutants may falsify test results with these toxicants.

III. CHOICE OF TEST CONDITIONS IN MICROALGAL TOXICITY TESTS

With an appropriate test organism, the next step is to design the microalgal tests in a way that ensures the best physiological conditions so that growth or the target physiological processes are optimized. Such a design guarantees that only the variable of interest, the toxic substance being tested, is responsible for the observed results. The characteristics of the test substance must also be considered since toxicity results may be difficult to interpret if unforeseen abiotic interactions occur between the toxicant and the test medium. The design of the toxicity test becomes crucial to the usefulness of the final results and depends on 1) the optimization of culture conditions, and 2) an understanding of toxicant characteristics.

A. Optimization of Culture Conditions

A variety of factors must be considered for optimizing the culture conditions of the test organisms. The most important factors which are selected on the basis of each organism's requirements for optimal growth include: culture type, growth medium, nutrient concentrations, light conditions, temperature, and pH.

The most commonly used culture type is batch culture which is used to maintain stock cultures of test organisms and to grow the bioassay test cultures.[2,11] Batch cultures are cultures growing in containers with a limited volume of medium, in which pH, CO_2 supply, essential nutrient availability and irradiance change more or less dramatically during growth. Problems associated with growth in batch cultures have been discussed in the context of basic algal physiology[31] and in the context of bioassays.[11] Other types of cultures (e.g., semicontinuous, continuous) reduce some of these problems but are expensive and difficult to set up and maintain, and are less commonly used for routine toxicity testing.[2]

The number of different growth media used in algal cultures is quite small given the variety of environmental conditions (e.g., conductivity, pH, major ion, and essential nutrient concentrations) of natural freshwaters. Culture collections generally employ only a few types of defined culture media[31] and use only one kind for each Class of algae. The simplest freshwater media include Beijerinck, Bold basal, Bozniak community, Chu 10, Rodhe VIII, Volvox, and Waris.[33] Media developed more recently (e.g., Fraquil[34]) generally incorporate a buffer such as Tris or PIPES, and an organic chelating agent such as EDTA.

The growth medium formulation is crucial for the proper assessment of toxicity. A medium formulation (designated ESW) was proposed to replace the U.S. EPA formula[3] because it permitted better growth for two out of three common microalgal test species, *Cryptomonas* and *Selenastrum*; there was no difference with *Chlamydomonas*.[35] Some differences in toxicity among the species for some contaminants were detected in ESW which were not apparent in tests with the standard medium.

The nutrient composition and the relative concentrations of the major (e.g., calcium, magnesium, potassium, sodium, bicarbonate, chloride, nitrate, phosphate, sulfate) and minor nutrients (e.g., trace metals) in the growth medium are crucial to achieving balanced algal growth. Since the requirements for major and trace elements differ among algae, it is important to appreciate the impact of the selected medium on the growth of the test organism. The Resource Ratio Gradient model of nutrient competition among algae is based on differences in affinity for nutrients, such as those of *Cyclotella meneghinina* and *Asterionella formosa* for silica and phosphorus.[36] Phosphate concentrations of the order used in many culture media are much higher than expected from natural waters and in some cases may even be inhibitory to some algae.[37] The sodium levels in some media (e.g., BG-11) may also be much higher than is appropriate for most freshwater algae.

Although major nutrient concentrations in culture media, in particular N, P, and Si, are several orders of magnitude higher than in natural waters, the duration of exponential growth in batch cultures is brief, usually 3 to 8 days, in most standard media, unless the population density at the start of growth is very low, <100 cells/mL. With inocula sufficient to give initial densities of >1000 cells/mL, which are in the range reported by most investigators[37] and recommended by standard protocols,[3,7] essential nutrients may become depleted as population density increases and self-shading leads to light limitation.

Such inadvertent depletion of nutrients and induction of light limitation is likely to modify the response of the test organism by altering the cellular composition of the microalgae. Therefore, the physiological state of bioassay cultures must be standardized in order to avoid the confounding effects of nutrient status, (e.g., differences in cell P, N, or C metabolism) on algal responses to toxic substances. The P state of the culture is an important factor controlling the extent of metal toxicity (e.g., aluminum, copper, vanadium)[39-41] as well as the toxicity of contaminated sediments to algal photosynthesis.[42] Phosphorus-loaded Cu-resistant cells of *Anabaena variabilis* could grow in higher Cu concentrations than phosphate-starved cells.[43]

But, since phytoplankton growth in many natural waters is, in fact, periodically nutrient-limited (usually by phosphorus in freshwaters), it is important to assess the effects of toxicants under simulations of such natural conditions. There are a number of examples where deliberate limitation by an essential nutrient is used as a condition during toxicity tests.[40,41] With an awareness of the double stress imposed on the test species, it is possible to obtain a realistic assessment of toxicity. The major difference between these studies and many routine tests in the literature is that the imposition of more than one stress is intentional and the appropriate conditions are used.

Phosphorus limitation has been investigated as a simultaneous stress during metal toxicity tests. Three strains of *Scenedesmus acutus* which were differentially sensitive to copper were tested in short-term photosynthesis bioassays.[40] Copper tolerance in all three strains was enhanced when the cells were P-sufficient or P-loaded (compared to the P-deficient state). The presence of polyphosphate reserves in the cells of these non-P-limited cultures appeared to be relatively unimportant as a Cu-tolerance mechanism because the two Cu-tolerant strains showed less dependence on cellular polyphosphate than did the Cu-sensitive strain. Similar results were reported with P-loaded and P-deficient cultures of *Anabaena variabilis* after exposure to copper.[43] The toxicity of substances such as arsenate and vanadate, which are chemically similar to phosphate and which compete for transport pathways, can be quite variable if the phosphate state of the test organisms is not considered.[41]

Nitrogen limitation has been reported from a variety of freshwater habitats, especially those associated with waste sites. The interaction between this nutrient stress and other toxicants (e.g.,

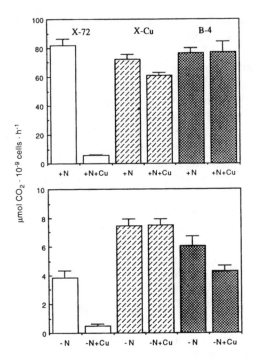

Figure 19.2 Effects of N limitation on copper toxicity to photosynthesis in three strains of *Scenedesmus.* Photosynthesis with 50 mM Cu as a percentage of -Cu control: X-72 N-sufficient = 6.9%; X-72 N-deficient = 13.2%; X-Cu N-sufficient = 83.9%; X-Cu N-deficient = 100%; B-4 N-sufficient = 100%; B-4 N-deficient = 70.8%. Vertical lines show standard error. (From Nalewajko, C. and Olaveson, M.M., *Can. J. Bot.,* 73: 1295, 1995. With permission.)

metal stress) indicates that the responses of the test organisms are not always predictable and are governed by the physiological characteristics of the algae. Figure 19.2 illustrates the results of the dual stresses of nitrogen limitation and elevated copper on three strains of *Scenedesmus acutus* known to have different mechanisms of dealing with copper toxicity.[44] In species that rely on metal exclusion as the tolerance mechanism, N-starvation may enhance extracellular fibril production and hence increase metal binding.[19]

Shuter[45] proposed a model to explain changes in cellular carbon distribution as a function of nutrients and irradiance. Nutrient deficiency enhances reserve carbon accumulation. Conversely at low irradiances, less carbon is directed into reserve products but relatively more carbon goes into the photosynthetic apparatus. Clearly, both nutrient concentrations and light levels can significantly alter the physiological responses of microalgae, and possibly affect the outcome of toxicity tests, where sufficient attention was not given to these factors.

Algae differ in their irradiance requirements for optimal photosynthesis and growth. According to Ryther,[46] these differences exist at the Division (or Class) level. Dinoflagellates (Dinophyceae) were characterized by highest irradiance requirements for optimal photosynthesis, Chlorophyta (Chlorophyceae) by the lowest, and diatoms (Bacillariophyceae) were intermediate in their requirements. Optimal irradiance levels are not fixed for a given species but change as a function of irradiance during growth.[47] Generally, pigment concentration in algae tends to increase as growth irradiance is decreased, and the P-I (photosynthesis-irradiance) curve parameters (e.g., P_{max}, I_k) change also. This acclimation to irradiance during growth has been documented in a number of species, but the molecular basis differs. According to Prezelin,[47] in some algae acclimation to low irradiance involves changes in the size of the photosynthetic unit (PSU) but in others, PSU density increases instead.

Low temperature in some species mimics low irradiance acclimation.[48] Temperatures for optimal growth also vary among algae, as does the range over which growth can occur.[30] It is logical to assume that algae are adapted to the thermal regime prevalent in their natural habitat. Algae from north temperate waters exhibit optimal growth rates in the 18 to 26°C range, and grow over a sufficiently wide range of temperatures to be considered as eurythermal. Among eight species of diatoms isolated from a variety of thermal regimes, species from low-temperature environments had a narrower range of temperatures for growth, and higher growth rates at low temperatures, than species from more temperate regimes.[49]

It is well established that many algal species have preferences for the pH conditions of their environment (and consequently their growth medium).[50-52] If the optimal pH range is exceeded, the algal species will respond directly to the pH conditions as well as any added toxicants. But pH can have indirect effects as well. In unbuffered media, the pH rises leading to CO_2 limitation particularly in unstirred cultures.[7,11,31] Likewise, in very acidic culture media (pH < 4), CO_2 limitation can occur due to depletion of the bicarbonate-carbonate buffer system.[51] In effect, then, these CO_2- or light-limited cultures are already stressed and will respond differently from optimally growing, exponential phase cultures when a second stress, the toxic substance, is imposed. Olaveson[53] has shown that in CO_2-limited batch cultures of *Euglena mutabilis*, the addition of copper (at a nominal concentration of 10 mg/L) causes significant depression in the growth rate to 48% of the control without added copper. However, if the stress caused by CO_2 limitation is alleviated by aeration, the addition of the same concentration of copper leads to only a slight depression in growth rate, to approximately 90% of the control. The significance of the CO_2 limitation to the outcome of this toxicity test is demonstrated by the increase in growth rates in response to aeration in the control cultures (0.57 div/day vs. 0.17 div/day in the nonaerated control). Nyholm and Kallqvist[11] reviewed the methods used in algal toxicity tests and concluded that excessive biomass and poor gas exchange conditions were used in many tests. These resulted in high pH and CO_2 limitation during the tests. In addition, algae isolated from soft-water lakes may be inhibited not only by higher pHs but also may be sensitive to the higher bicarbonate levels present above pH 6.[29]

B. Consideration of Toxicant Characteristics

The importance of the interaction between various abiotic factors and the test substances is frequently overlooked in toxicity tests[17] in spite of ample evidence that physical–chemical interactions can have a significant impact on the outcome of toxicity tests. The abiotic factors most likely to affect toxicant behavior depend on the toxicant; for heavy metals, these include medium composition (anion, cation, major nutrient, trace elements, and chelating agent concentration), pH, light, and temperature.[54,55]

The quantities of major cations (e.g., calcium, magnesium) in the medium must be considered, particularly in metal toxicity tests because they can affect metal (Cu, Ni) toxicity.[56] High concentrations of Ca and Mg compete with metals for binding sites on the cell surface which may explain why many metals are more toxic in soft waters than in hard waters.[57]

Studies with metal mixtures demonstrated the complicated interactions among metals which may alter results of toxicity tests. A study of the effects of metal mixtures on growth of *Selenastrum capricornutum* and *Chlorella stigmatophora* showed a synergistic interaction between Cu and Mn, and antagonistic reactions between Mn and Pb, and between Cu and Pb.[57] Other studies show similar complex interactions.[58] When mixtures of 10 metals were tested with the cyanobacterium, *Anabaena*, and the algae, *Ankistrodesmus, Chlorella, Scenedesmus,* and *Navicula*, results[59] were difficult to interpret.

When *Chlorella pyrenoidosa* was cultured without the chelating agent EDTA, it was found that the iron concentration in the medium affected Cu toxicity.[60] Copper toxicity to *Chlorella pyrenoidosa* ranged from 1 to 8 mg/L depending on Fe and chelating agent concentrations.[61] In bioassays with wastewater, total toxicity and metal toxicity to algae can be assessed separately by addition of sufficient EDTA to bind all the metals.[62]

The master variable, pH, has an important role in controlling metal toxicity as a result of its influence on metal speciation[63,64] as well as on metal adsorption by surfaces.[65] Natural river sediments adsorb more metal at increasing pH;[66] in contrast, adsorption of organics decreased with increasing pH.[67] Within the pH tolerance of *Scenedesmus acutus*, copper was less toxic at lower pH;[68] similar results were reported for Cd and Zn but Pb was more toxic at lower pH.[18]

Changes in pH during toxicity tests must be monitored because such changes can alter the responses and lead to misleading assessments of the toxicity of the test substance. In studies where both pH and other toxicants, e.g., a variety of metals[50,56,69] were examined, changes in pH were important determinants of growth and other physiological responses. In addition, changes in pH could alter the physical–chemical characteristics of the toxicant (e.g., speciation of metals) being tested. If the pH is inappropriate, problems with precipitate formation can occur when the toxicant is added, confounding the results. Frequent monitoring of pH is very important during toxicity tests at least until the test protocol is worked out and the physiological/growth behavior of the test organism is well established.[69] Differential toxicity of dissolved cadmium and copper to *Scenedesmus quadricauda* at pH 5 and 8 suggest the importance of metal speciation to toxicity responses.[70]

Light and temperature can modify xenobiotic chemistry and biodisposability, and thus differentially affect various species. Interactions of these two factors have been detected in the case of atrazine, where toxicity to the growth of a marine green flagellate was dependent on both, but only light intensity played a role in atrazine toxicity to a diatom.[17] The importance of temperature to the responses of microalgae to surfactants has also been documented.[71]

In using these microalgal toxicity tests, it is important to establish a protocol which gives reliable results. The development of such a protocol requires consideration of the method of toxicant addition and the duration of exposure. The method of toxicant addition is an important consideration when abiotic factors in the medium are likely to interfere with the toxic substance. Tests of contaminated sediments should involve direct exposure to the sediment material rather than the use of altered derivatives (e.g., elutriates).[2,42,72] In tests with contaminated sediments from Lake Ontario, elutriates stimulated Pmax (light-saturated photosynthesis rate) of phytoplankton, but direct exposure to sediments was strongly inhibitory.[72] The authors showed that phosphate availability was an important factor in these experiments, and concluded that, in elutriates, phosphate availability and metal toxicity to algae were sufficiently different from those in solid-phase experiments to nullify the utility of elutriate use in bioassays.[73]

The timing of toxicant addition is also important to consider. Addition prior to the introduction of the test organism allows sufficient time for the toxicant to equilibrate with the culture medium. For substances which have low solubility in water, this can be an important step in conducting a meaningful test since these substances are equilibrated in natural waters. When metal stock solutions are kept at acidic pH to maintain solubility, pH adjustments are often required before the test organisms are introduced. In some instances (e.g., with H_2O_2),[74] it has been shown that the addition of the toxicant 24 h after inoculating the test organism, *Synechococcus* sp. gave significantly different results than if the organism and the toxicant were added simultaneously.

Another major consideration is exposure time. The cultures have to be exposed to the toxicant for a sufficient period to give a reliable response. However, it is also important to select an appropriate period for monitoring toxicity (see Section IV). For instance, the use of closed containers which prevent the loss of CO_2 or O_2 during photosynthetic assays cannot be used for periods in excess of several hours, especially with dense cultures because photosynthesis may become carbon-limited. The change in photosynthetic or respiration rate will then be due to nutrient limitation rather than the test contaminant. One compromise is to pre-expose the test cultures to the toxic substance for a suitable period (e.g., 24 h) and then measure photosynthetic rates over a short-term (1 to 4 h) period. Such an approach has given responses comparable to longer-term (e.g., 5-day) growth bioassays.[40,56,68,75,76]

IV. ASSESSMENT OF TOXICITY RESPONSES

After selection of test organism and experimental conditions, the third consideration is the choice of an appropriate and convenient method to assess the toxicity response. This usually means choosing between an assessment of the organism's overall or integrated response to the toxicant (e.g., effect on growth) and an assessment of a more specific or targeted response to the toxicant (e.g., effect on photosynthesis, respiration, or other physiological/biochemical process). Growth measurements (as biomass or rate) are the most frequently used methods of toxicity assessment.[1]

A. Biomass

The majority of toxicity tests (e.g., algistatic tests) involve the exposure of the cultures to a standardized time period (usually 4 to 7 days) which allows for several generations of cells to experience the toxicant's effects.[6,77,78] The algal biomass or standing crop is determined at the beginning of the test and again at the end of the exposure period. The toxicity response is quantified as the change in biomass or standing crop over the exposure period. The use of standing crop, measured directly (as population density) or indirectly (as chlorophyll concentration, dry weight, optical density, packed cell volume) is a convenient approach in toxicity tests but suffers from several shortcomings.

One concern is that the pollutant may affect chlorophyll and growth differently. As well, growth may be inhibited by the pollutant but at the same time chlorophyll concentration per cell during the test may increase as a result of light limitation brought on by self-shading as population density in the culture increases. However, a comparison of chlorophyll, cell counts, optical density, and direct biovolume/biomass measurements found O.D., total carbon, and chlorophyll fluorescence to be equally good estimators of biomass.[78] In theory, all above measurements are equally acceptable during balanced growth, that is, when all variables change at the same rate during growth and the cultures are growing optimally. This is unlikely to occur when the algae are responding to toxic substances that typically have differential effects on growth rate (see Section IV.B) and cell size and behavior (see Section IV.C).

B. Growth Rate

In theory, algae growing in batch cultures exhibit five phases of growth: lag, exponential, declining growth rate, stationary, and death. However, a lag phase is not necessarily present if exponentially growing cells are used as an inoculum and there is no "medium shock," and the death phase occurs only after a very long stationary phase.[31]

Inhibition of growth rate by toxic substances has been the response reported most often, but in fact growth curves may be modified in other ways. For example, the duration of the lag phase may be increased,[58] or the population density in the stationary phase (the carrying capacity) may be decreased.[17] Longer lag phases have been documented most frequently in the case of heavy metals, e.g., in *Selenastrum capricornutum* in response to Cu[79] and Cu, Zn, and Cd.[58]

The particular effect on growth is dependent on the phase in the cell division cycle affected by the toxic substance. Cell weight and volume increase in the G1 phase; DNA replication occurs in the S phase, and mitosis in G2, hence inhibition of a particular phase will have different effects on the population growth curve as measured by cell numbers compared to biomass or biovolume.[80]

Most toxic effects on algae, however, are based on measurements of the EC_{50}, or the concentration of the substance in the medium that results in a 50% decrease of growth rate. This concentration is either estimated graphically from plots of growth rates at different concentrations of the toxic substance,[7] or by means of the probit method of curve fitting.[81] Other investigators have proposed alternative data analyses and found most give results as good as the probit method provided a suitable range of toxicant concentrations is used in the experimental design and compared to the appropriate control.[70,81-84]

Table 19.1   Differential Responses in Sensitivity Between Growth and Photosynthesis in Algae Pre-Exposed for 24 h to Nickel

Algal Species	EC$_{50}$ Value	
	Growth	Photosynthesis
Euglena gracilis	0.1 mM	0.5 mM
Euglena mutabilis	2.3 mM	>10.0 mM
Scenedesmus acutus (UTEX 72)	0.49 μM	0.62 μM
Scenedesmus acutus (Cu-tolerant)	10.0 μM	4.72 μM
Scenedesmus acutus (Ni-tolerant)	13.8 μM	11.4 μM
Scenedesmus acutus (B4)	21.7 μM	5.4 μM

Data from Olaveson, M.M., unpublished data, 1996; also Christensen, E.R., Scherfig, J., and Dixon, P.S., *Water Res.*, 13, 79, 1979.

C. Physiological Parameters

The use of changes in physiological processes (e.g., photosynthesis, respiration, nutrient uptake, etc.) to assess toxicity in lieu of growth is more common in research than in routine toxicity testing. However, the increasing need for obtaining rapid yet reliable results makes this approach increasingly attractive. The use of a targeted physiological process as a proxy for growth responses is appealing when large numbers of substances must be assessed. Such an approach also allows for more controlled conditions to be used and the opportunity to evaluate a large number of microalgal species in a relatively short period of time. The main drawback is the effort involved in establishing that the process chosen for assessment is indeed a target of toxicity. Clearly if the process used in the toxicity test is not affected by the toxic substance, the test results will be unreliable and misleading. The effort required to establish the validity of a particular physiological process as a reliable measure of toxicity has limited the use of this approach until recently.

1. Photosynthesis

With photosynthetic microalgal species, changes in photosynthetic responses can be a reliable indicator of toxicity. The methodologies for measuring photosynthesis are well established (e.g., ^{14}C-tracer methods, O$_2$ electrode or titration methods).[85] Photosynthesis has proven to be a reliable process for measurement of effects of a number of toxicants including metals,[75,86] surfactants,[76,87,88] organic compounds, and pesticides.[89-91]

The faster and logistically simpler photosynthesis assays can be used to predict effects of a given toxic substance on growth, but the relationship between the two variables must be first established for each species and each pollutant. In *Selenastrum capricornutum* exposed to copper, surfactants, and two organic compounds (phenol and anisole), 5-day growth EC$_{50}$ values were consistently lower than EC$_{50}$ values for photosynthesis.[87] Although the measured EC$_{50}$ values from 5-day-long growth experiments were lower than from photosynthesis experiments, there was a constant relationship between the values related to the duration of the assay. A number of researchers have shown that photosynthetic responses are reliable predictors of toxicity effects on growth. A comparison of nickel toxicity based on both growth and photosynthetic rate assays in two acidophilic *Euglena* species[53] and in several metal-tolerant strains of *Scenedesmus acutus*[75] is shown in Table 19.1. Among the Ni-pre-exposed cultures of four *Scenedesmus acutus* strains, EC$_{50}$s conformed to the expected pattern based on relative tolerances (e.g., from least to most tolerant: UTEX, Cu-Tol, Ni-Tol, B4) with one exception in which the photosynthesis and growth results were reversed.[75] It is likely that larger amounts of the toxic substance penetrate the cells during the longer growth experiments than during the shorter photosynthesis experiments. This may account for findings that short-term photosynthesis was a less sensitive effect parameter than growth over 3 to 4 days in assessment of herbicide and metal toxicity.[92-94] Similarly, vanadium was less toxic to photosynthesis than to growth in a number of different species of algae and cyanobacteria.[41]

A variation of this method uses the photosynthetic rate–light intensity (P-I) relationship to assess the effects of contaminants on algal photosynthesis. In this way, potential toxicity to both light-limited and light-saturated photosynthesis can be assessed.[47] This is particularly important to know for aquatic habitats of varying light conditions (e.g., turbulent waters, dredging projects, high turbidity waters). Values of P_{max}, the light-saturated photosynthetic rate, from P-I curve assays were used in tests of sediment toxicity in the Laurentian Great Lakes, Mackenzie River Delta, and in prairie dugouts treated with $CuSO_4$ over long periods of time.[2,42,72,73,93,94]

2. Respiration

Another process that can be used in toxicity tests is respiration. Several studies with microalgae have shown that respiration was more sensitive than photosynthesis to copper toxicity.[44,95] The effects of toxicants on respiration has been less well studied than photosynthesis in microalgae because of the complications with the most commonly used methods. The methods (e.g., O_2 electrode, Winkler titrations, ATP changes, electron transport chain [ETC] indicators) are well established[85,96] and have been used to assess water quality conditions (e.g., BOD tests).[6] It is important however to be aware of potential interferences with these methods. For instance, the O_2 electrode method cannot be used in cultures containing elevated metal concentrations. The Winkler titration method can be subject to unpredictable precipitate formation when used with some toxicants or with elevated concentrations of toxicants such as copper. A number of recent studies have followed changes in energy charge (as indicated by changes in ATP/ADP/AMP ratios) as an estimation of respiration responses.[17] The use of ETC indicators (INT or CTC methods[96]) is more difficult to assess in photosynthetic organisms than in nonphotosynthetic organisms due to interference with chlorophyll.

3. Nutrient Uptake

Other processes which have been used sporadically for toxicity assessment include nutrient (e.g., PO_4 and N) uptake. The methods used involve monitoring changes in chemical concentrations[58] or the use of radioactive tracers such as ^{32}P-PO_4, ^{15}N-NH_4 or ^{15}N-NO_3.[40,91] These processes have been used less frequently than growth or photosynthesis but offer considerable promise as rapid and reliable techniques when better calibrations with other processes are available.[70]

D. Other Parameters

Depending on the characteristics of the test organism, a variety of parameters and processes may be suitable for use in measurements of toxicity responses.[6,17,59,97] These include activity of various enzymes, cell motility, flagellar activity, loss and regeneration of flagella, cell size changes, and formation of teratogenic cells. While changes in these characteristics may be reliable indicators of toxicity, calibration with growth and more commonly used physiological indicators (e.g., photosynthesis) needs to be carried out.

E. Need for Cross-Calibration

The EC_{50} values reported for physiological processes are quite variable when compared with growth data. The reasons for this variability are complicated but may be related to the exposure time and to the nature of the toxic substance. The nature of the toxicant determines the site of action, so the process chosen for assessment must be the one most sensitive to the toxicant, otherwise there will be a significant discrepancy in response when compared with growth. Comparing several processes, whenever possible, is useful in increasing our understanding of the underlying mechanisms of toxicity.

With photosynthesis and respiration, the EC_{50} values were usually lower than the EC_{50} values for growth. This pattern is much more variable when considering other physiological processes. The EC_{50} values for nutrient uptake are not invariably higher than for growth. For example, EC_{50} values for P-uptake were much smaller than for growth inhibition with 3,5-dichlorophenol and potassium dichromate in *Selenastrum*.[98] Similarly, phosphate uptake was affected more than growth by Cu in three strains of *Scenedesmus acutus*.[44] However, phosphate assimilation in phytoplankton was inhibited at much higher herbicide concentrations than photosynthesis, and ammonium assimilation was inhibited at intermediate concentrations.[91]

Differential responses to cadmium between photosynthesis and phosphate uptake by two diatoms have been linked to mechanisms of action.[99] Therefore, appreciation of the underlying mechanisms of toxicity is often necessary in choosing the most appropriate process for testing. Some herbicides are known to act on photosynthesis, while others target processes such as phosphate uptake by uncoupling oxidative phosphorylation or N metabolism by altering protein synthesis.[91] In three strains of *Scenedesmus acutus* exposed to toxic Cu levels, the order of EC_{50}s, in decreasing sequence was photosynthesis, respiration, growth rate, and phosphate uptake; the results were not related to experimental duration.[56] Clearly, knowledge of the site of toxic action is a major determinant of the most sensitive process and should be considered when choosing the best method of assessment.

V. IMPORTANCE OF AN ECOPHYSIOLOGICAL APPROACH

The validity of differences among microalgal species in their responses to a toxic substance are difficult to confirm as real if ecological/physiological factors are not considered or there is no indication they have been considered. The literature is replete with examples of differential sensitivities among microalgal species to a variety of contaminants.[5] For example, an extensive survey of 118 algal species showed significant differential responses of photosynthesis to cadmium, copper, and zinc additions.[86] Likewise, a survey of a broad range of toxicants with two commonly used microalgae, *Microcystis aeruginosa* and *Scenedesmus quadricauda*, showed that just within these two species there was considerable variation in response depending on the chemical.[100] These findings are no different than have been reported for many other groups of organisms (e.g., invertebrates, mammals). However, more criticism is directed toward the microalgal tests because there is the expectation that all microalgal species are similar and should respond similarly to added toxicants. The above discussion with its stress on the taxonomic, ecological, and physiological differences among the microalgae should alleviate some of these concerns and place the use of microalgae in a more realistic perspective with regard to their potential in toxicity assessment. In fact, the variety of responses shown by the microalgae suggest that this group is very promising for toxicity tests. Such diversity in responses, however, can only be appreciated once we are confident that the differences are legitimate. This can only be accomplished by critically reviewing the results of toxicity testing to date and incorporating the results of research studies examining the responses to toxic substances under a variety of conditions with the goal of understanding the underlying mechanisms.

A number of recent reviews have listed a variety of taxa and their growth and physiological responses to various toxicants (e.g., metals,[70] surfactants,[88] and miscellaneous pollutants[90]). The differential sensitivities illustrated are usually accepted as reported because it is impossible to evaluate the experimental design used in the studies or because it is not the focus of the review to assess methodologies. Often there is insufficient information presented about the culture conditions to properly determine the validity of the test results.

Weiss and Helms[77] concluded, on the basis of round robin tests, that within-laboratory microalgal toxicity tests with *Selenastrum* had high precision, but large differences were apparent among laboratories. The coefficient of variation averaged for the eight laboratories varied from 3 to nearly 240%. All used the same culture of *Selenastrum* and a standard inoculum. The large variability was probably the result of differences in temperature, irradiance, aeration, and possibly the use of a different period in the culture growth for calculation of growth rates.

These publications indicate that, for some contaminants at least, large differences have been documented in responses among different algal genera, and in some cases, also among species of the same genus. The existence of these differences and their magnitude needs to be carefully examined because in some instances physical and chemical conditions during the assays were sub-optimal and may have affected the results.[11]

The most convincing reports of differences in growth responses to pollutants among species have four important characteristics. First, they include actual growth rates of the algae in the control treatments in growth tests (or in stock cultures in tests based on other physiological parameters). The provision of growth rates allows assessment of whether any factors were sub-optimal by comparison with published values.[17] Growth rates for most microalgal species are in the range of 1 to 2 div/day.[31] Second, to avoid complications from abiotic factors the same medium and the same physical conditions should be used in the tests. Third, actual mechanisms responsible for the differences in responses should be identified. This is best assessed by testing the responses of growth and of several physiological processes. Fourth, the significance of the observed differences should be statistically tested. When multispecies tests are conducted, a variety of multivariate analyses may be required.[10]

To date, there have been few studies[41,101] which meet all of these criteria. The study by Nalewajko et al.[41] indicated significant interspecific differences among a variety of microalgal species in their responses to vanadium (see Figure 19.3 and Table 19.2). It is interesting to note that the Cyanobacteria were most sensitive and the diatoms (Bacillariophyceae) were least sensitive; the Chlorophyceae were intermediate in their overall response, but within this group there was a range of responses suggesting that algal responses cannot be predicted at the Class level. By including all of the criteria listed above, this study provides reliable evidence of true differences in responses among species and allows comparisons with other species in terms of growth and several physiological processes. While such studies are time-consuming, they are essential in order to detect toxicity reliably and avoid the pitfalls discussed in this chapter. Also such a comprehensive approach would alleviate much of the frustration surrounding our inability to compare toxicity data derived not only under variable conditions but, more important, with different methods of assessment.[77]

The above discussion suggests that the use of cultures pre-exposed to the toxicant, then followed by short-term toxicity assays, has many advantages including significantly shorter test duration compared to growth tests. This method can also be less labor intensive although the personnel conducting the tests may require more extensive training than is necessary for the routine growth tests.

VI. CONCLUSIONS

It is necessary to develop ecologically meaningful, yet cost-effective, toxicity tests for assessing the impact of potential toxicants on the microalgae in natural ecosystems. As the major primary producers in freshwater ecosystems, microalgae play a pivotal role in the functioning of a healthy ecosystem. These organisms can provide an early indication of potential problems from contamination with a variety of toxicants.[1] From the discussion above, the following recommendations can be made:

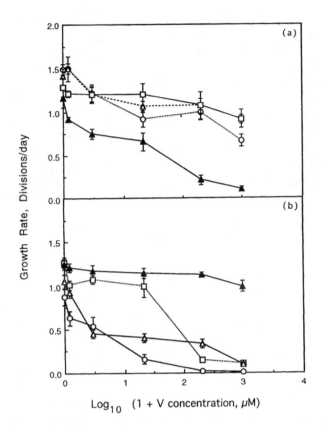

Figure 19.3 Growth rates in six species of algae and two species of cyanobacteria at a series of vanadium concentrations in Chu 10 (with 57.5 mM P). (From Nalewajko, C., Lee, K., and Olaveson, M., Responses of freshwater algae to inhibitory vanadium concentrations: the role of phosphorus, *J. Phycol.*, 31, 332, 1995. With permission.) a) *Anabaena* —□—, *Chlorella* -----○-----, *Dictyosphaerium* -----△-----, *Kirschneriella* —▲—; b) *Synechococcus* —▲—, *Ankistrodesmus* —△—, *Scenedesmus obliquus* -----□-----, *Diatoma* —○— . Each point represents the mean of four replicates — SE.

Table 19.2 Pairwise Comparison of Slopes from Linear Regressions of Growth Rate of P-sufficient Cultures of Eight Species Against Vanadium Concentration (see Figure 19.3) based on the Tukey-Kramer Test

Group	Species	Slope	t^2	Species Code	1	2	3	4	5	6	7	8
Cyanobacteria	*Anabaena*	−0.100	0.303	1	—	NS	S	NS	NS	S	S	S
	Synechococcus	−0.068	0.431	2		—	S	NS	NS	S	S	S
Chlorophyta	*Ankistrodesmus*	−0.263	0.743	3			—	S	S	NS	S	S
	Chlorella	−0.248	0.753	4				—	NS	S	S	S
	Dictyosphaerium	−0.163	0.561	5					—	S	S	S
	Kirschneriella	−0.314	0.867	6						—	NS	S
	Scenedesmus	−0.381	0.838	7							—	S
Bacillariophyta	*Diatoma*	−0.271	0.786	8								—

Note: Each species was subjected to 6 vanadium concentrations (0, 0.2, 2, 20, 200, and 1000 (μM); each treatment was conducted in quadruplicate. S = significant at P < 0.05, NS = not significant. (From Nalewajko, C., Lee, K., and Olaveson, M., Responses of freshwater algae to inhibitory vanadium concentrations: the role of phosphorus, *J. Phycol.*, 31, 332, 1995. With permission.)

1. Use the dominant microalgal species reported from the waters to be affected by the toxicant and, whenever possible, use more than one species. The ultimate test of toxicity to the primary producers of freshwater ecosystems is the use of multispecies toxicity tests using natural phytoplankton communities. Such tests are becoming more common,[101-104] but more careful assessment and monitoring of test conditions are required in order to obtain reliable and reproducible results. Reproducibility is more difficult to achieve in mixed algal populations especially when using natural site waters.

2. Ensure that the selected species exhibit maximal growth in the selected medium by using optimal environmental conditions. By optimizing growth conditions and reporting growth rates, it is possible to assess the test conditions and determine if the results are a true toxicity assessment.

3. Consider pre-exposing the test organism to the contaminant, then measuring photosynthesis or alternative physiological processes as a proxy for the more time-consuming and problem-prone biomass or growth rate measurements. Since algal physiological responses seem to be dependent on the characteristics of the toxicant,[91] such physiologically based toxicity tests should give effective and efficient assessments of toxicity.

4. If the receiving waters are likely to be N- or P-limited, include tests with both nutrient deficient and sufficient cultures. Other unusual physical–chemical characteristics (e.g., extremes of light, temperature, pH) likely to be present in the natural environment exposed to the toxicant should also be considered in microalgal toxicity tests.

REFERENCES

1. Klaine, S.J. and Lewis, M.A., Algal and plant toxicity testing, in *Handbook of Ecotoxicology,* Hoffman, D.J., Rattner, B.A., Burton, G.A., and Cairns, J., Lewis Publ., Boca Raton, FL, 1994, chap. 8.

2. Nalewajko, C., Use of algal bioassays in polluted environments, in *Phycotalk*, Vol. 1, Kumar, H.D. Ed., Rastogi and Company, Subhash Bazar, Meerut-250 002, India, 1989, chap. 2.

3. US Environmental Protection Agency, The *Selenastrum capricornutum* Printz Algal Assay Bottle Test, EPA-600/9-78-018, 126 pp., 1978.

4. Parrish, P.R., Acute toxicity tests, in *Fundamentals of Aquatic Toxicology*, Rand, G.M. and Petrocelli, S.R., Eds., Hemisphere Publishing Corp., Washington, 1985, chap. 2.

5. Lewis, M.A., Are laboratory-derived toxicity data for freshwater algae worth the effort?, *Environ. Toxicol. Chem.*, 9, 1279, 1990.

6. Matulova, D., The application of algae as test organisms, *Arch. Protistenkd.* 139, 279, 1991.

7. Environment Canada, Biological test method: Growth inhibition test using the freshwater alga *Selenastrum capricornutum*, Report EPS 1/RM/25, Ottawa, Ontario, 41 pp., 1992.

8. Radetski, C.M., Férard, J.F., and Blaise, C., A semistatic microplate-based phytotoxicity test, *Environ. Toxicol. Chem.*, 14, 299, 1995.

9. Blanck, H., Wallin, G., and Wangberg, S., Species-dependent variation in algal sensitivity to chemical compounds, *Ecotoxicology and Environmental Safety,* 8, 339, 1984.

10. Wangberg, S. and Blanck, H., Multivariate patterns of algal sensitivity to chemicals in relation to phylogeny, *Ecotoxicology and Environmental Safety,* 16, 72, 1988.

11. Nyholm, N. and Kallqvist, T., Methods for growth inhibition toxicity tests with freshwater algae, *Environmental Toxicology and Chemistry,* 8, 689, 1989.

12. Larsen, D.P., DeNoyelles, F., Stay, F., and Shiroyama, T., Comparisons of single species microcosm and experimental pond responses to atrazine exposure, *Environ. Toxicol. Chem.,* 5, 179, 1986.

13. Chiaudani, G. and Vighi, M., The use of *Selenastrum capricornutum* batch cultures in toxicity studies, *Mitt. Internat. Verein. Limnol.*, 21, 316, 1978.

14. Bold, H. C. and Wynne, M. J., *Introduction to the Algae*, Second Edition, Prentice-Hall, Englewood Cliffs, NJ, 1985.

15. Vymazal, J., *Algae and Element Cycling in Wetlands*, Lewis Publ., Boca Raton, FL, 689 pp. 1995.

16. Hawkins, P.R. and Griffiths, D.J., Uptake and retention of copper by four species of marine phytoplankton, *Botanica Marina*, 25, 551, 1982.

17. Puiseux-Dao, S., "Phytoplankton model" in ecotoxicology, in *Aquatic Toxicology: Fundamental Concepts and Methodologies*, Vol. II, Boudou, A. and Ribeyere, F., Eds., CRC Press, Boca Raton, FL, 1989.

18. Gimmler, H., Treffny, B. and Zimmerman, U. The resistance of *Dunaliella acidophila* against heavy metals: The importance of the zeta potential, *J. Plant Physiol.* 138, 708, 1991.

19. Strycek, T., Acreman, J., Kerry, A., Leppard, G.G., Nermut, M.V., and Kushner, D.J., Extracellular fibril production by freshwater algae and cyanobacteria, *Microbial Ecology*, 25, 53, 1992.

20. Reynolds, C.S., Phytoplankton assemblages and their periodicity in stratifying lake systems, *Holarctic Ecology*, 3, 141, 1980.

21. Reynolds, C.S. Phytoplankton periodicity: Its motivation, mechanisms and manipulation, *Fiftieth Annual Report of the Freshwater Biological Association*, Ambelside, 60, 1982.

22. Willen, E., A simplified method of phytoplankton counting, *Br. Phycol. J.*, 11, 265, 1976.

23. Willen, E., Phytoplankton in Lake Hjalmaren, *Acta Universitatis Upsaliensis*, 376, 18 pp., 1976.

24. Droop, M.R., Heterotrophy of carbon, in *Algal Physiology and Biochemistry,* Stewart, W.D.P., Ed., Univ. of California Press, Berkeley, 1974, chap. 19.

25. Kessler, E., Comparative physiology, biochemistry and taxonomy of *Chlorella* (Chlorophyceae), *Plant Sys. Evol.*, 125, 125, 1976.

26. Wood, A.M. and Leatham, T., The species concept in phytoplankton ecology, *J. Phycol.*, 28, 723, 1992.

27. Takishima, Y., Shimara, J., Ugawa, Y., and Sugawara, H., *Guide to the World Data Center in Micro-organisms with a List of Culture Collections in the World,* WFCC World Data Center on Microorganisms, 1989, 249 pp.

28. Lindstrom, K. and Rodhe, W., Selenium as a micronutrient for the dinoflagellate *Peridinium cinctum* fa. *westii*, *Mitt. Internat. Verein. Limnol.*, 21, 168, 1978.

29. Wehr, J.D. and Brown, L.M., Selenium requirement of a bloom-forming planktonic alga from softwater and acidified lakes, *Can. J. Fish. Aquat. Sci.,* 42, 1783, 1985.

30. Gensemer, R. W., The effects of pH and aluminum on the growth of the acidophilic diatom *Asterionella ralfsii* var. *americana*, *Limnol. Oceanogr.*, 36, 123, 1991.

31. Fogg, G.E. and Thake, B., *Algal Cultures and Phytoplankton Ecology*, University of Wisconsin Press, Madison, 1987, 269 pp.

32. Starr, R.C. and Zeikus, J.A., UTEX — The Culture Collection of Algae at the University of Texas at Austin, *J. Phycol.* (suppl.), 29, 106 pp., 1993.

33. Nichols, H.W., Growth media — freshwater, in *Handbook of Phycological Methods: Culture Methods and Growth Measurements,* Stein, J.R., Ed., Cambridge University Press, Cambridge, 1973, chap. 1.

34. Morel, F.M.M., Westall, J.C., Reuter, J.G., and Chaplick, J.P., Description of the algal growth media Aquil and Fraquil, *Mass. Inst. Technol. R.M. Parsons Lab. Tech. Note* 16, 1975.

35. Lauth, J.R., Cherry, D.S., and Cairns, J. Jr., A single reconstituted water formulation for the culture and toxicity testing of algae, invertebrates, and fish, *Environmental Auditor*, 1, 209, 1990.

36. Tilman, D. and Kilham, S.S., Phosphate and silicate growth and uptake kinetics of the diatoms *Asterionella formosa* and *Cyclotella meneghiniana* in batch and semi-continuous culture. *J. Phycol.* 12, 375, 1976.

37. Rodhe, W., Environmental requirements of freshwater plankton algae, *Symbolae Botan. Upsaliensis*, 10, 1948, 149 pp.

38. Winters, K., O'Donnell, R., Batterton, J.C., and van Baalen, C., Water-soluble components of four fuel oils: Chemical characterization and effects on growth of microalgae, *Marine Biology,* 36, 269, 1976.

39. Greger, M., Tillberg, J. and Johansson, M. Aluminium effects on *Scenedesmus obtusiusculus* with different phosphorus status. ll. Growth, photosynthesis and pH, *Physiologia Plantarum*, 84, 202, 1992.

40. Twiss, M.R. and Nalewajko, C., Influence of phosphorus nutrition on copper toxicity to three strains of *Scenedesmus acutus* (Chlorophyceae), *J. Phycol.*, 28, 291, 1992.

41. Nalewajko, C., Lee, K., and Olaveson, M., Responses of freshwater algae to inhibitory vanadium concentrations: the role of phosphorus, *J. Phycol.*, 31, 332, 1995.

42. Nalewajko, C. and Prepas, E., Responses of phytoplankton photosynthesis and phosphorus kinetics to resuspended sediments in copper sulfate-treated ponds, *J. Envir. Quality*, 25, 80, 1996.

43. Hashemi, F., Leppard, G.G., and Kushner, D.J., Copper resistance in *Anabaena variabilis:* Effects of phosphate nutrition and polyphosphate bodies, *Microbial Ecology*, 27, 159, 1994.

44. Nalewajko, C. and Olaveson, M.M., Differential responses of growth, photosynthesis, respiration, and phosphate uptake to copper in copper-tolerant and copper-intolerant strains of *Scenedesmus acutus* (Chlorophyceae), *Can. J. Bot.*, 73: 1295, 1995.

45. Shuter, B., A model of physiological adaptation in unicellular algae, *J. Theoret. Bot.*, 78, 519, 1979.

46. Ryther, J.H., Photosynthesis in the oceans as a function of light intensity, *Limnol. Ocean.* 1, 61, 1956.
47. Prezelin, B.B., Light reactions in photosynthesis, in Physiological bases of phytoplankton ecology, Platt, T., Ed., *Can. Bull. Fish. Aquat. Sci.*, 210, 1981.
48. Steeman-Nielsen, E. and Hansen, V.K., Light adaptation in marine phytoplankton populations and its interrelation with temperature, *Physiol. Pl.*, 12, 353, 1959.
49. Suzuki, Y. and Takahashi, M., Growth responses of several diatom species isolated from various environments to temperature, *J. Phycol.*, 31, 880, 1995.
50. Peterson, H.G. and Healey, F.P., Metal toxicity to algae: A highly pH-dependent phenomenon, *Can. J. Fish. Aquat. Sci.*, 41, 974, 1984.
51. Olaveson, M.M. and Stokes, P.M., Responses of the acidophilic alga *Euglena mutabilis* (Euglenophyceae) to carbon enrichment at pH 3, *J. Phycol.* 25, 529, 1989.
52. Wee, J.L., Millie, D.F., and Walton, S.P., A statistical characterization of growth among clones of *Synura petersenii* (Synurophyceae), *J. Phycol.*, 27, 570, 1991.
53. Olaveson, M.M., unpublished data, 1996.
54. Rai, L.C., Gaur, J.P., and Kumar, H.D., Phycology and heavy metal pollution, *Biol. Rev.*, 56, 99, 1981.
55. De Filippis, L.E. and Pallaghy, C.K., Heavy metals: sources and biological effects, in *Algae and Water Pollution,* Rai, L.C., Gaur, J.P. and Soeder, C.J., Eds., E. Schweizerbart'sche Verlagsbuchhandlung, Stuttgart, Germany, 1994, Chapter 2.
56. Rai, L.C., Rai, P.K., and Mallick, N., Regulation of heavy metal toxicity in acid-tolerant *Chlorella*:: physiological and biochemical approaches, *Environmental and Experimental Botany,* 36: 99, 1996
57. Christensen, E.R., Scherfig, J., and Dixon, P.S., Effects of manganese, copper, and lead on *Selenastrum capricornutum* and *Chlorella stigmatophora*, *Water Res.*, 13, 79, 1979.
58. Bartlett, L., Rabe, F.W., and Funk, W.H., Effects of copper, zinc, and cadmium on *Selenastrum capricornutum*, *Water Res.*, 8, 179, 1974.
59. Wong, P.T.S., Chau, Y.K., Kramar, O., and Bengert, G.A., Structure-toxicity relationship of tin compounds on algae, *Can. J. Fish. Aquat. Sci.*, 39, 483, 1982.
60. Steemann-Nielsen, E., and Wium-Andersen, S., Copper ions as poison in the sea and fresh water, *Mar. Biol.*, 6, 93, 1970.
61. Fitzgerald, G.P. and Faust, S.L., Factors affecting the algicidal and algistatic properties of copper, *Appl. Microbiol.*, 11, 345, 1963.
62. Wong, S.L., Wainwright, J.F., and Pimenta, J., Quantification of total and metal toxicity in wastewater using algal bioassays, *Aquatic Toxicology,* 31, 57, 1995
63. Campbell, P.G.C. and Stokes, P.M., Acidification and toxicity of metals to aquatic biota, *Can. J. Fish. Aquat. Sci.*, 42, 2034, 1985.
64. Harrison, G.I., Campbell, P.G.C., and Tessier, A., Effects of pH changes on zinc uptake of *Chlamydomonas variabilis* grown in batch culture, *Can. J. Fish. Aquat. Sci.*, 43, 687, 1986.
65. Bourg, A.C.M., Role of freshwater/seawater mixing on trace metal adsorption phenomena, in *Trace Metals in Sea Water*, Wong, C.S., Burton, J.D., Boyle, E., Bruland, W., and Goldberg, E., Eds., Plenum Press, New York, 1983, 195.
66. Scanferlato, V.S. and Cairns, J., Effect of sediment-associated copper on ecological structure and function of aquatic microcosms, *Aquat. Toxicol.*, 18, 23, 1990.
67. Schwarzenbach, R.P. and Westall, J., Sorption of hydrophobic trace organics in groundwater systems, in *Proc. Sem. Degradation, Retention, and Dispersion of Pollutants in Groundwater,* International Assoc. Water Pollut. Research Control, 1984, 39.
68. Nalewajko, C., Colman, B., and Olaveson, M., Physiological characteristics of three *Scenedesmus* strains: effects of pH on growth, photosynthesis, respiration and copper tolerance, *Envir. Exper. Botany*, 37, 11, 1997.
69. Lustigman, B., Lee, L.H., and Khalil, A., Effects of nickel and pH on the growth of *Chlorella vulgaris*, *Bull. Environ. Contamin. Toxicol.*, 55, 73, 1995.
70. Peterson, H.G. and Nyholm, N., Algal bioassays for metal toxicity identification, *Water Poll. Res. J. Canada*, 28, 129, 1993.
71. Lewis, M.A. and Hamm, B.G., Environmental modification of the photosynthetic response of lake plankton to surfactants and significance to laboratory-field comparison, *Water Res.*, 20, 1575, 1986.
72. Nalewajko, C., Ewing, C., and Mudroch, A., Effects of contaminated sediments and standard elutriates on primary production in Lake Ontario, *Water, Air, and Soil Pollution,* 45, 275, 1989.

73. Nalewajko, C., Ewing, C., and Mudroch, A., A comparison of the effects of sediments and standard elutriates on phosphorus kinetics in lakewater, *Water, Air, and Soil Pollution,* 44, 119, 1989.

74. Dupouy, D., Couter, A., Croute, F., Murat, M., and Planel, H., Sensitivity of *Synechococcus liordus* to hydrogen peroxide, *Envir. Exper. Botany,* 25, 339, 1985.

75. Jin, X., Nalewajko, C., and Kushner, D.J., Comparative studies of nickel toxicity to growth and photosynthesis in nickel-resistant and sensitive strains of *Scenedesmus acutus* f. *alternans* (Chlorophyceae), *Microbial Ecology,* 31, 103, 1996.

76. Yamane, A.N., Okada, M., and Sudo, R., The growth inhibition of planktonic algae due to surfactants used in washing agents, *Water Res.,* 18, 1101, 1984.

77. Weiss, C.M. and Helms, R.W., *The Interlaboratory Precision Test: an Eight Laboratory Evaluation of the Provisional Algal Assay Procedure Bottle Test,* Department of Environmental Sciences and Engineering, School of Public health, University of North Carolina at Chapel Hill, EPA-Water Quality Office Project-16010DQT, 1971.

78. Walsh, G.E., Deans, C.H., and McLaughlin, L.L., Comparison of the EC_{50}s of algal toxicity tests calculated by four methods, *Environ. Toxicol. Chemistry,* 6, 767, 1987.

79. Kuwabara, J.S. and Leland, H.V., Adaptation of *Selenastrum capricornutum* (Chlorophyceae) to copper, *Environ. Toxicol. Chem.,* 5, 197, 1986.

80. Chisholm, S.W. Temporal patterns of cell division in unicellular algae, in Physiological Bases of Phytoplankton Ecology, Platt, T., Ed., *Can. Bull. Fish. Aquat. Sci.* 210, 150, 1981.

81. Newman, M.C., *Quantitative Methods in Aquatic Ecotoxicology,* Lewis Publishers, Boca Raton, FL, 1995.

82. Christensen, E.R. and Nyholm, N., Ecotoxicological assays with algae: Weibull dose- response curves, *Environ. Sci. Technol.,* 18, 713, 1984.

83. Bruce, R.D. and Versteeg, D.J., A statistical procedure for modeling continuous toxicity data, *Environ. Toxicol. Chemistry,* 11, 1485, 1992.

84. Nyholm, N., Settergren, P., and Kusk, K.O., Statistical treatment of data from microbial toxicity tests, *Environ. Toxicol. Chemistry,* 11, 157, 1992.

85. Vollenweider, R.A., *A Manual on Methods for Measuring Primary Production in Aquatic Environments,* Blackwell Scientific Pub., London, 225 pp. 1974.

86. Takamura, N., Kasai, F., and Watanabe, M.M., Effects of Cu, Cd, and Zn on photosynthesis of freshwater benthic algae, *J. Appl. Phycol.,* 1, 39, 1989.

87. Nyholm, N. and Damgaard, B. M., A comparison of the algal growth inhibition toxicity test method with the short term ^{14}C-assimilation test, *Chemosphere,* 21, 671, 1990.

88. Nalewajko, C., Effects of surfactants on algae, in *Algae and Water Pollution,* Rai, L.C., Gaur, J.P. and Soeder, C.J., Eds., E. Schweizerbart'sche Verlagsbuchhandlung, Stuttgart, Germany, 1994, Chapter 7.

89. Hsiao, S.I., Effects of crude oils on the growth of arctic marine phytoplankton, *Environ. Pollut.,* 17, 93, 1978.

90. Nalewajko, C. and Dunstall, T.G., Miscellaneous Pollutants: thermal effluents, halogens, organochlorines, radionuclides, in *Algae and Water Pollution,* Rai, L.C., Gaur, J.P. and Soeder, C.J., Eds., E. Schweizerbart'sche Verlagsbuchhandlung, Stuttgart, Germany, 1994, Chapter 9.

91. Brown, L.S. and Lean, D.R.S., Toxicity of selected pesticides to lake phytoplankton measured using photosynthetic inhibition compared to maximal uptake rates of phosphate and ammonium, *Environ. Toxicol. Chemistry,* 14, 93, 1995.

92. Turbak, S.C., Olson, S.B., and McFeters, G.A., Comparisons of algal systems for detecting waterborne herbicides and metals, *Water Research,* 20, 91, 1986

93. Nalewajko, C., Effects of cadmium and metal-contaminated sediments on photosynthesis, heterotrophy, and phosphate uptake in Mackenzie River Delta phytoplankton, *Chemosphere,* 30, 1401, 1995

94. Nalewajko, C. and Prepas, E., Responses of phytoplankton photosynthesis and phosphorus kinetics to resuspended sediments in copper-sulfate treated ponds, *J. Envir. Quality,* 25, 80, 1996.

95. McBrien, D.C.H. and Hassall, K.A., The effect of toxic doses of copper upon respiration, photosynthesis and growth of *Chlorella vulgaris, Physiol. Plant.,* 20, 113, 1967.

96. Kenner, R.A. and Ahmed, S.I., Measurements of electron transport activities in marine phytoplankton, *Mar. Biol.,* 33, 119, 1975.

97. Rai, L.C., Husaini, Y., and Mallick, N., Physiological and biochemical responses of *Nostoc linckia* to combined effects of aluminum, fluoride and acidification, *Environmental and Experimental Botany,* 36: 1, 1996

98. Nyholm, N., Toxic effects on algal phosphate uptake, *Environ. Toxicol. Chemistry,* 10, 581, 1991.

99. Conway, H.L. and Williams, S.C., Sorption of cadmium and its effect on growth and the utilization of inorganic carbon and phosphorus of two freshwater diatoms, *J. Fish. Res. Board Can.,* 36, 579, 1979.

100. Bringmann, G. and Kuhn, R., Testing of substances for their toxicity threshold: Model organisms *Microcystis* (Diplocystis) *aeruginosa* and *Scenedesmus quadricauda*, *Mitt. Internat. Verein. Limnol.,* 21, 275, 1978.

101. Stauber, J.L. and Florence, T.M., Mechanism of toxicity of ionic copper and copper complexes to algae, *Mar. Biol.* 94, 511, 1987.

102. Lundy, P., Wurster, C.F., and Rowland, R.G., A two-species marine algal bioassay for detecting aquatic toxicity of chemical pollutants, *Water Res.,* 18, 187, 1984.

103. Kusk, K.O. and Nyholm, N., Evaluation of a phytoplankton toxicity test for water pollution assessment and control, *Arch. Environ. Contam. Toxicol.,* 20, 375, 1991.

104. Juttner, I., Peither, A., Lay, J.P., Kettrup, A., and Ormerod, S.J., An outdoor mesocosm study to assess ecotoxicological effects of atrazine on a natural plankton community, *Arch. Environ. Contamin. Toxicol.,* 29, 435, 1995.

CHAPTER 20

Development and First Validation of a "Stock-Culture Free" Algal Microbiotest: The Algaltoxkit

G. Persoone

CONTENTS

I. COMPLEXITY AND VARIABILITY OF ALGAL TOXICITY TESTS

Of all the short-term toxicity tests practiced to date for notification, screening, or monitoring purposes, bioassays with microalgae are among the most demanding from the point of view of infrastructure requirements, equipment, materials, and labor. In addition, performance of algal tests requires extensive know-how and good skills in microbiological and phycological technology. Consequently, it is no surprise that tests with microalgae are performed less frequently than acute bioassays with invertebrates or fish, and at much greater expense.

One of the consequences of the low popularity of algal toxicity tests is that it took until the adoption of the so-called "7th amendment" in 1992, to impose an algal assay at the "base set level," for preparation of the notification dossiers which are required by the Commission of the European Communities for the marketing of new chemicals.

In their critical reviews on toxicity tests with algae, Nyholm and Källqvist,[1] and Hörnström[2] address in detail the numerous problems associated with the performance of freshwater algal assays, with particular emphasis on the large variabilities and their potential causes, and with recommendations for increasing the (intra- and interlaboratory) precision of tests with microalgae.

II. DEVELOPMENT OF MICROBIOTESTS WITH ALGAE

In view of the complexity of the conventional algal assays, attempts were made in a few laboratories in different countries to develop alternative toxicity tests with microalgae; the major objective of all these studies was to make the tests less time-consuming and more user-friendly, and to reduce the bench space and the amount of materials needed.

The leading laboratory in these endeavors was, without any doubt, the Centre Saint Laurent of Environment Canada in Montreal, which in the early 1980s had already developed a miniaturized algal toxicity test in 96-well microtiter plates.[3] The efforts of Dr. Blaise and collaborators in refining, validating, and applying their new algal microbiotest, and promoting its use at the international level, eventually resulted in more than 20 publications from the Saint Laurent Centre on microplate algal tests, and more than 60 papers on application of this new technique worldwide. In 1992 the algal microbiotest was formally endorsed by Environment Canada as the recommended procedure for growth inhibition tests using the freshwater alga *Selenastrum capricornutum*.[4] The topic of microplate-based toxicity tests with microalgae is covered in detail in the review by Blaise in Chapter 18 of this volume.

Since algal microplate tests were not developed as a replacement technology for the regulatory tests with microalgae presently required by international organizations such as the Organization for Economic Cooperation and Development (OECD), International Standards Organization (ISO), or the European Union (EU), these new tests do not follow the testing procedures described in the standard protocols. For example, the microplate algal technique only prescribes determination of the algal numbers by counting (with a hemocytometer or an electronic particle counter), or by measuring absorbance (in a microplate photometer) at the end of the 72-h incubation period and not daily. In fact, neither the enumeration technique nor the measurement of optical density would be applicable to the microtiter plate technology on a daily basis. The volume of algal suspension needed for daily countings is indeed out of proportion to the volume of the microcups; in addition these cups are too shallow to give sensitive and precise optical density readings with a conventional microplate photometer, at least during the first days of the incubation period. Consequently, despite their numerous advantages, microplate algal microbiotests cannot be used in place of the standard algal assays required by international organizations for regulatory testing.

At the University of Ghent in Belgium, the Laboratory for Biological Research in Aquatic Pollution (LABRAP) has tried for many years to develop low-cost and user-friendly microbiotests with invertebrates for routine screening purposes. These endeavors gradually led to a new concept for invertebrate microbiotests, called Toxkits.[5] Besides the advantages inherent in small-scale tests, Toxkits have the added advantage of being totally independent of the culturing and maintenance of live stocks of the test species. The kits contain the biological material in a dormant form; these "cryptobiotic" stages can be stored for long periods of time without losing viability and hatched "on demand." More detailed information on this new generation of microbiotests can be found in Chapter 30.

In view of the urgent need for a cost-effective microbiotest which would be acceptable as an alternative to standard algal assays for regulatory testing, LABRAP has initiated research on the development of a Toxkit with microalgae, with the following objectives in mind:

1. To develop a new algal toxicity test (analogous to the Toxkits with invertebrate species) which would be independent of culturing of live stocks of microalgae
2. To miniaturize the testing procedure into a low-cost microbiotest
3. To design a rapid procedure for determination of the algal biomass, with acceptable sensitivity and precision

III. DEVELOPMENT OF A "STOCK-CULTURE-FREE" ALGAL MICROBIOTEST

The first objective, namely freedom from stock culturing the selected test species, started with the experimental comparison of the major storage techniques described in scientific literature, e.g., cultivation on agar,[6,7] immobilization in natural or synthetic polymers,[8-11] or cryopreservation.[12] Eventually the immobilization technique using alginate beads was given preference. This technique, which is practiced currently for fundamental as well as for applied purposes, not only appeared to be the best suited but has even already been tried for toxicity testing.[13]

The methods for immobilization and de-immobilization of microalgae described in the scientific literature were gradually refined and optimized in LABRAP for application with the most currently used algal test species: *Selenastrum capricornutum* (now renamed *Raphidocelis subcapitata*). Extensive experimental research revealed that after several months of storage in an appropriate medium in the refrigerator (at 4°C) in darkness, microalgae de-immobilized from alginate beads and resumed growth immediately. Growth rates after 72 h were found to be as high as those from algae taken from live stock cultures in the exponential phase of growth.[14]

Concurrently with the research on immobilization, storage, and de-immobilization of *Selenastrum capricornutum*, the technological aspects of the new algal test were investigated in depth. In view of the constraints inherent in the use of multiwell plates as test containers, extensive experimentation has been carried out with 10 mL glass tubes as test vials. The tubes were filled to approximately two thirds volume with the algae/toxicant suspensions, closed with a rubber stopper, and placed horizontally under a battery of light tubes on a tilting device (to keep the algae in continuous suspension). The algal biomass was determined daily during the 3-d exposure period, by optical density measurement of the algal suspensions in a colorimeter at 670 nm.

This simple and rapid new testing procedure was quite attractive for several reasons. Next to bypassing stock culturing (by using microalgae de-immobilized from alginate beads), the new microbiotest also required little bench space and equipment, and the daily scoring of the optical density (OD) in the glass tubes took less than 10 min per test.

The algal microbiotest in tubes was first tried out on a number of pure compounds. This study revealed that the 72-h effect levels correlated rather well with those obtained with standard algal test in flasks.[14] However, when the new technique was applied to real world samples, the results from the two types of tests differed markedly. This is probably because the closed test tubes (placed horizontally on the tilting device) preclude all gas exchange with the ambient environment. In addition, the test tube microbiotest unfortunately also suffered from a second major drawback, namely the low precision of the 0- and 24-h OD measurements. The optical density of algal suspensions containing 1.10^4 cells/mL (the start concentrations imposed in most standard tests), measured in test tubes of 16-mm diameter, indeed appeared to be at the sensitivity threshold of the measuring equipment; consequently the first reliable OD figures could only be obtained after 48-h of algal growth.

Eventually the test tube technique was abandoned, and alternative test containers were searched for, which, like cotton-plugged Erlenmeyer flasks, would allow for sufficient gas exchange but also allow rapid and precise daily measurement of algal growth. This challenging search ultimately led to the selection of spectrophotometric cells with 10-cm pathlength. Like in the glass tube technique, the 10-cm cells could serve as test containers, yet also allowed OD measurement directly in the cells with the aid of a conventional colorimeter or spectrophotometer provided with a holder for the long cells.

IV. DEVELOPMENT OF THE ALGALTOXKIT MICROBIOTEST

A new algal microbiotest was subsequently worked out in LABRAP, based on long cells, and with the following characteristics: 1) the bioassay is performed in 18 disposable polystyrene long

Figure 20.1   10-cm cell (long cell) with hinging lid, as test container for the Algaltoxkit microbiotest.

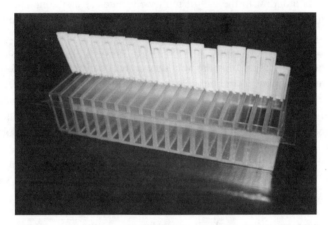

Figure 20.2   Eighteen long cells in the half-open position, in transparent holding tray, as test containers for one complete Algaltoxkit microbiotest.

cells; 2) each test comprises one control and 5 toxicant dilutions, each of which is prepared in triplicate; 3) all long cells are provided with a lid which hinges centrally (Figure 20.1). This special lid, on one hand, allows the cells to be closed for shaking (to homogenize the algal suspension prior to OD measurement), and on the other, to keep them half open during the incubation period, to allow for gas exchange. After inoculation with the algae/toxicant dilutions, the cells are placed in a transparent holding tray (Figure 20.2), and the open side is covered by an "anti-evaporation strip." Like the cotton plug in the flask test, this strip has the triple role of precluding contamination, minimizing evaporation, yet allowing for sufficient gas exchange with the ambient environment.

The new long cell technique was first tried to determine the variability of algal growth in the 18 cells contained in one holding tray. Each cell is inoculated with 25 mL algal suspension (1.10^4 cells/mL) obtained from alginate beads. The holding trays are placed in a temperature-controlled room at 24°C in front of a panel of light tubes (white type fluorescent lamps) providing 8.000 lux* at the surface of the holding tray.

Repeated experiments revealed excellent agreement of the algal growth in the individual cells, as shown by Figure 20.3; the variation coefficient of the ODs in the 18 cells after 72 h incubation

* Measured with a spherical collector.

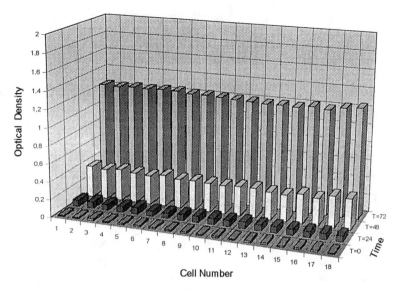

Figure 20.3   Growth of microalgae in the 18 long cells of one holding tray, after 24-h, 48-h, and 72-h incubation at 24°C and 8.000 lux illumination.

was as low as 4%! During the subsequent experiments with the long cell technique, it was also found that the growth rate of (de-immobilized) microalgae did not change significantly after several months of storage of the algal beads. This confirmed the preliminary findings reported above. The variation coefficient of the mean algal growth in the three control cells, for a total of 64 individual toxicity tests, was only 16%.

Eventually, the long cell technique was modeled into an Algaltoxkit, similar to the Toxkit tests used for invertebrate species. Each kit contains enough materials to perform two complete assays: 1) two holding trays with 18 long cells in each, 2) two tubes with algal beads in a specific storage medium, 3) two extra long cells (one for calibration of the measuring equipment, the second for the OD determination of the concentrated algal suspension), 4) a vial with matrix dissolving medium, and 5) five glass bottles with concentrated algal culturing medium. The algal growth medium has exactly the same composition as the medium recommended in OECD Guideline 201 for toxicity testing with microalgae[15] and is prepared by simply pouring the contents of the bottles with concentrated culturing medium into a calibrated flask and filling to the 1-liter mark with deionized or distilled water.

The detailed Standard Operational Procedure manual of the new algal microbiotest indicates that the preparation of the test starts with dissolving the alginate matrix of the beads. The algae are set free after approximately 15 min shaking and are first cleared from the dissolved matrix by centrifugation. After resuspension and rinsing in deionized water, they are centrifuged a second time and resuspended in the same tube in algal culturing medium. The exact number of algae is then determined by OD, with conversion of the OD into algal numbers with the aid of an (algal batch specific) OD/N regression sheet included in the kit. Subsequently, the concentrated algal suspension is diluted to exactly 1.10^6 cells/mL. One mL of this algal suspension will ultimately be added to the toxicant dilutions, prepared in 100 mL calibrated flasks in order to obtain the 1.10^4 cell/mL concentration prescribed in all standard algal tests to start growth tests.

After the transfer of the algae/toxicant dilutions into the long cells, the latter are placed in their holding tray in the half-open position and covered with the anti-evaporation strip. The trays are incubated for 3 d under standard temperature and light conditions as prescribed in national/international guidelines. Optical density is measured daily in each long cell, and the data are recorded on the sheets provided in the Toxkit.

Extensive application of the former operational procedure in LABRAP revealed that after 3 d incubation, the growth of the microalgae in the long cells exceeded the minimum multiplication factor (16 times) imposed by OECD and ISO guidelines. This growth was achieved even without continuous shaking of the long cells. In fact, experiments were performed specifically to determine the influence of intermittent (hand) shaking of the cells several times per day. These trials revealed that one shaking per day (at the time of OD measurement) was sufficient to reach the 16 times multiplication factor. Furthermore, the reduction of the number of shakings did not influence the outcome of the algal toxicity tests proper to any significant extent. These findings confirm earlier observations by Thellen et al.,[16] who, for the sake of test simplification, questioned the absolute requirement of "active" gas exchange by continuous shaking.

V. FIRST VALIDATION OF THE ALGALTOXKIT MICROBIOTEST

The Algaltoxkit happened to be finalized in LABRAP almost concurrently with two new invertebrate Toxkits (the Daphtoxkit *magna* and the Daphtoxkit *pulex*). As a result, comparison experiments on the sensitivity of all three new Toxkits versus that of conventional standard assays with the same test species could be performed concurrently and on the same chemicals and natural samples. For the data and comments on the validation study with the Daphtoxkit tests, we refer the reader to Chapter 30.

The Algaltoxkit first validation was based on the following 13 inorganic and organic chemicals: sodium chloride, potassium chloride, calcium chloride, mercuric chloride, cadmium chloride, zinc chloride, lead chloride, copper sulfate, nickel sulfate, magnesium sulfate, manganese sulfate, potassium dichromate, and Aminex, a commercial formulation of the pesticide 2,4,D, manufactured and distributed by the Belgian company Pratex. The new microbiotest was also applied to four industrial effluents from the textile industry. Each Algaltoxkit test was performed in parallel to the standard algal tests in Erlenmeyer flasks, following the prescriptions of OECD Guideline 201. For reasons of convenience, the standard assay in flasks was also started with microalgae de-immobilized from (the same batch) of algal beads. Consequently, the comparison of the algal microbiotest with the standard assay specifically reflects the influence of the materials used, such as the size and configuration of the test vessels (polystyrene cells versus glass Erlenmeyers) and the different test volumes (25 mL versus 100 mL), on the impact of the toxicants on the microalgae. In order to check the reliability and accuracy of OD as a good parameter for algal growth, measurements were made daily in all the long cells and all the Erlenmeyer flasks by OD readings in a colorimeter at 670 nm as well as by cell number counting with the aid of an electronic particle counter.

The 72-h EC_{50}s were calculated (as EbC_{50}s) according to the area under the curve method recommended by the OECD. The results are presented graphically and in order of decreasing toxicity in Figure 20.4. This figure, clearly shows the excellent concordance between the long cell test and the conventional flask test. For 5 of the 13 compounds, the difference between the 72-h EC_{50}s of the classical test and the new microbiotest was less than 10%, and for 4 chemicals it was less than 20%; for 3 compounds a difference between 20 and 30% was noted, and only for one compound was the difference higher than 30% (but less than 40%). The very good correlation between the effect data obtained with the two types of assays is even more obvious when looking at the linear regression in Figure 20.5, for which the r value is as high as 0.987 for the 13 data pairs.

The 72-h EC_{50} data for the four effluents of the textile industry are given in Table 20.1, in Toxic Units (1 TU = $1/EC_{50} \times 100$). This table presents the effect levels as obtained by OD measurement versus those determined by algal counts. As with the pure chemicals, the data of the Algaltoxkit correspond quite well with those obtained with the standard flask test. In addition, the table also shows the good correspondence between the effect levels obtained by OD readings versus cell counts. It may be underlined that for the OD readings in the long cells, the interference of the color of the effluent samples could be corrected by zero-calibrations of the photometer through a specific procedure outlined in the manual of the Algaltoxkit.

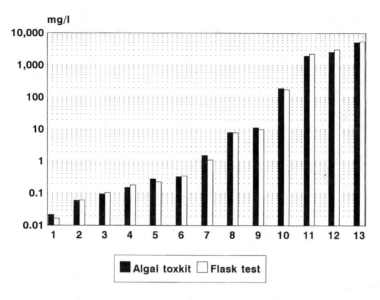

Figure 20.4   Comparison of 72-h EC$_{50}$s of the Algaltoxkit test and of the standard algal test in flasks for 13 pure compounds. The effect data (in mg/L) are ranked in order of decreasing toxicity. 1 = mercuric chloride; 2 = cadmium chloride; 3 = copper sulfate; 4 = zinc chloride; 5 = nickel sulfate; 6 = potassium dichromate; 7 = lead chloride; 8 = magnesium sulfate; 9 = manganese sulfate; 10 = Aminex; 11 = sodium chloride; 12 = potassium chloride; 13 = calcium chloride.

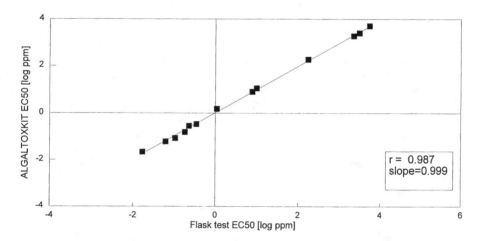

Figure 20.5   Linear regression for 13 data pairs of 72-h EC$_{50}$s of the Algaltoxkit test and the standard algal test in flasks.

VI. PRECISION OF THE ALGALTOXKIT MICROBIOTEST

The precision of the long cell algal assay technique has been determined with the reference toxicant potassium dichromate. Repeated assays have been performed over a period of five months and with microalgae from two different batches of algal beads. The average 72-h EC$_{50}$ for 10 toxicity tests was 0.33 mg/L, the range 0.29 to 0.38 mg/L, and the variation coefficient 8.1%, clearly showing the very high degree of standardization which has been achieved with this particular Toxkit. Although the mean value for the reference toxicant (0.33 mg/L) is slightly below the mean 72-h EC$_{50}$ (0.53 mg/L) reported in the ISO *Standard for Algal Growth Inhibition*,[17] for an intercalibration test

Table 20.1   72-h EC_{50}s for Four Effluents of the Textile Industry, as Determined with the Standard Algal Assay in Erlenmeyer Flasks, and with the Algaltoxkit Technology in Long Cells

Sample	Standard Algal Test in Flasks		Algaltoxkit in Long Cells	
	Optical Density Measurement	Algal Cell Counting	Optical Density Measurement	Algal Cell Counting
1	3.8	4.5	3.1	3.1
2	1.5	1.0	1.0	1.2
3	1.7	1.9	2.3	4.5
4	5	6.2	2.7	3.3

Note: Measurement of algal growth was made both by optical density readings and by algal cell countings. All data are expressed in Toxic Units.

performed among 16 laboratories, the average sensitivity of microalgae de-immobilized from alginate beads is situated in the 0.20 to 0.75 mg/L toxicity range of the former multi-lab calibration exercise.

VII. THE ALGALTOXKIT MICROBIOTEST: POTENTIAL AND LIMITATIONS

The major characteristics of the long cell algal microbiotest have been outlined above along with the description of the technological development of this new Toxkit assay. The advantages of the Algaltoxkit test can be briefly recapitulated as follows:

1. The new algal assay uses microalgae which are immobilized in alginate beads in which they survive for several months without losing their viability. The major practical consequence of storage of live test biota in an immobilized form is that performance of the test, like for all Toxkits, becomes independent of culturing/maintenance of the test species. An extra advantage of the Algaltoxkit test over the Toxkits with invertebrates is that the algal assays can be started immediately after de-immobilization of the algae, i.e., within a half hour of adding the matrix-dissolving medium to the tube with the algal beads.
2. The algal species selected for the Algaltoxkit is the unicellular chlorophyte *Selenastrum capricornutum* (renamed *Raphidocelis subcapitata*). This is the most commonly used species for toxicity testing with microalgae and is the recommended species in most national and international guidelines.
3. The new algal microbiotest makes use of disposable long cell vials which can be used for rapid and direct determination of algal growth by OD measurement in a colorimeter or spectrophotometer provided with a holder for 10-cm cells.
4. The long cells in their holding tray occupy only one third of the bench space needed for one standard test in flasks. Consequently the number of assays that can be performed concurrently can be tripled.
5. The first validation results indicate that the sensitivity of the Algaltoxkit test is the same as that of the standard assay performed in Erlenmeyer flasks and that the repeatability of this microbiotest is excellent.
6. The cost for performing Algaltoxkit tests is substantially lower than that of any conventional assay with microalgae. On the basis of the figure provided by Toussaint et al.[18] and taking into account all cost factors (equipment, materials, and labor), it was calculated that an Algaltoxkit test can be performed at a cost of only 25% of that of a standard algal assay.
7. Last but not least, test species, test medium, and testing protocol of the Algaltoxkit have been selected to follow the prescriptions of international guidelines such as OECD and ISO, making the Algaltoxkit an acceptable alternative for regulatory testing.

It may be mentioned at this stage that the Algaltoxkit (like the Daphtoxkits) is now available commercially, in a version containing the materials to perform two complete 72-h assays. In order to also keep the equipment costs for this new microbiotest as low as possible, the incubator for invertebrate Toxkit tests has been provided with a lateral illumination system. The (low-cost)

Algaltoxkit incubator allows the performance of two algal tests concurrently. Finally, thanks to the collaboration with an English company, a low-cost long cell colorimeter has been developed specifically for OD readings of the algal suspensions at 670 nm.

The attractive features of this appealing and cost-effective new algal microbiotest, should nevertheless be put in perspective with the limitations of the long cell testing procedure. For example, it is imperative that additional validation on pure compounds as well as on natural samples be performed to determine the importance of the limitations of the Algaltoxkit. There are indeed inherent differences between the conventional algal assay and the long cell technology such as the difference in testing volume (a factor of 1 to 4) and the much larger surface-to-volume ratio of the long cell test. This may lead to differences in adsorption and/or evaporation rates of particular toxicants. The importance of this factor should, however, not be overestimated, as demonstrated by Blanck et al.[19] for tests in polystyrene multiwells. Furthermore, continuous agitation of the test containers, which is highly recommended in most standard protocols to keep the algae in continuous suspension, is technically harder to put into practice with long cells than with Erlenmeyer flasks. Yet, as already discussed above, the necessity of shaking to achieve sufficient algal growth may be questioned. Finally the difference in the evaporation rate of the test medium in long cells versus in cotton-plugged flasks is one of the additional factors which still needs to be assessed quantitatively.

In conclusion, on the basis of all information available to date, and keeping the former questions in mind, it appears that the new Algaltoxkit test is a very attractive alternative to the (complex and time-consuming) standard algal assays in flasks. Intensive validation research is continuing in LABRAP and is also in progress in several laboratories in different countries to establish the full potential of this promising new Toxkit test.

REFERENCES

1. Nyholm, N. and Källqvist, T., Methods for growth inhibition toxicity tests with freshwater algae, *Environ. Toxicol. Chem.,* 8, 689, 1989.
2. Hörnström, E., Toxicity tests with algae — A discussion on the batch method, *Ecotox. and Environm. Safety,* 20, 343, 1990.
3. Blaise, C., Legault, R., and Birmingham, N., A simple microassay technique for measuring algal growth inhibition (EC$_{50}$s) in aquatic toxicity studies, in *Proc. 9th Ann. Aq. Tox. Workshop., Can. Techn. Rep. Fish. Aq. Sci.,* 1163, 1, 1982.
4. Environment Canada, Biological Test Method: Growth Inhibition Test using the freshwater alga *Selenastrum capricornutum*, Report EPS 1/RM/25, 42p., 1992.
5. Persoone, G., Cyst-based toxicity tests. I. A promising new tool for rapid and cost-effective toxicity screening of chemicals and effluents, *Zeitschr. f. Angew. Zool.,* 78, 2, 235, 1991.
6. Lukavsky, J., Controlled cultivation of algae on agar plates, *Arch. Hydrobiol. Suppl.* 46. Algalogical Studies 10, 90, 1974.
7. Lukavsky, J., The evaluation of algal growth potential by cultivation on solid media, *Water Res.,* 17, 5, 549, 1983.
8. Bucke, C., Immobilized cells, *Phil. Trans. R. Soc. Lond.,* B300, 369, 1983.
9. Vojtisek, V. and Jirku, V., Immobilized cells, *Folia Microbiol.,* 28, 309, 1983.
10. Dainty, A.L., Goulding, K.H., Robinson, P.K., Simpkins, I., and Trevan, M.D., Stability of alginate immobilized algal cells, *Biotechnol. and Bioengineering,* XXVIII, 210, 1986.
11. Robinson, P.K., Mak, A.L., and Trevan, M.D., Immobilized algae: a review, *Process Biochem.,* 122, 1986.
12. Sharp-McGrath, M., Daggett, P.M., and Dilivorth, S., Freeze drying of algae: Chlorophyta and Chrys-ophyta, *J. Phycol.,* 14, 521, 1978.
13. Bozeman J., Koopman, B., and Bitton, G., Toxicity testing using immobilized algae, *Aquat. Toxicol.,* 14, 345, 1989.
14. Amparado, R., Development and application of a cost-effective algal growth inhibition test with the green alga *Selenastrum capricornutum* (Printz), Ph.D. Thesis, University of Ghent, 215p. 1995.

15. OECD (Organization for Economic Cooperation and Development). Algal growth inhibition test. N° 21, OECD Guidelines for Testing of Chemicals, Paris, 1984.
16. Thellen, C., Blaise, C., Roy, Y., and Hickey, C., Round robin testing with the *Selenastrum capricornutum* microplate toxicity assay, *Hydrobiologia,* 188/189, 259, 1989.
17. ISO (International Standards Organization). Water quality — algal growth inhibition test, No. 8692, ISO, 1987.
18. Toussaint, M.W., Shedd, T.R., Van der Schalie, W.H., and Leather, G.R., A comparison of standard toxicity tests with rapid screening toxicity tests, *Environ. Toxicol. Chem.,* 14, 5, 907, 1995.
19. Blanck, H., Wallen, G., and Wangberg, S.A., Species dependent variation in algal sensitivity to chemical compounds, *Ecotox. and Environm. Safety,* 8, 339, 1984.

Protozoa

CHAPTER 21

Ciliated Protozoa as Test Organisms in Toxicity Assessments

Guy L. Gilron and Denis H. Lynn

CONTENTS

This chapter is dedicated to the late Dr. Jacques Berger of the University of Toronto, a renowned ciliate protozoologist who pioneered Canadian research on the effects of oil pollution on microbial systems.

I. INTRODUCTION

Ciliated protozoans (Kingdom Protista, Phylum Ciliophora) are ubiquitous unicellular eukaryotes: they inhabit soils, sediments, and marine and freshwater aquatic environments. Ciliates are important organisms in the transfer and transformation of nutrients in ecological food chains.[1,2] In the aquatic environment, ciliates are an integral part of the zooplankton community in that they feed predominantly on bacteria and small phytoplankton[3,4] and are responsible for the transfer of energy from the microbial food web to larger metazoan zooplankton.[1] In soil and sediment, ciliates feed primarily on bacteria and organic detritus.[5] The prevalence and conspicuousness of this group and their importance in trophic processes make them particularly appropriate as organisms used for the assessment of water quality.[6,7]

Recent advances in the assessment of ecotoxicity have focused on: 1) microscale testing; 2) more rapid bioassessment techniques; and 3) more sensitive indicator parameters of water quality (i.e., sublethal vs. lethal effects). The potential for using ciliates to evaluate water quality was recognized some years ago[8,9] by several investigators who recommended further work in this area. More recently, researchers have focused increasingly on ciliates as test and/or indicator organisms for the assessment of eutrophic and contaminated media, since they represent a neglected trophic level in most bioassay batteries and are sensitive to a broad range of toxicants in the natural environment.[10]

Three diverse technical approaches to utilizing ciliates in environmental health assessment are: 1) field population studies; 2) laboratory studies using "pure" chemicals of environmental concern; and 3) toxicity tests/bioassays. The first of these approaches is beyond the scope of this chapter but is discussed in some detail by Lynn and Gilron,[10] Cairns,[11] and Henebry and Cairns.[12] The second approach is also reviewed in Lynn and Gilron,[10] but has also been reviewed in other publications.[13] The third approach, namely, using ciliates as test organisms in toxicity assessments of contaminated media (i.e., bioassays), is the subject of this chapter.

As opposed to a comprehensive review of all environmental studies in which ciliates were used to evaluate toxic effects of chemicals in general, this chapter focuses on the literature dealing specifically with ciliate bioassay techniques, which have been to varying degrees, validated and calibrated, and which may have the potential to be used as standard or routine test methods. Several other reviews in the literature have provided a state-of-science perspective on this field.[14-16] This review will concentrate on literature subsequent to the Persoone and Dive[14] review of this field.

II. CILIATES AS TEST ORGANISMS IN BIOASSAYS

With the proliferation of new and innovative bioassay technologies for many different organismal groups, particularly with microscale tests, ciliates have also recently become test organisms in bioassays. There are several important reasons for this:

1. Ciliates are important as recyclers and remineralizers of organic material in terrestrial and aquatic ecosystems
2. Ciliates are relatively easy to culture in the laboratory
3. Ciliates have very rapid generation (population growth) rates

Ciliate bioassays have utilized a variety of test organisms; studies which have proposed these bioassays have also tested a broad range of chemical toxicants. Moreover, these studies have proposed a variety of different biological parameters as test endpoints.

Some methods have been developed as part of research aimed at determining the toxicity of a specific class of chemicals (e.g., the need for a surrogate test for mycotoxin poisons of food[17]), while others have been developed as standardized methods to be used for a broad range of toxicants and toxicant mixtures.[18-20] To date, none of these has attained the status of a standard regulatory test, although several have possibilities in this regard.

A. Ciliate Species Used in Bioassays

A relatively small number of ciliate species (representing only a half dozen genera) have been used as test organisms in the bioassay methods proposed to date. This can be attributed to the fact that relatively few species of ciliates are routinely cultured in protozoology research laboratories. Nonetheless, the species used for the majority of methods are common and ubiquitous, and their biology, ecology, and physiology are well understood. These species, therefore, are appropriate as test organisms in these bioassays. A review of the recent literature indicates that species from five genera have been used in most bioassay methods (see Table 21.1): *Tetrahymena* spp. (19 studies); *Colpidium* spp. (7 studies); *Paramecium* spp. (7 studies); *Colpoda* spp. (4 studies); and *Euplotes* spp. (3 studies). Although the use of a very limited number of genera and species is reasonable, future studies need to evaluate test methods by implementing a comparison of a method's utility among different unrelated species to understand the variability of specific bioassay methods. Only a few studies have addressed this issue.[21-23]

B. Toxicants Evaluated Using Ciliate Bioassays

Although ciliates have been used as bioassay test organisms relatively recently, the bioassays have been used to evaluate biological responses to a broad range of potential environmental contaminants. The majority of studies have evaluated the effects of single chemicals or prepared formulations of chemical mixtures (i.e., chemicals tested were reagent grade). A large proportion of the studies (greater than 50%) have evaluated metals (22 studies), organic chemicals (10 studies), and mycotoxins (7 studies). Other potential contaminants have also been evaluated, but fewer studies have been reported. These include: pesticides (3 studies), petroleum wastes (3 studies), fly ash (3 studies), and industrial chemicals (2 studies) (see Table 21.1 for references).

Five recent studies have investigated effects of contaminant mixtures such as effluents (e.g., sewage, industrial, mining) and contaminated soil/sediment leachates. These studies have developed tests which have the potential to be used as standard regulatory methods, since they have been shown to be effective for a variety of contaminated media.

C. Biological Parameters Used to Evaluate Ciliate Response

A number of biological parameters have been used as test endpoints in the evaluation of ciliate toxicity. These parameters reflect the diversity of biological responses which can be indicative of the effects of chemicals and toxicant mixtures on ciliated protozoa. The tests generally fall under the following categories of biological parameters: growth inhibition, survival, respiration, chemosensory behavior, mutagenicity, and ingestion rate. The evaluation presented below is organized according to the biological parameters used as endpoints in these test methods.

III. AN EVALUATION OF CURRENT CILIATE BIOASSAY METHODS

In order to implement a comparative analysis of test methods, a framework was developed and applied to categories of bioassay methods based on the biological parameter tested (Table 21.1). Our approach in this chapter is to evaluate test types with respect to critical aspects of toxicity test method implementation, namely: their cost, simplicity, speed, toxicant specificity, comparability with other bioassays, and any progress toward test method standardization. The framework used to conduct this evaluation is presented in Table 21.2.

A. Growth Inhibition

The largest number of bioassay methods developed to date have used growth inhibition as a test endpoint. These studies have taken advantage of the short generation time of ciliates and the usefulness of growth as a sensitive biological characteristic, the dynamics of which can reflect the effects of environmental stressors on ciliates.

Table 21.1   Summary of Bioassay Methods Using Ciliates as Test Organisms

Test Response	Test Species	Toxicants Tested	Reference
Growth Inhibition			
Growth inhibition	*Balanion sp.*	Cu	24
Growth inhibition	*Colpidium campylum*	Co, F, Ni, Se, As, Mo, Cr, Mn, Li, Hg, Zn, Cd, Cu, Pb, CN, Phenol, Ethanol.	25
Growth inhibition	*Colpidium campylum*	Effluents, leachates, elutriates, porewater.	19
Growth inhibition	*Colpidium campylum*	Some toxic substances found in drinking water.	26
Growth inhibition	*Colpidium campylum*	Potassium dichromate, cadmium sulfate, $NaBO_2$, $HgCl_2$, $ZnCl_2$, acrylnitril, aniline, cetrimide, T.P.B.S., phenol.	27
Growth inhibition	*Colpidium campylum*	Mycotoxins, *Penicillium roqueforti* toxin and metabolites, and fatty acids.	28
Growth inhibition	*Colpoda aspera*	Benzethonium chloride, benzalkonium chloride, didecyl dimethyl ammonium chloride, methyl dodecyl xykilene benzyl ammonium chloride+. (Invert soaps)	29
Growth inhibition	*Colpoda cucullus*	Cd, Cu, Pb, Zn.	30
Growth inhibition	*Colpoda inflata*	Cd, Cu, Zn.	31
Growth inhibition	*Colpoda inflata*	Mining effluent.	32
Growth inhibition	*Colpoda steinii*	Soil with sewage sludge and sewage sludge and Cd, Cu, Cr, Ni, Pb, Zn.	33
Growth inhibition	*Euplotes vannus*	Tributyltin and other metals, and petroleum hydrocarbons found in sediment, dibutyltin.	34
Growth inhibition	*Tetrahymena pyriformis*	Atrazine, bromacil, diuron, methyl parathion, lindane, 3,4-dichloroaniline, pentachlorophenol, cadmium, copper, and volatile 1,2-dichloropropane.	35
Growth inhibition	*Tetrahymena pyriformis*	Fungicides: Diphenyl, o-phenylphenol, 2-(4-thiazoyl) benzimidazole.	36
Growth inhibition	*Tetrahymena pyriformis*	Mycotoxins (Many)	17
Growth inhibition	*Tetrahymena thermophila*	Mycotoxins: Patulin, roquefortine, diacetoxyscirpenol, T-2 toxin.	37
Growth inhibition	*Uronema marinum*	Hg, Pb, Zn.	38
Growth inhibition	*Uronema parduczi*	Be, Cd, Ni, Hg, Pb, Ag, Cu, phenols, cresols, toluenes, benzenes, aldehydes.	39
Growth inhibition (and survival)	*Colpoda cucullus* *Blepharisma undulans*	Herbicides: Chlorex, MCPA, Dichlorprop, Matrigon. Fungicide: Benlate. Insecticide: Sumicidin.	40
Growth inhibition (and survival)	*Oxytricha fallax*	Pb.	41
Growth inhibition, (and survival, oxygen consumption, cell volume, weight, protein content).	*Tetrahymena pyriformis*	Cd.	42
Growth inhibition (and survival)	*Tetrahymena pyriformis*	Heavy metals: Hg, Zn, Pb, Co, Cd.	43
Growth inhibition, (and non-synchronous division, survival).	*Tetrahymena pyriformis* *Colpidium campylum*	Mycotoxins: Diacetoxyscirpenol, Tricothecin.	44
Growth inhibition, (and survival, oxygen consumption, locomotion).	*Tetrahymena pyriformis*	Shale oil retort water.	45
Survival			
Survival	*Aspidisca cicada* *Blepharisma americanum* *Colpidium campylum* *Euplotes affinis* *Euplotes patella* *Paramecium caudatum* *Uronema nigricans*	Cd, Cu, Hg, Zn.	22

Table 21.1 (continued) Summary of Bioassay Methods Using Ciliates as Test Organisms

Test Response	Test Species	Toxicants Tested	Reference
Survival	*Tetrahymena thermophila*	Cd, Cu.	46
Survival	*Vorticella convallaria*	Pb, Hg, Zn.	47
Survival, (and abnormal division, locomotion).	*Paramecium primaurelia*	Pesticide: Cartap.	48
Respiration			
Respiration (rate of oxygen uptake).	*Tetrahymena pyriformis*	Hg, Cd, Zn, Cu, Pb. Phenol, ammonium-N, Parathion, cyanide, pentachlorophenol.	18
Respiration (rate of oxygen uptake).	*Tetrahymena pyriformis*	Raw and treated wastewater effluent.	49
Respiration	*Tetrahymena pyriformis*	Phenol.	50
Respiration	*Tetrahymena pyriformis*	Shale oil retort water.	45
Chemotaxis			
Chemotaxis	*Tetrahymena pyriformis*	Cd, phenol, naphthalene.	51
Chemotaxis	*Tetrahymena pyriformis*	Phenol, naphthalene.	52
Chemotaxis	*Tetrahymena sp.* *Paranophrys sp.* *Miamiensis avidus*	Cd, Cu.	21
Chemotaxis (swimming rate - mm/sec.).	*Tetrahymena pyriformis*	Se, Va, Zr.	53
Chemotaxis	*Tetrahymena thermophila*	European Inventory of Existing Chemicals.	54
Chemotaxis	*Tetrahymena vorax*	NaCl, Cd, industrial effluents.	20
Chemotaxis	*Tetrahymena thermophila* *Colpidium campylum* *Glaucoma chattoni* *Tetrahymena corlissi* *Tetrahymena vorax*	NaCl, Cd, guaiacol, pulp and paper effluents.	23
Chemotaxis (and growth inhibition).	*Tetrahymena thermophila*	Industrial chemicals: aniline, 2,4-dinitroaniline, 4-nitroaniline, 3-hydroxyaniline, 4-hydroxyaniline.	55
Mutagenicity			
Mutagenicity (growth rate and survival of offspring).	*Paramecium tetraurelia*	Coal fly ash.	56
Mutagenicity (fraction of nonviable offspring) (and survival).	*Paramecium tetraurelia*	Nickel dust and iron-nickel dust.	57
Mutagenicity (abnormal cell replication and fraction of nonviable offspring).	*Paramecium tetraurelia*	Nickel dust: Nickel subsulfide, nickel monosulfide, 4-nitroquinoline-N-oxide.	58
Mutagenicity (survival when later exposed to UV light and DNA mutation affecting growth rate).	*Paramecium tetraurelia*	Suspected carcinogenic polycyclic aromatic hydrocarbons.	59
Mutagenicity (inhibition of synchronous division) (and survival).	*Tetrahymena pyriformis*	Mycotoxins: Tricothecenes, Patulin, Ochratoxin A, Zearalenone, aflatoxins B1, B2, G1, G2., Paracelsin.	60
Ingestion Rate			
Ingestion rate	*Euplotes mutabilis*	Cd, Cu, Hg, Ni.	61
Ingestion rate	*Paramecium aurelia*	Cd,Cu,Hg, phenols, naphthol, chlorpyrifos.	62
Miscellaneous			
Disruption of macroscopic bioconvective patterns.	*Tetrahymena pyriformis*	Cd.	63

TABLE 22.2 Framework for Comparative Analysis of Ciliate Bioassays

General Description	What does the test actually measure?
Cost	What costs are associated with: equipment? chemicals? technician time?
Simplicity	Does implementing the standard operating procedure (SOP) require special expertise? Is the procedure complex?
Speed	What is the duration of exposure to organisms? How long is analytical time?
Toxicant Specificity	Is the test specific to a particular group of toxicants or a broad range of toxicants?
Comparability with Other Bioassays	Have the data been compared to other lethal or sublethal bioassays? How do they compare?
Progress Toward Standardization	Have various test species been considered and tested with the method? Have different reference toxicants been tested? Is the test method appropriate as a universal method, or is it toxicant-specific (e.g., sensitive to organics but not to heavy metals)? Has an interlaboratory (round-robin) comparison been conducted? Has the test been compared to other standard or regulatory tests in order to evaluate its sensitivity?

General Description — These bioassays generally measure the population growth rate of the test species in response to a gradient of test concentrations. This approach is similar to that used for algal growth inhibition tests.[64,65] These tests evaluate sublethal effects, although several also utilize survival as an endpoint.[40] Some tests, in addition to growth inhibition, have included other sublethal endpoints, such as cell volume, protein content, and inhibition of synchronous division.[42,44,60]

Cost — In most cases, the cost of implementing this type of test is relatively low. However, some of the test methods require (or recommend) the use of electronic particle counters to estimate cell density (initial inocula and final counts). These counters can substantially increase the cost of the test unless an electronic particle counter is a standard piece of equipment in the laboratory. Manual counting using microscopy and counting chambers (i.e., hemocytometers, microwell plates, Palmer, or Sedgewick-Rafter cells) can be used to estimate cell density. However, costs associated with technician time need to be considered.

Simplicity — These types of tests are relatively easy to learn, implement, and interpret, which makes them quite attractive as standard tests. No complex techniques are required, and the results are simple to interpret.

Speed — Exposure duration for the tests can vary from 1 to 7 d. However, the majority of bioassays measuring growth rate take between 24 and 48 h. Analytical times can also vary from one to several hours, depending on whether particle counters are used.

Toxicant Specificity — Since a large number of studies using these methods has been reported, each of which has focused on a particular toxicant group, there is a large information base on a broad range of toxicants (e.g., heavy metals, pesticides, mycotoxins) and contaminant mixtures (e.g., sewage sludge, industrial effluents, and sediment leachates).

Comparability with Other Bioassays — With the exception of pesticides, this approach can be as sensitive, or more sensitive than standard *Daphnia* and fish bioassays.

Progress Toward Standardization — Over many years, Dive and colleagues[19,25-27] have developed a short-term bioassay technique based on a change in the population growth of *Colpidium campylum* over a 24-h period. It is relatively simple to implement and has a broad range of application for single toxicants and contaminant mixtures, such as effluents. The technique has been proposed as a standard test, and recent intercalibration data and technical review[66] demonstrate that it can be easily standardized.

Examples of test methods that have used growth inhibition as a test endpoint for aquatic assessments are described in Dive et al.,[19] Forge et al.,[33] and Janssen et al.[30] Pratt et al.[31] have recently proposed a standard test method for the evaluation of solid-phase media (i.e., soils and sediments) using the soil ciliate, *Colpoda inflata*. This method has also recently been applied to mining effluents.[32]

B. Survival

Survival, a common endpoint in many standard bioassays, has also been used as a parameter in ciliate bioassays (Table 21.1). In the majority of methods where survival has been used, it is often evaluated with other sublethal effects (e.g., locomotion, morphological abnormalities).

General Description — These tests specifically evaluate cell death, defined by the cessation of swimming activity. The approach to these tests is similar to that used in other lethality tests, where the reported endpoint is an LC_{10} or LC_{50}.[40]

Cost — These tests utilize standard lab equipment, such as standard glassware, fixatives, and microscopes; therefore, costs to implement the test are quite low. Those tests which also monitor other endpoints can have higher costs.

Simplicity — These tests are also relatively easy to learn, implement, and interpret, because no complex techniques are required, and the results are easy to interpret.

Speed — Relative to other approaches, these tests are not usually rapid: exposure times vary between 20 and 96 h and analysis at test completion can take between 3 and 5 h, depending on technician familiarity.

Toxicant Specificity — Tests using survival as the endpoint have been applied to heavy metals and pesticides. To date, no test specifically focusing on survival as the main endpoint has been implemented using chemical mixtures (i.e., effluents or leachates).

Comparability with Other Bioassays — These test methods with ciliates have been shown to be more sensitive (except for lead) compared with fish bioassay data. It has been reported that water hardness is critical in these evaluations.[43] Compared with *Daphnia magna* acute lethality results, ciliates are generally *less* sensitive. This observation, however, depends on the species of ciliate used in the test.[22]

Progress Toward Standardization — A recent method evaluating the effects of several pesticides on ciliates has proposed a new acute lethality methodology described in Schreiber and Brink.[40] Several standardization aspects were considered. An evaluation of different species (2 ciliates and 1 flagellate) was implemented, and two different methods of endpoint determination were also evaluated. The current test method appears to be specific to pesticides; however, only comparisons to literature values for other tests were undertaken. No reference toxicants were proposed, and an interlaboratory comparison has not yet been conducted.

Examples of test methods that have used survival as a test endpoint are described in Carter and Cameron,[43] Madoni et al.,[22] and Schreiber and Brink.[40]

C. Respiration

Cell respiration is a fundamental biological parameter, crucial to ciliate growth and viability. Nonetheless, relatively few studies have used cell respiration (or oxygen consumption/uptake rate) as a parameter to evaluate toxicant effects in ciliate bioassays.[18,45,49,50]

General Description — These tests are very rapid monitoring tests designed to measure changes in oxygen uptake rate or oxygen consumption as a function of toxicant concentration.

Cost — Due to the use of some elaborate laboratory equipment, such as temperature-controlled water baths, oxygen probes, and electronic recorders,[49] the infrastructural costs associated with equipment can be high, in addition to the incremental cost of training and technician time.

Simplicity — Although a relatively complex apparatus is used in the implementation of this test, the method is simple, once familiarity with the equipment is established.

Speed — The greatest advantage of this test is that the exposure duration is very short. One method[18] requires only a 10-min exposure. Analytical time, however, can still take as long as several hours.

Toxicant Specificity — A variety of chemical toxicants, primarily metals and organics, have been tested using the method.[18] The method has also been tested with raw and treated wastewater effluent[49] and synthetic fuel products, such as phenol[50] and shale oil retort water.[45] Based on a comparative evaluation, the method has been recommended for monitoring industrial effluents.[49]

Comparability with Other Bioassays — The method was determined to be at least as sensitive as the standard acute fish bioassay.

Progress Toward Standardization — In the two studies which used this approach by Slabbert and colleagues,[18,49] the method presented serves as a good standard operating procedure. However, crucial aspects of a standardization evaluation, such as a species comparison, an interlaboratory comparison, and comparisons to a battery of other sublethal tests to evaluate sensitivity, have not yet been conducted.

D. Chemosensory Behavior

Researchers have begun to evaluate new sublethal parameters that can indicate early (i.e., sublethal) toxicant effects on ciliates. The movement of ciliates toward or away from chemicals (i.e., chemosensory behavior) is a well-studied physiological response in ciliates,[67,68] which has been successfully applied in recent toxicity studies.

General Description — These types of tests measure chemosensory behavior or the inhibition of chemosensory behavior as the biological endpoint.

Cost — The test method is generally inexpensive to conduct. As discussed above (see *Cost* description in *Growth Inhibition* section), the cost of implementing this type of test will vary depending on whether electronic particle counting or manual counting is used for cell enumeration.

Simplicity — The test procedures are relatively simple to follow, and the data generated are simple to interpret.

Speed — The time required for conducting the test is one of its most attractive features with regard to method standardization. The exposure duration for these tests is quite short, generally between 15 and 30 min; analysis time may take between 1 and 5 h.

Toxicant Specificity — Several single toxicants have been tested using this approach, including heavy metals, phenol, and naphthalene.[21,51,52] In addition, some recent studies have investigated the use of this approach with toxicant mixtures, in particular, industrial effluents,[20,69] and specifically, pulp and paper mill effluents.[23]

Comparability with Other Bioassays — These tests have demonstrated comparability to fish and daphnid bioassays, especially for heavy metals, and have correlated quite well.[20,21] A recent study, which evaluated a chemotactic method using pulp and paper effluents, reported that this method is less sensitive than other standard sublethal tests, including fathead minnows, daphnids, rotifers, and algae.[23]

Progress Toward Standardization — Berk and colleagues[21,51,52] have developed a chemotaxis inhibition bioassay which, due to its practicability and speed, has promise for method standardization. It is relatively simple to implement and has a broad range of application for single toxicants and contaminant mixtures such as effluents. A modification of this technique has been proposed as a standard test in Canada,[20] and recent species comparison, technical refinement, interlaboratory calibration data, and technical review[23] demonstrate that it can be standardized with additional development.

Examples of this type of test include the pioneering work in this area by Berk and colleagues[21,51] (see Chapter 22). Their test method measured inhibition of chemotaxis. Gilron et al.[20,23] have modified this chemotactic model and applied it with a T-maze apparatus[68] and developed a toxicity endpoint (index) based on the direct measurement of chemotaxis as opposed to chemotactic inhibition.

E. Mutagenicity

Toxicants can have mutagenic effects. Mutations in ciliates caused by toxicants have been brought to expression through induction of autogamy.

General Description — Ciliates in the same cell cycle stage are exposed briefly to a toxicant and then established as clones. Following clonal growth, autogamy or self-fertilization is induced by starvation to bring any potential mutations to full expression. Exautogamous cells are randomly established as clones and are scored as *viable, detrimental,* or *lethal.* Morphological variations are also recorded.

Cost — Standard laboratory equipment, such as dissecting microscopes and chemicals, is required. Costs associated with training and technician time may be relatively high.

Simplicity — The test method is quite specialized in that it requires a familiarity with ciliate genetic techniques.

Speed — The bioassay takes at least several days, while the technical time is several hours.

Toxicant Specificity — The method has been evaluated using only a few specific toxicants, namely, components of nickel dust and coal fly ash.

Comparability with Other Bioassays — The test method proposed by Smith-Sonneborn and colleagues[56,57,59] showed good correlation with the Ames mutagenicity test.

Progress Toward Standardization — No specific information on standardization of the test method has been provided. The specialized knowledge of ciliate genetic techniques required for this bioassay makes its consideration as a routine, standardized test unlikely.

F. Ingestion Rate

The rate at which ciliates metabolize nutrients from their environment is correlated highly with cell growth and reproduction.[1] Two very recent studies have reported the effect of toxicants on ingestion rates of ciliates using fluorescent latex microspheres.[61,62]

General Description — In these methods, ciliates are exposed to toxicants and are provided with fluorescent latex microspheres for ingestion. Ingestion rate is measured either by quantifying fluorescence intensity or by calculating the number of particles ingested.

Cost — The requirements of a fluorescence microscope and fluorescent latex microspheres substantially elevate the cost of this bioassay; moreover, technician time may also be high.

Simplicity — The optimization of microsphere concentrations and microsphere size requires some specialized technical expertise.

Speed — Exposure duration is very short, between 10 and 30 min, although manipulation of images and analysis can take up to several hours.

Toxicant Specificity — To date, ingestion rate methods have been evaluated with heavy metals and organics.

Comparability with Other Bioassays — A recent study[62] concluded that the NOEC of ingestion rates in *Paramecium aurelia* compared favorably with the LC_{50} for reproduction tests using *Paramecium*, the rotifer, *Brachionus plicatilis*, and the cladoceran, *Ceriodaphnia dubia*.

Progress Toward Standardization — No specific information on standardization of these types of test methods has yet been provided.

IV. STUDIES USING CILIATES TO EVALUATE ECOLOGICAL EFFECTS

Other studies, which have not directly aimed at the use of ciliates in single-species tests, are also worthy of note, since they have demonstrated the impact of contaminants on the trophic and behavioral interactions of ciliates with their environment.

Berk et al.[70] investigated changes in mercury toxicity to *Uronema nigricans*, a marine ciliate, upon ingestion of mercury-laden *Pseudomonas* bacteria. Ciliates fed the mercury-laden bacteria were less sensitive to mercury poisoning than those fed the control bacteria. Berk and Colwell[71] investigated the transfer and accumulation of mercury in a three-component estuarine food web comprising bacteria (*Vibrio* spp., *Pseudomonas* spp.), ciliates (*Uronema nigricans*), and copepods (*Eurytemora affinis*). After establishing significant feeding of predator on prey, mercury transfer from one trophic level to another was examined. One of the most significant findings of the study was that the mercury was found to bioaccumulate most significantly in the ciliates. Lawrence et al.,[72] using a two-stage, nitrogen-limited chemostat, investigated the effects of cadmium on a microbial food chain, the prey, *Chlamydomonas reinhardtii*, and the predator, *Tetrahymena vorax*. An ecological approach to testing the behavioral effects of toxicants on a well-studied predator–prey system was also reported in Doucet and Maly.[73]

An increasing number of these types of studies have surfaced recently, in light of new approaches to evaluating ecologically relevant toxicity, rather than species-specific biological responses. Although microcosm studies[74] are still relatively expensive to conduct on a routine basis, microcosm studies yield valuable information on toxicant-induced changes to microbial processes in which ciliates play an important trophic role.

V. FUTURE DIRECTIONS

Ciliates have demonstrated their utility and relevance as test organisms in toxicity assessments, particularly in the last 10 to 15 years. Based on the momentum achieved in this field, and given the strong information base that has been established in the literature, the most likely future direction in this area will involve the development, validation, and establishment of standardized test methods to be used in comprehensive toxicity test batteries (also see Chapter 23, this volume). The inclusion of ciliates as test organisms in such batteries will serve a number of useful purposes, including the addition of a traditionally untested, but ecologically important, trophic group; test results that can be obtained relatively rapidly and economically; and new and varied sublethal biological endpoints.

In a recent review of whole-organism bioassays in Canada, Environment Canada[75] recommended the development of new tests using organisms that have not traditionally been used in test batteries for evaluating effluents and wastewater effects in surface waters. Two proposed test methods using ciliates were put forward as possible methods for further development with an aim toward standardization: the growth inhibition test method, developed and proposed by Dive et al.,[19] and the chemotactic response test method, developed and proposed by Gilron et al.[20] Furthermore, the American Public Health Association (APHA), in its upcoming 20th edition, is currently reviewing and considering the publication of several new test methods using ciliated protozoans as test organisms for different environmental media (i.e., freshwater, soils).

ACKNOWLEDGMENTS

We would like to acknowledge Ms. Joanne Charbonneau (Price Waterhouse) and Mr. Tim Fletcher (Ontario Ministry of the Environment) for their assistance with literature search and review aspects of this chapter. Funding was partially provided by the Research Advisory Committee of the Ontario Ministry of the Environment (Contract awarded to BOREALIS Environmental Consulting Inc. [1990]) and the Environmental Innovation Program (Environment Canada) (Contract awarded to Microvita Consulting [1993]).

REFERENCES

1. Fenchel, T., *Ecology of Protozoa, The Biology of Free-Living Phagotrophic Protists,* Springer-Verlag, Berlin, 1987.
2. Porter, K.G., Sherr, E. B., Sherr, B. F., Pace, M., and Sanders, R. W., Protozoa in planktonic food webs, *Journal of Protozoology,* 32, 409, 1985.
3. Pace, M. L. and Orcutt, J. D. Jr., The relative importance of protozoans, rotifers, and crustaceans in a freshwater zooplankton community, *Limnology and Oceanography,* 26, 822, 1981.
4. Sherr, E.B. and Sherr, B.F., High rates of consumption of bacteria by pelagic ciliates, *Nature,* 325, 710, 1987.
5. Clarholm, M., Interactions of bacteria, protozoa and plants leading to mineralization of soil nitrogen, *Soil Biology and Biochemistry,* 17, 181, 1985.
6. Bick, H., Autokologische und saprobiologische Untersuchungen an Susswasserciliaten, *Hydrobiologie,* 31, 17, 1968.
7. Sladecek, V., System of water quality from the biological point of view, *Arch. Hydrobiol. Beih. Ergebn. Limnol.,* 7, 1, 1973.
8. Antipa, G.A., Use of commensal protozoa as biological indicators of water quality and pollution, *Transactions of the American Microscopical Society,* 96(4), 482, 1977.
9. Cairns, John, Jr., Protozoans (Protozoa), in *Pollution Ecology of Freshwater Invertebrates,* C.W. Hart, Jr., and S.L.H. Fuller, Eds., Academic Press, New York, 1974, 1-28.
10. Lynn, D.H. and Gilron, G.L., A brief review of approaches using ciliated protists to assess aquatic ecosystem health, *Journal of Aquatic Ecosystem Health,* 1, 263, 1992.
11. Cairns, J., Jr. Zooperiphyton (especially Protozoa) as indicators of water quality, *Transactions of the American Microscopical Society,* 97(1), 44, 1978.
12. Henebry, M. S. and Cairns, J. Jr., Monitoring of stream pollution using protozoan communities on artificial substrates, *Transactions of the American Microscopical Society,* 99(2), 151, 1980.

13. Nilsson, J., *Tetrahymena* in cytotoxicology: with special reference to effects of heavy metals and selected drugs, *European Journal of Protistology*, 25, 2, 1989.

14. Persoone, G. and Dive, D., Toxicity tests on ciliates — a short review, *Ecotoxicology and Environmental Safety*, 2, 105, 1978.

15. Curds, C.R., Pelagic protists and pollution: a review of the past decade, *Annales de L'institut Oceanographique*, 58, 117, 1982.

16. Parker, J. G., Ciliated protozoa in marine pollution studies, *Ecotoxicology and Environmental Safety*, 7, 172, 1983.

17. Nishie, K., Cutler, H. G., and Cole, R. J., Toxicity of trichothecenes, moniliformin, zearalenone/ol, griseofulvin, patulin, PR toxin, and rubratoxin B on protozoan *Tetrahymena pyriformis*, *Research Communications in Chemical Pathology and Pharmacology*, 65(2), 197, 1989.

18. Slabbert, J. L. and Morgan, W. S. G., A bioassay technique using *Tetrahymena pyriformis* for the rapid assessment of toxicants in water, *Water Research*, 16, 517, 1982.

19. Dive, D., Blaise, C., and Le Du, A., Standard protocol proposal for undertaking the *Colpidium campylum* ciliate protozoan growth inhibition test, *Angewandte Zoologie*, 1, 79, 1991.

20. Gilron, G. L., Lynn, D. H., Hattie, S., and Holtze, K. E., Development and validation of a new, rapid and economical surrogate bioassay for industrial contaminants, Final Report prepared for the Ontario Ministry of the Environment, 46 pp., 1991.

21. Berk, S.G., Gunderson, J.H., and Derk, L.A., Effects of cadmium and copper on chemotaxis of marine and freshwater ciliates, *Bulletin of Environmental Contamination and Toxicology*, 34, 897, 1985.

22. Madoni, P., Esteban, G., and Gorbi, G., Acute toxicity of cadmium, copper, mercury, and zinc to ciliates from activated sludge plants, *Bulletin of Environmental Contamination and Toxicology*, 49, 900, 1992.

23. Gilron, G. L., Lynn, D. H., and Broadfoot, J., Further development of a sublethal bioassay for pulp and paper mill effluents using ciliated protozoans, Final Report prepared for Environment Canada, 1996, Environment Canada, Ottawa, ON.

24. Stoecker, D. K., Sunda, W. G., and Davis, L. H., Effects of copper and zinc on two planktonic ciliates, *Marine Biology*, 92, 21, 1986.

25. Dive, D. and Leclerc, H., Utilisation du protozoaire *Colpidium campylum* pour la mesure de la toxicité et de l'accumulation des micropollutants: analyse critique et applications, *Environmental Pollution*, 14, 169, 1977.

26. Dive, D. and Leclerc, H., Standardized test method using protozoa for measuring water pollutant toxicity, *Progress in Water Technology*, 7(2), 67, 1975.

27. Dive, D., Robert, S., Angrand, E., Bel, C., Bonnemain, H., Brun, L., Demarque, Y., Le Du, A., El Bouhouti, R., Fourmaux, M.N., Guery, L., Hanssens, O., and Murat, M., A bioassay using the measurement of the growth inhibition of a ciliate protozoan: *Colpidium campylum* Stokes, *Hydrobiologia*, 188/189, 181, 1989.

28. Dive, D., Moreau, S., and Cacan, M., Use of a ciliate protozoan for fungal toxin studies, *Bulletin of Environmental Contamination and Toxicology*, 489, 1978.

29. Kakiichi, N., Saito, M., Yayamoto, T., Kamata, S., Ito, O., Hayashi, M., Komine, K., Otsuka, H., and Uchida, K., Toxicity test for invert soap against ciliate *Colpoda aspera*, *Animal Science Technology (Japan)*, 62(12), 1113, 1991.

30. Janssen, M. P. M., Oosterhoff, C., Heijmans, G. J. S. M., and Van der Voet, H., The toxicity of metal salts and the population growth of the ciliate *Colpoda cucculus*, *Bulletin of Environmental Contamination and Toxicology*, 54, 597, 1995.

31. Pratt, J.R., Mochan, D., and Xu, Z., Rapid toxicity estimation using soil ciliates: sensitivity and bioavailability, *Bull. Environ. Contamin. Toxicol.*, 58, 387, 1997.

32. Bowers, N., Pratt, J.R., Beeson, D., and Lewis, M., Comparative evaluation of soil toxicity using lettuce seeds and soil ciliates, *Environ. Toxicol. Chem.*, 16, 207, 1997.

33. Forge, T. A., Berrow, M. L., Darbyshire, J. F., and Warren, A., Protozoan bioassays of soil amended with sewage sludge and heavy metals, using the common soil ciliate *Colpoda steinii*, *Biology and Fertility of Soils*, 16, 282, 1993.

34. Stebbing, A. R. D., Soria, S., Burt, G. R., and Cleary, J. J., Water quality bioassays in two Bermudan harbours using the ciliate *Euplotes vannus*, in relation to tributyltin distribution, *Journal of Experimental Marine Biology Ecology*, 138, 159, 1990.

35. Schäfer, H., Hettler, H., Fritsche, U., Pitzen, G., Röderer, G., and Wenzel, A., Biotests using unicellular algae and ciliates for predicting long-term effects of toxicants, *Ecotoxicology and Environmental Safety,* 27, 64, 1994.

36. Otsuka, K., Yoshikawa, H., Sugitani, A., and Kawa, M., Effect of Diphenyl, o-Phenylphenol and 2-(4-Thiazoyl)benzimidazole on growth of *Tetrahymena pyriformis, Bulletin of Environmental Contamination and Toxicology,* 41, 282, 1988.

37. Benitez, L., Martin-Gonzalez, A., Gilardi, P., Soto, T., Rodriguez de Lecea, J., and Gutierrez, J. C., The ciliated protozoan *Tetrahymena thermophila* as a biosensor to detect mycotoxins, *Letters in Applied Microbiology,* 19, 489, 1994.

38. Parker, J. G., Toxic effects of heavy metals upon cultures of *Uronema marinum* (Ciliophora: Uronematidae), *Marine Biology,* 54, 17, 1979.

39. Bringmann, V. and Kühn, R., Bestimmung der biologischen schedwirkung wassergefährdender stoffe gegen protozoen II. Bakterienfressende ciliaten, *Z. Wasser Abwasser Forsch.,* 1, 26, 1980.

40. Schreiber, B. and Brink, N., Pesticide toxicity using protozoans as test organisms, *Biology and Fertility of Soils,* 7, 289, 1989.

41. Nasreen, A. and Khan, M. A., Responses of the ciliate *Oxytricha fallax* Stein to lead acetate, *Indian Journal of Experimental Biology,* 17(9), 982, 1979.

42. Houba, C., Remacle, J., and De Parmentier, F., Influence of cadmium on *Tetrahymena pyriformis* in axenic culture, *European Journal of Applied Microbiology and Biotechnology,* 11, 179, 1981.

43. Carter, J.W. and Cameron, I.L., Toxicity bioassay of heavy metals in water using *Tetrahymena pyriformis, Water Research,* 7, 951, 1973.

44. Bijl, J., Dive, D., and Van Peteghem, C., Comparison of some bioassay methods for mycotoxin studies, *Environmental Pollution Series A,* 26, 173, 1981.

45. Schultz, T. W., Dumont, J. N., and Kyte, L. M., Cytotoxicity of synthetic fuel products on *Tetrahymena pyriformis.* II. Shale Oil Retort Water, *Journal of Protozoology,* 25(4), 502, 1978.

46. Schlenk, D. and Moore, C. T., Effect of pH and time on the acute toxicity of copper sulfate to the ciliate protozoan *Tetrahymena thermophila, Bulletin of Environmental Contamination and Toxicology,* 53, 800, 1994.

47. Sartory, D. P. and Lloyd, B. J., The toxic effects of selected heavy metals on unadapted populations of *Vorticella convallaria* var *similis, Water Research,* 10, 1123, 1976.

48. Komala, Z., Paramecium bioassay test in studies on cartap, *Bulletin of Environmental Contamination and Toxicology,* 28, 660, 1982.

49. Slabbert, J. L., Smith, R., and Morgan, W. S. G., Application of a *Tetrahymena pyriformis* bioassay system for the rapid detection of toxic substances in wastewaters, *Water SA,* 9(3), 81, 1983.

50. Schultz, T. W. and Dumont, J. N., Cytotoxicity of synthetic fuel products on *Tetrahymena pyriformis.* I. Phenol, *Journal of Protozoology,* 24(1): 164, 1977.

51. Roberts, R. O. and Berk, S. G., Development of a protozoan chemoattraction bioassay for evaluating toxicity of aquatic pollutants, *Toxicity Assessment,* 5, 279, 1990.

52. Berk, S.G., Mills, B.A., Stewart, K.C., Ting, R.S., and Roberts, R.O., Reversal of phenol and naphthalene effects on ciliate chemoattraction, *Bulletin of Environmental Contamination and Toxicology,* 44, 181, 1990.

53. Bovee, E. C. and O'Brien, T. L., Some effects of selenium, vanadium and zirconium on the swimming rate of *Tetrahymena pyriformis*: A bioassay study, *Univ. of Kansas Science Bulletin,* 52(4), 39, 1982.

54. Pauli, W., Berger, S., Schmitz, S., and Jaskulka, L., Chemosensory responses of ciliates: A sensitive end point in xenobiotic hazard assessment, *Environmental Toxicology and Water Quality,* 9, 341, 1994.

55. Pauli, W. and Berger, S., Chemosensory and electrophysiological responses in toxicity assessment: Investigations with a ciliated protozoan, *Bulletin of Environmental Contamination and Toxicology,* 49, 892, 1992.

56. Smith-Sonneborn, J., Fisher, G. L., Palizzi, R. A., and Herr, C., Mutagenicity of coal fly ash: A new bioassay for mutagenic potential in a particle feeding ciliate, *Environmental Mutagenesis,* 3, 239, 1981.

57. Smith-Sonneborn, J., Leibovitz, B., Donathan, R., and Fisher, G. L., Bioassay of environmental nickel dusts in a particle feeding ciliate, *Environmental Mutagenesis,* 8, 621, 1986.

58. Smith-Sonneborn, J., Use of a ciliated protozoan as a model system to detect toxic and carcinogenic agents, in *In Vitro Toxicity Testing of Environmental Agents: Current and Future Possibilities,* Kolber, A. R., Wong, T. K., Grant, L. D., DeWoskin, R. S., and Hughes, T. J., Eds., Plenum Press, New York, 1983, 113.

59. Smith-Sonneborn, J., Palizzi, R. A., McCann, E. A., and Fisher, G. A., Bioassay of genotoxic effects of environmental particles in a feeding ciliate, *Environmental Health Perspectives*, 51, 205, 1983.

60. Bijl, J. P., Rousseau, D. M., Dive, D. G., and Van Peteghem, C. H., Potentials of a synchronized culture of *Tetrahymena pyriformis* for toxicity studies of mycotoxins, *J. Assoc. Off. Anal. Chem.*, 71 (21), 282, 1988.

61. Al-Rasheid, K. A. L. and Sleigh, M. A., The effects of heavy metals on the feeding rate of *Euplotes mutabilis* (Tuffrau, 1960), *European Journal of Protistology*, 30, 270, 1994.

62. Juchelka, C. M. and Snell, T. W., Rapid toxicity assessment using ingestion rate of cladocerans and ciliates, *Archives of Environmental Contamination and Toxicology*, 28, 508, 1995.

63. Noever, D. A. and Matsos, H. C., A bioassay for monitoring cadmium based on bioconvective patterns, *Journal of Environmental Science and Health*, A26(2), 273, 1991.

64. Greene, J.C., Bartels, C.L., Warren-Hicks, W.J., Parkhurst, B.R., Linder, G.L., Peterson, S.A., and Miller, W.E., Protocol for short term toxicity screening of hazardous waste sites, EPA 3-88-029, 102 pp, 1989.

65. Environment Canada, Biological test method: growth inhibition test using the freshwater alga, *Selenastrum capricornutum*. Conservation and Protection, Ottawa, ON. EPS Report 1/RM/25, 42 pp., 1992.

66. Dive, D., Blaise, C., Robert, S., Le Du, A., Bermingham, N., Cardin, R., Kwan, A., Legault, R., Mac Carthy, L., Moul, D., and Veilleux, L., Canadian workshop on the *Colpidium campylum* ciliate protozoan growth inhibition test, *Angewandte Zoologie*, 1, 49, 1990.

67. Hellung-Larsen, P., Leick, V., Tommerup, N., and Kronborg, D., Chemotaxis in *Tetrahymena*, *European Journal of Protistology*, 25, 229, 1990.

68. Van Houten, J., Martel, E., and Kasch, T., Kinetic analysis of chemokinesis of *Paramecium*, *Journal of Protozoology*, 29, 226, 1982.

69. Roberts, R. O. and Berk, S. G., Effect of copper, herbicides and a mixed effluent on chemoattraction inhibition of *Tetrahymena pyriformis*, *Environmental Toxicology and Water Quality*, 8, 73, 1993.

70. Berk, S.G., Mills, A.L., Hendricks, D.L., and Colwell, R.R., Effects of ingesting mercury-containing bacteria on mercury tolerance and growth rates of ciliates, *Microbial Ecology*, 4, 319, 1978.

71. Berk, S.G. and Colwell, R.R., Transfer of mercury through a marine microbial food web, *Journal of Experimental Marine Biology and Ecology*, 52, 157, 1981.

72. Lawrence, S. G., Holoka, M. H., and Hamilton, R. D., Effects of cadmium on a microbial food chain, *Chlamydomonas reinhardtii* and *Tetrahymena vorax*, *The Science of the Total Environment*, 381, 1987/1988.

73. Doucet, C. M. and Maly, E. J., Effect of copper on the interaction between the predator *Didinium nasutum* and its prey *Paramecium caudatum*, *Canadian Journal of Fisheries and Aquatic Sciences*, 47, 1122, 1990.

74. Taub, F., Standard aquatic microcosms, *Environmental Science and Technology*, 23, 1064, 1989.

75. Keddy, C., Greene, J. C., and Bonnell, M. A., A review of whole organism bioassays for assessing the quality of soil, freshwater sediment and freshwater in Canada, Ecosystem Conservation Directorate, Evaluation and Interpretation Branch, Scientific Series No. 198, 1994.

CHAPTER 22

Development of a Protozoan Chemoattraction Inhibition Assay for Evaluating Toxicity of Aquatic Pollutants

Sharon G. Berk and Robert O. Roberts

CONTENTS

I. BACKGROUND AND TERMINOLOGY

    Protozoa are attractive organisms for toxicological studies, because they have short life cycles, are in intimate contact with the aquatic environment, and are relatively easy to culture and handle.[1] Effects of pollutants on protozoan populations may have significant impacts on ecosystems, because this group of organisms functions in several ecological processes. They may be a significant link in the food web, and their grazing activities can influence nutrient regeneration or control microbial populations.[2]

    The ciliated protozoan *Tetrahymena*, in particular, is well suited for laboratory toxicity testing. It can be easily cultured axenically (i.e., in the absence of bacteria or other organisms), and results can be obtained relatively quickly. The genus *Tetrahymena* is probably the most well-studied genus of all protists, and therefore, a large database is available on its physiology, biochemistry, genetics, morphology, behavior, and ecology. It is also found in many freshwater environments worldwide and is therefore a good representative ciliate for toxicity testing. *Tetrahymena* has been used by several investigators as a toxicity test organism.[3-8] However, very few studies have considered using chemosensory responses of ciliates as toxicity endpoints.

The term *chemotaxis* has been used rather loosely, even in our own initial studies to refer to behavior in a chemical gradient.[9,10] However, the term more specifically means a spatial chemosensory response in which there is an orientation of cells toward or away from a chemotactic stimulus.[11] Studies which look closely at the orientation of cells before they begin to move or while cells are in a semisolid medium can determine whether cells display a true chemotaxis. By observing immobilized cells in gelatin in the presence of a chemoattractant gradient, Leick has shown that the ciliate *Tetrahymena thermophila* displays chemotaxis.[12]

Chemokinesis, a temporal chemosensory response, includes orthokinesis and klinokinesis.[11] The term *orthokinesis* is used to denote an alteration (increase or decrease) in speed of movement, and *klinokinesis* denotes a change in the frequency of changing direction. Changes in swimming speed (orthokinesis) have been observed for ciliates under certain conditions.[13]

Regardless of the exact mechanism for chemoattraction, it is important for an organism to chemically locate compounds which signal food or mating types. Avoidance of unfavorable habitats is also important to the survival of populations. Many studies have been conducted to determine compounds that serve as attractants, with fewer studies on repellent compounds.[11]

One of the earliest studies demonstrating chemoattraction and repulsion was conducted using the ciliate *Paramecium caudatum*.[14] In that study drops of buffers of different pH values were added to a suspension of ciliates, and the pattern of swimming was determined by long-exposure photographs of the suspension at a relatively low magnification. Nakatani also studied responses of *Paramecium* to chemicals in capillaries.[15,16] Both of these approaches are semiquantitative, however, Van Houten et al. developed two methods for quantifying chemotaxis in *P. aurelia*.[17] One was a T-maze in which the ciliates were placed in a central tube and allowed to respond to a test chemical in one arm or to the control solution in the other arm of the T. An index of chemotaxis, I_{CHE}, was determined by: $I_{CHE} = T/T+C$, where T = number of ciliates in the test solution, and C = number of ciliates in the control solution. $I_{CHE} > 0.05$ indicated attraction to the test chemical, and $I_{CHE} < 0.05$ indicated repulsion from the chemical. The second method used a countercurrent flow design in which test chemicals and control solutions were of different densities and flowed past each other in a common tube. The ciliates could freely swim into and out of each solution. Solutions were collected separately and examined for numbers of ciliates. An index of repulsion or attraction was calculated. These tests ran for 30 to 40 min.

Levandowsky et al. used a flat capillary assay to study chemosensory responses of ciliates to amino acids.[18] Glass capillaries with rectangular cross-sections were filled with test compounds or a starvation medium and deployed in 3 mL of a suspension of *Tetrahymena thermophila*. Control and experimental capillaries were placed within the same ciliate sample. This design, however, yielded great variability in numbers of ciliates in replicate capillaries, and therefore the test required many replicates and the adoption of a sign test as a conservative measure of statistical significance.

Another capillary design was developed by Leick and Helle and consisted of an inner and outer compartment joined by capillaries.[19] The ciliates swam horizontally from outer to inner compartments if they were attracted to the inner chemical and became trapped in the inner compartment as they swam upward. Pauli et al. used this design to determine avoidance of *T. thermophila* to 43 industrial compounds.[8] A decrease in cell migration to the inner compartment indicated repulsion due to toxicants in the inner compartment.

Another quantitative assay for chemokinesis in *Tetrahymena* was developed by Koppelhus et al.[20] It consisted of a two-phase design in which the ciliate suspension was layered onto 2 to 5% metrizamide containing a chemoattractant. Optical density of cells in the lower phase could be read every 2 min automatically, and six cuvettes could be monitored in parallel. This design may be applicable to the study of inhibition of attraction due to toxicants, although a spectrophotometer set-up to accommodate this approach may be more costly than some of the simpler earlier test designs. For further comparisons of ciliate toxicity tests, including chemotactic tests, see Chapter 21 of this text as well as Chapter 23.[21]

Most tests regarding toxic pollutants and chemosensory responses examine avoidance reactions. Although avoidance of toxic chemicals is important to the survival of protozoa, there may be many circumstances, such as nonpoint source pollution, in which the populations are bathed in pollutants with no gradient of the pollutant to aid in avoidance responses. With this in mind we began investigations to determine whether immersion of ciliates in sublethal concentrations of chemicals could interfere with their ability to arrive at a chemoattractant. Such an inhibition may ultimately affect their populations in a polluted environment. The approach we took is one of testing inhibition of chemoattraction rather than avoidance of toxic chemicals. This chapter describes early and more recent studies by our laboratory group on developing a chemoattraction inhibition assay with ciliates.

II. CHEMOATTRACTION INHIBITION TESTING

The ecological significance of bacterial chemoreception inhibition was reported by Mitchell et al. in the early 1970s.[22-24] In their studies bacterial attraction to nutrients and prey organisms, determined by capillary methods, was inhibited by sublethal concentrations of phenol, toluene, crude oil, and ethanol.[22,24] Capillaries were filled with a test chemical, sealed on one end, and deployed into a bacterial suspension on a microscope slide. Bacteria were enumerated by plating the capillary contents onto sea water nutrient agar. Although average numbers attracted were reported, no statistical treatment of data was given, and therefore, the variability in replicates for this approach is not known.

In the late 1970s, one of the current authors (Berk) adopted the approach of Mitchell's group for use with ciliates, using capillaries in a Lucite® chamber developed by Palleroni.[25] However, ciliates tended to gather at the edges where the sides and bottom meet and in micro-crevices caused by drilling out the Lucite to form the chambers. Therefore subsequent work used troughs of several designs described below.

In the late 1980s, Levandowsky et al. studied inhibition of chemoreception in *Tetrahymena* using capillary techniques.[26] Results of this work are compared with our test results with metals described below.

Our early studies used one freshwater species and three marine species of ciliates in a capillary assay procedure in which glass troughs were made by cutting 1 dram vials in half.[9] Troughs were acid-washed, rinsed, and coated with a silicone compound to reduce the possibility of metals binding to the glass. In general, ciliates were suspended in various concentrations of cadmium and copper, made in 10X Osterhout's medium[27]: 18 mM NaCl, 0.031 mM KCl, 0.09 mM CaCl$_2$, 0.42 mM MgCl$_2$ · 6H$_2$O, 0.16 mM MgSO$_4$ · 7H$_2$O, or in filter-sterilized sea water for the marine species, then placed into the glass troughs to which 5 μL capillary tubes containing the attractant were added. The attractant, 0.15% yeast extract, was mixed with the same heavy metal concentration as that in which the ciliates were suspended to avoid the possibility of the ciliates escaping to a metal-free capillary and accumulating there, thereby appearing to be unaffected by the metal toxicants. Controls consisted of only yeast extract in the capillary and only Osterhout's medium or sea water for the ciliate suspension. Ciliates were exposed to the metals for 15 min, after which the capillaries were removed, contents were expelled, and ciliates attracted to the yeast extract were enumerated by direct microscopy after fixation with Lugol's iodine solution.

These studies showed that both marine and freshwater ciliates could be significantly inhibited in chemoattraction responses by the presence of sublethal concentrations of cadmium and copper. The freshwater species was more affected by cadmium than were the marine species, and conversely, the marine species were more affected by copper than were the freshwater species. An apparent stimulation of chemoattraction occurred from exposure of marine species to high sublethal concentrations of copper. Such stimulation most likely was a result of increased swimming speed observed for these organisms exposed to such copper concentrations. Later work added another testing aspect, one that determines changes in random movement due to exposure to toxicants.[28]

A. Standardization of Test Parameters

Using a marine and a freshwater species of ciliate, Berk and Mills examined several factors which may affect results of the chemoattraction inhibition test.[10] These included age of cultures, cell density during testing, axenic vs. bacteria-fed cultures, and starvation period prior to testing. For these tests, the percent inhibition of chemoattraction upon exposure to cadmium was determined. In addition, attraction, i.e., numbers of cells attracted into the capillaries, was examined for the various test parameters.

Two-day-old and seven-day-old cultures were used to compare sensitivity based on culture age. Results showed that the seven-day-old cultures of the marine species were inhibited by cadmium to a greater degree than were the younger marine cultures; however, inhibition of chemoattraction of the freshwater species, *Tetrahymena* sp., was not different for these two culture ages. No other parameter tested had an effect on the percent inhibition of chemoattraction to yeast extract by either species.

Although the percent inhibition of chemoattraction did not change with many of the parameters listed above, the *attraction* of the controls to the yeast extract did significantly change by different population densities and starvation periods (age of cultures and axenic vs. bacterized cultures were not examined in that portion of the study). For example, a twofold higher density resulted in a threefold increase in marine ciliates attracted to the capillaries; and the same increase in density resulted in fewer *Tetrahymena* attracted. Likewise, longer starvation periods resulted in more *Tetrahymena* attracted than at shorter starvation periods, with attraction fluctuating for the marine species under different starvation periods. The fact that chemoattraction can change with time after starvation and with cell densities underscores the need to run controls alongside the exposed cells for every trial throughout a day. In order to further standardize the parameters of the ciliate chemoattraction test for a revised test method Roberts studied the above-mentioned parameters as well as several others, including trough design.[29]

1. Trough Design

Five different trough types were tested for consistency of numbers of ciliates attracted to yeast extract. These included glass Petri dishes, polystyrene Petri dishes, glass troughs made from vials as in the first tests described above, plastic troughs made from aspirator bulbs, and blown glass troughs. The most consistent results from replicate to replicate were obtained using the blown glass toughs, plastic troughs, or the cut vials; however, the vials were harder to clean at the place where the bottom of the vial and sides met. The plastic troughs appeared to increase toxicity when phenol was used, as some toxic chemical or mixture may have been extracted from the plastic. Polystyrene Petri dishes also caused the test organisms to become immobilized and adhere to the bottom of the dishes, a phenomenon also observed by Levandowsky et al.[18] Glass Petri dishes resulted in a wide variance among replicates, and data were not normally distributed. Levandowsky et al. also experienced problems in obtaining normal distribution of data when using Petri dishes.[18] Therefore, the fabricated glass troughs 4.5 cm × 1 cm with a 2-mL capacity were used for all subsequent experiments.[28] They were set up in sets of four replicates attached to a microscope slide with double-sided tape (Figure 22.1).

2. Starvation Period

Unlike the 1- and 72-h starvation period of Berk and Mills, Roberts used a range of starvation periods from 3 to 30 h.[29] Cell divisions continue to occur for a while after ciliates are removed from their growth medium, however, divisions no longer occurred after 3 h. Results in Figure 22.2

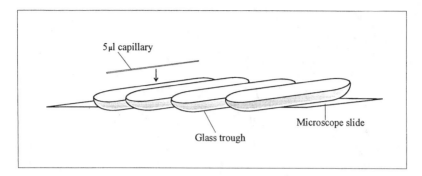

Figure 22.1 Blown glass troughs, 4.5 cm × 1 cm, for chemoattraction tests. Capillaries (5 μL) containing test solutions were placed into ciliate suspensions in each trough.

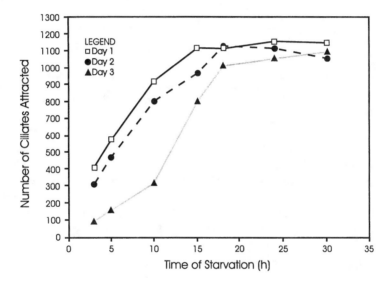

Figure 22.2 The effect of starvation period on number of ciliates attracted to yeast extract in capillary tubes. Each line is a separate culture used on a different day, and each point is the mean of four replicates. (From Roberts, R. O., Master's thesis, Tennessee Technological University, 1990.)

show that chemoattraction increased almost linearly up to 18 h, with no significant differences in numbers attracted between 18 and 30 h. Figure 22.2 also shows that the range in numbers attracted from day to day was smaller after 18-h of starvation, i.e., that the results were more consistent. The 18-h starvation period also works well for logistical reasons. Cells can be starved overnight, and the attraction test can be initiated the next day. This schedule fits well into an 8-h work day, which may be the routine schedule for lab technicians in consulting firms or industry. Koppelhus et al. studied chemoattraction of *T. thermophila* starved for 16 h, 3 d, 5 d, and 7 d, concluding that the 16-h starvation produced the greatest response.[30]

3. Population Density During Starvation and During Testing

The number of ciliates attracted to yeast extract after starvation at four different population densities was determined. Using densities of 10,000, 25,000, 50,000, and 100,000 ciliates/mL, Roberts found that significantly fewer ciliates were attracted as the population density during starvation increased.[29] This contrasts similar tests with *T. thermophila* attracted to proteose peptone, in which there was no difference in chemosensory activity when 25,000, 50,000, or 100,000 cells/mL were used.[30]

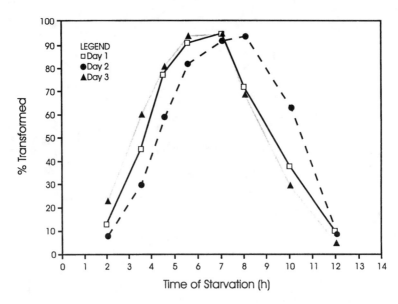

Figure 22.3   The effect of starvation period on the percent of the ciliate population in the transformed morphology (theronts). Each line represents the mean of three separate experiments performed on different days. (From Roberts, R. O., Master's thesis, Tennessee Technological University, 1990.)

The density of cells in the trough during the test also affects the numbers of cells attracted, as shown earlier by Berk and Mills.[10] Using 25, 50, and 100 ciliates per 25 µL, Roberts found the numbers attracted increased significantly with higher cell concentrations.[29] However, if the density is too high (500 per 25 µL or 750 to 1000 per 25 µL) fewer cells may be attracted to the capillaries.[10,19] Possibly physical interactions occur to affect behavior or the attractant gradient may be dispersed by the swimming of a high number of cells in the troughs.

4. Morphological Transformations

Within the first few hours of starvation *Tetrahymena* transforms from the pear-shaped feeding cell to an elongated faster swimming form called a theront. This form not only swims faster, but swims straighter, turning less frequently.[31,32] It is thought to be a dispersal form which may increase the chance of ciliates finding new sources of food.[32,33] After a period of time without added food, however, they revert back to their original pear-shaped form and swimming behavior.[29] Figure 22.3 shows the percent transformed cells in a suspension after various times of starvation. The maximum number of theronts occurred around 5 to 9 h after starvation, with less than 10% of cells in that form by 12 h. Therefore, the 18-h starvation period suggested above was also good for avoiding high numbers of transformed cells in the suspensions. This phenomenon is important for those using *Tetrahymena* in their studies, as preliminary observations (Roberts, unpublished) indicate that theronts were not as attracted to the yeast extract as the trophonts (feeding forms). Results in Figure 22.3 may explain why Koppelhus et al. did not observe increases in swimming speed of the *T. thermophila* which were tested only after 16 h of starvation.[30]

5. Osmolality

Osmolality is a measure of the total dissolved molecules and/or ions per volume of liquid. Addition of a test chemical or effluent to the ciliate suspension may increase the osmolality, and this may affect the behavior of the ciliates. One hour prior to determining chemoattraction, ciliate suspensions were treated to yield a range of osmolalities from 2 to 100 mOsm.[29] This was accomplished by using distilled water, sucrose, or salts. Results showed that shifts in osmolality from 12 (the control salts solution in which ciliates were always suspended for the tests) to 50 mOsm had no effect on chemoattraction to the yeast extract. Significant effects began above 50 mOsm using sucrose and above 75 mOsm using salts. Therefore, as long as the addition of a pollutant does not result in a solution above 50 mOsm, there should be no concern for changes in osmolality influencing the outcome of attraction. We recommend that osmolality be monitored if possible at various stages of the test preparation.

B. Revised Procedure

A revised protocol of the procedure of Berk and Mills was established with several additions and changes. A MOPS-buffered saline solution (MBSS) was used in place of the Osterhout's medium. MOPS was used because it has a low binding affinity for metals, and it does not diffuse across cell membranes.[34] Blanks were added to the test protocol to determine whether aquatic pollutants may increase or decrease the movement of cells into the capillaries. Blanks consisted of only the control solution (MBSS) in the capillary. A control test would consist of ciliates suspended in MBSS, with the capillary containing only MBSS. The experimental test for effect on movement would consist of the pollutant-exposed ciliates in the troughs with only the MBSS in the capillary. The 15-min exposure period was extended to 1 h or 5 h for longer-term comparisons. The 15-min exposure was too short to allow many tests to be run at once, since the short exposure did not allow enough time to deploy, remove, and empty capillaries on many troughs. The longer exposure facilitated processing of more replicates and pollutant concentrations at once. The total exposure time of the revised protocol was 1 h and 15 min (1-h exposure in tubes, then placement into troughs with capillaries for 15 min).

Cells from 4-day-old cultures were harvested, washed by centrifugation in MBSS, and allowed to starve at a density of 400 cells per 25 µL for 18 h. Cells were exposed by mixing test pollutants (diluted or made up in MBSS) with the starved cell suspensions to give a final ciliate density of 80 to 100 cells per 25 µL and the desired final pollutant concentration. The step-by-step revised protocol of Roberts and Berk is as follows:[28]

1. Expose 18-h starved ciliates to each toxicant concentration 1 h or 5 h (or both if comparing effects of exposure times)
2. Add 1 mL of control (nonexposed) ciliate suspensions to two sets of four replicate troughs
3. Add one 5 µL capillary filled with yeast extract (0.15%) or yeast extract and toxicant for exposed ciliates to each of four troughs, and one 5 µL blank capillary containing only MBSS (or MBSS with toxicant for exposed ciliates) to each of the remaining four troughs. Add capillaries at 15-s intervals or longer intervals if necessary for easy processing of many capillaries
4. Repeat steps 2 and 3 for toxicant-exposed ciliates
5. Remove each capillary 15 min after its addition to the trough, expel capillary contents onto a glass slide and add a drop of Lugol's iodine or fixative to enumerate ciliates

Slides with drops of capillary contents can be stored in a humidity chamber (small plastic boxes with damp cloths or pipette tip boxes with water below) to prevent evaporation before enumeration can be carried out.

Data are expressed as the % inhibition of chemoattraction by the following equation:

$$\% \ Inhibition = 100 \left[\frac{C - T}{C} \right]$$

where C is the number of control ciliates attracted to the yeast extract, and T is the number of toxicant-exposed ciliates attracted. EC_{50}s can be determined by probit analysis.

C. Reversibility and Recovery from Effects

The chemoattraction inhibition test described above can be used with additional treatments to determine whether the relatively short exposures to toxicants can be diminished or eliminated by removing the test organisms from the toxicants. Comparisons of 1-h exposures with longer exposures can show whether the organisms can recover from (or appear to adapt to) the initial 1-h exposures.

Berk et al., found that the 15-min exposure of *Tetrahymena* to 1 ppm cadmium could be nullified by washing the ciliates after the short exposure, however, the effect of 0.25 ppm copper could not be altered by such treatment.[9] Washing the marine ciliates after exposures to cadmium and copper removed the inhibitory effect, indicating that the metals did not cause an irreversible effect after the 15-min exposure. Inhibitory effects of phenol and naphthalene were actually reversed for *Tetrahymena*, i.e., after exposure and washing, the ciliates actually had a stronger chemoattraction than before exposure.[35] Movement tests indicated that there was no effect of these chemicals on movement that could account for the reversal. Perhaps a resistance or compensating mechanism was stimulated, leaving the ciliates with a better ability to follow the attractant gradient after the short exposure and subsequent removal of the toxicants. Inhibitory effects of hydrocarbons and ethanol on bacterial chemotaxis were also nullified by washing.[22,24]

When comparisons between 1-h and 5-h exposures were made for chemoattraction inhibition using *Tetrahymena pyriformis*, Roberts and Berk found that for cadmium, inhibitory effects increased with the longer exposure period except for the highest concentration tested, 100 μg/L.[28] Increased effect with time is characteristic of heavy-metal exposure. This was true for the Microtox assay, *Daphnia* and ciliates.[3,36,37] In contrast, the numbers of ciliates attracted to yeast extract after a 5-h exposure to phenol were higher than those attracted after 1 h, indicating adaptation to the chemical. Complete recovery to control levels occurred after the 5-h exposure to 100 mg/L. This trend was also true for the herbicides, Roundup® and 2,4-D. An increase in the EC_{50} concentration (indicating less sensitivity) between 5- and 15-min exposures also was observed with the Microtox assay testing phenol.[38] Naphthalene inhibition of the ciliate chemoattraction was not changed by the longer exposure periods.[28]

III. COMPARISON OF RESULTS WITH OTHER ASSAYS

During development of the chemoattraction assay Roberts and Berk also compared the results with those of a 24-h mortality test on the same organisms.[28,39] Results showed that the 1- or 5-h chemoattraction inhibition tests always gave lower values for the EC_{50} than for the LC_{50}. Prior to running chemoattraction inhibition tests, mortality tests were carried out to ascertain that the exposure time of the chemosensory test did not result in mortality, i.e., that the concentrations were sublethal for the exposure period. In comparison with other standard toxicity tests, the chemoattraction inhibition assay with cadmium and copper was more sensitive than other tests, except for those with algae and *Daphnia*. The EC_{50} values of cadmium using the Microtox and ATP-TOX assays were 368-fold and 25-fold higher, respectively, than those of the chemoattraction inhibition assay.[40] Levandowsky et al. also looked at chemosensory inhibition of *T. pyriformis* exposed to heavy metals and reported that the lowest concentration of cadmium which inhibited chemosensory responses was 0.01 mM (>1 mg/L); and this concentration is much higher than that which caused inhibition in our *T. pyriformis* test.[26] Reasons for the discrepancy are not clear, although Levandowsky's test design is slightly different from ours, as explained previously.

Similar results were observed for copper. Table 22.1 shows comparisons of the *T. pyriformis* chemoattraction inhibition assay and mortality assay with other standard tests for copper and the herbicides Roundup and 2,4-D. The chemoattraction inhibition assay EC_{50}s were similar to the 96-h LC_{50}s of the algae and *Daphnia* tests. However, as discussed above, a longer exposure to copper would probably have resulted in a lower EC_{50} value for the chemoattraction test.

Table 22.1 Comparison of *T. pyriformis* EC_{50} and LC_{50} Concentrations (mg/L) to Those of More Standardized Tests[a] for Six Chemicals and Substances

Test	Cd	Cu	Naph.	Phenol	Roundup	2,4-D
Microtox[b]	18.8	2.3	—	31	—	31
Microtox[c]	416	24.9	—	39.5	17.5	61
Microtox[d]	25	0.42	—	28	—	107
Algae[d]	0.041	0.043	—	—	—	96
Daphnia	0.041[d]	0.064[d]	8.6[e]	12[e]	25.5[f]	>240[d]
Fathead minnow	4.39[g]	0.070[h]	7.9[i]	28[j]	—	320[k]
T. pyriformis[l] chemoattraction assay	0.058	0.045	3	172	27	158
T. pyriformis[m] mortality test	0.112	0.132	14	642	1750	>500

[a] Other values obtained from the literature (mg/L).
[b] Reference 40 (15-min EC_{50})
[c] Reference 44 (15-min EC_{50})
[d] Reference 45
[e] Reference 43 (48-h LC_{50})
[f] Reference 46 (96-h LC_{50})
[g] Reference 47 (96-h LC_{50})
[h] Reference 48 (96-h LC_{50})
[i] Reference 49 (96-h LC_{50})
[j] Reference 50 (96-h LC_{50})
[k] Reference 51 (96-h LC_{50})
[l] Reference 29 (1-h EC_{50})
[m] Reference 29 (24-h LC_{50})

With respect to the organic chemicals the chemoattraction inhibition test was more sensitive for phenol (1 h EC_{50} = 151 mg/L; 5 h EC_{50} = 245 mg/L) than three of four assays reported by Dutka and Kwan, where EC_{50} values were ≥ 300 mg/L.[41] Ciliate inhibitory ranges for naphthalene (2.2 to 3.55 mg/L) were comparable to 48 h LC_{50} values for *Daphnia*.[42,43] Table 22.1 shows that the chemoattraction inhibition assay was similar to the Microtox and algae assays in its sensitivity to Roundup, but more sensitive to 2,4-D than the fathead minnow or *Daphnia* assays, and less sensitive to 2,4-D than Microtox or algal assays.

In spite of the comparisons above, it is important to bear in mind the fact that each test considers different endpoints, and the exposure periods of most tests are different from those of the chemoattraction test. Pauli et al. have also mentioned that their chemosensory assay with *Tetrahymena thermophila* had low intercorrelation with recommended tests, indicating that the chemosensory response measures interactions not measured by the standard tests.[8] Nevertheless, a chemosensory assay is a sensitive and ecologically significant endpoint that may have several targets for toxic chemicals. The overall effect of ciliates becoming inhibited from reaching an attractant may have important implications for their populations in nature. Since these organisms play significant roles in the ecosystem, especially in nutrient regeneration and in the base of food webs, an inability to locate their own food source may ultimately affect populations higher in the food web.

IV. DISCUSSION AND FUTURE IMPROVEMENTS

The chemoattraction inhibition assay has been developed using single toxic chemicals, however, its usefulness with effluents was also examined.[28,29] One effluent was from a metal-plating industry, and exposure of *T. pyriformis* to it gave a 1 h EC_{50} of 15.35% and a 5 h E_{50} of 6.95%, consistent with the trend of increasing toxicity with increasing time of metal exposure. The other effluent was from a paper mill, and although the effluent caused chemosensory inhibition at high concentrations, additional experiments demonstrated that the ciliates were attracted to the effluent. This reveals a shortcoming of the current protocol and the need to add another aspect to the protocol, i.e., testing a pollutant for its ability to serve as an attractant prior to running the chemoattraction inhibition test with yeast extract.

Another suggestion for improvement includes standardization of the starvation temperature and surface-to-volume ratio during starvation, as Koppelhus et al. found that the lower the O_2 tension, the greater the response of the *T. thermophila* to the attractant.[30] Automated enumeration of cells in the capillary may also improve the speed of gathering data. Perhaps electronic cell counters, microscopic image analysis, or even flow-cytometry may facilitate data collection.

Some aspects of the test approach can be flexible to accommodate different needs. For example, if effluents are tested, the natural receiving waters upstream of the effluent source can be used to wash and test ciliates, provided there is no attraction to or mortality from the natural waters. Various exposure periods can be run as long as the actual inhibition portion of the test is run for 15 min. If capillaries remain in the ciliate suspension longer than 15 min, the gradient of the attractant may be disrupted by the swimming cells, or the gradient may be too diffuse. Also, cells may begin to leave the capillary.

The test could be applied to the marine or estuarine environment if appropriate test species can be found. Berk et al. and Berk and Mills have used some marine species and found them to respond differently from *T. pyriformis*.[9,10] Other species to be considered are cyst-forming ciliates such as *Colpoda*, which may be good candidates for their potential use in assay kits. Stocks of cysts from one large population could be used to start new test cultures, and test results from such stocks may prove even more consistent than those using cultures which need to be transferred on a regular schedule.

The small standard error values for tests run on several different days indicate good reproducibility of this assay for most toxicants tested. Although the apparatus is simple compared with other test chambers described previously for ciliate chemotaxis studies, such as the T-maze, it accommodates the same applications as the other tests, and the relatively inexpensive troughs are easy to clean. The Palleroni chambers we tested were difficult to clean, since they were not smooth on the microscopic level, and any minute debris, although sterile, appeared to attract clusters of ciliates such that numbers of ciliates in capillaries were quite variable. Perhaps other more complex test chambers may also be more difficult to clean, and they may pose similar problems.

Development of the chemoattraction inhibition assay is not meant to replace other toxicity tests, but rather to complement the battery of standard tests. Inclusion of a ciliated protozoan fills a taxonomic gap in the battery of tests which includes prokaryotes, single-celled algae, fish, and invertebrates such as daphnids. Alone this test could serve as a rapid screening test. The relatively short-term chemoattraction inhibition assay may be used to gather data for predicting the outcome of longer-term exposures on protozoan communities. The applications of the chemoattraction inhibition assay can be determined only after more extensive trials are run to acquire a greater database.

REFERENCES

1. Cairns, J. Jr., Protozoans (Protozoa), in *Pollution Ecology of Freshwater Invertebrates*, Hart, C. W., Jr., and Fuller, S. L. H., Eds., Academic Press, New York, 1974, 1.
2. Johannes, R. E., Influence of marine protozoa on nutrient regeneration, *Limnol. Oceanogr.*, 10, 434, 1965.
3. Carter, J. W. and Cameron, I. L., Toxicity bioassay of heavy metals in water using *Tetrahymena pyriformis, Water Res.*, 7, 951, 1973.
4. Slabbert, J. L. and Maree, J. P., Evaluation of interactive toxic effects of chemicals in water using a *Tetrahymena pyriformis* toxicity screening test, *Water S.A.*, 12, 57, 1986.
5. Cooley, N. R., Keltner, J. M. Jr., and Forester, J., Mirex and Aroclor 1254: effect on and accumulation by *Tetrahymena pyriformis*–Strain W, *J. Protozool.*, 19, 636, 1972.
6. Yoshioka, Y., Ose, Y., and Sato, T., Testing for the toxicity of chemicals with *Tetrahymena pyriformis*, *Sci. Total Environ.*, 43, 149, 1987.

7. Schultz, T.W., Wayatt, N.L., and Lin, D.T., Structure–toxicity relationships for nonpolar narcotics: a comparison of data from the *Tetrahymena, Photobacterium* and *Pimephales* systems, *Bull. Environ. Contam. Toxicol.,* 44, 67, 1990.

8. Pauli, W., Berger, S., Schmitz, S., and Jaskulka, L., Chemosensory responses of ciliates: a sensitive end point in xenobiotic hazard assessment, *Environ. Tox. Water Qual.,* 9, 341, 1994.

9. Berk, S. G., Gunderson, J. H., and Derk, L. A., Effects of cadmium and copper on chemotaxis of marine and freshwater ciliates, *Bull. Environ. Contam. Toxicol.,* 34, 897, 1985.

10. Berk, S. G. and Mills B. A., Factors affecting chemotaxis inhibition tests with protozoa, *J. Testing Eval.,* 14, 140, 1986.

11. Leick, V. and Hellung-Larson, P., Chemosensory behavior of *Tetrahymena, BioEssays,* 14, 466, 1992.

12. Leick, V., Gliding *Tetrahymena thermophila*: oriented chemokinesis in a ciliate, *Eur. J. Protistol.,* 23, 354, 1988.

13. Lapidus, R. and Levandowsky, M., *Models of Biological Growth and Spread,* Springer-Verlag, Berlin, 1980.

14. Dryl, S., Contributions to mechanism of chemotactic response in *Paramecium caudatum, Animal Behavior,* 11, 393, 1963.

15. Nakatani, I., Chemotactic response of *Paramecium caudatum, J. Fac. Sci. Hokkaido Univ.,* 16, 553, 1968.

16. Nakatani, I, Effects of various chemicals on the behavior of *Paramecium caudatum, J. Fac. Sci. Hokkaido Univ.,* 17, 401, 1970.

17. Van Houten, J., Hansma, H., and Kung, C., Two quantitative assays for chemotaxis in *Paramecium, J. Comp. Physiol.,* 104, 211, 1975.

18. Levandowsky, M., Cheng T., Kehr A., Kim J., Gardner L., Silvern L., Tsang L., Lai G., Chung C., and Prakash E., Chemosensory responses to amino acids and certain amines by the ciliate *Tetrahymena*: a flat capillary assay, *Biol. Bull.,* 167, 322, 1984.

19. Leick, V. and Helle, J., A quantitative assay for ciliate chemotaxis, *Anal. Biochem.,* 135, 466, 1983.

20. Koppelhus, U., Hellung-Larsen, P., and Leick, V., An improved quantitative assay for chemokinesis in *Tetrahymena, Biol. Bull.,* 187, 8, 1994.

21. Gilron, G. and Lynn, D., Ciliated protozoa as test organisms in toxicity assessments, *Microscale Testing in Aquatic Toxicology — Advances, Techniques, and Practice,* Wells, P.G., Lee, K., and Blaise, C., CRC Press, Boca Raton, FL, Chapter 21.

22. Mitchell, R., Fogel, S., and Chet, I., Bacterial chemoreception: an important ecological phenomenon inhibited by hydrocarbons, *Water Research,* 6, 1137, 1972.

23. Walsh, F. and Mitchell, R., Inhibition of inter-microbial predation by chlorinated hydrocarbons, *Nature,* 249, 673, 1974.

24. Chet, I., Fogel, S., and Mitchell, R., Chemical detection of microbial prey by bacterial predators, *J. Bacteriology,* 106, 863, 1971.

25. Palleroni, N.J., Chamber for bacterial chemotaxis experiments, *Appl. Environ. Microbiol.,* 32, 729, 1976.

26. Levandowsky, M., Azcue, D., and Phung, P., Effects of heavy metals on chemosensory responses and phagocytosis in the ciliate, *Tetrahymena,* Abstr. 88th Annual meeting of the American Society for Microbiology, p. 307, 1988.

27. Taylor, C. V. and Strickland A. G. R., Some factors in the excystment of dried cysts of *Colpoda cucullus, Arch. Protistink.,* 86, 181, 1935.

28. Roberts, R. O. and Berk, S. G., Development of a protozoan chemoattraction bioassay for evaluating toxicity of aquatic pollutants, *Tox. Assess.,* 5, 279, 1990.

29. Roberts, R. O., Development of a protozoan chemoattraction bioassay for screening toxicity of aquatic pollutants, MS thesis, Tennessee Technological University, 1990.

30. Koppelhus, U., Hellung-Larson P., and Leik V., Physiological parameters affecting the chemosensory response of *Tetrahymena, Biol. Bull.,* 187, 1, 1994.

31. Lynn, D. H., The life cycle of the histophagous ciliate *Tetrahymena corlissi, J. Protozool.,* 32, 409, 1975.

32. Nelson, E. M. and Debault L. E., Transformation in *Tetrahymena pyriformis*: description of an inducible phenotype, *J. Protozool.,* 25, 113, 1978.

33. Nelson, E. M., Transformation in *Tetrahymena thermophila, Develop. Biol.,* 66, 17, 1978.

34. Gueffroy, D. E., *Buffers: A Guide for the Preparation and Use of Buffers in Biological Systems*, Calbiochem, San Diego, CA, 1981.

35. Berk, S. G., Mills, B. A., Stewart, K.C., Ting, R.S., and Roberts, R.O., Reversal of phenol and naphthalene effects on ciliate chemoattraction, *Bull. Environ. Contam. Toxicol.*, 44, 181, 1990.

36. Greene, J. C., Miller, W. E., Debacon, M. K., Long, M. A., and Bartels, C. L., A comparison of three microbial assay procedures for measuring toxicity of chemical residues, *Arch. Environ. Contam. Toxicol.*, 14, 659, 1985.

37. Khangarot B. S. and Ray P. K., Correlation between heavy metal acute toxicity values in *Daphnia magna* and fish, *Bull. Environ. Contam Toxicol.*, 27, 722, 1987.

38. Dutka, B. J., Nyholm, N., and Peterson, J., Comparison of several microbial screening tests, *Water Res.*, 17, 1363, 1983.

39. Roberts, R. O. and Berk, S. G., Effect of copper, herbicides, and a mixed effluent on chemoattraction of *Tetrahymena pyriformis*, *Environ. Tox. Water Qual.*, 8, 73, 1993.

40. Xu, H., and Dutka, B. J., ATP-TOX system — a new, rapid sensitive bacterial toxicity screening system based on the determination of ATP, *Tox. Assess.*, 2, 149, 1987.

41. Dutka, B. J. and Kwan, K. K., Battery of screening tests approach applied to sediment extracts, *Tox. Assess.*, 3, 303, 1988.

42. Crider, J. Y., Wilhur, J., and Harmon, H. J., Effects of naphthalene on the hemoglobin concentration and oxygen uptake of *Daphnia magna*, *Bull. Environ. Contam. Toxicol.*, 28, 52, 1982.

43. LeBlanc, G. A., Acute toxicity of priority pollutants to the water flea (*Daphnia magna*), *Bull. Environ. Contam. Toxicol.*, 24, 684, 1980.

44. McFeters, G. A., Bond, P. J., Olson, S. B., and Tchan, Y. T., A comparison of microbial bioassays for the detection of aquatic toxicants, *Water Res.*, 17, 1757, 1983.

45. Miller, W. E., Peterson, S. A., Greene, J. C., and Callahan, C. A., Comparative toxicology of laboratory organisms for assessing hazardous waste sites, *J. Environ. Qual.*, 14, 569, 1985.

46. Servizi, J. A., Gordon, R. W., and Martens, D. W., Acute toxicity of Garlon 4 and Roundup herbicides to salmon, *Daphnia*, and trout, *Bull. Environ. Contam. Toxicol.*, 39, 15, 1987.

47. Sherman, R. E., Gloss, S. P., and Lion, L. W., A comparison of toxicity tests conducted in the laboratory and in experimental ponds using cadmium and the fathead minnow (*Pimephales promelas*), *Water Res.*, 21, 317, 1987.

48. Norberg, T. J. and Mount, D. I., A new fathead minnow (*Pimephales promelas*) subchronic toxicity test, *Environ. Toxicol. Chem.*, 4, 711, 1985.

49. DeGrave, G. M., Overcast, R. L., and Bergman, H. L., Effects of naphthalene and benzene on the fathead minnow and rainbow trout, *Arch. Environ. Contam. Toxicol.*, 11, 487, 1982.

50. Phipps, G. L., Holcombe, G. W., and Fiandt, J. T., Acute toxicity of phenol and substituted phenols to the fathead minnow, *Bull. Environ. Contam. Toxicol.*, 26, 585, 1981.

51. Alexander, H. C., Gersich, F. M. and Mayes, M. A., Acute toxicity of four phenoxy herbicides to aquatic organisms, *Bull. Environ. Contam. Toxicol.*, 35, 314, 1985.

CHAPTER 23

Ciliate Microbiotest Applications: Metal Contaminants in Water and Soil

James R. Pratt, Daria G. Mochan, and Nancy J. Bowers

CONTENTS

I. INTRODUCTION

Methods for evaluating adverse effects of toxic substances rely on having test organisms of known size and age (usually juvenile) available for testing. Culturing test organisms often consumes much of the research effort leaving less time available for field evaluations of contamination. Additionally, taking healthy organisms to the field for onsite testing is often problematic. The goal of the research described below was to evaluate one microscale test, a growth test using ciliate protozoa,[1] for use in a battery of tests approach to contaminant and site evaluation. Microscale tests using ciliates satisfy two needs: the selected test organism reduces the need for long-term culturing and is representative of small, rapidly responding bacterivorous species in soils and freshwaters.

Protozoa are common and abundant in freshwater, soils, and litters. They may account for as much as 70% of nonbacterial respiration in soils.[2] Soil protozoa are adapted to drying, and many can grow rapidly from cysts when the soil is wetted. Colpodid ciliates can excyst in minutes and grow to large population sizes quickly. A few species of colpodid ciliates appear to be ubiquitous in soils and litters, and they are also considered bioindicators of enrichment in fresh waters.[3]

The studies described below illustrate the application of rapid testing to studies of contaminant bioavailability and effect using single toxic metals and metal mixtures derived from contaminated sites.

II. METHODS AND MATERIALS

A. Test Organism

The test species was *Colpoda inflata*, a common ciliate (length 40 to 60 μm) obtained as dry cysts (ATCC 30917) from the American Type Culture Collection (Rockville, MD, USA). Cultures were developed from stored cysts and were maintained for periods of 2 to 7 d. Culturing was done in 10% Sonneborn's *Paramecium* medium, a medium based on cerophyll and inorganic nutrients. Sonneborn's medium was prepared by boiling 2.5 g cerophyll (cereal leaves) for 5 min in 1 L of distilled, deionized water. After boiling, the mixture was filtered (#1 Whatman filter paper), the volume was adjusted to 1 L with distilled water, and 0.5 g Na_2HPO_4 was added. The full-strength medium was diluted prior to use with distilled, deionized water. A minimal salts medium[4] was also used and consisted of 6 mg KCl, 4 mg $CaHPO_4$, and 2 mg $MgSO_4$ in 1 L of distilled, deionized water. All media were autoclaved in small aliquots (50 mL) prior to use. Ciliates were fed a nonpathogenic strain of *Klebsiella pneumoniae* (ATCC 27889).

B. Test Procedure

Toxicity tests were conducted in sterile, 24-well, polystyrene tissue culture plates (Costar, Inc., Acton, MA, USA) following the method of Pratt et al.[1] described briefly below. Sterile medium was dispensed into wells and then amended with toxicant from stock solutions of reagent-grade chemicals prepared in sterile distilled, deionized water or from soil extracts prepared as described below. Each experiment consisted of four replicates of five test concentrations (arranged logarithmically) plus controls.

After medium and toxicant were dispensed into test wells, ciliates were added from log-phase cultures (48 to 96 h old) along with food bacteria. A constant volume of well-mixed culture was added to each well. This volume corresponded to approximately 100 cells/mL and was less than 10% of the total test volume in a given well. Previous research has shown that normal growth under test conditions results in cell densities of 400 to 1000 cells/mL in 24 h. After 24 h, subsamples were removed from each test well and enumerated using a direct counting technique as follows. Each well was thoroughly mixed with a micropipettor, and a 20 μL subsample was removed and transferred to a clean microscope slide by distributing the 20 μL as 3 or 4 drops. All of these drops were then immediately scanned at low magnification (40×) on the stage of a stereomicroscope to search for and enumerate active cells. Active cells are always moving and can be easily distinguished from bacterial aggregates or cysts. Dead cells are not present, because they lyse quickly as they lose the ability to osmoregulate. For a given well, the 20 μL subsampling technique was repeated a minimum of three times to assure that an accurate estimate of the population in the well had been obtained. Typically, a 20 μL sample had up to 20 cells. If repeat counts of subsamples varied by more than 30%, subsampling continued until population estimates stabilized. This procedure was repeated for each well, and the mean of subsample estimates for a given replicate was used to obtain cell densities used in later analyses.

C. Experiments

Bioavailability of toxic metals (cadmium, copper, and zinc) under different levels of organic matter was assessed by comparing dose-responses using the minimal salts medium and in 5% and 10% Sonneborn's medium as diluents. As prepared, the dissolved organic carbon concentration of 10% Sonneborn's medium was approximately 40 mg/L.

Toxicity of the soil elutriates was determined in a similar manner, but the maximum concentration of elutriate that could be tested was 82%. This concentration reflected dilution of the elutriate by the addition of ciliates and food in culture medium. Toxicity was assessed by first determining which elutriates showed toxicity in the ciliate growth test (using control and 100% elutriate screening tests) and then by conducting dose-response experiments to determine growth inhibiting concentrations of each toxic soil elutriate.

D. Soil Elutriates

Soil samples were collected in the vicinity of an abandoned precious metal and base metal mining district in central Colorado. Waste rock piles from mining operations had the potential to generate acids and contaminate soils. Three areas representing a gradient of contamination were sampled by collecting six subsamples from seven to ten sites in each area. Each set of six subsamples was composited, mixed, and then used for a variety of analyses. Aqueous elutriates of the composited samples were prepared using a 1:4 (v/v) mixture of sieved soil and deionized water.[5] Extraction was carried out for 1 h on an automatic shaker table. Extracts were settled for 1 h and then filtered (0.45 μm) prior to testing. Elutriates were analyzed for nine metals (arsenic, cadmium, copper, iron, lead, mercury, selenium, silver, and zinc) using standard EPA test methods.[6] Elutriate pH ranged from 2.1 to 8.4. Free metal concentrations of elutriates were estimated from total metal concentrations and pH, and toxicity assessments were done using estimated free metal concentrations.[7]

The experiments reported here include selected results for elutriates from three areas designated simply as areas A, B, and C. Area A appeared to be the least acid-producing and was located in a fluvial tailing area at the bottom of a small gulch. Area B was intermediate in impact severity. Area C was clearly acid-producing with an unvegetated area down gradient from the rock pile.

E. Data Analysis

Data were examined by characterizing the median tolerance limit for population growth, defined as the concentration corresponding to a 50% inhibition of growth (IG_{50}) relative to controls during the 24-h test period. These analyses were done by regressing cell number on log dose and then using inverse prediction[8] to estimate the IG_{50} from the control response (control mean = 100%). For these analyses, only the linear portion of the dose-response relationship was used; data for any test concentrations above the lowest concentration producing no growth were ignored, as were any test concentrations below the lowest concentration producing growth equivalent to controls.

III. RESULTS

A. Bioavailability

Dissolved and particulate organic carbon in natural waters can bind soluble metals and ameliorate toxicity, although not all metals are subject to significant binding by organic ligands. Experiments showed that the toxicity of cadmium and copper were reduced incrementally by the presence of dissolved organic materials in the test media (Table 23.1), but the toxicity of zinc was

Table 23.1   Effects of Toxic Metals on Growth of *Colpoda inflata* Cells and Cysts in a Single Test Series

Metal	10% Sonneborn's IG_{50}	5% Sonneborn's IG_{50}	Minimal Salts IG_{50}	Acute range
Cd	75	28	25	1–28,000
	(74–76)	(26–30)	(23–28)	
Cu	575	148	30	17–10,200
	(527–583)	(142–153)	(27–69)	
Zn	161	72	119	51–88,900
	(150–171)	(63–80)	(95–151)	

Note: The IG_{50}, the contaminant concentration inhibiting growth by 50%, was estimated by inverse prediction from the dose-response relationship for experiments conducted in organic (Sonneborn's) and inorganic (minimal salts) media. Data are means with ranges of test results shown in parentheses. All values are µg/L. Acute ranges based on water quality criteria (data from USEPA, *Quality Criteria for Water,* USEPA, Office of Water, Washington, DC, 1986).

not as affected by the organic media. Typically, the entire dose-response pattern was shifted as the concentration of dissolved organic carbon changed (Figure 23.1). Additionally, the within-treatment variability was reduced in the minimal salts medium, although dose-response relationships were strong regardless of the test medium. Short-term growth responses determined as 24 h IG_{50}s were in the lower portion of the range of acute toxicities for standard test species (Table 23.1).

B. Soil Contaminants

Elutriates from contaminated soils strongly affected ciliate growth (Figure 23.2, Table 23.2), and elutriates from those soils expected to be most toxic often completely inhibited ciliate growth. There were statistically strong relationships between growth responses and elutriate chemistry. Follow-up studies on the relationship between elutriate components showed that pH, copper, and cadmium contributed most strongly to growth inhibition. These components explained over 80% of the variability in elutriate effects on growth (Table 23.3). Elutriate pHs were generally low, and metals were predominantly in the free ion form (Table 23.4). The extreme toxicity values for some elutriates apparently precluded determining effects of other metals that should have contributed to measured toxicity (e.g., zinc). Free metal composition of nontoxic elutriates overlapped ranges of toxic elutriates, suggesting again that pH (much lower in the toxic samples) was the most significant factor contributing to the measured toxic effect.

IV. DISCUSSION

Microscale tests have great potential to influence the practice of ecotoxicology.[9] Success in applying microscale tests to research and field problem-solving depends on the sensitivity of such tests and the similarity of responses to those observed in more traditional tests. Our experiments with different media show that the ciliate growth test can respond to differential metal bioavailability. The binding of toxic metals to constituents of the organic growth medium clearly shows reductions in metal toxicity. A variety of naturally occurring organic compounds can bind soluble metals and reduce their bioavailability,[10] although some small-molecular-weight organic fractions may actually enhance metal uptake.[11] The order of toxicity (Cd>Cu>Zn) corresponds to expected toxicity in standard test species, although the precise components of natural waters that affect metal toxicity are many. Metals may be bound by a variety of organic and inorganic ligands, so differences between the metal binding capacity of test media and natural waters may be significant, suggesting that site-specific evaluations of metal toxicity may be of increasing importance.[12] Differences in

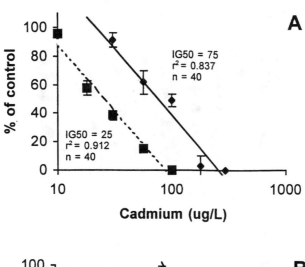

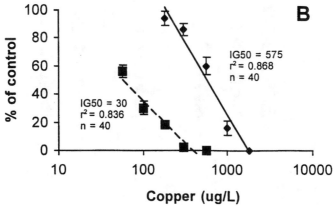

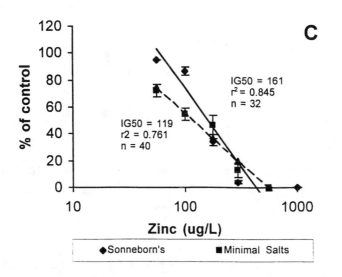

Figure 23.1   Differential effects of toxic metals on growth of *Colpoda inflata* reflecting differences in bioavailability in inorganic, minimal salts (MS–dashed line) and organic (10% Sonneborn's–solid line) test media: A) cadmium toxicity, B) copper toxicity, C) zinc toxicity. Error bars are standard deviations.

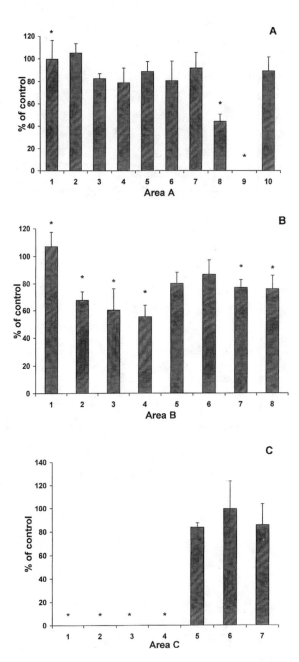

Figure 23.2   Effects of soil elutriates from three contaminated sites on the growth of *Colpoda inflata*: A) Area A (some toxicity at 2 of 10 sites), B) Area B (limited toxicity at 5 of 8 sites), C) Area C (high toxicity at 4 of 7 sites). Error bars are standard deviations. Asterisks identify treatments differing from controls (p<0.05).

toxicity among test media with differing organic carbon composition suggest that microscale tests seem to be capable of distinguishing metal-binding capacities of different waters.

The complexity of the interactions among toxicants and naturally occurring compounds (including ligands) is further represented by measurements of the toxicity of complex mixtures such as the soil elutriate tests described above. The observed toxicity (and, in some cases, lack of toxicity)

Table 23.2   Results of Stepwise Regression Analysis of 16
Colpoda inflata Toxicity Test Results Using Soil
Elutriate pH and Estimated Free Metal Concentrations

Endpoint	Variable Added	F	p	Model r^2
Ciliate growth	pH	60.25	0.0001	0.7237
	Copper	5.36	0.0304	0.7778
	Cadmium	4.43	0.0169	0.8318

Table 23.3   Predicted Elutriate IG_{50}s (% Elutriate
Concentration at which Ciliate Growth was
Reduced by 50%) Based on Ciliate Growth Tests

Elutriate	IG_{50}	95% Confidence Limits	r^2
A-8	75.0	65.1–90.6	0.7838
A-9	0.096	0.054–0.173	0.9072
C-2	0.049	0.018–0.130	0.7702
C-2	19.3	11.3–32.6	0.8070
C-3	11.5	4.76–34.4	0.6564
C-4	11.6	4.14–43.4	0.5853

Note: Growth response was regressed against the natural logarithm of the elutriate dilution, and the coefficient of determination for each test is shown.

Table 23.4   Estimated Free Metal Ion Concentrations and pH at Elutriate
Concentrations that Resulted in 50% Inhibition of Ciliate Growth (IG_{50})

Elutriate	Cadmium	Copper	Iron	Lead	Zinc	pH
A-8	45.3	13.5	0.00	1267	10567	5.8
A-9	5.73	0.13	0.00	4.60	754	5.4
C-1	0.55	9.17	186	1.90	94.5	2.2
C-2	1.90	258	31.5	38.0	1338	3.4
C-3	20.7	162	1.86	23.0	3005	2.7
C-4	16.4	349	81.2	23.4	3448	3.1
Nontoxic elutriates	0–20	0–155	0	0–3008	217–33135	6.7–8.0

Note: The ranges of metal ion concentrations in elutriates shown to be nontoxic (n = 14)
are shown for comparison. Metal values are in µg/L.

could not be predicted solely by the chemical composition of the elutriates and the strong effect
of low pH. Our assessments of elutriates from contaminated soils did not include a full analysis
of elutriate chemical composition, but it seems clear that interactions among toxicants and inorganic
and organic ligands could have significantly affected the speciation of the potentially toxic constituents.[13,14] The significant effect of pH on toxicity was expected[15] based on direct effects of low pH
and increases in free (toxic) metal ion concentrations. Although the complex mixtures had toxic
effects different than might have been anticipated given the comparatively high metal concentrations
in elutriates and although our estimates of free metal concentrations ignored the presence of other
factors affecting toxicity, free metal concentrations were acceptable predictors of additional mixture
toxicity beyond pH effects.

Microscale tests using eukaryotic microbes provide sensitive and rapid means of evaluating
toxicity. Several characteristics of eukaryotic microbes like ciliate protozoa make them ideal toxicity
test organisms. Eukaryotic microbes are ubiquitous in ecosystems and play major roles in primary
production (algae) and decomposition (protozoa, some fungi). Many eukaryotic microbes have
adapted to the high environmental variability in ecosystems by forming resistant stages, usually

resting cysts, and these allow the organisms to remain dormant during adverse conditions (e.g., drying) and then to exploit acceptable conditions rapidly. Growth rates of eukaryotic microbes can be high (3 to 4 doublings per day). Although many of the most opportunistic species are poor competitors,[2] the ability of these organisms to grow quickly from resting stages can be exploited for toxicological purposes.[16] Features of test organism biology make microscale tests rapid and inexpensive, since the high cost of maintaining organism cultures is reduced. Additionally, appropriate sensitivity to toxicants is retained. Additionally, opportunistic or pioneer species have been shown to be more sensitive to toxic chemicals than the slower growing species that comprise the remainder of communities.[17,18]

ACKNOWLEDGMENTS

The research reported here was supported, in part, by grant no. DAMD17-95-1-5068 from the U.S. Army Medical Research and Development Command, Fort Detrick, MD. We gratefully acknowledge the assistance of D. Beeson and M. Lewis of the S.M. Stoller Corporation for the preparation and analysis of soil elutriates.

REFERENCES

1. Pratt, J. R., Mochan, D. G., and Xu, Z., Rapid toxicity evaluation using soil ciliates: sensitivity and bioavailability, *Bull. Environ. Contam. Toxicol.,* 58, 387, 1997.
2. Foissner, W., Soil protozoa: fundamental problems, ecological significance, adaptations in ciliates and testaceans, bioindicators, and guide to the literature, in *Progress in Protistology,* Vol. 2, Corliss, J.O. and Patterson, D.J., Eds., Biopress Ltd., Bristol, United Kingdom, 1987.
3. Foissner, W., Blatterer, H., Berger, H., and Kohmann, F., Taxonomische und ökologisches Revision der Ciliaten des Saprobiensystem, Band I: Cyrtophorida, Oligotrichida, Hydrotrichida, Colpodea. Informationsnberichte der Bayerische Landesamt für Wasserwirtschaft, München, 1991.
4. Prescott, D. M. and James, T. W., Culturing of *Amoeba proteus* on *Tetrahymena, Exp. Cell Res.,* 8, 256, 1955.
5. Burton, G. A., Jr., Assessing the toxicity of freshwater sediments, *Environ. Toxicol. Chem.,* 10, 1585, 1991.
6. USEPA, *Methods for the Chemical Analysis of Water and Wastewater,* EPA 600/4-79-020, National Technical Information Service, Washington, DC, 1985.
7. Baes, C. F., Jr. and Mesmer, R. E., *The Hydrolysis of Cations,* John Wiley & Sons, New York, 1976.
8. Sokal, R. R. and Rohlf, F. J., *Biometry,* 2nd ed., W. C. Saunders Publishers, New York, 1981.
9. Blaise, C., Microbiotests in aquatic ecotoxicology: characteristics, utility, and prospects, *Environ. Toxicol. Water Qual.,* 6, 145, 1991.
10. Bitton, G. and Freihoffer, V., Influence of extracellular polysaccharides on the toxicity of copper and cadmium toward *Klebsiella aerogenes, Microb. Ecol.,* 4, 119, 1978.
11. Giesy, J. P., Jr., Laversee, G. J., and Williams, D. R., Effect of naturally occurring aquatic organic fractions on cadmium toxicity to *Simocephalus serrulatus* (Daphnidae) and *Gambusia affinis* (Poeciliidae), *Water Res.,* 11, 1013, 1977.
12. Hall, J. C., Raider, R. L., and Grafton, J. A., EPA's heavy metals criteria: strategies for obtaining reasonable limitations, *Water Env. Technol.,* March 1992, 60, 1992.
13. Leppard, G. G., Ed., *Trace Element Speciation of Surface Waters and its Ecological Implications,* Plenum Press, New York, 1983.
14. Stumm, W. and Morgan, J. J., *Aquatic Chemistry: Chemical Equilibria and Rates in Natural Waters,* 3rd ed. John Wiley & Sons, New York, 1996.
15. Schubauer-Berigan, M. K., Dierkes, J. R., Monson, P. D., and Ankley, G. T., pH-dependent toxicity of Cd, Ni, Pb, and Zn to *Ceriodaphnia dubia, Pimephales promelas, Hyalella azteca,* and *Lumbriculus variegatus, Environ. Toxicol. Chem.,* 12, 1261, 1993.

16. Forge, T. A., Berrow, M. L., Darbyshire, J. F., and Warren, A., Protozoan bioassays of soil amended with sewage sludge and heavy metals using the common soil ciliate *Colpoda steinii, Biol. Fertil. Soils,* 16, 282, 1993.
17. Ruthven, J. A. and Cairns, J., Jr., The response of freshwater protozoan artificial communities to metals, *J. Protozool.,* 20, 127, 1972.
18. Hart, K. M. and Cairns, J., Jr., The maintenance of structural integrity in freshwater protozoan communities under stress, *Hydrobiologia,* 108, 171, 1984.
19. USEPA, *Quality Criteria for Water,* USEPA, Office of Water, Washington, DC, 1986.

Invertebrates

CHAPTER 24

Utility and Practical Considerations for *In Vitro* Developmental Toxicity Testing: Hydra and FETAX Assays

E. Marshall Johnson

CONTENTS

ABSTRACT

The utility of *in vitro* developmental toxicity assays can be tested only by comparing their outcomes with those of standardized *in vivo* developmental toxicity studies. There are many potential ontogenetic systems that merit exploration for such application. In their development, each must be manipulated by the investigators to devise a means whereby the system can reliably recapitulate useful outcomes of standard tests. Failure to locate *in vitro* endpoints that recapitulate a useful aspect of standard *in vivo* studies can be a major impediment to *in vitro* assay development. An *in vitro* assay's reliability can be no greater than that of the *in vivo* studies with which it is compared. Just as one would not compare results of two-year and one-year carcinogenesis studies, one must exercise equal care in identifying a gold standard for validating *in vitro* developmental toxicity studies.

I. INTRODUCTION

This chapter is limited to a discussion of the usefulness of *in vitro* developmental screening tests. No attempt was made to describe the value of *in vitro* assays to explore normal and abnormal developmental mechanisms, a topic of considerable potential. Interested readers are directed to recent publications at the leading edge of these mechanistic type studies, e.g., Kochhar and Satre[1] and Flint et al.[2,3]

To appreciate the advantages and limitations of *in vitro* developmental toxicity methods as screening tests, it is necessary to first review the practical use and limitations of standard *in vivo* safety evaluations. The data from standard tests are the starting points for cross-species extrapolation to humans. Therefore, *in vivo* tests provide the gold standard against which the validity of *in vitro* test data, i.e., ability to predict or recapitulate *in vivo* data, can be determined. Just as *in vitro* tests must be performed and interpreted in a uniform manner, so too must the *in vivo* tests. The quality, quantity, and uniformity of interpretation of *in vivo* test data must be consistent, lest the gold standard itself become rate limiting in our understanding of the limitations and attributes of *in vitro* tests[4] and their appropriate use.

A. Goals of *In Vivo* Tests

The goals of *in vivo* developmental toxicity safety evaluations are threefold, at least. The first, and perhaps most obvious, is to identify a *permissible exposure level* (PEL) for industrial, and to a lesser extent, clinical settings. In industrial settings the PEL is established so that, using engineering or industrial hygiene methods, exposure is controlled at or below a level considered unlikely to cause harm in human pregnancies. In clinical settings, the goal is to provide information so the clinician can weigh risks to the conceptus against benefits to the woman. A second goal of developmental toxicity safety evaluations is to provide data supporting *individualized advice and counseling* subsequent to either a protracted or episodic exposure during pregnancy. A third goal is to provide a part of the necessary information to a jury deliberating causation in *toxic tort litigation*.

For the first goal mentioned, the margin-of-safety (MOS) approach, which is currently being refined through benchmark-dose statistical calculations,[5,6] has served society quite well. We have no examples where the no-observed-adverse-effect-level (NOAEL) of standard developmental toxicity studies made in relevant animals, when divided by a 100-fold MOS, fails to provide a sufficiently small human PEL. As more information became available about effect levels of known human teratogenic exposure intensities, it was possible to retrospectively examine MOS magnitudes for adequacy. Such an analysis was made recently[7] by comparing standard animal test data for four chemicals to the surmised lowest toxic exposure producing malformations in human pregnancies. The four known human teratogens examined all had margins of exposure less than 10, even though the data compared were animal NOAEL and human lowest-observed-adverse-effect-level (LOAEL). Considering that the actual animal NOAEL probably was higher than the one reported in the literature, i.e., between the NOAEL and the LOAEL, and that the human NOAEL was lower than the clinical human LOAEL used in the analysis, the actual margin-of-safety (margin-of-exposure in this context) had to be even lower. The conclusion from this is that a 100-fold margin of safety is adequate, and actually might be made somewhat smaller, when deriving the human PEL.

The second goal, individualized counseling, is not met adequately by contemporary *in vivo* data. For the present, when counseling individuals about specific exposures, use of animal data is limited to population-based risk estimations or rote application of the standard 100-fold MOS.

There is no means available to support individualized, probabilistic statements. The attending physicians are advised to weigh the risk and benefit of pharmaceutical therapy during pregnancy,[8] but there is no means whereby such can be achieved in a defensibly quantifiable manner for an individual pregnancy. Human epidemiological studies have served admirably even though after the fact to assess population risks, e.g., thalidomide, DES, and methyl mercury, but actually are of limited utility in detecting chemical exposures hazardous to embryos. This is due in large part to their design, which is inconsistent with the threshold concept applicable to developmental toxicology.[9] That is, epidemiologic studies often do not have adequate exposure data available and therefore must group all "exposed" pregnancies together for statistical comparison with an "unexposed" group. Take the example of aspirin, which is a potent teratogen in all animal species tested — but only at exposure levels high enough to also produce maternal toxicity. At submaternal toxic exposures, salicylate does not cause malformations in experimental animals. The epidemiologic surveys of aspirin intake all have the same flaw. All "exposed" are lumped together. All those who took 2 or 3 aspirin are grouped with those few who took 15 or 20 per day. The very large "n" of the former masks the very small "n" of the few high-exposed pregnancies. Surely salicylate is a human teratogen but such will only be detected when only the high-exposure group is compared with low or nonexposed pregnancies. On the other side of the coin, epidemiologic studies provide false positives simply because of statistical chance. This has a large downside because, unfortunately, in the face of the uncertainties and warnings, advice provided to pregnant women can be more hazardous to the conceptus than a maternal *de minimus* exposure to a teratogenic drug or industrial chemical.[10-12]

The third goal, toxic tort resolution on a plausible scientific basis, is adequately served by *in vivo* data, provided certain caveats are discussed in a relevant manner and understood by those passing judgment. In this situation, as well as in the former two, it is essential to avoid simplistic and misleading terms such as "teratogen" or "non-teratogen." Karnofsky[13] elegantly stated many years ago that almost any chemical could perturb *in utero* development provided the exposure was sufficiently intense and was present during vulnerable stages of embryonic differentiation and organogenesis. This factor is often difficult to place into a reasonable perspective and can become markedly confusing at several levels. Primary among these is that the standard animal developmental toxicology testing protocol is not really a test of teratogenicity[14] because it employs an agent-exposure duration that is overly long for efficient production of live-term terata. Protracted, intense maternal exposure can mask teratogenesis behind embryo/fetal resorption, or constitute an inadequate test because prolonged exposure of sufficient intensity to produce terata may be precluded by intervening maternal death. A second confusing factor is the large number of developmentally toxic scenarios in animals *vis-a-vis* the relatively few agent exposures recognized as human developmental toxicants. The conclusion from this discrepancy is not that animal data are irrelevant to humans. Instead, it reflects that standard animal tests require testing at maternally toxic exposure levels, but such intense exposures rarely occur in humans.

B. Role of *In Vitro* Tests

The possible roles that exist for *in vitro* developmental toxicity tests to achieve the three goals outlined above are:

1. Identify human exposure levels unlikely to perturb ontogenesis
2. Provide information to support individualized counseling
3. Provide data for evaluation of hazard magnitude for specific exposure scenarios to facilitate informed jury decisions

If one explores the means and extent to which standard testing meets these goals, one could develop a realistic perspective on how it, perhaps, could be augmented by *in vitro* tests.

II. STANDARD *IN VIVO* TESTING

The standard animal developmental toxicity test protocol employs four groups of pregnant animals.[15] The first is a sham-treated control; the remaining three constitute the low-, mid-, and high-exposure groups. The test agent is administered from day 7 of pregnancy, which marks the onset of organogenesis, through about day 17 in rats (or about day 21 in rabbits), which marks the end of major organogenesis. The presumed pregnant females are monitored by cage-side inspection, measurement of feed and water intake, and daily weight gain. They are killed just prior to anticipated natural delivery. The products of conception are examined by standardized means to detect any morphologic deviation from normal. The intensity of the three exposure levels of standard *in vivo* tests are defined, at least in part. The lowest exposure level must produce no agent-relatable effects on either the mother or the offspring. The highest exposure level must produce signs of systemic toxicity in the mother, but no more than 10% maternal deaths. The amount of agent administered to the mid-exposure group is based on several factors, usually derived from a pilot or range-finding experiment made in pregnant animals or based on anticipated human exposure to an agent coming onto the market.

The outcome derived from this experiment is the developmental toxicity no observed-adverse-effect-level (NOAEL), which when divided by 100 is generally regarded as a human exposure level unlikely to cause embryo/fetal harm (i.e., the PEL). In the event that a higher (usually the highest) exposure level results in a number of malformed young, an MOS of 1000 may be applied. In most instances, the extra tenfold factor is reserved for studies lacking a clear NOAEL. Its utilization is deemed necessary to estimate the NOAEL from the lowest-observed-adverse-effect level (LOAEL). The major exceptions to this 100-fold MOS generalization are Proposition 65 in California and, on occasion, at least one U.S. federal regulatory agency that sometimes applies an MOS of 1000 to the animal NOAEL. In the latter situation, consideration of the larger MOS can come up during sponsor/agency discussions if frank and severe terata were produced in the high-exposure animal pregnancies. The logic of this is difficult to resolve, in that the standard test is not a test of teratogenicity. Also, this "teratogen" label, without careful consideration of other factors, carries the connotation that agents lacking this type of developmental finding are "non-teratogens." This would be less than prudent because the standard test is not designed to elicit live terata. To produce such, short pulse-type exposures are needed. The exposure level may kill the pregnant female if continued but, when experienced for only two or three days, the mother can recover; however, the embryo's regulatory and self-repair mechanisms are overwhelmed short of embryo/fetal death.[16]

III. *IN VITRO* DEVELOPMENTAL SCREENING

A. Identifying the Animal NOAEL

Since the animal NOAEL provides the starting point for establishing the human PEL, an *in vitro* assay capable of predicting the animal NOAEL would be of great value and is a very worthwhile experimental goal.

Within individual species (perhaps only for specific animal strains[17]), *in vitro* prediction of the animal *in vivo* NOAEL may be at least a partially attainable goal. Between species it is unlikely to be reliable because the NOAEL of a developmental toxicity study made in rats does not reliably, or even usually, predict the NOAEL of a similar study made in another species, e.g., rabbits. This fact is recognized by regulatory agencies. Therefore, developmental toxicity tests must be made in at least two species and the NOAEL of the most sensitive becomes the starting point for MOS determination of the PEL. Even though the probabilities of predicting the *in vivo* NOAEL by an *in vitro* assay are not high, some progress has been made and, perhaps, might be achieved eventually, at least, within chemical classes and/or applicable to some animal species.

This type of approach is exemplified by the chick embryo retina system of Daston et al.[18] The technique involves use of explanted and dissociated neural retinas of day 6.5 chick embryos placed into rotating suspension culture for eight days. During this period the cells reaggregate and differentiate to a clearly assayable degree. The striking potential of the assay is evident by the observation that there was 71 and 89% concordance of effect levels (within at least an order of magnitude) between *in vitro* studies and studies using rats and mice, respectively. Because of this, there is some optimism that (after further experimentation) the data may predict toxic maternal serum levels. If such could be established, this may represent the first step toward an *in vitro* assay capable of replacing some animal testing in the all important arena of safety evaluation directed toward identifying the NOAEL from which the human PEL is derived, i.e., risk estimation.

B. Individualized Counseling

For individualized counseling, it perhaps is not surprising that the highly manipulable *in vitro* methods can have greater utility than tests made in intact pregnant animals. An outstanding example is provided by the studies reported in recent years from the laboratory of Dr. Norman W. Klein of Storrs, Connecticut.[19] A 1984 report[20] described how postimplantation rat embryos were grown in culture media containing blood serum from individual women who were on diverse pharmacologic therapies for epilepsy. The sera from women receiving phenytoin, valproic acid, or carbamazepine produced more incidences of abnormal embryogenesis than did sera of untreated pregnancies or those receiving phenobarbital. Even though it was not possible in this study to detect a direct relationship between malformation rate and serum parent drug concentration, it is a notable and intriguing study. At least one confounder was identified in this study that may assist in understanding the absence of a clear dose-response relationship between drug concentration and developmental effects. Nutritional supplementation overcame the teratogenic effects of some sera — a factor of unquantified magnitude that could have complicated direct drug-level and effect relationships. Another confounder was reported subsequently[21] when, in addition to nutritional factors, sera of women with prior histories of spontaneous abortion were found lacking in normal ability to support *in vitro* rat embryo development.

C. Litigation Support

For use in toxic tort litigation, *in vitro* assays presently are of little use and may be counterproductive of reasonable decisions. Their lack of utility seems to reside primarily in the fact that *in vitro* studies can produce adverse effects on almost any parameter by mechanisms irrelevant to humans, due to factors as simple as altered pH, accumulation of by-products, or (and most often) use of irrelevant agent concentrations *in vitro*.

IV. PRESCREENING TESTS FOR DEVELOPMENTAL SELECTIVITY

A fourth and somewhat less pressing use of *in vitro* tests is detecting agents that have a predilection to interfere with progressive *in utero* ontogenesis at exposure levels lower than those needed to perturb maternal homeostasis.[22] Even though agents of this type could be used safely because the developmental toxicity NOAEL would drive the regulatory scene, they represent the more insidious developmental toxicants. Fortunately, agents of this type are rather rare.[23] However, even when used with an adequate MOS, they tend to be driven from the marketplace for reasons unrelated to science.

A. Developmental Selectivity, the Hydra Assay

The degree of developmental selectivity can be quantified by considering the adult-to-developmental toxicity NOAELs as an adult/developmental (A/D) ratio. This has application for identifying those agents with selective developmental toxicity from among a group of congeners or industrially acceptable alternatives, as well as mixtures[24] and large groups of chemicals,[25] and eliminating them from further research and development.[26] The A/D ratio can be identified *in vitro* by a short-term test employing the freshwater hydroyoan *Hydra attenuata*.[26,27] This genus marks a crossing point of two important factors. It is the phylogenetically lowest animal form composed of complex organs and tissues, and it is also the highest animal capable of total whole-body regeneration from its dissociated cells. In this assay, adult hydra are dissociated into their component cells which, when randomly packed together by centrifugation, can achieve a remarkably wide spectrum of developmental events characteristic of any real embryo. This is an essential attribute for a developmental toxicity prescreening system because abnormal development is achieved by mechanisms as diverse as, for example, the long sequence of uncoupled oxidative phosphorylation leading to reduced ATP levels causing microtubular dysfunction manifested as failed chromosomal separation and reduced cell division, or as subtle as an improperly crosslinked extracellular matrix leading to faulty directional cell migration and differentiation.

The primary factor limiting wider application of this type of system is that consistent precision is needed, but frequently is lacking, in evaluating the standard developmental toxicity test data with which *in vitro* outcomes can be compared to test the latter's degree of validity. The greatest single confounder to ascertaining the validity, or lack thereof, of an *in vitro* assay is the adequacy and consistent uniformity of the standard tests — i.e., the gold standard — with which the *in vitro* assay determinations are compared and contrasted. A second confounder to ascertaining validity is the care with which the comparisons are made. These factors are exemplified by the consideration of the Hydra Assay's utility.

Determining a chemical's predilection to produce developmental toxicity in the presence or absence of maternal systemic toxicity, i.e., the size of the A/D ratio, is remarkably easy. This coelenterate's responses to graded concentrations of test agent are very clear, even stereotyped. The first and essential step in devising the assay was to sift through hydra's time-dependent sequence of responses to toxicant exposure levels to empirically identify endpoint assays such that division of the adult NOAEL by the developmental NOAEL, would result in proportions similar to those of standard tests evaluated by standard means.[28] Now that the endpoints and procedures have been standardized for the Hydra Assay, interlaboratory recapitulations of ratios are virtually identical. Problems such as technical errors in fabricating the artificial "embryo" produce an obviously spurious effect at a uniform time, and the rare problems of mold or bacterial contamination due to faulty feeding methods also are fixed both in nature and time. Thus, they do not confuse data interpretation, and the assay is useful for previously untested material[29] and retrospectively identified thalidomide as the markedly selective developmental toxicant it is.[30]

B. Cross-Species Extrapolation

Cross-species comparisons of A/D ratios can be difficult and confusing if they are made incautiously.[31] There are two major confounders, and each is related to the *in vivo* evaluation. The primary confounder is use of nonstandard exposure durations in two different species to calculate the animal adult and developmental NOAELs. This problem is not difficult to avoid but has led to some confusion, primarily with the adult (A) component of the *in vivo* ratio.

The most consistent method for both developmental and adult NOAEL determination is the mode used by national and international regulatory agency personnel in face-to-face encounters with manufacturers of specific chemicals. On these occasions, reliance solely on statistically significant differences of effects between exposed and control levels is not considered the NOAEL for either development or perturbed maternal homeostasis. More often each NOAEL is based on inspection of the exposure/response relationship; less often it is based on statistical pair-wise trend analysis. Since the A/D ratio must be based on data from standard animal tests, this test is an essential prelude to the *in vivo* A/D calculations necessary for validation studies. Because it is possible for reasonable persons to reasonably disagree, exact consensus cannot always be attained regarding the *in vivo* NOAEL. To overcome this potential flaw in the gold standard, there might be situations where a range of possible ratios is the best a standard animal study can provide for *in vitro* assay validation. These instances tend to be rare, but do occur. Much more typical are databases where agreement on NOAEL can be attained. For example, even under the difficult circumstances of glycol ether developmental effects (which involved groups as diverse as NIOSH, EPA, manufacturers, and a consultant for each), the issues of both developmental and maternal toxicity NOAELs were resolved, and in a manner that avoided simplistic acceptance of statistical significance levels from studies employing different-sized groups of animals.

Failure to use such methods of expert judgment when employing animal test data to explore levels of *in vitro* test accuracy can lead to significant differences in the conclusions regarding *in vitro* data validity and, therefore, utility. For instance, one published report on cross-species comparison of A/D ratio[32] purported that the ratios differed between three animal species: rat, mouse, and hamster. The investigators relied on statistical significance to derive the adult and developmental NOAELs, even though each study had an obvious dose-related incidence of effect at a lower exposure level and, in several instances, the cited study had failed to identify a NOAEL when the data were analyzed by either statistical significance level or dose-response relationship.

The A/D concept grew from standard animal tests employing exposure throughout the period of major organogenesis. When transient or more pulse-type exposures are employed, the A/D ratio will be a different number.[33] The NOAEL of a pulse-exposure in pregnant mammals tends to be about an order of magnitude larger than that of a standard test. If continued, that intensity of exposure would lead to maternal incapacitation or death. At present, it is not known if data of pulse-type exposures could be recapitulated *in vitro*, but it may be a topic worthy of experimental testing. Unfortunately, when the definition of the animal A/D ratio (derived from standard tests) is redefined to include pulse-type data and the interspecies ratios are compared, an incorrect conclusion can be reached regarding the ratio's reliability.[34] It is as illogical to compare such apples and oranges[35] as it would be to rank agents regarding carcinogenicity by comparing data from a 90-day study with one spanning two or three years of exposure and ascertainment.

An even more common confounder of adequate validity examination exists in differences in experimental designs.[36,37] This can be illustrated by exploring the manner in which maternal effects are reported.

The factor that most frequently drives the maternal NOAEL is decrement of anticipated maternal weight gain. Significant differences exist in how effects on this parameter are reported in the literature, and this can lead to confusion regarding selective developmental toxicity.[38,39] Ascertainment of maternal weight effects over the entire interval between initial maternal exposure and near-term sacrifice fails to take into account rebound of weight gain after exposure ceases, or even during the later stages of exposure. An initial decrement of maternal weight gain early in treatment (when the embryos are most vulnerable) can become masked by greater than control weight gain over the next several days. This phenomenon was illustrated by the study of Solomon et al.[40] In this standard developmental toxicity study of methyl methacrylate in rats, maternal weight gain was reported over several gestational intervals in the control, 99, 304, 1178, and 2028 mg/kg/day dose groups. Treatment began on pregnancy day 6 and ended on day 16; autopsy was on day 20. When maternal weight gain of treated and control dams was measured between days 6 and 20, only the gain of 2028 mg/kg/day dose group was statistically significantly lower than the controls. When maternal weight gain was reported between days 6 and 16, there was a statistically significant difference between the control and both the 1178 and 2028 mg/kg/day dose groups. When maternal weight gain as a measure of maternal toxicity was reported between days 6 and 8, all four dose groups were statistically significantly lower than the concurrent controls. From these data, it is obvious that clear definition of how the maternal effect levels are ascertained is of great importance when seeking a reliable and standardized gold standard. Failure to exercise adequate caution for such a basic methodologic factor will erroneously lead to the conclusion[41] that the phenomenon of developmental selectivity is not somewhat constant between mammalian species studied by means of the standard developmental toxicity experimental design.

C. The FETAX Assay

Another screening type assay that merits mention in this context employs blastula stages of the *Xenopus* frog[42] and is called the FETAX (Frog Embryo Teratogenesis Assay). This has a potential that may exceed that of hydra "embryos" because the frog embryo has many organs morphologically similar to mammals and therefore may one day be able to identify specific target organs of individual agents. Hydra is incapable of providing information about the organs most likely to be the primary targets of an agent, but *Xenopus* merits exploration for this attribute. As presently developed, the test attempts to rank chemical agents on their teratogenetic potential based on the difference (ratio) between the chemical concentration causing 50% embryonic lethality and that producing 50% abnormal embryos. As presently constructed, it is necessary to appreciate the confounders that proportionality of two 50% effect levels may give one relationship, whereas, lower- or higher-effect levels, say 10%–0% or 80%–80%, will provide different conclusions unless the two dose-response curves happen to be parallel. This is not necessarily an insurmountable difficulty and perhaps could be resolved by careful comparisons with mammalian *in vivo* animal data and selection of a goal other than differentiating between "teratogens" and "non-teratogens."

The FETAX assay also needs a better definition of potency. There are many criteria whereby developmental toxicant potencies can be compared:

- It produces live abnormal young
- It produces *in utero* death
- It produces severe abnormalities
- It produces developmental abnormalities in a high percentage of exposed pregnancies
- It produces developmental abnormalities at a low-exposure level
- It produces developmental abnormalities in numerous species
- It produces developmental abnormalities at exposure levels below those overtly toxic to the mother
- It produces developmental abnormalities only in the exposure range also toxic to the mother, but humans frequently experience exposure of this intensity

If this assay were modified by some means, it may be able also to provide a ranking index of developmental selectivity as is provided by Hydra. If such were achieved, it would be the most useful test at the present time.

V. DISCUSSION

Clearly, Karnofsky's "law" must be taken into account for developmental toxicology. Assays that purport separation of teratogens from non-teratogens[43,44] are incompatible with experience that began at the very onset of experimental developmental toxicology. Use of the results of assays that purport to identify such capabilities may be hazardous.

There are several lines of *in vitro* developmental toxicity investigation with the potential to actually replace some *in vivo* animal testing, and one is optimistic that a degree of such will be achieved eventually. The factors that presently limit full use of prioritization-type prescreening tests are not the biology involved. Tests of this type involve the basic factors of target organ toxicity applied to the pregnant animal. It is important to consider that embryos — whether they be mice, rabbits, or humans — achieve organogenesis by essentially the same mechanisms. The embryo evidently is more dependent on the timely achievement of such largely undiscovered developmental phenomena than is the adult, e.g., timely crosslinking of collagen, and these are the most likely causes of larger A/D ratios. Pharmacokinetic differences in *in vivo* systems are unlikely to be overcome in closed *in vitro* tests. This difficult hurdle may preclude absolute uniformity of A/D ratios between species, but the effects have only a rather small impact on the A/D ratio in all but rare instances.

By far, the largest problem for validating *in vitro* developmental prescreening techniques is the absence of uniformity in the application of the *in vivo* gold standard. Published *in vivo* studies frequently provide inadequate data on maternal toxicity, but such can be obtained from the original study reports. Last, but not least, we must move beyond comparisons of apples and oranges and also test *in vitro* validity on the basis of the dose-response relationship, lest the statistical limiting power of "n" of standard tests preclude employment of *in vitro* assays for the utility they possess.

REFERENCES

1. Kochhar, D. M. and Satre, M. A., Retinoids and fetal malformations, in *Dietary Factors and Birth Defects*, Sharma, R. P., Ed., Pacific Division, AAAS, San Francisco, 134, 1993.
2. Flint, O. P., Masters, B. A., Palmoski, M., Wasserman, J. A., and Durham, S. K., Antiviral nucleoside analogs (FddA, FddI, ddA, and AZT): comparative toxicity in rat embryo beating heart cell cultures, *in vitro Toxicol.*, 6, 201, 1993.
3. Flint, O. P., Weiss, A., and Durham, S. K., Anti-AIDS nucleoside analogs (ddI, d4T, ddC, and AZT): comparative *in vitro* neurotoxicity study using the micromass culture technique, *in vitro Toxicol.*, 6, 221, 1993.
4. Jensen, M., Newman, L. M., and Johnson, E. M., Problems in validation of *in vitro* developmental toxicity assays, *Fund. and Applied Toxicol.*, 13, 863, 1989.
5. Crump, K. S., A new method for determining allowable daily intakes, *Fund. and Appl. Toxicol.*, 4, 854, 1984.
6. Barnes, D. G. and Dourson, M., Reference dose (RfD): description and use in health risk assessments, *Regulatory Toxicol. and Pharm.*, 8, 471, 1988.
7. Newman, L. M., Johnson, E. M., and Staples, R. E., Assessment of the effectiveness of animal developmental toxicity testing for human safety, *Repro. Toxicol.*, 7, 359, 1993.
8. PDR, *Physicians' Desk Reference*, Medical Economics Data, Montvale, New Jersey, 1993.
9. Johnson, E. M., The relevance of developmental toxicity thresholds to epidemiologic detection of human teratogenic scenarios: an illustration employing acetylsalicylic acid, Manuscript in preparation, 1997.
10. Johnson, E. M., False positives/false negatives in developmental toxicology and teratology, *Teratology*, 34, 361, 1986.
11. Trichopoulos, D., Zavitsanos, X., Koutis, C., and Drogari, P., The victims of Chernobyl in Greece: induced abortions after the accident, *Br. Med. J.*, 295, 1100, 1987.
12. Koren, G., Bologa, M., Long, D., Feldman, Y. and Shear, N. H., Perception of teratogenic risk by pregnant women exposed to drugs and chemicals during the first trimester, *Am. J. Obstet. Gynecol.*, 160, 1190, 1987.
13. Karnofsky, D. A., Mechanisms of action of certain growth-inhibiting drugs, in *Teratology: Principles and Techniques*, Wilson, J. G. and Warkany, J., Eds., University of Chicago Press, Chicago, 185, 1965.
14. Johnson, E. M. and Christian, M. S., When is a teratology study not an evaluation of teratogenicity, *J. Am. Col. Tox.*, 3, 431, 1984.
15. U.S. EPA, Guidelines for Developmental Toxicity Risk Assessment, *Federal Register*, 58, 63798, 1991.
16. Murphy, M. L., A comparison of the teratogenic effects of five polyfunctional alkylating agents on the rat fetus, *Pediatrics*, 23, 231, 1959.
17. Kalter, H., Dose-response studies with genetically homogeneous lines of mice as a teratology testing and risk-assessment procedure, *Teratology*, 24, 79, 1981.
18. Daston, G. P., Baines, D., and Yonker, J. E., Chick embryo neural retina cell culture as a screen for developmental toxicity, *Tox. and Appl. Pharmacol.*, 109, 352, 1991.
19. Chatot, C. L., Klein, N. W., Piatek, J., and Pierro, L. J., Successful culture of rat embryos in human serum: use in the detection of teratogens, *Science*, 207, 1471, 1980.

20. Chatot, C. L., Klein, N. W., Clapper, M. L., Resor, S. R., Singer, W. D., Russman, B. S., Holmes, G. L., Mattson, R. H., and Cramer, J. A., Human serum teratogenicity studied by rat embryo culture: epilepsy, anticonvulsant drugs, and nutrition, *Epilepsia*, 25, 205, 1984.

21. Ferrari, D. A., Gilles, P. A., Klein, N. W., Nadler, D., Weeks, B. S., Lammi-Keefe, C. J., and Killman, R. E., Rat embryo development on human sera is related to numbers of previous spontaneous abortions and nutritional factors, *Am. J. Obstet. Gynecol.* 170, 228, 1994.

22. Johnson, E. M, Screening for teratogenic potential: are we asking the proper question, *Teratology*, 21, 259, 1980.

23. Johnson, E. M., A tier system for developmental toxicity evaluations based on considerations of exposure and effect relationships, *Teratology*, 35, 405, 1987.

24. Fu, L-J., Johnson, E. M., and Newman, L. M., Prediction of the developmental toxicity hazard potential of halogenated drinking water disinfection by-products tested by the *in vitro* Hydra assay, *Reg. Toxicol. and Pharmacol.*, 11, 1, 1990.

25. Newman, L. M., Giacobbe, R. L., Fu, L-J., and Johnson, E. M., Developmental toxicity evaluation of several cosmetic ingredients in the Hydra assay, *J. Am. Col. Toxicol.*, 9, 361, 1990.

26. Johnson, E. M., Christian, M. S., Dansky, L., and Gabel, B. E. G., Use of the adult developmental relationship in prescreening for developmental hazards, *Tera. Carcino. and Mutagen.*, 7, 273, 1987.

27. Johnson, E. M., Gorman, R. M., Gabel, B. E. G., and George, M. E., The *Hydra attenuata* system for detection of teratogenic hazards, *Tera. Carcino. and Mutagen.*, 2, 263, 1982.

28. Johnson, E. M. and Gabel, B. E. G., An artificial "embryo" for detection of abnormal developmental biology, *Fund. and Appl. Toxicol.*, 3, 243, 1983.

29. Fu, L.-J., Johnson, E. M., and Newman, L. M., Prediction of the developmental toxicity hazard potential of halogenated drinking water disinfection by-products tested by the *in vitro Hydra* Assay, *Reg. Toxicol. and Pharm.* 11, 1, 1990.

30. Newman, L. M., Johnson, E. M., Giacobbe, R., and Zang, R-W., The developmental toxicity hazard potential of thalidomide evaluated in *Hydra* with and without MFO activation, *Teratology*, 43, 462, 1991.

31. Jensen, M., Newman, L. M., and Johnson, E. M., Re: problems in validation of *in vitro* developmental toxicity assays, *Fund. and Applied Toxicol.*, 13, 863, 1989.

32. Rogers, J. M., Barbee, B., Burkhead, L. M., Rushin, E. A., and Kavlock, R. J., The mouse teratogen dinocap has lower A/D ratio and is not teratogenic in the rat and hamster, *Teratology*, 37, 553, 1988.

33. Welsch, F., CIIT Activities, *Chemical Industry Institute of Toxicology*, 6, 3, 1986.

34. Faustman, E. M., Short-term tests for teratogens, *Mutation Res.*, 205, 355, 1988.

35. Johnson, E. M. and Newman, L. M., Difficulties in identifying selective developmental toxicity, *Fund. and Applied Tox.*, 18, 635, 1992.

36. Setzer, R. W. and Rogers, J. M., Assessing developmental hazard: the validity of the A/D ratio, *Teratology*, 44, 653, 1991.

37. Johnson, E. M. and Newman, L. M., Selective developmental toxicity: misuse of the concept via mis-application of a mis-defined A/D ratio, *Teratology*, 46, 103, 1992.

38. Johnson, E. M., False positives/false negatives in developmental toxicology and teratology, *Teratology*, 34, 361, 1986.

39. Price, C. J., Kimmel, C. A., Tyl, R. W., and Marr, M. C., The developmental toxicity of ethylene glycol in rats and mice, *Toxicol. Appl. Pharmacol.*, 81, 113, 1985.

40. Solomon, H. M., McLaughlin, J. E., Swenson, R. E., Hagan, J. V., Wanner, F. J., O'Hara, G. P., and Krivanek, N. D., Methyl methacrylate: inhalation developmental toxicity study in rats, *Teratology*, 48, 115, 1993.

41. Rogers, J. M., Comparison of maternal and fetal toxic dose responses in mammals, *Tera. Carcino. and Mutagen.*, 7, 297, 1987.

42. Bantle, J. A., Fort, D. J., Rayburn, J. R., DeYoung, D. J., and Bush, S. J., Further validation of fetax: evaluation of the developmental toxicity of five known mammalian teratogens and non-teratogens, *Drug and Chem. Toxicol.*, 13, 267, 1990.

43. Smith, M. K., Kimmel, G. L., Kochhar, D. M., Shepard, T. H., Spielberg, S. P., and Wilson, J. G., A selection of candidate compounds for *in vitro* teratogenesis test validation, *Tera. Carcino. and Mutagen.*, 3, 461, 1983.

44. Jensen, M., Newman, L. M., and Johnson, E. M., Re: problems in validation of *in vitro* developmental toxicity assays, *Fund. and Applied Toxicol.*, 13, 863, 1989.

CHAPTER 25

Microscale Bioassays for Corals

M. Branton

CONTENTS

I. INTRODUCTION

Coral reef ecosystems worldwide are being degraded or destroyed at an unprecedented rate.[1,2] The decline of these ecosystems is attributed to anthropogenic impacts from increased rates of sedimentation, point and nonpoint source pollution, and physical habitat destruction which, alone or combined with natural stressors, may limit the ability of reef ecosystems to withstand and recover from stress events.[3-7] This may result in habitat loss, mortalities of reef inhabitants, alterations in community structure and an overall reduction in the biodiversity of the reef. When there is a loss of reef-building, hard corals to other reef organisms, such as soft coral, algae and zoanthids, the integrity of the base structure of the ecosystem may be compromised.

A suite of sensitive, standardized tests and criteria for monitoring the state of, and assessing risk to, coral reefs is necessary for the development of long-term management plans, and to facilitate rapid response to threats and prevention of further damage.[4,7,8] There are many well-established techniques used to measure and evaluate deterioration in the reef environment once it has caused individual and community changes. These include histopathological studies,[9,10] the monitoring of live coral cover and species diversity,[11-15] measuring productivity and calcification,[16-18] and the assessment of coral reproduction and recruitment patterns.[19-22] Although these provide vital information about the dynamics of disturbed reef communities, it is also crucial to develop and improve upon methods used to detect stress effects before deterioration has impacts at the community or ecosystem level.[23] A number of small-scale bioassays have been, and are being, developed for corals to meet this need. This chapter describes the advantages and disadvantages of using corals as biomonitoring organisms, and reviews several bioassays that show potential for being sensitive to sublethal stress in corals both in the laboratory and in the field.

II. CORAL BIOLOGY AND ECOLOGY

Corals are classified in the Phylum Cnidaria (formerly known as the Phylum Coelenterata) along with hydroids, jellyfish, and sea anemones.[24] The corals referred to in this chapter are hermatypic (reef building) corals of the Class Anthozoa, commonly known as hard corals. Hard corals and coralline algae are the primary sources of calcium carbonate which form the limestone structure of coral reefs. Reef-building corals derive their nutrition both from carnivory and from their symbiotic algae, zooxanthellae (*Symbiodinium sp.*), which live in the coral gastrodermis. Zooxanthellae provide their host animal with carbon compounds, an energy source used for metabolism, growth, and reproduction, and they aid in the secretion of calcium carbonate by lowering carbon dioxide levels in the coral tissues. Corals reproduce sexually, utilizing both internal and external fertilization, and asexually, through budding. Most tropical coral species grow colonially after larvae (planulae) settle onto the substrate. New corals grow into the calcium carbonate skeleton laid down by previous generations and then secrete their own exoskeleton forming either branching or massive colonies.[24-26]

Given their role as the primary builders of coral reefs, corals are an important group of biomonitoring organisms for aquatic bioassays and toxicity studies.[27] Both advantages and limitations of the use of corals in such studies are reviewed here. Corals are sedentary animals which are abundant in reef ecosystems. This facilitates coral collection, although the ease of sampling is dependent upon the target coral species, with the removal of pieces of massive corals generally requiring more exertion than that of branching corals. Selective sampling of corals should have low impact on remaining coral populations.[28] However even the conservative collection of corals in a system that is degraded may result in the amplification of stress effects that are already present. A coral colony of either branching or massive species can be fragmented into individual units for replicates thus minimizing genetic variability in test organisms. These units can then be re-attached in the field or stabilized in laboratory aquaria. The more closely experimental conditions mimic the native environment, including appropriate light, oxygen levels, sea water nutrient levels and pH, and water flow, the more successfully they will adapt to the laboratory environment. Regardless of holding conditions, some coral species will adjust better than others. Atkinson et al.[29] provide basic guidelines for the maintenance of coral in aquaria.

The selection of appropriate coral species for toxicity assays and monitoring studies requires the consideration of several factors. Accurate coral identification is critical and may be hampered by difficulties encountered in their identification to the species level.[30-34] Inter- and intraspecies variability in susceptibility to different stressors, the ability of corals to adapt to microniches, and the acclimatization of corals to changes in environmental conditions make it difficult to ensure that a measured response reflects exposure to and uptake of stressors.[7,35-38] Among the known variables that influence sensitivity are coral form, (e.g., branching or massive), polyp size, orientation, mucus production, and method of reproduction.[3,8,16,39]

III. PROMISING MICROSCALE ASSAYS FOR CORALS

Several small-scale, rapid bioassays exist or are being developed for use with corals in water quality and reef monitoring studies. These include buoyant weighing, quantification of zooxanthellae

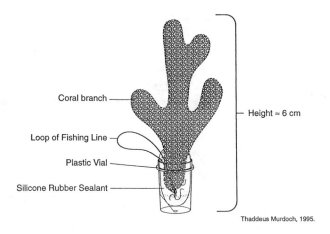

Thaddeus Murdoch, 1995.

Figure 25.1   A coral implanted in a clear plastic cup with silicone rubber sealant.

density and photosynthetic pigment, measurement of photosynthesis:respiration ratios,[40-43] monitoring the synthesis and accumulation of stress (heat shock) proteins, determination of metal loads in coral tissue and skeleton, and the observation of behavioral responses such as mesenterial filament extrusion and polyp expansion.[8,44-48] With the exception of metal analysis, these assays are nonstress-specific indicators of the general state of the coral. The monitoring of zooxanthellae density, chlorophyll concentration, photosynthesis:respiration ratios, coral growth, and behavioral responses has been conducted with both chronic and acute exposures to stressors such as heat, ultraviolet radiation, oil pollution, drilling fluids, and heavy metals.[3,39, 47-52] Heat shock proteins have primarily been studied in corals exposed to thermal stress in acute studies, but show potential for use with other stressors and may be useful in chronic exposures.[35,51-55] Metal load assays are metal specific and may be of value where metal contamination is suspected or known.[27,28,50] Small-scale bioassays used for all of these indicators, excluding photosynthesis:respiration rates and behavioral responses will be reviewed here.

A. Buoyant Weighing

Growth rates are frequently measured to assess the health of corals.[17,56,57] Growth is a useful parameter because it reflects the overall physiological condition of the coral. A number of methods have been devised to monitor growth in the field, but they require long measurement intervals or sacrificing the coral.[58-61] A more sensitive and rapid way to determine growth rates that allows repeated measurements is buoyant weighing. This technique has been used by a number of researchers[62-65] and recently refined by Davies.[66,67] Buoyant weighing is a robust and proven technique that requires only an accurate balance (precision = ±0.001) and can detect growth in branching species in as little as 24 hours for branching corals[66] or two to three days for massive species.[61,62] This technique may be most useful for laboratory assessments because of the possible added stress of handling corals maintained in the field during collection and transportation for shore or vessel-based weight measurements (S.R. Smith, personal communication).

1. Techniques

Growth rates of small pieces (nubbins) of branching corals and small diameter cores taken from the surface of massive corals are determined by estimating buoyant weight in sea water over time. Changes in the buoyant weight of the coral pieces indicate the rate of coral skeletal growth in terms of the amount of mass added between measurement intervals. Coral fragments are secured in a plastic cup or PVC tubing using silicone (Figure 25.1). A loop of monofilament fishing line is used

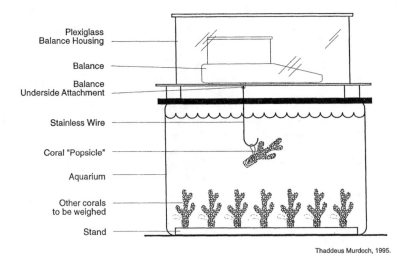

Thaddeus Murdoch, 1995.

Figure 25.2   Coral buoyant-weight apparatus.

to attach the coral to a balance (Figure 25.2). Cores of massive corals can be suspended in weighing trays if they are too large for plastic cups.[62] Corals can be maintained either in aquaria or in the field while they regenerate tissue damaged in the fragmentation process. Recovery takes approximately three days for branching coral and three weeks for massive coral after which they are weighed while still submerged, with corrections made for sea water and coral skeleton density.[66,67] Measurements should be taken at the same time each day. Prior to weighing, it is important to ensure that no water wets the aerial portion of the wire and that any debris or air bubbles are removed from the coral units.[66] In a study correlating lesion regeneration and coral growth, Meesters et al.[16] determined no significant handling effect for repeated weight measurements. Corals may be placed back in the field or maintained in the laboratory and exposed to the stressor of interest. If corals are ultimately sacrificed, growth rates can be standardized against the live surface area using either the wax method[68] or the aluminum foil method.[69]

B. Zooxanthellae Density Determination

When corals lose their zooxanthellae, or their zooxanthellae have reduced amounts of photosynthetic pigment, some physiological functions of the coral may be impaired or, in extreme cases, mortality may result.[70,71] These losses are critical for reef-building corals as zooxanthellae provide the energy needed for calcification, regeneration, and the ability to withstand further stresses.[47] Zooxanthellae also give corals their color, and their partial or total loss results in the phenomenon known as coral bleaching. The mechanism and conditions that bring about the expulsion of zooxanthellae are not fully understood;[72] however their loss is associated with a number of stressors including elevated temperatures, sedimentation, fluxes in salinity, exposure to chemicals, and aerial exposure.[8,39,49] There may be a loss of zooxanthellae or a diminishment of photosynthetic pigment *in vivo* without a noticeable change in color; therefore it is important to quantify algal density and chlorophyll concentrations to assess bleaching. Glynn[39] reviewed the difficulties that may be encountered in developing standardized methods for assessing the loss of zooxanthellae based on the complexity of the relationship between zooxanthellae and their hosts,[73] and the variety of species of zooxanthellae and associated pigments. Nonetheless, the monitoring of zooxanthellae loss has been used widely to indicate general stress in corals both in the laboratory and *in situ*.[48,74-76]

1. Techniques

There are several methods used to quantify zooxanthellae density. The first and most common is to remove coral tissue using an air brush or Water Pik.[77] The resulting blastate is homogenized, using a hand-held or motorized Dounce homogenizer, and centrifuged to obtain a zooxanthellae pellet. The pellet is resuspended in a known volume of fresh sea water and fixative, and the algal cells are counted with a hemocytometer, using several replicates for precision. Zooxanthellae density can be correlated with live coral surface area, using either the aluminum foil[68] or wax method,[69] or to total coral protein.[39] The loss of pigment per unit zooxanthella is determined by measuring chlorophyll concentration. Coral tissue is incubated for 24 hours in 90% acetone for extraction of chlorophyll a, chlorophyll c, and carotenoids. The chlorophyll b of corals is not measured because it is indistinguishable from that found in the filamentous green algae that is present in corals.[78] The amount of each pigment present is determined by measuring absorption with a spectrophotometer and using equations to convert absorption to chlorophyll concentrations[79,80] which are normalized to surface area as previously described.

C. Stress Proteins

As part of the cellular stress response cells may synthesize and accumulate heat shock proteins (hsp), also known as stress proteins, when exposed to physical and chemical stressors.[81-83] These proteins enhance cellular tolerance to further environmental insults and assist in repairing damaged cellular components.[52] Heat shock proteins are widely used as indicators of general stress and are induced by a variety of stressors including thermal stress, heavy metals, xenobiotics, and salinity stress.[52] It has been suggested that subsets of these proteins, including metallothioneins and hemeoxygenase,[84] may be linked to specific stressors such as heavy metals, oxidants, and DNA damage.[52,85] Recently, hsp synthesis has been demonstrated in corals exposed to sublethal, acute thermal shocks.[35,51,52,86] Monitoring biochemical changes may permit the detection of sublethal stress effects in corals as these reactions are often fundamental and are extremely rapid, minutes or hours, as compared to days to months for changes at the individual or community level.

1. Techniques

To assay coral for hsp, tissue is removed from the coral skeleton using a high-pressure jet of air from an air brush and immediately placed on ice. Protease inhibitors are added to the resulting blastate which is then homogenized using a hand-held or motorized Dounce homogenizer, and centrifuged to remove zooxanthellae from the host supernatant. The protein in the supernatant is separated by molecular weight using sodium dodecyl sulfide-polyacrylamide gel electrophoresis (SDS-PAGE).[87] Following electrophoresis, gels are either stained with Coomassie blue R250 or blotted to nitrocellulose for protein transfer.[88] Specific protein bands can be detected using autoradiography[51] or Western blotting[89] followed either by the development of the blot in nitroblue tetrazolium with 5-bromo-4-chloro-3-indol phosphate[52] or the application of an enhanced chemiluminescent agent to the blot prior to exposure to X-ray film.[35] The density of protein bands may be quantified using densitometry, then relative concentrations and frequency of the occurrence of bands can be compared across treatments.[35,51,52]

The use of the cellular stress response to indicate sublethal stress in corals is promising, however there are a number of factors that must be addressed before this can be established. One of the difficulties in using hsps in a dose-response relationship with environmental stress is that the speed and transience of hsp synthesis may be specific to different hsps and may vary among species. The range of hsps that are induced in corals and appropriate measurement intervals need to be determined to ensure that the detection of hsps, including their presence or absence, and the concentration of protein bands reflect exposure to adverse environmental conditions.[35] Furthermore, although the relative concentration of protein bands can be estimated, the techniques described are semiquantitative.[52] As such, it may be more appropriate to use the frequency of the occurrence of hsps rather than estimations of their protein concentration to determine exposure to stress.

D. Metal Uptake

While most small-scale, rapid bioassays developed for corals employ general stress indicators, work has been conducted examining the potential use of corals for monitoring metal pollution. Corals may be exposed to metals from mining and dredging operations, drilling muds, contaminated river water, and effluent discharges.[28,90,91] Several studies have examined metal loads in coral skeletal material, however there is no consensus on the extent to which such accumulation reflects bioavailable metals.[28,50,90,92] At sites with high metal contamination, it is possible to monitor environmental metal concentrations using skeletal analysis, while at less polluted sites, metal loads in the skeleton may not be adequate to be measured reliably.[28] Alternatively, metal loads in coral soft tissue can be determined and, according to comparative studies, they better reflect the presence of bioavailable metals in the environment. [27,28,92,93] The measurement of metals in coral soft tissue should be used in preference to skeletal material.

1. Techniques

After collection from the field, or exposure in the laboratory, corals must be washed and treated immediately, dried or frozen. McConchie and Harriott[28] recommend the immediate preparation of tissue and skeleton after collection because other methods, such as oven drying whole corals, have been found to change metal partitioning between tissue and skeletal material due to mucus production. The separation of tissue (organic) from skeletal (inorganic) material can be achieved by digesting the coral first in hydrogen peroxide to obtain the organic matter, followed by the digestion of remaining skeleton in nitric acid.[28,94] Alternatively, coral tissue can be removed immediately to avoid excess mucus production, using a Water Pik or air brush, then filtered and oven dried prior to analysis.[50] Trace metal concentrations in the resulting digest may be determined with spectrophotometry[94] or voltammetry.[28]

If corals are to be used as biomonitors of heavy metal pollution, further research is required into intraspecies sensitivity to different metals and mechanism(s) of metal partitioning. The role that zooxanthellae may have in metal transfer and or retention is unclear, and the extent to which corals may actively exclude some metals is unknown.[28,94] Developing an understanding of these issues is necessary to assess the extent to which metals accumulated in corals reflect environmental concentrations accurately.

IV. RESEARCH CHALLENGES WITH CORAL BIOASSAYS

Corals are important organisms for microscale ecotoxicological studies on the health of coral reefs due to their key role in the ecosystem. The techniques described are practical for water quality monitoring programs and toxicity studies. They can be applied under basic field conditions and can be used to analyze several stress responses from the same organism if tissue is aliquoted and appropriately treated for each bioassay. Coral growth rates, zooxanthellae density, and photosynthetic pigment concentration are proven indicators that reflect the overall physiological condition of corals and are useful in detecting sublethal stress. Monitoring the cellular stress response and measuring metal loads in coral tissue may provide mechanisms for the identification of specific stressors in the reef environment. However, the use of these assays is limited by our current understanding of processes and the development of appropriate protocols. Continued application and refinement of these procedures will contribute to their improved accuracy and reliability. The use of bioassays for the assessment of anthropogenic stress effects requires an understanding of natural fluctuations in the environment and the acclimation of corals to associated stressors. Due to the variability in coral response and acclimation to different stressors, there is a need for sensitivity mapping of corals both by species and by region. The identification and application of a suite of specific and nonspecific sublethal indicators will be invaluable in coral reef management and monitoring programs.

ACKNOWLEDGMENTS

This review was written with the support of the Bermuda Biological Station for Research (BBSR) and the School for Resource and Environmental Studies (SRES) of Dalhousie University. Thanks are due the Benthic Ecology Research Program of BBSR for permission to use their illustrations for the buoyant weighing procedure. I would also like to thank Dr. Robbie Smith and two anonymous reviewers for extremely useful remarks. Finally, my appreciation goes to Dr. Peter Wells of Environment Canada and SRES for helpful comments and guidance.

REFERENCES

1. Grigg, R.W., The international coral reef initiative: conservation and effective management of marine resources, *Coral Reefs*, 13, 197, 1994.
2. World Conservation Union, *Reefs at risk: A programme of action,* IUCN, Gland, 1993.
3. Peters, E.C., Gassman, N.J., Firman, J.C., Richmonds, R.H., and Power, E.A., Ecotoxicology of tropical marine systems, *Environ. Tox. Chem.,* 16, 1, 1997.
4. Crosby, M.P., Drake, S.F., Eakin, C.M., Fanning, N.B., Paterson, A., Taylor, P.R., and Wilson, J., The United States coral reef initiative: An overview of the first steps, *Coral Reefs*, 14, 1, 1995.
5. Done, T., Remediation of degraded coral reefs: the need for broad focus, *Mar. Poll. Bull.*, 30, 686, 1995.
6. Richmond, R.H., Coral reefs: Present problems and future concerns resulting from anthropogenic disturbance, *Amer. Zool.*, 33, 524, 1993.
7. Brown, B.E. and Howard, L.S., Assessing the effects of "stress" on reef corals, *Adv. Mar. Biol.*, 22, 1, 1985.
8. Done, T.J., Ecological criteria for evaluating coral reefs and their implications for managers and researchers, *Coral Reefs,* 14, 183, 1995.
9. Glynn, P.W., Szmant, A.M., Corcoran, E.F., and Cofer-Shabica, S.V., Condition of coral reef cnidarians from the Northern Florida reef tract: pesticides, heavy metals and histopathological examination, *Mar. Poll. Bull.,* 20(11), 568, 1989.
10. Peters, E.C., Meyers, P.A, Yevich, P.P., and Blake, N.J., Bioaccumulation and histopathological effects of oil on a stony coral, *Mar. Poll. Bull.,* 12, 333, 1981.
11. Dodge, R.E., Baca, B.J., Knap, A.H., Snedaker, S.C., and Sleeter, T.D., The effects of oil and chemically dispersed oil in tropical ecosystems: 10 years of monitoring experimental sites, *MSRC Technical Report 95-014*. Marine Spill Response Corporation, Washington, DC, 1995.
12. Guzmán, H.M., Burns, K.A., and Jackson, J.B.C., Injury, regeneration, and growth of Caribbean reef corals after a major oil spill in Panama, *Mar. Ecol. Prog. Ser.*, 105, 231, 1994.
13. Guzmán, H.M., Jackson, J.B.C., and Weil, E., Short-term ecological consequences of a major oil spill on Panamanian subtidal reef corals, *Coral Reefs*, 10, 1, 1991.
14. Bak, R.P.M., Effects of chronic oil pollution on a Caribbean coral reef, *Mar. Poll. Bull.,* 18, 534, 1987.
15. Hudson, J.H. and Robin D.M., Effects of offshore drilling on Phillipine reef corals, *Bull. Mar. Sci.* 32, 890, 1982.
16. Meesters, E.H., Noordeloos, M., and Bak, R.P.M., Damage and regeneration: links to growth in the reef-building coral *Monastrea annularis*, *Mar. Ecol. Prog. Ser.,* 112, 119, 1994.

17. Goreau, T. J. and MacFarlane, A. H., Reduced growth rates of *Monastrea annularis* following the 1987-1988 coral bleaching event, *Coral Reefs*, 8, 211, 1990.

18. Dodge, R.E., Knap, A.H., Wyers, S.C., Frith, H.R., Sleeter, T.D., and Smith, S.R., 1985, The effect of dispersed oil on the calcification rate of the reef-building coral *Diploria strigosa*, *Proc. 5th Int. Coral Reef Symp.* 6, 453, 1985.

19. Guzmán, H.M. and Holst, I. Effects of chronic oil-sediment pollution on the reproduction of the Caribbean reef coral *Sidastrea siderea*, *Mar. Poll. Bull.*, 26, 176, 1993.

20. Richmond, R.H., Effects of coastal runoff on coral reproduction, in *Proceedings, Colloquium on global aspects of coral reefs: health, hazards and history,* Miami Fl, USA, 360, 1993.

21. Atkinson, S. and Atkinson, M.J., Detection of estradiol-17 during a mass coral spawn. *Coral Reefs*, 11, 33, 1992.

22. Rinkevich, B, The contribution of photosynthetic products to coral reproduction, *Mar. Biol.*, 101, 259, 1989.

23. Brown, B.E., Assessing environmental impacts on coral reefs in *Proceedings of the 6th international coral reef symposium, Vol. 1,* Townsville, Australia, 71, 1988.

24. Barnes, R.S.K., Calow, P., and Olive, P.J.W., *The invertebrates: a new synthesis*, 2nd ed., Blackwell Scientific Publications, Oxford, 1993.

25. Humann, P., *Reef Coral Identification,* New World Publications Inc., Jacksonville, 1993.

26. Sebens, K. P., Biodiversity of coral reefs: What are we losing and why?, *Amer. Zool.*, 34, 115, 1994.

27. Hanna, R. G. and Muir, G. L., Red Sea corals as biomonitors of trace metal pollution, *Envir. Mon. Assess.,* 14, 211, 1990.

28. McConchie, D. and Harriott, V. J., The partitioning of metals between tissue and skeletal parts of corals: Application in pollution monitoring, in *Proceedings of the 7th International Coral Reef Symposium,* Vol. 1, Richmond, R. H. ed., University of Guam Press, Mangilao, 97, 1992.

29. Atkinson, M.J., Carlson, B., and Crow, G.L., Coral growth in high-nutrient, low-pH seawater: a case study of corals cultured at the Waikiki Aquarium, Honolulu, Hawaii, *Coral Reefs*, 12, 215, 1995.

30. Weil, E. and Knowlton, N., A multi-character analysis of the Caribbean coral *Monastrea annularis* (Ellis and Solander, 1786) and its two sibling species, *M. faveolata* (Ellis and Solander, 1786) and *M. franksi* (Gregory, 1895), *Bull. Mar. Sci.*, 55, 151, 1994.

31. Van Veghel, M.L.J. and Bak, R.P.M., Intraspecific variation of a dominant Caribbean reef building coral, *Monastrea annularis*: genetic, behavioral and morphometric aspects, *Mar. Ecol. Prog. Ser.,* 92, 255, 1993.

32. Willis, B.L., Species concepts in extant scleractinian corals: considerations based on reproductive biology and genotypic population structures, *Syst. Bot,* 15, 136, 1990.

33. Lang, J., Whatever works: the variable importance of skeletal and of non-skeletal characters in scleractinian taxonomy, *Paleontogr. Amer.*, 54, 18, 1984.

34. Foster, A.B., Phenotypic plasticity in the reef corals *Monastrea annularis* (Ellis and Solander) and *Siderastrea siderea, Bull. Mar. Sci.* 30, 678, 1979.

35. Branton, M.A., Coral reef monitoring: The use of heat shock proteins as biomarkers of thermal stress in the scleractinian coral *Madracis mirabilis*, M.E.S. thesis, Dalhousie University, Halifax, Nova Scotia, Canada, 1997.

36. Shick, J.M., Lesser, M.P., Dunlanp, W.C., Stochaj, W.R., Chalker, B.E., and WuWon, J, Depth-dependent responses to solar ultraviolet radiation and oxidative stress in the zooxanthellate coral *Acropora microphthalma, Mar. Biol.*, 122, 41, 1995.

37. Glynn, P.W., Widespread coral mortality and the 1981/82 El Niño warming event, *Environ. Conserv.*, 11, 133, 1984.

38. Hoeksema B.W., Control of bleaching in mushroom coral populations (Scleractinia: Fungiidae) in the Java Sea: stress tolerance and interference by life history strategy, *Mar. Ecol. Prog. Ser.*, 74, 225, 1991.

39. Glynn, P.W., Coral reef bleaching: ecological perspectives, *Coral Reefs*, 12, 1, 1993.

40. Fitt, W.K. and Warner, M.B., Bleaching patterns of four species of Caribbean reef corals, *Biol. Bull.,* 189, 298,1995.

41. Porter, J.W., Fitt, W.K., Spero, H.J., Rogers, C.S., and White, M.W., Bleaching in reef corals: physiological and stable isotopic responses, *Proc. Nat. Acad. Sci.*, 86, 9342, 1989.

42. Cook, C.B. and Knap, A.H., Effects of crude oil and chemical dispersant on photosynthesis in the brain coral *Diploria strigosa, Mar. Biol.*, 78, 21, 1983.

43. Wyers, S.C., Frith, H.R., Dodge, R.E., Smith, S.R., Knap, A.H., and Sleeter, T.D., Behavioral effects of chemically dispersed oil and subsequent recovery in *Diploria strigosa* (Dana), *Mar. Ecol.*, 7, 23, 1985

44. Loya, Y. and Rinkevich, B., Effects of oil pollution on coral reef communities, *Mar. Ecol. Prog. Ser.,* 3, 167, 1980.

45. Krone, M.A. and Biggs, D.B., Sublethal metabolic responses of the hermatypic coral *Madracis decactis* exposed to drilling fluids enriched with ferrochrome lignosulfate, *Proceedings, Symposium on research on environmental fate and effects of drilling fluids and cuttings, Vol. 2,* Lake Buena Vista, FL., USA, 1044, 1980.

46. Thompson, J.H. and Bright, T.J., Effects of an offshore drilling fluid on selected corals. *Proceedings, Symposium on research on environmental fate and effects of drilling fluids and cuttings, Vol.2,* Lake Buena Vista, FL., USA, 1079, 1980.

47. Meesters, E. H., Noordeloos, M., and Bak, R. P. M., Effects of coral bleaching on tissue regeneration and colony survival, *Mar. Ecol. Prog. Ser.*, 96, 189, 1993.

48. Muscatine L., Grossman, D., and Doino, J., Release of symbiotic algae by tropical sea anemones and corals after cold shock, *Mar. Ecol. Prog. Ser.,* 77, 233, 1991.

49. Jokiel, P. L. and Coles, S. L., Response on Hawaiian and other Indo-Pacific reef corals to elevated temperature, *Coral Reefs*, 8, 155, 1990.

50. Harland, A.D. and Brown, B.E., Metal tolerance in the scleractinian coral *Porites lutea*. Mar. Poll. Bull., 20, 353, 1989.

51. Black, N., A., Voellmy, R., and Szmant, M., Heat shock protein induction in *Monastrea faveolata* and *Aiptasia pallida*, exposed to elevated temperatures, *Biol. Bull.*, 188, 234, 1995.

52. Hayes, R. L. and King, C., M., Induction of 70-kDa heat shock protein in scleractinian corals by elevated temperature: significance for coral bleaching, *Mol. Mar. Biol. Biotech.*, 4, 36, 1995.

53. Sanders, B.M., Pascoe, V.M., Nakagawa, P.A., and Martin, L.S., Persistence of the heat-shock response over time in a common *Mytilus* mussel, *Mol. Mar. Biol. Biotech.,* 1, 1, 47, 1993.

54. Sanders, B.M., Stress proteins: Potential as multitiered biomarkers, in *Biomarkers of environmental contamination*, McCarthy J. F. and Shugart, L. R., Eds., Lewis Publishers, Boca Raton, 457, 1990.

55. Miller, D., Brown, B., Sharp, V., and Nganro, N., Changes in the expression of soluble proteins extracted from the symbiotic anemone *Anemonia viridis* accompany bleaching induced by hyperthermia and metal stressors, *J. Therm. Biol.* 17 (4/5), 217, 1992.

56. Yap, H. T., Alino, P. M., and Gomez, E. D., Trends in growth and mortality of three coral species (Anthozoa: Scleractinia), including effects of transplantation, *Mar. Ecol. Prog. Ser.*, 83, 91, 1994.

57. Tomascik, T. and Sander, F., Effects of eutrophication on reef building corals: growth rates of the reef building coral *Monastrea annularis, Mar. Biol.*, 87, 143, 1985.

58. Vago, R., Vago, E., Achituv, M., Ben-Zion, M., and Dubinsky, Z., A non-destructive method for monitoring coral growth affected by anthropogenic and natural long term changes, *Bull. Mar. Sci.,* 55(1), 126, 1994.

59. Gladfelter, E. H., Monahan, R. K., and Gladfelter, W. B., Growth rates of five reef-building corals in the Northeastern Caribbean, *Bull. Mar. Sci.*, 28, 728, 1978.

60. Buddemeier, R. W. and Kinzie, R. A., Coral growth, *Oceanogr. Mar. Biol. Ann. Rev.*, 14, 183, 1976.

61. Dodge, R.E., Knap, A.H., Wyers, S.C., Frith, H.R., Sleeter, T.D., and Smith, S.R., The effect of dispersed oil on the calcification rate of the reef-building coral *Diploria strigosa*, *Proceedings of the 5th international coral reef symposium,* 6, 453, 1985.

62. Dodge, R. E., Wyers, S. C., Frith, H. R., Knap, A. H., Smith, S. R., Cook, C. V., and Sleeter, T.D., Coral calcification rates by the buoyant weighing technique: effects of alizarin staining, *J. Exp. Mar. Bio. Ecol.,* 75, 217, 1984.

63. Jokiel, R. L., Maragos, J. E., and Franzisket, L., Coral growth: buoyant weighing technique, *Monogr. Oceanogr. Method.*, 5, 529, 1978.

64. Bak, R. P. M., The growth of coral colonies and the importance of crustose coralline algae and burrowing sponges in relation with carbonate accumulation, *Neth. J. Sea Res.*, 10, 285, 1976.

65. Bak, R. P. M., Coral weight increment *in situ*. A new method to determine coral growth, *Mar. Biol.*, 20, 45, 1973.

66. Davies, P. S., A rapid method for assessing growth rates of corals in relation to water pollution, *Mar. Poll. Bull.*, 21(7), 346, 1990.

67. Davies, P. S., Short-term growth measurements of corals using an accurate buoyant weighing technique, *Mar. Biol.*, 101, 389, 1989.

68. Glynn, P.W. and Croz, S.L., Experimental evidence for high temperature stress as the cause of El-Niño coincident coral mortality, *Coral Reefs* 8, 155, 1990.

69. Marsh, J.A., Primary productivity of reef-building calcareous red algae, *Ecology,* 51, 255.

70. Williams, E. H. and Bunkley-Williams, L., The world-wide coral reef bleaching cycle and related sources of coral mortality, *Atoll Research Bulletin*, 335, 1, 1990.

71. Hoegh-Guldberg, O. and Smith, G. J., Light, salinity and temperature and the population density, metabolism and export of zooxanthellae from *Stylophora pistillata* and *Seriatopora hystrix* (abstract), in *Proceedings of the Sixth International Coral Reef Symposium*, Vol. 1, Townsville, Australia, 1988.

72. Gates, R. D., Baghdasarian, G., and Muscatine, L., Temperature stress causes host cell detachment in symbiotic cnidarians: implications for coral bleaching, *Biol. Bull.,* 182, 324, 1992.

73. Buddemeier, R. W. and Fautin, D. G., Coral bleaching as an adaptive mechanism: a testable hypothesis, *Bioscience*, 43(5), 320, 1993.

74. Hayes, R. L. and Bush, P. G., Microscopic observations of recovery in the reef-building scleractinian coral, *Monastrea annularis*, after bleaching on a Cayman reef, *Coral Reefs,* 8, 203, 1990.

75. Szmant, A. M. and Gassman, N. J., The effects of prolonged "bleaching" on the tissue biomass and reproduction of the reef coral *Monastrea annularis*, *Coral Reefs,* 8, 217, 1990.

76. Harland, A. D. and Brown, B. E., Metal tolerance in the scleractinian coral *Porites lutea*, *Mar. Poll. Bull.,* 20, 353, 1989.

77. Johannes R. E. and Wiebe, W. J., Methods for determination of tissue biomass and composition, *Limnol. Oceanogr.*, 15, 822, 1970.

78. Kleppel, G.S., Dodge, R.E., and Reese, C.J., Changes in pigmentation associated with the bleaching of stony corals, *Limnol. Oceanogr.*, 34, 1331, 1989.

79. Jeffrey, S. W. and Haxo, F. T., Photosynthetic pigments of symbiotic dinoflagellates (zooxanthellae) from corals and clams, *Biol. Bull.,* 135, 149, 1968.

80. Jeffrey, S.W. and Humphrey, G.F., New spectrophotometric equations for determining chlorophylls *a, b, c*1 and *c*2 in higher plants, algae and natural phytoplankton. *Biochem. Physiol. Pfl.,* 167, 191, 1975.

81. Parsell, D. and Lindquist, S., Heat shock proteins and stress tolerance, in *The biology of heat shock proteins and molecular chaperones*, Cold Spring Laboratory Press, Plainview, N.Y., 1994.

82. Morimoto, R. I., Tissières, A., and Georgopoulos, C., Progress and perspectives on the biology of heat shock proteins and molecular chaperones, in *The Biology of Heat Shock Proteins and Molecular Chaperones*, Morimoto et al., Eds., Cold Spring Laboratory Press, Plainview, N.Y., 1994.

83. Stegeman, J., Brouwer, M., Di Giulio, R., Forlin, L., Fowler, B., Sanders, B., and Van Veld, P., Molecular responses to environmental contamination: Enzyme and protein systems as indicators of chemical exposure and effect, in *Biomarkers: Biochemical, Physiological and Histological Markers of Anthropogenic Stress,* Huggett, R. J. et al., Eds., Lewis Press, Chelsea, Michigan, 1992.

84. Hightower, L.E., A brief perspective on the heat-shock response and stress proteins, *Mar. Env. Res.,* 35, 79, 1993.

85. Bradley, B.P., Are stress proteins indicators of exposure or effect? *Mar. Env. Res.* 35, 85, 1993.

86. Bythell, J. C., Sharp, V. A., Miller, D., and Brown, B. E., A novel environmentally regulated 33 kDa protein from tropical and temperate cnidarian zooxanthellae, *J. Therm. Biol.,* 20(1/2), 15, 1995.

87. Laemmli, U. K., Cleavage of structural proteins during the assembly of the head of bacteriophage T4, *Nature*, 227, 680, 1970.

88. Towbin, H., Staehelin, T., and Gordon, J., Electrophoretic transfer of proteins from polyacrylamide gels to nitrocellulose sheets: procedure and some applications, *Biochem.*, 16, 4350, 1979.

89. MacRae, T.H., Langdon, C.M., and Freeman, J.A., Spatial distribution of posttranslationally modified tubulins in polarized cells of developing *Artemia, Cell. Motil. Cytoskel.* 18, 189, 1991.

90. Howard, L.S. and Brown, B.E., Heavy metals and reef corals, *Oceanogr. Mar. Biol. Ann. Rev.,* 22, 195, 1984.

91. Brown, B.E. and Holley, M.C., Metal levels associated with tin dredging and smelting and their effect upon intertidal reef flats at Ko Phuket, Thailand, *Coral Reefs,* 1, 131, 1982.

92. Shen, G.T., Boyle, E.A., and Lea, D.W., Cadmium in coral as a tracer of historic upwelling and historic fallout, *Nature*, 328, 794, 1987.

93. Bell, P. R. F., Grenfield, P. F., Hawker, D., and Connell, D., The impact of waste discharges on coral reef regions, *Wat. Sci. Tech.*, 21, 121, 1989.

94. Rainbow, P.S., Phillips, D.J.H., and Depledge, M.H., The significance of trace metal concentration in marine invertebrates. A need for laboratory investigation of accumulation strategies. *Mar. Poll. Bull.*, 21, 321, 1990.

95. Glynn, P.W., Szmant, A.M., Corcoran, E.F., and Cofer-Shabica, S.V., Condition of coral reef cnidarians from the Northern Florida reef tract: pesticides, heavy metals and histopathological examination, *Mar. Poll. Bull.*, 20, 568, 1989.

CHAPTER 26

The Use of Larvae and Small Species of Polychaetes in Marine Toxicological Testing

Donald J. Reish

CONTENTS

I. INTRODUCTION

Polychaetes are an important component of the marine environment especially in the benthos where they comprise 35 to 50% of both the total macroscopic species and specimen population.[1] Larval stages are pelagic with a free-swimming period lasting from a day to over a month. They may or may not feed during this period. Not all polychaetes have a free-swimming stage; others may lack a trochophore phase and are incubated by their parent and leave as juvenile worms. For purposes of this review, small polychaetes are defined as measuring less than 1.0 cm in length. They are found living among mussel beds, algal fronds, algal holdfasts, and other similar associations as well as in pore water between sediment particles.

Toxicological tests with polychaetes were initiated 35 years ago but have been used extensively only in the past 15 years.[2,3] Tests have measured the effects of toxicants in the water, sediment, and interstitial water (pore water). Toxicological tests have been conducted with over 40 species from 19 families, the majority of which were adults measuring greater than 1.0 cm in length.[3] The effect of toxicants on polychaetes in general has been reviewed in the past[3-7] but not specifically for small polychaetes or larvae suitable for microscale assays. Refer also to Chapter 27.

II. TOXICOLOGICAL TESTS

Tests with small polychaetes or larvae have been employed to a limited extent. Acute tests using the trochophore larval stage (e.g., *Capitella capitata*[8]) or juveniles (e.g., *Neanthes arenaceodentata*[9]) are generally of a 96-hour duration. Larval tests are generally conducted up to a 96-h period prior to the onset of metamorphosis and settlement.[10] The number of survivors per test concentration is determined, the LC_{50} is calculated, and the time period specified (i.e., 96-h LC_{50}). Chronic tests last for a longer period of time (7 to 28 days) and begin with either juveniles or immature adults (e.g., *Ophryotrocha diadema*[6]). Toxicity is also expressed as an LC_{50} for the test period. Reproductive tests frequently begin with the newly settled larvae and continue through the egg-laying stage (e.g., *Capitella capitata*[8]). Life cycle tests extend the reproductive test through the settlement of the F_1 generation (e.g., *Ophryotrocha diadema*[6]). In both the reproductive and life cycle tests, the results are expressed as the number of eggs or larvae produced within each test concentration and statistically compared to the control.[6]

Tests are usually conducted in an aqueous medium to which a toxicant is added (e.g., mercuric chloride). Other types of tests, which may be acute or chronic and include reproductive measures, include the pore water test (e.g., *Dinophilus gyrociliatus*[11]) or sediment test measuring growth as determined by the increase in body weight (e.g., *Neanthes arenaceodentata*[9]). Manuals outlining testing procedures have not yet been written specifically for larvae or small polychaetes, but test protocols have been published for polychaetes in general and would be helpful for these types of worms.[6,12-15]

A. Toxicants Tested

Table 26.1 summarizes the types of toxicants that have been tested with larvae or small species of polychaetes. Metals have been the most frequently tested class of toxicant probably because of the ease of testing with metal salts in an aqueous solution. Other toxicants tested include: detergents which are generally those used in the clean-up of oil spills; petroleum hydrocarbons, especially the water-soluble fractions of natural or refined products; pesticides; contaminated sedimental and other media substances such as pore water or organic compounds.

B. Test Species

A total of 14 species in 11 families of polychaete larvae or small species have been used in marine ecotoxicology (Table 26.1). Members of the genus *Ophryotrocha* are the most commonly used species. These species are small and easily cultured under laboratory conditions.[16] Tests begin with newly hatched juveniles possessing about four setigerous segments.[13] These species are convenient for life cycle tests since they require only about 35 days to complete one generation. *Dinophilus gyrociliatus* is a minute polychaete which completes its life cycle within 10 days or less. It has been used in pore water tests[11] as well as in studying the effects of metals or detergents. Tests with the trochophore larval stage of *Capitella capitata* are easy to initiate since a large number can be extracted from females incubating the developing embryos.[8] Reproductive or life cycle tests can be started with the trochophores of this species. While the adult of *Neanthes arenaceodentata* measures 3 to 6 cm in length, the juveniles recently emerged from the parent's tube (a male in this species) are less than 1.0 cm in length and have been used in comparisons with juvenile and adult sensitivity (see below). In addition, a sublethal test has been described using this species which measures the increase in body weight above the pretest weight and compares growth to that of a control.[11]

Table 26.1 The Effect of Chemicals and Environmental Samples on Polychaete Larvae and Small Species (less than 1 cm)

Family — Species	Metals	Petroleum	Detergent	Sediment	Pesticides	Other	References
Capitellidae							
Capitella capita	X		X		X		6,8,10,17,21-23
Cirratulidae							
Cirriformia luxuriosa	X						6
Ctenodrilidae							
Ctenodrilus serratus	X	X				X	6,24-26,43
Dinophilidae							
Dinophilus gyrociliatus	X		X		X	X	11,27,28,35
Dorvilleidae							
Ophryotrocha diadema	X	X	X			X	6,24,27,28,29-31,43
Ophryotrocha labronica	X		X				25,30,33,36
Ophryotrocha puerilis		X					6
Nereidae							
Nereis diversicolor	X					X	38-40
Neanthes arenaceodentata	X	X		X		X	6,41,42,45,46
Polynoidae							
Halosydna johnsoni	X						6
Sabellariidae							
Sabellaria spinulosa			X				18
Serpulidae							
Pomatoceros triqueter						X	19
Spionidae							
Streblospio benedicti					X		20
Spirorbinidae							
Dexiospira brasiliensis	X						6

Ctenodrilus serratus is another minute species which has a short life cycle. It reproduces asexually by transverse division with each segment of the 6- to 10-segmented worm forming a new individual.[6] This species has been used by different investigators to measure the effects of metals, petroleum hydrocarbons, and drilling mud on its life history. The remaining species in Table 26.1, with the exception of *Nereis diversicolor*, have been used only once in toxicity experiments. Summarizing, only five small species or larvae have been used several times as test animals by more than one worker. Five species of polychaetes have been used in which the test is initiated with trochophore larvae: *Capitella capitata*,[8,10,17] *Halosydna johnsoni*,[6] *Sabellaria spinulosa*,[18] *Pomatoceros triqueter*,[19] and *Streblospio benedicti*.[20]

III. EFFECT OF TOXICANTS

A. Metals

Table 26.2 summarizes the known data on the effect of metals and other toxicants on different life stages of polychaetes. Mercury and copper are the most toxic metals regardless of the species or stage used. The toxicity of cadmium, chromium, and zinc was similar with variations in relative toxicity measured with the different species.

The trochophore stage was tested with only two species, *Capitella capitata* and *Halosydna johnsoni*, and they were more sensitive than the adults in all tests but one. With some metals the difference was slight (e.g., copper with both species), but with mercury there was one order of magnitude difference between the two life stages.

Dinophilus gyrociliatus was the most sensitive species to these metals with the exception of mercury; however, the figure reported in Table 26.2 was for a 48-h testing period rather than 96 h. Furthermore, the test procedures in this case were different than for the other tests with this species and other species. It is possible that *D. gyrociliatus* would be the most sensitive species to mercury under the same testing procedures. The other small species, *Ophryotrocha diadema* and *Ctenodrilus serratus*, ranked next in sensitivity. The adults of the larger species, *Capitella capitata*, *Neanthes arenaceodentata*, and *Halosydna johnsoni*, were the most tolerant.

The effect of metals on reproduction has been studied using different species.[8,10,21,43,46,47] Abnormal larvae, especially two-tailed metatrochophores, were induced in the F_1 and F_2 generations of *Capitella capitata* exposed to chromium and zinc.[8] Similar observations were noted in the same species exposed to detergents.[22] Table 26.3 presents the results on the effects of mercury on two small species as an example of this type of investigation. Some generalities from Table 26.3 applicable to the effects of other metals on reproduction in polychaetes are: 1) survival at low concentrations of metals is similar to control; 2) as the concentration of the metal is increased, the survival rate decreases to zero; 3) reproduction (i.e., fecundity), as measured by the number of offspring produced, is frequently enhanced compared to controls at low concentrations of the metal; and 4) reproductive rate decreases to zero as the concentration of the metal is increased.

B. Petroleum Hydrocarbons

Crude oil and most refined petroleum products are a complex mixture of countless, largely organic, compounds. When an oil spill occurs in the ocean, the volatile components are largely dissipated into the air in a few days or less. The water soluble fraction persists in the water and is the component which is utilized in toxicity testing.[48] Table 26.2 summarizes the toxicological data for both crude and refined oils to juvenile or small adults. The water-soluble fraction of refined oil is more toxic than the comparable fraction in crude oil to all species tested. *Ophryotrocha diadema* is the most sensitive to crude oil, and *O. puerilis* to refined oil.

Table 26.2   Comparison of the Effect of Selected Toxicants on Different Life Stages of Polychaetes* (Data as 96-h LC_{50} in mg/L)

	Trochophore	Juvenile	Adult
Cadmium			
Capitella capitata	0.2	—	7.5
Ctenodrilus serratus	—	4.3	—
Dinophilus gyrociliatus	—	0.8	—
Neanthes arenaceodendata	—	12.5	12.0
Ophryotrocha diadema	—	4.2	—
Ophryotrocha labronica	—	4.0**	—
Chromium			
Capitella capitata	8.0	—	5.0
Ctenodrilus serratus	—	4.3	—
Dinophilus gyrociliatus	—	0.8,4.0**	—
Halosydna johnsoni	1.2	—	4.7
Neanthes arenaceodentata	—	1.0	1.0
Ophryotrocha diadema	—	7.5	—
Copper			
Capitella capitata	0.18	—	0.2
Ctenodrilus serratus	—	0.3	—
Cirriformia luxuriosa	—	0.35	0.9
Dexiospira brasiliensis	—	—	1.0
Dinophilus gyrociliatus	—	0.03	—
Halosydna johnsoni	0.1	—	0.15
Neanthes arenaceodentata	—	0.3	0.3
Ophryotrocha diadema	—	0.16	—
Mercury			
Capitella capitata	0.01	—	0.1
Ctenodrilus serratus	—	0.04	—
Dinophilus gyrociliatus	—	0.2**	—
Halosydna johnsoni	0.03	—	0.3
Neanthes arenaceodentata	—	0.1	0.2
Ophryotrocha diadema	—	0.09	—
Zinc			
Capitella capitata	0.9	—	1.8
Ctenodrilus serratus	—	7.1	—
Dexiospira brasiliensis	—	—	3.0
Dinophilus gyrociliatus	—	0.2	—
Halosydna johnsoni	2.2	—	6.0
Neanthes arenaceodendata	—	0.9	1.8
Ophryotrocha diadema	—	1.4	—
Detergent (BP1002)			
Sabellaria spinulosa	2.5	—	—
Crude Oil***			
Capitella capitata	—	—	12.0, >19.8
Ctenodrilus serratus	—	>19.8	—
Neanthes arenaceodentata	—	15.2	12.5
Ophryotrocha diadema	—	12.9	—
Ophryotrocha puerilis	—	17.2	—
Refined Fuel Oil***			
Capitella capitata	—	—	>8.7
Ctenodrilus serratus	—	4.1	—
Neanthes arenaceodentata	—	4.6	2.7
Ophryotrocha diadema	—	2.9	—
Ophryotrocha puerilis	—	2.2	—

*Data from References 6, 18, 24, 35, and 42-45
**48-h LC_{50}
***Water-soluble fraction in mg/L

Table 26.3   The Effect of Mercury on the Survival and Reproduction of Two Small Species of Polychaetes (Based on 40 specimens per concentration)

	Ctenodrilus serratus Population at:		Ophryotrocha diadema Population at:	
	4 days	21 days	4 days	21 days
Control	40	211	40	74
0.001 mg/L	40	208	40	11
0.005	40	242	40	120
0.01	40	207	40	69
0.05	15	41*	35	58
0.1	0	0*	16	16*
0.5	0	0*	0	0

* 0.05 level of significance for reproductive suppression

Data adapted from Reish, D. J. and Carr, R. S., *Mar. Pollut. Bull.,* 9, 24, 1978.

In a series of studies involving developing *Neanthes arenaceodentata* Rossi and associates[42,44,45] found the number of zygotes decreased with increasing sublethal concentrations of the water-soluble fraction of refined oil, but the survival rate of those that reached the embryonic stage was the same at all concentrations tested. The nonfeeding pre-21-segmented stage was more tolerant than juveniles which had commenced feeding. Apparently the greater toxicity of oils to the early feeding stage is the result of additional toxicant entering the organism via the digestive tract during the intake of food. It is possible that this hypothesis could be tested by comparing the results with nonfeeding trochophores and recently metamorphosed, feeding juveniles (i.e., *Capitella capitata*).

The effect of crude and refined oil on reproduction was studied using the same two small species of polychaetes reported in Table 26.3.[43] The results were similar. The population size at 28 days was inversely related to the concentration; however, in contrast to the experiments with mercury, there was reproductive enhancement in only one of the four experiments.

C. Detergents and Dispersants

Research on the effects of detergents on marine organisms was stimulated by the use of these chemicals following the oil spill of the *Torrey Canyon* in 1967. Studies have been conducted on the larvae of *Sabellaria spinulosa*,[18] *Ophryotrocha labronica*,[33] and different developmental stages of *Capitella capitata*[10,21,22] (Table 26.1). The results of the experiments with *C. capitata* are given in Table 26.4. Trochophores were exposed to polyethylene-glycol fatty acid to determine survival and duration of pelagic life.[10] Survival was inversely related to test concentration, however the duration of pelagic life of the larvae was directly related to the amount of detergent present. The duration of pelagic life for normal larvae is two days, but it was increased to 12 days at the highest concentration tested. Of interest is that the effect of this detergent on laboratory- and field-collected trochophores was the same. A second experiment[10] determined the effect of this detergent on reproduction, effect on the trochophore and metatrochophore stages, and metamorphosis to the benthic stage (Table 26.4). As noted in Table 26.4, the detergent affected each phase in the life history of this species. The number of benthic worms of the F_1 generation originating from 25 pairs was reduced from 153 in the control to 25 at the highest concentration tested.

D. Contaminated Sediments

Many toxicants adhere to the surface of sediment and other marine particles and are ultimately deposited on the ocean floor especially in bays, harbors, and shallow offshore water. Initial studies

Table 26.4   Relationship of the Concentration of a Detergent to Reproduction, Development, and Larval Survival in *Capitella capitata*

Concentration (mg/L)	No. of Females Developing Ovarian Tissue	No. of Females Laying Eggs (mean)	No. of Eggs (mean)	No. of Trochophores (mean)	No. of Free Trochophores (mean)	No. of Metatrochophores (mean)	No. of Benthic Worms (mean)
Control	21	19	285	250	183	167	153
0.01	21	17	285	227	176	158	133
0.1	19	14	288	214	165	144	114
1	19	13	315	217	159	108	77
10	17	10	278	161	133	78	77
100	11	6	238	124	102	51	25

Data reduced from Reference 10 based on an initial number of 25 males and 25 females.

measured the effect of contaminated sediment on adult bivalves, crustaceans, fish, and polychaetes. Later studies investigated the sublethal effects on the life history of *Capitella capitata*[17] and growth, as measured by increase in body weight (dry), on juvenile *Neanthes arenaceodentata*.[9,49-51] In tests with the latter species, juvenile worms of 2 to 3 weeks post-emergence (age from time of fertilization is 5 to 6 weeks) are taken from laboratory cultures and placed in one-liter beakers containing 2 cm of sediment and sea water. Tests are generally run for 20 days, after which the worms are removed, counted, allowed to depurate for one day, then dried and weighed. Body weight is compared to the preweight of randomly selected worms. Growth is expressed as the mg of dry weight per worm per day. The use of this measure of growth makes it possible to compare the results of different studies.

The rationale behind this test is that if an animal is able to grow, as measured by body weight, then presumably those sediments are not toxic to the organism. For example, experiments have shown that the sediment from some regions of Puget Sound are toxic[51] but not those present in the area surrounding the outfall sewer of Orange County, California.[52]

IV. DISCUSSION

Research has been preformed investigating the effects of many different toxicants on many different species of polychaetes.[3] This review has focused on the methodologies for evaluating effects on larvae and adults measuring less than 1.0 cm in length, truly the microscale toxicity procedures with polychaetes. Some general statements can be made based on this review, regardless of the type of toxicant, allowing for the fact that exceptions occur and the number of published accounts are limited: 1) small species of polychaetes are more sensitive to a toxicant than larger species; 2) larval stages are more sensitive than the adults of the same species; 3) all sublethal concentrations of a toxicant have quantifiable effects on all stages in the life history of the worm — that is, the measured effect increases with an increasing sublethal concentration; and 4) the difference in the concentration of a toxicant between the 96-h LC_{50} and a measurable sublethal effect may be slight or large depending upon the toxicant.

This review indicates the importance of conducting toxicological tests with larvae or small adult species of polychaetes. *Dinophilus gyrociliatus* and various species of *Ophryotrocha* offer many advantages as toxicological test organisms. All of these species can be cultured easily in the laboratory and require only about one hour per week per species to maintain. These species are hardy enough to be transported by overnight express to other laboratories with few or no fatalities.

The practice of relying on field-collected species to produce larvae in the laboratory is tenuous and unreliable at best. Many species of polychaetes have a short reproductive period and spawn in the water column. Usually the trochophore stage requires feeding upon phytoplankton. In contrast, species such as *Capitella capitata, Neanthes arenaceodentata, Dinophilus gyrociliatus,* and the three species of *Ophryotrocha* lay their eggs in tubes or capsules and the embryos do not require care or food. If larval testing is contemplated, the species listed above are recommended. An alternate selection might include members of the families Spionidae or Spirorbidae, which incubate their embryos. By following these recommendations, a new laboratory or one with limited experience in testing polychaetes might achieve meaningful results in a relatively short period of time.

As indicated in this review, microscale toxicological studies with polychaetes have included lethality, growth, and reproduction. Their small size and ease in culturing makes them convenient animals from which to obtain rapid results. What direction should the use of these animals take in the future? We need to develop culture techniques for additional species so that polychaetes will be available for use in all marine geographical localities. The effects of toxicants at the molecular and genetic level is largely an untapped field of inquiry. The development of this area of investigation will require a multidisciplinary approach. A good working hypothesis is that some changes are occurring at the cellular or subcellular level before death of the test organism takes place. Development of this line of research offers promise for achieving meaningful answers in a short period of time.

REFERENCES

1. Knox, G. A., The role of polychaetes in benthic soft-bottom communities, in *Essays in Honor of Dr. Olga Hartman,* Reish, D. J. and Fauchald, K., Eds., Allan Hancock Foundation, University of Southern California, Los Angeles, 1977, 547.

2. Reish, D. J. and Barnard, J. L., Field toxicity tests in marine waters utilizing the polychaetous annelid *Capitella capitata* (Fabricius), *Pacific. Nat.,* 1 (21), 1. 1960.

3. Reish, D. J. and Gerlinger, T. V., A review of the toxicological studies with polychaetous annelids, *Bull. Mar. Sci.,* 60, 584, 1997.

4. Oshida, P. S., Mearns, A. J., Reish, D. J. and Word, C. S., The effects of hexavalent and trivalent chromium on *Neanthes arenaceodentata* (Polychaeta: Annelida), *So. Calif. Coastal Water Res. Project, Tech. Mem. No.* 225, El Segundo, Calif., 1976.

5. Bryan, G. W., Pollution due to heavy metals and their compounds, in *Marine Ecology,* Kinne, O. Ed., John Wiley & Sons, London. 5, 1289, 1984.

6. Reish, D. J., The effect of different pollutants on ecologically important polychaete worms, *U.S. Environmental Protection Agency, Environmental Research Laboratory,* Narragansett, R. I. EPA 600/3-80-053, 1980.

7. Reish, D. J., Marine ecotoxicological tests with polychaetous annelids, in *Ecotoxicological Testing for the Marine Environment,* Persoone, G., Jaspers, E., and Claus, C., Eds., State Univ. Ghent and Inst. Mar. Sci. Res., Belgium, 1984, 427.

8. Reish, D. J., Effects of chromium on the life history of *Capitella capitata,* in *Physiological Responses of Marine Biota to Pollutants,* Vernberg, F. J., Calabrese, A., Thurberg, F. P., and Vernberg, W. B., Eds. Academic Press, New York, 1977, 190.

9. Dillon, T. M., Moore, D. W., and Gibson, A. B., Development of a chronic sublethal bioassay for evaluating contaminated sediment with the marine polychaete worm *Nereis arenaceodentata, Environ. Toxicol. Chem.,* 12, 589, 1993.

10. Bellan, G., Reish, D. J., and Foret, J. P., The sublethal effects of detergent on the reproduction, development, and settlement in the polychaetous annelid *Capitella capitata, Mar. Biol.,* 14, 183, 1972.

11. Carr, R. S., Williams, J. W., and Fragata, C. T. B., Development and evaluation of a novel marine sediment pore water toxicity test with the polychaete *Dinophilus gyrociliatus, Environ. Toxicol. Chem.,* 8, 533, 1989.

12. American Public Health Association, Annelida, in *Standard Methods,* Amer. Public Assoc., Amer. Water Works Assoc. and Water Environ. Assoc., Washington, DC, 1995, 8.

13. American Society for Testing and Materials, Standard guide for conducting acute, chronic and life cycle aquatic toxicity tests with polychaetous annelids, *Amer. Society for Testing and Materials,* Philadelphia, Vol 11.05, Guide No. 1562, 1995.

14. American Society for Testing and Materials, Standard guide for conducting sediment toxicity tests with polychaetous annelids, *Amer. Society for Testing and Materials,* Philadelphia, Vol. 11.05, Guide No. 1611, 1995.

15. Reish, D. J., Use of polychaetous annelids as test organisms for marine bioassay experiments, in *Aquatic Invertebrate Bioassays,* Buikema, A. L., Jr. and Cairns, J., Jr., Eds., Amer. Soc. Test. Materials, Special Tech. Publ. No. 715, 1980, 140.

16. Akesson, B., Morphology and life cycle of *Ophryotrocha diadema,* a new species from California, *Ophelia,* 15, 23.

17. Chapman, P. and Fink, R., Effects of Puget Sound sediments and the effluents on the life cycle of *Capitella capitata, Bull. Environ. Contam. Toxicol.,* 33, 451, 1984.

18. Wilson, D. P., Long-term effects of low concentrations of an oil-spill remover ("detergent"): Studies with the larvae of *Sabellaria spinulosa, J. Mar. Biol. Assoc. U.K.,* 48, 177, 1968.

19. Kloeckner, K., Rosenthal, H., and Willfuehr, J., Invertebrate bioassays with North Sea water samples. 1. Structural effect on embryos and larvae of serpulids, oysters and sea urchins, *Helgoländer wiss. Meersunters.,* 39, 1, 1985.

20. Chandler, G. T. and Scott, G. I., Effects of sediment bound endosulfan on survival, reproduction and larval settlement of meiobenthic polychaetes and copepods, *Environ. Toxicol. Chem.,* 10, 375, 1991.

21. Bellan, G., Reish, D. J., and Foret, J-P, Action toxique d'un détergent sur le cycle de développement de la polychète *Capitella capitata* (Fab.), *C. R. Acad. Sci., Paris,* 272, 2476, 1971.

22. Foret, J-P, Étude des effets à long terme de quelques détergents sur la séquence du développement de la polychètes sédentaire *Capitella capitata* (Fabricius), *Téthys*, 6, 751, 1975.

23. Hill, S. D. and Nelson, L., Lindane (1, 2, 3,4, 5, 6 - hexachlorocyclohexane) affects metamorphosis and settlement of larvae of *Capitella* I (Annelida, Polychaeta), *Biol. Bull.*, 183, 376, 1992.

24. Carr, R. S. and Reish, D. J., The effect of petroleum hydrocarbons on the survival and life history of polychaetous annelids, in *Fate and effects of petroleum hydrocarbons in marine organisms and ecosystems,* Wolfe, D. A., Ed., Pergamon Press, New York, 1977, 168.

25. Neff, J. M., Carr, R. S., and McCulloch W. L., Acute toxicity of a used chrome lignosulphonate drilling mud to several species of marine invertebrates, *Mar. Environ. Res.*, 4, 251, 1981.

26. Petrich, S. M. and Reish, D. J., Effects of aluminum and nickel on survival and reproduction in polychaetous annelids, *Bull. Environ. Contam. Toxicol.*, 23, 698, 1979.

27. Akesson, B., Bioassay studies with polychaetes of the genus *Ophryotrocha* as test animals, in *Sublethal effects of toxic chemicals on aquatic animals,* Koeman, O. J. M. and Strik, J. J. T. W. A., Eds., Elsevier Scientific Publ., Amsterdam, 1975, 121.

28. Carr, R. S., Curran, M. D., and Mazurkiewicz, M., Evaluation of the archiannelid *Dinophilus gyrociliatus* for use in short-term life cycle tests, *Environ. Toxicol. Chem.*, 5, 703, 1986.

29. Adema, D. M. M. and Vink, G. J., A comparative study of the toxicity of 1, 1, 2-trichloroethene dieldrin, pentachlorophenol, and 3, 4 dichloroaniline for marine and freshwater organisms, *Chemosphere*, 10, 533, 1981.

30. Akesson, B. and Ehrenstrom, F., Avoidance reactions of dorvilleid polychaetes when exposed to chemical- contaminated sediments, in *Ecotoxicological Testing for the Marine Environment,* Persoone, G., Jaspers, E., and Claus, C., Eds., State Univ. Ghent and Inst. Mar. Sci. Res., Belgium, 1978, 2, 3.

31. Parker, J. G., The effects of selected chemicals and water quality on the marine polychaete *Ophryotrocha diadema, Water Res.*, 18, 865, 1984.

32. Akesson, B. and Costlow, J. D., Effects of temperature and salinity on the life cycle of *Ophryotrocha diadema* (Polychaeta, Dorvilleidae), *Ophelia,* 17, 215, 1978.

33. Akesson, B., *Ophryotrocha labronica* as a test animal for the study of marine pollution, *Helgoländer wiss. Meeresunters.*, 20, 29, 1970.

34. Bogaert, T., Samoiloff, M., and Pulak, R., Development of a toxicity test for the determination of mutagenic activity in the marine environment-teratogenesis in a marine polychaete *Ophryotrocha labronica, Can. Tech. Rept. Fish. Aquatic Sciences*, Halifax, N. S., (1368) p. 395, 1985.

35. Roed, K. H., The effects of interacting salinity, cadmium and mercury on population growth of an archiannelid, *Dinophilus gyrociliatus, Sarsia*, 64, 245, 1979.

36. Roed, K. H., Effects of salinity and cadmium on reproduction and growth during three successive generations of *Ophryotrocha labronica* (Polychaeta), *Helgoländer wiss. Meeresunters*, 33, 47, 1980.

37. Saliba, L. J. and Ahsanullah, M., Acclimation and tolerance of *Artemia salina* and *Ophryotrocha labronica* to copper sulfate, *Mar. Biol.*, 23, 297, 1973.

38. Ozoh, P. T. E., Egg volume regulation in *Hediste diversicolor* eggs-a rapid bioassay, *Water, Air and Soil Pollution*, 49, 231, 1990.

39. Ozoh, P. T. E., The importance of adult *Hediste diversicolor* in managing heavy metal pollution in shores and estuaries, *Environ. Monit. Assess.*, 21, 165, 1992.

40. Ozoh, P. T. E. and Jones, N. V., The effects of copper ions on embryogenesis in the polychaete *Hediste (Nereis) diversicolor, Mar. Environ. Res.*, 24, 255, 1990.

41. Pastorok, R. A. and Becker, D. S., Comparative sensitivity of sediment toxicity bioassays at three superfund sites in Puget Sound, in *Aquatic Toxicity and Risk Assessment*, Landis, W. G. and van der Schalie, W. H., Eds., Amer. Soc. Testing and Materials, Philadelphia, STP No. 1096, 1990, 123.

42. Rossi, S. S., Anderson, J. W., and Ward, G. S., Toxicity of water-soluble fractions of four oils on the polychaetous annelids *, Neanthes arenaceodentata* and *Capitella capitata, Environ. Pollut.*, 10, 9, 1976.

43. Reish, D. J. and Carr, R. S., The effect of heavy metals on the survival, reproduction, development and life cycles for two species of polychaetous annelids, *Mar. Pollut. Bull.*, 9, 24, 1978.

44. Rossi, S. S. and Anderson, J. W., Toxicity of water soluble fractions of No. 2 fuel oil and South Louisiana crude oil on selected stages in the life history of the polychaete, *Neanthes arenaceodentata, Bull. Environ. Contam. Toxicol.*, 16, 18, 1976.

45. Rossi, S. S. and Anderson, J. W., Effects of No. 2 fuel oil on growth and reproduction in *Neanthes arenaceodentata* (Polychaeta: Annelida), *Water, Air and Soil Pollut.*, 9, 155, 1977.

46. Oshida, P. S., Word, L. S., and Mearns, A. J., Effects of hexavalent and trivalent chromium on the reproduction of *Neanthes arenaceodentata* (Polychaeta), *Mar. Environ. Res.*, 5, 41, 1981.

47. Reish, D. J. and Gerlinger, T.V., The effects of cadmium, lead and zinc on survival and reproduction in the polychaetous annelid *Neanthes arenaceodentata* Nereididae, in *Proceedings of the First International Polychaete Conference,* P. A. Hutchings, Ed., Linnean Soc. New South Wales, 1984, 383.

48. Anderson, J. W., Neff, J. M., Cox, B. A., Tatem, H. E., and Hightower, G. M., Characteristics of dispersions and water-soluble extracts of crude and refined oils and their toxicity to estuarine crustaceans and fish, *Mar. Biol.*, 27, 75, 1974.

49. Dillon, T. M., Moore, D. W., and Reish, D. J. A 28 day bioassay with the marine polychaete *Nereis (Neanthes) arenaceodentata,* in *Environmental Toxicology and Risk* Assessment, Third Vol., Hughes, J. S. Biddinger, G. R., and Mones, E., Eds., Amer. Soc. Testing Materials, Philadelphia, 1995, 201.

51. Johns, D. M., Pastorok, R. A., and Ginn, T. C., A sublethal sediment toxicity test using juvenile *Neanthes* sp. Polychaeta Nereidae, in *Aquatic Toxicology and Risk Assessment*, 14th Vol., Mayes, M. A. and Barron, M. G., Eds., Amer. Soc. Testing and Materials. Philadelphia, 1991, 180.

52. Gerlinger, T. V., Reish, D. J., and Fanizza, M., Survival and growth of juvenile *Neanthes arenaceodentata* (Annelida: Polychaeta) in marine sediment taken from the vicinity of an ocean outfall, *Bull. So. Calif. Acad. Sci.*, 94, 65, 1995.

CHAPTER 27

Development of a Canadian Marine Toxicity Test for Whole Sediments Using Cultured Spionid Polychaetes

Patricia Pocklington and Kenneth G. Doe

CONTENTS

I. INTRODUCTION

Sediment toxicity tests are used routinely within Canada and elsewhere to determine and monitor the adverse biological effects of substances that might be harmful to indigenous aquatic life in the environment (water and sediment). The Marine Environmental Quality Advisory Group of Environment Canada and Canada's Inter-Governmental Aquatic Toxicity Group (IGATG) recommended that a battery of sediment testing methods be developed for routine use in evaluating and managing contaminated sediment in the marine environment. This battery of test methods is to include the use of marine and/or estuarine organisms representing different trophic levels. To fulfill this objective, projects were supported by the Method Development and Application Section and the Office of Marine Disposal of Environment Canada to evaluate indigenous Canadian marine annelids, specifically members of the class Polychaeta, phylum Annelida, and appraise their usefulness as bioassay test organisms. To this end, polychaetes from the Atlantic, Pacific, and Arctic coasts of Canada waters were evaluated.

A. Polychaetes in Ecotoxicology

Following previous reviews (Reish 1984; Pocklington and Wells 1992), a literature review was conducted in 1993 to determine which species of polychaetes had been used, or were currently being used, as bioassay organisms for marine sediment toxicity experiments (Arenicola Marine 1993). This search generated a list of over 25 species (see Chapter 26). The literature indicated that many species of varying size and at different stages in their life cycle could be used, the latter because the short life span of many polychaete species allows increased emphasis on sublethal endpoints such as growth, development, and reproduction. It was also noted that both wild and cultured specimens were tested. With the development of sublethal tests such as growth, the small size of juvenile or young polychaetes used in ecotoxicology testing necessitates the miniaturization of experimental setups. Through these modifications the use of polychaete tests has contributed to advances in microscale aquatic toxicology procedures (refer also to Chapter 26).

Using the above information, lists of species known to inhabit Canadian waters (Banse and Hobson 1974; Hobson and Banse 1981; and Pocklington 1989) were examined to determine if the species which had been found useful as bioassay test organisms occurred in Canada or had Canadian counterparts. The possibility that Canadian counterparts could be found was high as many families and genera of polychaetes have worldwide distribution. Species which could be found in Atlantic, Pacific, and Arctic coastal environments and which were the same or closely related to currently used species were chosen for further evaluation. Fourteen species were chosen and assessed: *Arenicola cristata, Armandia brevis, Boccardia proboscidea, Capitella capitata, Ctenodrilus serratus, Dinophilus gyrociliatus, Glycinde picta, Mediomastus ambiseta, Neanthes arenaceodentata, Nereis (Neanthes) virens, Nereis diversicolor, Nephtys* spp. (juveniles), *Ophryotrocha diadema,* and *Polydora cornuta.* In addition, some families, not previously used as bioassay test organisms (e.g., representatives of the family Maldanidae) were also considered as possible test organisms because of their abundance in the field. Based on the numerical results of the evaluation, the animals were either rejected as possible test species or were recommended for further work. Those species recommended for further work were sought in sufficient numbers in the field. The ability to collect sufficient numbers of organisms in the wild for the required tests was found to be the single most important criterion.

Ten species of polychaetes were tested for survival and robustness in the laboratory. The Atlantic species included a spionid, *Polydora cornuta;* a nereid *Nereis (Neanthes) diversicolor;* a maldanid, *Clymenella torquata; and* a nephtyid, *Nephtys neotena.* The Pacific species included a spionid, *Boccardia proboscidea*; a nereid, *Hediste (?) limnicola*; a maldanid, *Axiothella rubrocincta*; and a nephtyid, *Nephtys caecoides.* The species used from the Arctic were the spionid, *Polydora quadrilobata* and the sabellid, *Euchone papillosa.*

They were tested to: 1) determine their ability to survive in the laboratory under a variety of environmental conditions (noncontaminant effects tests); and 2) determine their sensitivity to contaminated sediments and to a reference toxicant. Of the Atlantic species: *P. cornuta* and *C. torquata* could be held in the laboratory and proved to be the most sensitive to the test sediments and the reference toxicant; *N. neotena* did not perform well in the survival testing and so was not tested further; *N. diversicolor* showed little treatment response (it was neither sensitive to contaminated sediments nor to the reference toxicant). Of the Pacific species, *B. proboscidea* and *A. rubrocincta* survived in the laboratory and proved to be sensitive to the test sediments and the reference toxicant. One Arctic species, *E. papillosa*, was determined to be unsuitable but the other,

P. quadrilobata, belonging to the same family as *P. cornuta* and *B. proboscidea,* was thought to be suitable. Because of a need for a reliable source of test animals, attempts were made to culture the selected species and test the cultured animals for robustness and sensitivity. This required the development of methods for culturing the selected species along with choosing a suitable endpoint for the bioassay.

B. Literature Review of Bioassay Endpoints

Originally, lethal tests were used as a model, where the endpoint was average mortality measured over a period of time (Chapman et al. 1985; Craig and Caunter 1990; Jop 1989; Pesch and Hoffman 1983; Pesch et al. 1991).

It was decided that a more relevant approach for the development of regulatory guidelines would be to develop chronic sublethal sediment tests which could identify harmful contaminants before they reached lethal concentrations. It was found that a variety of simple sublethal tests have been used successfully, for example:

1. Modification of behavior such as reduced burrowing or aversion to burrowing in contaminated sediments (Johns et al. 1985; McLeese et al. 1982; Pesch and Hoffman 1983)
2. Effect on growth where animals displayed reduced growth when exposed to contaminated sediments (Chapman and Fink 1984; Chapman et al. 1985; Johns et al. 1985; Moore and Dillon 1993; Pastorok and Becker 1990; Zajac and Whitlach 1988)
3. Effect on fecundity as measured by change in production of eggs or larvae (Carr and Curran 1986; Carr et al. 1989; Grassle and Grassle 1984; Johns et al. 1991; Jop 1989; Kaag et al. 1994; Long et al. 1990; Pesch et al. 1991; Reish and Carr 1978; Zajac and Whitlach 1988)
4. Production of abnormal larvae (Long and Chapman 1985; Reish et al. 1974; Walsh et al. 1986)

More complex tests (i.e., those requiring a greater level of expertise/instrumentation), such as measurement of metabolites (Carr and Neff 1982), measurement of enzymes such as MFO (Lee and Singer 1980), examination of chromosomes for mutagenic effects (Pesch et al. 1987 and references therein), were considered too specialized for routine sediment testing. Only those tests with easily interpretable sublethal responses (i.e., modification of behavior, reduced growth, or reduced fecundity) were given further consideration. The endpoints finally selected for study were survival and growth, and for these endpoints, animals of a similar size and age were required.

II. METHODS

A. Test Animals for Sublethal Tests

Because of the need for a reliable supply of animals of the correct size (i.e., at a size for which growth over time could be easily measured) for bioassay testing, the next phase in the development of the test method was to culture the promising species. Protocols for conducting culturing experiments were developed for: 1) two maldanids, *Clymenella torquata* and *Axiothella rubrocincta,* and 2) three spionids, *P. cornuta, B. proboscidea,* and *P. quadrilobata.* It was found that some of the species could be cultured successfully in the laboratory (Pocklington et al. 1995a).

Five species were initially chosen for culturing. Of these, two could be induced to spawn and produce large numbers of larvae on demand. These were the spionids *Polydora cornuta* from the Atlantic and *Boccardia proboscidea* from the Pacific. The Arctic spionid, *Polydora quadrilobata,* and the two maldanids were held in the laboratory for extended periods of time and subjected to a variety of feeding and temperature regimes, but despite a high individual survival rate, these species could not be induced to produce large numbers of larvae.

TABLE 27.1 Chemical Analysis and Grain Size of Sediments Used in 20-Day NCETs[4] and CETs[4]

Sediment	Contaminants in Test Sediments (μg/g)					Grain Size (corrected for % moisture content)		
	Cd	Pb	PAHs	PCBs	TOC	% Gravel	% Sand	% Fines
Conrad's Beach[1]	0.05	3.08	<0.01	<0.01	0.36	0.1	95.2	4.1
Walton Beach[2]	0.03	8.94	0.08	<0.01	1.09	11.5	42.7	45.8
Silica flour	<0.3	8.4	—	—	0.8	<0.1	96.9	3.1
Silica sand 00	<0.3	5.0	—	—	0.8	<0.1	99.5	0.5
Silica sand 0	<0.3	2.7	—	—	3.7	<0.1	99.6	0.4
Silica sand 1	<0.3	1.3	—	—	1.2	3.0	96.5	0.6
Silica sand 2	<0.3	1.3	—	—	0.4	1.6	97.7	0.7
Tufts Cove[3]	3.57	242.0	57.56	0.46	2.66	13.2	72.1	20.5
Vancouver Shipyards[3]	1.63	43.3	19.38	0.06	2.98	2.6	46.6	49.4
False Creek[3]	0.80	39.9	39.91	0.07	1.01	7.4	36.8	55.8
Kaolin	<2.0	<10.0	—	—	0.0	0.0	0.44	99.56
Witty's Beach[1]	<2.0	<10.0	—	—	0.06	13.4	85.4	1.2
Witty: Walton blend	<2.0	<10.0	—	—	0.54	15.8	63.5	18
Witty: Kaolin blend	<2.0	<10.0	—	—	0.07	8.4	71.2	20.4
Ocean disposal criteria	0.6	500.0	2.5	0.1	—	—	—	—

[1]Control sediment from animal collection sites (Conrad's Beach for *P. cornuta*; Witty's Beach for *B. proboscidea*.
[2]Fine-grained, clean reference sediment.
[3]Test sediment (contaminated).
[4]Noncontaminant Effect Tests, Contaminant Effect Tests.

B. Noncontaminant and Contaminant Effect Tests (NCETs and CETs)

Adult *P. cornuta* and *B. proboscidea* were cultured at a temperature of 20 to 23°C, a salinity of 30 ± 2 ppt, and fed a 1:1 mixture of dried ground *Enteromorpha* sp. and a commercial fish food TetraMarin™ (ET). The young produced by these methods were allowed to develop to the juvenile stage (two to three weeks after emergence from the egg capsule). At this stage, the animals have a completely benthic existence, and sediment manipulation for deposit feeding and tube building exposes them directly to the sediments. The young produced in the culturing experiments were assessed for sensitivity using both survival and growth as endpoints (Pocklington et al. 1995 b,c). All cultured species were tested at: 1) three salinities (5 ± 2 parts per thousand (ppt), 15 ± 2 ppt, and 28 ± 2 ppt, and 2) two temperatures (10 ± 2°C and 20 ± 2°C). In experiments using the Atlantic species, test animals were exposed to several sediments of different grain size: a control sediment from the animals' habitat (Conrad's Beach, N.S.); a fine-grained reference sediment (Walton Beach, N.S.) considered to be noncontaminated, as all potential contaminants are below Ocean Disposal Guideline levels of Environment Canada; a 50:50 combination of the two; and a series of five commercially available silica sands ranging from silica flour to coarse sand (results of contaminant analyses and grain sizes of these are presented in Table 27.1). For the Pacific species, the test animals were exposed to the control sediment from the animals' habitat (Witty's Beach, B.C.), the reference sediment (defined above), calcined kaolin, and 50:50 mixtures of these sediments.

While being tested at a temperature of 20°C and a salinity of 30 ppt, the animals were offered three different food rations: *Enteromorpha* sp., TetraMarin, and a 50:50 combination of the two (ET). Once the best food was established, the quantity was varied to determine the feeding rate which would give the best growth and survival.

Based upon information in the literature (e.g., Johns et al. 1990; ASTM 1994), a test period of 20 days was chosen. But due to the rapid growth and short life-span of some of the test animals and because of the concern for loss of weight should spawning occur during the test, a test length of 14 days was also used.

To determine sensitivity, the animals were tested at what was considered the optimum temperature (20°C) and salinity (30 ppt) for growth and survival, and they were exposed to the contaminated sediments for both 14 and 20 days. The fine-grained reference sediment was from Walton Beach, N.S. The contaminated sediments came from Vancouver shipyards and False Creek, B.C., and Tuft's Cove, N.S. A grain-size and chemical profile of these sediments is given in Table 27.1.

The reference toxicant was cadmium chloride, following Johns et al. (1990) and ASTM (1994). The animals were exposed to a logarithmic series of concentrations of the reference toxicant to determine 96-h LC_{50}s in water-only exposures. The results were expressed as $CdCl_2$ concentration (nominal value).

III. RESULTS

A. Tests with *Polydora cornuta*

Results of the 14- and 20-day noncontaminant effects of salinity, temperature, diet, grain size, and food ration on survival and growth of cultured *P. cornuta* are summarized in Tables 27.2, 27.3, and 27.4. When *P. cornuta* was held at 20°C and fed 5 mg of a 1-to-1 mixture of ET per worm the mean survival in three test salinities (5 ppt, 15 ppt, and 28 ppt) ranged from 90 to 100% (Table 27.2). Growth was better at the two higher salinities (Table 27.2). Response to quantity of food and low temperature was predictable as the young grew best and had better survival at higher temperatures and larger rations (Tables 27.2 and 27.3). Growth rates (mg/worm/day) were similar in both 14- and 20-day tests when fed the same ration size (Table 27.3). Grain size and perhaps other physic-ochemical characteristics, on the other hand, had a significant effect on the growth of young *P. cornuta*, although survival was apparently unaffected by these variables (Table 27.4).

The growth of *P. cornuta* held in the fine-grained reference sediment (Walton Beach) and in all of the contaminated sediments was significantly lower than that of the corresponding group held in control sediment from the animals' habitat (Table 27.5). The mean % survival of animals held in False Creek sediment was significantly lower than survival in the control sediment, and no animals (0%) survived in the Tuft's Cove sediment. All animals held in Vancouver shipyard sediment survived (Table 27.5).

Data showing survival of *Polydora cornuta* in various concentrations of the reference toxicant, are given in Table 27.6. The 96-h LC_{50}s calculated for this species were 17.9, 19.4, and 8.5 mg $CdCl_2$/L, respectively, based upon nominal concentrations at the start of the test.

B. Tests with *Boccardia proboscidea*

As found with *P. cornuta*, optimal holding conditions for this species were a salinity of 28 ppt, a temperature of 20°C, and a feeding rate of 5 mg/worm of a 1:1 mixture of ET. At lower salinities, growth was lower; at lower temperatures, growth was lower (Table 27.7) and when fed a different ration (only *Enteromorpha* or only TetraMarin) growth was lower (Table 27.8). Results of experiments testing *B. proboscidea* sensitivity to grain size in the 14-day test indicated that survival and growth were best in the control sediment (Witty's Beach). Animals also showed reduced growth and survival in some fine-grained sediments (50:50 blend of Witty's Beach:kaolin). They also showed reduced growth in pure clay (calcined kaolin). For 20-day tests, survival was best in the control sediment, but growth was best in the fine-grained reference sediment. However, there was no significant difference (Table 27.8).

For 14-day tests on contaminated sediments, survival was greatest in the control, and significantly ($p \leq 0.05$) lower in the fine-grained reference sediment and in the sediment from Vancouver Shipyards (Table 27.9). None of the polychaetes exposed to the Tuft's Cove sediment survived (Table 27.9). Growth, on the other hand, was not significantly different in the test sediments than in the control sediment (Table 27.9). For 20-day tests, survival was highest in the control sediment, and significantly lower in the three test sediments. All animals died in Tuft's Cove sediment as in

Table 27.2   Effects of Temperature, Salinity, and Diet on Survival and Growth of *Polydora cornuta* in 20-Day Tests with Weekly Renewal

Treatment	Temperature (°C)	Salinity (ppt)	Diet[1]	Survival after 20 Days (%)	Mean Dry Weight per Worm (mg)[2]	Mean Growth Rate[5]
Conrad's Beach[4]	20	28	ET	100	4.37 ± 1.33	0.21
Conrad's Beach	10	28	ET	72[3]	2.07 ± 0.54[3]	0.095
Conrad's Beach	20	15	ET	95	3.98 ± 0.81	0.19
Conrad's Beach	20	5	ET	90	1.63 ± 0.53[3]	0.073
Conrad's Beach	20	28	E	95	2.08 ± 0.2[3]	0.095
Conrad's Beach	20	28	T	95	2.99 ± 1.21[3]	0.14

[1] Ration size = 5 mg of each diet per worm, fed 3 times per week. E = *Enteromorpha*; T = TetraMarin™; ET = 50:50 mixture of *Enteromorpha* and TetraMarin
[2] Initial mean dry weight per worm (day 0) = 0.18 ± 0.1mg.
[3] Significantly different from the control treatment.
[4] This treatment was used as the control treatment.
[5] Growth weight = mg dry weight increase/worm/day.

Table 27.3   Effects of Ration Size, Test Duration, and Water Renewal on Survival and Growth of *Polydora cornuta*

Treatment	Ration Size (mg/worm)	Test Duration	Water Renewal	Survival (%)	Mean Dry Weight per Worm (mg)[1]	Mean Growth Rate[5]
Conrad's Beach	1.25	20	Yes	88[4]	1.71 ± 0.71	0.072
Conrad's Beach[2]	5	20	Yes	100	2.94 ± 1.70	0.133
Conrad's Beach	5	20	No	92	5.63 ± 3.76	0.27
Conrad's Beach	20	20	Yes	80	7.32 ± 3.16[4]	0.35
Conrad's Beach[3]	5	14	Yes	92	2.08 ± 0.80	0.13
Conrad's Beach	5	14	No	84	2.39 ± 0.76	0.15

Note: All tests conducted at 20 ± 2°C and 28 ± 2 ppt salinity. Food was a 50:50 mix of dried *Enteromorpha* and TetraMarin

[1] Initial mean dry weight per worm (day 0) = 0.28 ± 0.9mg.
[2] Control treatment for 20-day tests.
[3] Control treatment for 14-day tests.
[4] Significantly different from control treatment.
[5] Growth rate = mg dry weight increase/worm/day.

Table 27.4   Effects of Grain Size on Survival and Growth of *P. cornuta* in 20-Day Tests

Treatment	Survival (%)	Mean Dry Weight per Worm	Mean Growth Rate[3]
Conrad's Beach	96	7.02 ± 1.52	0.31
Conrad's Beach:Walton mixture	96	5.01 ± 1.16[2]	0.21
Walton Beach	88	3.17 ± 0.99[2]	0.12
Silica sand No. 2	85	1.93 ± 0.5[2]	0.059
Silica sand No. 1	90	2.75 ± 1.4[2]	0.10
Silica sand No. 0	95	4.70 ± 1.13[2]	0.20
Silica sand No. 00	85	3.72 ± 1.92[2]	0.15
Silica flour	100	2.97 ± 0.93[2]	0.11

Note: All treatments conducted at 20°C, 28 ± 2 ppt salinity, and fed 5 mg of a 50:50 mix of dried *Enteromorpha*:TetraMarin per worm, three times per week.

[1] Initial mean dry weight per worm (day 0) = 0.76 ± 0.33 mg.
[2] Significantly different from control treatment (Conrad's Beach).
[3] Growth rate = mg dry weight increase/worm/day.

Table 27.5   20-day Effect of Test Sediments on Survival and Growth of Cultured *Polydora cornuta*

Treatment	Mean % Survival	Mean Dry Weight per Worm[1]	Mean Growth Rate[3]
Control	96	7.02 ± 1.52	0.31
Reference	88	3.17 ± 0.99[2]	0.12
Vancouver Shipyards, Vancouver, BC	100	3.68 ± 0.63[2]	0.15
False Creek, Vancouver, BC	76[2]	2.35 ± 0.82[2]	0.080
Tuft's Cove, Halifax, NS	0[2]	0	0

Note: Temperature 20°C; salinity = 28 ppt; food = 5mg/worm of ET mixture.

[1] Initial average weight of worms = 0.76 ± 0.33 mg.
[2] Significantly different from control response.
[3] Growth rate = mg dry weight increase/worm/day.

Table 27.6   Sensitivity of *Polydora cornuta* to Reference Toxicant $CdCl_2$

Concentration (mg/L as $CdCl_2$)	Concentration (mg/L as Cd)	%Mortality[1] Test 1	%Mortality[1] Test 2	%Mortality[1] Test 3
320	196.2	100	—	—
180	110.4	100	—	—
100	61.3	100	100	—
56	34.3	100	100	100
32	19.6	100	40	100
10	6.1	0	0	60
5.6	3.4	0	40	40
3.2	1.96	—	0	0
Control	**0.0**	20[2]	**0**	**0**

[1] 96-h LC_{50} = Test 1: 17.9 mg/L as $CdCl_2$ (95% confidence limit = 10.0 to 32.0 mg/L).
  Test 2: 19.4 mg/L as $CdCl_2$ (95% confidence limit = 12.2 to 30.9 mg/L).
  Test 3: 8.48 mg/L as $CdCl_2$ (95% confidence limit = 4.87 to 14.11 mg/L).
[2] I of 5 test animals was missing at test termination, and presumed dead.

Table 27.7   Effects of Temperature and Salinity on Survival and Growth of *B. proboscidea* in 20-Day Tests

Treatment	Temperature (°C)	Salinity (ppt)	Survival (%)	Mean Dry Weight per Worm (mg)	Mean Growth Rate[4]
Witty's Beach[1]	20	28	100 ± 0	2.7 ± 0.36	0.12
Witty's Beach	10	28	92 ± 11	1.73 ± 0.35	0.073
Witty's Beach	20	15	92 ± 11	1.20 ± 0.46[3]	0.047
Witty's Beach[2]	20	28	100 ± 0	0.46 ± 0.22	0.023
Witty's Beach	20	5	0 ± 0[3]	0	0

Note: Feeding was 5 mg per worm of *Enteromorpha*:TetraMarin mixture, fed three times per week.

[1] Used as control treatment for temperature and 15 ppt salinity experiments. Initial mean dry weight of worms for these experiments (day 0) = 0.27 mg.
[2] Used as control treatment for salinity experiments. Initial mean dry weight of worms for these experiments (day 0) = 0.007 mg.
[3] Significantly different from control treatment.
[4] Growth rate = mg increase in dry weight/worm/day.

the 14-day tests. Growth was statistically different from the control in Vancouver shipyards sediment, and was not statistically different from the control sediment in False Creek.

When comparing length of test on both survival and growth of *B. proboscidea*, it was found that survival was similar for both test durations. Growth rates were lower in 20-day tests when compared to 14-day tests (Tables 27.8 and 27.9).

Table 27.8   Effects of Diet and Grain Size on the Survival and Growth of *Boccardia proboscidea*

Treatment	14-Day Test			20-Day Test		
	Survival (%)	Mean Dry Weight per Worm (mg)	Mean Growth Rate[5]	Survival (%)	Mean Dry Weight per Worm (mg)	Mean Growth Rate[5]
Witty's Beach, ET[2]	100 ± 0	1.69 ± 1.12	0.12	100 ± 0	1.32 ± 0.16	0.062
Witty's Beach, E	84 ± 9[4]	0.65 ± 0.17[4]	0.04	88 ± 11	0.48 ± 0.08[4]	0.02
Witty's Beach, T	88 ± 18	1.20 ± 0.68	0.08	84 ± 26	0.96 ± 0.44	0.044
Witty's Beach, ET[3]	100 ± 0	2.7 ± 0.36	0.19	100 ± 0	1.32 ± 0.16	0.062
Walton Beach, ET	77 ±17[4]	1.80 ± 0.85	0.12	92 ± 18	1.95 ± 0.66	0.094
Witty:Walton, ET	92 ± 11	1.22 ± 0.79	0.08	92 ± 18	1.67 ± 0.28	0.080
Witty:kaolin blend, ET	72 ±11[4]	1.37 ± 0.10[4]	0.09	—	—	—
Kaolin, ET	88 ± 27	0.78 ± 0.12[4]	0.05	—	—	—

Note: All tests conducted at 20 ± 2°C, 28 ± 2 ppt salinity. Ration size was 5 mg of each diet per worm, fed three times per week.

[1] Initial mean dry weight of worms (day 0) = 0.08 mg.
[2] Control treatment for diet tests. ET = 50:50 mix of *Enteromorpha* and TetraMarin; E = *Enteromorpha*; T = Tetra-Marin.
[3] Control treatment for grain-size effects tests.
[4] Significantly different from control treatment.
[5] Growth rate = mg increase in dry weight/worm/day.

Table 27.9   14- and 20-day Effect of Test Sediments on Survival and Growth of Cultured *Boccardia proboscidea*

Treatment	Survival (%)	Mean Dry Weight per Worm[1]	Mean Growth Rate[4]	Survival (%)	Mean Dry Weight per Worm[1]	Mean Growth Rate[4]
Duration	14-DAY	14-DAY	14-DAY	20-DAY	20-DAY	14-DAY
Control	100	2.17 ± 0.64	0.13	100	1.76 ± 0.22	0.069
Reference	64 ± 26[2]	3.96 ± 2.04	0.26	—	—	—
Vancouver Shipyards	68 ± 23[2]	2.77 ± 1.37	0.17	72 ± 27[2]	1.42 ± 0.85[3]	0.052
False Creek	88 ± 18	2.60 ± 1.11	0.16	88 ± 11[2]	2.00 ± 0.68	0.081
Tuft's Cove	0[2]	0	0	0[2]	0	0

Note: Temperature 20°C; salinity 28 ppt; food 5mg/worm of ET mixture.

[1] Initial mean dry weight per worm = 0.38 mg.
[2] Survival in sediment was significantly less than that in the control (P = 0.05).
[3] Mean dry weight was significantly less than that in the same sediment in the 14- day test (P = 0.05).
[4] Growth rate = mg increase in weight/worm/day.

IV. DISCUSSION

A. Selection of Species

1. Atlantic Species — Polydora cornuta

After trials using several species of polychaetes which could be collected in reasonable numbers from coastal and estuarine waters on the Atlantic Coast of Canada, only one species — *Polydora cornuta* — was considered to be useful as a bioassay test organism. Following experiments to determine survival and sensitivity of this species under laboratory conditions, optimum environmental conditions for it were found to be: temperature of 20 ± 2°C; salinity of 28 ± 2 ppt; food ration of a 50:50 mixture of finely ground dried *Enteromorpha* sp. and TetraMarin at a feeding rate of 5 mg per worm three times a week. *P. cornuta* was found to be sensitive to sediments considered by Environment Canada to be contaminated, and also to cadmium, the reference toxicant, as $CdCl_2$.

In addition, it was found that *P. cornuta* could be cultured in the laboratory, and generation time for this species was found to be as short as 28 days; as of December 1996, ongoing cultures have been maintained for more than 29 months. It was found that when using as test organisms young juveniles which have been cultured in the laboratory (two to three weeks post-hatch), and using either survival or growth as an endpoint, this species was reasonably resilient to environmental variables such as salinity (≥ 15 ppt). Grain size and perhaps other physicochemical sediment characteristics, on the other hand, had a significantly deleterious effect on the growth of young *P. cornuta*, although survival was apparently unaffected by these variables.

The sensitivity of these juvenile polychaetes was assessed using sediments from False Creek and Vancouver shipyard in B.C. and Tuft's Cove in Halifax Harbour, N.S., which are considered by Environment Canada to be contaminated because the levels of polycyclic aromatic hydrocarbons (PAHs) and some metals in the sediments exceed Canada's Ocean Disposal Guidelines (Environment Canada 1995), as does the level of polychlorinated biphenyls (PCBs) in the sediments from Tuft's Cove (Table 27.1). The growth of *P. cornuta* held in all of these contaminated sediments was significantly lower than that of the corresponding group held in clean sediment from the animals' habitat. The mean % survival of animals held in False Creek sediment was significantly lower than survival in the control sediment, and no animals (0%) survived in the Tuft's Cove sediment. All animals held in Vancouver shipyard sediment survived.

Sensitivity was also assessed using various concentrations of a reference toxicant, $CdCl_2$. The LC_{50}s calculated for this species were 17.9, 19.4, and 8.5 mg $CdCl_2$/L based upon nominal concentrations at the start of the test.

Marked gains in mean dry weights (seven- to ninefold in a 14-day test and 11 to 25 times in a 20-day test), together with consistently high survival rates found during both 14-day and 20-day noncontaminant effects tests, showed that the proposed test conditions were suitable for this species. Growth rates were similar for 14- and 20-day tests.

2. Pacific species — Boccardia proboscidea

After trials using several species of polychaetes which could be collected in reasonable numbers from coastal and estuarine waters on the Pacific Coast of Canada, one species — *Boccardia proboscidea* — was considered to be useful as a bioassay test organism because of its sensitivity to contaminated sediments and to the reference toxicant cadmium (as $CdCl_2$). Following experiments to determine survival and sensitivity of this species under laboratory conditions, optimum environmental conditions were found to be: temperature of $20 \pm 2°C$; salinity of 28 ± 2 ppt; food ration of a 50:50 mixture of finely ground dried *Enteromorpha sp.* and TetraMarin at a feeding rate of 5 mg per worm three times a week.

In addition, *Boccardia proboscidea* could be cultured in the laboratory. When using as test organisms young juveniles which have been cultured in the laboratory (two to three weeks post-hatch), and using either survival or growth as an endpoint, this species was reasonably resilient to noncontaminant effect variables such as temperature. Grain size and perhaps other physicochemical sediment characteristics, on the other hand, had a significantly deleterious effect on the growth and survival of young *Boccardia proboscidea*.

The sensitivity of these juvenile polychaetes was assessed using sediments as described for *P. cornuta*. Survival in the 14-day test was greatest in the control sediment followed by survival in sediment collected from False Creek and Vancouver shipyards. None of the polychaetes exposed to the Tuft's Cove sediment survived. Growth, on the other hand, was slightly greater but was not significantly different in the test sediments than in the control sediments. In all cases the total organic carbon content of these sediments was greater in the test sediments than in the control sediment and may have provided additional available nutrition to the growing polychaetes.

The results of the 20-day contaminant trials were similar to the 14-day contaminant effects trials. Again survival was highest in the clean sediment collected from the animals' habitat, followed by False Creek and Vancouver shipyards. None of the polychaetes exposed to Tuft's Cove sediment for 20 days survived. Survival of polychaetes exposed to Vancouver shipyards sediment for 20 days was found to be significantly less in comparison with Witty's Beach control sediment. Growth was not significantly different in False Creek sediment from control sediment, but was significantly lower in Vancouver shipyard sediment. Growth rates for this species were generally one half or less for 20-day tests as compared with 14-day tests. The reason for this is not known. Fourteen-day tests are recommended for this species.

Based on these results, *B. proboscidea* was considered a suitable candidate for the sediment toxicity test method. The usefulness of the growth endpoint for this species needs further research and evaluation.

V. OVERVIEW AND RECOMMENDATIONS

It was found that bioassay tests can be miniaturized to use juvenile polychaetes as small as 1 mM in length (<1 mg dry weight) as test organisms. Two species of polychaetes, *Polydora cornuta* and *Boccardia proboscidea*, of the family Spionidae, have shown a resilience to some noncontaminant effects. Both species could be cultured in the laboratory, with production of same-aged juveniles on demand. To date, the survival response of cultured individuals of both species has been somewhat similar. Based on survival, *P. cornuta* has shown greater sensitivity to False Creek and Tuft's Cove sediments than to sediments from Vancouver shipyards. Growth was significantly lower than the control in all of the test sediments (including the reference sediment). Curiously, *B. proboscidea* showed greater sensitivity to Vancouver shipyards sediment than to False Creek sediment based on survival, and growth was a more variable endpoint. Like *P. cornuta*, *B. proboscidea* showed greatest sensitivity to the Tuft's Cove sediments. Results of tests using both species suggests that acceptable results may be obtained in a 14- rather than a 20-day test. This could lead to a reduction in cost of conducting the test compared to 20-day tests. Growth does not appear to be a particularly sensitive endpoint for *B. proboscidea*.

At the present time both *P. cornuta* and *B. proboscidea* might be used as bioassay test organisms for whole sediment destined for marine disposal. They are widespread in distribution, easy to find, readily available at most times of the year, easy to maintain in the laboratory, easy to culture, of suitable size, and have been shown in the preceding experiments to be somewhat tolerant of environmental change and are sensitive to sediments considered to be contaminated. Furthermore, both species are as sensitive to a reference toxicant ($CdCl_2$) as are other polychaete species for which information is available in the scientific literature.

More work is required, however, before the test method can be completed and implemented for sediment assessment. Improvements in culture techniques, including determination of the optimum culture temperature, would make the production of same-aged juveniles more reliable and possibly lower the variability in mean dry weight of the animals. A better estimation of the effects of differing grain sizes, test water salinity, sediment ammonia, and organic carbon content on the survival and growth of the test organisms is critical to the interpretation of test results. Finally, a comparison of the sensitivity of these polychaetes to that of the standardized test using *Neanthes arenaceodentata* (ASTM 1994; Johns et al. 1991) could address the relative performance and sensitivity of these test methods. A round-robin test would demonstrate that other laboratories could culture the test animals and conduct the test with a defined degree of precision.

REFERENCES

Arenicola Marine, *Development of Environment Canada Procedures for Aquatic Toxicity Testing using Poly-chaetes: Selection of Suitable Species. Phase 1. Initial Selection of Promising Species,* Status Report prepared for Marine Guidelines Implementation Specialist, Office of Waste Management, Environment Canada, 20p, 1993.

Arenicola Marine, *Development of Environment Canada's Procedure for Aquatic Toxicity Testing Using Polychaetes: Selection of Suitable Species,* May 1994, Final Report for Environment Canada, 27 p, Dartmouth, N.S., 1994a.

Arenicola Marine, *Experimental Design For 1994/95 Culturing Trials and Tests with Multiple Species of Marine or Estuarine Polychaetes,* June 1994, Report for McLeay Associates, Vancouver B.C., 19 p. Dartmouth, N.S., 1994b.

ASTM, *Standard Guide for Conducting Sediment Toxicity Tests with Marine and Estuarine Polychaetous Annelids,* E1611-94, 24 p., American Society for Testing and Materials, Philadelphia, PA, 1994.

Banse, K. and Hobson, K.D., *Benthic Errantiate of British Columbia and Washington, Bull. Fisheries Research Board of Canada,* 111p., 1974.

Carr, R.S. and Curran, M.D., Evaluation of the archiannelid *Dinophilus gyrociliatus* for use in short-term life-cycle toxicity tests, *Environ. Toxicol. Chem.,* 5, 703, 1986.

Carr, R.S. and Neff, J.M., Field assessment of biochemical stress indices for the sandworm *Neanthes virens* (Sars), *Marine Environmental Research,* 14, 267, 1984.

Carr, R.S., Williams, J.W., and Fragata, C.T.B., Development and evaluation of a novel marine sediment pore water toxicity test with the polychaete *Dinophilus gyrociliatus, Environ. Toxicol. Chem.,* 8, 533, 1989.

Chapman, P.M., Dexter, R.N., Kocan, R.M., and Long, E.R., An overview of biological effects testing in puget sound, washington: methods, results and implications, in *Aquatic Toxicology and Hazard Assessment: Seventh Symposium,* ASTM STP 845, Cardwell, R.D., Purdy, R., and Bahner, R.C., Eds., American Society for Testing and Materials, Philadelphia, PA, 1985, 344.

Chapman, P.M. and Fink, R., Effects of Puget Sound sediments and their elutriates on the life cycle of *Capitella capitata, Bull. Environ. Contam. Toxicol.,* 33, 451, 1984.

Craig, N.C.D. and Caunter, J.E., The effects of polydimethylsiloxane (PDMS) in sediment on the polychaete worm *Nereis diversicolor, Chemosphere,* 21(60), 751, 1990.

Environment Canada, *Users Guide to the Application Form for Ocean Disposal,* Report EPS 1/MA/1, 1995.

Grassle, J.P. and Grassle, J.F., Sibling species in the marine pollution indicator *Capitella* (Polychaeta), *Science,* 192, 267, 1976.

Grassle, J.P. and Grassle, J.F., The utility of studying the effects of pollutants on single species populations in benthos of mesocosms and coastal ecosystems, in *Concepts in Marine Pollution Measurements,* White, H.H., Ed., University of Maryland Sea Grant, College Park, MD, 1984, 621.

Hobson, K.D. and Banse, K., *Sedentariate and Archiannelid Polychaetes of British Columbia and Washington, Canadian Bulletin of Fisheries and Aquatic Sciences,* 209:144 p. 1981.

Huybers, A.L., Doe, K.G., Wade, S.J., and Wohlgeschaffen, G.D., Results of sensitivity tests with four polychaete species, Report to Arenicola Marine, Dartmouth Nova Scotia, January, 1994, 5 p., Environment Canada, Atlantic Region, 1994.

Johns, D.M. and Ginn, T.C., Test demonstration of a 10-day *Neanthes* acute toxicity bioassay, 16 p., August 1989, Report prepared by PTI Environmental Services for U.S. Army Corps of Engineers, Seattle, WA, 1989.

Johns, D.M., Ginn, T.C., and Ciammaichella, R., *Neanthes* long-term exposure experiment: further evaluation of the relationship between juvenile growth and reproductive success, 19 p., July 1991, Report EPA 910/9-91-026 (EPA Contract 68-D8-0085), prepared by PTI Environmental Services for U.S. Environmental Protection Agency, Seattle, WA, 1991.

Johns, D.M. and Ginn, T.C., Interlaboratory comparison of *Neanthes* 20-day sediment bioassay, Abstract, SETAC Platform Session #372, 13th Annual Meeting, Soc. Environ. Toxicol. Chem., Nov. 8-12, 1992, Cincinnati, OH, 1992.

Johns, D.M., Ginn, T.C., and Reish, D.J., The effect of contaminated sediments on the growth of the polychaete *Neanthes arenaceodentata, Bull. Mar. Sci.,* 48, 589, 1991.

Johns, D.M., Gutjahr-Gobell, R., and Schauer, P., Use of bioenergetics to investigate the impact of dredged material on benthic species: a laboratory study with polychaetes and Black Rock Harbour Material, 75 p., Final report prepared by PTI Environmental Services for U.S. Army Corps of Engineers and U.S. Environmental Protection Agency, Washington, DC, 1985.

Johns, D.M., Ginn, T.C., and Reish, D.J., Protocol for juvenile *Neanthes* sediment bioassay, 17 p., June 1990, Report EPA 910/9-90-011 (EPA Contract 68-D8-0085), prepared by PTI Environmental Services for U.S. Environmental Protection Agency, Seattle, WA, 1990.

Johns, D.M. and Ginn, T.C., *Neanthes* long-term exposure experiments: the relationship between juvenile growth and reproductive success, 15 p., Report (EPA Contract 910/9-90-010) prepared by PTI Environmental Services for U.S. Environmental Protection Agency, Seattle, WA, 1990.

Jop, K.M., Acute and rapid-chronic toxicity of hexavalent chromium to five marine species, in *Aquatic Toxicology and Hazard Assessment*, Vol. 12, ASTM STP 1027, Cowill, U.M. and Williams, L.R., Eds., American Society for Testing and Materials, Philadelphia, PA, 1989, 251.

Kaag, N.H.B.M., Foekema, E.M., and Bowmer, C.T., A new approach for testing contaminated marine sediments: fertilization success of lugworms following parental exposure, *J. Aquat. Ecosystem Health,* 3(3), 177, 1994.

Lee, R.F. and Singer, S.C., Detoxifying enzyme systems in marine polychaetes: increases in activity after exposure to aromatic hydrocarbons, *Rapports et Procès-verbaux des Réunions, Conseil International Pour L'Exploration de la Mer,* 179, 29, 1980.

Long, E.R. and Chapman, P.M., A sediment quality triad: measures of sediment contamination, toxicity and infaunal community composition in Puget Sound, *Mar. Poll. Bull.,* 16 (10), 405, 1985.

Long, E.R., Buchman, M.F., Bay, S.M., Breteler, R.J., Carr, R.S., Chapman, P.M., Hose, J.E., Lissner, A.L., Scott, J., and Wolfe, D.A., Comparative evaluation of five toxicity tests with sediments from San Francisco Bay and Tomales Bay, California, *Environ. Toxicol. Chem.,* 9, 1193, 1990.

McLeese, D.W., Burridge, L.E., and Van Dinter, J., Toxicities of five organochlorine compounds in water and sediment to *Nereis virens, Bull. Environ. Contam. Toxicol.,* 28, 216, 1982.

Moore, D.W. and Dillon, T.M., The relationship between growth and reproduction in the marine polychaete *Nereis (Neanthes) arenaceodentata* (Moore): implications for chronic sublethal sediment bioassays, *J. Exp. Mar. Biol Ecol.,* 173, 231, 1993.

Pastorok, R.A. and Becker, D.S., Comparative sensitivity of sediment toxicity bioassays at three superfund sites in Puget Sound, in *Aquatic Toxicology and Risk Assessment: Thirteenth Volume*, Landis, W.G. and van der Schalie, W.H., Eds., ASTM STP 1096, American Society for Testing and Materials, Philadelphia, PA, 1990, 123.

Pesch, C.E. and Hoffman, G.L., Interlaboratory comparison of a 28-day toxicity test with the polychaete *Neanthes arenceodentata*, in *Aquatic Toxicology and Hazard Assessment: Sixth Symposium*, Bishop, W.E., Cardwell, R.D., and Heidolph, B.B., Eds., ASTM STP 802, American Society for Testing and Materials, Philadelphia, PA, 1983, 482.

Pesch, G.G., Mueller, C., Pesch, C.E., Malcolm, A.R., Rogerson, P.F., Munns, W.R., Gardner, G.R., Heltshe, J., Lee, T.C., and Senecal, A., Sister chromatid exchange in a marine polychaete exposed to a contaminated harbour sediment, in *Short-Term Bioassays in the Analysis of Complex Environmental Mixtures V.* Sandhu, S.S., DeMarini, D.M., Mass, M.J., Moore, M.M., and Mumford, J., Eds., *Environ. Sci. Res. Ser.,* Vol. 36, Plenum Press, 1987, 237.

Pesch, C.E., Munns, W.R., Jr., and Gutjahr-Gobell, R., Effects of a contaminated sediment on life history traits and population growth rate of *Neanthes arenaceodentata* (Polychaeta: Nereidae) in the laboratory, *Environ. Toxicol. Chem.,* 10, 805, 1991.

Pocklington, P., *Polychaetes of Eastern Canada: An Illustrated Key to the Polychaetes of Eastern Canada Including the Eastern Arctic,* a manuscript report funded by the Ocean Dumping Control Act Research Fund, National Museums of Canada and the Department of Fisheries and Oceans in Mont Joli, Québec. 274 p., 1989 [Unpublished manuscript].

Pocklington, P., Vigerstad, T., Doe, K., Yee, S., Scroggins, R., and Chevrier, A., *Test Method Development for Sediment Toxicity using Multiple Species of Polychaetes found in Canadian Waters. Preliminary Study,* Abstract, Presented at First SETAC World Congress, March 1993, Lisbon, Portugal.

Pocklington, P., Doe, K., Yee, S., Vigerstad, T., Chevrier, A., McLeay, D., and Pocklington, M.E., Preliminary results of test development for sediment toxicity using multiple species of Atlantic, Pacific and Arctic polychaetes, in Westlake, G.F., Parrott, J. L., and Niimi, A.J., Eds., *Proceedings of the 21st Annual Aquatic Toxicity Workshop*: October 3–5, 1994, Sarnia, Ontario. Canadian Technical Report of Fisheries and Aquatic Science No. 2050, 33-36, 1995a.

Pocklington, P., Doe, K., Huybers, A., Wade, S., Lee, D., *Development of Whole Sediment Bioassay Using the Marine/Estuarine Polychaetes Polydora cornuta Bosc 1802 and Boccardia proboscidea Hartman 1940,* Second SETAC World Congress (16th Annual Meeting) Global Environmental Protection: Science, Politics and Common Sense, 5-9 November 1995, Vancouver B.C. (Abstract) pg. 259, 1995b.

Pocklington, P., Doe, K., Pocklington, M. E., Huybers, A., Wade, S., Lee, D., Yee, S., Fennel, M., and McLeay, D., *Development of Environment Canada's Growth and Survival Test for Sediment Toxicity Using Cultured Polychaete Worms,* Method Development and Application Section, Environment Canada, Ottawa, 56 p., 1995c.

Pocklington, P. and Wells, P.G., Polychaetes: key taxa for marine environmental quality monitoring, *Mar. Poll. Bull.,* 24(12), 593, 1992.

Reish, D.J., Marine ecotoxicological tests with polychaetous annelids, in *Ecotoxicological Testing for the Marine Environment*, Persoone, G., Jaspers, E., and Claus, C., Eds., State University of Ghent and the Institute of Marine Scientific Research, Bredene, Belgium, Vol. 1, 1984, 427.

Reish, D.J., The use of larvae and small species of polychaetes in marine toxicological testing, in *Microscale Testing in Aquatic Toxicology: Advances, Techniques, and Practice,* Wells, P.G., Lee, K., and Blaise, C., Eds., CRC Press, Boca Raton, FL, 1998, 383.

Reish, D.J. and Carr, R.S., The effect of heavy metals on the survival, reproduction, development and life-cycle for two species of polychaetous annelids, *Mar. Pollut. Bull.,* 9, 24, 1978.

Reish, D.J., Piltz, F., Martin, J.M., and Word, J.Q., Induction of abnormal polychaete larvae by heavy metals, *Mar. Pollut. Bull.,* 5(8), 125, 1974.

Walsh, G.E, Louie, M.K., McLaughlin, L.M., and Lores, E.M., Lugworm (*Arenicola cristata*) larvae in toxicity tests: survival and development when exposed to organotins, *Environ. Toxicol. Chem.,* 5, 749, 1986.

Zajac, R.N. and Whitlach, R.B., Population ecology of the polychaete *Nephtys incisa* in Long Island Sound and the effects of disturbance, *Estuaries,* 11 (2), 117, 1988.

CHAPTER 28

Microscale Toxicity Testing With Rotifers

Terry W. Snell and Colin R. Janssen

CONTENTS

Abstract

ABSTRACT

Toxicity tests are one of the essential tools for evaluating the effects of anthropogenic stress in aquatic ecosystems. Methods using microscale organisms have improved the speed, simplicity, and reduced the cost of making toxicity measurements. The size, ecology, and life cycle of rotifers makes them well suited for microscale assays and toxicity assessment. In the last five years the use of rotifers in routine toxicity assessment has dramatically increased, due primarily to the widespread availability of rotifer cysts and standardized protocols for performing tests. Development of new endpoints for rapid toxicity measurements are described, as well as their application to evaluating effluents, sediment and soil elutriates, and sediment pore waters. Most work has used 24-h acute tests with brachionid rotifers, but a standardized 2-d population growth test and a 7-d full life cycle test are available for estimating chronic toxicity. The chronic toxicity of several compounds has been evaluated and acute/chronic ratios of 0.9 to 33 reported. Behavioral assays of swimming are described that are based on computer tracking of rotifer movement. Methods for measuring ingestion rates are described using fluorescent microspheres and image analysis of single rotifers. Several substrates that are metabolized to fluorescent products have been used to measure *in vivo* enzyme activity. The reduction of enzyme activity with increasing toxicant exposure has been quantified in single rotifers with image analysis or in populations with a fluorometer. Results from changes in rotifer stress protein gene expression as a result of toxicant exposure are summarized. The need to relate biological effects at one level of organization to those at higher levels is discussed. Understanding how effects in population level processes like growth, predation, and competition translate into changes in community structure is especially critical. Several areas where rotifers could contribute to the understanding of how toxicants modify aquatic ecosystems are described.

I. INTRODUCTION

As the impact of environmental pollution and public awareness about it has increased, the need to monitor and control the release of toxic substances has grown. Measuring biological responses in test organisms is a key element in monitoring pollution effects. Conventional aquatic toxicity tests such as the internationally accepted methods with algae, cladocerans, and fish[1] have well-recognized limitations and are not sufficient in number or type for routine toxicity testing.[2-4] Some major drawbacks of conventional fish toxicity tests are: 1) problems in culturing test organisms — availability and response variability, 2) the need for large test volumes, 3) large amount of bench space required, and 4) the high costs of executing the tests. This has promoted the development of microbiotests which are rapid, small-volume toxicity tests with microscale organisms.[2,5] Although they are designed to be rapid, inexpensive, and simple, most of these screening assays attempt to retain high ecological relevancy. Recently, some advances in cell biology have been adapted for use in microbiotests, improving test speed, cost, and sensitivity.[6] Because of this, microbiotests can be used to assess toxic effects with more species. Risk assessments based on the toxic responses of several phylogenetically diverse species are more reliable for predicting real-world risks.

In this chapter, we describe how one group of microscopic animals, the rotifers, can be used for aquatic toxicity testing on a microscale. We describe features of their biology that enable them to be confined and manipulated in microvolumes and how this can be done while making ecologically relevant measurements.

II. ASPECTS OF ROTIFER BIOLOGY

A. Ecology

Rotifers play an important role in the ecological processes of many aquatic communities.[7,8] As suspension feeders, many species of planktonic rotifers influence algal species composition through selective grazing. Individual rotifers typically clear 1 to 10 μL/animal/h,[9,10] so that the rotifer zooplankton can collectively filter very large water volumes daily. They consume >10 times their body weight of dry mass per day, assimilating 20 to 80% of the energy.[11] Since rotifer populations can reach densities of >10/mL, rotifers make substantial amounts of biomass available to higher trophic levels.[12] Indeed, in oligotrophic and mixotrophic lakes, rotifers are probably the most important link between the pico- and nanoplankton and the macrozooplankton.[13] In eutrophic lakes, the importance of rotifers in the microbial loop is probably diminished as protozoans account for more of the bactivory.[14] Rotifers also often compete with cladocerans and copepods for phytoplankton in the 2 to 18 μm size range. Along with crustaceans, rotifers contribute substantially to nutrient recycling,[15,16] further influencing phytoplankton composition.

Because they reproduce rapidly, rotifers can account for >50% of zooplankton production.[17] This biomass is utilized by several predators including other rotifers like *Asplanchna*,[18] copepods,[19] insect larvae,[20] and fish.[21] Rotifers are often the first food of many larval fish, a fact that has led to their extensive use in aquaculture.[22,23]

B. Phylogeny

Rotifers are members of the animal phylum Rotifera, one of several phyla of lower invertebrates.[24] Recent phylogenetic analyses suggest that the closest relatives of the Rotifera are the Acanthocephala and Nematoda.[25,26] Rotifers are phylogenetically distinct from the other major zooplankton groups. Other major zooplankters such as the Cladocera and Copepoda are sister lineages in the phylum Arthropoda, whereas ciliates are classified in the kingdom Protista. This phylogenetic distinction suggests that the response of rotifers to xenobiotics may be quite different at the whole-organism level from that of cladocerans, copepods, and ciliates. Consequently, water quality criteria based on the responses of cladocerans or fish may or may not be protective of rotifers. As rotifers play a major functional role and represent an important element of biological diversity in aquatic environments, it could be argued that rotifers should be considered separately in the analysis of ecotoxicological effects.

C. Distribution

The phylum Rotifera is comprised of about 3000 described species, most of which are freshwater zooplankters in the class Monogononta.[26] These species are found in all major habitats from soils, mosses, and sediments, to freshwater and marine zooplankton.[24] Many species have broad geographic distributions; some are even pan-global. It is possible, therefore, to select species as ecotoxicological models that represent a particular habitat worldwide. For example, the planktonic species *Brachionus calyciflorus* is found in lentic freshwater habitats on all continents, and it is an ecologically significant component of many natural water bodies.[26] Thus, ecotoxicological results can be globally compared with high ecological relevancy.

D. Life Cycle

The rotifer life cycle is based on cyclical parthenogenesis where asexual reproduction predominates with occasional periods of sexual reproduction[24] (Figure 28.1). Asexual females produce genetically identical daughters via mitosis and are employed exclusively in ecotoxicology. For toxicity testing, these females are obtained from stock cultures or by hatching dormant embryos called cysts.[27] It is these cysts, which are described below, that give rotifers and other taxa producing dormant stages a special advantage in ecotoxicological studies.[3,28,29] Cysts are the product of sexual reproduction between ephemeral males and sexual females. To our knowledge, male rotifers have not been used in ecotoxicological studies, but there has been some recent work on the response of sexual female reproduction to toxicants.[30] Snell and Carmona[30] found that the rate of sexual female production, the initial step in sexual reproduction, is more reduced by pentachlorophenol, chlorpyrifos, cadmium, and naphthol than the rate of asexual female production. Fertilization rate and males produced per female were unaffected by the four chemicals at the concentrations tested.

E. Rotifer Size and Its Consequences

Rotifers are among the smallest metazoans, ranging in size from 50 to 1500 μm.[24] Their small size provides several advantages for use in ecotoxicological investigations. For example, typical exposure volumes for effluent or sediment pore water tests with rotifers range from 300 μL to a

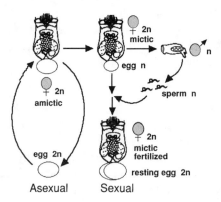

Figure 28.1 The rotifer life cycle.

few mLs depending on type of test. The ability to use small sample volumes can considerably reduce costs when working with pore water, solid waste elutriates, or when conducting toxicity identification evaluations.[31] Small size also permits a relatively large number of rotifers to be exposed in each treatment, increasing the power of statistical tests. Although small, rotifers are large enough so that single individuals can be easily transferred using standard plastic micropipets at 10 to 20× magnification using a dissecting microscope.

Rotifers also have one of the shortest generation times of any animal so that population growth rates can be measured in just a few days.[32,33] The population dynamics of rotifers are similar to bacteria, and some microbiological techniques are applicable to rotifers, such as growth in chemostats.[34] Test chambers developed for the mass microbiological market like multiwell microplates are well suited for rotifer experiments. This is important since many automated instruments are based on the microplate format.

The small size of rotifers allows them to be grown in continuous culture in chemostats. In fact, they are the only metazoan that can be grown in such cultures.[34] Other zooplankters like cladocerans have high enough growth rates for continuous cultures. However, they have a relatively long time to first reproduction, resulting in long residence times and time delays. These life cycle traits require very large culture vessels with small flow rates, all of which combine to prevent successful continuous culture. Chemostat culture of rotifers permits different kinds of experimental designs for toxicity analyses such as bioaccumulation studies and long-term, low level continuous exposures.

For example, the application of chemostats in toxicity studies is described for *B. calyciflorus*.[35] In these cultures, there were no detectable adverse effects at 0.02 mM and 0.06 mM lead after a 25-day exposure. The lethal effects of lead were detected at 0.6 mM and sublethal effects at 0.2 mM. At 0.2 mM, rotifers were washed out at a rate equal to the dilution rate, indicating cessation of growth, but no excess mortality. The use of continuous cultures allowed for accurate determination of lethal and sublethal thresholds of lead toxicity in exposures lasting 10 to 12 generations.

III. ROTIFERS IN AQUATIC TOXICOLOGY

A. History

Although rotifers have been used for experimental toxicity studies since the 1970s, it has been only in the last five years that they have been used routinely in toxicity testing.[27] The vast majority of these studies have used species in the genus *Brachionus*. This strong reliance on *Brachionus* has been more a matter of convenience rather than being based on any scientific rationale. It is not known how sensitive brachionids are to environmental contaminants as compared to other rotifer species.

The Brachionidae is certainly a large family that is distributed widely and abundant in many water bodies. However, other rotifer families, like the Asplanchnidae, Euchlanidae, Lecanidae, and Synchaetidae as well as the bdelloids, also have many species. The evaluation of the comparative sensitivity of some species in these families should therefore be recognized as an important research need.

Most studies on rotifers in aquatic toxicology have used mortality as an endpoint in one- or two-day exposures. A standardized protocol has been approved by ASTM[36] for estimating acute toxicity using the freshwater *B. calyciflorus* or the marine *B. plicatilis*. In addition to the rotifer acute tests, a variety of toxicity tests based on sublethal endpoints has also been developed.[27]

Perhaps the most significant recent advance in the use of rotifers in toxicity assessment is the controlled production of cysts.[28,29] Rotifer cysts, which can be stored for years, can be hatched in about 24 h by placing the dried cysts in water at 25°C in light. The neonates emerge synchronously and can be used in a variety of toxicity tests. As emphasized by several authors, eliminating the need to culture test animals results in significant savings in time, simplicity, and overall cost, while increasing the standardization and reproducibility of toxicity tests.[3] All labs can use the same strain and start a test with animals in the same physiological condition.[2,5] Rotifers and a few species of anostracan crustaceans are the only animals for which cysts currently are available commercially for toxicity testing. Recent advances in the controlled production of cladoceran ephippia suggest that these resting stages also will be available soon, according to G. Persoone of Belgium.

B. Recent Advances in Rotifer Toxicity Testing

The use of rotifers in ecotoxicity tests has increased in the past five years (Figure 28.2). Although rotifers have been used frequently in fundamental toxicological work, most test methods have not progressed past initial development. Only the acute (mortality) toxicity test with the freshwater rotifer *Brachionus calyciflorus* and, to a lesser extent its marine counterpart *B. plicatilis*, has been extensively evaluated as a routine toxicity screening tool. A review of recent advances in ecotoxicological research with rotifers, organized according to endpoints, is given below.

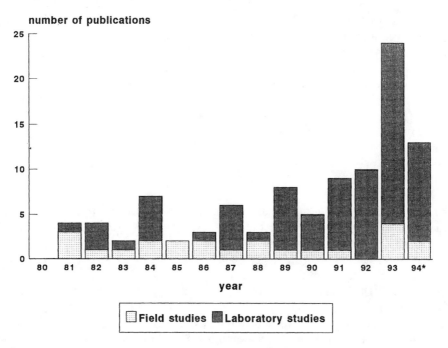

Figure 28.2 Temporal trends in the use of rotifers for ecotoxicological purposes. (Data from Poltox I database (Silverplate International). (*): period 1 January 1994 -31 March 1995.)

1. Mortality Assays

Standard protocols for performing a 24-h LC_{50} test with *B. rubens* (freshwater) and *B. plicatilis* (marine) were first described in 1989.[28,29] The freshwater species was, for reasons of convenience, replaced by *B. calyciflorus*.[37] These rotifer assays, which use dormant stages to obtain test organisms, have been incorporated in toxicity testing kits (Toxkits) and commercialized as the Rotoxkit F and Rotoxkit M tests.[3] An extensive review of the methodology, advantages, and limitations of these and other cyst-based toxicity tests is given in Chapter 30 in this volume.[38]

Several authors have reported on the use of acute rotifer tests in ecotoxicological studies. Acute (cyst-based) toxicity tests with *B. calyciflorus* and *B. plicatilis* were used as part of a battery of screening assays to determine the toxicity of the 50 priority chemicals of the Multicentre Evaluation of *in vitro* Cytotoxicity (MEIC) program in Europe. In a series of papers, Calleja and co-workers reported on the predictive power of these screening tests for evaluating human toxicity of various classes of chemicals.[39-42] Persoone et al.[31] have assessed the potential use and sensitivity of the *B. calyciflorus* test in a toxicity monitoring study of 398 environmental samples including effluents, sediment pore waters, solid waste elutriates, and monitoring wells. The rotifer assay was at least as sensitive as the acute *Daphnia magna* test for 62% of the samples. The Microtox® test was in 68% of the cases more sensitive than *B. calyciflorus*. However, this study also demonstrated that, for most samples, the rotifer assay gave nonredundant data. Similar evaluations of the freshwater rotifer test for screening environmental wastes have been reported.[43,44]

Recently, Keddy et al.[45] critically evaluated various whole-organism bioassays for the assessment of soil, freshwater sediment, and freshwater quality for their application in the Canadian National Contaminated Sites Remediation Program.[46] Of the 25 toxicity tests considered for freshwater quality assessment, three tests (the 72-h algal growth inhibition test with *Selenastrum capricornutum*, the 48-h survival test with *Daphnia magna,* and the 5- to 15-min bacterial test with *Photobacterium phosphoreum,* now named *Vibrio fischeri*) were selected for the recommended screening battery. The 24-h acute toxicity test with *B. calyciflorus* was suggested for the augmented screening battery. It was further recommended that a definitive testing battery should consist of the three above-mentioned screening tests plus an acute fish test. Also recommended was inclusion of the acute rotifer assay in the augmented definitive battery.

An in-depth analysis of the potential use of microbiotests was performed by Willemsen et al.[5] In their conclusions, these authors considered both the marine and freshwater rotifer test as practical and easy tests, exceptionally well documented and completely standardized. In comparison with the freshwater microbiotest with *Thamnocephalus platyurus,* the average sensitivity of the freshwater rotifer test for chemicals and environmental samples scored somewhat lower. A similar critical evaluation of the scientific and economic aspects of various conventional and alternative toxicity assays, including the *B. calyciflorus* test, is given by de Zwart.[47]

2. Reproduction Assays

A number of short-term tests to estimate chronic toxicity with rotifers, primarily *B. calyciflorus*, have been described.[32,33,48] In the simplest of these tests, the number of rotifers in a population in log-phase growth is counted at two time points and population growth rate (r) is estimated. For example, Janssen[48] and Janssen et al.[33] compared the sensitivity of a 72-h *B. calyciflorus* reproductive test used to estimate r with that of the 24-h LC_{50} test for copper, pentachlorophenol (PCP), 3,4 dichloroaniline (DCA), and lindane. The resulting acute to chronic ratios (A/C) were 1.2 and 12.4 for copper and 3,4 dichloroaniline, respectively. Using a 48-h toxicant exposure, a similar comparison was made by Snell and Moffat[32] for 11 chemicals. The A/C ratios ranged from 0.9 (2,4 dichlorophenoxyacetic acid) to 33 (chlorpyrifos). Such ratios are within acceptable limits as found for other species.

Because of their short generation time and lifespan, rotifers are ideally suited for life cycle toxicity testing. Under ecologically relevant conditions (25°C), *B. calyciflorus* typically produces, depending on the food availability, between 5 and 18 offspring during its lifespan which ranges from 5 to 7 days.[33,48] In life cycle toxicity tests, isolated neonates (<2 h old) are exposed to the test medium containing algal food in multiwell plates. Every 12 or 24 hours, the number of attached eggs, offspring, and mortality is recorded. Test medium is renewed daily, and the assay is terminated when the last rotifer dies, usually after 6 to 10 days. From these observations, survival and fertility tables are constructed and the following demographic parameters calculated: intrinsic rate of natural increase (r_m), net-reproductive rate, generation time, and life expectancy. Based on a series of life table experiments with Cu, PCP, DCA, and lindane, these authors showed that 95% of r_m calculated from an entire life table was reached after only 4 days and concluded that a life cycle test with rotifers can be conducted in one work week without loss of information or sensitivity. Comparison of the threshold toxicity levels obtained for the above-mentioned chemicals in life cycle tests with those of 3-day population growth tests showed that both tests have similar sensitivity.[33] The latter test, which only requires the initial setup and the final scoring of the number of rotifers after 2 or 3 days, could thus be considered a simple alternative to the more labor-intensive life cycle tests. Using life table techniques, the effect of methyl parathion, diazinon, and endosulfan on the demographic parameters of *B. calyciflorus* also was studied by Fernandez-Casalderrey et al.,[49-51] again demonstrating the usefulness of rotifers for ecotoxicological investigations of the entire life cycle.

3. *Behavioral–Physiological Assays*

A number of rapid toxicity tests based on behavioral endpoints such as swimming and feeding has been described.[27] The swimming behavior of *B. calyciflorus* exposed to Cu, PCP, DCA, and lindane, for periods ranging from 5 min to 5 h was studied by Janssen et al.[52] For the 3 of the 4 chemicals tested, the 3-h EC_{50}s based on swimming activity were lower than the 24-h LC_{50}s obtained in mortality tests. This assay, based on visual observation of individual rotifers swimming over a grid, is labor intensive and requires at least 3 h to complete. Recent developments in computer-aided video image capture have made it possible to acquire these data automatically. Charoy et al.[53] and Charoy and Janssen,[54] for example, have used computer-assisted data acquisition for tracking the swimming behavior of *B. calyciflorus* exposed to Cu, PCP, and DCA. After exposures of 5 min to 6 h, juvenile rotifers were placed individually in a microscale chamber and their swimming path recorded and digitized using a PC. From these data the average swimming speed, sinuosity, and the duration of swimming were calculated. In general, results similar to those of Janssen et al.[52] were obtained. Because rotifer swimming behavior responds rapidly and is a sensitive indicator of toxicity, automated data acquisition should expand the usefulness of behavioral endpoints in toxicity assessment.

Ingestion is an ecologically important behavior that has direct effects on growth and reproduction. The feeding activity of rotifers can easily be assessed by allowing a small population to feed on a known quantity of unicellular algae for a specified time and then quantifying the number of algal cells remaining at the end of the test. A 5-h toxicity screening test based on the feeding of *B. calyciflorus* was described by Janssen,[48] Janssen et al.,[55] and Ferrando et al.[56] Similar techniques were used to assess the effects of methyl parathion on the ingestion rate of *B. calyciflorus* and *Daphnia magna*.[51]

Juchelka and Snell[57,58] used fluorescently labeled 2 μm microspheres rather than algae to assess the effects of toxicants on the ingestion rate of *B. calyciflorus* and *B. plicatilis*. Since these rotifers are relatively nonselective suspension feeders,[59] they ingest more than a hundred of these microspheres in five minutes. Ingestion rate was quantified using image analysis by measuring the intensity of fluorescence in the gut of a single rotifer and dividing by the intensity of a single microsphere. For the 10 chemicals tested, these authors found that the ratio of NOECs based on 48-h reproductive rate (r) to those of the 1-h ingestion assay ranged from 0.13 to 1.0.

4. Biochemical and Molecular Assays

The use of *in vivo* enzyme activity as an endpoint in toxicity tests with invertebrates was first introduced for rapid toxicity screening with *Daphnia magna*.[6] This type of assay uses commercially available substrates that are initially nonfluorescent, but yield fluorescent products when cleaved by endogenous enzymes within the test animals. These probes of intracellular enzyme activity have been used as sublethal endpoints to assess toxicity in *B. calyciflorus*[60] and *B. plicatilis*.[61] *In vivo* esterase activity has proven especially useful, and it decreases in a concentration-dependent manner with toxicant exposure. The size of rotifers has allowed these enzyme reactions to be quantified in small rotifer populations in 200 μL using a 96-well microplate and an automated fluorometer or in single rotifers using an image analysis system.

Based on the rationale that the expression of certain classes of genes might be increased or decreased upon toxicant exposure, genes coding for stress proteins have proven useful for assessing toxicity in some aquatic species.[62] The expression of these genes in the rotifer *B. plicatilis* has been investigated[63,64] and techniques for visualizing stress protein abundance on Western blots with antibodies have been described. The abundance of a 58 kD stress protein (SP58) was increased fourfold upon exposure to a sublethal concentration of copper. The maximum SP58 increase occurred at a concentration of 5% of the copper 24-h LC$_{50}$, suggesting that stress protein abundance may be quite a sensitive endpoint. Although a similar response was detected for tributyl tin, several other toxicants had no effect on SP58 abundance. Antibody quantitation of stress protein abundance in rotifers needs more research before it can be widely applied for toxicity assessment. This technique, however, has good potential for automated microscale testing with existing instruments capable of making densitometric measurements.

Success in using SP gene mRNA abundance as an indicator of stress in *B. plicatilis* also has been reported.[64] The mRNA abundance was detected in dot blots using oligonucleotide probes generated using the polymerase chain reaction. This approach has potential for identifying particular genes involved in stressor-specific responses that could be useful as biomarkers.

5. Microcosm Assays

The small size of zooplankton and phytoplankton make it possible to construct a "realistic" aquatic community on a microscale in the laboratory. A three-liter standard laboratory microcosm was described by Taub[65] which is composed of rotifers (the bdelloid *Philodina*), algae, and crustaceans, all obtained from laboratory cultures. These microcosms are small enough so that sufficient replicates can be set up to allow a rigorous design and statistical analysis of the data. The microscale nature of this gnotobiotic community also allows more plant and animal species to be included in the microcosm than is possible with larger organisms. Exposure periods are typically 60 days, during which population density of each species is measured at regular intervals. This experimental system has the advantage of maintaining a considerable fraction of the ecological complexity of aquatic communities, such as competitive and predator-prey interactions. These interactions could well be more sensitive and/or more ecologically relevant measures of toxicity than traditional endpoints used in single-species tests, as stated many times by John Cairns and colleagues.[66] A similar microcosm has been described by Sugiura[67] that uses several species of bacteria, ciliates, algae, and the rotifers *Philodina* and *Lepadella*.

IV. RELATING EFFECTS ON DIFFERENT LEVELS OF BIOLOGICAL ORGANIZATION

Although the ultimate goal of ecotoxicological testing is to predict the ecological effects of chemicals and other stressors, currently methods are inadequate to reach that goal.[66,68-70] The scientific basis for the development and application of toxicity tests is limited because most basic scientific questions about the extrapolation from one level of biological organization to another remain unanswered. As yet, insufficient research effort has been directed toward understanding how environmental stress affects biological systems at different levels and how adverse effects at one level affect higher levels.[71] Because of their size and life cycle, rotifers can be useful as models to study the effects of xenobiotics at multiple levels of biological organization.

Some examples of research on the effects of xenobiotics on different levels of organization have been reported. Relationships between different levels like swimming and feeding behavior, and individual life history parameters and population dynamics of *B. calyciflorus* exposed to several toxicants were investigated by Janssen et al.[55] and Janssen.[48] A comparison of toxicant concentration thresholds found for the various test endpoints is given in Table 28.1. The relationship between ingestion rate and reproduction was also studied by Juchelka and Snell.[57] Using the same species, they examined the effect of chemicals on feeding behavior (ingestion rate) and how this subsequently affected reproduction at the population level. In the ingestion test, rotifers were exposed to pentachlorophenol, chromium, chlorpyrifos, or naphthol for 30 min at 25°C. Ingestion rate was estimated by feeding rotifers fluorescently labeled beads and quantifying with image analysis gut fluorescence intensity in single females. The reproductive test exposed rotifers for 48 h and population growth rate (r) was measured. For three of the four toxicants, the lowest observed effect concentration (LOEC) was identical for ingestion rate and reproductive tests. For naphthol, the ingestion LOEC (1.2 mg/L) was actually lower than the reproductive LOEC (2.4 mg/L). These results suggest that the rotifer 30-min ingestion test is a good predictor of the 48-h reproduction test LOEC, pointing to the strong relationship between ingestion and reproduction.

Rotifers and other small zooplankters like cladocerans and copepods are ideal for the experimental analysis of the relationship between endpoints. Future research along these lines, for example, how changes in stress protein gene expression or esterase enzyme activity affect reproduction, should be encouraged. A mechanistic understanding of toxicity effects requires such knowledge and will allow for the development of more appropriate biomarkers.

V. FUTURE APPLICATIONS

In most freshwater planktonic habitats, several rotifer species are present as part of the herbivore guild. In temperate and tropical lakes, there are typically 5 to 10 rotifer species, depending on the season.[72] This author further reported that the number of planktonic rotifer species resident in a lake over an extended period of time may be as high as 20 to 35, with the percentage of the most abundant species (dominance) ranging from 40 to 70%. Two symptoms of rotifer communities under stress have been noted: decrease in the number of species and an increase in dominance. A comparison of the differential sensitivities of rotifer species might help explain the patterns of dominance that are observed in natural rotifer assemblages exposed to anthropogenic impacts.[73-77]

An unexplored area of aquatic toxicology is whether species interactions like predator–prey relationships are disrupted at lower toxicant concentrations than single species survival and reproduction. This is important to know since water quality criteria are based exclusively on single species tests. The predator–prey interaction between the rotifers *Brachionus calyciflorus* and *Asplanchna girodi*[18] is an excellent model for examining the effect of toxicants above the single species level. The existing database of toxicant effects on the reproductive rate of *B. calyciflorus*[32,33] permit single species effects to be compared to predator–prey interactions. It would be an important result if the predator–prey interaction was disrupted at toxicant concentrations below that affecting reproduction in either species. The small size of rotifers allows for the experimental analysis of this type of problem quickly and inexpensively with rigorous experimental designs.

The use of rotifer behavior for toxicity assessment is becoming easier with computer-assisted data acquisition. Motion analysis systems for tracking swimming behavior in many animals simultaneously

Table 28.1　Comparison of Threshold Toxicity Levels for Cu and Pentachlorophenol (PCP) Obtained with *Brachionus calyciflorus* in Toxicity Tests with Various Endpoints

Criterion	Acute Test	Swimming Activity Test			Feeding Activity Test			Short Chronic Test		Life Table Studies		Population Studies	
Exposure period	Mortality 24 hours	Swimming activity 3 hours			Ingestion 5 hours			Population growth 3 days		Population growth 10 days		Carrying capacity 28 days	
Endpoint	LC_{50}	EC_{50}	LOEC	NOEC	EC_{50}	LOEC	NOEC	LOEC	NOEC	LOEC	NOEC	LOEC	NOEC
Cu (µg/L)	26	15	12	6	32	20	12	5	2.5	5	2.5	10	2.5
PCP (mg/L)	0.92	2.1	1.0	0.5	1.85	1.0	0.5	0.8	0.4	0.4	0.2	0.4	0.1

Data from references 33, 48, 52, 55, and 56. See text for explanation.

are now available.[78] Calculations of swimming speed, direction, turning frequency, and other parameters can be made automatically in a few seconds in 500 μL in multiwell plates. Image analysis also is finding applications in aquatic toxicology and has been used to quantitate the intensity of fluorescent signals in individual rotifers. Fluorescent markers have been used to measure the effects of toxicants on ingestion rate and *in vivo* enzyme activity in single individuals. Such applications are likely to expand as image analysis systems continue to decrease in price and software improves for automated image recognition.

As aquatic toxicology matures as a science, more emphasis is being placed on understanding the mechanisms of toxicity. As microscopic animals, rotifers are excellent experimental systems for investigating the relationship between molecular and cellular events and survival and reproduction at the whole-organism level. Better understanding of the effects of xenobiotics on biological systems will permit the identification of enzymes playing a regulatory role in the stress response. Similarly, regulation of the expression of specific genes could be implicated in the response to toxicant exposure. Knowledge of these stress responses will permit the design of specific molecular probes to track these endpoints and to use them in toxicity assessment. Automated measurements on small animals will surely play a central role in these advances in measuring toxicity.

REFERENCES

1. OECD (Organization for Economic Cooperation and Development), OECD guidelines for testing chemicals, OECD, Paris, 1993.
2. Blaise, C., Microbiotests in aquatic toxicology: characteristics, utility and prospects, *Toxicity Assessment*, 6, 145, 1991.
3. Persoone, G., Cyst-based toxicity tests: I. A promising new tool for rapid and cost-effective toxicity screening of chemicals and effluents, *Zeitschr. für Angew. Zool.*, 78, 235, 1991.
4. Dutka, B.J., Methods for toxicological analysis of waters, wastewaters and sediments. National Water Research Institute, Canada Centre for Inland Waters, Burlington, Ontario, Unpubl. man., 1991.
5. Willemsen, A., Vaal, M. A., and de Zwart, D., *Microbiotests as tools for environmental monitoring.* Rijksinstituut voor Volksgezondheid en Milieuhygiëne, Bilthoven, The Netherlands, 1995.
6. Janssen, C. R. and Persoone, G., Rapid toxicity screening tests for aquatic biota: I. Methodology and experiments with *Daphnia magna*, *Environ. Toxicol. Chem.*, 12, 711, 1993.
7. Pace, M. L. and Orcutt, J. D., The relative importance of protozoans, rotifers and crustaceans in a freshwater zooplankton community, *Limnol. Oceanogr.*, 26, 822, 1981.
8. Bogdan, K. G. and Gilbert, J.J., Seasonal patterns of feeding by natural populations of *Keratella, Polyarthra* and *Bosmina*: clearance rates, selectivities and contributions to community grazing, *Limnol. Oceanogr.*, 27, 918, 1982.
9. Wallace, R. L. and Starkweather, P. L., Clearance rates of sessile rotifers: *in vitro* determinations, *Hydrobiologia*, 121, 139, 1985.
10. Bogdan, K. G. and Gilbert, J. J., Quantitative comparison of food niches in some freshwater zooplankton, *Oecologia*, 72, 331, 1987.
11. Starkweather, P. L., Rotifera, in *Animal Energetics* Vol. 1: Protozoa through Insecta, Pandian, T. J. and Vernberg, F. J., Eds., Academic Press, Orlando, FL, 1987.
12. Edmondson, W. T., Secondary production, *Mitteilen International Verein. Limnologie*, 20, 229, 1974.
13. Arndt, H., Rotifers as predators on components of the microbial web (bacteria, heterotrophic flagellates, ciliates) — a review, *Hydrobiologia*, 255/256, 231, 1993.
14. Stockner, J. G. and Porter, K. G., Microbial food webs in freshwater planktonic ecosystems, in *Complex Interactions in Lake Communities*, Carpenter, S.R., Ed., Springer, New York, 1988.
15. Makarewicz, J.C. and Likens, G. E., Structure and function of the zooplankton community of Mirror Lake, New Hampshire, *Ecol. Monogr.*, 49,109, 1979.
16. Ejsmont-Karabin, J., Ammonia nitrogen and inorganic phosphorus excretion by the planktonic rotifers, *Hydrobiologia*, 104, 231, 1983.
17. Herzig, A., The analysis of planktonic rotifer populations: a plea for long-term investigations, *Hydrobiologia*, 147, 163, 1987.

18. Gilbert, J. J. and Stemberger, R. S., Prey capture in the rotifer *Asplanchna girodi*, *Inter. Ver. Theor. Ang. Limnol. Verh.*, 22, 2997, 1985.

19. Williamson, C.E., Invertebrate predation on planktonic rotifers, *Hydrobiologia*, 104, 385, 1983.

20. Moore, M. V. and Gilbert, J. J., Age-specific *Chaoborus* predation on rotifer prey, *Freshwater Biol.*, 17, 223, 1987.

21. Hewitt, D. P. and George, D. G., The population dynamics of *Keratella cochlearis* in a hypereutrophic tarn and the possible impact of predation by young roach, *Hydrobiologia*, 147, 221,1987.

22. Lubzens, E., Tandler, A., and Minkoff, G., Rotifers as food in aquaculture, *Hydrobiologia*, 186/187, 387, 1989.

23. Hoff, F. H. and Snell, T. W., *Plankton Culture Manual*, 3rd edition, Florida Aqua Farms, Dade City, Florida, 1993.

24. Wallace, R. L. and Snell, T. W., Rotifera, in *Ecology and Classification of North American Freshwater Invertebrates*, New York, Academic Press, 1991, 187.

25. Lorenzen, S., Phylogenetic aspects of pseudocoelomate evolution, in *The Origins and Relationships of Lower Invertebrates*, Clarendon Press, Oxford, UK, 1985, 210.

26. Nogrady, T., Wallace, R. L., and Snell, T. W., *Rotifera. Volume 1: Biology, ecology and systematics*, SPB Academic Publishing, The Hague, Belgium, 1993.

27. Snell, T. W. and Janssen, C. R., Rotifers in ecotoxicology: A review. *Hydrobiologia*, 313/314, 231,1995.

28. Snell, T.W. and Persoone, G., Acute toxicity bioassays using rotifers. I. A test for brackish and marine environments with *Brachionus plicatilis*, *Aquat. Tox.*, 14, 65, 1989.

29. Snell, T.W., Persoone, G., Acute toxicity bioassays using rotifers. II. A freshwater test with *Brachionus rubens*, *Aquat. Tox.*, 14, 81, 1989.

30. Snell, T. W. and Carmona, M. J., Comparative toxicant sensitivity of sexual and asexual reproduction in the rotifer *Brachionus calyciflorus*, *Environ. Tox. Chem.*, 14, 415, 1995.

31. Persoone G., Goyvaerts, M.P., Janssen, C.R., De Coen, W., and Van Steertegem, M., Cost-effective acute hazard monitoring of polluted waters and waste dumps with the aid of Toxkits, *Commission of the European Communities*, Contract ACE 89/BE2/D3, 1993.

32. Snell, T.W. and Moffat, B.D., A two-day life-cycle test with *Brachionus calyciflorus*, *Environ. Tox. Chem.*, 11, 1249, 1992.

33. Janssen, C.R., Persoone, G., and Snell, T.W., Cyst-based toxicity tests: VIII. Short chronic toxicity tests with the freshwater rotifer *Brachionus calyciflorus*, *Aquat. Tox.*, 28, 243, 1994.

34. Walz, N., *Plankton Regulation Dynamics*, Springer-Verlag, Berlin, 1993.

35. Seale, D. B. and Boraas, M. E., Application of rotifer continuous cultures to ecotoxicology, in *Plankton Regulation Dynamics*, N. Walz, Ed., Springer-Verlag, Berlin, 1993, 243.

36. ASTM, Standard guide for acute toxicity tests with the rotifer *Brachionus, Annual Book of ASTM Standards*, Vol. 11.04, E 1440, American Society for Testing and Materials, Philadelphia, PA, 1991.

37. Snell T.W., Moffat B.D., Janssen C.R., and Persoone, G., Acute toxicity tests using rotifers. IV. Effects of cyst age, temperature and salinity on the sensitivity of *Brachionus calyciflorus*, *Ecotox. Environ. Saf.*, 21, 308, 1991.

38. Persoone, G., Development and vaalidation of Toxkit microbiotests with invertebrates, in particular crustaceans, in *Microscale Testing in Aquatic Toxicology: Advances, Techniques, and Practice*, Wells, P.G., Lee, K., and Blaise, C., Eds., CRC Press, Boca Raton, FL, Chapter 30, 1998.

39. Calleja, M., Persoone, G., and Geladi, P., The predictive potential of a battery of ecotoxicological tests for human acute toxicity as evaluated with the first 50 MEIC chemicals, *Atla*, 21, 330, 1993.

40. Calleja, M., Persoone, G., and Geladi, P., Comparative acute toxicity of the first 50 Multicentre Evaluation of *in vitro* Cytotoxicity (MEIC) chemicals to aquatic non-vertebrates, *Arch. Environ. Contamin. Toxicol.*, 26, 69, 1994.

41. Calleja, M., Persoone, G., and Geladi, P., Human acute toxicity prediction of the first 50 Multicentre Evaluation of *in vitro* Cytotoxicity (MEIC) chemicals by a battery of ecotoxicological tests and physical-chemical properties, *Food and Chemical Toxicology Journal*, 32, 173, 1994.

42. Calleja, M., Persoone, G., and Geladi, P., QSAR models for predicting the acute toxicity of selected organic chemicals with diverse structures to aquatic non-vertebrates and humans, in *SAR and QSAR, Environmental Research*, 2, 193, 1993.

43. Van der Wielen, C., Persoone, G., Goyvaerts, M.P., Neven, B., and Quaghebeur, D., Toxicity of the effluents of three pharmaceutical companies as assessed with a battery of tests, *Tribune de l'eau*, Cebedeau, 43 n° 564, Etudes et Mémoires, 19, 1993.

44. Latif, M., Persoone, G., Janssen, C.R., De Coen, W., and Svardal, K., Cost-effective toxicity testing of waste waters in Austria with conventional and cost-effective bioassays, *Ecotox. Environ. Saf.*, 32, 139,1995.

45. Keddy, C.J., Greene, J.C., and Barnell, M.A., Review of whole-organism bioassays: soil, freshwater, sediment and freshwater assessment in Canada, *Ecotox. Environ. Saf.*, 30, 221, 1995.

46. CCME (Canadian Council of Ministers of the Environment), Subcommittee on environmental quality criteria for contaminated sites, the National contaminated sites remediation program, Eco-health branch, Environment Canada, Ottawa, Ontario, *Report CCME*, EPC-CS34, 1991.

47. de Zwart, D., *Monitoring water quality in the future*, Volume 3: Biomonitoring, National Institute of Public Health and Environmental Protection (RIVM), Bilthoven, The Netherlands, 1995.

48. Janssen, C.R., The use of sub-lethal criteria in toxicity tests with the freshwater rotifer *Brachionus calyciflorus* (Pallas), Ph.D. thesis, University of Ghent, Belgium, 1992.

49. Fernandez-Casalderrey, A., Ferrando, M.D., and Andreu-Moliner, E., Demographic parameters of *Brachionus calyciflorus* Pallas (Rotifera) exposed to sublethal endosulfan concentrations, *Hydrobiologia*, 226, 103, 1991.

50. Fernandez-Casalderrey, A., Ferrando, M.D., and Andreu-Moliner, E., Effect of sublethal diazinon concentrations on the demographic parameters of *Brachionus calyciflorus* Pallas (Rotifera), *Bull. Environ. Contam. Toxicol.*, 48, 202, 1992.

51. Fernandez-Casalderrey, A., Ferrando, M.D., and Andreu-Moliner, E., Chronic toxicity of methylparathion to the rotifer *Brachionus calyciflorus* fed on *Nannochloris oculata* and *Chlorella pyrenoidosa*, *Hydrobiologia*, 255/256, 41, 1993.

52. Janssen, C.R., Ferrando, M.D., Andreu-Moliner, E., and Persoone, G., Ecotoxicological studies with the freshwater rotifer *Brachionus calyciflorus*: IV. Rotifer behavior as a sensitive and sublethal indicator of toxic stress, *Ecotox. Environ. Saf.*, 28, 244, 1994.

53. Charoy, C., Janssen, C.R., Persoone, G., and Clément, P., The swimming behavior of *Brachionus calyciflorus* under toxic stress: I. The use of automated trajectometry for determining sublethal effects of chemicals, *Aquat. Tox.*, 32, 271, 1995.

54. Charoy, C. and Janssen, C.R., The swimming behavior of *Brachionus calyciflorus* under toxic stress: II. Exposure period, *Environ. Tox. Chem.*, 1997, submitted.

55. Janssen, C.R., Rodrigo, M.D.F., and Persoone, G., Ecotoxicological studies with the freshwater rotifer *Brachionus calyciflorus*. I. Conceptual framework and applications, *Hydrobiologia*, 255/256, 21, 1993.

56. Ferrando, M.D., Janssen, C.R., Andreu-Moliner, E., and Persoone, G., Ecotoxicological studies with the freshwater rotifer *Brachionus calyciflorus*: III. The effects of chemicals on the feeding behavior, *Ecotox. Environ. Saf.*, 26, 1, 1993.

57. Juchelka, C.M. and Snell, T.W., Using rotifer ingestion rates for rapid toxicity assessment, *Archiv. Environ. Contam. Toxicol.*, 26, 549, 1994.

58. Juchelka, C.M. and Snell, T.W., Rapid toxicity assessment using ingestion rate of cladocerans and ciliates, *Arch. Environ. Contam. Toxicol.*, 28, 508, 1995.

59. Starkweather, P. L., Aspects of feeding behavior and trophic ecology of suspension-feeding rotifers, *Hydrobiologia*, 73, 63, 1980.

60. Burbank, S.E. and Snell. T.W., Rapid toxicity assessment using esterase biomarkers in *Brachionus calyciflorus* (Rotifera), *Env. Tox. Water Qual.*, 9, 171, 1994.

61. Moffat, B. D. and Snell, T. W., Rapid toxicity assessment using an *in vivo* enzyme test for *Brachionus plicatilis* (Rotifera), *Ecotox. Environ. Safety*, 30, 47, 1995.

62. Sanders, B. M., Stress proteins in aquatic organisms: An environmental perspective, *Crit. Rev. Toxicol.*, 23, 49, 1993.

63. Cochrane, B. J., Irby, R. B., and Snell, T. W., Effects of copper and tributyl tin on stress protein abundance in the rotifer *Brachionus plicatilis.*, *Comp. Biochem. Physiol.*, 98C, 385, 1991.

64. Cochrane, B.J., DeLama, Y.M., and Snell, T.W., The polymerase chain reaction as a tool for developing stress protein probes, *Env. Tox. Chem.*, 13, 1221, 1994.

65. Taub, F., Standardized aquatic microcosms — development and testing. in *Aquatic Ecotoxicology* vol. II, Boudor, A. and Ribeyre, F., Eds., CRC Press, Boca Raton, FL, 1989, 47.

66. Cairns, J. Jr. and Pratt, J. R., The scientific basis of bioassays, in *Environmental Bioassay Techniques and Their Application*, Munawar, M., Dixon, G., Mayfield, C.I., Reynoldson, T., and Sadar, M.H., Eds., Kluwer Academic Publishers, London, 1989, 5.

67. Sugiura, K., A multispecies laboratory microcosm for screening ecotoxicological impacts of chemicals, *Envir. Tox. Chem.*, 11, 1217, 1992.

68. Calow, P., The choice and implementation of environmental bioassays, *Hydrobiologia*, 188/189, 61-64, 1989.

69. Geisy, J.P. and Graney, R. L., Recent developments in and comparisons of acute and chronic bioassays and bioindicators, in *Environmental Bioassay Techniques and Their Application*, Munawar, M., Dixon, G., Mayfield, C.I., Reynoldson, T., and Sadar, M.H., Eds., Kluwer Academic Publishers, London, 1989, 21.

70. Maltby, L. and Calow, P., The application of bioassays in the resolution of environmental problems: past, present and future, in *Environmental Bioassay Techniques and Their Application*, Munawar, M., Dixon, G., Mayfield, C.I., Reynoldson, T., and Sadar, M.H., Eds., Kluwer Academic Publishers, London, 1989, 65.

71. Janssen, C.R. and Persoone, G., Routine aquatic toxicity testing: some new approaches, in *Biological Indicators for Environmental Monitoring*, Bonotto, S., Nobili, R., and Revoltella, R., Eds., *Serano Symposia Review* 27, 1991, 195.

72. Green, J., Diversity and dominance in planktonic rotifers, *Hydrobiologia*, 255/256, 345, 1993.

73. Yan, N. D. and Geiling, W., Elevated planktonic rotifer biomass in acidified metal-contaminated lakes near Sudbury, Ontario, *Hydrobiologia*, 120, 199, 1985.

74. MacIsaac, H. J., Hutchinson, T. C., and Kellar, W., Analysis of planktonic rotifer assemblages from Sudbury, Ontario, area lakes of varying chemical composition, *Can. J. Fish. Aquat. Sci.*, 44, 1692, 1987.

75. Siegfried, C. A., Bloomfield, J. A., and Sutherland, J. W., Planktonic rotifer community structure in Adirondack, New York, USA lakes in relation to acidity, trophic status and related water quality characteristics, *Hydrobiologia*, 175, 33, 1989.

76. Havens, K. E., Pelagic food web structure in acidic Adirondack Mountain, New York, lakes of varying humic content, *Can. J. Fish. Aquat. Sci.*, 50, 2688,1993.

77. Havens, K. E., Structural and functional responses of a freshwater plankton community to acute copper stress, *Environ. Pollut.*, 86, 259, 1994.

78. Paffenhofer, G.A., Strickler, J.R., Lewis, K.D., and Richman, S., Motion behavior of nauplii and early copepodid stages of marine planktonic copepods, *J. Plankton Res.*, 18, 1699, 1996.

CHAPTER 29

Recent Advances in Microscale Toxicity Testing with Marine Mollusks

John W. Hunt, Brian S. Anderson, and Bryn M. Phillips

CONTENTS

I. RECENT ADVANCES IN MOLLUSKAN TOXICITY TESTING

A. Introduction

Of the wide variety of ecologically and economically important mollusks, a few common species have been used during the past 60 years as sensitive indicators of water quality. Recent reviews have described the species, protocols, sensitivity, and precision of various toxicity testing methods using mollusks.[1,2] The most commonly used methods measure toxicant effects on embryo-larval development, which is among the most sensitive of short-term test endpoints.[1,3] While standardized methods for such tests have been available for many years,[4,5] tests using mollusks are only now being approved for use in the United States for regulating effluent discharges as part of the U.S. National Pollutant Discharge and Elimination System (NPDES) permitting system. The test protocol for red abalone embryo-larval development[6] has been authorized for NPDES permit monitoring in California,[7] and this test and a similar one for marine mussels have been included in the new U.S. Environmental Protection Agency (EPA) manual for toxicity testing with west coast species,[8] making them likely candidates for inclusion in upcoming federal rules governing permit monitoring. Much of the supporting data used by regulatory agencies to characterize the precision of these test methods have come from tests using small-volume containers, as will be discussed in this chapter.

Recent advances in test methodology have demonstrated the utility of conducting molluskan toxicity tests in small volume containers. While testing in small volumes reduces sample requirements, eliminates subsampling bias in endpoint measurement, and provides logistical advantages, few studies have examined possible artifacts related to small volume testing or have documented its widespread use. In this chapter, we identify studies that demonstrate the various applications of small volume toxicity testing with marine mollusks, and examine the comparability of toxicity test results obtained in large-volume (e.g., 200 to 400 mL) and small-volume (e.g., 3 to 10 mL) test containers.

B. Exposure Systems Used in Molluskan Toxicity Testing

There are no "standard" test exposure systems in molluskan toxicity testing. The American Society for Testing and Materials (ASTM) standard guide for conducting toxicity tests with mollusks[9] specifies only that test chambers are usually of 1 to 2 L capacity, but it also cites early studies using test solution volumes as low as 10 mL and 30 mL.[10,11] It is likely, however, that test solution volumes of 200 mL to 1000 mL were used most often until very recently, requiring most investigators to subsample larvae for endpoint analysis, as described in the ASTM standard guide.[9] Cherr et al.[12] described small volume tests with embryos of the mussel *Mytilus californianus* and discussed the advantages of eliminating subsampling variability associated with larger containers. They demonstrated comparable sodium azide EC_{50} values derived from tests in large (400 mL) and small (3 mL) test solution volumes. Using small volumes for molluskan toxicity testing has rapidly gained acceptance, and it is probable that a majority of laboratories now conduct tests in 5 mL to 10 mL volumes, as evidenced by reports discussed below.

1. Methodological Adaptations and Advantages of Small Volume Testing

Many recent studies have been conducted using test containers holding 10 mL or less of test solution (Table 29.1). Glass scintillation vials, shell vials, multiwell plates, and other small transparent containers eliminate the need to transfer or subsample test organisms because all exposed embryos can be observed through the bottom of the test chamber with an inverted microscope. Bias associated with subsampling and the potential loss of fragmented or abnormal larvae is eliminated. Inverted microscopes are useful for this purpose in that they allow samples to be viewed from beneath, where preserved larvae and other material settle. This is especially important with small larvae such as those of many bivalve and echinoderm species. Screw-capped scintillation vials are convenient for analyzing larvae and allow efficient archiving and rechecking of samples for subsequent quality assurance purposes. Small test chambers allow conservation of test solution, preservatives, and cleaning solutions such as solvents or acids.

C. Applications of Small Volume Testing with Marine Mollusks

It is likely that marine mollusk toxicity tests are most commonly used to evaluate wastewater effluents and receiving waters. For these purposes, small volume tests allow efficient use of laboratory space, elimination of subsampling bias, and convenient archiving, as discussed above. However, there are additional applications for which these tests are particularly well suited. The ability to conduct tests with small sample volumes has facilitated their use in toxicity identification evaluations (TIEs) and assessments of sediment pore water, applications for which large volumes of test solution are difficult to obtain.

Table 29.1   Species and Applications for Small Volume Toxicity Testing with Marine Mollusks

Species	Volume (mL)	Container Type	Exposure Duration	Application	Reference
Oyster (*C. gigas*)	7	Lab Tek Slides	48 h	Pore water	26
Oyster (*C. gigas*)	10	Glass scint. vial	48 h	Complex effluents	27, 31
Oyster (*C. gigas*)	20	Glass scint. vial	48 h	Pore water	28
Oyster (*C. gigas*)	10	Glass scint. vial	48 h	Effluents (TIE)	13
Mussel (*M. edulis*)	5	Glass scint. vial	48 h	Pore water (TIE)	18
Mussel (*M. edulis*)	10	Glass scint. vial	48 h	Complex effluents	27, 31
Mussel (*M. edulis*)	10	Glass scint. vial	48 h	Sediment elutriate	19, 29
Mussel (*M. edulis*)	10	Glass scint. vial	48 h	Pore water	This paper
Mussel (*M. californianus*)	3	Lab Tek Slides	48 h	Sodium azide	12
Abalone (*H. rufescens*)	10	Glass scint. vial	48 h	Pore water	24, 25
Clam (*M. lateralis*)	10	Glass scint. vial	24 h	Effluents (TIE)	13, 17
Clam (*M. lateralis*)	10	Glass scint. vial	48 h	Metals, phenol	38

1. Toxicity Identification Evaluations

Recent developments in toxicity testing have included fractionation of complex toxicant mixtures to identify compounds responsible for observed toxicity.[13-16] In toxicity identification evaluation (TIE) procedures, many sample fractions are difficult to produce in large volumes. As a consequence, nearly all TIEs involving marine mollusks have been conducted using small volumes (5 to 10 mL).

U.S. EPA efforts to provide guidance for marine TIEs have included development of procedures using mollusks, especially the oyster *Crassostrea gigas* and coot clam *Mulinia lateralis*,[13,17] which has been used in both effluent and sediment TIE development. *Mulinia* has demonstrated acceptable tolerance to sodium thiosulfate, EDTA, and methanol at concentrations used in TIE procedures and has an operational pH range of 8.0 to 8.8, somewhat narrower than that demonstrated by other TIE test organisms.[13]

Studies have also been conducted to evaluate the tolerances of additional west coast test organisms to TIE manipulations. Preliminary results of studies funded by the San Francisco Bay Regional Water Quality Control Board, the U.S. EPA, and the San Francisco Bay Area Dischargers Association indicate that both red abalone and bay mussels can be used successfully in Phase 1 TIEs.[18-20] Considerations in using mollusk larvae in TIEs include sensitivity to ammonia and potential sensitivity to both sodium thiosulfate and the hypersaline brines used for sample salinity adjustments.

Both mussel and abalone larvae are relatively sensitive to ammonia, which can complicate the interpretation of TIE results.[21] A no observed effect concentration (NOEC) of 0.08 mg/L un-ionized ammonia has been calculated for oyster (*Crassostrea gigas*) larvae,[22] and our data for abalone larvae indicate a NOEC of 0.05 mg/L un-ionized ammonia. Abalone have been shown to be sensitive to ammonia concentrations found in effluent samples,[21] where concentrations can range from one to two orders of magnitude higher than these NOEC values.

A number of laboratories have indicated that brines used for salinity adjustment have caused toxicity to abalone larvae,[18,20] and a side-by-side test found abalone larvae adversely affected by a hypersaline brine mixture that was nontoxic to mussel larvae.[20] Hypersaline brines are generally prepared by freezing or evaporating natural sea water. The sea water source appears to be an important factor in brine quality, but brines from identical sea water sources do not always produce similar results at different laboratories. Communications with various investigators indicate that while one laboratory may generate acceptable brine with sea water from one source and toxic brine with sea water from another source, a different laboratory using sea water from the same sources may obtain exactly opposite results. Investigations into brine toxicity have generally been unsuccessful, and laboratories tend to search for acceptable sea water sources and then use them exclusively.

Abalone and mussel larvae have been adversely affected by sodium thiosulfate at concentrations used for TIE manipulations in one study[20] but have shown acceptable tolerance to the compound in other TIE investigations.[18] Both of the mollusks appear to be amenable to TIE testing of carbon column filtrates and extracts, allowing selective removal or addition of nonpolar organic compounds.[19,20] Mussel larval tests have demonstrated that toxicity detected in water extracts (elutriates) from San Francisco Bay sediments could be successfully removed by sample filtration through C-18 columns.[19]

2. Sediment and Pore Water

Another common application for small volume testing is in the toxicity assessment of sediments and pore water. Pore water is generally extracted from sediment in relatively small volumes by squeezing or centrifugation.[23] A number of studies have employed small volume pore water tests with abalone (*Haliotis rufescens*[24,25]), oysters (*Crassostrea gigas*[26,27]), and mussels (*Mytilus spp.*[18,28]). Small volume mussel tests have also been conducted with sediment elutriates.[19,29] The clam *M. lateralis* has been used successfully in solid-phase sediment toxicity tests using relatively small test chambers (150 mL crystallization dishes). The test incorporates juvenile clam growth inhibition as an endpoint that may provide greater sensitivity than is available in commonly used solid-phase tests measuring adult amphipod survival.[30]

Small volume pore water tests have shown consistent dose-response relationships, correlations with sediment trace metal concentrations, and good agreement among responses from different invertebrate species.[25] In a study sponsored by the California State Water Resources Control Board, pore water samples from 19 sites in an industrialized harbor area were tested at three concentrations (100%, 50%, and 25% pore water) using protocols for red abalone embryo–larval development, sea urchin embryo–larval development, and sea urchin fertilization. As would be expected of dose-response relationships, tests with all three species consistently showed decreasing toxicity with increasing pore water dilution. The tests also demonstrated significant relationships between toxicity and sediment pollutant concentrations. After excluding samples in which ammonia was found at toxic concentrations, data from red abalone pore water tests correlated significantly with sediment concentrations of copper, lead, and zinc. In the determination of significant pore water toxicity, test results from the three invertebrate protocols were in agreement 74% of the time: all tests indicated significant toxicity in 63% of samples and a lack of significant toxicity in 11% of samples.[25]

D. Small Volume Test Precision

The Washington State Department of Ecology has sponsored large interlaboratory studies to evaluate and determine the precision of a number of toxicity test methods under consideration for NPDES permit monitoring. Among the tests evaluated were the embryo/larval tests using the oyster *Crassostrea gigas* and the mussel *Mytilus spp.* Initially, these tests were to be conducted in beakers with 100 mL of test solution, as indicated in the protocol appended to the study report. However, during the course of project planning, a number of participants suggested the use of small-volume containers, and the bivalve tests for the study were conducted in 22 mL glass scintillation vials with 10 mL of test solution. This change necessitated the purchase of inverted microscopes at many participating laboratories, but participants have indicated that this investment was more than compensated for by the greater efficiency of analyzing all organisms within the test containers. Prior to the interlaboratory trials, participants expressed concerns regarding the use of micropipets to transfer embryos in very small volumes (100 µL). However, none reported problems in this regard, and experience at our laboratory has supported the use of micropipets for handling bivalve and echinoderm embryos. Abalone embryos are larger, however, and care should be taken to choose pipet tips with a relatively large aperture, so that the 250 µm embryos can be pipetted undamaged.

The Washington study found that *Mytilus* tests conducted by five laboratories during seven test periods were successfully completed within quality assurance guidelines 96% of the time; the success rate for tests with *C. gigas* was 81%. Test precision, as indicated by coefficients of variation (CV = standard deviation ÷ mean) were 21% and 29% for interlaboratory effluent tests with *Mytilus* and *C. gigas*, respectively. Intralaboratory precision estimates were similar: 19% for *Mytilus* and 25% for *C. gigas*. [27,31] These tests used lyophilized bleached kraft mill effluent, which allowed split samples to be reconstituted after extended storage without appreciable loss of effluent toxicity. Tests using cadmium as a reference toxicant produced results that were more variable than those from the lyophilized effluent tests, with interlaboratory CVs of 44% and 59% and intralaboratory CVs of 36% and 54% for *Mytilus* and *C. gigas*, respectively. A similar level of intralaboratory precision (CV = 31%) was obtained from thirteen *Mytilus* embryo-larval tests with cadmium chloride conducted at our laboratory over a two-year period using 10 mL volumes in glass scintillation vials. These precision estimates compare favorably with published estimates for the precision of other established toxicological and chemical methods[1,32-34] and indicate an acceptable level of performance for embryo–larval tests in small volumes.

With regard to this study and other work using west coast mussels, it should be noted that mussel taxonomy has been modified in recent years to identify bay mussels from the west coast of North America as *Mytilus galloprovincialis* (Baja California to San Francisco) and *Mytilus trossulus* (Monterey to Alaska). These mussels have previously been considered to be *M. edulis* and may be difficult to distinguish without electrophoretic characterization.[8] Differences in toxicant sensitivity among these species have not been demonstrated, and west coast bay mussels used in toxicity studies are now commonly referred to simply by genus.

II. EVALUATION OF THE EFFECTS OF TEST CHAMBER SIZE

As mentioned previously, Cherr et al.[12] exposed *Mytilus californianus* embryos to sodium azide in large and small containers. The large containers were 600 mL beakers holding 400 mL of test solution, while the small containers were Lab-Tek® (Nunc, Inc., Naperville, IL.) mini-polystyrene chambers adhered to glass microscope slides, each holding 3 mL of test solution. Three paired tests produced a mean EC_{50} of 25.5 mg/L in large volumes and 26.2 mg/L in small volumes, indicating that no significant artifacts were introduced by the small containers. We have conducted a number of toxicity tests with red abalone (*Haliotis rufescens*) embryos in large (200 mL) and small (10 mL) containers and have seen a similar level of agreement. Eleven tests in 10 mL volumes produced a mean zinc EC_{50} of 44.5 ± 5.7 µg/L, while eight tests in 200 mL volumes produced a mean zinc EC_{50} of 43.8 ± 9.0 µg/L, again indicating the comparability of results from tests conducted in large and small containers.

Described below are further experiments to investigate possible effects of container size, using two additional toxicants, copper chloride and the organic biocide sodium pentachlorophenate. Both of these compounds are toxic to red abalone embryos at relatively low concentrations[1] that could be affected by differential sorption to container surfaces. Because small volume containers have larger surface area to volume ratios, we investigated whether small containers might adsorb a proportionately greater toxicant mass, resulting in lower test solution concentrations and reduced toxicity.

A. Methods for Container Comparisons

1. Experimental Design

Two experiments were conducted, one each with copper chloride and sodium pentachlorophenate. Each experiment consisted of four concurrent red abalone 48-h embryo–larval toxicity tests.

Each of the four tests was conducted in one of the following types of test container: 600 mL borosilicate glass beakers with 200 mL of test solution, 30 mL borosilicate glass scintillation vials with 10 mL of test solution, 250 mL polypropylene plastic beakers with 200 mL of test solution, or 30 mL high-density polyethylene (HDPE) plastic vials with 10 mL of test solution. The ratios of wet surface area (cm^2) to test solution volume (mL) for the respective containers were 2.21, 5.42, 1.68, and 4.84, giving the small vials about two-and-a-half times more surface area per volume than the large beakers. There were five replicates of each container treatment at each toxicant concentration.

Each test container was inoculated with an equal density of abalone embryos (10 embryos/mL). Nominal test concentrations for all container types were 0, 2, 4, 8, 16, and 32 µg/L copper, and 0, 10, 20, 40, 80, and 160 µg/L pentachlorophenate. Test solutions at each concentration were delivered to all container treatments from the same volumetric mixing flasks. Copper test solutions were measured at the beginning and end of the exposure period, and from the container surfaces at the end of the experiment, as described below.

2. Test Protocol

The red abalone embryo–larval test followed a protocol[6] similar to that used in many bivalve tests (for example, ASTM[9]). Adult abalone were induced to spawn by exposure to dilute hydrogen peroxide in Tris-buffered sea water.[35] Males and females were spawned in separate containers to isolate eggs and sperm, which were then mixed at appropriate densities. The resulting embryos were rinsed, concentrated into a beaker, and delivered into test containers to provide a final embryo density of 10/mL. Embryos were exposed to static test solutions (without renewal) for 48 h. Water quality parameters of dissolved oxygen, pH, temperature, and salinity were measured in each concentration at the beginning and end of each test to verify compliance with quality assurance criteria.

All test containers were cleaned prior to testing using the following procedure: triple rinse with tap water, triple rinse with hexane solvent, triple rinse with deionized water, 24-h soak in 2N hydrochloric acid, triple rinse with deionized water, 24-h soak in deionized water, triple rinse with distilled water, dry in a clean drying oven.

During the 48-h exposure period, the embryos developed into trochophore and then veliger larvae. At the end of the exposure, larvae were prepared for observation in one of three ways. Larvae in the large glass or plastic beakers were poured onto a 37 µm mesh screen, then rinsed into scintillation vials where they were preserved in 3% buffered formalin. Larvae in small polyethylene vials were poured directly into glass scintillation vials, where they were similarly preserved. Larvae tested in glass scintillation vials were not transferred, but were preserved by addition of formalin directly into the vials. An inverted compound microscope was used to analyze all larvae through the glass bottom of the scintillation vials. We analyzed all of the approximately 100 larvae from each small test container, and a random subsample of 100 larvae from each large container. Endpoint data were reported as the percentage of larvae counted that had developed normally.

3. Sampling and Analysis of Copper Concentrations

At the beginning and end of the experiment, copper concentrations were measured in test solutions from all four test container types at the three highest nominal concentrations (8, 16, and 32 µg/L). Test solutions with nominal concentrations of 0, 2, and 4 µg/L were not measured because we had less confidence in the reliability of calculations based on measurements nearer the method detection limit (0.4 µg/L). Copper concentrations in acid rinsates of test container surfaces were also measured at these test concentrations. Pentachlorophenate concentrations were not measured because of logistical and funding constraints.

Initial copper concentrations were measured in samples taken directly from the mixing flask that was used to deliver test solution to the toxicity test containers. Final copper concentrations were measured in samples taken from additional replicates (with embryos) of each container type at the three concentrations measured (8, 16, and 32 µg/L). These samples were preserved in high-density polyethylene vials with 1% by volume 14 N quartz-distilled nitric acid.

The following procedure was used to measure copper adsorbed to test container surfaces during testing. At the end of the test period, the entire contents of a test container from each treatment at each concentration was discarded, and the test container was manually centrifuged in a sling for two minutes to remove any remaining test solution. Centrifuge speeds ranged from 90 to 120 rpm for large beakers and 120 to 150 rpm for small vials. All containers were visually inspected for consistency of droplet removal. Estimates of the remaining water volume ranged from 0.1 to 0.3 mL per container. Test containers were then rinsed by slowly swirling 2N nitric acid to contact all inner surfaces at least three times. Ten mL of acid was used to rinse the small containers, and 30 mL was used to rinse large containers. The acid rinsate was then poured into high-density polyethylene vials for chemical analysis. All copper concentrations were analyzed using a Perkin Elmer Zeeman/3030 graphite furnace atomic absorption spectrophotometer with a method detection limit of 0.4 µg/L. All measured concentrations were corrected by subtraction of appropriate matrix blanks. Analytical precision, as indicated by replicate measurements of 32 µg/L samples, was 3.8% (mean = 31.9, S.D. = 1.2 µg/L). Total copper was measured, and there was no attempt to distinguish "free" copper ions from complexed forms.

4. Data Analysis

Toxicity data were used to compare the effects observed in tests with each of the four different container types. EC_{50} data for each toxicant and test container type were calculated from nominal concentrations using the trimmed Spearman–Karber method.[36] A two-way analysis of variance (ANOVA) was used to determine the statistical significance of differences between concentrations and container types. Proportion data (percent normally developed larvae) were normalized by transformation to the arcsine square root prior to performing the analysis of variance.[37]

Chemistry data were used to calculate changes in copper concentration during the exposure period and to determine the mass of copper removed from container surfaces after testing. The calculated reduction in test solution copper concentration (µg/L) due to adsorption to container surfaces was calculated by dividing the concentration of copper in the acid rinsate (µg/L) by the volume of acid (L) to get the mass of copper from the container surface (µg), and then dividing this mass by the original test solution volume (L). This value was then compared to changes between initial and final copper concentrations as measured directly from the test solutions.

B. Results and Discussion of Container Comparisons

1. Effects of Container Type on Toxicant Concentration

Comparison of initial and final measured test solution copper concentrations from the 48-h experiment indicated that the magnitude and direction of change was variable among treatments and concentrations. Measured concentrations were unchanged in 2 of the 12 treatments. In 6 of the 12 treatments, concentrations decreased over time (by an average of 4.3 µg/L; Table 29.2). In the 4 remaining treatments, measured copper concentrations increased (by an average of 1.4 µg/L). While copper loss to container walls may have contributed to decreases in test solution concentrations in some cases, as discussed below, other factors must have affected measured concentrations. These factors may include measurement error, uptake by test larvae, precipitation from solution, and sample or test solution contamination, either from container surfaces or through exposure to air or laboratory apparatus. Some measurements are available to evaluate the possible contribution of some of these factors. Analytical precision was 3.8%, based on replicate analyses of a single 32 µg/L test solution sample (n = 5). While matrix blank values were subtracted from copper

Table 29.2   Measured Copper Concentrations from Test Solutions and Acid Rinsates of Container Surfaces in Small Volume Toxicity Tests with Red Abalone, *Haliotus rufescens*.

Nominal Cu Concentration (µg/L)	Initial Cu Concentration (µg/L)	Final Cu Concentration (µg/L)	Change in Cu Concentration During 48-h Exposure (µg/L)	(%)	Rinsate Cu per Test Sol. Vol. (µg/L)	Rinsate Cu as a Percentage of: Initial Cu Concentration	Cu Lost Over Exposure Period
Plastic Vial							
8	6.4	6.9	0.5	8%	1.6	25%	nc
16	12.0	12.0	0.0	0%	3.2	27%	nc
32	33.0	28.0	-5.0	-15%	2.5	8%	50%
Plastic Beaker							
8	6.4	6.3	-0.1	-2%	1.4	23%	nc
16	12.0	12.0	0.0	0%	2.2	19%	nc
32	33.0	27.0	-6.0	-18%	3.6	11%	59%
Glass Vial							
8	6.4	7.4	1.0	16%	nd	nc	nc
16	12.0	13.0	1.0	8%	1.0	8%	nc
32	33.0	27.0	-6.0	-18%	1.5	5%	25%
Glass Beaker							
8	6.4	2.6	-3.8	-59%	0.6	9%	15%
16	12.0	15.0	3.0	25%	0.8	7%	nc
32	33.0	28.0	-5.0	-15%	1.4	4%	27%

Note: nc indicates value could not be calculated because measured rinsate copper was greater than any measured decrease in test solution copper concentration.

nd indicates copper not detected. Method detection limit = 0.4 µg/L for all samples. Precision of replicate measurements: CV = 3.8% (n = 5).

measurements to yield the values given in Table 29.2, the blank values indicate that contaminant copper was present and may have contributed to the observed variability. These blanks included laboratory sea water (no copper detected), fresh nitric acid (0.75 µg/L), and nitric acid rinsates from each type of test container prior to exposure to test solutions (0.4 to 1.2 µg/L). Many potential factors affecting measured copper concentrations in test solutions were not investigated, and some variability remains unexplained.

Analysis of copper recovered from container surfaces did not support the hypothesis that container size and surface-to-volume ratio contributed to loss of toxicant to container walls. Copper was detected in all but one of the acid solutions used to rinse container surfaces after the test ($n = 12$; Table 29.2). Calculations indicated that copper recovered in the rinsate accounted for 4 to 27% of the initial copper in test solutions. This mass of copper is one to two orders of magnitude greater than could be accounted for by test solution droplets left behind on container surfaces after centrifugation. Of those samples where test solution chemistry indicated a decline in concentration over the test period, copper recovered from container surfaces was sufficient to account for 15 to 59% of the decrease (Table 29.2). Container material appeared to be a greater factor than container size in determining the amount of copper recovered from container surfaces. Significantly more copper was recovered from plastic surfaces than from glass (ANOVA, $\alpha = .01$; Table 29.2). Large and small glass containers had strikingly similar levels of container surface copper on a mass per test solution volume basis, and there was similar agreement between values for large and small plastic containers (Table 29.2). For the purposes of this experiment, it was assumed that the different recovery rates were the result of differential copper sorption to the glass and plastic surfaces. It must be kept in mind, however, that differential rates of copper removal during acid rinsing may also be a factor in determining the mass of copper recovered, although we assumed that a triple acid rinse would be effective with both materials.

2. Effects of Container Type on Toxicity

For both copper and pentachlorophenate, there were significant differences in rates of normal embryo–larval development among tests conducted in different types of containers (ANOVA: $\alpha < .0001$). Contributing to this statistical significance was the low between-replicate variability characteristic of the test, with a mean between-replicate CV of 10%. The differences among tests are indicated by the EC_{50} values for the respective dose-response curves (Table 29.3, Figures 29.1 and 29.2). In all cases, tests conducted in small vials resulted in lower EC_{50}s than tests in beakers, indicating greater toxicity measured in small containers. In two of three available comparisons, the differences in EC_{50}s were statistically significant, as indicated by the lack of overlapping 95% confidence intervals. Data were only available for three of the four pentachlorophenate container treatments, since the test conducted in plastic vials produced 100% abnormally developed larvae at all concentrations, including the control. This was probably due to uniform container contamination, which could not be explained by re-examination of cleaning and handling procedures that were the same for all containers. Many types of materials are known to release toxic substances that may contaminate test solutions, such as plasticizer compounds present in plastic and polyvinyl chloride containers. It is possible that exposure to acid or hexane during washing may have allowed the release of toxic compounds from the HDPE vials. The presence of such compounds may also be a factor contributing to the differential toxicity observed among other treatments in this experiment.

The observed differences in toxicity associated with different container types did not appear to be the result of differential loss of toxicant to container surfaces. The results were counter to the original hypothesis that toxicity would be decreased in smaller containers due to their greater surface-to-volume ratio and proportionately greater sorption of toxicant from test solutions. In both the trace metal and organic experiments, greater toxicity was observed in the smaller containers (Table 29.3). The trend was similar with both glass and plastic containers.

Table 29.3   Surface Area-to-Volume Ratios, Change in Copper Concentrations, and Concentrations Causing Abnormal Development in 50% of Test Larvae (EC$_{50}$s) for Tests Conducted in Different Types of Test Containers

Container Type	Surface Area to Volume Ratio	Mean Change* in Cu Concentration (%)	Copper Chloride (µg Cu/L)		Sodium Pentachlorophenate (µg/L)	
			EC$_{50}$	(C.I.)	EC$_{50}$	(C.I.)
Plastic vials	4.84	−2.3%	9.4	(8.3–10.6)	na	na
Plastic beakers	1.68	−6.7%	13.4	(11.7–15.3)	54.3	(47.6–61.8)
Glass vials	5.42	2.0%	9.5	(8.4–10.8)	49.8	(45.5–54.4)
Glass beakers	2.21	−1.6%	10.6	(9.9–11.5)	65.0	(57.2–73.9)

Note: C.I. is 95% confidence interval for EC$_{50}$ estimate.
na indicates that data were not available for this treatment.

* Average of values from all concentrations given in Table 29.2.

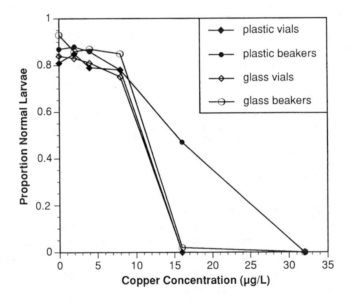

Figure 29.1   Dose-response curves for normal development of abalone larvae exposed to copper chloride in four types of test containers.

In the experiment with copper, the greatest difference in response was at 16 µg/L, where about 50% of the larvae developed abnormally in the large plastic beakers, as compared with 100% abnormal larvae at that concentration in the other three container treatments (Figure 29.1). There was little difference between treatments in measured final test solution copper at this test concentration (16 µg/L; Table 29.2). All water quality measurements were within normal ranges. Dissolved oxygen averaged 89% saturation, with low variability among measurements. Test solution ammonia concentrations were not measured, but embryo densities were the same in all containers.

In the pentachlorophenate experiment, the dose-response curve was steepest for the test in small glass vials, with 100% abnormal larvae at the 80 µg/L concentration, as compared to about 60% abnormal larvae at that concentration in the two beaker treatments (Figure 29.2). There were more abnormal larvae in the glass vial controls compared to the other tests, but larval response in the low pentachlorophenate concentrations was similar among all treatments. Organic chemical concentrations were not measured, so there are no chemical data with which to evaluate the observed differences in effects. Measured water quality parameters were all within normal ranges.

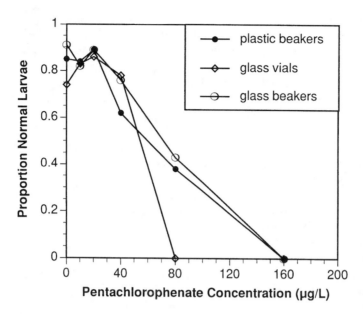

Figure 29.2   Dose-response curves for normal development of abalone larvae exposed to sodium pentachlorophenate in four types of test containers.

The toxicity data indicate that test container size may have been more important than test container material in determining the final test result. The copper EC_{50} values were similar for tests in glass and plastic vials, and similarly higher for tests in beakers of both materials. This contrasts with the copper recovery data suggesting that container material was more important than size in determining loss of copper to sorption. In both experiments there were differences in larval development between test container treatments, but these differences did not appear to be the result of toxicant sorption related to container size. Further experimentation would be necessary to investigate alternative causes for the observed differences in toxicity among container types.

III. SUMMARY

A large number of laboratories throughout the world are engaged in toxicity testing for permit monitoring activities and other environmental assessments. Many of these toxicity evaluations use a number of molluskan species and protocols, most involving embryo–larval development. The published literature on pollution studies with marine mollusks has been recently reviewed,[1] but it is likely that the majority of recent data from mollusk testing remains relatively inaccessible in client reports and other formats that are not widely circulated. Recent developments have focused on new applications for existing protocols. Primary among these new applications are assessments of sediment pore water toxicity and identification of individual toxicants in complex mixtures through TIEs. Both of these applications are greatly facilitated through conducting tests in small volumes.

Small-volume toxicity tests using mollusk embryos have demonstrated adequate precision for regulatory purposes, as indicated by available intralaboratory data and by the more comprehensive interlaboratory variability study sponsored by the State of Washington.[27,31] Variability among mollusk tests was shown to be lower than that observed in tests with echinoderms and kelp, and compared favorably with published values for other protocols.[1,32-34]

The side-by-side experiments presented in this chapter indicate that test container type may affect the results of toxicity tests, though the causes of observed differences were unclear. In tests with both trace metal and organic compounds, greater toxicity was observed in smaller containers compared to larger containers with the same toxicant concentrations. This result was contrary to what would be expected if the greater surface area to volume ratio of smaller containers were allowing relatively greater loss of toxicant through sorption to container surfaces. Measured copper concentrations from test solutions and acid rinsates of container surfaces indicated that the loss of copper to container surfaces was probably influenced more by container material than by container size (Table 29.2).

Test solution volume and container type should be considered as possible factors affecting comparability between results obtained from different exposure systems. This may be especially true of the more hydrophobic compounds. However, data from multiple bivalve tests, as presented by Cherr et al.,[12] and multiple abalone tests, as presented in this chapter, indicate that for some toxicants, there may be little appreciable difference in EC_{50} values from tests conducted in small volumes (3 to 10 mL) as compared to larger volumes (200 to 400 mL).

Since practical laboratory exposure systems are designed for maximum precision, sensitivity and convenience, rather than in an attempt to mimic field conditions, and since there has been minimal previous test container standardization, no advantage is gained by using large test solution volumes. Conducting toxicity tests with marine mollusks in small volume containers increases laboratory efficiency, eliminates subsampling and handling bias, produces an acceptable level of precision, and increases the feasibility of environmental assessments, especially those involving sediment pore water, TIEs, and other applications where large sample volumes are difficult to obtain.

ACKNOWLEDGMENTS

We appreciate the efforts of our co-workers in conducting the experiments presented. We thank Mark Slezak, Witold Piekarski, and Michelle Hester for assistance with the toxicity tests, and Jon Goetzl for chemical analysis of trace metal concentrations in test solutions. The manuscript was improved through the thoughtful suggestions of three anonymous reviewers.

REFERENCES

1. Hunt, J. W. and Anderson, B. S., From research to routine: a review of toxicity testing with marine molluscs, in *Environmental Toxicology and Risk Assessment, ASTM STP 1179*, Landis, W. G., Hughes, J. S., and Lewis, M. A., Eds., Amer. Soc. Test. Materials, Philadelphia, PA., 1993, 320.
2. Calabrese, A., Ecotoxicological testing with marine molluscs, in *Ecotoxicological Testing for the Marine Environment*, Vol. 1, Persoone, G., Jaspers, E., and Claus, C., Eds., State University of Ghent and Institute for Marine Scientific Research, Bredene, Belgium, 1, 455, 1984.
3. Conroy, P. T., Hunt, J. W., and Anderson, B. S., Validation of a short-term toxicity test endpoint by comparison with longer-term effects on larval red abalone *Haliotis rufescens*, *Environ. Toxicol. Chem.*, 15, 1245, 1996.
4. Woelke, C. E., *Development of a Receiving Water Quality Bioassay Criterion Based on the 48-h Pacific Oyster (Crassostrea gigas) Embryo*, Technical Report No. 9, Washington Department of Fisheries, Olympia, WA, 1972.
5. APHA, *Standard Methods for the Examination of Water and Wastewater*, 16th Ed., American Public Health Association, Washington, DC, 1985.
6. Anderson, B. S., Hunt, J. W., Turpen, S. L., Coulon, A. R., Martin, M., McKeown, D. L., and Palmer, F. H., Procedures manual for conducting toxicity tests developed by the Marine Bioassay Project, Report 90-10WQ, California State Water Resources Control Board, Sacramento, CA, 1990.
7. California State Water Resources Control Board, Water Quality Control Plan for Ocean Waters of California, California State Water Resources Control Board, Sacramento, CA, 1990.

8. Chapman, G. A., Denton, D. L., and Lazorchak, J. M., Short-term methods for estimating the chronic toxicity of effluents and receiving water to west coast marine and estuarine organisms, Report EPA/600/R-95-136, U.S. Environmental Protection Agency, Washington, DC, 1995.

9. ASTM, *Annual Book of ASTM Standards*, Vol. 11.04, American Society for Testing and Materials, Philadelphia, PA. 1993.

10. Byrne, C. J. and Calder, J. A., Effect of water soluble fractions of crude, refined, and waste oils on the embryonic and larval stages of the Quahog clam, *Mercenaria sp.*, *Mar. Biol.*, 40, 225, 1977.

11. Hidu, H., Roosenburg, W. H., Drobeck, K. G., McErlean, A. J., and Mihursky, J. A., Thermal tolerance of oyster larvae, *Crassostrea virginica* Gmelin, as related to power plant operation, *Proceedings of the National Shellfisheries Association*, 64, 102, 1974.

12. Cherr, G. N., Shoffner-McGee, J., and Shenker, J. M, Methods for assessing fertilization and embryonic/larval development in toxicity tests using the California mussel (*Mytilus californianus*). *Environ. Toxicol. Chem.* 9, 1137, 1990.

13. Burgess, R. M., Ho, K. T., Morrison, G. E., Chapman, G., and Denton, D. L., Marine toxicity identification evaluation (TIE): phase I guidance document. Report EPA 600/R-95/054, Office of Research and Development, U.S. Environmental Protection Agency, Washington, DC 20460, 1996, Draft.

14. Mount, D. I., Methods for aquatic toxicity identification evaluations: Phase III toxicity confirmation procedures, Technical Report EPA/600/3-88/036, U.S. Environmental Protection Agency, Environmental Research Laboratory, Duluth, MN, 1988.

15. Mount, D. I., and Anderson-Carnahan, L., Methods for aquatic toxicity identification evaluations: Phase I toxicity characterization procedures, Technical Report EPA/600/3-88/034, U.S. Environmental Protection Agency, Environmental Research Laboratory, Duluth, MN, 1988.

16. Mount, D. I., and Anderson-Carnahan, L., Methods for aquatic toxicity identification evaluations: Phase II toxicity identification procedures, Technical Report EPA/600/3-88/035, U.S. Environmental Protection Agency, Environmental Research Laboratory, Duluth, MN, 1988.

17. Pelletier, M. C., Kuhn-Hines, A., Burgess, R., Morrison, G., and Ho, K., Utility of the bivalve *Mulinia lateralis* as a marine TIE species, presented at Society of Environmental Toxicology and Chemistry 15th Annual Meeting, Denver, October 30 to November 3, 1994, SETAC, Pensacola, FL.

18. Hansen, S.R. and Associates, Development and application of estuarine sediment toxicity identification evaluations, Report, S. R. Hansen and Associates, Concord, CA, 1996.

19. Kline, K. F., Griffen, D. M., and Fisler, M. W., Preliminary TIE assessment of a region-wide toxicity observed in sediments from San Francisco Bay, presented at Society of Environmental Toxicology and Chemistry, Houston, November, 1993.

20. Edge, D., Newton, F., Cherr, G., Schipper, L., and Anderson, J., A marine TIE: when Phase I becomes a research project, presented at Society of Environmental Toxicology and Chemistry, Vancouver, B. C., November, 1995.

21. Griffen, D. M., Kline, K. F., and Targgart, L. S., Development of Phase I TIE methods for the giant kelp, *Macrocystis pyrifera* and red abalone, *Haliotis rufescens*, presented at Society of Environmental Toxicology and Chemistry, Houston, November, 1993.

22. USEPA, Refinements of current PSDDA bioassays, final report summary, Technical Report EPA/910/R-9-93-014a, U.S. Environmental Protection Agency, Water Division, Surface Waters Branch, Region 10, Seattle, WA, 1993.

23. Carr, R. S., and Chapman, D. C., Comparison of methods for conducting marine and estuarine sediment porewater toxicity tests — extraction, storage, and handling techniques, *Arch. Environ. Contam. Toxicol.*, 28, 69, 1995.

24. Sapudar, R.A., Wilson, C.J., Reid, M.L., Long, E.R., Stephenson, M., Puckett, H.M., Fairey, R.A., Hunt, J.W., Anderson, B.S., Holstad, D.R., Newman, J.W., and Birosik, S., Sediment chemistry and toxicity in the vicinity of the Los Angeles and Long Beach Harbors, Final Draft Report, California State Water Resources Control Board, Sacramento, CA, 1994.

25. Anderson, B. S., Hunt, J. W., and McNulty, H. R., Evaluation of three marine toxicity test protocols for sediment interstitial water, presented at Society of Environmental Toxicology and Chemistry (SETAC), Houston, November, 1993, SETAC, Pensacola, FL.

26. Flegal, A. R., Riseborough, R. W., Anderson, B.S., Hunt, J.W., Anderson, S., Oliver, J., Stephenson, M., and Packard, R, San Francisco Estuary pilot regional monitoring program: sediment studies, Final Report, California Regional Water Quality Control Board, San Francisco Bay Region, Oakland, CA, July, 1994.

27. Pastorok, R. A., Anderson, J. W., Butcher, M.K., and Sexton, J. E., West coast marine species chronic protocol variability study, Final Report, Washington Department of Ecology, Industrial Section, Olympia, WA, 1994.

28. Anderson, S. L., Knezovich, J. P., Jelinski, J., and Steichen, D. J., The utility of pore water toxicity testing for development of site-specific marine sediment quality objectives for metals, Report LBL-37615 UC-000, Lawrence Berkeley National Laboratory, University of California, Berkeley, CA, 1995.

29. San Francisco Estuary Institute, San Francisco estuary regional monitoring program for trace substances, Annual Report, San Francisco Estuary Institute, Richmond, CA, 1994.

30. Burgess, R. M. and Morrison, G. E., A short exposure, sublethal, sediment toxicity test using the marine bivalve *Mulinia lateralis*: statistical design and comparative sensitivity, *Environ. Toxicol. Chem.* 13, 571, 1994.

31. Bioscience Monitoring Advisory Board, West coast marine species chronic protocol variability study, Report BSAB#1, Washington Department of Ecology, Industrial Section, Olympia, WA, 1994.

32. Rue, W. J., Fava, J. A., and Grothe, D. R., A review of inter- and intralaboratory effluent toxicity test method variability, in *Aquatic Toxicology and Hazard Assessment*, 10th Volume, Adams, W. J., Chapman, G. A., and Landis, W. G., Eds., ASTM Special Technical Publication (STP) 971, Amer. Soc. Test. Materials, Philadelphia, PA, 1988.

33. Morrison, G. E., Torello, E., Comeleo, R., Walsh, R., Kuhn, A., Burgess, R., Tagliabue, M., and Greene, W., Intralaboratory precision of saltwater short-term chronic toxicity tests, *Water Pollut. Control Feder.* 61, 1707, 1989.

34. Parkhurst, B.R., Warren-Hicks, W., and Noel, L. E., Performance characteristics of effluent toxicity tests: summarization and evaluation of data, *Environ. Toxicol. Chem.* 11, 771, 1992.

35. Morse, D. E., Duncan, H., Hooker, N., and Morse, A., Hydrogen peroxide induces spawning in molluscs, with activation of prostaglandin endoperoxide synthetase, *Science*, 196, 298, 1977.

36. Hamilton, M.A., Russo, R.C., and Thurston, R.V., Trimmed Spearman-Karber method for estimating median lethal concentrations in toxicity bioassays, *Environ. Sci. and Technol.*, 11, 714, 1977.

37. Zar, J. H., *Biostatistical Analysis*, 2nd Edition. Prentice-Hall Inc., New Jersey, 1984.

38. Morrison, G. E. and Petrocelli, E., Suitability of *Mulinia lateralis* as a euryhaline toxicity test species, Proceedings of the Seventeenth Annual Aquatic Toxicity Workshop: Nov. 5-7, 1990, Vancouver, B.C. Vol. 1, p. 337, Can. Tech. Rept. Fish. Aquat. Sci. No. 1774, Feb. 1991.

CHAPTER 30

Development and Validation of Toxkit Microbiotests with Invertebrates, in Particular Crustaceans

G. Persoone

CONTENTS

I. INTRODUCTION

Between the end of the 1970s and the mid 1980s, three comprehensive reviews on toxicity tests with invertebrates were published by Murphy,[1] Buikema and Cairns,[2] and Persoone et al.,[3] and an update paper on predictive toxicity testing on aquatic biota was written 10 years later by Maltby and Calow.[4] The latter review revealed that in fact bioassay practice had not changed much since the former four reviews. Invertebrates were still much more commonly used test organisms than fish; the majority of the tests were short-term bioassays; and from the 18 taxonomic groups of invertebrates on which effect data have been published, cladocerans and in particular *Daphnia* were the most widely used test animals. In the recent *Handbook of Ecotoxicology*, the review on freshwater invertebrate toxicity tests by Persoone and Janssen[5] indicates that *Daphnia* tests are in fact currently the only type of freshwater invertebrate bioassays that are formally endorsed by international organizations (e.g., the OECD and ISO) and that tests with waterfleas are, together with fish, the most used assays for regulatory testing worldwide.

With regard to the marine environment, a variety of invertebrate test species including many crustaceans, is currently used in the framework of international conventions (such as PARCOM or

MARPOL), with an increasing tendency for the mysid *Mysidopsis bahia* and the calanoid copepod *Acartia tonsa* as preferred test organisms.[3]

Whatever method or species is recommended, all tests have the same drawbacks enumerated by Snell and Janssen in their Introduction to Chapter 28 in this volume, namely: problems in culturing test organisms, availability and response variability, the need for large test volumes, large amount of bench space required, and high cost of executing the tests.

These bottlenecks, which become critical whenever the need arises for routine testing, have fostered the development of so-called "microbiotests," a generic term introduced by Blaise[6] to characterize all microscale toxicity tests.

II. DEVELOPMENT OF THE FIRST CYST-BASED MICROBIOTESTS

The first success in developing a standard test which was independent of the burden of culturing/maintenance of live stocks of test organisms was made by Vanhaecke et al.[7] in the Laboratory for Biological Research in Aquatic Pollution (LABRAP) at the University of Ghent in Belgium. An acute toxicity microbiotest, the ARC-test, with nauplii from the brine shrimp *Artemia salina* (renamed *Artemia franciscana*) was worked out, departing from "cryptobiotic" eggs (cysts). As underlined by Van Steertegem and Persoone,[8] the use of "dormant" stages of selected aquatic invertebrate species as starting biological material for toxicity tests has numerous advantages. Besides bypassing the problem of recruitment and maintenance of live stocks of test organisms in good health and in sufficient numbers, this new approach also provides test biota with uniform characteristics and eliminates differences in sensitivity resulting from variations in culturing conditions of live stocks.

The first cyst-based toxicity test, based on the physiologically and ecologically quite special marine anostracan crustacean *Artemia*, triggered the development at LABRAP of additional microbiotests based on the dormant egg concept. Yet, this noble objective rapidly appeared to be a road with a large number of hurdles. Contrary to brine shrimp cysts, which can be harvested in large amounts in natural (saline) sites and which are commercialized for aquaculture purposes, no dormant eggs of any other appropriate candidate species seem to be available in nature in numbers large enough to satisfy the needs for toxicity testing. Consequently, the development of other cyst-based microbiotests had to face and solve all biological and technological problems inherent in the controlled production, the controlled storage, and the controlled hatching of the cryptobiotic stages of the candidate species.

It took until 1989 before two additional cyst-based toxicity tests were completed, based on a freshwater rotifer (*Brachionus rubens*, later replaced by *Brachionus calyciflorus*) and a marine rotifer (*Brachionus plicatilis*), respectively.[9,10] For details on the numerous toxicity studies already performed with these two microbiotests, we refer the reader to Chapter 28 by Snell and Janssen on toxicity testing with rotifers in this volume.

The substantial time and cost-saving of the new generation of "stock-culturing free" microscale tests was exemplified for the first time by a large factorial experiment study dealing with the influence of experimental conditions on the sensitivity of the test organisms.[11] The latter investigations encompassed more than 300 replicated bioassays and were based on two marine cyst-producing test species (*A. franciscana* and *B. plicatilis*) and the conventional *D. magna* acute assay. The factorial study confirmed the species and chemical specificity of toxic effects and again drew attention to the important influence of environmental conditions on toxic thresholds, which were found to vary by up to two orders of magnitude.

In order to better standardize the promising new microbiotest technology and to reduce costs and increase user-friendliness, LABRAP endeavored to miniaturize the cyst-based tests into "Toxkits." Besides the biological material in the form of dormant eggs, Toxkits were provided with a standard reconstituted freshwater for cyst hatching and preparation of the toxicant dilution series, and with multiwell plates as convenient miniature test vessels. Toxkit protocols were elaborated and upgraded at LABRAP to achieve adequate intralaboratory precision of the Artoxkit M and the Rotoxkit F and M.[12]

III. VALIDATION OF THE FIRST CYST-BASED MICROBIOTESTS

There was considerable international interest in the new microbiotests, but it was necessary to determine whether the Toxkits were satisfactory from the point of view of practicability and accuracy. Hence, in 1989 an intercalibration exercise was launched almost concurrently in Europe, the U.S., and Canada. Toxkits were sent out to approximately 200 laboratories that had expressed an interest in performing the Toxkit assays on the reference chemical copper sulfate. The data and information provided by 183 laboratories who pledged their results were analyzed in detail, and the conclusions of the exercise were published.[13] This extensive bioanalytical intercalibration study revealed that 60 to 80% of the participants were able to successfully complete the tests; repeatability varied from one Toxkit test to the other and ranged from as low as 25% up to 68%. Thanks to the comments and input of the participants, the research teams at LABRAP were able to address the variability problems, leading to commercialization of improved Toxkit versions at the end of 1991.

Taking into account the need for a freshwater microbiotest with a pelagic crustacean, research had in the meantime been initiated at LABRAP on a cyst-based toxicity test with the freshwater anostracan *Streptocephalus proboscideus*.[14] The Streptoxkit F assay which was eventually developed was basically analogous to the Artoxkit M test, except for the freshwater hatching and test medium.

Due to sponsorship by the European Communities and two major environmental organizations in Flanders, Belgium, a three-year microbiotest validation study was undertaken by LABRAP dealing with acute hazard monitoring of waste dumps, ground waters, river sediments, and industrial effluents. The test battery was composed of three microbiotests with biota belonging to different phylogenetic groups: a bacterial assay (the Microtox), a rotifer test (the Rotoxkit F), and a crustacean microbiotest (the Streptoxkit F). For reasons of higher sensitivity and more reliable hatching, the latter test was replaced toward the end of the study by the Thamnotoxkit F as indicated below. Acute tests were also carried out in parallel on *D. magna* on a restricted number of samples.

The goals and objectives of this extensive research effort were:

1. To evaluate the potential and the limitations of the new cost-effective Toxkit microbiotests for detection and quantification of the acute hazard of contaminated environments
2. To determine the applicability of the new microbiotests for routine toxicity monitoring of solid and liquid wastes
3. To compare the sensitivity of the new bioassay technologies with that of conventional toxicity tests
4. To establish the relationship between the toxic effects and their chemical causes

Eventually, close to 350 natural samples originating from 24 waste dumps, 23 monitoring wells, 16 purification plants, 36 river stretches, and 25 industries were analyzed ecotoxicologically and chemically. All facts and figures of this ACE biomonitoring study are reported in full detail in a 600-page final report to the EEC.[15]

The ACE validation study revealed that a small battery of low-cost microbiotests appeared capable of detecting and quantifying the acute hazard of complex environmental samples. A signal of acute hazard was obtained in virtually all toxic samples, even with only two microbiotests, namely the bacterial and the crustacean assays. In turn, a more extended battery appeared to be necessary to determine the intensity of the hazard. Yet, for an ecologically meaningful hazard evaluation of contaminated sites, the authors also emphasized the need to enlarge the test battery with low-cost assays based on species from other important phylogenetic groups, primarily microalgae. In addition,

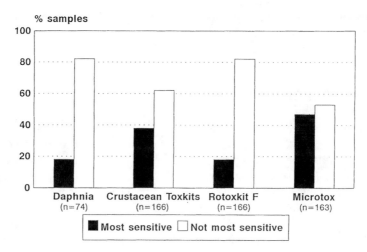

Figure 30.1   Intensity of the effect signal given by the individual bioassays for all of the toxic samples of the ACE study.

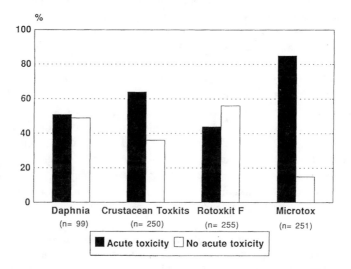

Figure 30.2   Toxicity detection potential of the individual bioassays for the toxic samples of the ACE study.

the ACE validation study confirmed that microbiotests are particularly suited for routine toxicity screening. Besides their low cost, microscale assays enable many samples to be handled per day and require only a small amount of bench space.

The sensitivity comparison of the bacterial, rotifer, and crustacean microbiotests with the conventional *D. magna* acute immobilization tests confirmed the need for a battery of tests. Figure 30.1 shows the toxicity detection potential of the individual bioassays for all of the samples found toxic to one or several of the test organisms used. All toxic effects were converted to Toxic Units (1 TU = $1/L(E)C_{50} \times 100$), and ranked in five toxicity classes. Figure 30.1 clearly reveals that the *D. magna* assays only detected a toxic impact in half of the toxic samples, whereas the crustacean and rotifer microbiotests were able to detect toxicity in 60 and 40%, respectively, and the Microtox test in more than 80% of the toxic samples. If, however, the intensity of the signal is taken into consideration instead of toxicity detection potential, the outcome is quite different. Indeed, as shown in Figure 30.2, each individual test species was only found to be the most sensitive in less than half of the cases.

Table 30.1   Correlation of the Toxicity Findings for the Most Toxic Samples of the Five Surveys of the ACE Study, with Their Chemical Causes

Survey	Degree of Toxicity	Number of Samples	Positive Correlation	No Correlation
Solid waste leachates	Extremely toxic	6	6	0
	Very toxic	2	2	0
Monitoring wells	Extremely toxic	4	2	2
	Very toxic	15	7	8
Sludge leachates	Extremely toxic	0	0	0
	Very toxic	9	2	7
Sediment pore waters	Extremely toxic	0	0	0
	Very toxic	1	1	0
Industrial effluents	Extremely toxic	3	3	0
	Very toxic	7	4	3
TOTAL		**47**	**27**	**20**

As far as the relationship between effects and causes is concerned, the efforts to earmark the chemical(s) responsible for the toxic impact on the test biota were only partially successful. Table 30.1, which summarizes the correlations for the most toxic samples of the five surveys, reveals that even in cases of severe acute toxicity, the chemical origin of the toxicity could often not be traced. Only for 27 out of 47 very toxic (>10 TU) to extremely toxic (>100 TU) samples could the toxic impact be attributed to contaminants present in concentration(s) higher than the toxicity thresholds for the test species used. As emphasized in the ACE Final Report, the lack of explanations for chemical causes versus toxic effects may be due to several reasons:

1. Despite the considerable budget allocated to the chemical part of the study, the number of inorganic and organic compounds that were analyzed was limited. Consequently, some of the toxic effects found were probably due to chemicals that had not been analyzed.
2. In many samples, various contaminants were present in concentrations below the toxic thresholds, and it is very hard (if not impossible) to extrapolate from this information whether or not the sum total of these chemicals may or may not affect the biota.

The rather poor cause-to-effect findings of the ACE study confirmed analogous poor correlations reported by Lambolez et al.[16] in a study on the environmental risks of industrial waste disposal. Consequently, one of the recommendations made by the authors of the ACE study was that financial resources for chemical investigations would be used much better if the suggestion made by Blaise (personal communication), "Why look for (chemical) causes, if there are no (toxic) effects," would be followed. According to Blaise, there is no reason to perform extensive chemical analyses on samples that are not toxic. Chemistry should instead focus on the restricted number of toxic samples to find out the cause of toxicity, through in-depth chemical analysis.

IV. EVALUATION AND APPLICATION OF CYST-BASED MICROBIOTESTS

During the early 1990s, Toxkit microbiotests were gradually incorporated into various monitoring studies of wastewaters,[17] landfills,[18] and river sediments.[19,20] More recently the Rotoxkit F and the Thamnotoxkit F were selected by Environment Canada for inclusion in a battery of five microbiotests for cost-effective aquatic-effects monitoring in the mining industry.[21] Each of the former studies confirmed the usefulness of the new microbiotests which, besides low cost and user-friendliness, were found to be as sensitive as conventional toxicity tests.

Next to the application of cyst-based assays for the detection and quantification of the acute impact of environmental contaminants, the predictive potential of Toxkit microbiotests as first screening assays has also been explored in human toxicology. In the framework of the MEIC study (Multicenter Evaluation of *in vitro* Cytotoxicity), an initiative of the Scandinavian Society of Cell

Toxicology, the acute toxicity of 50 chemicals has been determined with several cyst-based tests and the *Daphnia* acute assay, and the results compared with literature data on human lethal doses.[22-27] This research revealed that the battery of aquatic tests used was at least as accurate as the conventional rat or mouse tests in predicting the acute hazard to humans. The predictive power was increased by combination of the microbiotests with physicochemical properties through either PLS regression or neural networks.

Toxkits, and more particularly the Rotoxkit F and the Thamnotoxkit F, have also been used for QSAR modeling of nonionic surfactants and proved to be an appealing tool.[28]

A substantial amount of toxicity data has gradually been generated by LABRAP on the sensitivity of the crustacean microbiotests with the fairy shrimp *Streptocephalus proboscideus* and *Thamnocephalus platyurus* on pure chemicals and on various types of environmental samples. In order to compare the sensitivity of the crustacean Toxkit assays with those from toxicity tests using the standard acute test with the cladoceran *D. magna*, regression equations were calculated from 146 data pairs taken from five different studies.[29] This statistical analysis revealed that correlation coefficients ranged from 0.84 to 0.92 and that in the majority of cases, effect ratios between the conventional tests and the microbiotests were within a factor of 2. Repeated tests at LABRAP showed that the precision of the crustacean microbiotests and hence their degree of standardization was most satisfactory. The coefficients of variation with the Thamnotoxkit for 10 repeated tests performed "in house" by the same operator on the reference chemical potassium dichromate, was as low as 13.5%.

The growing need for ecotoxicological testing and the active search for new standard methods which could be recommended in national guidelines for hazard assessment of contaminated environments gradually focused attention on the cyst-based assays in various countries. Shortly after the appearance on the market of the Rotoxkit F test, this microscale assay was submitted in Canada to an in-depth evaluation for possible recommendation for application in the assessment and remediation of contaminated sites, under the National Contaminated Sites Remediation Program. In their "Review of Whole Organism Bioassays: Soil, Freshwater Sediment and Freshwater Assessment in Canada," Keddy et al.[30] acknowledged the potential of the rotifer microbiotest, and a 24-h survival test with *B. calyciflorus* was recommended for augmenting the set of screening tests for freshwater quality assessment. In the meantime, the two assays with *B. calyciflorus* and *B. plicatilis* have been endorsed by the American Society for Testing and Materials for a new toxicity guideline.[31]

In Europe, two studies also focused attention on the Toxkits. An in-depth analysis was performed by the Dutch National Institute of Public Health and Environmental Protection RIVM.[32] Twenty-six aquatic microbiotests were reviewed and compared on the basis of the following characteristics: convenience, completeness of documentation, cost, reproducibility, level of standardization, influence of experimental conditions, sensitivity to environmental samples and sensitivity to single compounds. A battery of three microbiotests was selected, which should be able to detect most classes of toxicants at low concentrations with a high reliability. The proposed battery was composed of a bacterial test, an algal assay, and the Thamnotoxkit.

The second study[33] was co-funded by the European Communities and two Dutch Ministries. Volume 3 of the extensive report specifically focuses on biomonitoring and makes recommendations on the bioassays which should be taken into consideration for effluent toxicity monitoring and ambient toxicity testing.[33] One of the stringent conclusions of the latter review is that "there is an increasing demand for alternative tests which are rapid, user-friendly, and more cost-effective, without neglecting ecological realism and possibilities for extrapolation." Among the variety of testing procedures that were taken into consideration in the latter study, all the Toxkit microbiotests obtained high to very high scores for each of the three selection criteria "science, efficacy, and costs."

Because of their advantages, the freshwater rotifer and crustacean Toxkit tests are to date the subject of intensive practical evaluation studies in various countries, in view of incorporation into national guidelines for toxicity testing. Since cyst-based microbiotests have a relatively short existence, none of these assays has so far been formally endorsed in any country for regulatory

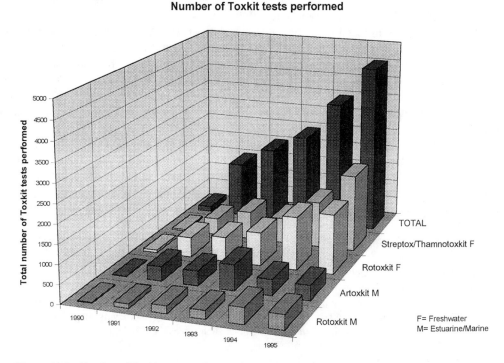

Figure 30.3   Number of Toxkit tests performed yearly since their commercialization at the end of 1990.

toxicity testing. Yet, the number of Toxkit users is increasing steadily, and more and more cyst-based assays are performed for research and/or monitoring purposes. Figure 30.3 shows the number of cyst-based assays that have been carried out yearly, revealing that by the end of 1996, more than 25,000 toxicity tests had already been performed with culture-free invertebrate microbiotests.

V. RECENT NEW DEVELOPMENTS IN TOXKIT MICROBIOTESTS WITH CRUSTACEANS: THE DAPHTOXKITS

The various qualities of the cyst-based Toxkit technology outlined above and the enthusiasm with which these microscale assays have been received by ecotoxicologists should not hide their limitations nor the reluctance in some jurisdictions to apply these low-cost and user-friendly tests. Both the rotifer and the crustacean microbiotest indeed suffer from the same major handicap, namely that these assays are not (yet) imposed by any national or international organization for the hazard evaluation of chemicals or wastes. This is due in the first place to the fact that these Toxkits do not make use of any of the test species imposed or recommended for regulatory testing. Currently all first-tier ecotoxicological evaluations are based mainly on the following three assays:

- An algal growth inhibition test
- An immobilization test with *Daphnia*
- A mortality test with fish

Although, as shown above, the crustacean Toxkit assay is as sensitive as the standard *Daphnia* test, it is clear that the latter bioassay, for which there is a very large database, will continue to be the workhorse in toxicity testing with invertebrates. Conscious of this fact, scientists at LABRAP wondered whether it might not be possible to develop a *Daphnia* and a microalgae microbiotest which would benefit from the same advantages of the other Toxkits, yet adhere to internationally accepted standard methods for toxicity testing.

In fact, research was initiated more than a decade ago at LABRAP on this subject, but it took more than 10 years of investigation before all the hurdles leading to fully standardized new Toxkit tests complying with accepted standards were overcome. This chapter is the first report on two Daphtoxkits (based on *D. magna* and *D. pulex*) and an Algaltoxkit (based on *Selenastrum capricornutum*, renamed *Raphidocelis subcapitata*) which were completed early in 1996. The characteristics of the Algaltoxkit assay, with facts and figures on this new microbiotest, are detailed in Chapter 20.

Like the other Toxkits with invertebrates, the Daphtoxkit *magna* and the Daphtoxkit *pulex* are based on dormant stages, namely cryptobiotic eggs of *D. magna* Straus and *D. pulex* Leydig, respectively. The dormant eggs are protected by a chitinous capsule (called an ephippium) and can be stored for long periods of time without losing their viability. When the ephippia are placed in specific environmental conditions and triggered, the eggs develop in 3 to 4 days into neonates which can then be used immediately for the toxicity tests.

Besides the advantages of the other Toxkits, the Daphtoxkits have two additional assets:

1. The test species are Daphnids which are used worldwide for toxicity testing and imposed or recommended by national as well as by international organizations for regulatory toxicity testing.
2. Both Daphtoxkit assays can be performed in accordance with the testing conditions prescribed by OECD Guideline 202 (*Daphnia* sp., Acute Immobilisation Test), and can be easily adapted to adhere to other national and international standard methods such as ISO, EEC, USEPA, DIN, AFNOR, and ASTM.

The strains of *D. magna* and *D. pulex* which are used for ephippia production have been selected by LABRAP from a large number of strains originating from various locations. The *D. magna* UZ94 and *D. pulex* LP94 strains were eventually chosen on the basis of a number of important characteristics — growth and reproduction, production, hatching and shelf life of the ephippia, and sensitivity of the neonates (hatched from the dormant eggs) to toxicants.

The principle of the Daphtoxkit testing procedure is the same as that of all cyst-based microbiotests, i.e., performance of 24-h (or 48-h) acute assays with neonates, less than 24 hours old, of either *D. magna* or *D. pulex* hatched from ephippia, and making use of specially designed multiwell plates as test vessels. In analogy with the conventional *Daphnia* tests, the effect criterion is immobility instead of mortality. Each Daphtoxkit contains all the disposable materials needed to perform six complete toxicity tests, either range finding or definitive. The only equipment needed is an incubator or a temperature-controlled room at 20°C, a small light table with a transparent stage, and conventional laboratory glassware.

VI. FIRST VALIDATION OF THE DAPHTOXKITS

After completion of the test procedure of the new *Daphnia* microbiotests and after selection of the materials to be included in these Toxkits, extensive in-house validation studies were started at LABRAP. The first objective was to determine the precision of the new assays and their sensitivity in comparison to the standard *D. magna* test, which makes use of neonates taken from laboratory stock cultures. Since each Daphtoxkit contains standard test materials, standard dilution water, and ephippia produced under strictly controlled conditions, the repeatability of these microscale tests was found to be as good as that of the previous Toxkits. The coefficients of variation for repeated tests were found to be below 30%, which can be considered as quite satisfactory in comparison to the intralaboratory variabilities reported in the literature.[34,35]

Table 30.2 Comparison of the 24- and 48-h EC$_{50}$s of the Daphtoxkit Tests with Those of the Conventional *Daphnia magna* Assay for 19 Inorganic and Organic Chemicals

| | Chemical compound | Daphnia magna | | | | Daphnia pulex | |
| | | Stock (ISO water) | | Ephippia (EPA water) | | Ephippia (EPA water) | |
		24 h	48 h	24 h	48 h	24 h	48 h
1	HgCl$_2$	0.029	0.020	0.023	0.013	0.022	0.010
2	PbCl$_2$	0.29	0.19	0.53	0.29	0.16	0.12
3	CuSO$_4$·5H$_2$O	0.51	0.34	0.18	0.12	0.040	0.026
4	K$_2$Cr$_2$O$_7$	0.61	0.24	0.28	0.11	0.71	0.30
5	NaPCP	1.09	0.88	0.74	0.53	0.75	0.39
6	CdCl$_2$·5H$_2$O	1.26	0.49	0.39	0.23	0.36	0.16
7	Lindane	1.58	0.90	0.96	0.71	1.11	0.87
8	3,4 DCA	2.90	2.09	3.53	2.47	9.10	2.11
9	TCA	14.05	9.96	14.29	7.74	11.38	8.26
10	Phenol	16.15	10.04	43.76	19.08	56	8.12
11	NiSO$_4$·6H$_2$O	20.02	16.63	24.90	10.0	24.36	14.34
12	SLS	21.84	17.36	15.95	13.22	9.87	7.82
13	ZnSO$_4$·7H$_2$O	28.66	15.70	15.95	6.98	7.13	2.26
14	FeSO$_4$·7H$_2$O	225	158	91	87	232.81	26.14
15	2,4 D	228	135	255	159	165	64.7
16	MnSO$_4$·H$_2$O	258	110	130	80	68.24	50.04
17	KCl	748	437	752	483	609	313
18	CaCl$_2$·2H$_2$O	3092	2192	2121	1398	2616	748
19	NaCl	6104	3094	4557	4365	2413	1341

Note: The effect data are expressed in mg/L and ranked in order of decreasing toxicity for the standard *Daphnia* test.

Table 30.3 Comparison of the 24- and 48-h EC$_{50}$s of the Daphtoxkit Tests with Those of the Conventional *Daphnia magna* Assay, for Nine Effluents from the Textile Industry

| | Sample | Daphnia magna | | | | Daphnia pulex | |
| | | Stock (ISO Water) | | Ephippia (EPA Water) | | Ephippia (EPA Water) | |
		24 h	48 h	24 h	48 h	24 h	48 h
1	9690001	0	1	0	0.3	1.32	9.12
2	9690007	3.09	4.76	3.92	5.58	7.05	12.85
3	910011	2.47	2.47	2.34	2.46	7.25	9.64
4	960018	2.57	2.64	1.68	2.25	2.65	3.07
5	9690020	0.4	1.37	1.16	1.31	1.81	9.11
6	920023/31-01	0	0	0	0.3	0.4	2.49
7	920023/1-02	0	0	0	0	0	1.75
8	92304/31-01	1.53	1.79	1.75	1.81	4.21	4.70
9	92304/1-02	1.29	1.37	1.19	1.41	2.89	5.00

Note: The effect data are expressed in Toxic Units.

Sensitivity comparisons between the Daphtoxkits and the conventional *D. magna* assay have been carried out so far with 19 inorganic and organic chemicals, and with 9 textile industry effluents (Tables 30.2 and 30.3).

Table 30.2 shows that for the majority of the chemicals the Daphtoxkit tests gave lower EC$_{50}$s than the conventional assay. For some compounds, however, the sensitivity of both assays was approximately the same and there were a few chemicals for which the Daphtoxkit neonates were less sensitive than neonates from stock cultures. The higher sensitivity of the ex-ephippia neonates for certain compounds is in fact not related to the test organisms as such, but is most probably due to the difference in water hardness of the media used in the conventional *D. magna* test versus the

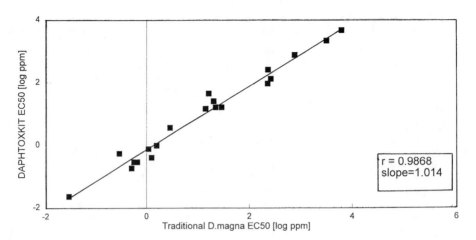

Figure 30.4   Comparison of the 24-h EC_{50}s of the Daphtoxkit *magna* test with those of the conventional *D. magna* assay. Linear regression for the data pairs of 19 inorganic and organic compounds.

microbiotests. Indeed, the artificial freshwater selected for the Daphtoxkits for hatching and for the toxicant dilutions, is the "EPA moderately hard water" which is also used in the other invertebrate Toxkits. The standard *D. magna* tests were performed in reconstituted freshwater made up with the ISO formula, which is much richer in calcium than the EPA medium; the former medium is also used at LABRAP for stock culturing of daphnids. Persoone et al.[11] quote seven literature sources which indicate the higher sensitivity of daphnids to chemicals, in particular metals, in soft versus hard waters. Experiments conducted at LABRAP with the reference chemical potassium dichromate with either stock neonates exposed to the toxicant in EPA water instead of ISO medium or alternatively with ex-ephippia neonates in ISO water instead of EPA water, confirmed the lower sensitivity of the daphnids in the harder ISO medium.

For more than half the pure chemicals tested, the 24-h and 48-h EC_{50}s of the conventional *D. magna* test did not differ from those of the Daphtoxkit *magna* assays by more than a factor of 1.5; for the other compounds, the EC_{50}s differed by factors ranging from 1.5 up to 3. For industrial effluents from the textile industry, the differences in the toxic signal between the two types of *D. magna* assays ranged from very low up to a factor of 3 after 24-h exposure. EC_{50}s after two days of exposure were the same for both tests or differed less than 30%. The excellent overall correlation between the data obtained with the Daphtoxkit *magna* and the conventional *Daphnia magna* test for pure compounds is visualized in Figure 30.4. The linear regression calculated for the nineteen 24-h EC_{50} data pairs indeed revealed an r value as high as 0.98.

Comparison of the *D. magna* and the *D. pulex* microbiotests indicated sensitivity differences from zero up to a factor of 3 for the 19 pure chemicals; for one compound (copper sulfate) the difference was surprisingly as high as a factor of 4. For the 24-h exposure period, *D. pulex* ex-ephippia neonates were more sensitive than stock *D. magna* neonates for 8 of the 19 chemicals; extension of the exposure to 48-h eventually increased the sensitivity of the former daphnid species leading to lower EC_{50}s for 13 chemicals. Table 30.3 shows that *D. pulex* was in all cases more sensitive to mixed wastes than *D. magna*, after both 24-h and 48-h exposure.

Lilius et al.[36] have addressed the question of sensitivity differences between *Daphnia* species. Departing from contradictory literature statements that *D. magna* is either more sensitive, equally sensitive, or less sensitive to toxicants than *D. pulex*, these investigators compared the 24-h effects of 30 inorganic and organic chemicals on both daphnids. The conclusion of the authors that there is no significant difference in the overall sensitivity of these two species as calculated from

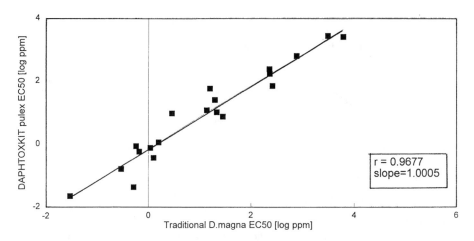

Figure 30.5 Comparison of the 24-h EC_{50}s of the Daphtoxkit *pulex* test with those of the conventional *D. magna* assay. Linear regression for the data pairs of 19 inorganic and organic compounds.

regression analysis of the EC_{50} data is in agreement with the findings reported above for pure chemicals with the Daphtoxkits. Indeed, Figure 30.5, which shows the linear regression for the 24-h EC_{50} data pairs of the Daphtoxkit *pulex* and the conventional *D. magna* test for the 19 inorganic and organic chemicals, reveals a very high concordance (r = 0.96). In turn, the first validation study on the new *Daphnia* microbiotests, however, also shows that *D. pulex* is more sensitive than *D. magna* for mixed wastes such as industrial effluents from the textile industry. It is clear that more comparisons are needed with natural samples from various sources before this conclusion can be generalized. Furthermore, the higher sensitivity of *D. pulex* after 48-h exposure has to be interpreted with caution. Indeed, during the elaboration of the sensitivity data set, it was repeatedly found that the percentage mortality of test organisms in the controls, although matching the test acceptability threshold, was higher after 48-h exposure with *D. pulex* than with *D. magna*. This may indicate that the smaller daphnid species is more stressed than *D. magna* by prolonged starvation; this raises the question whether it is (ecologically) sound to prolong acute tests with *D. pulex* beyond the 24-h exposure period.

The preliminary validation studies made at LABRAP on the relative sensitivity comparison of the Daphtoxkits versus the conventional *D. magna* assay, can be considered an extension of the recent investigations of Toussaint et al.[37] These scientists compared five inexpensive rapid toxicity tests with five standard aquatic acute tests on 11 chemicals and came to the conclusion that a test with lettuce and the rotifer (cyst-based) test ranked most similar in sensitivity to the standard tests. The information generated for the Daphtoxkits reveals that these two new microbiotests are not only as sensitive as the standard *D. magna* test, but that they have several additional advantages, not the least of which are the financial ones. On the basis of the figures provided by Toussaint et al.,[37] it was calculated that Daphtoxkit assays can be performed at a cost which is a small fraction of that of *Daphnia* assays departing from live stock cultures.

In conclusion, the first toxicity data provided in this paper on the two new crustacean microbiotests with daphnids need extensive additional intra- and interlaboratory validation before the value of these promising microscale bioassays will be fully evaluated. Yet, on the basis of the information available to date, it seems that the low-cost and user-friendly Daphtoxkits, in combination with the culture-independent Algaltoxkit, will be powerful new tools in many countries for routine toxicity screening of chemicals and biomonitoring of contaminated sites.

REFERENCES

1. Murphy, P.M., A manual for toxicity tests with freshwater microinvertebrates and a review of the effects of specific toxicants, Univ. of Wales. Inst. Science Techn., Cardiff, 249 pp., 1979.
2. Benfield, E.F. and Buikema, A.L., Jr., Synthesis of miscellaneous invertebrate toxicity tests, in *Aquatic Invertebrate Bioassays,* Buikema, A.L. and Cairns, J., Jr., Eds., ASTM Special Publication, 715, 1980, 174-187, ASTM Philadelphia, PA.
3. Persoone G., Jaspers, E., and Claus, C., Eds., Ecotoxicological testing for the marine environment, State Univ. Ghent and Inst. Mar. Scient. Res., Bredene, Belgium, Vol. 1. 798 p., Vol. 2. 588 p., 1984.
4. Maltby, L. and Calow, P., The application of bioassays in the resolution of environmental problems: past, presence and future, in *Environmental Bioassay Techniques and their Application,* Munawar, M., Dixon, G., Mayfield, C.I., Reynolds, T., and Sadar, M.H., Eds., Kluwer Academic Publishers, London, 1989, 65.
5. Persoone, G. and Janssen, C.R., Freshwater invertebrate toxicity tests, in *Handbook of Ecotoxicology,* Calow, P., Ed., Vol. 1, Blackwell Scientific Publications, 1993, 51.
6. Blaise, C., Microbiotests in aquatic toxicology: characteristics, utility and prospects, *Toxicity Assessment,* 6, 145-155, 1991.
7. Vanhaecke, P., Persoone, G., Claus, C., and Sorgeloos, P., Proposal for a short term toxicity test with *Artemia* nauplii, *Ecotoxicol. Environ. Safety,* 5, 382, 1981.
8. Van Steertegem, M. and Persoone, G., Cyst-based toxicity tests V: Development and critical evaluations of standardized toxicity tests with the brine shrimp *Artemia* (Anostraca, Crustacea), in *Progress in Standardization of Aquatic Toxicity Tests,* Soares, A.M.V.M. and Calow, P., Eds., Lewis Publishers, Boca Raton, FL, 1993, 81.
9. Snell, T.W. and Persoone, G., Acute toxicity bioassays using rotifers. I. A test for brackish and marine environments with *Brachionus plicatilis, Aquat. Toxicol.,* 14, 65, 1989.
10. Snell, T.W. and Persoone, G., Acute toxicity bioassays using rotifers. II. A freshwater test with *Brachionus rubens, Aquat. Toxicol.,* 14, 81, 1989.
11. Persoone G., Van de Vel, A., Van Steertegem, M., and De Nayer, B., Predictive value of laboratory tests with aquatic invertebrates: influence of experimental conditions, *Aquat. Toxicol.,* 14, 149, 1989.
12. Persoone, G., Cyst-based toxicity tests I. A promising new tool for rapid and cost-effective toxicity screening of chemicals and effluents, *Zeitschr. für Angew. Zool.,* 78, 235, 1991.
13. Persoone, G., Blaise, C., Snell, T., Janssen, C., and Van Steertegem, M., Cyst-based toxicity tests II. Report on an international intercalibration exercise with three cost-effective Toxkits, *Zeitschr. für Angew. Zool.,* 79 (1), 17, 1992/1993.
14. Centeno, M.D., Brendonck, L., and Persoone, G., Cyst-based toxicity tests. III. Development and standardization of an acute toxicity test with the freshwater anostracan crustacean *Streptocephalus proboscideus,* in *Progress in Standardization of Aquatic Toxicity Tests,* Soares, A.M.V.M. and Calow, P., Eds., Lewis Publishers, Boca Raton, FL, 37, 1992.
15. Persoone, G., Goyvaerts, M., Janssen, C., De Coen, W., and Vangheluwe, M., Cost-effective acute hazard monitoring of polluted waters and waste dumps with the aid of Toxkits, Final Report, Commission of the European Communities, Contract ACE 89/BE 2/D3, 600 p., 1993.
16. Lambolez, L., Vasseur, P., Ferard, J.F., and Gisbert, G., The environmental risks of industrial waste disposal: an experimental approach including acute and chronic toxicity studies, *Ecotox. Environ. Safety,* 28, 317-328, 1994.
17. Latif, M., Persoone, G., Janssen, C., De Coen, W., and Svardal, K., Toxicity evaluations of waste waters in Austria with conventional and cost-effective bioassays, *Ecotox. Environ. Safety,* 32, 139, 1995.
18. Clément, B., Persoone, G., Janssen, J., and Le Dû-Delepierre, A., Estimation of the hazard of landfills through toxicity testing of leachates. 1. Determination of leachate toxicity with a battery of acute tests, *Chemosphere,* 33 (11), 2303, 1996.
19. Vangheluwe, M., Janssen, C., Van Sprang, P., and Persoone, G., Sediment toxicity screening with cost-effective microbiotests and conventional assays: a comparative study, Abstract, Second SETAC World Congress, Vancouver, November 1995. Soc. Environ. Toxicol. Chem., Pensacola, FL.

20. Ugazio, G., Burdino, E., Crespi, M., Barbero, N., Garizio, M., Arru, G., and Congiu, A.M., Richerche ecotossicologiche eseguite con batterie di tests biologici e fitologici sui sedimenti prelevati in sequenza da 24 fiumi affluente del Po, Rendiconto del XII Convegno sulla patologia da Tossici ambientali ed occupazionali, 1-15, 1995.

21. Rodrigue, D., Mailhiot, K., Hynes, T.P., and Wilson, L.J., Aquatic effects monitoring in the mining industry: Review of appropriate technologies, in Hynes, T.P. and Blanchette, C.M., Eds., Proceedings of Sudbury '95 — *Mining and the Environment*, Vol. 2, 813, 1995.

22. Calleja, M.C. and Persoone, G., Cyst-based Toxicity Tests: IV. The potential of ecotoxicological tests for the prediction of acute toxicity in man as evaluated on the first ten chemicals of the MEIC programme, *ATLA*, 20, 396, 1992.

23. Calleja, M.C., Persoone, G., and Geladi, P., Comparative acute toxicity of the first 50 MEIC chemicals to aquatic non-vertebrates, *Arch. Environ. Contamin. Toxicol.*, 26, 69, 1994.

24. Calleja, M.C., Persoone, G., and Geladi, P., The predictive potential of a battery of ecotoxicological tests for human acute toxicity, as evaluated with the first 50 MEIC chemicals, *ATLA*, 21, 330, 1993.

25. Calleja, M.C., Geladi, P., and Persoone, G., QSAR models for predicting the acute toxicity of selected organic chemicals with diverse structures to aquatic non-vertebrates and humans, *SAR and QSAR in Environ. Res.*, 2, 3, 193, 1994.

26. Calleja, M.C., Geladi, P., and Persoone, G., Modeling of human acute toxicity from physico-chemical properties and non-vertebrate acute toxicity, of the 38 organic chemicals of the MEIC priority list by PLS regression and neural networks, *Fd. Chem. Toxic.*, 32, 10, 923, 1994.

27. Calleja, M.C., Persoone, G., and Geladi, P., Human acute toxicity prediction of the first 50 MEIC chemicals by a battery of ecotoxicological tests and physicochemical properties, *Fd. Chem. Toxic.*, 32, (2), 173, 1994.

28. Lindgren, A., Sjöström, M., and Wold, S., QSAR modeling of the toxicity of some technical non-ionic surfactants toward fairy shrimp, *Quantitative Structure Activity Relationships*, 15(3), 208, 1966.

29. Persoone, G., Janssen, C., and De Coen, W., Cyst-based toxicity tests X: Comparison of the sensitivity of the acute *Daphnia magna* test and two crustacean microbiotests for chemicals and wastes, *Chemosphere*, 29, 12, 2701, 1994.

30. Keddy, C.J., Green, J.C., and Bonnell, M.A., Review of whole organism bioassays: soil, freshwater sediment and freshwater assessment in Canada, *Ecotox. Environ. Safety*, 30, 221, 1995.

31. American Society for Testing and Materials, Standard Guide for Acute Toxicity Tests with the Rotifer *Brachionus*, ASTM Designation E1440-91, in *Annual Book of ASTM Standards*. section 11. Water and Environmental Technology. Volume 11.04. Pesticides, Resource Recovery, Hazardous Substances and Oil Spill Responses, Waste Management, Biological Effects, 1210-1216, 1996. 1992.

32. Willemsen, A., Vall, M.A., and de Zwart, D., Microbiotests as tools for environmental monitoring, Report from the National Institute of Public Health and Environmental Protection RIVM, N°9, 607042005, 39 p., 1995.

33. De Zwart, D., *Monitoring Water Quality in the Future*. Volume 3. Biomonitoring Report from the Ministry of Housing, Spatial Planning and the Environment, Zoetermeer, The Netherlands, 83 p., 1995.

34. Grothe, D.R. and Kimerle, R.A., Inter and intralaboratory variability in *Daphnia magna* effluent toxicity tests, *Environ. Toxicol. Chem.*, 4, 189, 1985.

35. Lewis, P.H. and Weber, C.I., A study on the reliability of *Daphnia* acute toxicity tests, in Aquatic Toxicology and Hazard Assessment, Seventh Symposium, Cardwell, R.D., Purdy, R., and Bahner, R.C., Eds., American Society for Testing and Materials. Philadelphia, PA, 77, 1985.

36. Lilius, H., Hästbacke, T., and Isomaa, B., A comparison of the toxicity of 30 reference chemicals to *Daphnia magna* and *Daphnia pulex*, *Environ. Toxicol. Chem.*, 14 (12), 2085, 1995.

37. Toussaint, M.W., Shedd, T.R., Van der Schalie, W.H., and Leather, G.R., A comparison of standard acute toxicity tests with rapid screening toxicity tests, *Environ. Toxicol. Chem.*, 14 (5), 907, 1995.

CHAPTER 31

Death by Mud: Amphipod Sediment Toxicity Tests

Peter M. Chapman

CONTENTS

ABSTRACT

A historical perspective on amphipod sediment toxicity tests is provided. Amphipod tests have been miniaturized, from 25-L aquaria to 1-L test vessels, and have become a major benchmark technique for sediment toxicity testing. The historic and projected development of these tests provides a model for the development of other invertebrate microbiotests. Techniques and practice for short-term acute and long-term, chronic and/or sublethal tests are reviewed. Advances and future directions are described and projected, respectively. Major future directions include: development and usage of full life-cycle chronic tests, usage with other tools to determine causative agents, and determination of toxic mechanisms.

I. INTRODUCTION

Amphipod sediment toxicity tests are not usually considered microbiotests (defined as small-scale biological and ecotoxicological tests, see Chapter 1). However, they have been miniaturized during the process of development (see Section II), and further miniaturization is possible (see Section IV). In addition, they are simple, cost-efficient, use relatively small animals and, most important, provide one of the best models for sediment toxicity methods development. Thus, their inclusion in this book is well warranted.

This chapter is not intended to be an authoritative review of all aspects of amphipod sediment toxicity testing. Rather, it provides both knowledgeable and novice readers with useful information on past, present, and anticipated future techniques, advances, and practice for amphipod sediment toxicity tests within a model of test development and application which could usefully be followed by microscale tests in general. The reader should also consult Chapters 38, 39, and 43.

II. HISTORICAL PERSPECTIVE

Amphipod sediment toxicity tests were primarily developed in response to U.S. regulatory requirements for evaluating the suitability of dredged material for in-water disposal. Specifically, the 1977 Ocean Dumping Regulations and Criteria[1] required testing with at least three species, consisting of a filter-feeder, a deposit-feeder, and a burrower. Agency scientists addressed this requirement by recommending that same year that the species selected "include a crustacean, an infaunal bivalve, and an infaunal polychaete" and stating that "infaunal amphipods seem to be among the most sensitive crustaceans and, for this reason, are among the preferred organisms for solid phase bioassays."[2] Two amphipod taxa were recommended: *Ampelisca* spp. and *Paraphoxus* spp. Testing of various species was to be conducted in aquaria under flow-through conditions, with test material deposited on top of the organisms.

Additional details of such required sediment toxicity tests were published two years later in 1979 with *Paraphoxus epistomus* as the sole crustacean in a suite of five different species tested in 25-L aquaria.[3] In 1982 Swartz et al.[4] published a key paper which described broad-scale testing of sediments at a marine hazardous waste disposal site with this species (now taxonomically reclassified as *Rhepoxynius abronius*) to determine in-place sediment toxicity. This objective differed from the original, namely classifying dredged material, and required testing large numbers of samples quickly. Logistical and cost considerations led to a miniaturization of test vessels (from 25-L aquaria to 1-L beakers) and to a change from flow-through to static test conditions. Comparisons of resident amphipod distributions and sediment toxicity indicated that this sediment toxicity test was ecologically relevant.[4]

In 1985 Swartz et al.[5] published a definitive method for determining sediment toxicity with *R. abronius* for a variety of applications; this publication provides an excellent model for general factors to be considered when developing a new sediment toxicity test method. This particular method involved 1-L beakers, static conditions, 10-d exposures, 5 replicate test beakers each containing 20 amphipods for each tested concentration of each material, and three endpoints: survival after 10 d, emergence during each of the 10 d, and reburial in clean sediment after 10 d. Similarly, that year Swartz et al.[6] suggested that "this amphipod is a useful test species for sediment bioassays used in research, monitoring, and regulatory programs," while Kemp et al.[7] described its life history and productivity. In the intervening 10 years, literally hundreds of papers have been published using infaunal amphipods for sediment testing, either alone or in combination with other species. The method also has been formalized for this and other amphipod species.[8-11]

The most recent manual for testing dredged material in U.S. waters[12] specifically identifies 20 candidate test species for determining the potential benthic impact of dredged-material disposal. Of the 20, seven are amphipods and five of the amphipods are among the seven recommended

Table 31.1   Some Amphipod Species Commonly Used in Sediment Testing[a]

Marine	Estuarine	Freshwater
Short-Term Tests[b]		
Ampelisca abdita[c]	*Eohaustorius estuarius*	*H. azteca*
Rhepoxynius abronius	*Hyalella azteca[e]*	
Leptocheirus plumulosus		
Grandidierella spp.		
Corophium spp.[c]		
Long-Term Tests[d]		
A. abdita[c]	*H. azteca[e]*	*H. azteca*
L. plumulosus		
Corophium spp.[c]		

[a] List is not exhaustive, merely indicative. Also see reference 66.
[b] 10-d exposures
[c] Tube builder, not a deposit feeder. All others burrowers and deposit feeders.
[d] Include full life-cycle exposures.
[e] Can be used in estuarine tests, provided salinities are relatively low.[67]

benchmark test organisms.[12] Two of the benchmark amphipod species, *Ampelisca abdita* and *Rhepoxynius abronius*, were recommended in the 1977 document. The other three benchmark amphipods are: *Leptocheirus plumulosus* (marine), *Eohaustorius estuarius* (estuarine), and *Hyalella azteca* (freshwater).

The above discussion refers to acute lethal testing. Procedures for longer-term (e.g., chronic) testing are in an earlier stage of development and are discussed under "Techniques and Practice."

III. TECHNIQUES AND PRACTICE

Infaunal amphipods are: sensitive to contaminants; readily available; as a group, tolerant of a wide range of grain sizes[13,14] and laboratory exposure conditions; and, ecologically relevant to most sites and areas. Acute toxicity test methods are well developed; longer-term test methods are being developed (Table 31.1). General guidance for conducting sediment toxicity tests including statistical data analysis is available.[15]

The limitations of amphipod sediment toxicity tests, and of sediment toxicity tests in general, include: altered sediment-borne contaminant/chemical bioavailability through manipulations; potential false positives or false negatives (due to effects from variations in factors such as interstitial salinity, grain size); and the inherent limitations of any laboratory test compared to the field environment. Advantages (similarly of general applicability) include relative primarily to the acute test: simplicity (cost and level of effort); extensive methods standardization and database; site specificity; interpretable results; and regulatory precedence.

A. Acute Tests

Current methodology for acute lethality testing[8-12] has the following generic features applicable to all amphipod sediment toxicity tests. Tests are static nonrenewal, although static renewal (minimum 48-h water changes), intermittent flow, or continuous flow tests may be used where it is necessary to maintain water quality parameters such as dissolved oxygen, or where ammonia is a suspected toxicant.[16] Test duration is 10 d, generally in 1-L aerated (trickle-flow) test vessels containing 775 mL of water and a 2 cm layer (175 mL volume) of sediment (smaller containers can be used, see Section IV). Testing is conducted in 1 to 5 replicate test vessels (5 provides an optimum balance between power and number of replicates[5]), ideally containing 20 individuals each, though as few as 10 individuals can be used. A negative control (clean sediment) and positive control (96-h survival with water-only exposure to a reference toxicant) are run with each test series. Controls for other variables (e.g., particle size, interstitial salinity) may also be appropriate in specific cases. The test acceptability criterion is ≥90% survival in the negative control.

Acute tests and techniques shown in Table 31.1, are commonly used but are not limited to those species. For example, in New Zealand, two local species (*Paracorophium excavatum* and *Proharpinia hurleyi*) are being used.[17] In Canada, a guidance manual[18] has been based on testing the suitability of six species of marine or estuarine amphipods (*Amphiporeia virginiana*, *Corophium volutator*, *Eohaustorius estuarius*, *E. washingtonianus*, *Foxiphalus xiximeus*, and *Rhepoxynius abronius*). Typically, the sensitivity of local species is benchmarked against *R. abronius* in side-by-side tests,[19] as this is the most suitable amphipod species currently available in the Northern Hemisphere for sediment toxicity tests.[18] Arguably, free-burrowing species (e.g., *R. abronius*) are more sensitive to contaminants in sediment interstitial water than tube-building species (e.g., *Corophium spinicorne*).[20] However, deposit feeders (e.g., *C. volutator*) are likely more susceptible to particle-bound contaminants due to sediment ingestion than are meiofaunal predators (e.g., *R. abronius*).[21]

B. Sublethal Tests

There are no standard methods for short- or long-term, sublethal tests. Such tests can involve behavioral[22,23] or other short-term responses but are most useful when they involve full life-cycle responses.[24]

As of early 1996, only three species were known to have been reasonably developed for full life-cycle testing with sediments: *A. abdita*,[25] *L. plumulosus*,[26,27] and *C. insidiosum*.[28] However, active research on other species is under way and both a 28-d partial life-cycle test and a 10-d sublethal growth test have been developed for *H. azteca*.[29-31] In Portugal (V. Quintino, University of Aveiro, personal communication), *C. multisetosus* is being cultured and its utility explored in 10-d acute lethality tests similar to the protocol described above, and in 28-d full life-cycle tests similar to the procedures described below. In South Africa, *Grandidierella* spp. may be appropriate for full life-cycle sediment tests.[32] In New Zealand, *Chaetocorophium* sp. is being used for experimental sublethal sediment toxicity tests.[33]

The following description is inclusive for *L. plumulosus* and *C. insidiosum*, which are tolerant of a wide range of salinities. These methods have been used to test the toxicity of DDT- and PCB-spiked sediments and field-collected sediments.[34]

Testing is conducted with five replicates for each sediment treatment and for the negative control, which consists of uncontaminated sediment. Each replicate contains 20 juvenile amphipods. A positive control currently consists of a reference toxicant water-only 96-h acute exposure; spiked sediment positive controls which match the test duration and endpoints have not been developed to the point that they can be routinely used.

Both amphipod species can be cultured in the laboratory using a 0.5 to 1.0 cm depth of relatively fine, clean sediment which is sieved to ≤250 μm, renewed every 6 to 8 weeks and covered with clean overlying sea water. Conditions are static renewal (50% renewal, three times per week). Feeding coincides with water renewal and consists of an algal concentrate (cell density 1×10^6 cells/mL) of *Skeletonema* and *Thalassiosira*, and of a dry food mixture called GORP (48% TetraMin® flaked fish food, 24% alfalfa, 24% wheat grass, 4% Neonovum®). Both species are maintained at 20°C; culture salinities are 29 ppt for *L. plumulosus* and 20 ppt for *C. insidiosum*. Overcrowding (population densities >1.5 cm^{-2}) decreases fecundity, and periodic thinning by sieving is required.

One gravid female, having dark-colored eggs in the mid-abdominal brood pouch, is isolated by sieving (1.0-mm mesh size for *L. plumulosus*, 0.5 mm for *C. insidiosum*) for each juvenile required in tests. *L. plumulosus* can be tested as newborns (≤24 h); *C. insidiosum* should be tested as 7-d-old juveniles due to a natural high initial incidence of mortality,[35] and their small size (<1 mm), which makes handling of younger juveniles very difficult.

Testing is conducted for 28 d with the same water change and feeding schedule as the cultures in 1-L glass aerated containers using a 2-cm layer of sediment and approximately 800 mL of overlying water. After the 28-d exposure period, the sediments are sieved (500 µm, then 250 µm mesh size), and the amphipods retained on the sieve are examined for survival, growth, and reproduction. During testing, containers are examined daily and observations on organism activity (e.g., burrowing) are used to assess their health qualitatively prior to test termination. Regular water quality measurements (DO, temperature, pH, salinity, ammonia) are conducted either daily or during water changes. Growth is determined as length in survivors measured from top of rostrum to last segment before the telson. A digital analysis system is optimal for such measurements: three live images of each amphipod are frozen on video for measurement and calculation of mean length. Fecundity is measured as mean number of offspring per surviving female in each treatment. The material remaining on the 250 µm screen is preserved in 70% ethanol with Rose bengal stain. Adults are sexed using an inverted microscope to examine gender-specific traits. For *L. plumulosus*, sexing is based on presence of eggs in the oviducts or brood pouch (females) and gnathopod morphology (males have a notched palm on the dactyl and stout terminal segments). For *C. insidiosum*, mature males have larger, more developed second antennae than females.

Three variables may have a strong influence on test outcome: source of amphipods, source of negative control sediment, and renewal of test sediment during test exposure. Close attention to these variables is required for a successful test.

Tests are considered acceptable if mean negative control survival is ≥75% and the reference toxicant LC_{50} is within 2 S.D. of the laboratory mean. LC_p (survival), EC_p (reproduction), and IC_p (growth) endpoints are calculated. NOEC (no observed effect concentration) and LOEC (lowest observed effect concentration) can be determined for spiked-sediment tests for endpoints with statistically significant effects and an indication of concentration-response trends but have higher uncertainty than the aforementioned point estimates.[36] The power of the test (i.e., an indication of the sensitivity of the experimental design within the natural variability of the system under study) should be on the order of 0.80 to be acceptable.

IV. ADVANCES AND FUTURE DIRECTIONS

Major advances which can be pursued or which are required with small-scale amphipod assays include: further miniaturization; spiked sediment controls; and isolating or excluding ammonia and hydrogen sulfide toxicity from more persistent contaminants. There are three major future directions for amphipod sediment toxicity tests: complete development and usage of full life-cycle chronic tests; integrated usage with other ecotoxicological tools to determine causative agents; and determination of mechanisms of amphipod toxicity. These major advances and directions are discussed below.

A. Further Miniaturization

Some freshwater amphipod sediment toxicity tests are routinely conducted in 300-mL containers with reference toxicity tests conducted in 30-mL containers.[11] Further miniaturization of these tests is possible and has been attempted by investigators who examined the possibility of using test containers as small as 100 mL for 10-d *R. abronius* acute lethality tests.[37] There were advantages (e.g., fewer amphipods required [5 per container], less test material required, less space required) but also disadvantages (e.g., loss of power due to the smaller n, more variability) to this miniaturization. More success was obtained in 10-d *H. azteca* growth tests, also using 5 amphipods per container.[31] With general standardization of the test to 1-L containers,[8] further work on miniaturization has not been widely pursued but could be worthwhile if the above disadvantages can be addressed.

B. Spiked Sediment Controls in Amphipod Assays

As previously noted, spiked sediment positive controls which match test durations (10 to 28 d) and endpoints (other than acute mortality, e.g., growth and reproduction) have not been developed yet to the point where they can be routinely used. At present there are cases where acute positive controls are used when amphipod tests are based on chronic response. Such comparisons (e.g., growth and reproductive success, as compared to survival) are inappropriate and probably do not reflect responses measured in the actual test. Positive controls serve to provide a measure of repeatability and reproducibility (precision), within a laboratory and between laboratories, respectively. Development of such controls is an essential part of test quality assurance/quality control and must include firm criteria for rejection/acceptance of positive control results.[38]

C. Ammonia and Hydrogen Sulfide Toxicity

Ammonia and/or hydrogen sulfide toxicity have been implicated in test failures and in apparently unexplained (or undesired) toxicity test results by some testing laboratories, permitees, and even regulators. However, ammonia and hydrogen sulfide can also be contaminants of concern to which toxicity test results are correctly responding.[16]

The following guidance is from EPA/USACE.[12] When chemical analyses indicate that ammonia is present in sediments at levels which may cause toxicity, and ammonia is not a contaminant of concern, ammonia in the interstitial water should be reduced to below 20 mg/L before adding test organisms. This can be done by sufficiently aerating the sample at saturation and replacing two volumes of water per day, with daily measurements until the ammonia concentrations are sufficiently reduced. One or more test containers are set up explicitly for measuring interstitial ammonia. Testing can be conducted by continuous flow or by replacing up to two volumes per day.

Daily measurement of ammonia during amphipod testing has not been reported in the literature. Such data are provided in Figure 31.1, for three marine sediments, C being the control. There is a pulse of ammonia release to overlying water for Sediment B, with maximal release occurring after test initiation and before termination. This pattern has also been found in some other marine sediments examined during routine testing (EVS Environment Consultants, unpublished data). Although the ammonia concentrations shown in Figure 31.1 are below those which are acutely toxic to amphipod species routinely used in testing,[12] this may not always be the case. Investigators who only measure ammonia at the beginning and end of testing should be suspicious where relatively high mortalities are associated with ammonia concentrations approaching lethal levels, because lethal levels may have been exceeded during intermediate test periods.

Hydrogen sulfide toxicity is not believed to be a problem if dissolved oxygen levels are maintained at the requisite concentrations in the overlying water.[12]

D. Full Life-Cycle Chronic Tests

Measures of short-term toxicity are considered to be inadequate to protect field populations compared to tests which measure "…abnormal development, reproductive failure, or growth of sensitive life stages."[39] Thus long-term tests are expected to be of increasing importance, possibly as part of a second test tier after short-term testing has been conducted. In this respect, full life-cycle tests are the most useful.[24]

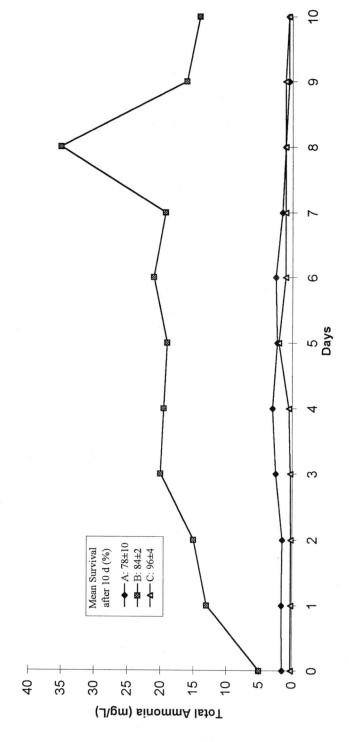

Figure 31.1 Daily ammonia measurements[66] in overlying water for three different sediments (A, B,C): 10-d *R. abronius* sediment toxicity tests.

Table 31.2   Proposed Decision Criteria: Sediment Toxicity Testing with Amphipods

Situation	Effects[a]		Possible Decision
	Growth	Reproduction	
1	No	No	Strong evidence that sediments are not chronically toxic. No further sediment toxicity testing needed; continue with chemical and other monitoring. If significant increases in chemical contamination occur, further testing may be needed.
2	Yes	No	Evidence for sediment chronic toxicity. Determine significance.
3	No	Yes	Evidence for sediment chronic toxicity. Determine significance.
4	Yes	Yes	Strong evidence that sediments are chronically toxic. Determine significance.

[a] Significant differences between test and reference stations.

However, there are no standard methods for full life-cycle chronic amphipod tests, and only three salt water species have, as previously described, been used to date in such testing. It is doubtful that these three species are suitable or appropriate for all possible test situations. Whether they are representative of the range of amphipod sensitivities to contaminants remains to be established. The lack of freshwater species for full life-cycle testing also needs to be addressed.

When new amphipod tests are developed, they need to be standardized, much as has occurred for 10-d acute lethality tests. As part of such standardization, controls need to be appropriate. Presently, controls are generally based on acute rather than chronic endpoints. (This is especially true for reference toxicants, which are conducted in water-only exposures rather than in the ideal which would be spiked-sediment exposures.) All control comparisons (i.e., positive and negative controls) should be, as previously noted, to the endpoints used in actual testing, i.e., not only survival but also for growth and fecundity. Further, the sensitivity of new tests needs to be compared with existing tests. This currently occurs in marine 10-d acute tests using the *R. abronius* test as a benchmark.[19] No similar standard has been proposed let alone developed for full life-cycle chronic tests.

Testing with full life-cycle tests will involve the general hypothesis that "sediments are not chronically toxic." This hypothesis is tested statistically by evaluating the specific hypothesis: "No differences can be detected between sampling sites," which involves comparing test sites to each other and to at least one reference site. Interpretation of test results is not as straightforward as for acute lethal tests in which organisms either live or die. Thus, *a priori* decision criteria are required.

A matrix of suggested generic decision criteria for amphipod sediment testing is provided in Table 31.2, related to the above hypothesis. Determining significance could involve determining the extent of observed effects, for instance retesting with the same or different organism(s). For instance, presuming that growth effects were determined, one approach could be to retest using the polychaete *Neanthes arenaceodentata* 20-d growth test,[40] 28-d test,[41] or full life-cycle test.[42] Determining the significance of any effects must involve assessing whether statistical significance could relate to ecological significance. Minimum levels of growth and reproduction which would be considered ecologically relevant would have to be defined *a priori* using best professional judgment and knowledge. Intensity of effect is not included in these decision criteria; however, clearly a situation where no reproduction occurred would engender more concern than one where there was a minor but statistically significant reduction in growth. Where effects are major as well as statistically significant, it may be necessary to determine the reason(s) for the observed effects, through such means as a sediment toxicity identification evaluation (TIE).[43] However, sediment TIEs are still in the early development phase.[44]

Population growth rate of amphipods will, in future, be used to interpret life-cycle responses. Work is currently under way integrating responses (e.g., reproduction, growth) into an amphipod population growth estimate and interpreting changes in rate relative to reference sediments.[45]

E. Integrated Usage of Amphipod Assays

Field-collected sediments returned to the laboratory for testing differ from the same sediments prior to collection. For instance, contaminant bioavailability may be affected by holding times, which remain to be fully standardized (only maximum holding times are presently specified).[12,46] Animals used in the laboratory are also not in the same physiological condition and health as those found in the field, even when they are the same species. For example: for field-collected test organisms, length of holding time affects sensitivity to toxicants;[47] H. azteca are sensitive to the presence of other species in test sediments and to sterilization of sediments,[48] and their recovery from test sediments can be variable;[49] field populations may have adapted to pollution and hence be more tolerant than laboratory test organisms;[50] and field populations and individuals are affected by competition, predation, and parasitism as well as by individual and mixed toxins.[51]

Early in the development of sediment toxicity tests, it was recognized that toxicity determined by these tests could not be directly linked to chemical measures of contamination.[52] It is now recognized that sediment toxicity tests do not rely on field validation,[53] but are best interpreted using a burden of evidence approach, ideally combining observational data with these experimental data.[54-56] Swartz et al.[57] have shown that: "Correlations between toxicity, contamination, and biology [e.g., the Sediment Quality Triad[58,59]] indicate that acute sediment toxicity to E. estuarius, R. abronius, or H. azteca in laboratory tests provides reliable evidence of biologically adverse sediment contamination in the field." Similar findings have been made throughout temperate regions[59] and in the Antarctic.[60]

Integrated usage has more often involved acute sediment toxicity test responses than chronic responses. New ways of integrating chronic tests involving multiple endpoints into Sediment Quality Triad analyses and into meaningful yet intelligible presentations are required. In particular, the relative importance of chronic versus acute responses should be considered. One approach has been to use a single representative test endpoint to represent all tests and endpoints;[61] other approaches need to be proposed and explored.

F. Mechanisms of Toxicity in Amphipod Assays

The mechanisms (i.e., reasons) by which toxicity is exhibited to certain contaminants by certain amphipods and not by others remain to be fully investigated but could provide two useful products. First, such research should provide an understanding of when (and when not) toxicity tests with certain amphipod species are relevant to investigating or monitoring particular sediment contaminants. Such information would be particularly useful in determining the utility of historic or proposed test species and possibly in explaining any anomalous results. Second, this research might allow for the use of differential taxa responses in multitaxa test batteries to provide indications of causative chemicals when toxicity is observed.[62] Specifically, sediments comprise complex mixtures of contaminants whose combined mode of toxic action may be additive, less than additive, or more than additive; understanding these interactive effects would allow for useful guidelines for screening sediments.[63]

The few amphipod studies of species sensitivity differences which have been done are instructive and could usefully be expanded and miniaturized. For instance, Reichert et al.[64] exposed R. abronius and E. washingtonianus to benzo[a]pyrene (BaP) in sediments. They found that the former exhibited toxicity but no tissue accumulation; E. washingtonianus exhibited the reverse pattern. This difference was determined to be due to the fact that R. abronius metabolized the BaP to toxic intermediates, whereas E. washingtonianus did not. This study confirmed the utility of R. abronius for assessing toxicity in polyaromatic hydrocarbon (PAH)-contaminated sediments. Similar work on differential sensitivities of marine infaunal amphipods has been conducted with tributyltin.[65]

V. CONCLUSIONS

Sediment toxicity tests with amphipods have increased in importance since their conception. The small size of these animals and of the test containers makes them a convenient test organism for marine microscale toxicology. The fact that they are generally abundant in the field and/or relatively easy to culture allows for relatively high replication and overall n per replicate, increasing statistical validity. Other advantages include the facts that they are taxonomically widespread (many species to a genus), have a short life cycle and high reproductive potential, and are generally sensitive indicators of pollution in both the field and the laboratory.[51] Further, despite their small size, they can be used for tissue body burden analyses, if necessary by pooling specimens within a replicate.

Acute sediment toxicity test methods are well developed and appropriately sensitive, and have been widely applied in research and regulatory programs. Long-term sediment toxicity test methods are under development but show promise; growth and reproduction are key endpoints related to protecting field populations by means of laboratory data.[24]

The utility of these tests is expected to only increase with further development. The detail involved in their initial development[5] and subsequent standardization[8] provides a model which should be emulated by other tests. A key to the ultimate utility of these relatively simple laboratory tests is the continued demonstration that they provide useful information applicable to the health of field populations of amphipods and other organisms.

ACKNOWLEDGMENTS

This chapter was written at the invitation of the editors. It was word-processed by Jackie Gelling.

REFERENCES

1. EPA, Ocean Dumping — final revisions of regulations and criteria, U.S. Environmental Protection Agency. *Federal Register* Part IV, Vol. 42, No. 7, Tuesday, 11 January 1977.
2. EPA/USACE, Environmental Protection Agency/Corps of Engineers Technical Committee on Criteria for Dredged and Fill Material, Ecological Evaluation of Proposed Discharge of Dredged Material into Ocean Waters, Environmental Effects Laboratory, U.S. Army Engineer Waterways Experiment Station, Vicksburg, MS, 1977.
3. Swartz, R. C., DeBen, W. A., and Cole, F. A., A bioassay for the toxicity of sediment to the marine macrobenthos, *J. Water Pollut. Control. Fed.*, 51, 944, 1979.
4. Swartz, R. C., DeBen, W. A., Sercu, K. A., and Lamberson, J. O., Sediment toxicity and the distribution of amphipods in Commencement Bay, Washington, USA, *Mar. Pollut. Bull.*, 13, 359, 1982.
5. Swartz, R. C., DeBen, W. A., Jones, J. K. P., Lamberson J. O., and Cole, F. A., Phoxocephalid amphipod bioassay for marine sediment toxicity, in *Aquatic Toxicology and Hazard Assessment: Seventh Symposium*, Cardwell, R. D., Purdy, R., and Bahner, R. C., Eds., STP 854, American Society for Testing and Materials, Philadelphia, PA, 1985, 284.
6. Swartz, R. C., Ditsworth, G. R., Schults, D. W., and Lamberson, J. O., Sediment toxicity to a marine infaunal amphipod: cadmium and its interaction with sewage sludge, *Mar. Environ. Res.*, 18, 133, 1985.
7. Kemp, P. F., Cole, F. A., and Swartz, R. C., Life history and productivity of the phoxocephalid amphipod *Rhepoxynius abronius* (Barnard), *J. Crustacean Biol.*, 5, 449, 1985.
8. ASTM, Standard guide for conducting 10-day static sediment toxicity tests with marine and estuarine amphipods, Annual Book of ASTM Standards, Vol. 11.04, Method E1367-92. American Society for Testing and Materials, Philadelphia, PA, 1994.
9. ASTM, Standard guide for conducting sediment toxicity tests with freshwater invertebrates. Annual Book of ASTM Standards, Vol. 11.04, Method E1383-94. American Society for Testing and Materials, Philadelphia, PA, 1994.

10. EPA, Methods for assessing the toxicity of sediment-associated contaminants with estuarine and marine amphipods, EPA/600/R-94/025, Washington, DC, 1994.

11. EPA, Methods for measuring the toxicity and bioaccumulation of sediment-associated contaminants with freshwater invertebrates, EPA/600/R-94/024, Washington, DC, 1994.

12. EPA/USACE, Evaluation of dredged material proposed for discharge in waters of the U.S.-testing manual, EPA-823-B-94-002, Washington, DC, 1995.

13. DeWitt, T. H., Ditsworth, G. R., and Swartz, R. C., Effects of natural sediment features on survival of the Phoxocephalid amphipod, *Rhepoxynius abronius, Mar. Environ. Res.,* 25, 99, 1988.

14. Suedel, B. C. and Rodgers, J. H. Jr., Responses of *Hyalella azteca* and *Chironomus tentans* to particle-size distribution and organic matter content of formulated and natural freshwater sediments, *Environ. Toxicol. Chem.,* 13, 1639, 1994.

15. Hill, I. R., Matthiessen, P., and Hemibach, F. (Eds.), Guidance document on sediment toxicity tests and bioassays for freshwater and marine environments, Workshop on Sediment Toxicity Assessment, SETAC (Society of Environmental Toxicology and Chemistry) Europe Special Technical Publication, 105 pp., 1993.

16. Kohn, N. P., Word, J. Q., Niyogi, D. K., Ross, L. T., Dillon, T., and Moore, D. W., Acute toxicity of ammonia to four species of marine amphipod, *Mar. Environ. Res.,* 38, 1, 1994.

17. Hickey, C. W. and Roper, D. S., Acute toxicity of cadmium to two species of infaunal marine amphipods (tube-dwelling and burrowing) from New Zealand, *Bull. Environ. Contam. Toxicol.,* 49, 165, 1992.

18. Environment Canada, Biological test method: acute test for sediment toxicity using marine or estuarine amphipods, Environmental Protection Report EPS VRM/26, Ottawa, ON, 1992.

19. Chapman, P. M., Marine sediment toxicity tests, in *Chemical and Biological Characterization of Sludges, Sediments, Dredge Spoils, and Drilling Muds,* Lichtenberg, J. J., Winter, F. A., Weber, C. I., and Fradkin, L., Eds., ASTM STP 976, American Society for Testing and Materials, Philadelphia, PA, 1988, 391.

20. Swartz, R. C., Schults, D. W., DeWitt, T. H., Ditsworth, G. R., and Lamberson, J. O., Toxicity of fluoranthene in sediment to marine amphipods: a test of the equilibrium partitioning approach to sediment quality criteria, *Environ. Toxicol. Chem.,* 9, 1071, 1990.

21. Caparis, M. E. and Rainbow, P. S., Accumulation of cadmium associated with sewage sludge by a marine amphipod crustacean, *Sci. Tot. Environ.,* 156, 191, 1994.

22. Oakden, J. M., Oliver, J. S., and Flegal, A. R., Behavioral responses of a phoxocephalid amphipod to organic enrichment and trace metals in sediment, *Mar. Ecol. Prog. Ser.,* 14, 253, 1984.

23. Pascoe, D., Kedwards, T. J., Maund, S. J., and Taylor, E. J., Laboratory and field evaluation of a behavioral bioassay — the *Gammarus pulex* (L.) precopula separation (GaPPS) test, *Water Res.,* 28, 369, 1994.

24. Chapman, P. M., Environmental quality criteria — what type should we be developing? *Environ. Sci. Technol.,* 25, 1353, 1991.

25. Redmond, M. S., Scott, K. J., Swartz, R. C., and Jones, J. K. P., Preliminary culture and life-cycle experiments with the benthic amphipod *Ampelisca abdita, Environ. Toxicol. Chem.,* 13, 1355, 1994.

26. DeWitt, T. H., Redmond, M. S., Sewall, J. E., and Swartz, R. C., Development of a chronic sediment toxicity test for marine benthic amphipods, Report prepared for the U.S. Environmental Protection Agency, Newport, OR, EPA Contract CR-8162999010, 1992.

27. McGee, B. L., Schlekat, C. E., and Reinharz, E., Assessing sublethal levels of sediment contamination using the estuarine amphipod *Leptocheirus plumulosus, Environ. Toxicol. Chem.,* 12, 577, 1993.

28. Lamberson, J. O., Redmond, M. S., Jones, J. K. P., Sewall, J. E., Chapman, J. W., Cole, F. A., Schultz, D. W., and Swartz, R. C., Development of an amphipod sediment toxicity test for Hawaii, Report prepared for the U.S. Environmental Protection Agency, Newport, OR. EPA Contract 68-C0-0051, 38pp., 1993.

29. Borgmann, U. and Munawar, M., A new standardized sediment bioassay protocol using the amphipod *Hyalella azteca* (Saussure), *Hydrobiologia,* 188/189, 425, 1989.

30. Borgmann, U. and Norwood, W. P., Spatial and temporal variability in toxicity of Hamilton Harbour sediments: evaluation of the *Hyalella azteca* 4-week chronic toxicity test, *J. Great Lakes Res.,* 19, 72, 1993.

31. Kubitz, J.A., Besser, J.M., and Giesy, J.P., A two-step experimental design for a sediment bioassay using growth of the amphipod *Hyalella azteca* for the test end point, *Environ. Toxicol. Chem.,* 15, 1783, 1996.

32. Connell, A. D. and Airey, D. D., Life-cycle bioassays using two estuarine amphipods, *Grandidierella lutosa* and *G. lignorum*, to determine detrimental levels of marine pollutants, *S. African J. Sci.*, 75, 313, 1979.

33. Hickey, C. W. and Martin, M. L., Relative sensitivity of five benthic invertebrate species to reference toxicants and resin-acid contaminated sediments, *Environ. Toxicol. Chem.*, 14, 1401, 1995.

34. Murdoch, M. H., Chapman, P. M., and Paine, M. D., Southern California damage assessment surface water injury: sediment, Unpublished report prepared by EVS Environment Consultants, N. Vancouver, B.C., 136pp. + appendices, 1994.

35. Nair, K. K. C. and Anger, K., Life cycle of *Corophium insidiosum* (Crustacea, Amphipoda) in laboratory culture, *Helgolander wiss. Meersunter.*, 32, 279, 1979.

36. Chapman, P. M., Caldwell, R. S., and Chapman, P. F., A warning: NOECs are inappropriate for regulatory use, *Environ. Toxicol. Chem.*, 15, 77, 1996.

37. Shields, A. R., Klopfer, D. C., and Karas, L. L., Miniaturization of Puget Sound dredged disposal analysis sediment bioassays, in *Abstract Book, Twelfth Annual Meeting of the Society of Environmental Toxicology and Chemistry*, Pensacola, FL, 157, 1991.

38. Murdoch, M. H., Stewart, J. V., Crane, J. L., McPherson, C. A., and Chapman, P. M., Reference toxicant test results — what do we do with them?, in *Abstract Book, Second World Congress of the Society of Environmental Toxicology and Chemistry*, SETAC, Pensacola, FL, 315, 1995.

39. Martin, M. and Richardson, B. J., A paradigm for integrated marine toxicity research? *Mar. Pollut. Bull.*, 30, 8, 1995.

40. Johns, D. M., Pastorok, R. A., and Ginn, T. C., A sublethal sediment toxicity test using juvenile *Neanthes* sp. (Polychaeta:Nereidae), in *Aquatic Toxicology and Hazard Assessment: Fourteenth Volume*, Mayes, M. A. and Barron, M. G., Eds., ASTM STP 1124, American Society for Testing and Materials, Philadelphia, PA, 1992, 280.

41. Dillon, T. M., Moore, D. W., and Reish, D. J., A 28-d sediment bioassay with the marine polychaete *Nereis* (*Neanthes*) *arenaceodentata*, in *Environmental Toxicology and Risk Assessment: Third Volume*, Hughes, J. S., Biddinger, G. R., and Mones, E., Eds., ASTM STP 1218, American Society for Testing and Materials, Philadelphia, PA, 1995, 201.

42. Dillon, T. M., Moore, D. W., and Gibson, A. B., Development of a chronic sublethal bioassay for evaluating contaminated sediment with the marine polychaete worm *Nereis* (*Neanthes*) *arenaceodentata*, *Environ. Toxicol. Chem.*, 12, 605, 1993.

43. Ankley, G. T., Schubauer-Berigan, M. K., and Hoke, R. A., Use of toxicity identification evaluation techniques to identify dredged material disposal options: a proposed approach, *Environ. Manage.*, 16, 1, 1992.

44. Ankley, G. T. and Schubauer-Berigan, M. K., An overview of the application of toxicity identification evaluation procedures for contaminated sediments, *J. Aquat. Ecosystem Health*, 4, 133, 1995.

45. DeWitt, T. H., Word, J. Q., Swartz, R. C., Scott, K. J., and Chapman, P. M., From death to extinction: interpreting existing and forthcoming amphipod sediment toxicity tests, in *Abstract Book, Second World Congress of the Society of Environmental Toxicology and Chemistry*, Pensacola, FL, 1995, 39.

46. Landrum, P. F., Eadie, B. J., and Faust, W. R., Variation in the bioavailability of polycyclic aromatic hydrocarbons to the amphipod *Diporeia* (spp.) with sediment aging, *Environ. Toxicol. Chem.*, 11, 1197, 1992.

47. Robinson, A. M., Lamberson, J. O., Cole, F. A., and Swartz, R. C., Effects of culture conditions on the sensitivity of a phoxocephalid amphipod, *Rhepoxynius abronius*, to cadmium in sediment, *Environ. Toxicol. Chem.*, 7, 953, 1988.

48. Day, K. E., Kirby, R. S., and Reynoldson, T. B., The effect of manipulations of freshwater sediments on responses of benthic invertebrates in whole-sediment toxicity tests, *Environ. Toxicol. Chem.*, 14, 1333, 1995.

49. Tomasovic, M. J., Dwyer, F. J., Greer, I. E., and Ingersoll, C. G., Recovery of known-age *Hyalella azteca* (Amphipoda) from sediment toxicity tests, *Environ. Toxicol. Chem.*, 14, 1177, 1995.

50. Hoare, K., Beaumont, A. R., and Davenport, J., Variation among populations in the resistance of *Mytilus edulis* embryos to copper: adaptation or pollution? *Mar. Ecol. Prog. Ser.*, 120, 155, 1995.

51. Conlan, K. E., Amphipod crustaceans and environmental disturbance: a review, *J. Nat. History*, 28, 519, 1994.

52. Swartz, R. C., Schults, D. W., Ditsworth, G. R., and DeBen, W. A., Toxicity of sewage sludge to *Rhepoxynius abronius*, a marine benthic amphipod, *Arch. Environ. Contam. Toxicol.,* 13, 207, 1984.

53. Chapman, P. M., Do sediment toxicity tests require field validation? *Environ. Toxicol. Chem.,* 14, 1451, 1995.

54. Swartz, R. C., Schults, D. W., Ditsworth, G. R., DeBen, W. A., and Cole, F. A., Sediment toxicity, contamination, and macrobenthic communities near a large sewage outfall, in *Validation and Predictability of Laboratory Methods for Assessing the Fate and Effects of Contaminants in Aquatic Ecosystems,* Boyle, T. P., Ed. ASTM STP 865, American Society for Testing and Materials, Philadelphia, PA, 1985, 152.

55. Swartz, R. C., Cole, F. A., Schults, D. W., and DeBen, W. A., Ecological changes in the Southern California Bight near a large sewage outfall: benthic conditions in 1980 and 1983, *Mar. Ecol. Prog. Ser.,* 31, 1, 1986.

56. Chapman, P. M., Extrapolating laboratory toxicity results to the field, *Environ. Toxicol. Chem.,* 14, 927, 1995.

57. Swartz, R. C., Cole, F. A., Lamberson, J. O., Ferraro, S. P., Schults, D. W., DeBen, W. A., Lee, H. II, and Ozretich, R. J., Sediment toxicity, contamination and amphipod abundance at a DDT- and dieldrin-contaminated site in San Francisco Bay, *Environ. Toxicol. Chem.,* 13, 949, 1994.

58. Chapman, P. M., The Sediment Quality Triad approach to determining pollution-induced degradation, *Sci. Tot. Environ.,* 97-8, 815, 1990.

59. Chapman, P. M., Presentation and interpretation of Sediment Quality Triad data, *Ecotoxicology,* 5, 327, 1996.

60. Lemhan, H.S., Kierst, K.A., Conlan, K.E., Slattery, P.N., Konar, B.H. and Oliver, J.S., Patterns of survival and behavior in Antarctic benthic invertebrates exposed to contaminated sediments: field and laboratory bioassay experiments, *J. Exptl. Mar. Biol. Ecol.,* 192, 233, 1996.

61. Chapman, P. M., Paine, M. D., Arthur, A. D., and Taylor, L. A., A triad study of sediment quality associated with a major, relatively untreated marine sewage discharge, *Mar. Pollut. Bull.,* 32, 47, 1996.

62. Reish, D. J., Effects of metals and organic compounds on survival and bioaccumulation in two species of marine gammaridean amphipod, together with a summary of toxicological research on this group, *J. Nat. History,* 27, 781, 1993.

63. Swartz, R. C., Kemp, P. F., Schults, D. W., and Lamberson, J. O., Effects of mixtures of sediment contaminants on the marine infaunal amphipod, *Rhepoxynius abronius, Environ. Toxicol. Chem.,* 7, 1013, 1988.

64. Reichert, W. L., Eberhart, B.-T. L., and Varanasi, U., Exposure of two species of deposit-feeding amphipods to sediment-associated [³H]Benzo[a]pyrene: uptake, metabolism and covalent binding to tissue macromolecules, *Aquat. Toxicol.,* 6, 45, 1985.

65. Meador, J. P. and Varanasi, U., Differential sensitivity of marine infaunal amphipods to tributyltin, *Mar. Biol.,* 116, 231, 1993.

66. APHA, *Standard Methods for the Examination of Water and Wastewater,* 18th Edition. American Public Health Association, Washington, DC, 1992.

67. Nebeker, A. V. and Miller, C. E., Use of the amphipod crustacean *Hyalella azteca* in freshwater and estuarine sediment toxicity tests, *Environ. Toxicol. Chem.,* 7, 1027, 1988.

CHAPTER 32

Physiological Dysfunction in Estuarine Mysids and Larval Decapods with Chronic Pesticide Exposure

Charles L. McKenney, Jr.

CONTENTS

I. ESTUARINE MYSIDS AND PESTICIDES

Adequate assessment of the biotic hazard of contaminants released into the marine environment must be established to detect and prevent ecological damage resulting from long-term exposure to low concentrations of contaminants. In the late 1970s, the estuarine mysid, *Mysidopsis bahia,* was proposed as a suitable species for laboratory exposures to determine the chronic effects of pollutants.[1] Methods were developed for both laboratory culturing[2] and entire life-cycle toxicity testing utilizing flow-through laboratory sea water systems.[3] Several years later, a review of toxicity-testing results with a variety of marine contaminants identified *M. bahia* as one of the most sensitive members of the estuarine community to pollutant stress.[4]

Even though lower levels of biological organization (cellular and organismal) respond to environmental stress long before higher levels (population and community) in most pollution situations, the ecological significance of biotic responses at lower levels may be difficult to interpret.[5,6] Aquatic ecologists have for several decades, however, examined the productive processes of communities and ecosystems by studying the underlying physiological responses of selected species to environmental variables through their life cycles.[7,8] Physiological approaches to ecological energetics may provide valuable tools for use as early-warning bioindicators both in laboratory studies of aquatic toxicology and in field monitoring of water quality.

A considerable body of evidence exists demonstrating the sensitivity of marine crustaceans to a wide variety of xenobiotics and the utility of decapod biology in ecotoxicological studies in the marine environment.[9-13] Considering the close phylogenetic relationship between crustaceans and the insect targets at which many pesticides are directed, it is not surprising that crustaceans are often more sensitive to pesticides than other marine organisms. For the majority of pesticides examined in life-cycle toxicity tests using *M. bahia*, sublethal reductions in growth and reproductive capacity have proven to be the most sensitive criteria for chronic biological effects.[1,14-16] Since both growth and egg production are restrained by physiologically available energy, the objectives of the more recent research described herein were to compare the dose-response relationships of growth and reproduction with individual rate functions of energy metabolism and combined metabolic indices for various life stages during life-cycle exposure in the laboratory.

A. Methods for Life-Cycle Exposures and Reproductive Measures

Newly released juvenile mysids (<24 h old, dry wts = 12-32 µg) were obtained from laboratory cultures using a recirculating sea water system similar to that described by Reitsema and Neff.[17] Juveniles were reared in cylindrical nylon baskets situated in triplicate exposure chambers in a flow-through exposure system as described by McKenney[18] (Figure 32.1). Throughout the study, all groups of mysids were fed freshly hatched *Artemia* nauplii (<24 h old) at densities (2 to 3 nauplii mL^{-1}) previously determined to maximize growth and reproduction.[19] Mysids were observed daily for survival. As juveniles matured, ovigerous females were individually paired with mature males in separate brood cups.[18] Daily records were maintained for day of release of first brood and number of young released in the first brood of isolated females. The study was terminated seven days after the mean day of release of the first brood by control mysids, since experience has shown that toxicant-induced delay is manifest within seven days. Reproductive responses include day of release of the first brood, number of young per female in the first brood, total number of young per replicate population, and percentage of reproductively active females.

B. Methods for Measurements of Energy Metabolism

For the various life stages (early juvenile, juvenile, late juvenile, early adult),[20] 8 to 10 mysids from each exposure condition were individually sealed in 5-mL all-glass syringes containing 1 to 3 mL (dependent on the size of the mysid) of fully oxygenated media at the appropriate exposure concentration. Syringes containing mysids were incubated in a 25°C water bath with control syringes without mysids. After 4 to 6 h, 75 to 100 µL of the fluid in the syringe was injected into a Radiometer BMS3 Mk2 Blood Micro System with attached PHM71 Mk2 Acid-Base Analyzer, and a direct measurement was recorded for partial pressure of oxygen.[20] In addition, 100 µL was injected into a micro-flow-through cap (Model 95-00-25) attached to an Orion* Ammonia Electrode (Model 95-10) connected to an Orion Research Microprocessor Ionalyzer (Model 901). A direct reading of ammonia–nitrogen concentration was recorded from the Ionalyzer as described in detail by McKenney and Matthews.[20] Each mysid was then removed from the syringe, rinsed briefly in

---

* Mention of commercial products does not imply endorsement by the U.S. Environmental Protection Agency.

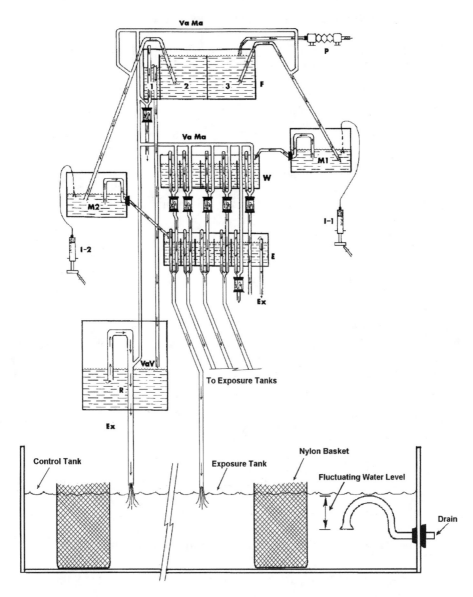

Figure 32.1 Exposure system used to continuously expose *Mysidopsis bahia* through an entire life cycle to pesticides.

distilled water, and placed in an oven at 60°C to dry for 48 h. Dry weights of individual mysids were subsequently measured to the nearest 0.1 μg on a Cahn 21 Automatic Electrobalance. Weight-specific respiration rates were determined using the equation described by McKenney.[21] Weight-specific ammonia excretion rates were calculated by the difference between ammonia–nitrogen concentrations in control syringes and syringes containing a mysid. O:N ratios were calculated as the ratios of atoms of oxygen consumed to atoms of nitrogen excreted. A theoretical O:N ratio of 7 closely approximates wholly protein catabolism by planktonic marine crustaceans.[22] A higher proportion of carbohydrates and/or lipids utilized as substrate by the organism increases the value of the O:N ratio.

C. Methods for Growth Measurements and Bioenergetic Terms

Several growth indices were calculated by the formulas of Winberg[23] and Winberg et al.[24] using dry weights of the various mysid life stages under each exposure condition. The mean specific rate of growth was calculated by the equation:

$$g_n = \frac{\left(\ln W_{n+1} - \ln W_n \right)}{t} \qquad (32.1)$$

where g_n = the mean specific rate of growth of a particular life stage n; W_n and W_{n+1} = the mean dry weight of life stages n and n+1, respectively; and t = the number of days between life stage n and life stage n+1. The mean specific rate of growth was incorporated into the formula:

$$G_n = \left(e^{g_n} - 1 \right) \qquad (32.2)$$

to quantify the fractional daily weight gain (G_n) for each life stage under the various exposure conditions. G_n was multiplied by the replicate dry weights to produce values of daily weight gain ($\mu g \ d^{-1}$) for each life stage-exposure condition combination.

Using daily weight gain values and weight-specific respiration values for each life stage and exposure concentration, daily production energy in joules and daily maintenance expenditure in joules were calculated for each life stage under the various exposure conditions by the formulas described by McKenney and Matthews.[20] Net growth efficiency (K_2)[23,24] was derived for each life stage-exposure condition by the formula:

$$K_{2_n} = \frac{P_n}{\left(P_n + R_n \right)} \qquad (32.3)$$

where K_{2_n} = the net growth efficiency for life stage n, P_n = the daily production in joules for life stage n, and R_n = the daily expenditure for maintenance in joules for life stage n. Net growth efficiency values represent the percentage of assimilated energy converted into growth of new tissue.

D. Relationships of Physiological Performance to Energy Metabolism

In separate studies, groups of *Mysidopsis bahia* were exposed through entire life-cycles to a range of concentrations of representatives of some of the major classes of pesticides: including endrin[25] (an organochlorine), thiobencarb[26] (a carbamate), fenthion[18,20] (an organophosphate insecticide), and DEF[27] (an organophosphate herbicide) (Table 32.1). Without exception, retarded juvenile growth rates and inhibited young production proved more sensitive to chronic pesticide exposure than direct lethality by the pesticide. For each pesticide, reductions in these measures of physiological performance among pesticide-exposed mysids were accompanied by alterations in their energy metabolism.

Exposure to pesticide concentrations producing significant mysid mortality resulted in elevated respiration rates of juveniles early in the exposure period. In general, the linear relationship that existed between respiration rates and pesticide concentrations in younger juveniles became a curvilinear relationship in older juveniles. This curvilinear dose-response in respiration continued into early adult stages in most cases. An interaction between mysid age and pesticide concentration modified growth rates in actively growing juvenile mysids. With some pesticides, reduced growth rates were seen predominantly in early juveniles, while for other pesticides, effects were manifest

Table 32.1   Summary of the Effects of Four Pesticides on *Mysidopsis bahia* Exposed Through an Entire Life Cycle in the Laboratory

Pesticide	Days of Exposure	Life Stage	Lowest Effect Concentration (μg liter^{-1})	Biological Response
Endrin	1	Early juvenile	0.12	↑ VO$_2$
	4	Juvenile	0.12	↓ Survival
			0.061	↓ Growth, ↓ K$_2$
	10	Late juvenile	0.061	↑ VO$_2$, ↓ Growth
	16	Early adult	0.030	↑ O:N
	20	Adult	0.061	↓ Survival
			0.030	↓ No. of Young
Thiobencarb	1	Early juvenile	76	↑ VO$_2$
	4	Juvenile	22	↓ Growth, ↓ K$_2$
	10	Late juvenile	22	↓ Growth
	16	Early adult	76	↑ O:N
	24	Adult	181	↓ Survival
			35	↓ No. of Young
Fenthion	1	Early juvenile	0.300	↑ VO$_2$
	4	Juvenile	0.300	↓ Survival
			0.166	↓ Growth
	9	Late juvenile	0.166	↑ VO$_2$, ↓ Growth
	16	Early adult	0.079	↑ VO$_2$, ↓ Growth
			0.016	↑ O:N
	20	Adult	0.079	↓ No. of Young
DEF	5	Juvenile	0.085	↑ O:N
	12	Late juvenile	0.140	↑ VO$_2$, ↓ Growth
			0.085	↑ O:N
	21	Adult	0.246	↓ Survival
			0.085	↑ VO$_2$, ↑ O:N, ↓ No. of Young

Note: VO$_2$ = weight-specific oxygen consumption rate; K$_2$ = net growth efficiency; O:N = atoms of oxygen consumed to atoms of nitrogen excreted.

in older juveniles and young adults. In either case, lower net growth efficiency values (K$_2$ values) suggested that retarded growth rates were a result of increased metabolic demand, reducing the amount of assimilated energy available for production of new tissue. Without exception, dose-response relationships were significantly stronger for the bioenergetic index (K$_2$) than for either of the individual physiological functions (respiration rate and growth rate) from which this index was derived. These results suggest that differences in the ratio of assimilated energy available for growth in mysids may serve as a useful indicator of physiological stress from pesticide exposure.

While exposure to pesticides modified ammonia excretion rates in mysids, dose-response relationships were stronger with the metabolic index, O:N ratio, than with either of the two individual rate functions (respiration rate and ammonia excretion rate) from which this index is derived. Furthermore, O:N ratios not only varied among mysid life stages, but were significantly influenced by a pesticide-life stage interaction. High O:N ratios indicative of greater lipid catabolism in developing juveniles were significantly lower in early adult stages, suggesting a shift toward more proteinaceous substrates during the maturation process. This pattern was modified among mysids reared in sublethal concentrations of all the pesticides studied. Higher O:N ratios during maturation of pesticide-exposed mysids indicate greater reliance on the more energy-rich lipid substrates to support the elevated rates of oxidative metabolism, resulting in fewer lipids being available for gamete production and reduced reproductive success.

Reductions in the physiological performance of an important estuarine zooplankter (*Mysidopsis bahia*) during low-level exposure to organic pesticides (several orders of magnitude lower than those levels causing direct fish mortality) were accompanied by alterations in the energy metabolism

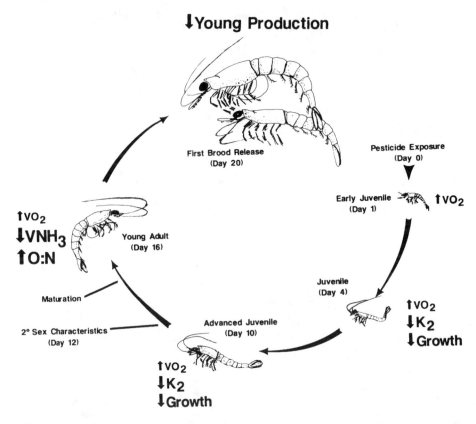

Figure 32.2   Generalized summary of the effects of four pesticides on survival, growth, reproduction, and energy metabolism of *Mysidopsis bahia* exposed through an entire life cycle in the laboratory. Terms are defined in the note for Table 32.1.

of exposed individuals. A generalized summary of these successive responses by *M. bahia* through an entire life cycle to chronic pesticide exposure is shown on Figure 32.2. Lower survival rates through the life cycle, retarded juvenile growth rates, and inhibited reproductive success would all contribute to lower secondary production rates in populations of this important estuarine prey species. Measurements of modifications in the energy metabolism of zooplankton exposed to xenobiotics, therefore, offer the potential of monitoring for, and predicting, ecological disruptions in the estuarine community under chronic, low-level pesticide stress.

Low-level exposure in the field from ground application of fenthion, an organophosphate insecticide used in mosquito control, produced both increased mortality and sublethal growth retardation of *Mysidopsis bahia* juveniles.[28] Significantly higher rates of oxygen consumption accompanied reduced weights of fenthion-exposed mysids 8 days after the field spray, suggesting bioenergetic imbalances. These field results confirm earlier laboratory studies in which short-term measurements of metabolic dysfunction predict altered production rates in mysid populations. Secondary production in nonreproducing zooplankton populations without continuous recruitment is the outcome of two opposing processes: increased weight of individuals in the population and decreased numbers of individuals in the population due to mortality.[29,30] Increased mortality and sublethal growth retardation of juvenile mysids following low-level exposure to fenthion, therefore, would result in reduced production in populations of this crustacean, which serves as an important link in the estuarine food chain between primary producers and commercially important fish utilizing the estuary as a nursery.[31,32]

II. DECAPOD LARVAE AND INSECT GROWTH REGULATORS

Insect growth regulators (IGRs) are an extensive group of insect hormones (most notably juvenile hormone [JH]) and their analogs which act as biological control agents in the management of insect pests by altering larval development during the critical and complex process of metamorphosis.[33,34] To determine whether these compounds affect estuarine biota, crustaceans, the dominant estuarine member of the same phylogenetic group containing insects (Arthropoda), are of particular relevance. In fact, compounds with JH activity in insects have been extracted from crustaceans, while JH and its analogs have been found to have various effects on crustacean reproduction.[35]

Given the mode of action of IGRs, it is not surprising that larval stages of insect pests are routinely used as bioassay organisms in testing the activity of juvenile hormone analogs as potential insecticides.[34,36] A series of pioneering studies by Costlow and colleagues[37-39] demonstrated that these compounds, intended for insects, are also toxic to larval stages of several species of estuarine crabs. Therefore, to thoroughly evaluate the potential environmental impact of these new insecticides on estuarine biota, analogous examination of their influence on the complete larval development and metamorphosis of other ecologically important estuarine crustaceans was needed. For this purpose, a series of studies were conducted to examine the effects of methoprene, a JH analog used in mosquito control, on developing larvae of two types of estuarine crustaceans with different developmental patterns. The grass shrimp, *Palaemonetes pugio*[40,41] (Figure 32.3), which has a variable number of larval stages through metamorphosis, was studied; as was the mud crab, *Rhithropanopeus harrisii*[42] (Figure 32.4), which has a fixed number of larval stages. The objective of these studies was to compare and contrast the influence of methoprene on the complete larval development and metamorphic success of these two types of crustaceans. Since JH is thought to control various aspects of metabolic homeostasis in insects,[34] a secondary objective was to examine growth and energy metabolism of grass shrimp larvae through the complete metamorphic process during exposure to methoprene.

A. Methods for Measuring Developmental Responses

Grass shrimp and mud crab larvae were obtained in the laboratory from previously collected ovigerous females. Larvae were reared through completion of metamorphosis in a range of nominal concentrations (0.1 to 1000.0 μg l^{-1}) of the single isomer (S)-methoprene (isopropyl [2E, 4E, 7S]-11-methoxy-3, 7, 11-trimethyl-2, 4-dodecadienoate) as contained in the formulation Altosid Liquid Larvicide (ALL). *P. pugio* larvae were also reared in the same concentrations of the isomeric mixture (R,S)-methoprene (isopropyl [2E,4E]-11-methoxy-3, 7, 11-trimethyl-2, 4-dodecadienoate). The exposure media were renewed daily for each exposure condition in each of three replicates (10 to 18 larvae per replicate) in either compartmented plastic boxes (each larva individually reared in an approximate liquid volume of 25 mL) or glass culture bowls (groups of 10 larvae reared in 10-cm culture bowls containing an approximate liquid volume of 50 mL). Daily observations were recorded on survival and molting frequency, after which larvae were fed ≤24-hr-old *Artemia* nauplii. Developmental rate was determined by the presence of cast exuvia. Observations for each individual larva were ended upon completion of larval development through metamorphosis.

B. Methods for Measuring Growth and Metabolic Responses

Newly released *P. pugio* larvae were randomly distributed in replicate beakers (10 larvae per beaker) contained within separate exposure tanks for each exposure concentration of the single isomer (S)-methoprene as contained in the formulation Altosid Liquid Larvicide (for a more detailed description, refer to Celestial and McKenney[42]). After the daily recording of larval mortality, larvae were fed an abundance of freshly hatched *Artemia* nauplii. Observations were complete when individual larvae completed metamorphosis to the postlarval form.

LIFE CYCLE OF <u>P</u>ALAEMONETES <u>P</u>UGIO

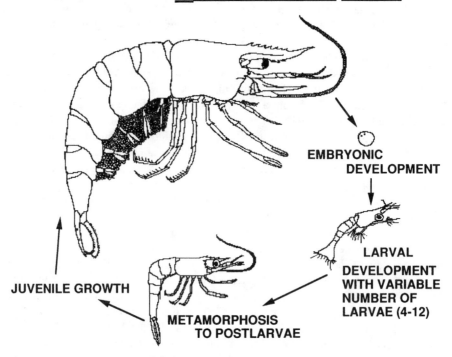

Figure 32.3 Life cycle of the grass shrimp, *Palaemonetes pugio*.

On days corresponding with the intermolt of various larval stages, 8 larval grass shrimp from each exposure condition were individually sealed in glass syringes with fully-oxygenated media. Oxygen consumption rates, ammonia excretion rates, and dry weights of the various larval stages under the different exposure regimes were determined using methods described previously. O:N ratios were calculated as the ratio of atoms of oxygen consumed to atoms of nitrogen excreted, and growth rates and K_2 values were calculated as previously described.

C. Comparative Developmental Responses of Crab and Shrimp Larvae

Larvae of grass shrimp[40] and mud crabs[42] exhibited differential rates of metamorphic success dependent on exposure concentrations of methoprene. No crab or shrimp larvae successfully completed metamorphosis over a 2- to 3-week period with exposure to 1000 µg methoprene L^{-1}. Exposure to 100 µg methoprene L^{-1} significantly reduced metamorphic success in crab larvae, but had no significant effect on grass shrimp larvae. Similar concentrations of methoprene are lethal to a number of pest insects including salt marsh mosquitoes,[36] while concentrations of this IGR which proved lethal to several species of fish and other nontarget organisms were two to three orders of magnitude higher.[36,43]

The mortality pattern during larval development of both crustaceans indicates that certain larval stages are more sensitive to methoprene toxicity than others and that stage-sensitivity varies between crabs and shrimp. Grass shrimp larvae were more sensitive to toxicity by dual isomeric methoprene formulation during the first two larval stages and the final larval stages.[40] Crab larvae appeared most resistant to methoprene toxicity during the second zoeal stage and most sensitive during the final premetamorphic larval stage, the megalopal stage.[42] This apparent stage-specific sensitivity to juvenile hormones and their analogs is seen during insect larval development[34] and accounts for

METAMORPHOSIS IN RHITHROPANOPEUS HARRISII

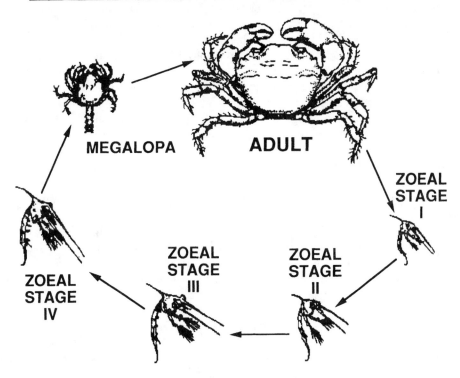

Figure 32.4   Metamorphosis in the mud crab, *Rhithropanopeus harrisii*, with larval development consisting of four zoeal stages and one megalopal stage.

the predominant use of the final premetamorphic larval instars of insects during efficacy testing of these compounds as insecticides.[34,37]

Developmental rates of crustacean larvae were modified by methoprene exposure dependent on concentration, species, and developmental stage. Methoprene exposure did not alter either the duration of total larval development or the total number of larval stages prior to metamorphosis in *P. pugio*.[40] Developmental rates of *R. harrisii* larvae were delayed through the zoeal stages with methoprene exposure, while those for megalopae were not affected.[42] These developmental observations, when coupled with the differential larval toxicity patterns to methoprene between these two types of crustaceans, suggest the possibility of different endocrine control systems for larval development and metamorphosis between these types of crustaceans. Larvae from the Natantia (inclusive of most shrimp) have a more labile developmental pattern with a variable number of larval stages prior to metamorphosis (suggestive of reduced hormonal control), while larvae of the Reptantia (inclusive of crabs) are considered to have a less flexible, more "stable developmental pattern" with a fixed number of larval stages (suggestive of greater hormonal control).[44,45]

D. Interactions Among Larval Stage, Growth, and Metabolism

Modifications in the energy metabolism of grass shrimp larvae occurred with exposure to methoprene concentrations which prevented successful development through metamorphosis.[41] Respiration rates of *P. pugio* larvae were significantly elevated as early as 24 h post exposure.

Changing O:N ratios suggest a shift from use of lipid as the major metabolic substrate during larval development to an increased usage of protein during premetamorphic larval stages. Similar enhanced reliance on protein catabolism just prior to completion of metamorphosis has been observed during the larval development of other marine crustaceans[46,47] and may represent a physiological prerequisite for successful metamorphosis. Lipid catabolism remained dominant, however, for premetamorphic larval *P. pugio* exposed to methoprene, as indicated by their significantly higher O:N ratios. Modifications of this premetamorphic pattern of energy utilization by exposure to methoprene could indicate an important physiological mechanism of toxicity of these substances to developing crustacean larvae.

The earliest and most sensitive response of grass shrimp larvae to methoprene exposure was growth retardation. Reduced net growth efficiency values suggest that retarded larval growth rates resulted from less assimilated energy being available for tissue production. Judged by elevated respiration rates of methoprene-exposed larvae, proportionally more of the physiologically available energy was channeled into energy required for metabolic maintenance. Given that JH is thought to also play a functional role in the regulation of insect energy metabolism,[34] these metabolic and bioenergetic responses of crustacean larvae to an insect growth regulator with juvenile-hormone activity may suggest similar endocrine control of these functions in this closely related group of organisms.

These findings support an analogous functional approach in the selection of appropriate testing procedures to evaluate potential environmental hazards from these new pesticides. They suggest that the use of crustacean larval testing procedures, most of which are microscale by necessity, would be a valuable adjunct in the assessment of insect growth regulators in the marine environment.

III. FUTURE DIRECTIONS

Studies on mysids and larval decapods reviewed in this chapter suggest that reproductive and developmental processes are those life functions most vulnerable to low-level, chronic pesticide exposure in pesticide-sensitive populations of marine crustaceans. Furthermore, these reductions in physiological performance among individuals in the population are accompanied by alterations in their energy metabolism. Experimental studies directed toward an increased understanding of these relationships among reproductive and developmental aberrations and metabolic alterations may provide sensitive biomonitoring tools of pesticide impact on production at a sensitive and critical link in estuarine trophodynamics.

REFERENCES

1. Nimmo, D. R., Bahner, L. H., Rigby, R. A., Sheppard, J. M., and Wilson, A. J., *Mysidopsis bahia*: An estuarine species suitable for life-cycle toxicity tests to determine the effects of a pollutant, in *Aquatic Toxicology and Hazard Evaluation*, ASTM STP 632, Mayer, F. L. and Hamelink, J. L., Eds., American Society for Testing and Materials, Philadelphia, 1977, 109.
2. Nimmo, D. R., Hamaker, T. L., and Sommers, C. A., Culturing the mysid (*Mysidopsis bahia*) in flowing sea water or a static system, in Bioassay Procedures for the Ocean Disposal Permit Program, EPA 600/9-78-010, U.S. Environmental Protection Agency, Cincinnati, OH, 1978, 59.
3. Nimmo, D. R., Hamaker, T. L., and Sommers, C. A., Entire life cycle toxicity test using mysids (*Mysidopsis bahia*) in flowing water, in *Bioassay Procedures for the Ocean Disposal Permit Program*, EPA 600/9-78-010, U.S. Environmental Protection Agency, Cincinnati, OH, 1978, 64.
4. Nimmo, D. R. and Hamaker, T. L., Mysids in toxicity testing- a review, *Hydrobiologia*, 93, 171, 1982.
5. Sastry, A. N. and Miller, D. C., Application of biochemical and physiological responses to water quality monitoring, in *Biological Monitoring of Marine Pollutants*, Vernberg, F. J., Calabrese, A., Thurberg, F. P., and Vernberg, W. B., Eds., Academic Press, New York, 1981, 265.

6. Capuzzo, J. M., Moore, M. N., and Widdows, J., Effects of toxic chemicals in the marine environment: predictions of impacts from laboratory studies, *Aquat. Toxicol.*, 11, 303, 1988.

7. Mann, K. H., The dynamics of aquatic ecosystems, in *Advances in Ecological Research*, Craig, J. B., Ed., Academic Press, London, 1969, 1.

8. Klekowski, R. Z. and Duncan, A., Physiological approach to ecological energetics, in *Methods for Ecological Bioenergetics*, Grodzinski, W., Klekowski, R. Z., and Duncan, A., Eds., Blackwell Scientific Publications, Oxford, 1975, 15.

9. Williams, A. B. and Duke, T. W., Crabs (Arthropoda: Crustacea: Decapoda: Brachyura), in *Pollution Ecology of Estuarine Invertebrates*, Hart, C. W., Jr. and Fuller, S. L. H., Eds., Academic Press, New York, 1979, Chap. 6.

10. Couch, J. A., Shrimps (Arthropoda: Crustacea: Penaeidae), in *Pollution Ecology of Estuarine Invertebrates*, Hart, C. W., Jr. and Fuller, S. L. H., Eds., Academic Press, New York, 1979, Chap. 7.

11. Epifanio, C. E., Larval Decapods (Arthropoda: Crustacea: Decapoda), in *Pollution Ecology of Estuarine Invertebrates*, Hart, C. W., Jr. and Fuller, S. L. H., Eds., Academic Press, New York, 1979, Chap. 8.

12. Gentile, J. H., Johns, D. M., Cardin, J. A. and Heltshe, J. F., Marine ecotoxicological testing with crustaceans, in *Ecotoxicological Testing for the Marine Environment*, Persoone, G., Jaspers, E., and Claus, C., Eds., State Univ. Ghent and Inst. Mar. Scient. Res., Bredene, Belgium, Vol. 1, 1984, 479.

13. Brown, T., Hagerdorn, C., Johnels, A., Lacy, G., Landa, D., Peakall, D., Pimentel, D., Soldan, T., and Veleminsky, J., Methods to assess toxic effects on ecosystems, in *Methods to Assess Adverse Effects of Pesticides on Non-target Organisms*, Tardiff, R. G., Ed., John Wiley & Sons Ltd, Chichester, 1992, Chap. 4.

14. Nimmo, D. R., Hamaker, T. L., Moore, J. C., and Sommers, C. A., Effect of diflubenzuron on an estuarine crustacean, *Bull. Environ. Contam. Toxicol.*, 22, 767, 1979.

15. Nimmo, D. R., Hamaker, T. L., Moore, J. C., and Wood, R. A., Acute and chronic effects of Dimilin on survival and reproduction of *Mysidopsis bahia*, in *Aquatic Toxicology*, Eaton, J. G., Parrish, P. R., and Hendricks, A. C., Eds., American Society for Testing and Materials, Philadelphia, 1980, 366.

16. Nimmo, D. R., Hamaker, T. L., Matthews, E., and Moore, J. C., Overview of the acute and chronic effects of first and second generation pesticides on an estuarine mysid, in *Biological Monitoring of Marine Pollutants*, Vernberg, F. J., Calabrese, A., Thurberg, F. P., and Vernberg, W. B., Eds., Academic Press, New York, 1981, 3.

17. Reitsema, L. A. and Neff, J. M., A recirculating artificial seawater system for the laboratory culture of *Mysidopsis almyra* (Crustacea: Pericaridea), *Estuaries*, 3, 321, 1980.

18. McKenney, C. L., Jr., Influence of the organophosphate insecticide fenthion on *Mysidopsis bahia* exposed during a complete life cycle. I. Survival, reproduction, and age-specific growth, *Dis. Aquat. Org.*, 1, 131, 1986.

19. McKenney, C.L., Jr., Optimization of environmental factors during the life cycle of *Mysidopsis bahia*, Environmental Research Brief, EPA/600/M-87/004, U.S. EPA, Cincinnati, 1987.

20. McKenney, C. L., Jr. and Matthews, E., Alterations in the energy metabolism of an estuarine mysid (*Mysidopsis bahia*) as indicators of stress from chronic pesticide exposure, *Mar. Environ. Res.*, 30, 1, 1990.

21. McKenney, C. L., Jr., Alterations in growth, reproduction, and energy metabolism of estuarine crustaceans as indicators of pollutant stress, in *Biological Monitoring of the Environment: A Manual of Methods*, Salanki, J., Jeffrey, D., and Hughes, G. M., Eds., International Union of Biological Sciences, CAB International, Paris, 1994, 111.

22. Mayzaud, P., Respiration and nitrogen excretion of zooplankton. II. Studies of metabolic characteristics of starved animals, *Mar. Biol.*, 21, 19, 1973.

23. Winberg, G. G., *Methods for the Estimation of Production of Aquatic Animals*, Academic Press, New York, 1971, 175.

24. Winberg, G. G., Patalas, K., Wright, J. C., Hillbricht-Ilkowska, A., Cooper, W. E., and Mann, K. H., Methods for calculating productivity, in *Secondary Productivity in Fresh Waters*, Edmondson, W. T. and Winberg, G. G., Eds., Blackwell Scientific Publications, Oxford, 1971, 296.

25. McKenney, C. L., Jr., Interrelationships between energy metabolism, growth dynamics, and reproduction during the life cycle of *Mysidopsis bahia* as influenced by sublethal endrin exposure, in *Physiological Mechanisms of Marine Pollutant Toxicity*, Vernberg, W. B., Calabrese, A., Thurberg, F. P., and Vernberg, F. J., Eds., Academic Press, New York, 1982, 447.

26. McKenney, C. L., Jr., Associations between physiological alterations and population changes in an estuarine mysid during chronic exposure to a pesticide, in *Marine Pollution and Physiology: Recent Advances*, Vernberg, F. J., Thurberg, F. P., Calabrese, A., and Vernberg, W. B., Eds., University of South Carolina Press, Columbia, 1985, 397.

27. McKenney, C. L., Jr., Hamaker, T. L., and Matthews, E., Changes in the physiological performance and energy metabolism of an estuarine mysid (*Mysidopsis bahia*) exposed in the laboratory through a complete life cycle to the defoliant DEF, *Aquat. Toxicol.*, 19, 123, 1991.

28. McKenney, C. L., Jr., Matthews, E., Lawrence, D. A., and Shirley, M. A., Effects of ground ULV application of fenthion on estuarine biota. IV. Lethal and sublethal responses of an estuarine mysid, *J. Fl. Anti-mosq. Assoc.*, 56, 72, 1985.

29. Parsons, T. R., Zooplankton production, in *Fundamentals of Aquatic Ecosystems*, Barnes, R. S. K. and Mann, K. H., Eds., Blackwell Scientific Publications, Oxford, 1980, 46.

30. Williams, R., An overview of secondary production in pelagic ecosystems, in *Flows of Energy and Materials in Marine Ecosystems*, Fasham, M. J. R., Ed., Plenum Press, New York, 1984, 361.

31. Odum, W. E., Pathways of energy flow in a south Florida estuary, University of Miami Sea Grant Program, Sea Grant Technical Bulletin No. 7, Miami, 1971.

32. Mauchline, J., The biology of mysids and euphausiids. Part 1. The biology of mysids, *Adv. Mar. Biol.*, 18, 1, 1980.

33. Staal, G. B., Insect control with growth regulators interfering with the endocrine system, *Ent. Exp. Appl.*, 31, 15, 1982.

34. Downer, R. G. H. and Laufer, H., *Endocrinology of Insects*, Alan R. Liss, Inc., New York, 1983.

35. Laufer, H., Ahl, J. S. B., and Sagi, A., The role of juvenile hormones in crustacean reproduction, *Amer. Zool.*, 33, 365, 1993.

36. Mian, L. S. and Mulla, M. S., Biological and environmental dynamics of insect growth regulators (IGRs) as used against Diptera of public health importance, *Res. Rev.*, 84, 27, 1982.

37. Costlow, J. D., Jr., The effect of juvenile hormone mimics on development of the mud crab, *Rhithropanopeus harrisii* (Gould), in *Physiological Responses of Marine Biota to Pollutants*, Vernberg, F. J., Calabrese, A., Thurberg, F. P. and Vernberg, W. B., Eds., Academic Press, New York, 1977, 211.

38. Christiansen, M. E., Costlow, J. D., Jr., and Monroe, R., Effects of juvenile hormone mimic ZR-515 (Altosid) on larval development of the mud crab *Rhithropanopeus harrisii* in various salinities and cyclic temperatures, *Mar. Biol.*, 39, 269, 1977.

39. Christiansen, M. E., Costlow, J. D., Jr., and Monroe, R., Effects of juvenile hormone mimic ZR-512 (Altozar) on larval development of the mud crab *Rhithropanopeus harrisii* at various cyclic temperatures, *Mar. Biol.*, 39, 281, 1977.

40. McKenney, C. L., Jr. and Matthews, E., Influence of an insect growth regulator on the larval development of an estuarine shrimp, *Environ. Pollut.*, 64, 169, 1990.

41. McKenney, C. L., Jr. and Celestial, D. M., Variations in larval growth and metabolism of an estuarine shrimp *Palaemonetes pugio* during toxicosis by an insect growth regulator, *Comp. Biochem. Physiol.*, 105C, 239, 1993.

42. Celestial, D. M. and McKenney, C. L., Jr., The influence of an insect growth regulator on the larval development of the mud crab *Rhithropanopeus harrisii*, *Environ. Pollut.*, 85, 169, 1994.

43. Lee, B. M. and Scott, G. I., Acute toxicity of temephos, fenoxycarb, diflubenzuron, and methoprene and *Bacillus thuringiensis* var. *israelensis* to the mummichog (*Fundulus heteroclitus*). *Bull. Environ. Contam. Toxicol.*, 43, 827, 1989.

44. Costlow, J. D., Jr., Metamorphosis in crustaceans, in *Metamorphosis: A Problem in Developmental Biology*, Etkin, W. and Gilbert, L. I., Eds., Appleton-Century-Crofts, Educational Division, Meredith Corp., New York, 1968, 3.

45. Gore, R. H., Molting and growth in decapod larvae, in *Crustacean Issues 2: Larval Growth*, Wenner, A. M., Ed., A. A. Balkema, Rotterdam, 1985, 1.

48. Capuzzo, J. M. and Lancaster, B. A., Some physiological and biochemical considerations of larval development in the American lobster, *Homarus americanus* Milne-Edwards, *J. Exp. Mar. Biol. Ecol.*, 40, 53, 1979.

49. Anger, K., Changes in respiration and biomass of spider crab (*Hyas araneus*) larvae during starvation, *Mar. Biol.*, 90, 261, 1986.

Teleosts and Amphibians

CHAPTER 33

Aquatic Testing with Early Life Stages of Killifish

Judith S. Weis and Peddrick Weis

CONTENTS

ABSTRACT

Early life stages of the mummichog (*Fundulus heteroclitus*) have been used in experimental laboratories for many years. They are readily available (in season) and easy to handle and maintain, and therefore have been used extensively in basic developmental biology and in microscale aquatic toxicology. Numerous studies have been reported on responses to contaminants including metals, pesticides and other organics, and mixtures, with parameters including abnormal development (especially in the craniofacial, cardiovascular, and skeletal systems), mortality, developmental retardation, hatching success and rate, and biochemical responses. In addition, recent studies have revealed that functional deficits in post-hatching larval behavior can be a subtle response to embryonic exposures. Small-scale aquatic toxicity procedures, such as those described here, can contribute much to testing pollution effects. In addition, research on this species has yielded much interesting and important information enhancing our understanding of the ways in which organisms respond to environmental perturbations.

I. THE MUMMICHOG

The common killifish or mummichog, *Fundulus heteroclitus*, is a small Cyprinodont fish which is very common in estuaries along the Atlantic coast from the Gulf of St. Lawrence region of Canada to northern Florida (Figure 33.1). It is a notoriously hardy species, tolerates a very wide range of temperatures and salinities, adapts well to the laboratory, and is found both in contaminated and pristine environments. In some highly contaminated environments it may be the only fish species present. *Fundulus* maintains a restricted home range[1] and is an important component of estuarine tidal marsh communities with intermediate trophic status, functioning as both predator and prey. It feeds primarily at high tide on the marsh surface and eats primarily small crustaceans and annelids. It is in turn eaten by birds, predatory fishes, and blue crabs.[2] *Fundulus* is one of the most common fish species to be studied in the laboratory and is one of the few to be studied in outer space.[3] Along the East Coast there are northern and southern subspecies which can be distinguished by a number of differences in gene frequencies[4] and egg morphology.[5]

As small-scale techniques in aquatic toxicology have developed, early life history stages of fish, including mummichogs, have proven to be extremely useful subjects for research, chemical screening, effluent testing and regulations, monitoring of aquatic contaminants, and environmental quality assessments. The reader can also refer to Chapters 4, 7, and 34 in this context.

II. USE IN AQUATIC TOXICOLOGY

In the past 25 years, as the science of environmental toxicology (including ecotoxicology and aquatic toxicology) has developed, *Fundulus* has been used in many studies of aquatic toxicity. The species has been used in many kinds of toxicity tests and has generally proved to be among the more resistant estuarine fish species to various chemicals. Acute bioassays have been done by Eisler and others for pesticides,[6,7] metals,[8-10] detergents,[11] oil dispersants, and other toxicant mixtures.[12,13]

Sublethal effects that have been examined in adult specimens include effects on chloride cells,[14] biochemical responses,[15] chemoreception,[16] liver enzyme function,[17,18] blood cells,[19] lateral line,[16] behavior,[20,21] histopathology,[19,22] and fin regeneration.[23,24] There have also been many studies focusing on bioaccumulation of various contaminants.[25-28]

III. EMBRYOS AND LARVAE

Embryos and larvae are often the most sensitive stages in an animal's life cycle. In addition, some toxicants exert their effects especially on developing systems and can produce structural or

Figure 33.1   The common killifish, or mummichog, *Fundulus heteroclitus*. The male is dark with light vertical bars; the female is light with faint dark bars. Bar = 1 cm. (From Weis, J.S. and Weis, P., *BioScience*, 39, 89, 1989. © American Institute of Biological Sciences. With permission.)

functional abnormalities in developing embryos. Fish eggs are useful for studies of developmental toxicology since they are numerous and their transparency allows them to be studied and examined throughout all stages of embryonic development. In standardized "early life stage" fish toxicity tests, organisms are exposed from the time of fertilization until some weeks after hatching, and hatching success, survival, and growth of the larvae are recorded. Such tests were not designed as tests for developmental toxicants (teratogens) in particular, but rather as general toxicity screening methods. Hatching success, a commonly used endpoint, can be affected by many unrelated factors, including embryonic death, gross abnormalities, or more subtle changes. On the other hand, there have been many studies particularly concerned with reproductive and developmental effects of environmental toxicants which have focused on the nature and degree of the abnormalities produced, and sometimes on the mechanisms of action of the teratogenic substances. In most studies, embryos are exposed to the toxicant after fertilization. However, in the natural environment, embryos could also be exposed via the yolk, synthesized during oogenesis in exposed females, and during the brief period between shedding of the eggs and sperm and the elevation of the chorion. In most but not all cases, the chorion acts as a barrier to partially protect the developing embryo from toxicants in the water.

Fundulus heteroclitus breeds, generally with a biweekly cycle, through the late spring and summer months. Spawning is concentrated during the spring tides, and eggs are normally deposited in algal mats or empty *Geukensia* shells in the intertidal zone.[29] Its eggs and sperm can be easily stripped from ripe adults during the breeding season, and fertilization and embryonic development can be readily studied. Although they can be bred in captivity,[30] in most studies field-collected fish have been used as a source of eggs and sperm.

There is a long history of using *Fundulus* embryos for laboratory studies of both normal and abnormal development, going back over a century.[31] The ease of obtaining and caring for the eggs, as well as their large size, made them attractive to investigators. In the early 1900s, Stockard experimented with the effects of lithium[32] and magnesium chloride[33] on *Fundulus* embryos. He described the production of embryos with a single large median cyclopic eye, fused nasal pits, and a protruding mouth, one of the classic teratology studies. Subsequently, he and others[34-36] showed that the mechanical removal of some anterior embryonic shield or the addition of some other chemicals (ethyl alcohol, butyric acid, acetone) could also produce cyclopia. Stockard also noted that the eggs were capable of developing out of the water in a moist atmosphere, that they were incapable of hatching outside the water, but that they could survive almost two weeks after the normal hatching time, and then hatch within moments of submersion in water.

IV. DEVELOPMENTAL TOXICOLOGY

Teratogens tend to be fairly nonspecific in the types of defects they produce. Many different substances can produce the same kinds of deformities, even though their modes of action may differ. Developmental processes that can be disrupted include cell or tissue differentiation, cell death (apoptosis), morphogenetic movements, tissue communication and interaction (e.g., induction), growth, and metabolism (respiration, absorption, excretion). In some cases the etiology of a developmental malformation may be traced back to a mutagenic event such as alterations in DNA sequence or cytogenetic changes, but in many cases developmental abnormalities are not associated with mutagenesis.

Fish embryos tend to become abnormal in specific ways.[37] The skeletal system appears to be the most sensitive, and flexures (scoliosis and lordosis) are very common, as well as stunting and partial twinning. The circulatory system is also sensitive, resulting in frequent observations of circulatory stasis, a failure of the heart tube to bend, and edema of the pericardial cavity, in response to many different toxicants. The developing optical system and brain are also sensitive, and many investigators report microphthalmia (small eyes), anophthalmia (missing eyes), as well as cyclopia (single median eye) or intermediate conditions of fusion of the optic cups. Another phenomenon frequently observed in exposed embryos is a general retardation of development, sometimes manifested as a developmental arrest. Such a reduction in developmental rate can result in more severe abnormalities, as the teratogens can act continuously on embryos during sensitive stages which are prolonged as a result of the retardation of development. However, this retardation is difficult to distinguish statistically from the normal developmental rate because both normal developmental rate and abnormal retardation have considerable variance.

In recent years, interest has reawakened in the use of eggs and larvae of *Fundulus* to investigate developmental responses to environmental toxicants, usually in static or static-renewal situations. This has been aided by Armstrong and Child's description of normal stages of development.[38] In addition to elucidating effects of various chemicals, these studies investigated interactions of toxicants and population differences in responses, as described below.

V. EMBRYONIC ABNORMALITIES IN MUMMICHOGS IN RESPONSE TO VARIOUS CHEMICALS

A. Pesticides

1. DDT

In 1949, Schreiman and Rugh[39] exposed embryos throughout development to 1 mg/L DDT. Embryos hatched with tails with an acute bend. When older embryos were treated, only mild lordosis was seen. Neuromuscular abnormalities were seen in spasmodic contractions of the musculature, poor swimming movements, or a lack of swimming ability. These effects were reversible if the embryos were placed back in clean water before the 16th day of development, a very important observation. Crawford and Guarino[40] exposed adult females to 0.1 µg/L for two 24-hr periods. Eggs from these females showed normal fertilization, but development was retarded relative to controls. Hatching was normal, however.

2. Carbaryl

Embryos exposed continuously to 10 mg/L experienced developmental arrest at stage 22–24 (four days/early organogenesis), depending on the time of initial exposure. If they were transferred to clean water after 4 days, they could resume development, but most developed circulatory abnormalities involving thin cardiac walls, failure of the heart tube to bend, decreased circulation, hemostasis, pericardial edema, and lack of blood pigment.[41] Since developmental arrest itself does not cause malformations,[42] in this case the period of arrest made them more vulnerable to the teratological effects of the insecticide. Pericardial edema, which is a commonly observed abnormality, is believed to be due to a fluid imbalance caused by the retarded circulation.[43]

3. Parathion

Embryos exposed to 1 or 10 mg/L exhibited developmental arrest at stage 25. When exposure was stopped after 3 days and embryos were returned to clean water, 50% of the embryos from 10 mg/L developed circulatory defects, while very few embryos were affected by 1 mg/L.[41]

B. Oil

The water-soluble fraction (WSF) of number 2 fuel oil at concentrations equivalent to 0.28 to 0.7 ppm total naphthalene caused decreased length, skeletal malformations, decreased yolk utilization,

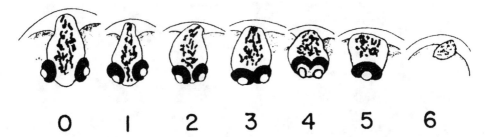

Figure 33.2   Heads of 7-day old mummichog embryos, demonstrating the craniofacial index (CFI) for ranking the severity of craniofacial anomalies: 0 = normal, 3 = synophthalmia, 5 = cyclopia, 6 = anophthalmia. The utility of the CFI is demonstrated in Figure 33.4. (From Weis, J.S., Weis, P., and Heber, M., Variation in response to methylmercury by killifish (*Fundulus heteroclitus*) embryos, in *Aquatic Toxicology and Hazard Assessment: Fifth Conference,* ASTM STP 766, Pearson, J.G., Foster, R., and Bishop, W., Eds., Amer. Soc. for Testing and Materials, Philadelphia, 1982, 109. © American Society for Testing and Materials. With permission.)

and decreased number of vertebrae.[44] Interactions of temperature and salinity with oil were noted. The toxicity of the oil was greater when temperature and salinity were suboptimal.[45] The day on which exposure was started and the total length of exposure were important in determining hatching success. When embryos of the related species *F. grandis* were exposed to concentrations of WSF corresponding to 1.1, 2.2, and 4.4 ppm, those in the lowest concentration hatched prematurely but were otherwise normal; those in the intermediate concentration showed some pathology of liver, kidney, lens, and epithelia; and those in the highest concentration failed to hatch.[46] Exposure to 50% of the WSF (corresponding to 5 ppm total hydrocarbons) produced 50% hatch and depressed heartbeat,[47] while 10% and 30% WSF stimulated hatching. Accelerated hatching was associated with reduced embryonic movements, which may have concentrated the hatching enzyme in a localized area. Thus, accelerated hatching may be a useful parameter in aquatic toxicity of oil.

C. Metals

1. Mercury

Weis and Weis[48,49] observed cyclopia and intermediate conditions in embryos exposed to mercuric chloride ($HgCl_2$) or methylmercury (meHg) (Figure 33.2). This condition was similar to that observed much earlier by Stockard.[32] The mechanism underlying the fusion of the optic cups is believed to be inadequate induction of the forebrain, which then permits the two optic vesicles to approach one another in the anterior midline of the embryo. Gastrulation was the critical period for the genesis of these defects; this corresponds to the time of induction of the forebrain. Cardiac malformations such as the "tube heart," and pericardial edema were also observed, as were skeletal malformations including bends and stunting.[48,49] Sharp and Neff[50] found the duration of exposure to mercuric chloride was important in the genesis of spinal curvature; thus, for this syndrome, no critical period was noted. Sharp and Neff[51] stated that when hatching success was the effect studied, all embryos exposed prior to gastrulation were equally sensitive; i.e., duration of exposure was not important. Thus, craniofacial, cardiovascular, and skeletal defects, the severity of which are the results of both concentration and duration of exposure, are useful parameters for mercury toxicity.

a. Interactions

Weis et al.[52] found that lowered salinity and decreased temperature increased the degree of abnormality in meHg-exposed embryos. The decreased temperature, which slowed down development, presumably prolonged the embryos' sensitive period during exposure to the teratogen.

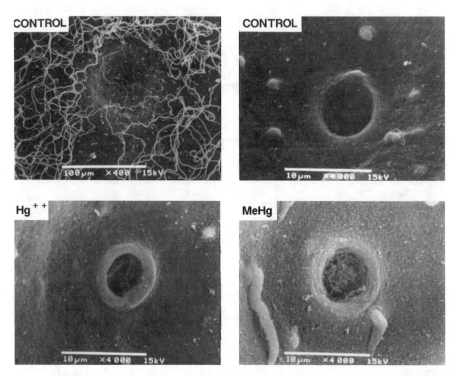

Figure 33.3   Micropyles of unfertilized mummichog eggs, showing controls at low and high magnification (top row) and eggs exposed to 1 mg/L Hg^{2+} and 1 mg/L methylmercury (meHg), both at high magnification. (From Khan, A.T. and Weis, J.S., *Environ. Biol. Fish,* 37, 323, 1993. With permission of Kluwer Academic Publishers.)

Presumably the same phenomenon occurred in the Linden et al.[45] study with oil. The presence of Zn or Cd reduced the magnitude of effects, perhaps because of competition for sites of entry. In this study an index was used to evaluate the severity of a defect in each embryo. Indices of severity of craniofacial, cardiovascular, and skeletal effects were devised to give a more quantitative evaluation, allowing for more detailed and precise analysis than simply reporting the percentage of embryos which were normal or abnormal.

b. Effects on Gametes

Mercury also affected gametes when they were exposed prior to fertilization. Khan and Weis[53-56] noted that mummichog sperm exposure to 50 µg/L meHg or $HgCl_2$ for one minute reduced fertilization success, while eggs needed to be exposed to higher concentrations or for longer periods of time before effects could be seen. After exposure of eggs prior to fertilization, some abnormalities were subsequently noted in the developing embryos. The mechanisms by which exposure of eggs to methylmercury or to inorganic mercury inhibited fertilization appeared to be different. While methylmercury triggered the cortical reaction making the egg refractory to sperm penetration, mercuric chloride caused a swelling of the lip of the micropyle (Figure 33.3).[57]

2. Lead

Exposure of embryos to 10 µg/L produced fish larvae which could not uncurl after hatching.[48] Such larvae would not be viable because their swimming ability was very limited.

3. Tributyltin (TBT)

Embryos exposed to 30 µg/L TBT showed a variety of optical malformations, including anophthalmia (no eyes), microphthalmia (small eyes), and rudimentary supernumerary eyes at the anterior end of the embryo. Cardiovascular problems (tube hearts), stunting, retardation, and inhibition of hatching were also seen.[58] However, responses were variable among batches produced by different females, with some batches being very resistant to the teratogenic effects but showing high mortality instead. This is probably due to genetic differences within wild stocks.

VI. BIOCHEMICAL RESPONSES

In some studies, the biochemical mechanisms underlying the above developmental toxicity studies have been investigated. Binder et al.[59] showed that PCB exposure induced the production of microsomal aryl hydrocarbon hydroxylase (AHH) in mummichog embryos. The amount of induction was much greater after hatching. Since this enzyme can convert PCBs to more damaging metabolites, this may explain why larvae are more sensitive than embryos. AHH activity, which in the adult is highest in the liver, could be detected in embryos before the appearance of the liver rudiment.

The metal-binding protein, metallothionein, can convey tolerance to various divalent cations. *Fundulus* embryos, however, do not synthesize these proteins until shortly before hatching, well after the genesis of abnormalities;[60] thus it is not the mechanism of tolerance for mummichog embryos.

VII. VARIATION IN RESPONSES AND POPULATION DIFFERENCES

Using the indices described previously, striking population differences were noted in susceptibility of embryos to meHg effects. These population differences were related to pollution history (as a local population adapted to specific pollutants) and suggest the utility of using embryo responses for monitoring. Embryos from a polluted environment (Piles Creek [PC] a tributary of the Arthur Kill in the highly industrialized and urban Newark Bay estuary) were much more resistant than those from a more pristine environment,[61] although there was considerable variation in resistance among batches of eggs from different females from the clean population from Long Island, (LI).[62] Some of the latter females produced eggs that were severely affected, while others produced eggs that were only mildly affected or unaffected by 50 µg/L. In contrast, almost all females from the polluted population produced resistant eggs (Figure 33.4). The variation seen in the clean population indicates that some of them could continue to survive and breed if that area became contaminated. Perry et al.[63] noted that embryos which were more resistant to the teratogenic effects were also more resistant to the cytogenetic effects of meHg. Possible mechanisms for the increased tolerance in the polluted population include more rapid development time (thus less exposure during critical periods), and a less permeable chorion, resulting in less meHg penetrating into the embryo.[64] These are strictly embryonic adaptations, and after hatching, larvae and juveniles of the two populations have comparable tolerance in acute tests.[54]

Embryos from the polluted Piles Creek site were less resistant to inorganic mercury than embryos from the clean LI population, however.[65] A concentration of 50 µg/L mercuric chloride was lethal to a large number of PC embryos but was much less toxic to LI embryos. This indicates that the tolerance is specific to the form of mercury. Khan and Weis[53-56] found similar relationships in toxicity of meHg and inorganic Hg to sperm and eggs in the two populations, with LI gametes being more affected by meHg, and PC gametes more affected by $HgCl_2$. Eggs from the Piles Creek population were much less salinity tolerant. Fertilization was unsuccessful at salinities approaching

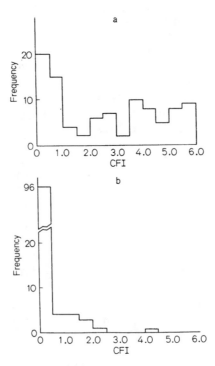

Figure 33.4   The mean craniofacial indices (CFIs) of individual clutches of embryos from (a) a reference population for mummichogs from eastern Long Island, NY, and (b) Piles Creek, a polluted estuary, resulting from exposure to 50 µg/L methylmercury. The CFI is illustrated in Figure 33.2. (From Weis, J.S., Weis, P., Heber, M., and Vaidya, S., Methylmercury tolerance of killifish (*Fundulus heteroclitus*) embryos from a polluted vs. non-polluted environment, *Mar. Biol.*, 65, 283, 1981. © Springer-Verlag New York, Inc. With permission.)

30‰, due to a triggering of the cortical reaction and raising of the fertilization membrane, thus preventing sperm from entering the micropyle.[66]

Fish from the Newark Bay area were also found to have increased resistance to dioxin, another contaminant found in this estuary. Embryos were unaffected by the same exposures (12 to 100 ng/L) which produced pericardial edema and hemorrhage (and death) in the reference population from Tuckerton, NJ.[67] Unlike the methylmercury situation, this resistance was not correlated with a decreased uptake of the TCDD, and the increased resistance was also manifested in adults from the Newark population. The observations indicate that responses of embryos could be used for monitoring for specific pollutants.

VIII. FUNCTIONAL DEFICIENCIES

In mammalian teratology, a field much more intensively studied than fish teratology, it has become clear that functional impairment is more subtle than the production of gross anatomical abnormalities. The field of "behavioral teratology" is the study of behavioral abnormalities after embryonic exposure to toxicants. Although there has been considerable work on effects of contaminants on morphological development of fish, little attention has been paid to behavioral responses following embryonic exposures to pollutants. A 1995 study[68] showed that larvae of the mummichog exposed as embryos to 5 or 10 µg/L meHg (concentrations below those which cause structural malformations) were less effective in capturing prey than were controls. When larvae were maintained in clean water, differences between prey capture rate of experimental and control larvae decreased over time until they were no longer significant. However, even at two weeks after hatching, larvae that had been exposed as embryos made more miscues (i.e., missed their prey on some attempts) than control larvae. Differences were also seen between controls from different populations. Fish from polluted Piles Creek, which have impaired predation as adults[69] but which develop more rapidly, had a higher larval prey capture rate as young larvae, although they exhibited more miscues than the LI reference population.[70] Differences were also seen in the response to meHg: Piles Creek fish, which previously had been found to be more resistant to teratogenic effects of higher concentrations of meHg, appeared to be more resistant than the reference population to these behavioral effects as well.[70]

Larvae from the clean LI population were also tested for swimming performance after embryonic and/or early larval exposure by chasing them with a glass rod. Swimming performance improved with larval age. Larvae which had been exposed to 10 µg/L meHg as embryos swam greater distances than controls, while those that were exposed only as larvae swam less than controls.[71] The effects of the embryonic exposure diminished over time, and the larval environmental conditions took precedence. Larvae were also tested with two predators, the grass shrimp, *Palaemonetes pugio*, and with year-old mummichogs. Those that had been exposed to mercury, either as embryos (which swam more than controls) or as larvae (which swam less than controls) were both more susceptible to predation than controls. The increased swimming of those exposed as embryos may have reflected greater activity, making them more likely to attract the predator's attention, while the slower swimming of those exposed as larvae could make them easier to capture. Since larval stages are particularly sensitive, and the ability to feed and to avoid predation are critical to survival, these subtle effects of embryonic and early larval exposure could have major impacts on larval success. These subtle effects make suitable endpoints for small-scale assays with mummichog embryos.

IX. DISCUSSION

Early life stages of the mummichog have been in use in experimental laboratories for a full century. Their ready availability (in season), ease of maintenance, and economy of scale have favored their use in basic developmental biology and in microscale aquatic toxicology. Numerous studies have been reported on exposures to metals, pesticides and other organics and mixtures, with parameters including abnormal development, mortality, retardation, hatching success, biochemical responses, and post-hatching behavior.

As a generally robust species, the mummichog will not be ideal for toxicity tests that require very sensitive responses. This is compensated for by the ease of handling and working with the gametes and embryos. In addition, new subtler endpoints (such as larval behavior) can be discovered in future studies. Small-scale aquatic toxicity procedures, such as described here, can contribute much to testing pollution effects. However, beyond the realm of "toxicity testing," we feel that research on this species has yielded much interesting and important information on the ways in which organisms respond to environmental perturbations. Much more basic biology and ecology of pollution responses remains to be learned. We recommend this species to experienced researchers and newcomers alike.

REFERENCES

1. Lotrich, V.A., Summer home range and movements of *Fundulus heteroclitus* (Pisces: Cyprinodontidae) in a tidal creek, *Ecology*, 56, 191, 1975.
2. Kneib, R.T., The effects of predation by wading birds (Ardeidae) and blue crabs (*Callinectes sapidus*) on the population size structure of the common mummichog, *Fundulus heteroclitus*, *Estuar. Coast. Shelf Sci.*, 14, 159, 1982.

3. Hoffman, R.B., Salinas, G.A., and Baky, A.A., Behavioral analysis of killifish exposed to weightlessness in the Apollo-Soyuz test project, *Aviat. Space Environ. Med.*, 48, 712, 1977.
4. Powers, D.A., Ropson, I., Brown, D.C., van Beneden, R., Cashon, R., Gonzalez-Villasenor, L.I., and DiMichele, L.A., Genetic variation in *Fundulus heteroclitus*: Geographic distribution, *Amer. Zool.*, 26, 131, 1986.
5. Able, K.W. and Felley, J.D., Geographical variation in *Fundulus heteroclitus*: Tests for concordance between egg and adult morphologies, *Amer. Zool.*, 26, 145, 1986.
6. Eisler, R., Factors affecting pesticide-induced toxicity in an estuarine fish, U.S. Fish Wildlife Serv. Tech. Pap., 45, 1, 1970.
7. Eisler, R., Acute toxicities of organochlorine and organophosphorus insecticides to estuarine fishes, U.S. Fish Wildl. Serv. Tech. Pap., 46, 1, 1970.
8. Eisler, R., Acute toxicity of zinc to the killifish *Fundulus heteroclitus, Chesapeake Sci.*, 8, 262, 1967.
9. Eisler, R., Cadmium poisoning in *Fundulus heteroclitus* (Pisces: Cyprinodontidae) and other marine organisms, *J. Fish. Res. Bd. Canada*, 28, 1225, 1971.
10. Eisler, R. and Henneky, R.J., Acute toxicities of Cd^{2+}, Cr^{+6}, Hg^{2+}, Ni^{2+}, and Zn^{2+} to estuarine macrofauna, *Arch. Environ. Contam. Toxicol.*, 6, 315, 1977.
11. Eisler, R., Some effects of a synthetic detergent on estuarine fishes, *Trans. Amer. Fish. Soc.*, 94, 26, 1965.
12. LaRoche, G., Eisler, R., and Tarzwell, C.M., Bioassay procedures for oil and oil dispersant toxicity evaluation, *J. Water Poll. Contr. Feder.*, 42, 1982, 1970.
13. Eisler, R. and Gardner, G.R., Acute toxicology to an estuarine teleost of mixtures of cadmium, copper, and zinc salts, *J. Fish Biol.*, 5, 131, 1973.
14. Crespo, S. and Karnaky Jr., K.J., Copper and zinc inhibit chloride transport across the opercular epithelium of seawater-adapted killifish, *Fundulus heteroclitus, J. Exper. Biol.*, 102, 337, 1983.
15. Burns, K.A., Microsomal mixed function oxidases in an estuarine fish, *Fundulus heteroclitus*, and their induction as a result of environmental contamination, *Comp. Biochem. Physiol. B. Comp. Biochem.*, 53, 443, 1976.
16. Gardner, G.R. and LaRoche, G., Copper-induced lesions in estuarine teleosts, *J. Fish. Res. Bd. Canada*, 30, 363, 1973.
17. Jackim, E., Hamlin, J.M., and Sonis, S., Effects of metal poisoning on five liver enzymes in the killifish (*Fundulus heteroclitus*), *J. Fish. Res. Bd. Canada*, 27, 383, 1970.
18. Pruell, D. and Engelhardt, F.R., Liver cadmium uptake, catalase inhibition and cadmium thionein production in the killifish (*Fundulus heteroclitus*) induced by experimental cadmium exposure, *Mar. Environ. Res.*, 3, 101, 1980.
19. Gardner, G.R. and Yevich, P.P., Histological and hematological responses of an estuarine teleost to cadmium, *J. Fish. Res. Bd. Canada*, 27, 2185, 1970.
20. Eisler, R., Behavioral responses of marine poikilotherms to pollutants. *Phil. Trans. Roy. Soc. London, B.*, 286, 507, 1979.
21. Weis, J.S. and Khan, A.A., Effects of mercury on the feeding behavior of the mummichog, *Fundulus heteroclitus,* from a polluted habitat, *Mar. Environ. Res.*, 30, 243, 1990.
22. DiMichele, L. and Taylor, M.H., Histopathological and physiological responses of *Fundulus heteroclitus* to naphthalene exposure, *J. Fish. Res. Bd. Canada*, 35, 1060, 1978.
23. Weis, P. and Weis, J.S., Effects of heavy metals on fin regeneration in the killifish, *Fundulus heteroclitus, Bull. Environ. Contam. Toxicol.* 16, 197, 1976.
24. Weis, P. and Weis, J.S., Effects of zinc on fin regeneration in the mummichog, *Fundulus heteroclitus*, and its interaction with methylmercury, *Fish. Bull. Natl. Mar. Fish. Serv.*, 78, 163, 1980.
25. Eisler, R., Radiocadmium exchange with seawater by *Fundulus heteroclitus* (L.)(Pisces: Cyprinodontidae), *J. Fish Biol.*, 6, 601, 1974.
26. Bennett, R.O. and Dooley, J.K., Copper uptake by two sympatric species of killifish *Fundulus heteroclitus* (L.) and *F. majalis* (Walbaum), *J. Fish Biol.*, 21, 381, 1982.
27. Chernoff, B. and Dooley, J.K., Heavy metals in relation to the biology of the mummichog, *Fundulus heteroclitus, J. Fish Biol.*, 14, 309, 1979.
28. Khan, A.T. and Weis, J.S., Bioaccumulation of heavy metals in two populations of mummichogs (*Fundulus heteroclitus*), *Bull. Environ. Contam. Toxicol.*, 51, 1, 1993.
29. Taylor, M.H., DiMichele, L., and Leach, G.J., Egg stranding in the life cycle of the mummichog, *Fundulus heteroclitus, Copeia*, 1977, 397, 1977.

30. Boyd, J.F. and Simmons, R.C., Continuous laboratory production of fertile *Fundulus heteroclitus* (Walbaum) eggs lacking chorionic fibrils, *J. Fish. Biol.,* 6, 389, 1974.

31. Atz, J.W., *Fundulus heteroclitus* in the laboratory: A history, *Amer. Zool.,* 26, 111, 1986.

32. Stockard, C.R., The development of *Fundulus heteroclitus* in solutions of lithium chloride, with appendix on its development in fresh water, *J. Exper. Zool.,* 3, 99, 1906.

33. Stockard, C.R., The development of artificially produced cyclopean fish — "the magnesium embryo," *J. Exper. Zool.,* 6, 285, 1909.

34. Stockard, C.R., The influence of external factors, chemical and physical, on the development of *Fundulus heteroclitus, J. Exper. Zool.,* 4, 165, 1907.

35. Werber, E.I., Experimental studies on the origin of monsters. I. An etiology and an analysis of the morphogenesis of monsters, *J. Exper. Zool.,* 21, 485, 1916.

36. Lewis, W.H., The experimental production of cyclopia in the fish embryo (*Fundulus heteroclitus*), *Anat. Rec.,* 3, 175, 1909.

37. Weis, J.S. and Weis, P., Pollutants as developmental toxicants in aquatic organisms, *Environ. Health Persp.,* 71, 77, 1987.

38. Armstrong, P.B. and Child, J.S., Stages in the normal development of *Fundulus heteroclitus, Biol. Bull.,* 128, 143, 1965.

39. Schreiman, E. and Rugh, R., Effect of DDT on functional development of larvae of *Rana pipiens* and *Fundulus heteroclitus, Proc. Soc. Exper. Biol. Med.,* 70, 431, 1949.

40. Crawford, R.B. and Guarino, A.M., Effects of DDT in *Fundulus*: studies on toxicity, fate, and reproduction, *Arch. Environ. Contam. Toxicol.,* 4, 334, 1976.

41. Weis, P. and Weis, J.S., Cardiac malformations induced by insecticides in embryos of the killifish, *Fundulus heteroclitus, Teratology,* 10, 263, 1974.

42. Laale, H.W. and McCallion, D.J., Reversible developmental arrest in the embryo of the zebrafish, *Brachydanio rerio, J. Exper. Zool.,* 167, 117, 1968.

43. Laale, H.W., Teratology and early fish development, *Amer. Zool.,* 21, 517, 1981.

44. Linden, O., Laughlin, R., Sharp, J.R., and Neff, J.M., The combined effect of salinity, temperature, and oil on the growth pattern of embryos of the killifish, *Fundulus heteroclitus Walbaum. Mar. Environ. Res.,* 3, 129, 1980.

45. Linden, O., Sharp, J.R., Laughlin, R., and Neff, J.M., Interactive effects of salinity, temperature, and chronic exposure to oil on the survival and developmental rate of embryos of the estuarine fish *Fundulus heteroclitus, Mar. Biol.,* 51, 101, 1979.

46. Ernst, V., Neff, J.M., and Anderson, J., The effects of the water soluble fraction of No. 2 fuel oil on the early development of the estuarine fish *Fundulus grandis* Baird and Girard, *Environ. Poll.,* 14, 25, 1977.

47. Anderson, J.W., Dixit, D.B., Ward, G.S., and Foster, R.S., Effects of petroleum hydrocarbons on the rate of heart beat and hatching success of estuarine fish embryos, in *Physiological Responses of Marine Biota to Pollutants,* Vernberg, F.J., Calabrese, A., Thurberg, F.P., and Vernberg, W.B., Eds., Academic Press, New York, 1977, 241.

48. Weis, J.S. and Weis, P., The effects of heavy metals on embryonic development of the killifish, *Fundulus heteroclitus, J. Fish Biol.,* 11, 49, 1977.

49. Weis, P. and Weis, J.S., Methylmercury teratogenesis in the killifish, *Fundulus heteroclitus, Teratology,* 16, 317, 1977.

50. Sharp, J.R. and Neff, J.M., Effects of the duration of exposure to mercuric chloride on the embryogenesis of the estuarine teleost, *Fundulus heteroclitus, Mar. Environ. Res.,* 3, 195, 1980.

51. Sharp, J.R. and Neff, J.M., Age-dependent response differences of *Fundulus heteroclitus* embryos following chronic exposure to mercury, in *Marine Pollution and Physiology: Recent Advances,* Vernberg, F.J., Thurberg, F.P., Calabrese, A., and Vernberg, W.B., Eds., University of South Carolina Press, Columbia, 1985, 281.

52. Weis, J.S., Weis, P., and Ricci, J., Effects of cadmium, zinc, salinity and temperature on the teratogenicity of methylmercury to the killifish, *Fundulus heteroclitus, Second Int. Symp. on Early Life History of Fishes, Rapp. P.-V. Réun. Cons. Int. Explor. Mer.,* 178, 64, 1981.

53. Khan, A.T., and Weis, J.S., Effects of methylmercury on sperm and egg viability of two populations of killifish, *Fundulus heteroclitus, Arch. Environ. Contam. Toxicol.,* 16, 499, 1987.

54. Khan, A.T. and Weis, J.S., Effects of mercuric chloride on eggs and juvenile viability in two populations of killifish, *Mar. Poll. Bull.,* 18, 504, 1987.

55. Khan, A.T. and Weis, J.S., Effects of methylmercury on egg and juvenile viability in two populations of *Fundulus heteroclitus, Environ. Res.,* 44, 272, 1987.
56. Khan, A.T. and Weis, J.S., Effects of mercuric chloride on sperm and egg viability in two populations of the mummichog *Fundulus heteroclitus, Environ. Poll.,* 48, 263, 1987.
57. Khan, A.T. and Weis, J.S., Differential effects of organic and inorganic mercury on the micropyle of eggs of *Fundulus heteroclitus, Environ. Biol. Fish,* 37, 323, 1993.
58. Weis, J.S., Weis, P., and Wang, F.S., Developmental effects of tributyltin on the fiddler crab, *Uca pugilator,* and the killifish, *Fundulus heteroclitus, Oceans '87 Proceedings Int. Organotin Symp.,* 4, 1456, 1987.
59. Binder, R.L., Stegeman, J.J., and Lech, J.J., Induction of cytochrome P-450-dependent monooxygenase systems in embryos and eleutheroembryos of the killifish, *Fundulus heteroclitus, Chem. Biol. Interactions,* 55, 185, 1985.
60. Weis, P., Metallothioneins and methylmercury tolerance in *Fundulus heteroclitus, Mar. Environ. Res.,* 14, 153, 1984.
61. Weis, J.S., Weis, P., Heber, M., and Vaidya, S., Methylmercury tolerance of killifish (*Fundulus heteroclitus*) embryos from a polluted vs. non-polluted environment, *Mar. Biol.,* 65, 283, 1981.
62. Weis, J.S., Weis, P., and Heber, M., Variation in response to methylmercury by killifish (*Fundulus heteroclitus*) embryos, in *Aquatic Toxicology and Hazard Assessment: Fifth Conference,* ASTM STP 766, Pearson, J.G., Foster, R., and Bishop, W., Eds., Amer. Soc. for Testing and Materials, Philadelphia, 1982, 109.
63. Perry, D., Weis, J.S., and Weis, P., Cytogenic effects of methylmercury in embryos of the killifish, *Fundulus heteroclitus, Arch. Environ. Contam. Toxicol.,* 17, 569, 1988.
64. Toppin, S.V., Heber, M., Weis, J.S., and Weis, P., Changes in reproductive biology and life history of *Fundulus heteroclitus* in a polluted environment, in *Pollution Physiology of Marine Organisms,* Vernberg, W., Calabrese, A., Thurberg, F., and Vernberg, F.J., Eds., University of South Carolina Press, Columbia, 1985, 171.
65. Weis, J.S., Weis, P., Heber, M., and Vaidya, S., Investigations into mechanisms of methylmercury tolerance in killifish (*Fundulus heteroclitus*) embryos, in *Physiological Mechanisms of Marine Pollutant Toxicity,* Vernberg, W., Calabrese, A., Thurberg, F., and Vernberg, F.J., Eds., Academic Press N.Y. 1982, 311.
66. Bush, C.P. and Weis, J.S., Effects of salinity on fertilization success in two populations of killifish, *Fundulus heteroclitus, Biol. Bull.,* 164, 406, 1983.
67. Prince, R. and Cooper, K.R., Comparisons of the effects of 2,3,7,8-tetrachlorodibenzo-p-dioxin on chemically impacted and nonimpacted subpopulations of *Fundulus heteroclitus.* 1. TCDD toxicity, *Environ. Toxicol. Chem.,* 14, 579, 1995.
68. Weis, J.S. and Weis, P., Effects of embryonic exposure to methylmercury on larval prey capture in the mummichog, *Fundulus heteroclitus, Environ. Toxicol. Chem.,* 14, 153, 1995.
69. Weis, J.S. and Khan, A.A., Reduction in prey capture ability and condition of mummichogs from a polluted habitat, *Trans. Amer. Fish. Soc.,* 120, 127, 1991.
70. Zhou, T., Scali, R., Weis, P., and Weis, J.S., Behavioral effects in mummichog larvae (*Fundulus heteroclitus*) following embryonic exposure to methylmercury, *Marine Environ. Res.,* 42, 45, 1996.
71. Weis, J.S. and Weis, P., Effects of embryonic and larval exposure to methylmercury on larval swimming performance and predator avoidance in the mummichog, *Fundulus heteroclitus, Can. J. Fish. Aquat. Sci.,* 52, 2168, 1995.
72. Weis, J.S. and Weis, P., Tolerance and stress in a polluted environment. The case of the mummichog, *BioScience,* 39, 89, 1989.

CHAPTER **34**

Genotoxicity in Fish Embryos

Armin Herbert and Peter-Diedrich Hansen

CONTENTS

I. INTRODUCTION

Genotoxicity is the potential of a physical or chemical agent to alter genetic information of infectious particles (viroids and viruses), cells, or individuals. Genetic information is the information encoded in nucleic acids, which is inherited from one to the next generation of infectious particles, cells, or individuals. In viroids and viruses, this can be ribonucleic acid (RNA), however, in some of them and in all living cells, it is deoxyribonucleic acid (DNA). In all living organisms, DNA is the primary information carrier — the central element of life processes. DNA is the information matrix for two processes, essential to life: replication and transcription. Replication is the central function during the process of heredity, whereas transcription is the central function for realization of inherited information, i.e., the programmed biosynthesis during ontogenesis.

The total of genetic information of a cell constitutes the genome. Thus genotoxicity may be seen as a toxicity-dependent alteration of the genome. Alterations can be defined on two structural levels. First, changes can be found in the primary structure of DNA, which is any molecular alteration in the linear biopolymer with its units sugar, phosphate, and the purine and pyrimidine bases of adenine, cytosine, guanine, and thymine. Second, a change of base sequence alters the encoded information. The primary lesion is usually called DNA damage, whereas the change of base sequence is called mutation. Mutations can result from DNA damage, mostly due to DNA repair errors, but also due to errors in the replication process. DNA repair systems have evolved in both prokaryotes and eukaryotes to maintain a high stability of the genetic information, however, they are not free of errors. Both DNA damage and mutations occur spontaneously or as exogenously triggered events. Mutations arising from errors during DNA repair occur with rates between 1 in 5000 to 500,000 primary lesions.[1] As a result, exogenous induction of DNA primary lesions (DNA damage) by genotoxic pollutants also increases mutation rates. However, DNA primary lesions appear to be reversible and not detectable after repair, whereas mutations accumulate with time due to constant rates of repair errors. Thus, due to the difference in reversibility of lesions on the primary and secondary level, the determination of exogenously induced DNA damage rates must *a priori* be considered as an underestimate of the actual induction of mutation rates.

Nevertheless, attempts have been made to develop and adapt methods to determine DNA damage in aquatic vertebrates and invertebrates, since easy-to-use mutation assays are presently not at hand for these organisms. The method of ^{32}P-postlabeling has successfully been used for the specific determination of DNA adducts.[2-4] However, the method is restricted to the detection of adducts only, and does not provide a sum parameter including other DNA lesions like strand breaks, depurinations, and depyrimidinations (AP sites), deaminations or forms of oxidative damage. Sensitivity toward alkaline conditions can be used instead as a sum parameter for various types of DNA damage, which is increased by many of these lesions and additionally by excision repair patches.[5] Various techniques are suitable to determine alkali sensitivity and have successfully been used with mammalian systems in radiation biology and human (mammalian) genetic toxicology. They include alkaline gradient centrifugation and the derived nucleoid sedimentation technique,[6-8] alkaline filter elution,[9-11] alkaline unwinding in different modifications,[12-18] alkaline viscometry,[19-21] and alkaline agarose gel electrophoresis and the derived single cell electrophoresis (COMET) assay.[22-26]

For investigations in aquatic invertebrates and fish, a protocol for an alkaline unwinding/hydroxylapatite "batch" elution assay was derived which was based on procedures published by Rydberg and Johanson,[13,27] Kanter and Schwartz,[14] and Cesarone, Bolognesi, and Santi[28] and was particularly adapted to the cytogenetic characteristics of aquatic vertebrate and invertebrate taxa.[29-33] The method has recently been miniaturized and offers the following advantages:

1. Wide spectrum of sensitivities providing a sum parameter for different types of DNA lesions
2. High cost-efficiency
3. Easy use
4. Rapidity (results in less than 1 day)
5. High sample throughput (up to 40 samples per day)
6. Portability for use in field studies, e.g., on research vessels and field stations
7. No demand for radioactively labeled or highly toxic chemicals (with the exception of formamide)

The DNA unwinding/hydroxylapatite (HAP) batch elution assay has been recently applied on a large scale to field and laboratory studies on genotoxicity in fish embryos, where investigations of DNA damage are still rare.

II. DNA UNWINDING/HYDROXYLAPATITE BATCH ELUTION ASSAY (PROTOCOL)

A. Principle

The alkaline DNA unwinding/hydroxylapatite (HAP) batch elution assay may be divided into three subsequent processes: the actual unwinding of sample DNA under controlled alkaline conditions,

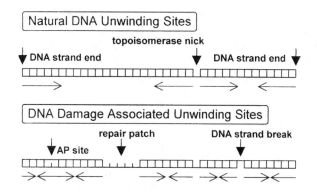

Figure 34.1   Theory of alkaline DNA unwinding to detect DNA lesions.

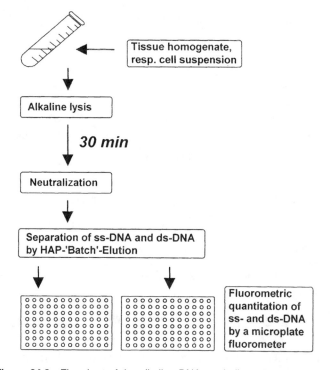

Figure 34.2   Flowchart of the alkaline DNA unwinding assay procedure.

the separation of single-stranded (ss) from double-stranded (ds) DNA by HAP batch elution, and the quantitation of ss- and ds-DNA fractions by fluorometry (Figures 34.1 and 34.2).

In the first step, a buffered alkaline solution is added to a tissue homogenate or cell suspension. Within a few seconds all hydrogen bonds of cell constituents are disrupted, including the hydrogen bonds between paired bases of DNA. A second comparatively slow process — the unwinding of the plektonemic DNA double helix into single strands — starts at the same time. After a defined period of time, this process is stopped by neutralization of the lysate. A detergent (SLS, see below) is added, and the sample is sonicated to prevent reconstitution of DNA-protein complexes and rejoining of complementary DNA strands. Single (ss)- and double (ds)-stranded DNA are separated by ion exchange chromatography with hydroxylapatite batches. DNA is quantified in both fractions by fluorometry. A decrease in the relative amount of double-stranded DNA (F) indicates DNA damage. The negative logarithm of F (–logF) is proportional to the relative average number of DNA unwinding points, if a random distribution of these points within the genome is assumed.[13] Natural DNA unwinding points are the DNA strand ends, topoisomerase nicks, and spontaneous DNA lesions. Many exogenously induced DNA lesions result directly or indirectly in additional alkali-sensitive unwinding points and increase the value of –logF.

B. Reagents

Homogenization buffer: 0.114 mol/L Tris HCl, 0.077 mol/L NaCl, pH 9.00 at 0°C (pH 8.54 at 23°C); store at 0°C or sterilize by autoclaving (to prevent bacterial growth)

Lysis buffer: 0.15 mol/L NaOH, 0.05 mol/L Na_2HPO_4, 1 mol/L NaCl; do not use old NaOH solutions and keep fresh ones air-tight

Neutralizing solution: 0.18 mol/L HCl, 5 mg/L phenol red

SLS solution: 0.5% sodium lauryl sarcosinate, 0.02 mol/L Na_2-EDTA, pH 7.0

Hydroxylapatite: (BioRad DNA-Grade Bio-Gel HTP); wash one volume hydroxylapatite (HAP) twice with ten volumes 0.01 mol/L phosphate buffer (see below), then suspend in two volumes of the same buffer. Heat the hydroxylapatite suspension for 5 min before use in a boiling water bath.

Formamide: (purification: 1L formamide + 10 g Norit A charcoal + 50 g mixed-bed ion exchanger, 2 h stirring at 4°C, filter through a double paper filter; the pH value of a 1:2 dilution with water should be below 7.5 and conductivity should be below 40 µS) Handle with care!

Potassium phosphate stock buffer: 1 mol/L (0.5 mol/L KH_2PO_4 plus 0.5 mol/L K_2HPO_4, do *not* adjust pH)

0.01 mol/L Phosphate buffer: dilute 1 mol/L phosphate stock buffer

Wash buffer: 0.01 mol/L potassium phosphate buffer, 20% formamide

Elution buffer I: 0.125 mol/L potassium phosphate buffer, 20% formamide

Elution buffer II: 0.5 mol/L potassium phosphate buffer, 20% formamide (use 1 mol/L phosphate stock buffer for all buffer preparations)

Bisbenzimide buffer: 0.05 mol/L potassium phosphate buffer, 2.2 mol/L NaCl, adjust pH 7.4 with 5 mol/L KOH

Bisbenzimide stock solution: 250 µmol/L bisbenzimide in distilled water; store cold (4°C) and dark; stable for at least 3 months

Bisbenzimide dilution: Bisbenzimide stock solution, diluted 1:100 with bisbenzimide buffer

C. Equipment

Nine items are required:

1. Tip sonicator
2. Desktop centrifuge for 1.5 mL reagent tubes
3. 1.5 mL reagent tubes (1 per sample in microtiter plate assay/3 per sample in cuvette assay)
   2.0 mL reagent tubes (1 per sample)
4. Reagent tube mixer or shaker
5. Eppendorf Multipette 4780 (50 µL, 250 µL, 500 µL)

6. Microplate fluorometer or conventional spectro-/filterfluorometer with wavelength set at 360/450 nm
7. Adjustable microliter pipettes 10 to 200 μL, 200 to 1000 μL
8. Glassware (beakers, etc.)
9. Microplates

D. Assay

1. Preparation and Lysis of Samples

All steps are carried out with samples placed on ice! (Temperature control is strictly required and temperatures approximate 0° to 2°C!)

Transfer fish embryos or tissue samples immediately into 5 volumes ice-cold (0°C) homogenization buffer and homogenize (preferentially with a Dounce glass homogenizer). Allow debris to settle for 15 sec.

Alternatively for invertebrate hemolymph: Place 1.2 mL hemolymph in 1.5 mL reagent tubes for 5 min on ice and allow hemocytes to settle; to increase the cell titer discard 0.6 mL of the upper supernatant and suspend cells in the remaining hemolymph; carry on immediately.

Transfer 0.2 mL of the homogenized sample or suspended hemocytes into a 2 mL reagent tube and leave it on ice. (Usually duplicates!)

To determine the reagent blank later on, do the same with 0.2 mL homogenization buffer and process it in the same way as the samples. (Usually duplicates!)

Add 0.5 mL ice-cold lysis buffer, and leave samples on ice for 30 min; avoid mechanical agitation and exposure to light.

IMPORTANT: Add lysis buffer and later neutralizing solution by forceful injection with the pipette *along the tube wall* and *never* directly into the sample to minimize bubbles and other DNA shearing forces.

To stop alkaline DNA unwinding, add 0.5 mL ice-cold neutralizing solution and 0.5 mL SLS solution (stored at room temperature).

To minimize renaturation of partially unwound DNA, immediately (if samples are placed on ice within 20 min after neutralization) sonicate samples for 10 sec (method of choice) or shear them 4 times through a 27G syringe needle.

After this step, samples can be stored or transported frozen at –20°C or below.

2. Hydroxylapatite Batch Elution to Separate Single (ss)- and Double-Stranded (ds) DNA

Preparation of reagent tubes: per sample and blank control, prepare and label one 1.5 mL reagent tube, which is used for the hydroxylapatite batch elution. (Two more are required for single (ss)- and double (ds)-stranded DNA fractions, if standard fluorometer cuvettes (3 mL) are used instead of microplates.)

Boil the hydroxylapatite suspension for 5 min (in a boiling water bath).

Transfer 0.2 mL of the hydroxylapatite suspension into a 1.5 mL reagent tube (one per sample and blank).

IMPORTANT: Assure in all following steps, that the hydroxylapatite is completely suspended by mixing.

Add 1.0 mL of the lysed samples and blank controls to the hydroxylapatite and suspend by repeated mixing (at least 3 times, at 3-min intervals; at low DNA concentrations more mixing, e.g., 6 times, and time, e.g., 15 to 20 min, increases binding efficiency). Centrifuge at 600 to 1200 × g for 2 to 3 sec; remove the supernatant by turning the reagent tube quickly upside-down (alternatively with a pipette or — for convenience — with a pipette tip connected to a vacuum pump) and discard it. If the DNA content of the samples is expected to be very low, this step may be repeated with the remaining lysate.

Add 1.25 mL wash buffer (preferably with a multistep pipette) and suspend the hydroxylapatite by mixing twice at an interval of 3 min. Centrifuge and discard the supernatant as described above.

Repeat washing a second time.

Add 1 mL elution buffer I and suspend the hydroxylapatite by mixing three times at intervals of 3 min. Centrifuge as above, but save a part of the supernatant with a multistep pipette and dispense four 100 μL aliquots into wells of a 96-well microtiter plate for the determination of the ss-DNA. (Alternatively, if a microtiter plate fluorometer is not available, save 0.4 mL of the supernatant and transfer it to a 1.5 mL reagent tube.) Discard remaining supernatant.

Repeat last step using elution buffer II, and keep aliquots of the supernatant for the determination of the ds-DNA.

3. Determination of DNA Concentrations in the ss- and ds-DNA Supernatants

Aliquots of the supernatants are heated in a 95°C waterbath for 5 min to transfer all DNA to single-stranded conformation to get comparable relative fluorescence intensities; then cool rapidly to room temperature (use 20°C water bath).

Prepare the totally required volume of bisbenzimide dilution: dilute bisbenzimide stock solution 1/100 v/v with bisbenzimide buffer.

Add 0.2 mL bisbenzimide dilution to each sample well and determine fluorescence at an excitation wavelength of 360 nm and emission of 450 nm. (Alternatively, if a microtiter plate fluorometer is not available, add 0.8 mL bisbenzimide dilution to each sample tube and determine fluorescence.)

Calculate DNA-dependent fluorescence by subtracting control blanks from sample fluorometer readings; distinguish between blanks obtained from the first and second elution step.

Calculate the fraction F of double-stranded(ds)-DNA by the formula:

$$F = \frac{ds}{(ss + ds)} \tag{34.1}$$

where ss is the relative fluorescence of the single-stranded(ss)-DNA fraction, and ds that for the double-stranded(ds)-DNA fraction.

The negative logarithm of F [–log F] is proportional to the relative average number of DNA unwinding points, and thus to the degree of fragmentation. The relative average number of exposure-related DNA lesions per unwinding unit (i.e., the average distance between unwinding points) can be calculated as follows:

$$n = \frac{-\log F_x}{-\log F_0} - 1 \qquad (34.2)$$

where the index x indicates values from polluted and 0 from unpolluted organisms.

4. Possible Causes of Troubles

Tissue or hemolymph samples have been prepared too slowly and insufficiently cooled: perform initial steps steadily and completely on ice to minimize activities of endogenous nucleases. The high pH (9.0) of the homogenization buffer should minimize nuclease activity in homogenates, however, in some cases the additional use of 20 mmoles/l Na_2EDTA may be considered.

pH changes of the lysis buffer or NaOH solutions after long storage: avoid CO_2 binding by regular preparation of new solutions and air-tight closed storage of aliquots. **NOTE:** Reaction conditions of the assay are defined by the given concentrations of solution constituents and *not* by adjustment to any pH value. The quality of the lysis buffer can, however, be controlled by occasional pH determinations to detect possible deviations from initial values after storage.

Use of hydroxylapatite of insufficient quality: only "spherical" hydroxylapatite is suitable, but still not all commercially available preparations will separate ss- and ds-DNA properly. Suitable preparations will sediment slowly after suspension in ten volumes 0.01 mol/L potassium phosphate buffer (about 5 to 10 min); crystal spikes around the crystallization core can be visualized by light or scanning electron microscopy and cannot be seen in preparations of insufficient quality. Avoid mechanical destruction of the crystals, e.g., by use of magnetic stirrers.

Binding of CO_2 to suspended hydroxylapatite: always boil hydroxylapatite solutions before use!

Insufficient amount of DNA binding to the hydroxylapatite: when low concentrations of DNA in lysates occur, the given time for loading the samples and the lysate volume of 1 mL may be insufficient. **ALTERNATIVE:** Extend time given for HAP–DNA binding to 15 or 20 min with additional vortexing. Then load the remaining lysate in a second step in the same way.

Insufficient quality of formamide and formamide-containing solutions: be aware of the instability of formamide; prepare formamide-containing solutions fresh at least every four weeks; check quality (see above) before preparation of solutions; clean formamide may be stored in aliquots at –70°C for longer periods. Handle formamide with care.

III. DNA DAMAGE IN FISH EMBRYOS (EXPERIMENTAL RESULTS)

T.H. Morgan, both an embryologist and founder of modern genetics, put embryology and genetics into context when he wrote a book about it.[34] Unfortunately, in his time "so little actual connection could be drawn between the two subjects."[35] Only in more recent decades, has it become known that about 10% of an estimated 50,000 to 100,000 genes of a higher eukaryotic organism[36] are functionally used by adults, but 90% are required for embryonic development.[37] This indicates that embryos may suffer from adverse health perturbations caused by genotoxic impacts with a risk which is an order of magnitude higher than that of adults. Disruption of embryonic development or development of malformations could occur and interfere with the viability and fertility of the embryo.

In fish embryos from the North Sea, malformations occur in spatial and temporal patterns with high incidences, i.e., with rates up to 90%.[38] In March 1993 the Department of Ecotoxicology of the Berlin University of Technology investigated, in cooperation with the Institut für Fischereiökologie of the Bundesforschungsanstalt für Fischerei, whether DNA damage also occurs at increased levels. For this purpose, embryos were collected at different stations along a sampling grid in the

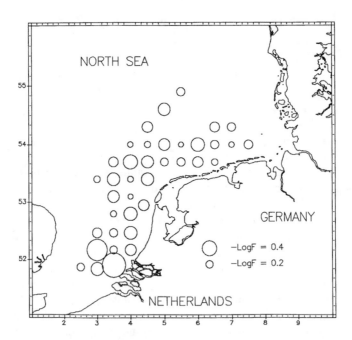

Figure 34.3   Spatial distribution of DNA damage indicated by increased –log F values in dab (*Limanda limanda*) embryos from the North Sea in March 1993. (–log F is indicated by diameter of circles.)

southern part of the North Sea. They were caught by means of a plankton net just below the sea surface. They were sorted by species. A set of 200 dab (*Limanda limanda*) embryos were pooled and homogenized in 500 μL of homogenization buffer. Duplicate samples were investigated by the unwinding assay, and –logF was determined according to the above given protocol. The procedure of Link and Wallace was applied to analyze the statistical significance of variation between sampling sites, which is suitable for the use of duplicates only, employing the range of data as measure of dispersion.[39-41] Results are summarized in Figure 34.3. As can be seen, samples showed a distinct spatial pattern in terms of DNA fragmentation, and from 38 sampling stations low values between 0.14 and 0.18 were found at 9 locations and high values above 0.30 at 13 locations with a maximum of 0.65 at the Rhine estuary.

From these numbers a careful estimate can be given of the number of DNA lesions per cell, if the assumption is made that the lowest observed –log F values represent embryos without DNA damage. Kanter and Schwartz estimated the average molecular weight of the alkaline unwinding unit in mammalian cells at 4.2×10^8 dalton.[14] Since –logF is inversely proportional to the average molecular weight,[13] a value of 5.8×10^8 dalton can be estimated for fish embryos from comparisons with mammalian and fish samples.[30] With a roughly estimated total of nuclear DNA in fish cells of 10^{12} dalton (20% of the mammalian genome),[42] the nuclear genome in the dab embryos contains approximately 1700 unwinding units. Taking these numbers into account and using formula (34.2), –logF values of 0.3 and 0.65 would represent DNA damage rates of 1900 lesions per cell with n = 1.14, and 6200 lesions per cell with n = 3.6.

It became clear from studies with mussels, that in some invertebrate species the average molecular weight of the alkaline unwinding unit is much lower than in many fish or mammalian cell and tissue types.[29-31,43] DNA fragmentation varies with tissue types and can vary with changing physiological conditions when genotoxic impact is not apparent.[29-31] From the above calculated numbers it is clear that the major part cannot be strand ends of chromosome-sized DNA molecules.[44] Spontaneous DNA lesions also cannot account for the majority of unwinding points.[45] However, it has theoretically been argued that the topoisomerase cleavage sites could constitute most of the

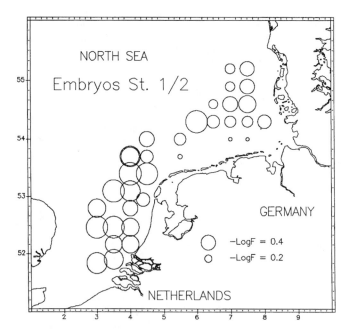

Figure 34.4   Spatial distribution of DNA damage indicated by increased –log F values in dab (*Limanda limanda*) embryos of developmental stages I and II from the North Sea in March 1995. (–log F is indicated by diameter of circles.)

unwinding points.[29-31] There are now strong experimental indications that topoisomerase action is part of transcriptional processes.[46-52] For the mussel *Mytilus edulis* it was recently demonstrated by investigation of naturally covalently DNA-binding proteins and inhibitor studies with camptothecin that topoisomerases may bind and cleave the DNA with a four times higher density in this organism as compared to mammalian cells (unpublished data).

Transcription related topoisomerase-like cleavages also occur in high densities in sea urchin embryos and vary with developmental stages.[53] A survey was made during a sea cruise in March 1995 in the southern North Sea similar to the one in 1993. Differently, however, dab embryos were sorted according to their developmental stage. The purpose of this investigation was to determine, whether DNA cleavages related to developmental physiology may interfere with the determination of DNA damage caused by genotoxic impact. In order to obtain enough DNA from the samples, embryos of the stages 1 and 2, as well as 3 and 4, respectively, were pooled. Developmental stages were classified according to Cameron et al.[54]

Data for DNA fragmentation are summarized in Figures 34.4 and 34.5. As in the study from 1993, lowest observed values for the DNA fragmentation parameter were about 0.1 and maxima at 0.6. Application of the Link and Wallace procedure to investigate the significance of observed differences indicates high significance with probabilities of error (P) clearly below 0.01, both for early and late developmental stages. On average the DNA fragmentation was found to be a little lower in the late stages (mean = 0.353), when compared to early stages (mean = 0.415). However, as the similar spatial patterns in Figures 34.4 and 34.5 suggest, DNA fragmentation highly correlates (r = 0.796) between early and late stages. In contrast to the study from 1993, however, embryos with a high degree of DNA damage were found almost all over the investigated area. Exceptions were sampling sites in the vicinity of the West and East Frisian Islands, where data suggest low or absent genotoxic impacts.

Careful comparisons between embryonic and adult dab can be drawn from investigations on DNA fragmentation in liver from adult dab (*Limanda limanda*) during sea cruises in the North Sea in 1993 and 1995. Liver was chosen for investigation in adult fish, because it serves as the main

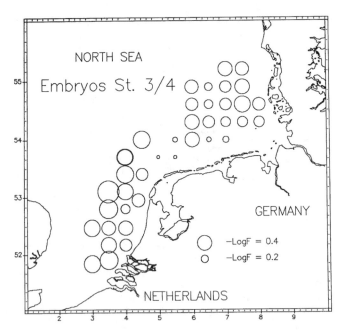

Figure 34.5   Spatial distribution of DNA damage indicated by increased –log F values in dab (*Limanda limanda*) embryos of developmental stages III and IV from the North Sea in March 1995. (–log F is indicated by diameter of circles.)

fat deposit in this taxon and DNA from this organ may be mostly affected by lipophilic genotoxic pollutants, when compared to other organs. Samples from adult fish in March 1993 showed a distinct spatial pattern in terms of DNA fragmentation, with peak values significantly increased over others. Analysis of variance indicated differences between sampling sites ($p < 0.0001$). In adult fish, the minimum for –logF was found to be 0.32 (SEM = 0.01). Maximum values ranged from 0.40 (SEM = 0.02) to 0.44 (0.06) at four sampling sites. In the study from January 1995, analysis of variance again indicated differences between sampling sites ($p < 0.0001$). This time, the means of –logF from 9 sampling sites ranged from 0.26 (SEM = 0.01) to 0.43 (0.02). Taking the lowest observed values as references, exogenously induced maximum DNA lesion rates can be estimated to be about 1000. This compares to peak values in embryos of 6000 and more, as has been calculated above.

In an accompanying series of experiments, the possible impact of solar UV radiation on DNA integrity in fish embryos was investigated. For this purpose, fish embryos were exposed to artificial sunlight generated by a sunshine simulation apparatus which was built at the Alfred-Wegener-Institut in Bremerhaven (Germany). Figures 34.6 through 34.8 summarize some experimental results, which are typical for a whole series. Figure 34.6 gives the range of data for control and exposed embryos, each pair collected at different sampling sites. Variation of exposure time is given in the figure legend. In all experiments, spectral characteristics of sunlight from Orlando, Florida, were simulated with 100% intact atmospheric ozone layer.

Figure 34.7 summarizes a time series experiment with dab embryos, and Figure 34.8 shows a parallel experiment with plaice (*Pleuronectes platessa*) embryos. Again, sunlight's spectral characteristics from Orlando were simulated, however, with 50% ozone depletion. In summary, there was no significant detectable radiation effect on the DNA integrity, either in embryos caught at sampling sites with low background fragmentation or with high background fragmentation. The reduction of –log F in dab embryos after a 10-h exposure may be caused by acute toxicity, i.e., on topoisomerase activity, which is similarly known from experiments with mussels (unpublished data).

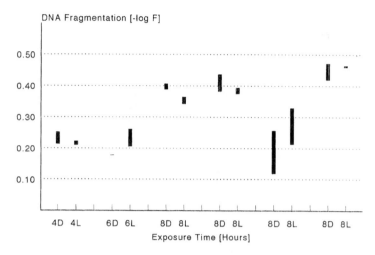

Figure 34.6   DNA fragmentation in dab (*Limanda limanda*) embryos exposed to simulated sunlight (X axis labels: number — hours of exposure/D — dark exposure [controls]/L — light exposure).

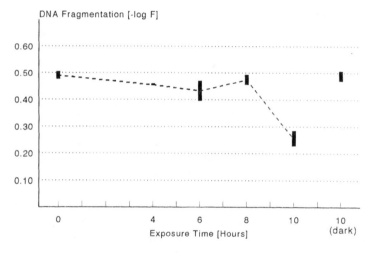

Figure 34.7   DNA fragmentation in dab (*Limanda limanda*) embryos exposed to simulated sunlight: time-course study.

IV. DISCUSSION AND PERSPECTIVES

Genotoxicity in general and genotoxicity in embryos, especially, is toxicity with some unique characteristics. First, genotoxicity may not be identified when it is accompanied by severe acute toxicity or coincides with acute toxic impact. On the other hand, severe genotoxic impact may not be readily recognizable when it is not associated with acute toxicity, and it will establish its adverse effects with a time delay, e.g., during carcinogenesis.[55,56] As other toxicants, genotoxic agents interact trivially with biological systems (cells, tissues, organs, individuals, populations, communities) on the molecular level. However, whereas many of the nongenotoxic toxicants form weak chemical interactions, following rules of receptor-ligand binding and the law of mass action, genotoxic agents change more or less directly the covalent structures of the target molecule, i.e., the DNA. As a result, effects are not fully reversible after removal of the genotoxic agent, even

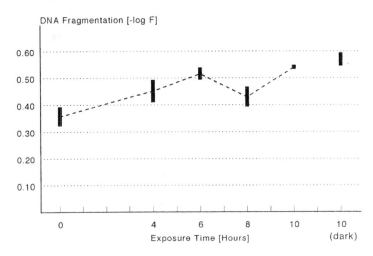

Figure 34.8   DNA fragmentation in plaice (*Pleuronectes platessa*) embryos exposed to simulated sunlight: time-course study.

despite sophisticated repair mechanisms,[1] and threshold doses appear not to exist.[57,58] In addition to these particular qualities of genotoxic impact, there appear to be increased genotoxic risks to fish embryos when compared to adult fish. First, taking into account that most, i.e., ~90%, of an individual's genes are realized during embryogenesis, the mere stochastic risk of being harmfully hit by genotoxic impact is about tenfold higher in embryos.[37] Second, the actual rate of DNA lesions appears to be higher in embryos as compared, for example, to liver DNA lesions in adult fish.

Similarly high incidences have also been reported in the Baltic Sea, where DNA damage has been investigated in cod (*Gadus morhua*) embryos and larvae by means of ^{32}P-postlabeling analysis.[59] This study reports 2.3 nmol of adducts/mol nucleotides in embryos and larvae from the Baltic Sea, compared to 0.12 nmol of adducts/mol nucleotides in samples from the Barents Sea (reference site). This dramatically increased rate of DNA damage in embryos and larvae from the Baltic Sea was associated with highly increased mortality and malformation incidence in *in vitro* fertilized cod embryos with parental fish from the same regions.[60]

The last example illustrates the possible consequences of genotoxic impact on embryos. It has been argued that complex adverse effects are more likely the result rather than a single distant endpoint of, for example, carcinogenesis in an individual's life.[33,61] These adverse effects can be summarized by the parameter of genetic fitness,[62,63] which is expected to be reduced during and after genotoxic impact. This could be expressed as reduced viability and fertility on the individual's level in the hierarchy of biological organization. Genetic fitness can also be defined on population and community (ecosystem) levels. On the population level, genotoxic impact could cause reduction of abundance or biomass production; on the community level there could be a reduction of species diversity, which links the term "loss of biological information" between community and molecular levels. Future research will require that investigations cover studies both at the molecular level on DNA damage and — in parallel — parameters of genetic fitness on the individual, population, and community levels. The alkaline DNA unwinding/hydroxylapatite batch elution assay, with its many advantages, could contribute to such an approach.

V. SUMMARY

DNA damage in fish embryos can be investigated by an alkaline DNA unwinding/hydroxylapatite batch elution, which provides a sum parameter for a wide spectrum of lesions with sensitivity to alkaline conditions. A detailed experimental protocol is given. Field studies on DNA damage in dab (*Limanda limanda*) embryos from the North Sea in 1993 and 1995 are summarized. They indicate extremely high genotoxic impacts at some sampling sites. Laboratory studies with fish embryos in a sunshine simulation apparatus do not cause UV-related DNA damage, which is detectable by the applied assay. Possible consequences of DNA damage in embryos on population and community levels are discussed. A combined approach to investigate DNA damage on the molecular level and genetic fitness on higher levels of biological organization is suggested.

ACKNOWLEDGMENTS

All sampling and laboratory work was carried out on board the FFS *Walther Herwig* of the German Bundesforschungsanstalt für Fischerei. Embryo sampling and characterization were supervised by Dr. V. Dethlefsen, Institut für Fischereiökologie of the Bundesforschungsanstalt für Fischerei, Cuxhaven, and Dr. H. von Westernhagen, Biologische Anstalt Helgoland, Hamburg. The sunshine simulation experiments were carried out by Dr. von Westernhagen, U. Hoge, and M. Kruse, the latter from the Alfred-Wegener-Institut, Bremerhaven, where the simulation apparatus was built. D. Warnecke, H.C. da Silva de Assis, and A. Sturm kindly assisted in laboratory work.

The investigations were funded as part of the project BIOMAR by the European Union.

REFERENCES

1. Loeb, L. A. and Cheng, K. C., Errors in DNA synthesis: A source of spontaneous mutations, *Mutation Research,* 238, 297, 1990.
2. Dunn, B. P., Black, J. J., and Maccubbin, A., ^{32}P-Postlabeling analysis of aromatic DNA adducts in fish from polluted areas, *Cancer Res.,* 47, 6543, 1987.
3. Kurelec, B., Chacko, M., and Gupta, R. C., Postlabeling analysis of carcinogen-DNA adducts in mussel, *Mytilus galloprovincialis, Mar. Environ. Res.,* 24, 317, 1988.
4. Varanasi, U., Reichert, W. L., and Stein, J. E., ^{32}P-Postlabeling analysis of DNA adducts in liver of wild English sole (*Parophrys vetulus*) and winter flounder (*Pseudopleuronectes americanus*), *Cancer Res.,* 49, 1171, 1989.
5. Kohn, K. W., The significance of DNA-damage assays in toxicity and carcinogenicity assessment, *Ann. New York Acad. Sci.,* 407, 106, 1983.
6. McGrath, R. A. and Williams, R. W., Reconstruction *in vivo* of irradiated *Escherichia coli* deoxyribonucleic acid; the rejoining of broken pieces, *Nature,* 212, 534, 1967.
7. Lett, J. T., Caldwell, I., Dean, C. J., and Alexander, P., Rejoining of X-ray induced breaks in the DNA of leukemia cells, *Nature,* 214, 790, 1967.
8. Cook, P. R. and Brazell, I. A., Detection and repair of single-strand breaks in nuclear DNA, *Nature,* 263, 679, 1976.
9. Kohn, K. W. and Grimek-Ewig, R. A., Alkaline elution analysis, a new approach to the study of DNA single-strand interruptions in cells, *Cancer Res.,* 33, 1849, 1973.
10. Petzold, G. L. and Swenberg, J. A., Detection of DNA damage induced *in vivo* following exposure of rats to carcinogens, *Cancer Res.,* 38, 1589, 1978.
11. Eastman, A. and Bresnick, E., A technique for the measurement of breakage and repair of DNA alkylated *in vivo, Chem.-Biol. Interactions,* 23, 369, 1978.
12. Ahnström, G. and Erixon, K., Radiation induced strand breakage in DNA from mammalian cells. Strand separation in alkaline solution, *Int. J. Radiat. Biol.,* 23, 285, 1973.
13. Rydberg, B., The rate of strand separation in alkali of DNA of irradiated mammalian cells, *Radiat. Res.,* 61, 274, 1975.
14. Kanter, P. M. and Schwartz, H. S., A hydroxylapatite batch assay for quantitation of cellular DNA damage, *Anal. Biochem.,* 97, 77, 1979.
15. Birnboim, H. C. and Jevcak, J. J., Fluorometric method for rapid detection of DNA strand breaks in human white blood cells produced by low doses of radiation, *Cancer Res.,* 41, 1889, 1981.

16. Kanter, P. M. and Schwartz, H. S., A fluorescence enhancement assay for cellular DNA damage, *Molec. Pharmacol.*, 22, 145, 1982.

17. Storer, R. D. and Conolly, R. B., An *in vivo* — *in vitro* alkaline DNA unwinding assay for hepatic DNA damage: Comparison with the alkaline sucrose gradient centrifugation technique, *Anal. Biochem.*, 142, 351, 1984.

18. Daniel, F. B., Haas, D. L., and Pyle, S. M., Quantitation of chemically induced DNA strand breaks in human cells via an alkaline unwinding assay, *Anal. Biochem.*, 144, 390, 1985.

19. Parodi, S., Carlo, P., Martelli, A., Taningher, M., Finollo, R., Pala, M., and Giaretti, W., A circular channel crucible oscillating viscosimeter. Detection of DNA damage induced *in vivo* by exceedingly small doses of dimethylnitrosamine, *J. Mol. Biol.*, 147, 501, 1981.

20. Brambilla, G., Carlo, P., Finollo, R., Bignone, F. A., Ledda, A., and Cajelli, E., Viscometric detection of liver DNA fragmentation in rats treated with minimal doses of chemical carcinogens, *Cancer Res.*, 43, 202, 1983.

21. Zahn, R. K., Zahn-Daimler, G., and Beyer, R., Highly protective alkalinization by ammonia vapor diffusion in viscosimetric DNA damage assessment, *Anal. Biochem.*, 168, 387, 1988.

22. Freeman, S. E., Blackett, A. D., Monteleone, D. C., Setlow, R. B., Sutherland, B. M., and Sutherland, J. C., Quantitation of radiation, chemical, or enzyme-induced single strand breaks in nonradioactive DNA by alkaline gel electrophoresis: Application to pyrimidine dimers, *Anal. Biochem.*, 158, 119, 1986.

23. Sutherland, J. C., Monteleone, D. C., Mugavero, J. H., and Trunk, J., Unidirectional pulsed-field electrophoresis of single- and double-stranded DNA in agarose gels: Analytical expressions relating mobility and molecular length and their application in the measurements of strand breaks, *Anal. Biochem.*, 162, 511, 1987.

24. Sutherland, J. C., Lin, B., Monteleone, D. C., Mugavero, J. H., Sutherland, B. M., and Trunk, J., Electronic imaging system for direct and rapid quantitation of fluorescence from electrophoretic gels: Application to ethidium bromide-stained DNA, *Anal. Biochem.*, 163, 446, 1987.

25. Singh, N. P., McCoy, M. T., Tice, R. R., and Schneider, E. L., A simple technique for quantitation of low levels of DNA damage in individual cells, *Exp. Cell Res.*, 175, 184, 1988.

26. Olive, P. L., Banáth, J. P., and Durand, R. E., Heterogeneity in radiation-induced DNA damage and repair in tumour and normal cells measured using the "Comet" assay, *Radiation Res.*, 122, 86, 1990.

27. Rydberg, B. and Johanson, K. J., Radiation-induced DNA strand breaks and their rejoining in crypt in villous cells of the small intestine of the mouse, *Radiat. Res.*, 64, 282, 1975.

28. Cesarone, C. F., Bolognesi, C., and Santi, L., Improved microfluorometric DNA determination in biological material using 33258 Hoechst, *Anal. Biochem.*, 100, 188, 1979

29. Herbert, A., Genotoxische und gen-physiologische Veränderungen bei *Mytilus galloprovincialis* und anderen wasserlebenden Tieren als Folge veränderter Umweltbedingungen und Schadstoffeinwirkung, Ph.D. Thesis, Johannes-Gutenberg-Universität, Mainz (Germany), 1987.

30. Herbert, A. and Zahn, R. K., Monitoring DNA damage in *Mytilus galloprovincialis* and other aquatic animals: I. Basic studies with a DNA unwinding technique, *Angew. Zoologie,* 76, 143, 1989.

31. Herbert, A. and Zahn, R. K., Monitoring DNA damage in *Mytilus galloprovincialis* and other aquatic animals: II. Pollution effects on DNA denaturation characteristics, *Angew. Zoologie,* 77, 13, 1990.

32. Herbert, A., Monitoring DNA damage in *Mytilus galloprovincialis* and other aquatic animals: III. A case study: DNA damage in fish from a Florida marsh, *Angew. Zoologie,* 77, 143, 1990.

33. Herbert, A. and Hansen, P.-D., Erfassung des erbgutverändenden Potentials von Gewässern durch Messung von DNS-Schäden mittels alkalischer Denaturierungsverfahren, in *Biologische Testverfahren*, Steinhäuser, K. G. and Hansen, P. — D., Eds., Gustav-Fischer-Verlag, Stuttgart/New York, 1992, 745.

34. Morgan, T. H., Embryology and genetics, Columbia University Press, New York, 1934.

35. Allen, G. E. and Thomas Hunt Morgan — The man and his science, Princeton University Press, Princeton, NJ, 1978, 300.

36. Ohno, S., The total number of genes in the mammalian genome, *Trends Genet.*, 2, 8, 1986.

37. Galau, A. G., Klein, W. H., Davis, M. M., Wold, B. J., Britton, R. J., and Davidson, E. H., Structural gene sets active in embryos and adult tissues of the sea urchin, *Cell,* 7, 487, 1976.

38. Cameron, P., Berg, J., Dethlefsen, V., and von Westernhagen, H., Developmental defects in pelagic embryos of several flatfish species in the southern North Sea, *Netherlands Journal of Sea Research*, 29, 239, 1992.

39. Kurtz, T. E., Link, R. F., Tukey, J. W., and Wallace, D. L., Short-cut multiple comparisons for balanced single and double classifications: part 1, results, *Technometrics*, 7, 95, 1965.

40. Kurtz, T. E., Link, R. F., Tukey, J. W., and Wallace, D. L., Short-cut multiple comparisons for balanced single and double classifications: part 2, derivations and approximations, *Biometrika*, 52, 485, 1965.

41. Sachs, L., Angewandte Statistik: statistische Methoden und ihre Anwendung, Springer Verlag, Berlin, 1978, 416.

42. Ohno, S., Protochordata, cyclostomata and pisces, in John, B., Bauer, H., Brow, S., Kayano, H., Levan, A., and White, M., Eds., *Animal Cytogenetics, Vol. 4,Chordata 1*, Berlin, 1974.

43. Accomando, R., Viarengo, A., Bordone, R., Taningher, M., Canesi, L., and Orunesu, M., A rapid method for detecting DNA strand breaks in *Mytilus galloprovincialis* Lam. induced by genotoxic xenobiotic chemicals, *Int. J. Biochem.*, 23, 227, 1991.

44. Huberman, J. A. and Riggs, A. D., Autoradiography of chromosomal DNA fibers from chinese hamster cells, *Proc. Natl. Acad. Sci. USA*, 55, 599, 1966.

45. Zahn, R. K., *Über physische Umweltrisiken*, Abhandlungen der Akademie der Wissenschaften und der Literatur, Mainz-Mathematisch-Naturwissenschaftliche Klasse, Franz-Steiner Verlag, Stuttgart, 1985.

46. Fleischmann, G., Pflugfelder, G., Steiner, E. K., Javaherian, K., Howard, G. C., Wang, J. C., and Elgin, S. C. R., *Drosophila* DNA topoisomerase I is associated with transcriptionally active regions of the genome, *Proc. Natl. Acad. Sci. USA*, 81, 6958, 1984.

47. Muller, M. T., Pfund, W. P., Mehta, V. B., and Trask, D. K., Eukaryotic type I topoisomerase is enriched in the nucleolus and catalytically active on ribosomal DNA, *EMBO J.*, 4, 1237, 1985.

48. Bonven, B. J., Gocke, E., and Westergaard, O., A high affinity topoisomerase I binding sequence is clustered at DNAase I hypersensitive sites in *Tetrahymena* R-chromatin, *Cell*, 41, 541, 1985.

49. Gilmour, D. S., Pflugfelder, G., Wang, J. C., and Lis, J. T., Topoisomerase I interacts with transcribed regions in *Drosophila* cells, *Cell*, 44, 401, 1986.

50. Stewart, A. F. and Schütz, G., Camptothecin-induced *in vivo* topoisomerase I cleavages in the transcriptionally active tyrosine aminotransferase gene, *Cell*, 50, 1109, 1987.

51. Zhang, H., Wang, J. C., and Liu, L. F., Involvement of DNA topoisomerase I in transcription of human ribosomal RNA genes. *Proc. Natl. Acad. Sci. USA*, 85, 1060, 1988.

52. Culotta, V. and Sollner-Webb, B., Sites of topoisomerase I action on *X. laevis* ribosomal chromatin: Transcriptionally active rDNA has an 200 bp repeating structure, *Cell*, 52, 585, 1988.

53. Case, S. T., Mongeon, R. L., and Baker, R. F., Single-stranded regions in DNA isolated from different developmental stages of the sea urchin, *Biochim. Biophys. Acta*, 349, 1, 1974.

54. Cameron, P., Berg, J., Dethlefsen, V., and von Westernhagen, H., Mißbildungen bei Fischembryonen der südlichen Nordsee, in *Warnsignale aus der Nordsee*, Lozan, J.L., Lenz, W., Rachor, E., Watermann, B., and von Westernhagen, H., Eds., Verlag Paul Parey, Berlin/Hamburg, 1990, 281.

55. Barrett, J. C. and Ts'o, P. O. P., Relationship between somatic mutation and neoplastic transformation, *Proc. Natl. Acad. Sci. USA*, 75, 3297, 1978.

56. Trosko, J. E. and Chang, C.-C., Genes, pollutants and human diseases, *Quarterly Reviews of Biophysics*, 11, 603, 1978.

57. Hagen, U., Genetische Wirkungen kleiner Strahlendosen, *Naturwissenschaften*, 74, 3, 1987.

58. Johannsen, F. R., Risk assessment of carcinogenic and noncarcinogenic chemicals, *CRC Crit. Rev. Toxicol.*, 20, 341, 1990.

59. Ericson, G., Åkerman, G., Liewenborg, B., and Balk, L., Comparison of DNA damage in the early life stages of cod, *Gadus morhua*, originating from Barents Sea and Baltic Sea, *Mar. Environ. Res.*, 42, 119, 1996.

60. Åkerman, G., Tjärnlund, U., Broman, D., Näf, C., Westin, L., and Balk, L., Comparison of reproductive success of cod, *Gadus morhua*, from the Barents Sea and Baltic Sea, *Mar. Environ. Res.*, 42,139, 1996.

61. Kurelec, B., The genotoxic disease syndrome, *Mar. Environ. Res.*, 35, 341, 1993.

62. Wilson, E. O. and Bossert, W.H., Population genetics, in *A Primer of Population Biology*, Sinauer Associates, Inc. Publisher, Connecticut, USA, 1971, Chap. 2.

63. Ayala, F.J. and Kiger, Jr., J.A., Natural selection, in *Modern Genetics*, The Benjamin/Cummings Publishing Company, Menlo Park, CA, 1980, Chap. 20.

CHAPTER 35

Genotoxicity Tests in Amphibians — A Review

V. Ferrier, L. Gauthier, C. Zoll-Moreux, and J. L'Haridon

CONTENTS

I. INTRODUCTION: *IN VITRO* AND *IN VIVO* GENOTOXICITY TESTS — PERSPECTIVES

Over the last decade, genotoxicity in inland waters (natural water, drinking water, industrial and urban wastewater) mostly has been evaluated using *in vitro* bioassays, the most widely employed being short-term bacterial tests such as those of Ames[1-2] and the SOS Chromotest.[3] Tests based on the observation of sister chromatid exchanges (SCE), chromosomal aberrations (CA), or micronuclei (MN) in the cells of various species of fish and mammals also have been employed, albeit to a lesser extent.

We will not present a detailed review of these tests here (instead, see Chapter 34; a good review is that of Godet and Vasseur[4]). Nevertheless, some of the features of the *in vitro* tests, especially those involving bacteria, are worth noting. The obvious advantages of the *in vitro* tests are that they are inexpensive, quick and convenient, and their procedures can be readily scaled down, standardized, and automated. For these reasons, such tests are in routine use, at least for primary screening purposes. They do, however, have certain drawbacks for applications in genetic ecotoxicology. They can only detect the genotoxic effects of low levels of micropollutants after concentration of samples or after some form of extraction process. Concentration and extraction are time-consuming and costly processes and do not recover all micropollutants (10 to 30% in the study of Wilcox et al.[5]). They also do not take into account the synergistic or antagonistic actions between different pollutants in a complex medium. Furthermore, even when applied to raw samples, they do not take into account bioavailability, bioaccumulation, excretion, and metabolism — processes which are characteristic of whole organisms, especially in the higher species. Moreover, in contrast to prokaryotic species, aquatic vertebrates can reveal the toxic effects of nonsoluble chemicals (that may damage gills and can also be absorbed by the digestive tract after being solubilized by digestive processes). Another problem is that sterilization of the aqueous samples before addition to the bacterial culture medium may remove pollutants adsorbed onto suspended matter in solution (sterilization by filtration) or may modify the structure of the pollutants (sterilization by gamma radiation). These processes pose problems for the representativeness of the *in vitro* data in relation to the true genotoxic impact of the aqueous contaminants on the environment.

In attempts to circumvent these difficulties, genotoxicity tests using whole aquatic organisms (*in vivo* tests) have been developed based on observation of SCE, CA, and MN in the cells of a wide variety of species. Some work has been carried out on flies and worms, although most of the studies over the last decade have been devoted to fish[6-7] and amphibians.

In amphibians, the genotoxicity tests so far developed are all based on the observation of micronuclei *in vivo*. In eukaryotes, chromosomal or genomic mutations can give rise to the formation of small intracytoplasmic masses of chromatin or micronuclei. They consist essentially of chromosomal fragments or whole chromosomes which have not migrated to poles during mitosis. They result either from chromosome breaks (clastogenic effects) or dysfunction of the spindle apparatus or centromere kinetochore complexes leading to the elimination of whole chromosomes (aneugenic effects). Exposure of organisms to clastogens, spindle poisons, and ionizing radiation will tend to increase the number of micronuclei in cells. The erythrocytes (RBCs) of lower vertebrates, such as amphibians, are nucleated and undergo cell division in circulating blood, especially during the larval stages (e.g., 10% of mitoses in the larvae of the urodele amphibian *Pleurodeles waltl*[8]). These cells are thus readily accessible to experimentation and observation.

The initial experiments were carried out over 10 years ago on tadpoles of two species of Anuran — *Rana catesbeiana*[9] and *Caudiverbera caudiverbera*[10] — after exposure to gamma radiation. A direct relationship was observed between the dose of radiation and the number of micronucleated RBCs.

II. THE AMPHIBIAN MICRONUCLEUS TEST

The most significant results to date have been obtained at the Centre de Biologie du Développement in Toulouse, France, where the amphibian micronucleus test has been validated and standardized.[11] This micronucleus test was first developed on the aquatic larvae of the newt *Pleurodeles waltl* (Pleurodele)[12] and the Axolotl *Ambystoma mexicanum*.[13] It has subsequently been extended to the tadpole of the African toad *Xenopus laevis*.[14,15]

A. Principle of the Micronucleus Test

The principle of the test is simple, and the detailed procedure for the newt is laid down in the French Standard AFNOR T90-325, where it is referred to as the newt micronucleus test[11] (or Jaylet test, after its inventor). Newt larvae are reared for 12 d either in the presence of known concentrations of one or more chemicals, or in raw or diluted samples of water (e.g., industrial or urban effluents, drinking water or samples taken during treatment, inland waters, aqueous extracts of industrial wastes or contaminated soils). The maximum toxic concentrations in the test medium compatible with survival over the duration of the test are determined in a preliminary acute toxicity study. At the end of the exposure period, blood samples are removed by cardiac puncture. Blood smears on microscope slides are then stained and examined under the light microscope. The levels of micronucleated RBCs per thousand (Ω) are counted and compared to levels observed in animals reared in pure water (negative control) or in water containing a reference genotoxic agent such as cyclo-phosphamide or BaP (positive control). If there is a statistically significant[16] elevation in the Ω of micronucleated RBCs with respect to the negative controls, the test medium is assumed to contain a genotoxic agent. The procedure for the African toad is similar to that for the newt except for a few minor changes (e.g., temperature of water, diet of test animals).

B. Studies with X-rays and Chemicals

Siboulet et al.[12] and Grinfeld[17] provided the first demonstration of the induction of micronuclei in the RBCs of newt larvae exposed to various doses of X-rays and a variety of chemical compounds. In this test system, relatively low doses (6 rads) were found to give rise to significant effects, and the dose-response curve was linear up to a dose of 150 rads (maximal response reached at 600 rads). Since these first studies, 78 chemical compounds have now been examined in the newt micronucleus test, of which 44 were found to be genotoxic. Fifteen substances have also been tested in the *Xenopus*, of which 9 have so far been found to be genotoxic.[4,15,18,28]

The compounds shown to be genotoxic in the newt or *Xenopus* belong to a variety of chemical classes: amines (aromatic, aliphatic), nitroso compounds, polycyclic aromatic hydrocarbons, N-, S-, O- mustards, aziridines, carbamates, organohalogenated derivatives, and heavy metals). It can be seen from the results summarized in Table 35.1 that the amphibian micronucleus test is highly sensitive. For certain worrisome contaminants of the environment, genotoxic effects were observed at low concentrations, within the range found in polluted waters.[29] Such compounds include BaP (0.0125 ppm in the newt), and methyl mercury or mercuric chloride (0.00125 ppm and 0.0025 ppm, respectively, in the African toad). A particularly interesting feature is that the larvae of amphibians have a high capacity for bioaccumulation of aquatic contaminants. For example, after exposure of newt larvae for 12 d, concentration factors of 1200 and 600 have been observed for CH_3HgCl and $HgCl_2$, respectively.[20] These results agree with those obtained by Grinfeld et al.[30] in the same species after a 12-h exposure to BaP (200-fold bioaccumulation). This capacity for bioaccumulation accounts, at least in part, for the ability of the amphibian micronucleus test to detect the genotoxic activity of micropollutants in natural raw water samples without the need for concentration (a normal procedure for the *in vitro* tests).

In most cases, dose-response relationships are observed (Figure 35.1).[18-28] Another advantage of the micronucleus amphibian test is that it accounts for antagonistic and synergistic effects, which are often encountered in complex mixtures. In a study of samples taken from a reservoir behind a river dam in France, where the water is considered to be rich in humic acids, Gauthier[31] observed a potentiating effect of this water on the genotoxicity of BaP. In contrast, addition of commercial humic acids to reconstituted water markedly reduced the genotoxicity of BaP (i.e., there was a masking action; Figure 35.2). In a similar study of an effluent enriched in metals (iron, chromium, zinc), Godet and Vasseur[4] investigated interactions between the three metals. They observed a genotoxicity effect of FeIII ($FeCl_3$) at 0.6 mg/L in the presence of CrVI ($K_2Cr_2O_7$) at 1 mg/L in the test medium. This indicated that iron, which is not generally regarded as an environmental hazard (French standard for discharge 5 mg/L), may become so when accompanied by hexavalent chromium.

There have been relatively few studies to date comparing the performance of the amphibian micronucleus test to that of the *in vitro* bacterial tests widely used in the evaluation of the geno-toxicity of pure substances (SOS Chromotest and Ames fluctuation test). For example, Lecurieux et al.[25] examined the activity of 7 chemicals (4-nitroquinoline 1-oxide, potassium bichromate, formaldehyde, sodium hypochlorite, benzo(a)pyrene, cyclophosphamide, and 2-naphthylamine) in all three tests. They found that: 1) none of the three tests detected genotoxicity in all compounds; 2) the Ames fluctuation test appeared to be most suited for detection of direct genotoxic effects; and 3) the newt micronucleus test was most sensitive for compounds with indirect genotoxic action. Godet and Vasseur[4] demonstrated the superiority of the newt micronucleus test over the Ames test

Table 35.1   Positive Genotoxicity Results Obtained with Some Compounds in *Pleurodeles waltl* and *Xenopus laevis*

A- *Pleurodeles waltl*	Concentrations (ppm)	References
acridine orange	0.2	24
benz(a) anthracene	0.187	22
benzo(a) pyrene	0.0125	18
bromodichloromethane	25	26
bromoform	2.5	34
caffeine	100	24
ε-caprolactam	100	21
captan	0.125	24
carbaryl	2.5	24
cyclophosphamide monohydrate	0.5	21
dibromoacetonitrile	0.00012	26
dibromochloromethane	25	26
dichloroacetonitrile	0.00025	26
dichloropropanone	0.000025	26
diethylsulfate	6	18
7,12-dimethyl-benz(a) anthracene	0.0125	22
epichlorhydrin	1	13
ethidium bromide	25	21
ethylene imine	0.5	24
ethylmethane sulfonate	10	13
ethylenedibromide	1	21
N-ethyl-N'-nitro-N-nitrosoguanidine	0.4	18
N-ethyl-N-nitrosourea	1	21
ferric chloride	0.6	4
hexamethylphosphoramide	30	21
mercuric chloride	0.012	20
3-methylcholanthrene	0.5	24
methylmercuric chloride	0.012	20
monobromoacetonitrile	0.00125	26
monochloramine	0.15	33
monochloroacetonitrile	0.000125	26
naphthalene	0.25	27
2-naphthylamine	0.0025	25
4-nitroquinoline 1-oxide	0.000002	25
N-nitrosoatrazine	7.5	23
N-nitrosocarbaryl	0.125	24
N-nitrosodiethanolamine	12.5	23
potassium dichromate	10	28
pyrene	0.175	21
sodium hypochlorite	0.525	33
o-toluidine	4	21
trichloroacetaldehyde hydrate	200	24
trichloroacetonitrile	0.0001	26
1-1-3-trichloropropanone	0.001	26
B-*Xenopus laevis*		
benzo(a) pyrene	0.05	15
cadmium chloride	2	28
cyclophosphamide monohydrate	5	15
mercuric chloride	0.00125	15
methyl benzimidazole carbamate	0.025	28
methyl mercuric chloride	0.0025	15
potassium dichromate	10	28

for detecting the genotoxicity of certain metals. In their study, mentioned above, only the newt micronucleus test detected the additive effect of chromium on the genotoxicity of iron, which produced negative results in the Ames test.

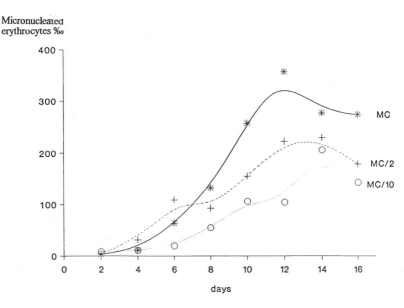

Figure 35.1   Genotoxic effects of benzo(a) pyrene in the newt *Pleurodeles waltl* in accordance with duration of treatment and concentration. MC is the no toxic maximal concentration (0.5 ppm) (From Jaylet, A., Deparis, P., Ferrier, V., Grinfeld, S., and Siboulet, R., *Mutation Res.*, 164, 245, 1986. With permission.)

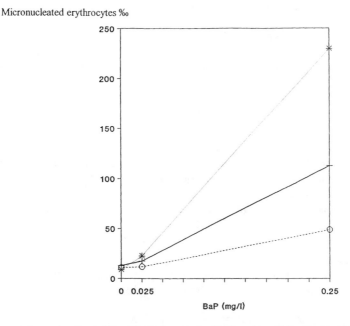

Figure 35.2   Synergistic and antagonistic effects revealed with the pleurodele micronucleus-test. * = dam water + DMSO + BaP; + = reconstituted water + DMSO + BaP; ○ = reconstituted water + DMSO + BaP + humic acids (Fluka) (From Gauthier, L., Rapport au Groupement d'Etudes Scientifiques du Test Micronoyau Triton (Gestmint), C.I.R.S.E.E., Ed., Le Pecq, France, 1992. With permission.)

Fernandez et al.[24] have compared the qualitative results for 47 chemicals and X-rays obtained with the newt micronucleus test to those of various short-term genotoxicity tests (Ames, CA, rodent micronucleus tests) and long-term carcinogenicity assays in rodents. Concordances between the newt micronucleus test and the other tests of 86% (Ames test), 83% (CA test), 71% (rodent micronucleus test), and 76% (rodent carcinogenicity assay) were found, which is indicative of the validity of the newt micronucleus test.

C. Studies *In Situ*

The high sensitivity of the amphibian micronucleus test means that it can be employed in ecotoxicology for the detection of genotoxicity in raw or diluted samples of complex aquatic media (e.g., stages in the production of drinking water, industrial and domestic wastes, inland waters).

1. Application to the Production of Drinking Water

After initial observations on the genotoxicity of certain samples of drinking water to newt larvae,[34] further experiments conducted in our laboratory in collaboration with private firms investigated the genotoxic impact of various processes, especially the chlorination used in the production of water for human consumption. It was shown that concentrations of chlorine of 0.125 ppm and 0.25 ppm (levels commonly encountered in drinking water) led to significant increases in the number of micronucleated RBCs in newt larvae.[35] Similar results were obtained with other substances used in the production of drinking water such as monochloramine (0.15 ppm chlorine) and even low concentrations of ozone under certain conditions. The genotoxic potential of certain by-products of chlorination produced by disinfection of water rich in organic matter (such as bromoform) has been identified in a study carried out by the Dutch Institute for Inland Water Management and Wastewater Treatment in Lelystad. Extracts removed from treated river water on an XAD4 column were found to be genotoxic in newt larvae. In this case, the amphibian micronucleus test was found to be 25-fold more sensitive than the Ames test.[36]

2. Application to Polluted Water

The amphibian micronucleus test shows most promise for *in vivo* evaluation of genotoxicity in polluted inland waters. In recent years, there has been increasing use of *in vivo* methods for testing industrial wastes, although relatively few species have so far been examined. The induction of chromosome aberrations (CA), sister chromatid exchanges (SCE), and micronuclei (MN) have been investigated (see review of Godet and Vasseur,[4] 15 cited references) in the following species: the mollusks *Mytilus galloprovincialis* and *M. edulis*, the fish *Umrea pygmea, Heterpneustes fossilis, Notobranchius rachowi,* and the amphibians *Pleurodeles waltl* and *Xenopus laevis*.

To date, the amphibian micronucleus test is the best validated *in vivo* genotoxicity test for industrial waste, both in terms of the number of effluents tested and the variety of industrial sites examined. Waste from more than 40 industrial sites, mostly in France, have been examined. They include chemical plants, metal extraction and fabrication industries, surface treatment works, paper mills, tanning factories, mines, urban and highway discharges. More than half of these sites produce genotoxic wastes.[4,28,37,38]

We will describe three typical examples of the application of the amphibian micronucleus test to industrial or urban polluted waters: effluents from tanning industries and coke works, and urban wastes.

a. Effluents from Tanning Factories

The water from the river Dadou (a tributary of the Tarn, France) sampled downstream from the central outflow from the tanning factories in the town of Graulhet was tested at two different times: in October 1985 and four years later in November 1989 after modernization of the tanning factories.[38] The contaminants of the river water stem from the substances used in the processing of animal hides and leather (sodium sulfide, solvents, natural and synthetic tannins, chromium salts, organic dyes, metal oxides). In the first test, samples of Dadou river water were found to be

Micronucleated
erythrocytes ‰

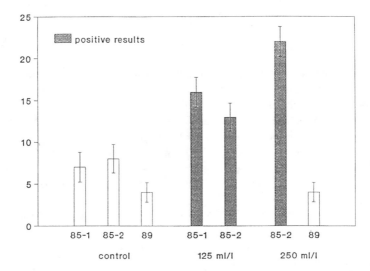

Figure 35.3   Induction of micronuclei in the newt *Pleurodeles waltl* by effluents from tanneries. (From Gauthier, L., Etude du pouvoir génotoxique des eaux de surface potables ou en cours de traitement, par la formation de micronoyau chez le triton *Pleurodeles waltl*, Thèse de Doctorat de l'Université Paul Sabatier, Toulouse, 1989. With permission.)

genotoxic in newt larvae at concentrations of 125 mL/L and 250 mL/L in the test medium (Figure 35.3) with a clear-cut dose-response relationship. In the tests carried out four years later after modernization of the factories (modification of the tanning process and treatment of wastewater prior to discharge), no genotoxic effects were observed at the doses used in the previous tests. This shows that the amphibian micronucleus test can be usefully employed to assess the efficiency of treatment of industrial effluents.

b. Effluent from a Coke Works

A coke works site, also in France, produces wastes rich in ammonia, phenols, chlorides, and HAP. The wastewater is subjected to biological treatment before discharge into the river, which markedly reduces the phenol index and the metal content, although the HAP and AOX contents are still elevated in the outflow. The newt micronucleus test was employed to evaluate the genotoxicity of the water before and after the biological treatment, and also to test the river water downstream of the discharge point.[28] The effluent prior to treatment was shown to be genotoxic, and some residual genotoxicity was also detected after treatment (Figure 35.4) despite the dilutions employed (up to 15.625 ml/L for untreated water samples). Dose-response relationships were observed for both the untreated and treated water samples. Similar results were obtained with the *Xenopus* micronucleus test.

c. Urban Effluents

Effluents were taken from the outflow of a wastewater treatment plant outside a French town of around 200,000 inhabitants. The effluent consists of both household and industrial (essentially smelting and metal working) wastewater. The *Xenopus* micronucleus test was used to evaluate genotoxicity in the outflow of the plant after biological treatment. Significant genotoxicity was observed even after a ¼ dilution (Figure 35.5). A dose-response relationship was observed, and similar results were obtained with the newt micronucleus test.[28]

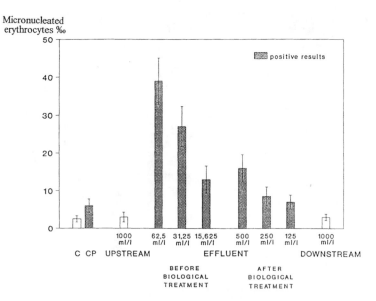

Figure 35.4   Genotoxic effects of an effluent from a cokery in the newt *Pleurodeles waltl.* C = negative controls; CF = positive controls (cyclophosphamide 2 ppm). (From Zoll-Moreux, C. and Ferrier, V., Etude comparative entre le test Jaylet (test micronoyau triton) et le test micronoyau xénope, Etudes Interagences, Agence de l'Eau et Ministère de l'Environnement, Paris, 1995 (in press). With permission.)

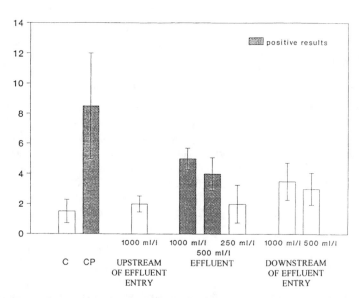

Figure 35.5   Genotoxic effects of an urban effluent in the toad *Xenopus laevis.* C = negative controls; CP = positive controls (cyclophosphamide 10 ppm) (From Zoll-Moreux, C. and Ferrier, V., Etude comparative entre le test Jaylet (test micronoyau triton) et le test micronoyau xénope, Etudes Interagences, Agence de l'Eau et Ministère de l'Environnement, Paris, 1995 (in press). With permission.)

3. Application to Industrial Wastes and Contaminated Soils

Part of ecotoxicity of industrial wastes and contaminated soils lies in the genotoxic impact on the aquatic environment (infiltration or runoff). Nearly all of the studies to date have employed *in vitro* bacterial tests, although in some recent trials, plant systems have been used (micronucleus and adduct tests). For such situations, we still do not have a validated *in vivo* method for evaluating genotoxicity based on an animal system.

A multicenter program initiated by the French Agency for the Environment and Energy Control (ADEME) is currently examining the amphibian micronucleus test for the evaluation of genotoxicity in aqueous extracts of various industrial wastes and contaminated soils. Extracts of soil contaminated with hydrocarbons (tanneries and coke works) obtained under standard conditions[39] have been shown to induce a significantly higher level of micronucleated RBCs in *Xenopus* larvae.[40] This opens new perspectives for the amphibian micronucleus test provided that the methods of aqueous extraction are optimized for the biological system employed. This test would thus appear to be well suited for evaluating the genotoxicity of water-extractable contaminants of solid matter.

4. Comparison of In Vivo Amphibian Micronucleus Test with Other In Vitro and In Vivo Genotoxicity Tests

Few studies comparing the performance (sensitivity, reliability, feasibility, cost) of the various currently available genotoxicity tests either *in vitro* or *in vivo* have been reported. One of the first studies was that of Van der Gaag[37] on the effluent from a treatment plant for wastewater from a petrochemical complex in Holland. Samples of wastewater were subjected to various genotoxicity tests: two *in vitro* bacterial tests (Ames and SOS Chromotest) and three *in vivo* tests (SCE in the fish *Notobranchus rachowi*, and the micronucleus tests in the mussel *Mytilus edulis* and the amphibian *Pleurodeles waltl*). The bacterial tests only gave positive results with extracts/concentrates obtained using XAD4 pH7/pH2 columns. Raw and diluted samples of the effluent, however, were found to be genotoxic in the three *in vivo* tests. Furthermore, the SCE-*Notobranchus* and newt micronucleus tests revealed marked residual genotoxicity in the final effluent of the XAD pH2 columns (end of extraction/concentration process; Figure 35.6). The authors estimated that 50% of the dissolved organic carbon and 30% of the AOX escaped the extraction/concentration process and were therefore thought to represent supports for the residual genotoxicity not detected by the *in vitro* tests. The *in vivo* tests, especially the newt micronucleus test, were considered to be more ecotoxicologically relevant than were the *in vitro* tests. In the same study, the authors also evaluated the feasibility and cost of the five tests (total duration of tests and hours of work involved). They found similar feasibility and costs, albeit with a slight advantage for the SCE-*Notobranchus* test in view of the shorter duration of exposure required (96 h).

Another study[4] compared the performance of the newt micronucleus and Ames tests at five sites in France: an urban water treatment plant (A), a surface treatment plant (B), a polystyrene production factory (C), paper pulp factory (D), and a plant for production of chlorinated derivatives (E). A wide variety of contaminants were produced including: metals (Cr, Fe, Zn) in effluent A, benzene derivatives (effluent B), chlorinated compounds (effluent D), and chloromethane and chloroethane (effluent E). The micronucleus test detected genotoxicity in effluents B (surface treatment) at concentrations of 2.5, 10, and 20%; effluent C (the polystyrene factory) at a concentration of 2.5%; and effluent E (chlorinated derivatives) at concentrations of 25 and 50%. It was negative with the effluent from the paper pulp factory. The Ames test was found to be less sensitive than the micronucleus test, as it only detected genotoxicity in effluents B (raw and freeze-dried samples) and D (organic extract).

Although definite conclusions about the respective performances of *in vitro* and *in vivo* genotoxicity tests cannot be drawn from the results of these two studies, they do provide support for

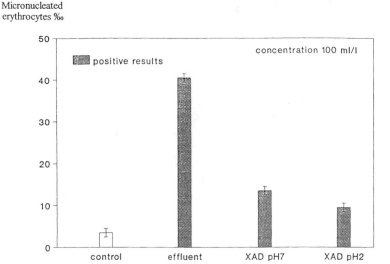

Micronucleated
erythrocytes ‰

Figure 35.6   Genotoxic effect of a petrochemical industry effluent in the newt *Pleurodeles waltl*. (From Gauthier, L., Etude du pouvoir génotoxique des eaux de surface potables ou en cours de traitement, par la formation de micronoyau chez le triton *Pleurodeles waltl*, Thèse de Doctorat de l'Université Paul Sabatier, Toulouse, 1989. With permission.)

1) using a battery of complementary tests (initial screening tests and second line tests) and 2) the specific value of *in vivo* tests, especially the amphibian micronucleus tests, which are more representative of true conditions of exposure to environmental hazards. Further development of such tests is desirable, such as that currently being undertaken by the French Water Authorities comparing the *Xenopus* micronucleus test with the *in vitro* Mutatox® bacterial test at 30 different industrial sites.

III. PRACTICALITY OF THE AMPHIBIAN MICRONUCLEUS TEST

The results for both pure substances and complex media described above demonstrate the usefulness of the amphibian micronucleus test for the evaluation of ecotoxicological hazards. Although it is a simple and convenient test, it does present certain drawbacks for routine use, such as the long duration of exposure of the animals to the water samples, the variability in results stemming from the procedure for reading the blood smears, and differences in sensitivity of the hatching of larvae used which has been demonstrated with the positive control substances. For example, with the newt test, Godet and Vasseur[4] observed a range of micronucleated RBCs from 0 to 49% with a 25 μg/L dose of BaP. Using BaP in the same test, Djomo[27] found a minimum effective concentration of 5 μg/L, whereas in the initial studies of Jaylet et al.[18] a value of 12.5 μg/L was established. These sources of variability can be circumvented by systematic use of positive controls and by the use of animals from the same hatching for a given test.

Another source of variability is the procedure for reading the blood smears. Godet and Vasseur[4] found an extreme case of an 80% variation around the mean level of micronucleated RBCs on three separate readings of the same blood smear (counting 1000 cells). This variability stems from the fact that the micronucleated RBCs are not distributed evenly over the blood smear. This source of variability can be reduced to some extent by counting a larger number of cells (2000 to 3000 cells per smear). Work is currently in progress on the automation of blood smear readings either by image analysis or flow cytometry. In view of the current complexity and cost of image analysis (cost of equipment and software, reliability of autofocusing systems for moving slides), flow cytometry shows the most promise. Interest in this technique has been shown in experiments carried out with the *Xenopus* micronucleus test at the Centre Saint Laurent (Montreal, Canada) by L'Haridon and Blaise.[41] However, the test in *Xenopus* larvae with automatic counting of micronucleated cells after exposure to BaP appears to be only half as sensitive as the manual test. Experiments on cell conditioning designed to improve sensitivity are in progress. If the sensitivity of the automated method reaches that of current manual procedures, the feasibility and the specificity of the amphibian micronucleus test will be considerably enhanced (counting several thousand cells per animal, there will be a reduction in time for reading blood smears). The test will then be applicable for routine screening and will broaden its scope and international dissemination. It should be noted that the development of automated methods of reading data is not specific to the amphibian micronucleus test, as it also applies to other micronucleus tests such as that of the mouse.[42]

The procedure for the amphibian micronucleus tests as laid down in the French Standard T90325 for the newt[11] indicates a duration of exposure of 12 d. This is considerably longer than the exposures required for other *in vivo* or *in vitro* genotoxicity tests (SCE test-*Notobranchus* 48 to 96 h, Ames test 48 to 63 h; SOS Chromotest 24 to 72 h), but is of the same magnitude as that used in the *Mytilus* micronucleus test. This is an obvious drawback for widespread routine use. In a study of the effect of reducing the duration of exposure on sensitivity, Zoll-Moreux[15] found that the level of micronucleated RBCs induced by 0.25 mg/L and 0.025 mg/L BaP for a given exposure (2, 4, 6, or 8 d) was increased if the animals were kept for a further 4 d in water lacking BaP. This effect was observed for exposures as short as 2 d. Such procedures merit further investigation, especially for tests carried out in complex media. Nevertheless, current data indicate that 12 d is an optimum duration of exposure. The amphibian micronucleus test is therefore well suited to investigating situations of chronic toxicity. Under such conditions, a duration of exposure of 12 d would not be considered as an obstacle to routine use.

IV. CONCLUSIONS

It is well recognized that no single bioassay can evaluate all situations involving genotoxic pollution, due to the wide variety of genotoxic agents and their biological targets, and the wide range of exposure conditions (duration, concentration, and bioavailability of contaminants). Batteries of complementary tests chosen on the basis of their sensitivity, relevance, feasibility, and cost are thus needed. The use of an aquatic vertebrate along with test systems using less evolved species would appear to be desirable. This review shows the high level of interest in the amphibian micronucleus test. Another point in favor of an amphibian system is that extensive laboratory studies are required before a new test is employed *in situ*. Although amphibians have a different distribution in freshwater to that of fish, they are widely used in laboratory studies in developmental genetics and molecular and developmental biology. Taken together, these considerations point to the interest in amphibians such as *Xenopus* in genetic ecotoxicology.

REFERENCES

1. Ames, B. N., MacCann, J., and Yamasaki, E., Methods for detecting carcinogens and mutagens with the *Salmonella* mammalian microsome mutagenicity test, *Mutation Res.*, 31, 347, 1975.
2. Maron, D. and Ames, B. N., Revised methods for the *Salmonella* mutagenicity test, *Mutation Res.*, 113, 173, 1983.
3. Quillardet, P., Huisman, O., D'Ari, R., and Hoffnung, M., SOS Chromotest, a direct assay of induction of an SOS function in *Escherichia coli* K-12 to measure genotoxicity, *Proc. Natl. Acad. Sci.*, 79, 5971, 1982.

4. Godet, F. and Vasseur, P., Evaluation de la génotoxicité des effluents; étude comparative des tests d'Ames et micronoyaux Triton, Etude Inter Agences n° 29, Agences de l'Eau et Ministère de l'Environnement, Paris, 50, 1994.

5. Wilcox, P., Van Hoff, P., and Van der Gaag, M. A., Isolation and characterization of mutagens from drinking water, in *Proc. of the XVth Annual Meeting of the European Environmental Mutagen Society*, Brussels, Belgium, 1986.

6. Chouroulinkov, I. and Jaylet, A., Contamination of aquatic systems and genetic effects in *Aquatic Ecotoxicology: Fundamental Concepts and Methodologies*, Boudou, A. and Ribeyres, F. S, Eds., CRC Press, Boca Raton, FL, 1984, 211.

7. Zoll, C., Ferrier, V., and Gauthier, L., Use of aquatic animals for monitoring genotoxicity in unconcentrated water samples, in *Mechanisms of Environmental Mutagenesis and Carcinogenesis*, Kappas, A., Ed., Plenum Press, New York, 1990, 233.

8. Deparis, P., Le sang circulant au cours de la croissance larvaire de *Pleurodeles waltlii* Michah (amphibien urodèle), *J. Physiol*, 66, 423, 1973.

9. Krauter, P. W., Anderson, S. L., and Harrison, F. L., Radiation induced micronuclei in peripheral erythrocytes of *Rana catesbeiana*: an aquatic animal model for *in vivo* genotoxicity studies, *Environ. Mol. Mutagen*, 10, 285, 1987.

10. Venegas, W., Hermosilla, I., Gavilan, J.F., Naveas, R., and Carrasco, P., Larval stages of the anuran amphibian *Caudiverbera caudiverbera*: a biological model for studies for genotoxic agents, *Biol. Soc. Biol. Conception*, 58, 171, 1987.

11. AFNOR, Essai des eaux: évaluation de la génotoxicité au moyen de larves de batraciens (*Pleurodeles waltl*), Association Française de Normalisation, Paris, T90-325, 1992.

12. Siboulet, R., Grindfeld, S., Deparis, P., and Jaylet, A., Micronuclei in red blood cells of the newt *Pleurodeles waltlii* Michah: induction with X-rays and chemicals, *Mutation Res.*, 125, 275, 1984.

13. Jaylet, A., Deparis, P. and Gaschignard, D., Induction of micronuclei in peripheral erythrocytes of axolotl larvae following *in vivo* exposure to mutagenic agents, *Mutagenesis*, 1, 211, 1986.

14. Van Hummelen, P., Zoll, C., Paulussen, J., Kirsch-Volders, M., and Jaylet, A., The micronucleus test in *Xenopus*: a new and simple *in vivo* technique for detection of mutagens in fresh water, *Mutagenesis*, 4, 12, 1989.

15. Zoll-Moreux, C., Conséquences de la contamination du milieu hydrique par des sels de mercure, du benzo(a) pyrène ou des pesticides organochlorés chez deux amphibiens *Pleurodeles waltl* et *Xenopus laevis*, Thèse de Doctorat de l'Université Paul Sabatier, Toulouse, 1991, 191.

16. MacGill, R., Tuckey, J. W., and Larsen, W. A., Variations of box plots, *Am. Statist.*, 32, 12, 1978.

17. Grinfeld, S., Recherches sur l'induction de micronoyaux par les rayons X et divers agents chimiques dans les globules rouges de larves de *Pleurodeles waltl*, Absorption, relargage et excrétion de l'un d'entre eux: le benzo(a)pyrène, Thèse de Doctorat de 3è cycle, Université Paul Sabatier, Toulouse, 1983.

18. Jaylet, A., Deparis, P., Ferrier, V., Grinfeld, S., and Siboulet, R., A new micronucleus test using peripheral blood erythrocytes of the newt *Pleurodeles waltl* to detect mutagens in fresh-water pollution, *Mutation Res.*, 164, 245, 1986.

19. Fernandez, M. and Jaylet, A., An antioxidant protects against the clastogenic effects of benzo(a) pyrene in the newt *in vivo*, *Mutagenesis,* 2, 293, 1987.

20. Zoll, C., Saouter, E., Boudou, A., Ribeyre, F., and Jaylet, A., Genotoxicity and bioaccumulation of methyl-mercury and mercuric chloride *in vivo* in the newt *Pleurodeles waltl*, *Mutagenesis*, 3, 337, 1988.

21. Fernandez, M., Gauthier, L., and Jaylet, A., Use of newt larvae for *in vivo* genotoxicity testing of water: results on 19 compounds evaluated by the micronucleus test, *Mutagenesis*, 4, 17, 1989.

22. Fernandez, M. and L'Haridon, J., Influence of lighting conditions on toxicity and genotoxicity of various PAH in the newt *in vivo*, *Mutation Res.*, 298, 31, 1992.

23. L'Haridon, J., Fernandez, M., Ferrier, V., and Bellan, J., Evaluation of the genotoxicity of N-nitrosoatrazine, N-nitrosodiethanolamine and their precursors *in vivo* using the newt micronucleus test, *Wat. Res.*, 27, 855, 1993.

24. Fernandez, M., L'Haridon, J., Gauthier, L., and Zoll-Moreux, C., Amphibian micronucleus test(s): a simple and reliable method for evaluating *in vivo* genotoxic effects of fresh water pollutants and radiations. Initial assessment, *Mutation Res.*, 292, 83, 1993.

25. Lecurieux, F., Marzin, D., and Erb, F., Comparison of three short-term assays: results on seven chemicals, *Mutation Res.*, 319, 223, 1993.

26. Lecurieux, F., Marzin, D., Brice, A., and Erb, F., Utilisation de trois tests de génotoxicité pour l'étude de l'activité génotoxique de composés organohalogénés, d'acides fulviques et d'échantillons d'eau (non concentrés) en cours de traitement de potabilisation, *Revue des Sciences de l'Eau*, (in press).

27. Djomo, J. E., Ferrier, V., Gauthier, L., Zoll-Moreux, C., and Marty, J., Amphibian micronucleus test *in vivo*: evaluation of the genotoxicity of some major polycyclic aromatic hydrocarbons found in a crude oil, *Mutagenesis*, 10, 223, 1995.

28. Zoll-Moreux, C. and Ferrier, V., Etude comparative entre le test Jaylet (test micronoyau triton) et le test micronoyau xénope, Etudes Interagences, Agence de l'Eau et Ministère de l'Environnement, Paris, (in press).

29. Giraud, M. and Guillet, H., Teneur en mercure des milieux naturels, in *La pollution par le mercure et ses dérivés*, Ministère de l'Environnement, Paris, 1972, 55.

30. Grinfeld, S., Jaylet, A., Siboulet, R., Deparis, P., and Chouroulinkov, I., Micronuclei in red blood cells of the newt *Pleurodeles waltl* after treatment with benzo(a)pyrene: dependence on dose, length of exposure, post-treatment time and uptake of the drug, *Environ. Mutagenesis*, 8, 41, 1986.

31. Gauthier, L., Rapport au Groupement d'Etudes Scientifiques du Test Micronoyau Triton (Gestmint), C.I.R.S.E.E., Ed., Le Pecq, France, 1992.

32. Godet, F., Vasseur, P. and Babut, M., Essais de génotoxicité *in vitro* et *in vivo* applicables à l'environnement hydrique, *Revue des Sciences de l'eau*, 6, 285, 1993.

33. Godet, F., Babut, M., Burnel, D., Veber, A.M., and Vasseur, P., The genotoxicity of iron and chromium in electroplating effluents, *Mutation Res.*, in press, 1997.

34. Jaylet, A., Gauthier, L., and Fernandez, M., Detection of mutagenicity in drinking water using a micronucleus test in the larval newt (*Pleurodeles waltl*), *Mutagenesis*, 2, 211, 1987.

35. Gauthier, L., Levy, Y., and Jaylet, A., Evaluation of the clastogenicity of water treated with sodium hypochlorite or monochloramine using a micronucleus test in the larval newt (*Pleurodeles waltl*), *Mutagenesis*, 4, 170, 1989.

36. Gauthier, L., Etude du pouvoir génotoxique des eaux de surface potables ou en cours de traitement, par la formation de micronoyau chez le triton *Pleurodeles waltl*, Thèse de Doctorat de l'Université Paul Sabatier, Toulouse, 1989.

37. Van der Gaag, M. A., Gauthier, L., Noordsij, A., Levi, Y. and Wrisberg, M. N., Methods to measuring genotoxins in waste water: evaluation with *in vivo* and *in vitro* tests, in *Genetic Toxicology of Complex Environmental Mixtures*, M.S. Waters et al., Eds., Plenum Press, New York, VI, 1990, 215.

38. Gauthier, L., Van der Gaag, M. A., L'Haridon, J., Ferrier, V., and Fernandez, M., *in vivo* detection of waste water and industrial effluent genotoxicity: use of the newt micronucleus test (Jaylet test), *Sci. Total Environ.*, 138, 249, 1993.

39. AFNOR, Déchets: Essai de lixiviation, Association Française de Normalisation, X31-210, 1992.

40. Bekaert, C., Evaluation du potentiel génotoxique de sols contaminés et de déchets à l'aide du test micronoyau *Xénope*, personal communication, 1995.

41. L'Haridon, J. and Blaise, C., Evaluation de la génotoxicité sur Xénope par dénombrement cytométrique des micronoyaux, Centre Saint-Laurent, Environnement Canada, Montréal, unpublished data, 1995.

42. Romagna, F. and Staniforth, C.D., The automated bone narrow micronucleus test, *Mutation Res*, 213, 91, 1989.

Multitrophic Assessment in Practice

CHAPTER 36

Marine and Estuarine Porewater Toxicity Testing

R. Scott Carr

CONTENTS

I. INTRODUCTION

Historically the standard approach for assessing the quality or potential toxicity of marine or estuarine sediments has been to expose macrobenthic organisms directly to the whole sediments for a specified period of time (usually 10 days) after which the survival of the test species is determined. This has been the standard approach for assessing the suitability of dredged material for different disposal options in the regulatory arena in the U.S. since the mid-1970s[1] and continues in an updated form[2,3] to be the most commonly employed sediment toxicity assessment method. There are a number of limitations, however, with the whole-sediment test approach.

Whole-sediment methods, as currently employed,[2,3] use adult macrobenthic organisms and the only endpoint measured is mortality. In addition, the standard amphipod test protocol underestimates the potential toxicity of contaminated sediments because the porewater, in which it has taken weeks or perhaps months for the hydrophobic contaminants to come to equilibrium, is at least partially diluted with fresh overlying water before the start of the exposure. These tests are useful when working with highly contaminated sediments or when only acute toxicity test data are needed. The combined insensitivity of the test species, life stage, endpoint, and the artificial manipulations of the test sediment prior to exposure, result in a test which has limited ability to predict the potential chronic impacts of moderately contaminated sediments.

Equilibrium-partitioning theory predicts that porewater is the controlling exposure medium in the toxicity of sediments to infaunal organisms.[4-6] Except for the sediment-associated contaminants which may be released in the gut of sediment-ingesting animals, only the contaminants that are solubilized in the interstitial water are bioavailable. Exposing organisms to porewater provides a direct measure of contaminant exposure that incorporates all the physical and chemical parameters (e.g., pH, salinity, grain size, TOC, AVS) which affect bioavailability. It is difficult to obtain large volumes of porewater for toxicity testing, therefore, microscale techniques are most appropriate for conducting tests with this matrix. This apparent limitation is more than offset by the fact that the porewater approach is amenable to testing with sensitive life-stages of sensitive species which provides a measure of chronic sublethal impacts, as opposed to the acute lethal endpoint used in the whole-sediment tests. The use of more sensitive species and life-stages in conjunction with the sediment equilibrated porewater matrix greatly increases the sensitivity of porewater tests for detecting contaminant effects.

Although some work has been done with freshwater porewater samples, most of the studies on porewater testing have focused on marine and estuarine species. Freshwater porewater studies have primarily employed the Microtox®* assay or the *Ceriodaphnia* sp. reproduction, feeding behavior, or mortality assays to detect toxicity. Because a larger variety of marine and estuarine species and endpoints have been used in porewater testing, this chapter will focus primarily on work done with these species. Much of the information reported here is, however, applicable to freshwater porewater testing as well.

II. POREWATER EXTRACTION AND HANDLING METHODS

A. Centrifugation

Centrifugation is the most commonly employed method for extracting porewater from sediments.[7-13] Various centrifugation speeds, times, and vessels composed of different materials have been employed by different investigators. Centrifuge bottles or tubes composed of polycarbonate, Teflon®, or glass are recommended for minimizing loss of soluble contaminants due to adsorption. In order to remove colloids from suspension, high speeds ($\geq 10,000$ g) are required,[14] but colloid removal may not be necessary or desirable in toxicity testing. Centrifugation speeds of 2000 to 5000 g for 10 to 60 min are more commonly employed.[9,13,15] Regardless of the speed used for the initial extraction, it may not be possible to remove all of the fine particulate material because of electrostatic repulsion, unless the supernatant is decanted and recentrifuged.[13] Centrifugation at different speeds does not appear to affect the chemistry of porewater.[16]

Another potential problem with centrifugation that has been observed with sediments containing high concentrations of organic contaminants (e.g., PAHs or PCBs) is that these hydrophobic contaminants can become concentrated in the supernatant and sometimes form an emulsion or a

* Registered trademark of AZUR Environmental, Carlsbad, California.

visible layer on the surface which makes retrieval of the porewater difficult.[13] Centrifugation has been recommended as the extraction method of choice for minimizing the loss of highly hydrophobic contaminants.[11,13,15] Standard centrifugation does not work well for sandy sediments because little porewater is obtained due to the limited compaction of coarse-grained sediments.[13] Inverted centrifugation techniques have also been employed with sandy sediments,[17] but this method requires the use of a filter which may increase the loss of hydrophobic compounds.

B. Squeezing

One of the most commonly employed methods for extracting porewater is squeezing sediments using either mechanical or pneumatic pressure. The earliest squeezers were made of stainless steel,[18] which caused trace metal contamination problems as noted by Presley.[19] These pressurized extractors evolved from all stainless steel to Teflon-coated stainless steel,[19] to Lucite®,[20] to all Teflon,[21] and most recently to polyvinyl chloride (PVC).[13] The results of recent studies indicate that if the extractor is composed of chemically inert materials to minimize leaching and adsorption of contaminants from and to the interior surfaces of the extractor, that the toxicity of the sample will not be affected.[13] For example, no difference was found in the toxicity of porewater obtained with Teflon and PVC pneumatic extractors.[13] Significant differences have been observed with different types of filters which are used with all squeeze extraction devices.[13] Glass fiber filters adsorbed the majority of hydrophobic contaminants such as DDT,[15] whereas 5 μm polyester filters appear to adsorb fewer contaminants than most of the commercially available filters that have been tested.[13] Pneumatic extraction can be used with sediments of all textures and is the most efficient method for processing large numbers of samples.[13]

C. Vacuum

Vacuum extraction methods have been used with field collected samples[22] and *in situ*.[23,24] These techniques generally involve the use of a syringe to supply the vacuum to a needle or filtered port which extracts the porewater from different depths depending on the design of the device. This extraction method may be particularly useful for sampling sediments with volatile organic compounds. The device developed by Winger and Lasier[22] is the most useful for toxicity testing purposes because it allows larger volumes to be collected than the other vacuum methods. In order to obtain sufficient porewater for toxicity testing within a reasonable length of time, numerous suction devices are usually employed simultaneously. One of the biggest advantages of the Winger and Lasier apparatus (which is composed of a plastic syringe, aquarium tubing, and a crushed glass air stone) is the low cost. Some of the problems with this method are that the air stones tend to clog rapidly with very fine silty sediments, and sandy sediments result in a large amount of particulate material in the porewater sample. The suspended particulates can be removed by an additional centrifugation, which is highly recommended when using the vacuum extraction method initially.

D. Passive Diffusion

Collection devices containing a dialysis membrane or filter, sometimes referred to as "peepers," have been used to obtain porewater by means of passive diffusion.[25-28] This method may have the greatest likelihood of maintaining the *in situ* conditions of the porewater, but this advantage is offset by other technical limitations. One of the primary disadvantages of this technique is that a week or more is required for equilibrium to be achieved. The other drawback, with regard to toxicity testing, is that only small volumes can be obtained with the passive samplers used to date, and a large ratio of sediment to porewater collected is required. These characteristics make this porewater collection method undesirable for routine toxicity testing applications, even with microscale testing procedures, particularly when large numbers of samples need to be processed rapidly.

E. Water Quality Adjustments

In sediment quality assessment surveys in bays and estuaries particularly, it is not uncommon to obtain porewater with a wide range of salinities, pH, dissolved oxygen levels, and varying concentrations of ammonia and sulfide. Sulfide concentration is the only one of these parameters that appears to be noticeably affected by the porewater extraction method. Centrifuge extraction tends to produce a sample with lower sulfide content than pneumatic extraction as more off-gassing of volatile compounds is likely to occur during centrifugation. The freezing/thawing of porewater also reduces the sulfide concentration. The other water quality parameters (pH, DO, ammonia, and salinity) do not differ noticeably among the different extraction techniques or after freezing and thawing. We recommend that all of these parameters be measured just prior to testing and that the salinities be adjusted for all samples to ±1‰ of the test salinity. We recommend that if the dissolved oxygen concentration of the sample is <80% saturation, the sample should be gently stirred on a stir plate until this level is achieved, before testing is commenced. Un-ionized ammonia concentrations are sometimes sufficiently elevated to be the primary toxicant in porewater, particularly in anaerobic sediments. We have observed that ammonia often co-varies with the concentration of other contaminants and believe that this is related to reduced bioturbation resulting from contaminant impacts on benthic organisms.[29] As the elevated ammonia concentration in contaminated sediments may either be a direct or indirect result of anthropogenic impacts, we believe it is important to include ammonia as part of the sediment quality assessment and not attempt to remove it from the sample prior to testing. Sediment porewater ammonia concentrations can increase significantly during storage, even when refrigerated. It is recommended, therefore, that porewater be extracted from sediments as soon as possible after collection.

Our standard procedure when testing porewater that has been stored frozen is to place the frozen sample in a refrigerator or incubator (~4°C) overnight to thaw slowly. On the following day, the samples are removed from the refrigerator and thawing completed at the test temperature. The salinity of each sample is then measured and adjusted to ±1‰ of the test salinity by the addition of hypersaline brine (~100‰) or ultrapure deionized water. It is necessary to adjust all samples to ±1‰ of the test salinity so that this variable does not confound the results of the test response. Salinity adjustments may alter the bioavailability of some contaminants (e.g., divalent metals) which may need to be considered in the interpretation of the toxicity test data. We have found that hypersaline brine prepared by slow evaporation of natural sea water is preferable to other methods (e.g., freezing sea water or brine prepared with artificial sea salts) which may produce artifactual toxicity due to trace nutrient limitations in sensitive tests such as fertilization or embryological development assays. It is important to add as little brine or deionized water as possible to minimize the dilution of the porewater sample. After the salinity adjustments have been made, the water quality parameters are all measured at the test temperature. The test is then either commenced immediately or the samples are held refrigerated overnight and then returned slowly to the test temperature before the test is started on the following day. A reconstituted brine control at the test salinity and a reference porewater sample that has been handled identically to the test samples should be included with every test series. We recommend that samples be tested within 48 h after thawing and kept refrigerated (except when water quality measurements are made and just prior to test initiation) because bacterial processes may alter the toxicity of samples held for longer periods, even when refrigerated.

F. Method Comparison Studies

Several extraction method comparison studies have been published.[11,13,27] Schults et al.[11] studied the loss of specific trace metal and organic contaminants due to adsorption among four different extraction procedures. They concluded that there was significant loss of several highly hydrophobic organic compounds with all the methods but that the centrifuge method was the most accurate and

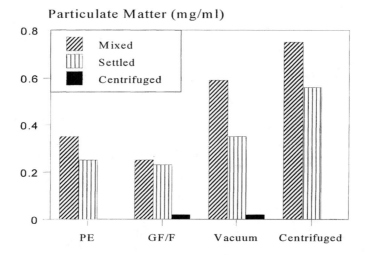

Figure 36.1   Mean weight of particulate matter (≥0.45 μm) remaining in porewater (22‰ salinity) after different
extraction methods and manipulations. The PE (5 μm polyethylene) and GF/F (0.7 μm glass
fiber filter) refer to the filters used in the pneumatic extraction device. The "mixed" porewater
samples were shaken vigorously and sampled immediately; the "settled" samples were allowed
to settle for approximately 1 h after shaking before they were sampled; the "centrifuged" samples
were centrifuged at 2000 g for 20 min after shaking before they were sampled.

precise for PAHs and PCBs. All of the samples in this study[11] were filtered through a GF/F glass
fiber filter after the initial extraction. Filtration undoubtedly affected the results of this study as
this type of filter has been shown to strongly adsorb highly hydrophobic organics.[15]

In a more recent comparison study, toxicity rather than the concentration of particular analytes
was used to discern differences among extraction techniques.[13] It was observed that the different
extraction methods (centrifugation, vacuum, and pneumatic) produced samples with a visible and
quantifiable difference in the amount of suspended particulate material (Figure 36.1). It was found
that this suspended particulate material could be significantly reduced or eliminated by an additional
centrifugation (2000 g for 20 min) regardless of the initial extraction method. It has been shown
that suspended particulate material can produce artifactual responses in microscale tests such as
the echinoderm fertilization assay,[13] so it is highly recommended that particulates be removed by
centrifugation (and not filtration) before toxicity tests are conducted.

Recent studies have shown that if the suspended particulate material is removed, the toxicity
of porewater is not affected by freezing and thawing (Table 36.1). Porewater was extracted by two
different methods and an aliquot of the sample was frozen and then thawed and compared with a
sample that had not been frozen but stored refrigerated overnight. These samples were not centrifuged after the initial extraction to remove particulate material and the difference between the
frozen vs. fresh comparison for centrifuge samples Nos. 12 and 13, particularly, is due to the higher
suspended particulate load as compared with the pneumatically extracted sample (Table 36.1).

An extensive series of comparisons with 33 different contaminated sediments using two different
assays with the sea urchin *Arbacia punctulata* showed that there were no statistically significant
differences between centrifuge and pneumatically extracted porewater samples for 79% of the
comparisons when a second centrifugation was performed after the initial extraction (Table 36.2).
Of the 14 samples that were different between the two methods, 57% were more toxic for the
centrifuge-extracted sample, and all of these samples had a visible oil sheen on the supernatant
after the initial extraction. Centrifugation may tend to concentrate oily contaminants (e.g., PAHs
and PCBs) which would explain the higher toxicity observed for these samples as compared with
the pneumatically extracted samples. Volatile organic compounds may be lost to a greater degree
during the more rigorous initial centrifugation extraction than by the pneumatic extraction and is

Table 36.1   Summary of Sediment Extraction and Storage Experiments with Contaminated Sediments

Sample No.	Endpoint	Centrifuged		Pneumatic	
		Fresh	Frozen	Fresh	Frozen
1	% Fertilization (100% PW)	81.6[1]	87.6	93.2	90.8
2	% Fertilization (100% PW)	98.4	82.0[2]	59.4[2]	96.0
3	% Fertilization (25% PW)	34.8[1]	16.8[1]	97.6	98.0
4	% Fertilization (100% PW)	0[1]	0[1]	97.0	93.8
5	% Fertilization (100% PW)	54.4[1]	35.6[1]	98.6	97.8
6	% Fertilization (100% PW)	2.2[1]	1.4[1]	26.2	16.8
7	% Normal development (25% PW)	0	0	0	0
8	% Normal development (100% PW)	98.0	93.6	98.0	97.6
9	% Normal development (100% PW)	95.4	90.0[1]	97.4	98.0
10	% Normal development (100% PW)	98.6	86.6[1,2]	99.0	97.8
11	% Normal development (100% PW)	96.0	96.8	98.2	97.0
12	% Normal development (100% PW)	88.0	35.0[1,2]	98.6	98.0
13	% Normal development (100% PW)	98.6	30.6[1,2]	99.0	83.6
14	EC_{50} (% Normal development)	51.8	30.0	41.4	48.6

Note: Porewater samples were not centrifuged prior to freezing after the initial extraction.

[1] Statistically significant difference between centrifuged vs. pneumatic for either the fresh or frozen samples compared independently (<0.01 using statistical and minimum detectable significance criteria).

[2] Statistically significant difference between fresh vs. frozen for either the centrifuged or pneumatically extracted samples compared independently (<0.01 using statistical and minimum detectable significance criteria).

From Carr R. S. and Chapman, D. C., *Archives of Environmental Contamination and Toxicology*, 28, 1995, Springer-Verlag New York. With permission.

a possible explanation for the higher toxicity displayed in pneumatically extracted samples compared with centrifuged samples. When the previously mentioned precautions are taken regarding extractor materials, different extraction methods (centrifugation, vacuum, and pneumatic) were found to produce samples exhibiting similar toxicity the majority of the time, with some exceptions for certain types of contaminants.[13]

The advantages and disadvantages for the three primary extraction methods utilized in toxicity testing are compared in Table 36.3.

III. SPECIES AMENABLE TO TESTING WITH POREWATER

Because of the difficulty in obtaining large volumes of porewater, organisms which require small volumes are most suitable for porewater toxicity testing. Microscale techniques are therefore the methods of choice when testing porewater. A wide variety of organisms have been used successfully in porewater tests. Minute species or larval forms are preferable not only for their small volume requirements but also because they tend to be the most sensitive species and life-stages, which increases their relevance when extrapolating from laboratory results to the field.

A. Microtox®

The Microtox test system, which measures changes in the photoluminescence of the marine bacterium *Vibrio fischeri,* formerly *Photobacterium phosphoreum,* as an endpoint, has been used more in fresh porewater studies[9,10,30] than with marine or estuarine porewaters.[31] Although the small sample size required by this technique is well suited for limited sample sizes, the sensitivity of the standard Microtox assay for porewater from freshwater, estuarine, or marine sediments is low

Table 36.2 Comparison of Centrifuge (-C) and Pneumatically Extracted Sediment Porewater Samples for the Sea Urchin Fertilization and Embryological Development Toxicity Tests

Sample No.	% Fertilization	% Normal Development	Sample No.	% Fertilization	% Normal Development
1	2.0 ± 1.2	0.0 ± 0.0	17-C	8.0 ± 6.2	0.0 ± 0.0
1-C	1.0 ± 1.2	0.4 ± 0.9	18	98.6 ± 1.3	91.8 ± 1.0
2	96.0 ± 3.0	45.6 ± 34.8	18-C	99.2 ± 1.3	93.6 ± 2.9
2-C	98.6 ± 1.7	9.0 ± 10.7*	19	55.0 ± 18.3	0.0 ± 0.0
3	94.8 ± 2.9	1.8 ± 2.5	19-C	76.8 ± 17.0	0.0 ± 0.0
3-C	98.4 ± 1.1	0.2 ± 0.4	20	55.4 ± 17.7	0.0 ± 0.0
4	95.6 ± 3.6	6.4 ± 6.5	20-C	78.4 ± 11.7*	0.0 ± 0.0
4-C	95.0 ± 3.0	0.6 ± 1.3	21	95.2 ± 2.5	0.2 ± 0.4
5	96.6 ± 1.3	4.0 ± 3.0	21-C	95.8 ± 2.8	0.6 ± 1.3
5-C	96.0 ± 2.5	0.0 ± 0.0	22	96.6 ± 2.2	0.0 ± 0.0
6	95.8 ± 2.8	29.8 ± 12.2	22-C	95.8 ± 6.7	0.0 ± 0.0
6-C	95.4 ± 3.8	0.4 ± 0.9**	23	11.4 ± 6.5	0.4 ± 0.9
7	93.8 ± 2.9	87.4 ± 4.2	23-C	3.4 ± 2.1	0.0 ± 0.0
7-C	83.8 ± 1.3	63.8 ± 14.6**	24	15.4 ± 6.4	0.2 ± 0.4
8	97.4 ± 1.9	77.4 ± 5.3	24-C	73.0 ± 8.3**	0.2 ± 0.4
8-C	16.8 ± 7.0**	49.4 ± 20.7**	25	51.8 ± 0.8	0.0 ± 0.0
9	98.8 ± 1.3	7.2 ± 8.4	25-C	57.0 ± 7.2	0.0 ± 0.0
9-C	35.0 ± 4.5**	1.2 ± 1.6	26	0.6 ± 0.6	0.0 ± 0.0
10	98.2 ± 1.3	0.0 ± 0.0	26-C	1.6 ± 2.1	0.0 ± 0.0
10-C	83.4 ± 4.4*	0.6 ± 0.9	27	0.2 ± 0.4	0.4 ± 0.9
11	91.6 ± 3.0	0.0 ± 0.0	27-C	0.6 ± 0.5	0.2 ± 0.5
11-C	90.0 ± 1.9	0.0 ± 0.0	28	93.0 ± 2.1	0.0 ± 0.0
12	0.4 ± 0.6	0.0 ± 0.0	28-C	96 ± 2.0	0.0 ± 0.0
12-C	0.4 ± 0.5	0.6 ± 1.3	29	8.6 ± 2.5	0.0 ± 0.0
13	0.4 ± 0.6	0.6 ± 1.3	29-C	15.6 ± 3.2	0.0 ± 0.0
13-C	73.6 ± 11.6**	0.0 ± 0.0	30	86.4 ± 9.5	0.0 ± 0.0
14	0.4 ± 0.6	0.0 ± 0.0	30-C	78.0 ± 11.3	0.0 ± 0.0
14-C	90.2 ± 2.3**	0.0 ± 0.0	31	28.2 ± 7.3	0.0 ± 0.0
15	1.4 ± 1.5	0.8 ± 1.3	31-C	8.4 ± 12.4*	0.0 ± 0.0
15-C	31.4 ± 12.0**	0.2 ± 0.4	32	86.8 ± 9.6	0.2 ± 0.4
16	0.8 ± 0.8	0.0 ± 0.0	32-C	30.8 ± 10.5**	0.0 ± 0.0
16-C	2.8 ± 2.5	0.0 ± 0.0	33	88.2 ± 5.1	0.0 ± 0.0
17	0.6 ± 0.5	0.0 ± 0.0	33-C	90.8 ± 5.8	0.0 ± 0.0

* Significant difference between centrifuged and pneumatically extracted samples (two-tailed Student's t test for equal variances, performed on arcsine–square root transformed data, $\alpha = 0.05$; minimum detectable significant difference {$\alpha = 0.05$} between means for fertilization test = 13).

**Significant difference between centrifuged and pneumatically extracted samples (two-tailed Student's t test for equal variances, performed on arcsine-square root transformed data, $\alpha = 0.01$; minimum detectable significant difference {$\alpha = 0.01$} between means for fertilization test = 16).

compared to other toxicity tests that have been directly compared.[30,31] Other modifications of the Microtox approach utilizing organic sediment extracts[32] or solid-phase exposures[31] appear to offer more promise for evaluating sediment quality, particularly of marine and estuarine sediments.

B. Algal Assays

No studies of marine algal assays with porewater have yet been published, but preliminary reports indicate that a zoospore germination and growth test with *Ulva fasciata* and *Ulva lactuca* has been used successfully in porewater tests.[33] Preliminary results suggest that the germination endpoint is as sensitive as some of the more sensitive embryological development assays. This test

Table 36.3   Advantages and Disadvantages of Different Porewater Extraction Techniques

Extractor type	Advantages	Disadvantages
Centrifuge	1. Minimizes contact with surfaces, compared to other methods thereby minimizing adsorptive loss of contaminants. 2. Majority of porewater can be extracted from clay sediments. 3. By recentrifuging the supernatant (or by centrifuging porewater extracted by other methods), most suspended solid particles of greater than 0.45 μm may be removed from marine/estuarine porewater.	1. Very little porewater can be collected from sandy sediments because sand does not compact easily. 2. Cleaning centrifuge tubes after porewater extraction from clay sediments is difficult and time consuming. Large centrifuge tubes are too expensive to be considered disposable. 3. Cannot easily be performed in the field. 4. Most expensive technique, in terms of initial equipment cost.
Vacuum	1. Least expensive method, in terms of equipment cost. 2. All materials are disposable, which minimizes cleanup costs and reduces possibility of cross sample contamination. 3. Can easily be performed in field.	1. Was least successful technique for removing particles of greater than 0.45 μm from porewater. 2. Extraction of porewater from clay samples may be prohibitively time consuming.
Pneumatic	1. Can extract majority of the porewater from sediment, regardless of grain size. Best method for maximizing the quantity of porewater obtained from a limited quantity of sediment for studies with a wide variety of sediment textures. 2. Was most successful method for removal of suspended solid particles of greater than 0.45 μm. 3. Can easily be performed in field.	1. Extraction of porewater from clay sediments takes longer than with the centrifugation method. 2. Requires more initial start-up costs than the suction method but in the long run may be more cost effective for large-scale programs. 3. More hydrophobic contaminants may be adsorbed by the filter and other surfaces as compared with the centrifugation methods.

From Carr R. S. and Chapman, D. C., *Archives of Environmental Contamination and Toxicology*, 28, 1995, Springer-Verlag New York. With permission.

appears to be particularly resistant to ammonia toxicity, which may be a useful characteristic in toxicity identification evaluations (TIEs). Many of the algal species used in microplate procedures as described in this volume[34] could easily be adapted for use with porewater samples.

C. Polychaetes

The first published report of porewater toxicity testing with a marine organism was with the polychaete *Dinophilus gyrociliatus*.[21] This minute cosmopolitan species attains a maximum length of approximately 1 mm, and a life cycle test can be conducted in 7 days.[35] Sexual dimorphism is conspicuous in *D. gyrociliatus*, with small eggs developing into dwarf males and large ones developing into females. Both male and female eggs are deposited in each egg capsule, with the total number of eggs per capsule ranging from 2 to 10, the majority being female. Before the siblings emerge from the egg capsule, the dwarf males copulate with the females and die soon after. Experiments are conducted with newly emerged juvenile females approximately 0.1 mm in length.[21,35] After 5 to 6 days at 20°C, the females begin to lay eggs. After 7 days the number of eggs produced per female is enumerated and is the most sensitive endpoint for this test. The reproductive test with *D. gyrociliatus* has proved to be very sensitive to a number of porewater samples.[36,37]

Other minute species such as *Ctenodrilus serratus* or *Ophryotocha* spp.[38,39] could be tested in small volumes but require exposures of up to 28 days in order to quantify the reproductive endpoint. This length of exposure would probably necessitate the use of a static renewal test design in order to maintain acceptable water quality over the duration of the test. Other species of polychaetes which are being evaluated in Canada for the sediment toxicity test may also be amenable for testing with porewater.[40] No other reports of porewater testing with polychaetes have been published to my knowledge.

D. Mollusks

One mollusk test that has been used successfully with porewater is the embryological development test using the abalone *Haliotis rufescens*.[41] Other more common embryological development tests with oysters (e.g., *Crassostrea* spp.[42] and clams (e.g., *Mercenaria mercenaria*[43]) could easily be adapted for use with porewater samples. Embryological development tests with bivalves are among the most sensitive assays used in effluent or water quality monitoring. As mentioned previously, it is important that suspended particulate material is removed, by centrifugation, prior to porewater testing to minimize artifactual responses with fertilization and embryological development assays, particularly.

E. Crustaceans

The nauplii of the benthic harpacticoid copepod *Longipedia americana* were used in a study of the impacts of offshore oil and gas exploration and production activities on the benthos.[37] Newly hatched nauplii from laboratory-reared animals were exposed under static conditions for 48 h (through the first molt) and were fed an algal mixture during the test. The laboratory cultured harpacticoid *Amphiascus tenuiremis*[44] has also been used in tests with porewater. It should be possible to use a wide variety of early life stages of other crustaceans, particularly species of decapods and penaeids that have been used in toxicity testing previously.

F. Echinoderms

The majority of the toxicity testing with marine and estuarine porewater has been conducted with sea urchin gametes and embryos.[13,36,37,45-53] The species that has been most routinely employed is the sea urchin *Arbacia punctulata,* but other species of sea urchin (e.g., *Strongylocentrotus* spp. and *Lytechinus* spp.) as well as the sand dollar (e.g., *Dendraster* spp.) have been used successfully in porewater tests. The most common tests are the fertilization test (also known as the sperm cell test) and the embryological development test. The fertilization test involves exposing sperm initially for 10 to 60 min followed by the addition of eggs. The eggs and sperm are incubated together for an additional 25 to 60 min before the test is fixed and the number of eggs exhibiting a fertilization membrane are enumerated. The embryological development test can either be initiated with fertilized eggs or embryos in the first division stages, or by adding the eggs followed shortly thereafter by the sperm and allowing the embryos to develop to the echinopluteus stage which, in control treatments, takes from 2 to 4 days depending on the species and the test temperature. The number of embryos developing normally is the standard endpoint but a number of different developmental abnormalities can also be enumerated and used as endpoints as well. A cytogenetic assay has also been conducted with sea urchin embryos exposed to porewater where the embryos are fixed and stained at the blastula stage and examined for chromosomal abnormalities following the procedures of Hose et al.[54]

G. Teleosts

Fish embryos and larvae of red drum *Sciaenops ocellatus* have been used successfully in porewater testing.[55] Short-term (48-h) survival tests were not as sensitive as tests with sea urchin embryos. Unusual developmental abnormalities have been observed with red drum after exposure to porewater from PAH-contaminated sediments (Carr, unpublished data) which may be useful for screening for teratogenic substances. Teleost species with a relatively short (1 to 3 days) embryological period are likely to be most amenable to testing with porewater under static conditions. Due to their high metabolic activity, it is difficult to maintain fish larvae in small volumes for very long once they have reached the feeding stage. Many of the commonly used marine toxicity testing

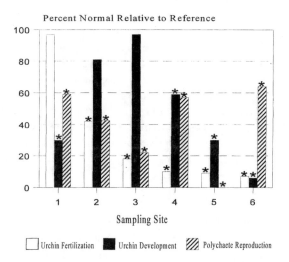

Figure 36.2   Comparison among three different toxicity tests (sea urchin fertilization and embryological development tests with *Arbacia punctulata* and polychaete reproduction test with *Dinophilus gyrociliatus*) for six porewater samples from contaminated sites. Asterisks represent a significant decrease ($\alpha \leq 0.01$) as compared with a reference porewater sample.

species (e.g., *Cyprinodon variegatus* or *Menidia menidia*) could be used in porewater testing using survival, hatching, and developmental abnormalities as endpoints.

H. Comparison of Species and Endpoints

Studies have been conducted to compare the relative sensitivities of some of the tests that have been described above to porewater extracted from field-collected sediments. The relative sensitivity of the sea urchin fertilization and embryological development tests and the polychaete reproduction test is shown in Figure 36.2 for porewater from sediments with different types of contamination. Statistically significant toxicity was observed for at least two tests for each of the six porewaters tested, but the relative sensitivity was different for different samples, indicating differences in contaminant-specific responses for the different tests. In general, the embryological development test tends to be more sensitive than the shorter-term fertilization assay, but this is not always the case, as some contaminants may be particularly spermatocidic.

Sediments from the vicinity of an offshore gas production platform provided an opportunity to compare the sea urchin embryological development test with the polychaete reproduction test and a short-term (48 h) survival test with nauplii of an indigenous benthic copepod *Longipedia americana* (Figure 36.3). There was excellent agreement among the three species tested, which indicates that the sea urchin assay, which is the most economical and convenient of the three assays to perform, could be used in this context as a reasonable surrogate for predicting the impacts of sediments with similar contaminants on sediment dwelling organisms.[37]

The standard 10-day amphipod survival test has been tested in side-by-side comparisons with the sea urchin fertilization and embryological development assays in numerous recent sediment quality assessment surveys (totaling more than 800 samples) in bays and estuaries along the Atlantic and Gulf coasts of the US. The results of these comparisons have demonstrated that the sea urchin porewater tests are always considerably more sensitive (by at least an order of magnitude on average) than the amphipod whole-sediment test, which is the most sensitive test used for regulatory testing of dredged material in the U.S. This difference in sensitivity is not entirely unexpected as the amphipod test measures acute toxicity with an adult benthic crustacean, whereas the sea urchin

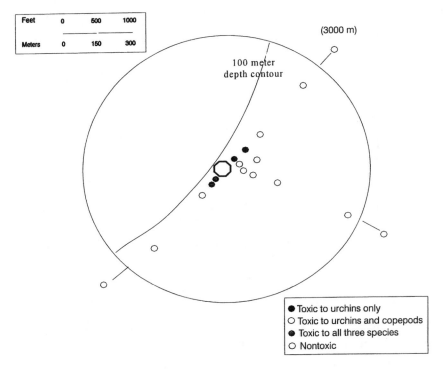

| Feet | 0 | 500 | 1000 |
| Meters | 0 | 150 | 300 |

(3000 m)

100 meter
depth contour

● Toxic to urchins only
○ Toxic to urchins and copepods
● Toxic to all three species
○ Nontoxic

Figure 36.3   Comparison among three different toxicity tests (sea urchin fertilization test with *Arbacia punctulata*, polychaete reproduction test with *Dinophilus gyrociliatus*, and survival of nauplii copepod *Longipedia americana*) for porewater samples collected in the vicinity of an offshore oil and gas production platform.

fertilization and embryological development tests with porewater provide measures of sublethal impacts on sensitive life stages of a sensitive organism.

Numerous recent sediment surveys have shown excellent agreement between toxicity predicted using sediment quality assessment guidelines[56] and toxicity observed with porewater toxicity tests.[57] A high degree of association has also been observed between sediment contaminant concentrations and sea urchin porewater toxicity tests.[29] Using the effects range median (ERM, the concentration of a particular chemical above which biological effects are likely to occur) values of Long et al.[56] for the 25 chemicals for which ERM values have been calculated, an ERM cumulative index can be calculated by dividing the sediment concentrations for each chemical by the corresponding ERM value and summing the quotients for each chemical for each station. The cumulative ERM index was calculated and compared with the observed toxicity for sediments collected in Tampa Bay, Florida[29] (Figure 36.4). In this study, 100% of the samples with a cumulative ERM index of >2.5 were highly toxic in the porewater assay. For sediments from this system, this cumulative ERM index could be employed with considerable confidence to predict potentially toxic sediments. Of course, it would be more cost effective to employ sensitive porewater toxicity tests to make these assessments directly.

IV. THE WAY FORWARD

A. Helpful Hints

Although different porewater extraction methods, in general, yield samples with similar toxicity when the recommended precautions are taken, certain types of contaminants may be preferentially

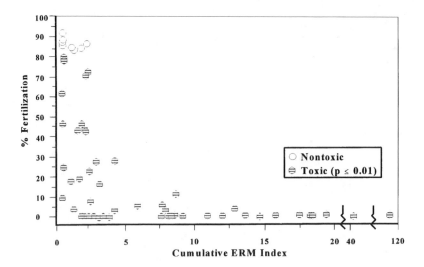

Figure 36.4   Relationship between sea urchin fertilization in 100% water quality adjusted porewater and cumulative ERM index for sediments from Tampa Bay, Florida. (From Carr, R.S. et al., *Environmental Contamination and Toxicology*, 15, 1996, SETAC Press. With permission.)

lost (e.g., volatile organics) or concentrated (e.g., oil and PCBs) by centrifugation or lost through adsorption (e.g., chlorinated organics) during filtration. Therefore, it is highly recommended that:

1. Only one extraction method should be used for all samples within a particular study to minimize the influence of this variable among samples.
2. If many different sediment texture types, including sandy sediments, are to be compared, pneumatic extraction is the most efficient method.
3. If the contaminants of concern are highly hydrophobic and the sediments are fine grained, centrifugation is the method of choice for minimizing the loss of contaminants due to adsorption.
4. Regardless of the method used to extract the porewater initially, centrifuge the sample prior to testing or freezing to remove suspended particulate material.
5. Do not use glass fiber filters to remove particulates unless you are also attempting to remove the hydrophobic organic compounds from solution as well.
6. Keep sediment cool (~4°C) after collection and extract the porewater as soon as possible after collection; composite samples should be gently rehomogenized prior to processing.
7. Composite all porewater collected from the same station and store subsamples of the composite sample so that spare samples are available for retesting or additional analyses.
8. For large sediment surveys, test as many samples as possible together to minimize between-test variability.
9. Extraction devices, glassware, and other nondisposable objects which come in contact with the porewater sample should be meticulously cleaned and great care taken to avoid contamination of the sample when using sensitive toxicity tests.

The biggest source of variability when conducting sensitive porewater toxicity tests, in my experience, is the quality of the test organisms. In order to minimize the effects of this variable, it is recommended that:

10. Laboratory-cultured organisms or organisms easily held in the laboratory (for the purpose of obtaining gametes or young) or locally collected species should be used in testing.
11. Ideally the test organism will be readily available during most, if not all, times of year.
12. A reference toxicant test, a reference porewater, a reconstituted brine control (if salinity adjustments were required) should always be run in conjunction with any test in order to monitor the quality of the test organisms and media.

B. Research Needs

Most of the recent sediment quality assessment studies in which porewater toxicity testing has been employed have not included chemical analysis of porewater contaminant concentrations. Other than a few spiked-sediment studies, the parameters which control equilibrium partitioning of contaminants in conjunction with toxicity tests have not been well studied. There is a definite need for studies to evaluate the influence of dissolved organic material (DOM) on the bioavailability of porewater-associated contaminants. By using the approaches developed for toxicity identification evaluation (TIE) studies, the influence of chemical and physical interactions among the sediment and porewater constituents on bioavailability could be ascertained, thereby allowing models, such as equilibrium partitioning theory, to be refined and improved.

The porewater toxicity test approach has been shown to be amenable with a wide variety of species. Many of the microscale methods described in this book could likely be adopted for testing porewater. While this review has focused on marine and estuarine test species, most of the general guidance is applicable to freshwater species as well. There is a need to evaluate more freshwater species, particularly embryo–larval life stages of benthic-dwelling organisms. The evaluation of new species should ideally include comparisons with other standardized sediment test methods. The fact that recent studies have observed excellent concordance between toxicity predicted by sediment quality guideline values and toxicity measured with pore toxicity tests[29,57] suggests that this approach has the potential to detect chronic benthic impacts as opposed to acute lethality, which is the endpoint used in standard whole-sediment tests used for regulatory testing in the U.S.

REFERENCES

1. U.S. Environmental Protection Agency/U.S. Army Corps of Engineers (USEPA/USACE), Ecological evaluation of proposed discharge of dredged material into ocean waters, Environmental Effects Laboratory, U.S. Army Engineer Waterways Experiment Station, Vicksburg, Mississippi, 1977.
2. U.S. Environmental Protection Agency/U.S. Army Corps of Engineers (USEPA/USACE), Evaluation of dredged material proposed for ocean disposal, EPA-503/8-91/001, 1991.
3. American Society for Testing and Materials (ASTM), Standard guide for conducting 10-day static sediment toxicity tests with marine and estuarine amphipods, ASTM, E1367-92, ASTM, Philadelphia, PA, 24 pp., 1993.
4. Adams, W. J., Kimerle, R. A. and Mosher, R. G., Aquatic safety assessment of chemicals sorbed to sediments, in *Aquatic Toxicology and Hazard Assessment: Seventh Symposium.* ASTM STP 854, Cardwell, R. D., Purdy, R., and Bahner, R. C., Eds., American Society for Testing and Materials, Philadelphia, PA, 429, 1985.
5. Di Toro, D. M., Zarba, C. S., Hansen, D. J., Berry, W. J., Swartz, R. C., Cowan, C. E., Pavlou, S. P., Allen, H. E., Thomas, N. A., and Paquin, P. R., Pre-draft technical basis for establishing sediment quality criteria for non-ionic organic chemicals using equilibrium partitioning, US Environmental Protection Agency, Office of Water, Washington, DC, 1991.
6. U.S. Environmental Protection Agency (USEPA), Technical basis for deriving sediment quality criteria for nonionic organic contaminants for the protection of benthic organisms by using equilibrium partitioning, EPA-822-R-93-011, USEPA Office of Water, Washington, DC, 62 pp., 1993.
7. Edmunds W. M. and Bath, A. H., Centrifuge extraction and chemical analysis of interstitial waters, *Environmental Science and Technology,* 10, 467, 1976.
8. Landrum P. F., Nihart S. R., Eadie B. J., and Herche L. R., Reduction in bioavailability of organic contaminants to the amphipod *Pontoporeia hoyi* by dissolved organic matter of sediment interstitial waters, *Environmental Toxicology and Chemistry,* 6, 11, 1987.
9. Giesy J. P., Graney R. L., Newsted J. L., Rosui C. J., Benda A., Kreis R. G., and Horvath F. J., Comparison of three sediment bioassay methods using Detroit River sediments, *Environmental Toxicology and Chemistry,* 7, 483,1988.

10. Giesy, J. P., Rosiu, C. J., Graney, R. L., and Henry, M. G., Benthic invertebrate bioassays with toxic sediment and porewater, *Environmental Toxicology and Chemistry*, 9, 233, 1990.

11. Schults, D. W., Ferraro, S. P., Smith, L. M., Roberts, F. A., and Poindexter, C. K., A comparison of methods for collecting interstitial water for trace organic compounds and metals analyses, *Water Research*, 26, 989, 1992.

12. Burgess, R. M., Schweitzer, K. A., McKinney, R. A., and Phelps, D. K., Contaminated marine sediments: water column and interstitial toxic effects, *Environmental Toxicology and Chemistry*, 12, 127, 1993.

13. Carr R. S., and Chapman, D. C., Comparison of methods for conducting marine and estuarine sediment porewater toxicity tests — Extraction, storage, and handling techniques, *Archives of Environmental Contamination and Toxicology*, 28, 69, 1995.

14. Chin, Y. and Gschwend, P. M., The abundance, distribution, and configuration of porewater organic colloids in recent sediment, *Geochimica Cosmochimica Acta*, 55, 1309, 1991.

15. Word, J. Q., Ward, J. A., Franklin, L. M., Cullinan, V. I., and Kiesser, S. L., Evaluation of the equilibrium partitioning theory for estimating the toxicity of the nonpolar organic compound DDT to the sediment dwelling amphipod *Rhepoxynius abronius,* Report prepared by Battelle Marine Research Laboratory, Sequim, Washington for U.S. Environmental Protection Agency, Criteria and Standards Division, Washington, DC, 60 pp., 1987.

16. Adams, D. D., Sediment porewater sampling, in *Handbook of Techniques for Aquatic Sediments Sampling*, Mudroch, A. and MacKnight, S. D., Eds., CRC Press, Inc., Boca Raton, Florida, Chap. 7, 1991.

17. Saager, P. M., Sweerts, J.-P., and Ellermeijer, H. J., A simple porewater sampler for coarse, sandy sediments with low porosity, *Limnological Oceanography*, 35, 747, 1988.

18. Reeburgh, W. S., An improved interstitial water sampler, *Limnological Oceanography*, 12, 163, 1967.

19. Presley, B. J., Brooks, R. R., and Kappel, H. M., A simple squeezer for removal of interstitial water from ocean sediments, *Journal of Marine Research*, 25, 355, 1967.

20. Bender, M., Martin, W., Hess, J., Sayles, F., Ball, L., and Lambert, C., A whole-core squeezer for interfacial porewater sampling, *Limnological Oceanography*, 32, 1214, 1987.

21. Carr, R. S., Williams, J. W., and Fragata, C. T. B., Development and evaluation of a novel marine sediment porewater toxicity test with the polychaete *Dinophilus gyrociliatus, Environmental Contamination and Toxicology*, 8, 533, 1989.

22. Winger, P. V. and Lasier, P. J., A vacuum-operated porewater extractor for estuarine and freshwater sediments, *Archives of Environmental Contamination and Toxicology*, 21, 321, 1991.

23. Sayles, F. L., Mangelsdorf, P. C., Jr., Wilson, T. R. S., and Hume, D. N., A sampler for the in situ collection of marine sedimentary porewaters, *Deep Sea Research*, 23, 259, 1976.

24. Knezovich, J. P. and Harrison, F. L., A new method for determining the concentrations of volatile organic compounds in sediment interstitial water, *Bulletin of Environmental Contamination and Toxicology*, 38, 937, 1987.

25. Hoepner, J., Design and use of a diffusion sampler for interstitial water from fine grained sediments, *Environmental Technology Letters,* 2, 187, 1981.

26. Carignan, R., Interstitial water sampling by dialysis: methodological notes, *Limnological Oceanography*, 29, 667, 1984.

27. Carignan, R., Rapin, F. and Tessier, A., Sediment porewater sampling for metal analysis: a comparison of techniques, *Geochimica Cosmochimica Acta*, 49, 2493, 1985.

28. Di Toro, D. M., Mahoney, J. D., Hansen, D. J., Scott, K. J., Hicks, M. B., Mayr, S. M., and Redmond, M.S., Toxicity of cadmium in sediments: the role of acid volatile sulfide, *Environmental Toxicology and Chemistry*, 9, 1487, 1990.

29. Carr, R. S., Long., E. R., Chapman, D. C., Thursby, G., Biedenbach, J. M., Windom, H., Sloane, G., and Wolfe, D. A., Toxicity assessment studies of contaminated sediments in Tampa Bay, Florida, *Environ. Toxicol. Chem.,* 15, 1218, 1996.

30. Ankley, G. T., Lodge, K., Call, D. J., Balcer, M. D., Brooke, L. T., Cook, P. M., Kreis, R. J., Jr., Carlson, A. R., Johnson, R. D., Niemi, G. J., Hoke, R. A., West, C. W., Giesy, J. P., Jones, P. D., and Fuying, Z. C., Integrated assessment of contaminated sediments in the lower Fox River and Green Bay, Wisconsin, *Ecotoxicology and Environmental Safety*, 23, 46, 1992.

31. Hoskin, S. J., Evaluation of a new sediment toxicity test: correlation with a sediment quality triad, Master of Science Thesis, University of Texas Health Science Center at Houston, Houston, Texas, 57 pp., 1993.

32. Long, E. R. and Markel, R., An evaluation of the extent and magnitude of biological effects associated with chemical contaminants in San Francisco Bay, California, NOAA Technical Memorandum, NOS ORCA 64, 86 pp, 1992.

33. Hooten, R., and Carr, R. S., Development and application of a marine sediment porewater toxicity test using *Ulva fasciata* and *U. lactuca* zoospores, *Environmental Toxicology and Chemistry,* in press, 1997.

34. Blaise, C., Férard, J.-F., and Vasseur, P., Microplate toxicity tests with microalgae: a review, in *Microscale Testing in Aquatic Toxicology: Advances, Techniques and Practice*, Wells, P. G., Lee, K., and Blaise, C., Eds., CRC Press, Boca Raton, FL, 1998, Chap. 18.

35. Carr, R. S., Curran, M. D., and Mazurkiewicz, M., Evaluation of the archiannelid *Dinophilus gyro-ciliatus* for use in short-term life-cycle toxicity tests, *Environmental Contamination and Toxicology*, 5, 703, 1986.

36. Carr, R. S. and Chapman, D. C., Comparison of solid-phase and porewater approaches for assessing the quality of marine and estuarine sediments, *Chemistry and Ecology*, 7, 19, 1992.

37. Carr, R. S., Chapman, D. C., Presley, B. J., Biedenbach, J. M., Robertson, L., Boothe, P., Kilada, R., Wade, T., and Montagna, P., Sediment porewater toxicity assessment studies in the vicinity of offshore oil and gas production platforms in the Gulf of Mexico, *Canadian Journal of Fisheries and Aquatic Sciences,* 53, 2618, 1996.

38. Reish, D. J. and Carr, R. S., The effect of heavy metals on the survival, reproduction, development and life cycles for two species of polychaetous annelids, *Marine Pollution Bulletin* 9, 24, 1978.

39. Carr, R. S. and Reish, D. J., The effect of petroleum hydrocarbons on the survival and life history of polychaetous annelids, in *Fate and Effects of Petroleum Hydrocarbons in Marine Ecosystems and Organisms*, D.A. Wolfe, Ed., Pergamon Press, New York, 166, 1977.

40. Arenicola Marine, Development of Environment Canada's procedure for aquatic toxicity testing using polychaetes: selection of suitable species, report prepared for Environment Canada, contract No. KA168-2-2236 by Arenicola Marine, Dartmouth, NS, Canada, 1992.

41. Hunt, J.W., Anderson, B. S., and Phillips, B. M., Recent advances in microscale toxicity testing with marine mollusks, in *Microscale Testing in Aquatic Toxicology — Advances, Techniques and Practice*, Wells, P. G., Lee, K., and Blaise, C., Eds., CRC Press, Boca Raton, FL, 1998, Chap. 29.

42. Long, E. R., Buchman, M. R., Bay, S. M., Breteler, R. J., Carr, R. S., Chapman, P. M., Hose, J. E., Lissner, A. L., Scott, J., and Wolfe, D. A., Comparative evaluation of five toxicity tests with sediments from San Francisco Bay and Tomales Bay, California, *Environmental Toxicology and Chemistry,* 9, 1193, 1990.

43. Laughlin, R. B., Jr., Gustafson, R. G., and Pendoley, P., Acute toxicity of tributyltin (TBT) to early life history stages of the hard shell clam, *Mercenaria mercenaria*, *Bulletin of Environmental Contamination and Toxicology*, 42, 352, 1989.

44. Green, A. S., Chandler, G. T., and Blood, E. R., Aqueous-, pre-water-, and sediment-phase cadmium: toxicity relationships for a meiobenthic copepod, *Environmental Toxicology and Chemistry*, 12, 1497, 1993.

45. Carr, R.S., Toxicity testing of sediments from Massachusetts Bay, Massachusetts. Final report submitted to Arthur D. Little, Inc., 5 pp. + 7 tables and 4 attachments, 1993.

46. Carr, R.S., Sediment quality assessment survey of the Galveston Bay System, Galveston Bay National Estuary Program report, GBNEP-30, 101 pp., 1993.

47. U.S. Fish and Wildlife Service (USFWS), Toxicity testing of sediments from Charleston Harbor, South Carolina and vicinity, Final report submitted to National Oceanic and Atmospheric Administration, 7 pp. + 16 tables and 4 attachments, 1993.

48. National Biological Service (NBS), Survey of sediment toxicity in Pensacola Bay and St. Andrew Bay, Florida, Final report submitted to National Oceanic and Atmospheric Administration, 12 pp. + 24 tables and 5 attachments, 1994.

49. National Biological Service (NBS), Toxicity testing of sediment from western Florida and coastal South Carolina and Georgia. Final report submitted to National Oceanic and Atmospheric Administration, 14 pp. + 35 tables and 4 attachments, 1995.

50. National Biological Service (NBS), Toxicity testing of sediment from Biscayne Bay, Florida and surrounding areas. Final report submitted to National Oceanic and Atmospheric Administration, 11 pp. + 17 tables, 11 figures, and 4 attachments, 1995.

51. Long, E. R., Carr, R. S., Thursby, G. A., and Wolfe, D.A., Sediment toxicity in Tampa Bay: Incidence, severity, and spatial extent, *Florida Scientist*, 58, 163, 1995.

52. Carr, R. S., Chapman, D. C., Howard, C. L., and Biedenbach, J. M., Sediment Quality Triad assessment survey in the Galveston Bay Texas system, *Ecotoxicology*, 5, 1, 1996.

53. D'Unger, C. D., Chapman, D. C., and Carr, R. S., Discharge of oilfield produced water in Nueces Bay, Texas: a case study, *Environmental Management*, 20, 143, 1996.

54. Hose, J. E., Puffer, H. W., Oshida, P. S., and Bay, S. M., Developmental and cytogenetic abnormalities induced in the purple sea urchin by environmental levels of benzo(a)pyrene, *Archives of Environmental Contamination and Toxicology*, 12, 319, 1983.

55. Roach, R. W., Carr, R. S., Howard, C. L., and Cain, B. W., An assessment of produced water impacts in Galveston Bay system, U.S. Fish and Wildlife Service Report, 61 pp., 1992.

56. Long, E. R., MacDonald, D. D., Smith, S. L., and Calder, F. D., Incidence of adverse biological effects within ranges of chemical concentrations in marine and estuarine sediments, *Environmental Management*, 19, 81, 1995.

57. MacDonald, D. D., Carr, R. S., Calder, F. D., Long, E. R., and Ingersoll, C. G., Development and evaluation of sediment quality guidelines for Florida coastal waters, *Ecotoxicology*, 5, 253, 1996.

CHAPTER **37**

Correspondence of a Microscale Toxicity Test to Responses to Toxicants in Natural Systems

John Cairns, Jr., B. R. Niederlehner, and Eric P. Smith

CONTENTS

I. INTRODUCTION

Scientists often use models to reduce the complexity of intractable problems and to increase their understanding of the natural world. These models are recreations of the system of interest, but they incorporate only the parts relevant to the problem at hand.[1] When evaluating the ecological risk posed by the use and disposal of chemicals, the dominant model is the single species toxicity test. Much information about toxicity can be obtained from traditional toxicity tests through observing the response of individual species to the chemical. However, many parts relevant to the problem of predicting ecological risk are excluded from this model. Interactions between species and between species and physical features of their environment are key determinants of ecological outcome. Examples include disease resistance, predation rate, replacement of sensitive species by more resistant ones, changes in communities of organisms used for food, avoidance, biologically mediated changes in toxicant fate, or chemically mediated habitat destruction. These interactions cannot be assessed in tests using an individual species, but are the mechanics by which communities are shaped and reshaped by toxicant stress. However, these interactions can be included in models for studying ecological risk by using community-level toxicity tests.

Interactions shape the community. However, there are too many to monitor individually, and it is very difficult to know which interactions will be relevant in any particular situation or time. It is more efficient to monitor community-level endpoints that integrate all these interactions over many species. Field survey methods that index the health of indigenous communities have been developed over a long period of experience.[2,3] Endpoints such as biotic diversity and trophic structure are considered good evidence of ecological health in field surveys. Because these endpoints correspond closely to societal goals, they are more relevant. These community-level endpoints avoid extrapolating from one endpoint that is easily measured to another qualitatively different endpoint with relevance to management goals. Yet, these community-level endpoints can be measured directly in easily replicated and controlled microscale model systems and then used to establish cause-and-effect relationships between chemical and adverse ecological effects at the community level.

II. THE ARTIFICIAL SUBSTRATE MICROCOSM TEST (AS–M)

Many microcosms have been designed to reduce relevant characteristics of communities to a manageable size (see review by Huckabee[4]). Among the more well-known microscale toxicity tests at the community-level are the standardized aquatic microcosm[5] (SAM), the mixed flask assay[6] (MFA), and the soil core microcosm.[7] The artificial substrate microcosm (AS–M) is another microcosm designed to incorporate increased complexity into evaluations of toxicity.[8,9]

The AS–M is similar in both approach and physical and temporal scale to conventional toxicity tests,[10] but the community-level endpoints monitored in AS–M are more comparable to those used to evaluate natural systems.[3] In this test, microbial communities are obtained by colonizing artificial substrates, small blocks (6 × 5 × 3.75 cm) of reticulated polyurethane foam (PF), in a reference system. The interstices of the artificial substrate are colonized by indigenous microbes — bacteria, fungi, algae, protozoans, and rotifers, with occasional ostracods, cladocerans, gastrotrichs, insects, and nematodes. These naturally derived communities are then collected and transported to the laboratory for use in designed experiments.

In a typical toxicity test, three replicates of five concentrations and a control are tested. The test monitors both functional and structural changes in transplanted microbial communities and colonization, i.e., the establishment of new communities on initially barren substrates. Many taxonomic groups are represented on the substrate and are included in functional and some structural measures. However, taxonomic identifications have been limited to motile unicellular eukaryotes, i.e., protozoans. This group is a particularly effective index of effects to the community because of its unique functional and phyletic diversity. What other single taxonomic group encompasses primary producers, decomposers, and several levels of consumers? The communities of organisms identified included all these groups — from primary producers (e.g., *Chlamydomonas*) to voracious predators (e.g., *Acineta*), with many bacteriovores and algivores in between.[11] Tests are monitored for 7 to 21 d. Comparisons between results after 7 d and 21 d suggest that, in most cases, conclusions are comparable.[12] Of the many endpoints that have been monitored in this system, taxonomic richness and composition of the protozoan component of newly developed communities have proven to be particularly consistent and are sensitive indicators of stress across many toxicant types.[13,14] Concentration-response curves that describe the loss of taxonomic richness over increasing concentration have been derived for over 30 toxicants. In addition, the Peoples Republic of China now routinely uses this system for monitoring surface water quality. Professor Shen Yun-fen of Academia Sinica of the Institute of Hydrobiology at Wuhan in the Peoples Republic of China has trained significant numbers of people to do this relatively quickly.[15]

III. VALIDATION OF MICROSCALE TOXICITY TESTS

There has always been a great deal of confusion about what is meant by validation of toxicity tests. The general concept must refer to truth testing: is a toxicity test able to accomplish what is expected of it? A recent editorial[16] implies that a field validation of toxicity test results depends upon finding identical responses in the field and laboratory. The expectation in this case may be to prove that a particular chemical is the cause of an impairment. Suter[17] lists the environmental analogs to Koch's postulates as: 1) the adverse effects are regularly associated with exposure to the toxicant, 2) indicators of exposure to the toxicant are found in adversely affected organisms, 3) toxic effects are seen when normal organisms are exposed to the toxicant, and 4) the *identical* indicators of exposure and effects are identified in the laboratory and the field.

In predictive applications, the validity of a toxicity test depends on the ability to use test results to predict environmental outcome in natural systems with a degree of precision adequate to management needs. If the predictions are precise enough to avoid expensive management errors, they are valid. If not, then expensive treatment facilities may be built without any resulting biological benefit or environmental damage that results in long-term loss of ecosystem services and amenities or remediation costs.

We have suggested that correlations of laboratory toxicity and field impairment are necessary, but are not sufficiently rigorous tests of validity for predictive methods.[18] A more demanding criterion for predictive validity requires that the costs of management decisions made on the basis of imperfect information not be too high. An operational definition of this concept would require that valid tests have predictive errors within a predefined range, ideally based on management costs, but tentatively set at 20% for the preliminary analyses.

Using this 20% predictive error as a criterion, we have tested the validity of the AS–M in two ways. First, we have compared predictions of taxonomic loss made from AS–M toxicity tests of effluents to observed ecological outcome in receiving systems. Second, we have used the AS–M to derive tolerance distributions that describe the range in sensitivity to a toxicant across diverse taxa and then compared these distributions to those observed in ecosystem experiments across a gradient of toxicant stress.

A. Predictions of Taxonomic Loss

When predictions of environmental effects based on the AS–M were compared to environmental outcome in five systems that received point source discharges, errors were <20% in four of five comparisons (Figure 37.1).[12] In the remaining case (field study #2), predictive accuracy was very poor. Both AS–M and traditional toxicity tests with *Ceriodaphnia* and fathead minnows predicted no instream toxicity, perhaps due to volatility or degradation of toxic components in the effluent. The mean error for all five predictions of environmental outcome was 18.4 ± 21.4%. The magnitude of these prediction errors was compared to those from more traditional tests used in the U.S. EPA's Complex Effluent Toxicity Testing Program.[19] Predictive errors varied from 2 to 60% in seven comparisons. Error was <20% in only one of seven comparisons, and the mean error was 28.1 ± 17.3%.

B. Tolerance Distributions

Although a toxicity test using a single species provides useful information about toxicity, it does not provide much information about how the thousands of other types of organisms likely to be affected by any environmental discharge will respond.[20] Yet, the goal of some protective efforts has been operationally defined as protecting 95% of the species.[21] Obviously, in order to do this, information about the distribution of tolerances of various organisms to a chemical is necessary.

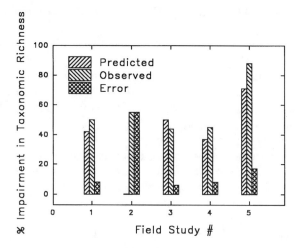

Figure 37.1   Comparison of predictions of environmental outcome based on the AS–M to contemporaneously observed field impact. (From Niederlehner, B. R. and Cairns, J., Jr., *Ecological Toxicity Testing: Scale, Complexity, and Relevance*, Cairns, J., Jr., and Niederlehner, B. R., Eds., CRC Press, Boca Raton, FL, 1995, 123. With permission.)

Often, this information is provided by gathering toxicity test data for many individual species. From the results of many population-level toxicity tests, some inferences about the distribution of tolerances can be made. Generally, there are a few very sensitive species, a few very tolerant species, and the great majority responds somewhere in between, like a Gaussian or bell curve. As such, the values are usually described with a log-normal or log-logistic distribution. The mean and standard deviation describing this distribution are used to extrapolate to a chemical concentration corresponding to the protection of 95% of the distribution.[22,23]

This extrapolation approach based on tolerance distributions is much more appealing than the use of arbitrary application factors, but it has a number of uncertainties and assumptions, both statistical and ecological.[17,24,25] The first of these assumptions is that the endpoints (usually NOECs) are random samples from the tolerance distribution. This is almost certainly not the case. Species used in laboratory tests are limited to those tolerant of laboratory conditions and amenable to human time scales. In addition, a previous assumption has been that more sensitive organisms make better toxicity test subjects. As such, organisms for which there are years of experience and well-developed test methods have been systematically selected for a lack of tolerance to organic pollution.

A second uncertainty concerns the appropriateness of the distribution used to describe the endpoint. While a log-normal or log-logistic distribution may be appropriate, there are usually not enough data to check for goodness-of-fit using a Komolgorov-Smirnov or other statistical test. If the underlying distribution is not normal or logistic, especially if it is not symmetric, the extrapolated estimates could be grossly in error. Stated simply, very little is known about the shape of the tolerance distribution for the species in an ecosystem.

Finally, there is the holist's question: will an ecosystem be protected if the chemical concentration is below that affecting most species individually? If an ecosystem is nothing more than the sum of the tested parts and if the few endpoints tested, usually reproduction or growth, are as sensitive as all other responses the organism is capable of making, then the extrapolations may be accurate.

Microscale community-level tests provide an alternative line of evidence in describing the distribution of tolerances of organisms to a chemical that address the first and third of these objections. Each artificial substrate has from 20 to 100 protozoan species. Because these communities are obtained from systems that have not been exposed to the test chemical, this community has not been selected for tolerance to the chemical. The species making up any community

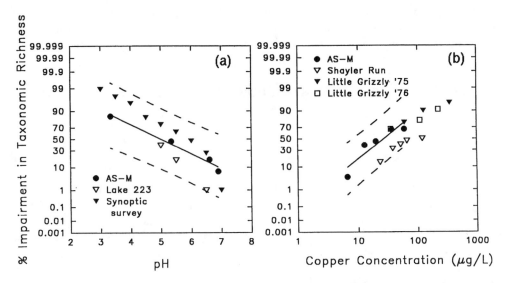

Figure 37.2   Comparison of tolerance distributions based on the AS–M to existing field data across gradients
of stress. Dashed lines are the 95% prediction interval for the AS–M based model. Data for field
studies with pH were taken from Schindler et al.[27] and Eilers et al.[28] Data for field studies with
copper were taken from Sheehan and Winner.[30] (From Niederlehner, B. R. and Cairns, J., Jr.,
Ecological Toxicity Testing: Scale, Complexity, and Relevance, Cairns, J., Jr., and Niederlehner,
B. R., Eds., CRC Press, Boca Raton, FL, 1995, 123. With permission.)

developed in absence of pollution stress are likely to be a random sample of tolerances. When the
loss in taxonomic richness of these communities is plotted over increasing chemical stress, the
concentration-response curve is analogous to a cumulative tolerance distribution. At each increase
in concentration, additional species are lost, in addition to all the species lost at lower concentrations.
In this approach, individual species are not tracked but integrated as a measure of taxonomic loss.

Niederlehner and Cairns[12] compared predictions of loss of taxonomic richness with chemical
stress from the AS–M to the published results of several ecosystem experiments. Loss of taxonomic
richness over pH gradients has been reported for the ecosystem experiment in Lake 223 in the
Experimental Lakes Area of Canada over the progress of its acidification.[26,27] In addition, Eilers
and colleagues[28] compiled tolerance distributions over pH for eight major taxonomic groups based
on more than 1000 observations of minimum pH tolerance from both field and laboratory studies.
Loss of taxonomic richness over copper gradients in natural systems has also been reported. The
Shayler Run study, in which copper was added to a second-order stream in southwest Ohio, remains
a classic ecosystem experiment.[29] Sheehan and Winner[30] reported loss of macroinvertebrate taxa
across the copper gradient in Shayler Run and reported the loss of macroinvertebrate richness in
Little Grizzly Creek, a second-order stream in the Sierra Nevada Mountains in California receiving
drainage from a copper mine.

The pH tolerance model based on the AS–M fell between the two field-based distributions
(Figure 37.2a). Impairment predicted from the AS–M was correlated to observed impact in the field
studies of pH (N = 11, p = 0.0002, r = 0.8945), and field observations fell within the 95% prediction
interval based on the AS–M. Predictions of environmental outcome based on the AS–M were within
20% for 10 of 11 cases. However, the AS–M overestimated impairment in Lake 223 at a pH of
5.5 by >20% (43% vs. 17%). The mean error of predictions of impairment compared to both field
data sets was 15.1 ± 6.2% (N = 11).

In the case of copper (Figure 37.2b), predicted impairments were correlated to observed outcome
in field studies (N = 11, p = 0.0068, r = 0.5758). The mean predictive error was 20.5 ± 16.6%. But
only 6 of 11 predictions were within 20%. The predictions from the AS–M consistently overestimated macroinvertebrate response in Shayler Run, perhaps partially due to the substantial difference

in hardness in the two systems (244 mg/L in Shayler Run vs. 55 mg/L in the AS–M). Better correspondence between distributions might be possible if chemical data for each system allowed the concentrations to be plotted as ionic copper rather than the total of the dissolved fraction.

IV. PERSISTENT OBSTACLES TO MICROSCALE COMMUNITY TOXICITY TESTS

A. Cost

It is sometimes stated that the cost of multispecies toxicity tests is excessive. At one time, this statement may have been true. Multispecies testing clearly can be more costly than single species testing when fish and macroinvertebrates are the focus of large-scale, long-term, multiple-endpoint studies. Artificial pond studies can cost millions of dollars. However, microscale community-level toxicity tests reduce the scale and the cost of the tests without proportionately reducing the complexity of the tests. The approximate cost of an AS–M is $3,321.[12] Based on a cost for short-term chronic tests with single species of $1,500 (generalized from price lists received in 1994), using three to seven of these tests to make a tolerance distribution based prediction of community-level effects would cost $4,500 to 10,500.

Perhaps more important, two types of costs are associated with choosing a toxicity test method: 1) the actual costs of the toxicity test and 2) the cost of making decisions based on that test. If a test is cheap but insufficiently accurate, no money will be saved by using it. If there are false positives, expensive treatment facilities might be built that yield no biological benefit. If there are false negatives, environmental damage will occur and both the ecosystem and public opinion may be expensive to remediate.

B. Standardization

It has been repeatedly suggested that multispecies tests are difficult to standardize and their results are too highly variable to be used in decision making. Toxicity tests are standardized to ensure repeatability, i.e., tests conducted one after another and reproducibility by different laboratories should yield similar conclusions. Because there is great variability in the structure and function of microbial communities, it is reasonable to question the repeatability of tests using these widely varying communities. However, concentration-response models describing the distribution of tolerances of organisms to toxic substances may be similar across varying community composition as long as a sufficiently large number of species are sampled and no source community is subjected to selective pressure from contamination in the natural system. When the replicability of the AS–M was compared to that for more traditional toxicity tests, the community-level test did slightly better than single species tests, with mean ratios of maximum to minimum effect levels of 2.4 and 3.0, respectively.[12] Because community-level test results are reasonably repeatable, there seems to be no need to recreate the same community over and over again for testing purposes. The relevant structural characteristic (i.e., the tolerance distribution) is "standard." While there is no question that different ecosystems respond differently to toxic stress, it may be that toxicant fate is the major source of such differences.

Standardization systematically eliminates the dynamics and complexities of ecosystems. In the most rigorous applications, relevant dynamics and complexities will be reintroduced into the heuristic model at some later stage. But if decisions have to be made early on in the process, over-standardization of toxicity testing at all scales may diminish the possibility of producing predictive models that correspond well to events in natural systems. Standard practices that retain sufficient flexibility to be modified for highly site-specific situations are a desirable quality-control mechanism. However, highly standardized tests displace the need for professional judgment from the stage of designing and carrying out an appropriate test to the stage where test information is applied to a specific environmental problem, extrapolating from standardized tests to a site-specific application of the results.

C. Uncharismatic Microfauna

One of the characteristics by which any ecological toxicity test is judged is its social relevance.[31-33] Does society in general care about this consequence? The farther a test endpoint is from obvious damage to a valued ecosystem component, the harder it is to use for regulatory purposes. While at least some portion of the population cares about the fate of bald eagles or sea otters, very few will react to local reductions in microbial diversity with a sigh or a shudder. Obviously, microscale tests with microorganisms face considerable challenges in demonstrating their social relevance. While scientists are aware of the complete dependence of human society on the ecosystem services provided by microorganisms (e.g., the breakdown of wastes, the regeneration of nutrients), the general view is likely to be that microorganisms are primitive, simple, uninteresting, and unimportant. Nonscientists are more likely to believe tests based on familiar organisms, such as fish or mice. Even scientists may think of microbial communities as interchangeable; one being no worse than any other. Microscale toxicity tests simply require a level of environmental and ecological literacy that, unfortunately, is rather rare.

V. CONCLUSIONS

Microscale community toxicity tests provide an opportunity to evaluate cause-and-effect relationships between chemicals and community-level endpoints. The results of such tests are reasonably repeatable. They have provided predictions of environmental outcome that were within 20% of observed field effects in 20 out of 27 comparisons. Of the failures to meet this criterion, six were false positives, suggesting a biological impairment where none was observed. One was a false negative, predicting no impairment where a substantial impairment was observed. In that case, traditional tests also failed to predict an adverse effect. While the 74% rate of success in predicting ecological outcome with reasonable accuracy is encouraging, the acceptance of such tests remains limited due to factors other than validity. Misinformation about cost, replicability, standardization, and the uncharismatic nature of microbes remain as obstacles to microscale community-level toxicity tests.

REFERENCES

1. O'Neill, R. V., DeAngelis, D. L., Waide, J. B., and Allen, T. F. H., *A Hierarchical Concept of Ecosystems*, Princeton University Press, Princeton, NJ, 1986.
2. Karr, J. R., Biological integrity: a long-neglected aspect of water resource management, *Ecol. Appl.* 1, 66, 1991.
3. Plafkin, J. L., Barbour, M. T., Porter, K. D., Gross, S. K., and Hughes, R. M., *Rapid Bioassessment Protocols for Use in Streams and Rivers: Benthic Macroinvertebrates and Fish*, EPA 444/4-89-001, USEPA, Washington, DC, 1989.
4. Huckabee, J. W., Evaluation of tests to predict chemical injury to ecosystems: microcosms, in *Methods for Estimating Risk of Chemical Injury: Human and Non-human Biota and Ecosystems, SCOPE*, Vouk, V. B., Butler, G. C., Hoel, D. G., and Peakall, D. B., Eds., John Wiley & Sons, New York, 1985, 637.
5. Taub, F. B., Measurement of pollution in standardized aquatic microcosms, in *Concepts in Marine Pollution Measurements*, White, H. H., Ed., University of Maryland Press, College Park, MD, 1984, 159.

6. Leffler, J. W., The use of self-selected, generic aquatic microcosms for pollution effects assessment, in *Concepts in Marine Pollution Measurements*, White, H. H., Ed., University of Maryland Press, College Park, MD, 1984, 139.

7. Van Voris, P., Tolle, D. A., Arthur, M. F., and Chesson, J., Terrestrial microcosms: validation, applications, and cost-benefit analysis, in *Multispecies Toxicity Testing*, Cairns, J., Jr., Ed., Pergamon Press, New York, 1985, 117.

8. Cairns, J., Jr., Pratt, J. R., and Niederlehner, B. R., A provisional multispecies toxicity test using indigenous organisms, *J. Test. Eval.*, 13, 316, 1985.

9. Pratt, J. R. and Bowers, N. J., A microcosm procedure for estimating ecological effects of chemicals and mixtures, *Toxicity Assess.*, 5, 189, 1990.

10. USEPA, *Short-term Methods for Measuring the Chronic Toxicity of Effluents and Receiving Waters to Freshwater Organisms*, 3rd ed., EPA-600/4-91-002, Washington, DC, 1991.

11. Pratt, J. R. and Cairns, J., Jr., Functional groups in the Protozoa: Roles in differing ecosystems, *J. Protozool.*, 32, 95, 1985.

12. Niederlehner, B. R. and Cairns, J., Jr., Naturally derived microbial communities as receptors in toxicity tests, in *Ecological Toxicity Testing: Scale, Complexity, and Relevance*, Cairns, J., Jr., and Niederlehner, B. R., Eds., CRC Press, Boca Raton, FL, 1995, 123.

13. Pratt, J. R. and Bowers, N. J., Variability of community metrics: detecting changes in structure and function, *Environ. Toxicol. Chem.*, 11, 451, 1992.

14. Niederlehner, B. R. and Cairns, J., Jr., Consistency and sensitivity of community-level endpoints in microcosm tests, *J. Aquat. Ecosystem Health*, 3, 93, 1994.

15. Shen, Y., Feng, W., Gu, M., Wang, S., Wu, J., and Tan, Y. *Monitoring of River Pollution: Evaluation of Water Pollution by Using PFU Microbial Communities in Hanjiang River*, Centre File 91-0176-02, Institute of Hydrobiology, The Chinese Academy of Science, China Architecture and Building Press, Wuhan, 1994.

16. Chapman, P. M., Do sediment toxicity tests require field validation?, *Environ. Toxicol. Chem.*, 14, 1451, 1995.

17. Suter, G. W., II, Retrospective risk assessment, in *Ecological Risk Assessment*, Suter, G. W. II, Ed., Lewis Publishers, Chelsea, MI, 1993, 311.

18. Cairns, J., Jr., and Smith, E. P., Developing a statistical support system for environmental hazard evaluation, *Hydrobiologia*, 184, 143, 1989.

19. Parkhurst, B. R., Marcus, M. D., and Noel, L. E., *Review of the Results of the Complex Effluent Toxicity Testing Program*, Utility Water Act Group Report, 1990.

20. Mayer, F. L., Jr., Deans, C. H., and Smith, A. G., *Inter-taxa Correlations for Toxicity to Aquatic Organisms*, USEPA 600/x-87/332, National Technical Information Service, Springfield, VA, 1987.

21. Stephan, C. E., Mount, D. I., Hansen, D. J., Gentile, J. H., Chapman, G. A., and Brungs, W. A., *Guidelines for deriving numerical national water quality criteria for the protection of aquatic organisms and their uses,* NTIS PB85-227049, Springfield, VA, 1985.

22. Wagner, C. and Lokke, H., Estimation of ecotoxicological protection levels from NOEC toxicity data, *Water Res.*, 25, 1237, 1990.

23. Aldenberg, T. and Slob, W., Confidence limits for hazardous concentrations based on logistically distributed NOEC toxicity data, *Ecotoxicol. Environ. Saf.*, 25, 48, 1993.

24. Smith, E. P. and Cairns, J., Jr., Extrapolation methods for setting ecological standards for water quality: statistical and ecological concerns, *Ecotoxicology*, 2, 203, 1993.

25. Forbes, T. L. and Forbes, V. W., A critique of the use of distribution-based extrapolation models in ecotoxicology, *Functional Ecol.*, 7, 249, 1993.

26. Schindler, D., Mills, K., Malley, D., Findlay, D., Shearer, J., Davies, I., Turner, M., Linsey, G., and Cruikshank, D., Long-term ecosystem stress: the effects of years of experimental acidification on a small lake, *Science*, 228, 1395, 1985.

27. Schindler, D. W., Dasian, S. E. M., and Hesslein, R. H., Losses of biota from American aquatic communities due to acid rain, *Environ. Monitor. Assess.*, 12, 269, 1989.

28. Eilers, J. M., Lien, G. J., and Berg, R. G., *Aquatic Organisms in Acidic Environments: A Literature Review*, Tech. Bull. No. 150, Department of Natural Resources, Madison, WI, 1984.

29. Geckler, J., Horning, W., Neiheisel, T., Pickering, Q., and Robinson, E., *Validity of Laboratory Tests for Predicting Copper Toxicity in Streams*, EPA/600/3-76-116, National Technical Information Service, Springfield, VA, 1976.

30. Sheehan, P. J. and Winner, R. W., Comparison of gradient studies in heavy-metal-polluted streams, in *Effects of Pollutants at the Ecosystem Level*, Sheehan, P. J., Miller, D. R., Butler, G. C., and Bourdeau, P., Eds., John Wiley & Sons, New York, 1984, 255.

31. Kelly, J. R. and Harwell, M. A., Indicators of ecosystem response and recovery, in *Ecotoxicology: Problems and Approaches*, Levin, S. A., Harwell, M. A., Kelly, J. R., and Kimball, K. D., Eds., Springer-Verlag, New York, 1989, 9.

32. Suter, G. W., II, Endpoints for regional ecological risk assessments, *Environ. Man.*, 14, 9, 1990.

33. Cairns, J., Jr., McCormick, P. V., and Niederlehner, B. R., A proposed framework for developing indicators of ecosystem health, *Hydrobiologia*, 263, 1, 1993.

CHAPTER 38

Role of Microbiotests in Contaminated Sediment Assessment Batteries

Philippe Ross

CONTENTS

I. INTRODUCTION

A. Test Batteries for Sediment Assessment

Environmental samples from areas of suspected pollution may contain complex mixtures of contaminants. In such cases, it is widely accepted that one cannot rely on a single bioassay with a "most sensitive species" to detect all potential hazards.[1-3] Different types of organisms may, due to fundamental differences in their biology, have varying sensitivities to different groups of toxic substances.[3] If the purpose of toxicity testing is to protect the environment from toxic insult, then a testing program must optimize its ability to detect pollution. For this reason, batteries of tests are now frequently used.[4-6] An ideal suite of bioassays should maximize the yield of nonredundant information.[7]

A battery of tests typically covers several trophic levels. The suite of tests should have good sensitivity to a broad range of toxic substances. For a test system to have value in a battery, it must have the characteristics of a good individual test, as well as compatibility with other components of the battery.[8] The battery approach for the assessment of contaminated sediments has been recommended in standardized forms by the governments of Canada,[9] the U.S.,[10] and the U.S.–Canadian International Joint Commission.[11]

B. Microbial Assays in Sediment Assessment Test Batteries

Designs for contaminated sediment screening surveys often call for large numbers of sampling sites, in order to ensure adequate coverage of potential "hot spots" and sufficient resolution for mapping procedures.[12] Thus it is desirable that sediment test batteries contain assays that are rapid and cost effective, and do not require large sample volumes. Microscale tests (including microbial tests), by the nature of the size and metabolism of the organisms used, often have these characteristics.[8] In virtually all aquatic ecosystems the most important trophic level in terms of energy flow, metabolic activity, system biogeochemistry, and nutrient cycling is found in the microbial community. A drastic alteration of bacterial processes will alter, more surely than any other impact, the fundamental character of life throughout the system. It is entirely appropriate to include a bacterial test system in any test battery designed to protect aquatic ecosystems.[13] This chapter discusses the use of microbial toxicity test systems in the assessment of contaminated sediments. It begins with brief discussions of indigenous, community-level enzyme activity tests and of liquid-phase tests on sediment extracts. An in-depth discussion of whole-sediment toxicity tests follows.

II. INDIGENOUS MICROBIAL ENZYME ACTIVITY

Microbial enzyme activity measured in whole sediment samples has been shown to be a meaningful indicator of contamination.[14,15] Microbial communities are useful monitors because they can respond rapidly to changes in environmental conditions and play a major role in ecosystem biogeochemistry.[16] Suites of microbial enzyme assays have been used as components of multi-trophic level bioassays in several studies.[12,14-16,18-25] Enzyme activities assayed in sediments include alkaline phosphatase, amylase, dehydrogenase, electron transport, p- galactosidase, p-glucosidase, and protease.[15]

Interpretation of enzyme assay results is difficult, however, for several reasons.[17] Measurement of enzyme activity can be a rapid process,[14] but determining the population densities of the various species that make up the bacterial community that produces those enzymes is a time-consuming endeavor which requires taxonomic expertise. Thus, spatial and temporal variations in components of sediment microbial populations usually cannot be compared to observations of enzyme activity.[17] Furthermore, low concentrations of toxicants can cause stimulation of enzyme activity,[14] due to nutrient co-occurrence, community adaptation, hormesis, or disruption of feedback mechanisms.[17] Finally, when contamination affects carbon or nitrogen substrates, it is possible that one enzyme system may be stimulated at the same time that another is inhibited.[17] Caution should be used in the interpretation of microbial enzyme activity data from field-collected sediments.

III. LIQUID-PHASE ASSAYS

A variety of bacterial species have been used in pure culture to assess the toxicity of sediment extracts (Table 38.1). Extraction methods used include mechanical removal of pore water (interstitial water), extraction by water (elutriation), and extraction by organic solvents.[17] Numerous studies have used such methods as components of batteries to assess sediment contamination in terms of toxicity or genotoxicity.[4,6,11,26-32] Of the eleven bacterial species listed in Table 38.1, the majority of sediment extract toxicity testing has been performed with the bacterial luminescence assay.

Table 38.1   Bacterial Species Used in Sediment Extract Assays

Species	Type of Assay	References
Vibrio fischeri, (formerly Photobacterium phosphoreum)	Microtox® Test; bioluminescence	51
Spirillum volutans	Motility inhibition test	27
Escherichia coli	SOS-Chromotest; genotoxicity	52
	TOXI-Chromotest	53
Nitrobacter sp.	Nitrogen metabolism	54
Azotobacter vinelandii	Nitrogen metabolism	55
Aeromonas hydrophila	Oxygen metabolism	56
Pseudomonas alcaligenes	Dehydrogenase activity test	57
Pseudomonas fluorescens	Fluorescence test	58
Salmonella typhimurium	Ames test; genotoxicity	59
Bacillus cereus	ECHA biocide monitor	29

A. Testing Sediment Extract Toxicity with *Vibrio fischeri*

The Microtox® Basic Test[33] measures toxicity by tracking decreases in the light output of *Vibrio fischeri,* formerly *Photobacterium phosphoreum,* a naturally occurring luminescent marine bacterium. The presence of toxic or bioreactive substances that disrupt or inhibit cellular metabolism will ultimately affect the cell's electron transport system, and can be quantified by measuring the change in luminescence of the test cell suspension. The strengths and weaknesses of toxicity testing with *V. fischeri* have been reviewed by Warren-Hicks et al.[34] and Ross.[13] The Microtox Basic Test has been used to assess the toxicity of bottom sediments by exposing bacteria to elutriates,[4,6,8] to extracted pore water,[5,31] or to organic solvent extracts.[35]

1. Cost and Efficiency Comparison

In a cost and efficiency comparison, Ross et al. examined a broad spectrum of existing bioassay methods in order to select a suite of bioassays to meeting criteria of sensitivity, rapidity, replicability, cost, and sample volume.[8] It was also stipulated that, as a group, the suite of assays should provide a broad range of biotypes and an acceptable degree of nonredundant information for testing sediment elutriates. A large number of tests were assessed from literature information and unpublished data. Rapidity was determined as the minimum time between sample receipt and the availability of results. Replicability was assessed by calculating coefficients of variation for replicate tests with reference toxicants. Cost was determined by applying a complex assessment of all resources required to perform a bioassay.[8] The Microtox test was found to be among the highest ranking tests in terms of exposure time, operator time expended, replicability, sample volume, and total cost.

The initial selection process resulted in the choice of four microscale biotest methods for sensitivity comparison: the Microtox test, an algal photosynthesis test (*Selenastrum capricornutum*) based on [14]carbon uptake,[36] a lettuce seed (*Latuca sativa*) root elongation test,[37] and a rotifer (*Brachionus calyciflorus*) survival assay.[38] Sensitivity was assessed by parallel tests with 10 reference toxicants (Table 38.2). The suite of four microscale tests exhibits a great deal of sensitivity. For each of the reference toxicants used in the study, at least one test in the battery was at or beyond the sensitivity level of the water flea *Daphnia magna* and the fathead minnow *Pimephales promelas,* two commonly used freshwater tests. The suite of tests also satisfies the group criterion for nonredundant information. Each biotest ranked as having the most sensitive response to at least one of the ten reference toxicants (Table 38.2). This indicates that the four tests are complementary, each having its own role in enhancing the overall sensitivity of the battery.[8] The bacterial test was relatively more sensitive to organic compounds than to metals, compared to the other three assays.

Table 38.2   Ranking of EC_{50}/IC_{50} Values of Four Microscale Toxicity Test Systems for Ten Reference Toxicants

	Microtox	Algae	Lettuce	Rotifer
Cadmium	4	1	3	2
Copper	3	1	4	2
Nickel	4	1	3	2
Zinc	3	2	4	1
$\sum$ ranks (inorganics)	14	5	14	7
Benzene	3	4	2	1
Methoxychlor	3	4	2	1
Pentachlorophenol	3	4	1	2
Phenanthrene	2	1	4	3
Phenol	1	3	4	2
Toluene	1	4	2	3
$\sum$ ranks (organics)	13	20	15	12
$\sum$ ranks (all chemicals)	27	25	29	19

Note: A value of 1 indicates the most sensitive response, while 4 indicates the least sensitive.

Adapted from Ross, P., Burnett, L., Kermode, C., and Timme, M., *Proceedings of the Seventeenth Annual Aquatic Toxicity Workshop*, Chapman, P., Bishay, F., Power, E., Hall, K., Harding, L., McLeay, D., Nassichuk, M., and Knapp, W., Eds., November 5-7, 1990, Vancouver, BC, Canada, Can. Tech. Rep. Fish. Aquat. Sci. No. 1774, 331, 1991.

IV. SOLID-PHASE ASSAYS

While liquid-phase microbial tests have yielded useful data, they do not expose the bacterial cells directly to sediment particles. Contaminant molecules bound to sediment particles are not in contact with the test organisms. In recent years the *V. fischeri* test has been modified to incorporate contact of the cells with the whole sediment sample, i.e., particles as well as pore water. Brouwer et al.[39] used a sediment contact assay in which *V. fischeri* cells were mixed with a slurry of whole-sediment sample and diluent medium. After exposure for 15 min, the bacteria were recovered by centrifugation. The light output of the cells was then measured in a photometer. Brouwer et al. used this procedure to screen sediments from 48 sites in Hamilton Harbor (Ontario, Canada).[39]

The direct-contact assay developed by AZUR Environmental (Carlsbad, CA, USA) is referred to as the Microtox Solid-Phase Test, or SPT.[40] The main difference between the SPT and the Brouwer et al.[39] procedure is that the SPT uses filtration, rather than centrifugation, to extract the bacteria after exposure.[41] Exposure times and conditions are also slightly different in the SPT protocol. The *V. fischeri* SPT has been proposed as a rapid-screening tool for sediment quality surveys.

Tay et al.[42] used the SPT to assess toxicity in samples from seven sites in Halifax Harbor (Nova Scotia, Canada), where contamination by heavy metals and organic compounds is known to exist. In this study, the Microtox SPT proved to be the most sensitive of five assays, including two amphipod tests (*Corophium volutator* and *Rhepoxynius abronius*), a polychaete test (*Neanthes* sp.), and the Microtox pore water assay. Hoskin[43] used the SPT to screen for toxicity at 34 sites in the Galveston Bay complex and found the test to be sensitive and in good agreement with a more costly Sediment Quality Triad[44,45] approach. Ross and Leitman[46] used the SPT as part of a multi-trophic battery of tests to monitor the effects of bridge runoff on a coastal salt marsh ecosystem, and Carlson and Morrison[47] used the assay to screen sewage sludge samples. In a study comparing three microbial sediment assays with toxicity tests for four benthic macroinvertebrate species and with benthic community structure metrics, Day[48] found the SPT to be significantly correlated ($p <$ 0.05) with most of the macroinvertebrate test endpoints and strongly correlated ($r = 0.830$) with alterations in benthic community structure.

Benton et al.[49] and Quintino et al.[50] have found SPT results to be strongly dependent on sediment granulometry. There is evidence that this phenomenon is related to increased retention of bacterial cells by silt and clay particles, relative to their retention by sand grains, although the phenomenon is both complex and sediment-dependent.[61] Ringwood et al. propose an array of options for correcting raw IC_{50} values for grain-size effects.[60]

V. CONCLUSIONS

It must be remembered that screening-level tests measuring acute toxicity, such as many of those described above, cannot replace more costly bioassays using locally relevant species and measuring chronic responses and full or partial life cycles. Screening tests should be viewed as operationally defined detection systems, used to identify problem areas where more resources and intensive effort should be focused. There is an inevitable trade-off between the value of direct environmental relevance and sensitivity to the measurement of chronic effects, both of which favor macroscale tests, and the capacity to evaluate large numbers of samples at a reasonable cost, which favors microscale tests. Microscale, including microbial, screening assays are increasing in acceptance, act as alternate tests to those using macroscale organisms such as many juvenile and adult invertebrates, and will continue to be used as an important component of multispecies and multitrophic level toxicity screening protocols for the assessment of contaminated sediments.

REFERENCES

1. Cairns, J., Jr., Multispecies toxicity testing, *Environ. Toxicol. Chem.,* 3, 1, 1984.
2. Cairns, J., Jr., The myth of the most sensitive species, *Bioscience,* 36, 670, 1986.
3. Cairns, J. Jr., and Neiderlehner, B.R., Problems associated with selecting the most sensitive species for toxicity testing, *Hydrobiologia,* 153, 87, 1987.
4. Burton, G. A., Jr., Stemmer, B. L., Winks, K. L., Ross, P. E., and Burnett, L. C., A multitrophic level evaluation of sediment toxicity in Waukegan and Indiana Harbors, *Environ. Toxicol. Chem.,* 8, 783, 1989.
5. Giesy, J. P. and Hoke, R. A., Freshwater sediment toxicity bioassessment: rationale for species selection and test design, *J. Great Lakes Res.,* 15, 539, 1989.
6. Ross, P. E. and Henebry, M. S., Use of four microbial tests to assess the ecotoxicological hazard of contaminated sediments, *Tox. Assess.,* 4, 1, 1989.
7. Ross, P. E. and Burton, G. A., Jr., Non-redundancy in sediment toxicity test batteries, Abstracts, 13th Annual Meeting, Society for Environmental Toxicology and Chemistry. Cincinnati, OH, 8-12 November, 1992.
8. Ross, P., Burnett, L., Kermode, C., and Timme, M., Miniaturizing a toxicity test battery for screening contaminated sediments, in *Proceedings of the Seventeenth Annual Aquatic Toxicity Workshop,* Chapman, P., Bishay, F., Power, E., Hall, K., Harding, L., McLeay, D., Nassichuk, M., and Knapp, W., Eds., November 5-7, 1990, Vancouver, BC, Canada, Can. Tech. Rep. Fish. Aquat. Sci. No. 1774, 331, 1991.
9. MacDonald, D. D., Smith, S. L., Wong, M. P., and Mudroch, P., The development of Canadian marine environmental quality guidelines, Environment Canada, Marine Environmental Quality Series, No. 1. Ottawa, ON, Canada. 121 pp. 1992.
10. USEPA (United States Environmental Protection Agency), Assessment and Remediation of Contaminated Sediments (ARCS) Program: Assessment Guidance Document, USEPA, Great Lakes National Program Office, Chicago, IL. Report No. EPA-905-B94-002, 247 pp., 1994.
11. IJC (US-Canada International Joint Commission), Procedures for the assessment of contaminated sediment problems in the Great Lakes, IJC, Windsor, ON, Canada, 140 pp., 1988.

12. Ross, P. E., Burton, G. A., Jr., Crecelius, E. A., Filkins, J. C., Giesy, J. P., Jr., Ingersoll, C. G., Landrum, P. F., Mac, M. J., Murphy, T. J., Rathbun, J. E., Smith, V. E., Tatem, H. E., and Taylor, R. W., Assessment of sediment contamination at Great Lakes Areas of Concern: the ARCS program Toxicity-Chemistry Work Group strategy, *Journal of Aquatic Ecosystem Health,* 1, 193, 1992.

13. Ross, P. E., The use of bacterial luminescence systems in aquatic toxicity testing, in *Ecotoxicology Monitoring,* Richardson, M., Ed., VCH Publishers, New York, NY, USA, 1993, Chap. 13.

14. Burton, G. A., Jr., and Stemmer, B. L., Evaluation of surrogate tests in toxicant impact assessment, *Tox. Assess.,* 3, 255, 1988.

15. Burton, G. A., Jr., Nimmo, D., Murphey, D., and Payne, F., Steam profile determinations using microbial activity assays and *Ceriodaphnia, Environ. Toxicol. Chem.,* 6, 505, 1987.

16. Griffiths, R. P., The importance of measuring microbial enzymatic functions when assessing and predicting long-term anthropogenic perturbations, *Mar. Poll. Bull.,* 13, 273, 1983.

17. Burton, G. A., Jr., Assessing the toxicity of freshwater sediments, *Environ. Toxicol. Chem.,* 10, 1585, 1991.

18. Buikema, A. L., Rutherford, C. L., and Cairns, J., Jr., Screening sediments for potential toxicity by *in vitro* enzyme inhibition, in *Contaminants and Sediments,* Vol. 1, Baker, R. A., Ed., Ann Arbor Science Publishers, Ann Arbor, MI, USA, 463, 1980.

19. Burton, G. A., Jr., Stemmer, B. L., and Winks, K. L., A multitrophic level evaluation of sediment toxicity in Waukegan and Indiana harbors, *Environ. Toxicol. Chem.,* 8, 1057, 1989.

20. Burton, G. A., Jr., Lazorchak, J. M., Waller, W. T., and Lanza, G. R., Arsenic toxicity changes in the presence of sediment, *Bull. Environ. Contam. Toxicol.,* 38, 491, 1987.

21. Griffiths, R. P., Caldwell, B. A., Broich, W. A., and Morita, R. Y., Long-term effects of crude oil on microbial processes in subarctic marine sediments, *Mar. Poll. Bull.,* 13, 273, 1982.

22. Stemmer, B. L., Burton, G. A., Jr., and Sasson-Brickson, G., Effect of sediment spatial variance and collection method on cladoceran toxicity and indigenous microbial activity determinations, *Environ. Toxicol. Chem.,* 9, 1035, 1990.

23. Trevors, J. T., Electron transport system activity in soil sediment and pure cultures, *CRC Crit. Rev. Microbiol.,* 11, 83, 1984.

24. Sayler, G. S., Puziss, M., and Silver, M., Alkaline phosphatase assay for freshwater sediments: application to perturbed sediment systems, *Appl. Environ. Microbiol.,* 38, 922, 1979.

25. Tabata, M., Osawa, K., Ohtakara, A., Nakabayashi, H., and Suzuki, S., Evaluation of river sediments by *in vitro* enzyme inhibition. *Bull. Environ. Contam. Toxicol.,* 44, 892, 1990.

26. Blaise, C., Microbiotests in ecotoxicology: characteristics, utility and prospects. *Environ. Toxicol. Wat. Qual.,* 6, 145, 1991.

27. Dutka, B., Method for determining acute toxicant activity in water, effluent and leachates using *Spirillum volutans, Tox. Assess.,* 1, 139, 1986.

28. Dutka, B., Priority setting of hazards in waters and sediments by proposed ranking scheme and battery of tests approach, *Germ. J. Appl. Zool.,* 75, 303, 1988.

29. Dutka, B., and Gorrie, J. F., Assessment of toxicant activity in sediments by the ECHA biocide monitor, *Environ. Pollut.,* 57, 1, 1989.

30. Ross, P. E., A summary of the sediment assessment strategy recommended by the International Joint Commission, in *Sediment Classification Methods Compendium,* United States Environmental Protection Agency. Report No. EPA 823-R-92-006, Washington, DC, USA, 1992, Chap. 12.

31. Giesy, J. P., Graney, R. L., Newsted, J. L., Rosiu, C. J., Benda, A., Kreis, R. G., Jr., and Horvath, F. J., Comparison of three sediment bioassay methods using Detroit River sediments, *Environ. Toxicol. Chem.,* 7, 483, 1988.

32. Dutka, B., Kwan, K., Rao, S., Jurkovic, A., McInnis, R., Palmateer, G., and Hawkins, B., Use of bioassays to evaluate river water and sediment quality, *Environ. Toxicol. Wat. Qual.,* 6, 309, 1991.

33. Bulich, A. A., Use of luminescent bacteria for determining toxicity in aquatic environments, in *Aquatic Toxicology,* ASTM STP 667, Marking, L. L., and Kimerle, R. A., Eds., American Society for Testing and Materials, Philadelphia, PA, 98, 1979.

34. Warren-Hicks, W., Parkhurst, B., and Baker, S., Jr., Eds., *Ecological Assessment of Hazardous Waste Sites,* EPA/600/3-89/013. U.S. Environmental Protection Agency, Environmental Research Laboratory, Corvallis, OR, 1989.

35. Carr, R. S., Long, E. R., Windom, H. L., Chapman, D., Thursby, C. G., Sloane, G. M., and Wolfe, D. A., Sediment quality assessment studies of Tampa Bay, Florida, *Environ. Toxicol. Chem.*, 15, 1218, 1996.

36. Ross, P. E., Jarry, V., and Sloterdijk, H., A rapid bioassay using the green alga *Selenastrum capricornutum* to screen for toxicity in St. Lawrence River sediment elutriates, ASTM STP 766, American Society for Testing and Materials, Philadelphia, PA, 179, 1988.

37. Wang, W., Root elongation method for toxicity testing of organic and inorganic pollutants, *Environ. Toxicol. Chem.*, 6, 409, 1987.

38. Snell, T. W., and Persoone, G., Acute toxicity bioassay using rotifers II. A freshwater test with *Brachionus rubens, Aquatic Toxicology,* 14, 81, 1989.

39. Brouwer, H., Murphy, T. P., and McArdle, L., A sediment-contact bioassay with *Photobacterium phosphoreum, Environ. Toxicol. Chem.,* 9, 1353, 1990.

40. Microbics, *Microtox® Users Manual,* Microbics Corporation, Carlsbad, CA, USA, 1992.

41. Tung, K. K., Scheiber, M. G., and Walbourn, C. C., The solid-phase assay: a new Microtox® test procedure, in *Proceedings of the Seventeenth Annual Aquatic Toxicity Workshop,* Chapman, P., Bishay, F., Power, E., Hall, K., Harding, L., McLeay, D., Nassichuk, M., and Knapp, W., Eds., November 5-7, 1990, Vancouver, BC, Canada, Can. Tech. Rep. Fish. Aquat. Sci. No. 1774, 495, 1991.

42. Tay, K.-L., Doe, K. G., and Wade, S. J., Sediment bioassessment in Halifax Harbor, *Environ. Toxicol. Chem.,* 11, 1567, 1992.

43. Hoskin, S. J., Evaluation of a new sediment toxicity test: correlation with a Sediment Quality Triad, M. S. thesis, University of Texas Science Center at Houston, School of Public Health. Houston, TX, 58 pp., 1993.

44. Chapman, P. M., Sediment Quality Triad Approach, in *Sediment Classification Methods Compendium,* United States Environmental Protection Agency, Report No. EPA 823-R-92-006. Washington, DC, 1992, Chap. 10.

45. Long, E. R. and Chapman, P. M., A sediment quality triad: measures of sediment contamination, toxicity and infaunal community composition in Puget Sound, *Mar. Poll. Bull.,* 16, 405, 1985.

46. Ross, P. E. and Leitman, P. A., Solid-phase testing of soil and sediment samples using *Vibrio fischeri.* Test design and data interpretation, in *Environmental Toxicology Assessment,* Richardson, M., Ed., Taylor & Francis, London, 65, 1995.

47. Carlson, C. and Morrison, G., Fractionation and toxicity of metals in sewage sludge, *Environ. Technol.,* 13, 751, 1992.

48. Day, K., Dutka, B., Kwan, K., Batista, N., Reynoldson, T., and Metcalfe-Smith. J., Correlations between solid-phase microbial screening assays, whole sediment toxicity tests with macroinvertebrates and *in situ* benthic community structure, *J. Great Lakes Res.,* 21, 192, 1995.

49. Benton, M., Mallott, M., Knight, S., Cooper, C., and Benson, W., Influence of sediment composition on apparent toxicity in a solid-phase test using bioluminescent bacteria, *Environ. Toxicol. Chem.,* 14, 411, 1995.

50. Quintino, V., Picado, A., Rodrigues, A., Mendonça, E., Costa, M., Bordalo Costa, M., Lindgaard-Jørgensen, P., and Pearson, T., Sediment chemistry — infaunal community structure in a southern European estuary related to solid-phase Microtox testing, *Neth. J. Aquat. Ecol.,* 29, 427, 1995.

51. Schiewe, M., Hawk, E., Actor, D., and Krahn, M., Use of bacterial luminescence assay to assess toxicity of contaminated marine sediments, *Can. J. Fish. Aquat. Sci.,* 42, 1244, 1985.

52. Fish, F., Lampert, I., Halachmi, A., Reisenfeld, G., and Herzberg, M., The SOS Chromotest kit: a rapid method for the detection of genotoxicity, *Tox. Assess.,* 2, 135, 1987.

53. Reinhartz, A., Lampert, I., Herzberg, M., and Fish, F., A new, short-term, sensitive, bacterial assay kit for the detection of toxicants, *Tox. Assess.,* 2, 193, 1987.

54. Williamson, K. and Johnson, D., A bacterial bioassay for the assessment of wastewater toxicity, *Water Res.,* 15, 383, 1981.

55. Tam, T.-Y. and Trevors, J., Toxicity of pentachlorophenol to *Azotobacter vinelandii, Bull. Environ. Contam. Toxicol.,* 27, 230, 1981.

56. Flemming, C. and Trevors, J., Copper toxicity in freshwater sediment and *Aeromonas hydrophila* cell suspensions measured using an O_2 electrode, *Tox. Assess.,* 4, 473, 1989.

57. Bitton, G., Khafif, T., Chataigner, N., Bastide, J., and Coste, C., A direct INT- dehydrogenase assay (DIDHA) for assessing chemical toxicity, *Tox. Assess.,* 1, 1, 1986.

58. Trevors, J., Mayfield, C., and Inniss, W., A rapid toxicity test using *Pseudomonas fluorescens, Bull. Environ. Contam. Toxicol.,* 26, 433, 1981.

59. Ames, B., McCann, J., and Yamasaki, E., Methods for detecting carcinogens and mutagens with the *Salmonella*/mammalian microsome mutagenicity test, *Mutation Res.,* 31, 347, 1975.

60. Ringwood, A. H., DeLorenzo, M. E., Ross, P. E., and Holland, A. F., Interpretation of Microtox® solid-phase toxicity tests: the effects of sediment composition, *Environ. Toxicol. Chem.,* 16, 6, 1135, 1997.

61. Cook, N. H. and P. G. Wells, Toxicity of Halifax Harbour sediments: An evaluation of the Microtox® solid-phase test, *Water Qual. Res. J. Canada,* 31, 4, 673, 1996.

PART THREE

Validation, Applications, and Training with Microscale Testing

CHAPTER **39**

The Influence of Particle Size, Ammonia, and Sulfide on Toxicity of Dredged Materials for Ocean Disposal

K. L. Tay, K. G. Doe, A. J. MacDonald, and K. Lee

CONTENTS

I. INTRODUCTION

Aquatic toxicity tests have been used by regulators and managers for environmental assessment and regulation since the 1960s.[1] However, early biological tests were designed to assess only the waterborne contaminants, particularly in the area of controlling and detecting pollutants discharged to the freshwater and marine environments through industrial and municipal wastewaters.[2-6] The development of simple and cost-effective sediment toxicity tests to assay field-collected sediments in the early 1980s has provided a new dimension to regulators and managers to better predict environmental impacts in both water column and sediment phases of the freshwater and marine ecosystems.[7-13]

Sediment bioassays are now used for development of sediment quality guidelines,[14-15] environmental assessment of contaminated sediments,[16-20] and for making pass/fail decisions in regulatory programs.[21-23] To ensure proper use and interpretation of the endpoints of each of the sediment bioassays, guidelines and protocols have been developed in the United States and Canada. However, these guidances have not adequately addressed the concerns raised recently regarding factors, other than sediment contaminants, that may be the major causes of toxicity in some sediments. These factors include particle size effects[24-26] and the toxicity effects of ammonia and sulfide.[27-35] Until the role of these factors as causative agents for measured toxicity is known, the relationship between persistent sediment contaminants and bioassay results cannot be properly understood. The significant influence of these factors on test results has raised the question whether the apparent toxicity as measured by a sediment bioassay should be used to develop sediment quality criteria or to make pass/fail regulatory decisions.

This chapter will focus on the interpretation of the influence of particle size effects, and ammonia and sulfide toxicity on microscale sediment bioassay results used in decision making under regulatory programs with special emphasis on evaluating the suitability of dredged materials for disposal in marine waters. Ocean disposal of dredged materials in Canada is regulated under the Canadian Environmental Protection Act (CEPA) Part VI by means of a permit system. Before a permit can be issued, physical and chemical testing of the dredged materials is required. Should chemical screening criteria be exceeded, the material can be further evaluated through the application of, *inter alia*, three sediment toxicity tests: Microtox* solid-phase test, amphipod lethality test, and the echinoid fertilization test.[36-38]

Under the Canadian Ocean Disposal Program, we have conducted a series of laboratory experiments and compiled bioassay results submitted to and generated by the program to address the confounding effects of particle size, ammonia, and sulfide on sediment toxicity. The results will be discussed with special reference to use of the bioassays in the regulatory context of the Canadian Ocean Disposal Program.

II. LITERATURE REVIEW

A. Particle Size Effects

The mechanisms of particle size effects on bioassay endpoints are quite complex. Due to the larger surface area and the greater binding capacities of small particles, sediments with a greater proportion of small particles may have relatively higher concentrations of contaminants, organic carbon, and water content.[24,39-41] Because of the poor understanding of the physical and chemical properties of fine-grained marine sediments and how these properties affect the health of benthic organisms, the particle size effects on bioassay endpoints have been considered as "super-variable," i.e., the other factors that co-occur with fine particle size may be the actual cause of toxicity.[24,42]

Toxicity as measured by the Microtox solid-phase test increases with decreasing sediment particle size. The relationship is most obvious in clean sediments. It has been postulated that this phenomenon is likely due to the fact that bacteria are lost through the filtering process of the method because of their adhesion to the small sediment particles.[25,43,44] In fact, the filtration losses have been demonstrated by Greene et al.[43] using traditional plate counting techniques. Benton et al.[25] used different mixtures of sand, silt and clay, and spiked formulated sediments to assess the influence of sediment composition on Microtox solid-phase toxicity. They concluded that particle size effects may have been masked by the observed toxicity of methyl parathion-spiked sediments. An alternative conclusion that could be drawn from the same data is that at the dose level tested there was essentially no effect of particle size on observed toxicity of spiked mixtures of sand:silt and sand:clay in the range of 10 to 50% fines.

To minimize the particle size effects, Greene et al.[43] proposed using reference samples of similar physical make-up to compare with test samples if the observed EC_{50}s are about 500 mg/kg or higher. In a draft guide under consideration by the American Society for Testing and Materials (ASTM) Subcommittee E47.03 it was recommended that for samples of EC_{50} greater than about 2500 mg/kg, data calculation must include comparison to the results obtained with a reference

* Registered Trademark of AZUR Environmental, Carlsbad, CA, USA

sediment of lower toxicity but of comparable physical characteristics.[45] However, it may not be practical to obtain reference samples representative of all sediment types and compositions.[25]

The particle size effects on the amphipod bioassay endpoints have also been demonstrated by many researchers.[16,24,26,42] However, the nature of these effects is probably not due to deficiency of the bioassay method but due to species-specific particle size tolerance. Benthic ecologists have long known that the distribution of a benthic species is affected by particle size of the sediment.[46] When the phoxocephalid amphipod, *Rhepoxynius abronius*, which has a realized niche of fine, well-sorted sands,[20,47-49] was exposed to a wide variety of marine sediments, it was found that some very fine and very coarse "clean" sediments were toxic to the organism. This effect was more pronounced when clean reference sediments were used. Mean *R. abronius* survival in fine, uncontaminated field sediment (≥80% silt-clay) can be 15% lower than survival in native sediment.[20,50] To address the possible effects of particle size, regression-based models have been developed for *R. abronius* to examine the extent of the grain size interference.[24] Also, other species of amphipod, such as *Ampelisca abdita* and *Eohaustorius estuarius*, which have survived well in fines were recommended as replacement species for toxicity tests when a fines content of a sediment sample exceeded 60%.[50]

Nipper et al.[51] found that the growth, not the survival, of the amphipod, *Grandidierella japonica*, an inhabitant of coastal southern California, was affected by variations in sediment grain size. The growth of another species of amphipod, *Chaetocorophium* cf. *lucasi*, an inhabitant of New Zealand estuarine muddy sediments, was also affected by sediments with different grain size distributions. However, the percent survival of this amphipod species was affected by the type of sand more than by the concentration of sand in the sediment samples.[26]

B. Ammonia and Sulfide

Ammonia can arise from both natural and anthropogenic sources and can also be formed in sediment during handling and storage. Ammonia exists in sediment in two forms, the un-ionized ammonia that is highly toxic to some marine organisms and the ammonia ion (NH_4^+) that is less toxic.[32] Ammonia measurements can be affected by many variables, especially salinity, pH, dissolved oxygen concentration, and temperature.[29,52,53] Jones-Lee et al.[32] suggested that ammonia is largely dissolved in the interstitial water because of the high ionic concentrations of sodium, calcium, and magnesium in the interstitial waters, the comparatively low concentration of ammonia, and the low sorption tendency of ammonia compared with calcium, magnesium, and sodium. Measurement methods for ammonia in pore water and sediment for toxicity study have not been standardized. In most recent studies, ammonia measurement in sediments was conducted in pore water.[29-31,54,55] A concentration range of 0.17 to 17 mg/L ammonia in sediment pore water was quite common, while concentrations as high as 430 mg/L had also been reported.[55] The effects of ammonia on measured toxicity of Microtox, amphipod, and sea urchin tests varied from one study to another. While Ankley et al.[28] reported that there was no correlation between pore water ammonia concentrations with the Microtox pore water endpoints, Bay et al.[56] showed that the toxic effects of 72 sediment samples collected from Southern California Bight on the amphipod, *A. abdita*, and sea urchin, *Strongylocentrotus purpuratus*, were correlated to ammonia concentration in virtually all cases. Another amphipod, *R. abronius*, showed an ammonia sensitivity of about 10 to 20 times less than many other "sensitive" test organisms commonly used for toxicity tests.[32]

For the amphipod, *A. abdita*, the no observable effects concentrations (NOEC) for un-ionized ammonia is 0.4 mg/L, and for the total ammonia is 30 mg/L.[57] The un-ionized ammonia NOEC and lowest observable effects concentrations (LOEC) for the sea urchin, *Arbacia punctulata*, used in fertilization assays are 0.4 and 0.8 mg/L, respectively.[31] The un-ionized ammonia NOECs for the fertilization assay with the sand dollar, *Dendraster excentricus*, and the sea urchin, *S. purpuratus*, are 0.21 and 0.17 mg/L, respectively.[35] However, at concentrations higher than the NOEC, the IC_{50} of these two species are quite different (16.5 mg/L for *D. excentricus* and 1.69 mg/L for *S. purpuratus*).

Carr et al.[31] indicated that although concentrations of un-ionized ammonia covaried with toxicity to sea urchin fertilization success in their study, these concentrations rarely equaled or exceeded the concentrations expected to cause toxicity. Thus, most of the observed toxicity in their study was not due to the effect of ammonia but more likely due to other factors. Bailey et al.[35] used the same approach to conclude that the toxicity measured in their fertilization test could not be caused by ammonia in their test samples.

Elemental sulfur is a common component of anaerobic marine sediments.[58] Concentrations of sulfides in pore water between 0.7 and 170 mg/L are common in sediments, while concentrations as high as 2693 mg/L have been reported.[34,55] Under anoxic conditions, sulfur is reduced to hydrogen sulfide and is highly toxic.[59] The toxicity is primarily caused by the undissociated hydrogen sulfide molecule that freely diffuses across membranes. This toxicity increases with increasing temperature and/or decreasing pH and dissolved oxygen.[34] The Microtox bioassay has been shown to be highly sensitive to elemental sulfur dissolved in several organic solvents.[58] A positive correlation was also reported between Microtox endpoints and sediment sulfide concentrations in a direct contact bioassay.[59] Based on a literature review, Sims and Moore[34] concluded that the comparison of reported exposure and effect concentrations suggests a strong potential for hydrogen sulfide toxicity in dredged material bioassays. However, due to the lack of standard measurement methods for sulfides in sediment, the difficulties encountered by researchers in making accurate measurements of sulfide concentrations in pore water and sediment, and the volatility of sulfide during toxicity tests, there is a general lack of literature data on the relationship between sediment sulfide and toxicity endpoints.

III. LABORATORY AND FIELD DATA COMPILATION

A. Particle Size Effects

1. Microtox Solid-Phase Assay

The standard protocol for the Microtox solid-phase assay was released by Microbics Corporation (now AZUR Environmental) in 1992.[60] Since its release, the method has been used extensively for comparison with the other sediment bioassays and has been adopted by regulators for the assessment of dredged materials for ocean disposal. To address the effects of particle size on Microtox solid-phase endpoints, we conducted two experiments using commercially available silica sands and kaolin clay (Table 39.1). The results showed good agreement with those reported by Greene et al.[43] and Benton et al.[25] that particle size effects on apparent toxicities were influenced by sediment fines. The relationship between clay and Microtox endpoints in this study was more pronounced than those between silt and Microtox endpoints. The inverse fines/endpoints relationship extended over a range of 0% to 97% (Figure 39.1). When a fixed amount of a marine sediment analytical reference sample (National Research Council of Canada Marine Sediment Reference Sample HS-6) for PAHs (polynuclear aromatic hydrocarbons) was added to a mixture of silica sand and kaolin clay, the particle size effects were less pronounced over the range of 40 to 100% fines.

To compare the results of the laboratory studies using commercially obtained media to the results of studies using natural sediments, 14 Microtox solid-phase studies with a total of 145 samples conducted using natural sediments from estuaries, reference ocean sites, and harbors were compiled. The results showed good agreement with the previous studies in the following areas: 1) the apparent toxicity (i.e., combined contaminant effect and particle size effect) increased with increased percentages of silt and clay; 2) the effects of particle size were less pronounced in harbor sediments (Figure 39.2). Benton et al.[25] indicated that from 25 to 50% silt or clay, the change in EC_{50} appeared negligible, and that sediment formulation of 50 to 75% silt or clay also showed no change in EC_{50}. Our data, which covered an extensive range of particle sizes, showed that particle

Table 39.1 Measured percent sand and fines (sum of silt and clay) and Microtox EC_{50} (mean ± standard deviation) of silica sand,[1] Kaolin clay,[2] mixtures of silica and Kaolin clay, and mixtures of silica sand, Kaolin clay and a fixed percent (40%) of Marine Reference Sample HS-6.[3] All Microtox tests were conducted in triplicate.

Test Samples	% Sand	% Fines	EC_{50} (mg/kg) (mean ± sd)
100% sand[1]	95	5	>97000
95% sand, 5% clay	92	10	28000 ± 1080
90% sand, 10% clay	86	11	13900 ± 1800
80% sand, 20% clay	76	23	5747 ± 399
60% sand, 40% clay	58	41.5	4353 ± 733
40% sand, 60% clay	39	62	2613 ± 92.9
20% sand, 80% clay	20.5	77.5	2267 ± 497
100% clay[2]	<1	100.5	1373 ± 32.1
0% sand, 60% clay, 40% HS-6[3]	1	100.5	512 ± 116
12% sand, 48% clay, 40% HS-6	14	88.5	693 ± 129
24% sand, 36% clay, 40% HS-6	24.5	76.5	904 ± 134
36% sand, 24% clay, 40% HS-6	39	62	976 ± 121
48% sand, 12% clay, 40% HS-6	48	50	781 ± 46.6
54% sand 6% clay, 40% HS-6	54	44	817 ± 65.8
57% sand, 3% clay, 40% HS-6	57	41	852 ± 80.5
60% sand, 0% clay, 40% HS-6	60	39	899 ± 18.8

[1] Sand = 1:1 mix of silica sand 0 (medium-fine sand) and silica flour (very fine sand containing 5% silt).
[2] Clay = Kaolin clay (3 to 4% silt, 96 to 97% clay, and <1% sand).
[3] HS-6 = National Research Council of Canada reference material containing 30 mg/kg total PAH (sum of 16 PAHs designated priority pollutants by the US EPA).

size effects were more pronounced at the coarse end of the size spectrum and that the effects were less pronounced when fines were more than 40%. In fact, the relationship between particle size and observed toxicity disappeared when EC_{50} was less than 6000 mg/kg.

Environment Canada's sediment bioassay interpretation guidance proposed that test sediments producing a 5-min EC_{50} < 1000 mg/kg are toxic. This criterion obviates the use of reference site samples which could be significantly affected by grain-size effects (Figure 39.2). However, when a fixed criterion is used for decision making, it is important to ensure that the same method used to generate the toxicity endpoints be used consistently. Over the years, the Microtox solid-phase test has undergone several methodology changes. When these methods were tested in our laboratory, it was quite evident that different methods may produce different results (Figures 39.3A and B).

The Environment Canada 1000 mg/kg EC_{50} criterion is a good choice since all the EC_{50} of the "clean" silica sand and the various types of clay (up to 100% clay) are all above the 1000 mg/kg level and below this level when PAH-contaminated sediment was added to the sand and clay mixtures (Figure 39.1). However, due to the increase of particle size effects on the coarser side of the size spectrum (<40% fines), a higher criterion may be needed for test sediment containing <40% fines. If reference site samples are used instead of using a fixed criterion, then particle size should be used as one of the selection criteria for reference sites, and particle size of the reference site sample should match that of the test sample.

2. Amphipod Bioassay

To assess particle size effects on amphipod endpoints, the Environment Canada Pacific Bioassay Laboratory used two china clays, southern bentonite and calcined kaolin, and five grades of silica sand to test two west coast amphipod species, *Eohaustorius washingtonianus* and *R. abronius*.[61] The

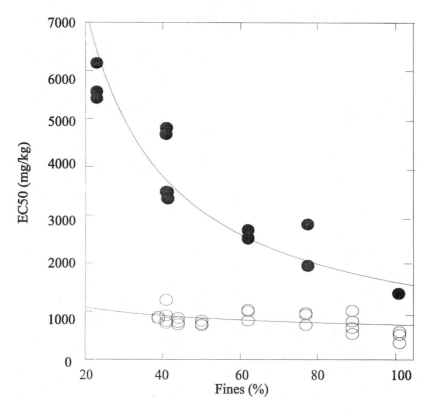

Figure 39.1   Effects of particle size on Microtox solid-phase EC_{50}s of clean sand/clay mixtures (●———●) and sand/clay mixtures with a fixed amount (40%) of National Research Council of Canada Marine Sediment Reference Sample HS-6 (○———○).

results showed that particle size effects on the amphipod are species specific. While the percent survival of *E. washingtonianus* was lowest at both ends of the grain size spectrum, *R. abronius* was only affected by high proportion of fines (Figures 39.4A and B).

The relationship between percent clay and the endpoints of the *R. abronius* test was less evident (r = −0.4, p = 0.02) in the four sets of bioassay data (n = 41 sediment samples) obtained under the Environment Canada Ocean Disposal Program. This relationship, however, becomes more pronounced (r = −0.6, p = 0.002, n = 24) when the two sets of "contaminated sediments" were removed indicating that contaminants in sediments may have masked the particle size effects.

Particle size effects were quite prominent (% clay versus % survival: r = −0.8, p = 0.01) in *E. washingtonianus* toxicity tests conducted by Environment Canada using thirteen sediment samples collected from reference ocean sites off the west coast of British Columbia (Figure 39.5). Since none of the sediment samples contained <10% fines, the effect of the coarser materials, as demonstrated by our laboratory experiments conducted using commercially available sand and clay, remains to be proven by tests using natural sediments.

In two studies (n = 19 sediment samples) conducted under the Ocean Disposal Program, *E. estuarius*, another west coast amphipod, was not affected by the particle size. No particle size effects were observed in nine studies (n = 93 sediment samples) using the Canadian east coast species, *Amphiporeia virginiana,* as the test organism. Another east coast species of amphipod, *Corophium volutator,* was also quite tolerant to a wide range of particle sizes (two studies with n = 21 sediment samples).

Environment Canada recommended all the above amphipod species as test organisms for the Ocean Disposal Program.[37] Since the response of amphipods to particle size is species specific, the

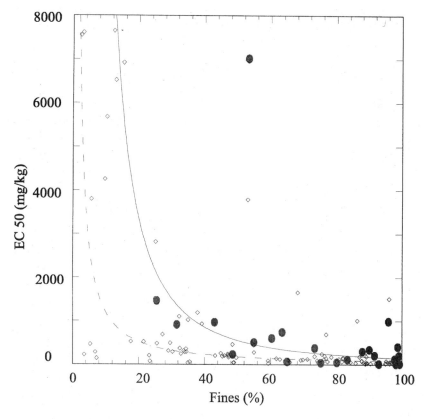

Figure 39.2   Effects of particle size on Microtox solid-phase EC_{50}s on harbor sediments (◊----◊) estuarine and clean reference sediments (● — ●).

sensitivity of each amphipod species to particle size should be taken into consideration in the selection of test organism before conducting the amphipod test. Based on our study, we concluded that the degree of sensitivity to particle size among the four amphipods can be ranked as follows: *E. washingtonianus* > *R. abronius* > *E. estuarius* = *A. virginiana* = *C. volutator*.

Both the Puget Sound Dredged Disposal Analysis (PSDDA) and the Washington Department of Ecology recommended that when testing marine sediments (>25‰ salinity) which exhibit high percentages of fines (i.e., >60%), *A. abdita* should be considered as a substitute species for *R. abronius* to avoid the particle size effects.[50] For the Canadian Ocean Disposal Program, for the same salinity, *A. virginiana,* which can be found on the east coast, is a good substitute because of its tolerance to the fines and its sensitivity to pollutants. For low salinity sediments (≤25‰), *E. estuarius* and *C. volutator* are recommended.

B. Ammonia and Sulfide

In our laboratory, we have adapted the standard method for the use of ammonia and sulfide ion specific electrode to measure the sediment ammonia and sulfide for the Microtox solid-phase assay and the amphipod test.[62] These methods eliminate the variability created by the different extraction methods for pore water and measure the total ammonia and sulfide concentrations in the sediments. For ammonia, results obtained from these methods are correlated significantly with the pore water measurements (Figure 39.6).

To assess the effects of ammonia and sulfide on the Microtox solid-phase endpoints, we examined a total of five sediment studies (n = 49 sediment samples) conducted recently under the

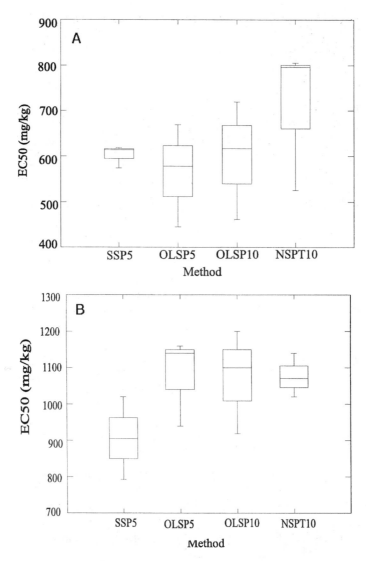

Figure 39.3   A) Box and whisker plots of four Microtox solid-phase procedures. Sample A is from an industrialized harbor. SSP5 = Small Sample Procedure (5-min EC_{50}), OLSP5 = Old Large Sample Procedure (5-min EC_{50}), OLSP10 = Old Large Sample Procedure (10-min EC_{50}), NSPT10 = New Solid Phase Test (10-min EC_{50}). Values represented are median, standard deviation, and range (n = 3). B) Box and whisker plots of four Microtox solid-phase procedures. Sample B is from an industrialized harbor. SSP5 = Small Sample Procedure (5-min EC_{50}), OLSP5 = Old Large Sample Procedure (5-min EC_{50}), OLSP10 = Old Large Sample Procedure (10-min EC_{50}), NSPT10 = New Solid Phase Test (10-min EC_{50}). Values represented are median, standard deviation, and range (n = 3).

Ocean Disposal Program. Spearman correlation coefficients between sediment sulfide and the Microtox endpoints (r = –0.6; p = 0.0001; Figure 39.7) were found to be better than those between ammonia and the endpoints (r = –0.4; p = 0.007). This is in agreement with previous studies indicating that Microtox bacteria are more sensitive to sulfide than ammonia.[58] The ammonia NOEC level for the Microtox solid-phase test is not available. Microtox tests conducted in our laboratory resulted in a 15-min EC_{50} of >10,000 mg/L as NH_4Cl (>2620 mg/L as ammonia-N) at a test pH of 8.0. Although the correlation between total ammonia in sediment and the Microtox endpoints in our data is very significant, all of the ammonia concentrations are below the EC_{50} level of

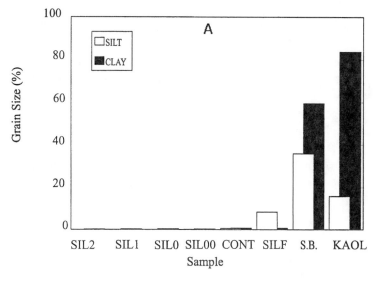

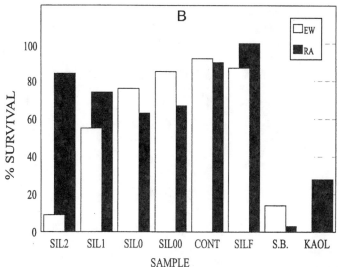

Figure 39.4 A) Particle size of clean sand, silt, and clay. SIL2 = Silica Sand No. 2, SIL1 = Silica Sand No. 1, SIL0 = Silica Sand No. 0, SIL00 = Silica Sand No. 00, CONT = Control Sediment, SILF = Silica Flour, SB = Southern Bentonite Clay, and KAOL = kaolin clay. B) Percent survival of two species of amphipods in clean sand, silt, and clay. EW = *E. washingtonianus*, RA = *R. abronius*, SIL2 = Silica Sand No. 2, SIL1 = Silica Sand No. 1, SIL0 = Silica Sand No. 0, SIL00 = Silica Sand No. 00, CONT = Control Sediment, SILF = Silica Flour, SB = Southern Bentonite Clay, and KAOL = kaolin clay. (Data from Lee, D. L., Ten-day Amphipod Sediment Bioassay — A Comparison of Particle Size Effects on Two Species of Amphipod, *Eohaustorius washingtonianus* and *Rhepoxynius abronius*. Environment Canada, Pacific and Yukon Region, unpublished data, 1995.)

>10,000 mg/L, indicating that most of the observed toxicity effects are probably not related to ammonia but due to other factors. No sulfide NOEC and LOEC for Microtox are available for our data interpretation.

Of the four amphipods, ammonia seemed to correlate better with the two *Eohaustorius* species (r = –0.57; p = 0.001; Figure 39.8) than *R. abronius* and *C. volutator*. An un-ionized ammonia NOEC of 0.4 mg/L has been reported by Mueller and Scott[57] for *A. abdita*. Since no ammonia NOEC is available for *Eohaustorius* species, it is not possible to conclude that the observed toxicity is related to ammonia, and there was no significant correlation between sediment sulfide and survival.

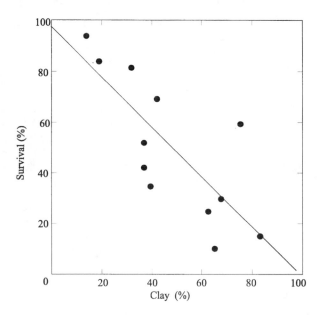

Figure 39.5   Relationship between clay content and the survival of *E. washingtonianus.*

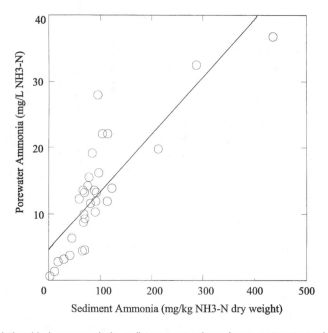

Figure 39.6   Relationship between whole sediment ammonia and pore water ammonia concentrations.

We also compiled two sets of bioassay data to assess the effects of pore water ammonia and sulfide on the echinoid fertilization test (*L. pictus*). Only pore water ammonia was found to be significantly correlated with the % fertilization endpoints (r = –0.7, p = 0.0001). Since the concentration range of the undissociated ammonia in the two sets of bioassay studies is 0.002 to 0.49 mg/L, which is well below the NOEC (0.66 mg/L) reported by our bioassay laboratory for *L. pictus*, it is likely that other factors were responsible for the observed toxicity effects (Figure 39.9).

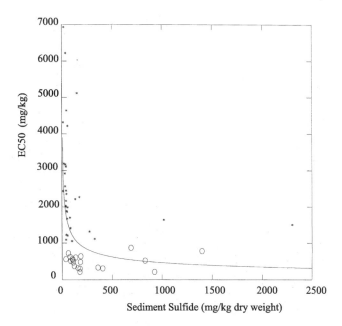

Figure 39.7   Relationship between Microtox solid-phase EC_{50} and sediment sulfide. Spearman correlation coefficient r = −0.6; p = 0.001. * = nontoxic, o = toxic (according to Canadian Ocean Disposal Guidelines).

IV. DISCUSSION

Sediment toxicity tests were developed using the following criteria: 1) test organisms were sensitive to contaminants, ecologically relevant, easily accessed, and maintained in laboratory; 2) tests exhibit good repeatability; 3) tests simulate *in situ* toxicity effects; 4) tests are simple and easy to run; and 5) cost effective. A test species that is sensitive to contaminants in sediments could also be quite sensitive to particle size distribution, ammonia, and sulfide in sediments. To ensure that the test results are interpreted properly, each method should address the effects of these factors on the endpoints and provide advice and recommendation for the proper use of the method to meet the study objectives.

The effects of ammonia and sulfide on the sediment toxicity endpoints are more difficult to control than the effects of particle size. Particle size effects can be avoided by careful selection of test species and reference sediment samples if necessary. However, when ammonia and sulfide are of concern in the test sediments, it will be the decision of the toxicologists on whether these variables should be eliminated or remain untouched. If the objectives of the test are to generate toxicity data for persistent chemicals, the elimination of the effects of ammonia and sulfide on the toxicity endpoints may be needed, since these factors can be controlled by sediment treatment methods.[63-65] However, if the objective is to assess the health of an ecosystem, to develop sediment quality criteria, or to make a regulatory decision on ocean disposal of dredged materials, the influence of these characteristics on sediment toxicity must be understood.

To avoid the effects of ammonia on the toxicity endpoints, the U.S. Environmental Protection Agency and the U.S. Army Corps of Engineers recommended a procedure in the amphipod test to reduce interstitial ammonia by flushing the overlying water in test chambers over a certain period of time before adding the animals.[63] However, this proposed process may result in reducing other soluble toxicants in the sediment. Other proposed attempts to remove ammonia from sediment

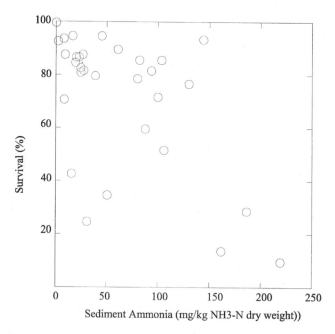

Figure 39.8   Relationship between percent survival of *E. washingtonianus* and sediment ammonia.

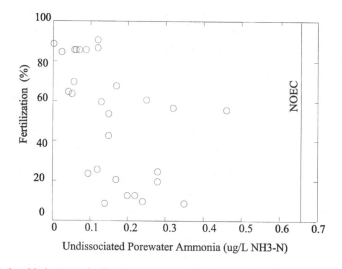

Figure 39.9   Relationship between fertilization success of *L. pictus* and undissociated pore water ammonia.

include introduction of aquatic plants, inducing microbial activities, modifying pH, and adding zeolite to sediment.[64,65]

For the assessment of dredged material for ocean disposal, chemical and biological treatment, or replacement of overlying water to remove ammonia and sulfide from test sediment may not be the best solution, since these processes may alter the chemical characteristics of the sediment. A better way to assess the effects of ammonia and sulfide of sediment is to calculate the correlations between all the measured factors and bioassay endpoints. If significant correlations exist between ammonia, sulfide, and bioassay endpoints, but the concentrations of these factors rarely equaled or exceeded the concentrations expected to cause toxicity according to established sediment quality guidelines, it would be reasonable to conclude that other factors are contributing to the toxicity.[31]

More information is needed to address the issues of ammonia and sulfide in sediment toxicity tests. Without a reliable database, the practical way to address this issue is to ensure that all potential factors of test sediment should be measured in all tests. Interpretation of bioassay endpoints will depend on the judgment of the decision makers, who should use all the available data for reaching a conclusion from the tests.

For regulatory programs such as the Ocean Disposal Program, proper data interpretation is critical in making pass/fail decisions on a disposal permit application. Ignoring the effects of particle size, ammonia, and sulfide on the bioassay endpoints could cause significant problems leading to wasted time and resources. Regulators and managers should be aware of this concern when using toxicity endpoints to make regulatory decisions.

In addition to the particle size effects and the toxic effects of ammonia and sulfide on sediment endpoints, data generated from our laboratory also showed that Eh (sediment oxidation-reduction potential) was highly correlated ($r = 0.6$; $p = 0.0001$; $n = 49$ sediment samples) with Microtox toxicity. The highly significant correlations of sediment sulfide versus Microtox toxicity ($r = -0.6$; $p = 0.0001$, $n = 49$) and total organic carbon versus Microtox toxicity ($r = 0.5$; $p = 0.0009$; $n = 49$) in these sediments indicated that the toxicity of highly reduced sediments may be related to the presence of sulfide and total organic carbon concentrations which are also related to the Eh levels. Thus, measurements of Eh,[66] total organic carbon, and other parameters that have potential effects on toxicity endpoints should also be considered in the list of parameters to be measured in any toxicity test.

V. FUTURE DIRECTIONS

In preparation of data interpretation protocols for regulatory use of sediment bioassays, special attention should be paid to the possible effects of particle size, ammonia, sulfide, and other factors such as sediment Eh, etc. More research is needed in establishing the cause-effect relationship of these factors with the various bioassay tests because they correlate as strongly as persistent contaminant concentrations with the endpoints of many bioassays. Also, more studies are required in the development and standardization of sediment TIEs (toxicity identification evaluations). There is a need to standardize the measurement methods for pore water and sediment ammonia and sulfide. There is also a need for a consensus on how to deal with "natural" toxicants like ammonia and sulfide. Should they be removed from the test sediments (e.g., by chemical or biological treatment, flushing, etc.) before testing or should they be counted as part of the toxic potential of sediment? The use of reference sediments (natural and artificial) requires further research, and the establishment of good reference sites for test sediments of different particle size, organic carbon content, ammonia, sulfides, Eh, etc., is needed for comparison purposes. Improved criteria (including biological performance criteria) for reference sediments is urgently required, especially when bioassay tests are used for making pass/fail decisions.

REFERENCES

1. Foster, R. B., Environmental Legislation, in *Fundamentals of Aquatic Toxicology*, Rand, G. M. and Petrocelli, S. R., Eds., Hemisphere Publishing Corporation, McGraw-Hill International Book Company, Montreal, 1985, Chap. 20, 587.
2. Sprague, J. B., Measurement of pollutant toxicity to fish, I. Bioassay methods for acute toxicity, *Water Res.*, 3, 793, 1969.
3. Sprague, J. B., Measurement of pollutant toxicity to fish, II. Utilizing and applying bioassay results, *Water Res.*, 4, 3, 1970.
4. Sprague, J. B., Measurement of pollutant toxicity to fish, III. Sublethal effects and "safe" concentrations, *Water Res.*, 5, 245, 1971.

5. Pessah, E. and Cornwall, G. M., Use of toxicity tests in regulating the quality of industrial wastes in Canada, *ASTM STP 707,* American Society for Testing and Materials, Philadelphia, Pa., 130, 1980.

6. Blaise, C., Sergy, G., Wells, P., Bermingham, N., and Van Coillie, R., Biological testing -Development, application, and trends in Canadian Environmental Protection Laboratories, *Tox. Assess.,* 3, 385, 1988.

7. Williams, L. G., Chapman, P. M., and Ginn T. C., A comparative evaluation of marine sediment toxicity using bacterial luminescence, oyster embryo and amphipod sediment bioassays, *Marine Environ. Res.,* 19, 225, 1986.

8. Pesch, G. G. and Pesch, C. E., Chromosome complement of the marine worm *Neanthes arenaceodentata* (Polychaeta: Annelida), *Can. J. Fish. Aquat. Sci.,* 37, 286, 1980.

9. Pesch, G. G. and Pesch, C. E., *Neanthes arenaceodentata* (Polychaeta: Annelida), a proposed cytogenetic model for marine genetic toxicology, *Can. J. Fish. Aquat. Sci.,* 37, 1225, 1980.

10. Pesch, G. G., Pesch, C. E., and Malcolm, A. R., *Neanthes arenaceodentata,* a cytogenetic model for marine genetic toxicology, *Aquat. Toxicol.,* 1, 301, 1981.

11. Chapman, P. M. and Morgan, J. D., Sediment bioassays with oyster larvae, *Bull. Environ. Contam. Toxicol.,* 31, 438, 1983.

12. Swartz, R. C., DeBen, W. A., Jones, J. K. P., Lamberson, J. O., and Cole, F. A., Phoxocephalid amphipod bioassay for marine sediment toxicity, in *Aquatic Toxicology and Hazard Assessment, Proceedings of the Seventh Annual Symposium, ASTM STP 854,* Cardwell, R. D., Purday, R., and Comotto-Bahner, R., Eds., American Society for Testing and Materials, Philadelphia, PA, 1985, 284.

13. Ingersoll, C., Sediment toxicity and bioaccumulation testing, *ASTM Standardization News,* April 1991, 28.

14. Chapman, P. M., Current approaches to developing sediment quality criteria, *Environ. Toxicol. Chem.,* 8, 589, 1989.

15. Long, E. R., MacDonald, D. D., Smith, S. L., and Calder, F. D., Incidence of adverse biological effects within ranges of chemical concentrations in marine and estuarine sediments, *Environ. Manag.,* 19, 81, 1995.

16. Becker, D. S., Bilyard, G. R., and Ginn, R. C., Comparisons between sediment bioassays and alterations of benthic macroinvertebrate assemblages at a marine superfund site: Commencement Bay, Washington, *Environ. Toxicol. Chem.,* 9, 669, 1990.

17. Chapman, P. M., Power, E. A., Dexter, R. N., and Andersen, H. B., Evaluation of effects associated with an oil platform, using the sediment quality triad, *Environ. Toxicol. Chem.,* 10, 407, 1991.

18. Lamberson, J. O., DeWitt, T. H., and Swartz, R. C., Assessment of sediment toxicity to marine benthos, in *Sediment Toxicity Assessment,* Burton, G. A., Eds., Lewis Publishers, Ann Arbor, 1992, Chap. 9.

19. Tay, K. L., Doe, K. G., Wade, S. J., Vaughan, D. A., Berrigan, R. E., and Moore, M. J., Sediment bioassessment in Halifax Harbour, *Environ. Toxicol. Chem.,* 11, 1567, 1992.

20. DeWitt, T. H., Swartz, R. C., and Lamberson, J. O., Measuring the acute toxicity of estuarine sediments, *Environ. Toxicol. Chem.,* 8, 1035, 1989.

21. Chapman, P. M., Sediment bioassay tests provide toxicity data necessary for assessment and regulation, *Proceedings of the Eleventh Annual Aquatic Toxicity Workshop, Nov. 13 -15, 1984,* Vancouver, B. C., Canada, Green, G. H., and Woodward, K. L., Eds., 1984, 178.

22. Chapman, P. M., Regulatory uses of aquatic toxicology (opportunities and pitfalls), *Proceedings of the Eighteen Annual Aquatic Toxicity Workshop,* Ottawa, Ontario, Sept. 30–Oct. 3, 1991.

23. Southerland, E., Karvitz, M., and Wall T., Management framework for contaminated sediments (The U.S. EPA Sediment Management Strategy), in *Sediment Toxicity Assessment,* Burton, G. A., Eds., Lewis Publishers, Ann Arbor, 1992, Chap. 15.

24. DeWitt, T. H., Ditsworth, G. R., and Swartz, R. C., Effects of natural sediment features on survival of the phoxocephalid amphipod, *Rhepoxynius abronius, Mar. Environ. Res.,* 25, 99, 1988.

25. Benton, M. J., Malott, M. L., Knight, S. S., Cooper, C. M., and Benson, W. H., Influence of sediment composition on apparent toxicity in a solid-phase test using bioluminescent bacteria, *Environ. Toxicol. Chem.,* 14, 411, 1995.

26. Nipper, M. G. and Roper, D. S., Growth of an amphipod and a bivalve in uncontaminated sediments: Implications for chronic toxicity assessments, *Mar. Pollut. Bull.,* 31, 424, 1995.

27. Tsai, C. F., Welch, J., Chang, K. Y. and Shaeffer, J., Bioassay of Baltimore Harbour sediments, *Estuaries,* 2, 141, 1979.

28. Ankley, G. T., Katko, A., and Arthur, J. W., Identification of ammonia as an important sediment-associated toxicant in the Lower Fox River and Green Bay, Wisconsin, *Environ. Toxicol. Chem.,* 9, 313, 1990.

29. Thompson, B. E., Bay, S. M., Anderson, J. W., Laughlin, J. D., Greenstein, D. J., and Tsukada, D. T., Chronic effects of contaminated sediments on the urchin *Lytechinus pictus, Environ. Toxicol. Chem.,* 8, 629, 1989.

30. Dillon, T. M., Moore, D. W., and Gibson, A. B., Development of a chronic sublethal bioassay for evaluating contaminated sediment with the marine polychaete worm *Nereis (Neanthes) arenaceodentata, Environ. Toxicol. Chem.,* 12, 589, 1993.

31. Carr, R. S., Long, E. R., Chapman, D. C., Thursby, G., Beidenbach, J.M., Windom, H. L., Sloane, G. M., and Wolfe, D. A., Toxicity assessment studies of contaminated sediments in Tampa Bay, Florida. *Environ. Toxicol. Chem.,* 15, 1218, 1996.

32. Jones-Lee, A. and Lee, G. F., Toxicity of ammonia in aquatic sediments and its implications for sediment quality evaluation and management, *Sediment Management Annual Review Meeting Minutes,* Tacoma, Washington, 1995, Appendix B.

33. Tay, K. L., Doe, K. G., MacDonald, A. J., and Lee, K., Monitoring of the Black Point Ocean Disposal Site, Saint John Harbour, New Brunswick, 1992–1994, Environment Canada, Atlantic Region, Internal Report, ISBN: 0-662-25655-7, Cat. no.: En40-214/9E, 133, 1997.

34. Sims, J. G., and Moore, D. W., Risk of pore water hydrogen sulfide toxicity in dredged material bioassays. U.S. Army Corps of Engineers, Waterways Experiment Station, Miscellaneous Paper D-95-4, 1995.

35. Bailey, H. C., Miller, J. L., Miller, M. J., and Dhaliwal, B. S., Application of toxicity identification procedures to the echinoderm fertilization assay to identify toxicity in a municipal effluent, *Environ. Toxicol. Chem.* 14, 2181, 1995.

36. Environment Canada, Biological test method: Toxicity test using luminescent bacteria (*Photobacterium phosphoreum*), Environmental Protection, Conservation and Protection, Report EPS 1/RM/24, 1992.

37. Environment Canada, Biological test method: Acute test for sediment toxicity using marine or estuarine amphipods, Environmental Protection, Conservation and Protection, Report EPS 1/RM/26, 1992.

38. Environment Canada, Biological test method: Fertilization assay using echinoids (sea urchins and sand dollars), Environmental Protection, Conservation and Protection, Report EPS 1/RM/27, 1992.

39. Marquenie, J. M., Bioavailability of micropollutants, *Environ. Technol. Lett.,* 6, 351, 1985.

40. Muir, D. C., Rawn, G. P., Townsend, B. E. Lockhart, W. L., and Greenhalgh, R., Bioconcentration of cypermethrin, deltamethrin, fenvalerate and permethrin by *Chironomus tentans* larvae in sediment and water, *Environ. Toxicol. Chem.,* 4, 51, 1985.

41. True, C. J. and Heyward, A. A., Relationships between Microtox test results, extraction methods, and physical and chemical compositions of marine sediment samples, *Tox. Assess.,* 5, 29, 1990.

42. Long, E. R., Buchman, M. F., Bay, S. M., Breteler, R. J., Carr, R. S., Chapman, P. M., Hose, J. E., Lissner, A. L., Scott, J., and Wolfe, D. A., Comparative evaluation of five toxicity tests with sediments from San Francisco Bay and Tomales Bay, California, *Environ. Toxicol. Chem.,* 9, 1193, 1990.

43. Greene, M. W., Bulich, A. A., and Underwood, S. R., Measurement of soil and sediment toxicity to bioluminescent bacteria when in direct contact for a fixed time period, *Proceedings 65th Annual Conference and Exposition of the Water Environment Federation,* New Orleans, LA, September 20–24, 1992.

44. Cook, N. H. and Wells, P. G., Toxicity of Halifax Harbour sediments: an evaluation of the Microtox® solid-phase test, *Water Qual. Res. J. Canada,* 31, 673, 1996.

45. ASTM standard guide for conducting whole sediment toxicity test with luminescent bacteria, submitted to ASTM Subcommittee E47.03, Draft #7, 9, 1995.

46. Gray, J. S., *The ecology of marine sediments,* Cambridge Studies in Modern Biology 2, Cambridge University Press, Cambridge, 1981, 45.

47. Oakden, J. M., Feeding and substrate preference in five species of Phoxocephalid amphipod from central California, *J. Crustacean Bio.,* 4, 233, 1984.

48. Kemp, P. F., Cole, F. A., and Swartz, R. C., Life history and productivity of the Phoxocephalid amphipod *Rhepoxynius abronius* (Barnard), *J. Crustacean Bio.,* 5, 449, 1985.

49. Slattery, P. N., Life histories of infaunal amphipods from subtidal sands of Monterey Bay, California, *J. Crustacean Bio.,* 5, 635, 1985.

50. Betts, B., Review of new scientific information and proposed modifications to the SMS Rule, Triennial Review, Sediment Management Standards (SMS Rule), Publication No. 95-318, Washington State Department of Ecology, 1995.

51. Nipper, M. G., Greenstein, D. J., and Bay, S. M., Short- and long-term sediment toxicity test methods with the amphipod, *Grandidierella japonica, Environ. Toxicol. Chem.,* 8, 1191, 1989.

52. Jones, R. A. and Lee, G. F., Evaluation of the elutriate test as a method of predicting contaminant release during open water disposal of dredged sediment and environmental impact of open water dredged material disposal, Vol. I: Discussion, Technical Report D78-45, US Army Corps of Engineers, WES, Vicksburg, MS, 1978.

53. Lee, G. F. and Jones, R. A., Water quality significance of contaminants associated with sediments: An overview, in *Fate and Effects of Sediment-Bound Chemicals in Aquatic Sediments,* Pergamon Press, Elmsford, NY, 1, 1987.

54. Sarda, N. and Burton, G. A., Ammonia variation in sediments: Spatial, temporal and method-related effects, *Environ. Toxicol. Chem.,* 14, 1499, 1995.

55. Gibson, A. B., Dillon, T. M., and Engler, R. M., Naturally occurring levels of ammonia and sulfide in pore water: An assessment of the literature, *Environmental Effects of Dredging Technical Notes, EEDP-01-34,* US Army Engineer Waterways Experimental Station, Vicksburg, MS, 1995.

56. Bay, S., Greenstein, D., Brown, J., and Jirik, A., Toxicity of sediments and interstitial waters from the Southern California Bight, presented at Second SETAC World Congress, Vancouver, November 5–9, 1995, 203.

57. Mueller, C. A. and Scott, J., The dynamics of ammonia in the 10-day sediment testing using the amphipod, *Ampelisca abdita,* presented at Second SETAC World Congress, Vancouver, November 5–9, 1995, 265.

58. Jacobs, M. W., Delfino, J. J., and Bitton, G., The toxicity of sulfur to Microtox from acetonitrile extracts of contaminated sediments, *Environ. Toxicol. Chem.,* 11, 1137, 1992.

59. Moller, A., Murphy, T., Thachuk, J., and Brouwer, H., Sulfide toxicity in anoxic sediments, *Proceedings of the Twentieth Annual Aquatic Toxicity Workshop,* Oct. 17–21, 1993, Quebec City, Quebec, 1994, 260.

60. Microbics Corporation, *Microtox manual. A toxicity testing handbook, Vol. 2 — Detailed protocols and Vol. 3 — Condensed protocols.* Carlsbad, CA, 1992.

61. Lee, D. L., Ten-day Amphipod Sediment Bioassay — A comparison of Particle Size Effects on Two Species of Amphipod, *Eohaustorius washingtonianus* and *Rhepoxynius abronius.* Environment Canada, Pacific and Yukon Region, unpublished data, 1995.

62. Environment Canada, Standard Operating Procedure #29: Measurement of Redox, Ammonia, and Sulfides in Sediments and Water, Environment Canada, Atlantic Region, 1995.

63. USEPA, EPA methods for assessing the toxicity of sediment-associated contaminants with estuarine and marine amphipods, EPA 600/R-94/025, June 1994.

64. Mount, D. R., Henke, C. E., Ingersoll, C. G., and Besser, J. M., Development of toxicant identification procedures for whole sediment toxicity tests, presented at Second SETAC World Congress, Vancouver, November 5–9, 1995, 9.

65. Cohn, N., Pinza, M., Ohlrogge, S., Ferguson, C., and Word, J., Manipulated and unmanipulated reduction of ammonia in sediment toxicity tests with the amphipod, *Rhepoxynius abronius,* presented at Second SETAC World Congress, Vancouver, November 5–9, 1995, 264.

66. Buckley, D. E. and Hargrave, B. T., Geochemical characteristics of surface sediments, in *Investigations of Marine Environmental Quality in Halifax Harbour,* Nicholl, H. B., Ed., Canadian Technical Report of Fisheries and Aquatic Sciences, No. 1693, 1989, 9.

CHAPTER 40

Oil Spill Treating Agents: Present Knowledge and Toxicity Testing Needs

Alan J. Mearns

CONTENTS

ABSTRACT

Data and information were compiled and reviewed on the toxicity of oil spill treating agents to aquatic organisms. Oil spill treating agents include solidifiers, herders, emulsion treating agents, dispersants, shoreline cleaners, and bioremediation products. Small-scale aquatic toxicity tests have been used both to screen products and to test their efficacy during actual oil spill responses. In general, treating agents are less toxic (96-h LC_{50} range: 8 to >500,000 ppm) than the oils they are intended to treat (96-h LC_{50} range: .2 to >1000 ppm). However, the toxicities of oils and some shoreline cleaners and bioremediation products overlap, raising cause for concern among response agencies. Furthermore, the ability to compare products is hampered by use of various test species and exposure conditions. It is recommended that response authorities, both national and international, continue to encourage standardized testing procedures. Manufacturers should be encouraged to conduct toxicity tests during product development as well as during product promotion. Finally, toxicologists should continue to develop tests that more accurately reflect the physical phases of treating agents (solid, dispersed, water accommodated) as well as realistic response conditions (i.e., pulse exposure) and also conduct a battery of tests to compare species sensitivities. Acceptance of new chemical and biological technologies require all three groups to work in concert.

I. INTRODUCTION

Oil spill treating agents are chemicals and other materials used to change the fate and behavior of oil and to reduce its impact on marine and freshwater life (Sergy et al., 1993; Walker et al., 1993; Walker et al., 1995a; SEA, Inc., 1995). Examples include dispersants, herders, de-emulsifiers, shoreline cleaners, and bioremediation materials. This article explores the use of freshwater and marine toxicity tests for evaluating the ecological safety of oil spill treating agents.

Biotechnology firms and the chemical and oil industries are producing an intriguing array of chemical, physical, and biological materials to help disperse, herd, contain, solidify, and degrade crude oil and refined products. The oil spill response community — industry and government — would like to use these products but are often reluctant to do so because of a lack of confidence in their effectiveness and concern about exacerbating the environmental effects of the oil spill. A major problem is that testing these materials barely keeps pace with their development, production, and promotion. Furthermore, the testing that is done — especially toxicity testing — is incomparable and difficult to interpret. As a result, local, state, provincial, and national agencies are now searching to find the best methods for confirming and comparing the effectiveness and safety of oil spill treating agents. Standardized microscale toxicity testing — using small containers and small marine and freshwater organisms — play, a pivotal role in that assessment.

This article reviews what has been accomplished in evaluating the toxicity of oil spill treating agents using small-scale testing procedures. It is especially directed to 1) manufacturers, to help guide them toward producing acceptable products, 2) toxicologists, who can benefit from understanding the decision challenges faced by the spill response managers, and 3) the spill response community itself, those whose job it is to decide what response tools to use under various spill response situations and who need to understand the scope and limits of toxicology testing. In this connection, it is hoped this chapter will be useful as an aid for teaching oil spill responders how to evaluate the value and utility of ecologically-relevant toxicity tests.

The chapter is not a critique of microscale toxicity test methods *per se*. Instead, it highlights properties of oil and oil spill treating agents, what toxicity testing methods have been used and recommended, how product toxicities compare with each other and with oil itself, and what challenges lie ahead in product evaluation and testing. Complete references are included for those interested in the mechanics and scales of the tests.

The chapter is based on a review of pertinent literature and on the experience of the author and his colleagues during spill responses, drills, and response planning activities. In addition to published and unpublished scientific literature, a primary source of information for this article is the National Contingency Plan (NCP) Product Schedule Database and an NCP Product Schedule Notebook published by the U.S. Environmental Protection Agency's Emergency Response Division (for a description, see Thomas et al., 1995). These documents are designed to provide responders, support personnel, and researchers with information to allow quick comparison of product data: composition, application rates, effectiveness, availability, storage time, and toxicity. The 1995 Product Schedule Database contains information on over 100 products authorized for potential use on oil discharges. Proprietary information is collected but is kept confidential.

This chapter does not review the many microscale tests available for determining the *effectiveness* of treating agents to control oil. Effectiveness testing is an essential part of the evaluation of new products and indeed uses many of the microscale tools (small containers, flasks, etc.) also used to evaluate their toxicity. For a review of effectiveness testing, see NRC 1989, Fingas et al. 1995, and Walker et al., 1993 and 1995.

II. OIL IMPACTS AND SPILL RESPONSE

Oil is a natural substance that freshwater, marine, and terrestrial systems have dealt with for eons. When exposure is low to moderate, such as near natural oil seeps, marine and freshwater ecosystems are capable of degrading, consuming, and metabolizing oil. However, natural systems are overwhelmed by large quantities. To understand the need for oil spill treating agents it is necessary to understand basic features of the fate, transformations, and effects of oil spilled at sea, in rivers, lakes, and streams. As shown in the upper part of Figure 40.1, oil is composed of many substances. The fate of these substances is largely determined by physical conditions and processes in the sea and freshwater environments (lower left circle, Figure 40.1), conditions that vary greatly among spills. The biological effects of the oil (lower right circle, Figure 40.1), including toxicity, depends on what mix of substances organisms are exposed to and how long they are exposed. The physical processes illustrated can serve both to enhance or mitigate toxicity.

For example, as a thick, viscous material, oil can smother or coat shoreline plants and animals, leading to death by asphyxiation or, in the case of birds and mammals, by hypothermia. Although short-lived, soluble volatile compounds — benzene, toluene, xylenes, and other aromatic hydrocarbons such as naphthalenes — are abundant in fresh oil and fuels and are the principal compounds causing narcosis, loss of equilibrium, and death to exposed marine and freshwater organisms (Anderson et al., 1974). Fortunately, early loss of these compounds — through "weathering" — can eventually turn much of the oil into a relatively innocuous material. Finally, oils and fuels, on the sea surface or mixed into the water column or sediments, are a source (through ingestion) of mutagenic and carcinogenic chemicals, most notably the higher-molecular-weight polycyclic aromatic hydrocarbons (PAHs) such as benzo-(a)-pyrene.

The steps in an oil spill response move from notification to containment, shoreline protection, and, if necessary, cleanup. At each stage, a variety of tools can come into play. Alternatives include "no action" (the oil is not a threat or is rapidly dispersing or weathering), mechanical or physical treatment, chemical treatment, and biological treatment (bioremediation). Since the 1989 Exxon Valdez oil spill (e.g., Wolfe et al., 1994), there has been a shift in approach for responding to spills, including a major move to pre-approve many new treatment products. While still relying on conventional response methods (booms, skimmers, pressure washing, etc.), responders are taking a more cautious approach with high-energy methods and looking to supplement these with what are perceived as more "environmentally sensitive" approaches, including the use of chemical and other treating agents. In the U.S. the decision to use a particular response method is at the discretion of 1) a unified command composed of federal and state on-scene coordinators (OSCs) and the spiller and 2) regional response teams (RRTs) composed of resource trustee agencies. The effectiveness and safety of new response methods, including treating agents, is often questioned. The central issue with toxicity testing is the extent to which products used to combat oil are, in fact, safe.

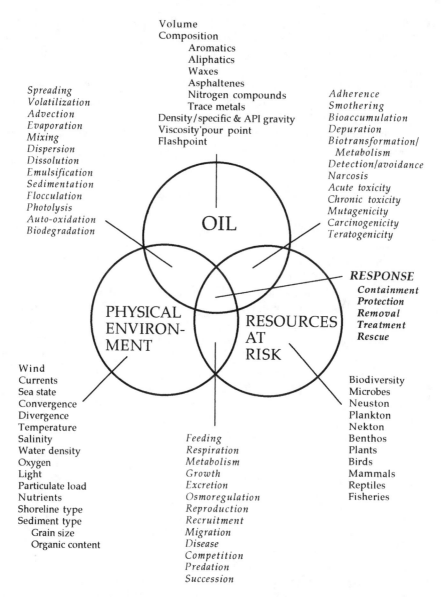

Volume
Composition
   Aromatics
   Aliphatics
   Waxes
   Asphaltenes
   Nitrogen compounds
   Trace metals
Density/specific & API gravity
Viscosity'pour point
Flashpoint

Spreading
Volatilization
Advection
Evaporation
Mixing
Dispersion
Dissolution
Emulsification
Sedimentation
Flocculation
Photolysis
Auto-oxidation
Biodegradation

Adherence
Smothering
Bioaccumulation
Depuration
Biotransformation/
* Metabolism*
Detection/avoidance
Narcosis
Acute toxicity
Chronic toxicity
Mutagenicity
Carcinogenicity
Teratogenicity

OIL

PHYSICAL ENVIRON-MENT

RESOURCES AT RISK

RESPONSE
Containment
Protection
Removal
Treatment
Rescue

Wind
Currents
Sea state
Convergence
Divergence
Temperature
Salinity
Water density
Oxygen
Light
Particulate load
Nutrients
Shoreline type
Sediment type
   Grain size
   Organic content

Feeding
Respiration
Metabolism
Growth
Excretion
Osmoregulation
Reproduction
Recruitment
Migration
Disease
Competition
Predation
Succession

Biodiversity
Microbes
Neuston
Plankton
Nekton
Benthos
Plants
Birds
Mammals
Reptiles
Fisheries

Figure 40.1   Interactions among properties and processes (italicized) of oil, aquatic environments, and biota that need to be considered in oil and oil spill treating agent toxicity testing.

III. OIL SPILL TREATING AGENTS

A. Classification

As indicated above, oil spill treating agents are chemicals and other materials applied to oil to reduce its impact on marine and freshwater life (Walker et al., 1993; Walker et al., 1995a; SEA, Inc., 1995). Specifically, treating agents are offered to minimize biological, ecological, and public health impacts from several forms of oil, including sheens, slicks and tar balls, soluble compounds, and the carcinogenic and mutagenic compounds such as PAHs. The conventional classification of oil spill treating agents is based on use, not chemical properties:

1. Dispersants
2. Bioremediation agents
3. Chemical oil spill treating agents
   Herding agents
   Emulsion treating agents
   Solidifiers
   Elasticity modifiers
   Shoreline cleaning agents
   Oxidation agents
4. Sorbents

Dispersants are ionic and nonionic surfactants that act upon contact to enhance the ability of oil on the water surface to break into small (<70 μm) droplets (NRC, 1989). Dispersants are delivered directly to floating oil by fine mist spraying from aircraft or boats and at concentrations on the order of about 10% of the spilled oil. The quickly formed oil droplets do not sink *per se* but mix into the upper water column to be diluted by prevailing currents and water motion. When appropriately used, dispersants can reduce the amount of oil that strands on shorelines and reduce potential contact of sea birds with oil slicks. Use of dispersants is limited to the first few hours or days of a spill after which emulsification (water in oil) inhibits dispersion. Dispersed oil is not recoverable and environmental concerns include the fate and toxicity of the dispersants as well as the oil.

Bioremediation products include oil-degrading microorganisms, nutrients, and oxygen-producing chemicals designed to accelerate the degradation of oil. Relative to an oil spill response (hours, days), biodegradation is a slow process (days, weeks, months). Only certain groups of hydrocarbons — namely, straight-chain alkanes and some PAHs — are degradable in this time frame: waxes, asphaltenes, and other recalcitrant compounds remain. However, some bioremediation products contain surfactants and emulsifiers that change the appearance and mobility of stranded oil, but these processes must be distinguished from true degradation. While there is good evidence that "microbial seeding" and nutrient amendments work in the laboratory and at some contained waste sites, oil-degrading microbes occur in all freshwater and marine environments and their effectiveness is often limited only by nutrients (Venosa et al., 1996) and/or available oxygen; indeed materials that release oxygen are among the products considered as treating agents (Mendelsshon et al., 1995). Environmental concerns include the toxicity of nutrients, pathogenicity of microbes, and toxicity of intermediate metabolites of PAHs.

Herding agents are liquids that have a higher spreading pressure than oil and work like "chemical booms" to compress oil slicks on the water surface, thus improving chances of recovery. They are by definition insoluble in water. Recommended doses are in the range of 1 to 15 L per mile of slick. Environmental concerns include the fate and effects of material not recovered or removed.

Emulsion treating agents are hydrophilic surfactants that destabilize water-in-oil emulsions and include both emulsion inhibitors and emulsion breakers. They can be effective at application rates on the order of 1% of the oil or less. They can help extend in time the window-of-opportunity to use dispersants or *in situ* (open water) burning, which are inhibited by emulsification. Since their main use is to support dispersion, they are not recoverable and environmental concerns focus on their fate and effects.

Solidifiers are polymers or granules that turn oil into a coherent mass, bonding it into carpet- or gel-like masses for improving recovery. Environmental concerns include the fate and effects of material not recovered or removed.

Elasticity modifiers currently include one liquid material that conveys viscoelastic properties to light oils, aiding in their physical removal. It increases the stickiness of oil, which could increase the effectiveness of skimming but also might be detrimental in marshes and to birds should the treated material get loose.

Shoreline treating agents include a wide variety of surface active liquids and surfactant-based solids designed to enhance removal of stranded oil by water washing. Some cause more dispersal than others (Fingas et al., 1995), making recovery of removed oil more or less difficult. Some include ingredients now common in cleaners and degreasers for public sale and consumption. Environmental concerns include the fate and effects of material not recovered or removed.

Only one *oxidizing agent* — a titanium oxide-based granule — has been offered to date. This product is claimed to promote direct chemical oxidation of oil. It is not to be confused with oxygen-producing agents (i.e., peroxides) used to enhance bioremediation (see above). Environmental concerns include the fate and effects of material not recovered or removed.

Sorbents are solid or fibrous materials in the form of pillows, pads, boom, or woven sheet used to absorb/adsorb oil on direct contact. During a response, sorbents are often deployed as passive collectors and left unattended. Their potential to be lost, and not biodegrade, indicates a need for testing (Sergy et al., 1995).

In situ burning of oil, either on the sea surface or on shorelines such as marshes, is also a new tool coming into play in spill response. Burning requires use of ignition fluids, but its principal effects are atmospheric contamination (a possible concern for wildlife as well as humans) and production of residues. Open-water burning is a highly desired response method with response niche overlapping that of dispersants.

Each of these classes of agents has several commonly used synonyms, can fill one or more uses or response niches in the spill response working environment, and has a particular window of opportunity when it is effective and outside of which it is not (Walker et al., 1995). Each has certain tradeoffs or environmental drawbacks. For example, dispersants can be used to stop the rapid spreading of oil on the water: they put the oil into the water column as small droplets. But these are effective only on certain commonly transported oils and only in the early hours of a spill before the oil weathers or emulsifies (water-in-oil emulsion) and most require some wind energy. By contrast, herding agents and elasticizers have the potential to help confine and collect oil at the surface, but work only in calm conditions, and must be used with recovery systems.

B. Composition

Oil spill treating agents are produced from both synthetic and natural materials. Unfortunately for the toxicologist, the specific composition and content of many commercial products is indeed proprietary. However, some information is provided in the 1995 NCP Product Schedule database. Chemical composition information was encountered for 24 dispersants and 59 other products including: 34 bioremediation; 2 shoreline cleaning agents; one each viscoelasticizer and solidifier; and 9 miscellaneous products. Twelve products were not classified.

More than 100 synthetic and natural chemicals and materials have been used as ingredients of oil spill treating agents. Many are food additives and therefore are presumably not toxic and otherwise biodegradable. However, others, such as some of the metals and organic chemicals, are also U.S. EPA "Priority Pollutants," and therefore should be the subject of ecological concern if they are to be introduced into the environment in significant quantities to treat oil.

The 59 nondispersant products listed in the NCP contained a total of 71 materials including 17 types of microorganisms (e.g., bacillus, salmonella, *Klebsiella pneumoniae*, pseudomonads, corynebacteria, enterobacteria, etc.); 11 trace metals (arsenic, barium, copper, chromium, lead, zinc, etc.), 15 types of nutrient additives (micronutrients, tricalcium phosphate, ammonium phosphate, etc.); 8 enzymes (oxidases, proteases, transferases, amylases, etc.); 5 organic chemicals (methyl ethyl ketone, d-limonene, polyisobutylene, etc.); 9 carriers (wheat, sugar, etc.); 1 soap; 10 other specific ingredients (extracts of kelp, aloe, calcium stearate, etc.) and 5 confidential ingredients (Table 40.1). Where reported, concentrations of the metals ranged up to 51 ppm (mg/dry kg) for zinc in one product. Chromium, lead, zinc, and barium were the most frequently reported, with concentrations above detection limits. However, there was no standard detection limit, and they ranged 2 to 3 orders of magnitude (e.g., from 0.01 to 10.0 ppm). One shoreline cleaner, tested recently at several oil spill sites, contains proprietary bacterial fermentation by-products in combination with a carrier solvent, d-limonene. The product-oil mixture is claimed to be stable and water insoluble. Chemical analysis indicated that the major component of the product is d-limonene (90 to 97% by weight). D-limonene, a dipertane, is a colorless liquid with a lemon-like odor and is used as a dispersing agent in oils, resins, paints, lacquers, varnishes, and floor waxes and furniture polishes. It is also used as a flavor and fragrance, a mosquito larvicide, and a freshwater herbicide and is effective in killing fleas (Hoff et al., 1994).

In summary, oil spill treating agents can contain a variety of microorganisms and organic and inorganic chemicals, many of which may be toxic. As a result, responders and toxicologists alike sense a need for direct, reliable, comparable data on both the effectiveness of products and on their toxicity, both alone and in the presence of oil.

IV. TOXICITY TESTING

There are two basic objectives for testing the toxicity of treating agents: 1) to compare and rank products so that responders can select among the most effective and safest, and 2) to confirm or document actual performance (effectiveness and safety) in the field during actual oil spill responses. The first objective can be met through use of standardized acute toxicity tests such as those regularly employed to monitor effluents and other materials. The second objective, where toxicity testing may be part of a larger environmental effects monitoring program, may be accomplished using standardized tests or site- and species-specific tests, including chronic or sublethal components as well as bioaccumulation.

A. Product Screening

Small-scale fish and invertebrate tests have been used to determine toxicities of chemicals under strict protocols for more than half a century (e.g., Jones, 1964). Standardized aquatic toxicity screening tests have been used to evaluate, compare, and select dispersants for several decades whereas their use with most other types of oil spill treating agents has begun only recently. However, existing and proposed tests are not necessarily standardized across types of agents (i.e., dispersants compared to bioremediation agents) or among countries. Table 40.2 summarizes 12 existing and proposed national tests in the U.S. (NETAC, 1993; Walker et al., 1995), Canada (Sergy et al., 1993 and 1995), and the United Kingdom (Anon. 1994a and 1994b; Rycroft et al., 1994). These involve 10 species of fishes and invertebrates (5 salt water and 5 freshwater), 5 exposure times (100 min, 6 h, 48 h, 96 h, and 7 d) and tests with or without one of several oils or fuels (No. 2 fuel, Prudhoe Bay crude oil, Kuwait crude oil). All are static or partial-replacement small-scale tests in the sense that they are done with small organisms in small containers. Testing details can be found in the cited papers. There are other tests used internationally: for dispersants alone, Pauwels and Clark (1993) list 17 tests required by 11 countries.

In the U.S., responsibility for conducting the toxicity tests rests with the product manufacturers or vendors. The results must be reported to the EPA if the product is to be listed in the NCP. In Canada and the U.K., national laboratories perform the tests and list or publicize only those products that pass predetermined criteria (i.e., LC_{50}s >1000 ppm or no more toxic than the oil).

In addition, manufacturers, vendors, industries, states, and academia have conducted many tests with nonstandard methods. Walker et al. (1993) found that vendors relied heavily on two test species, brine shrimp (*Artemia salina*) and top minnows (*Fundulus* spp.) and the recommendation was advanced that these should be the standard species. However, to be listed in the NCP, the U.S. EPA

Table 40.1   List of Materials Reported in 71 Nondispersant Oil Spill Treating Agents Listed in the National Contingency Plan (NCP) of the United States

Metals

Arsenic	Lead	Barium
Cadmium	Mercury	Selenium
Chromium	Nickel	Silver
Copper	Zinc	

Microorganisms

Acinetobacter sp.	Biological-hydrocarbon cleaner	*Pseudomonas stutzeri*
Acinetobacter johnsonii	Confidential microbes	*Pseudomonas flourescens*
Alcaligenes faecalis	*Coryneform sp.*	*Pseudomonas putida*
Bacillus sp.	*Enterobacter gergoviae*	Saprophytic microbes
Bacillus circulans	Facultative anaerobes	Ubiquitous hydrocarbon-oxidizing
Bacillus spore group 1	*Klebsiella pneumoniae*	microorganisms
Bacillus spore group 2		

Nutrients

Ammonium phosphate	NH_4	PO_4
Inorganic salts	Nitrogen	Potassium nitrate
Micronutrients	Oleophilic nutrient additive	Supplemental salts
Mineral nutrients	Phosphoric acid	Tricalcium phosphate
Monoammonium phosphate	Phosphorus	Urea
N-K-P 36-6-6		

Enzymes

Amylase	Fermentation enzymes	Oxidoreductases
Carbohydrases	Hydrolases	Protease
"Confidential" enzymes	Ligases	Transferases
Enzyme	Lipases	

Organic Chemicals

Block copolymer	"Confidential" solvent	Polyisobutylene
Calcium stearate	Methyl ethyl ketone	Solvent
Chlor hydro	Methyl urea	

Carriers

Aloe extract	Milk products	Sugar
Coated capsules	Prills	Wheat bran
Kelp extract		

Other

Biodegradable soap	"Confidential" additive	Emulsifier
BOD	"Confidential" surface active agent	Wetting agents
Buffers	Cyanide	

Note: Listed as provided to EPA by product vendors.

now requires acute toxicity data for silversides and mysids (marine) or fathead minnow and *Daphnia* (freshwater; see Table 40.2). Other species have been used for purposes other than screening but which may lead to better standardized tests in the future (such as the use of several invertebrates and algae with spiked or pulse bioassays for dispersants, Singer et al., 1991; Pace and Clark, 1993; Bragin et al., 1994; Pace et al., 1995). Indeed, one of the largest bodies of aquatic toxicology data is for dispersants, data which has been developed over the past 30 years using yet a different variety of European marine species (e.g., Rycroft et al., 1994).

Table 40.2   Comparison of Regulatory Standard Acute Toxicity Tests for Oil Spill Treating Agents from the U.S., Canada, and the U.K.

Agency	Treating Agent	Species	Life Stage	Medium	LC_{50}-Type	Product Alone	Product+Oil?	Reference
EPA	All (NCP)	Silverside	Juvenile	SW	96 h	5 dilutions	1:10, No. 2	1
EPA	All (NCP)	Mysid	Juvenile	SW	48 h	5 dilutions	1:10, No. 2	1
Environment	All	Rainbow trout	Juvenile	FW	96 h	100 mg/L	no	2
Canada	Biorem.	Rainbow trout	Juvenile	FW	96 h	10,000 mg/L	no	5
	Biorem.	Daphnia	Adult	FW	48 h	10,000 mg/L	no	5
NETAC/EPA	Biorem.	Silverside	Juvenile	SW	7 d	yes	PBCO	3
NETAC/EPA	Biorem.	Mysid	Juvenile	SW	7 d	yes	PBCO	3
NETAC/EPA	Biorem.	Fathead minnow	?	FW	7 d	yes	PBCO	3
NETAC/EPA	Biorem.	Daphnia	?	FW	7 d	yes	PBCO	3
MAFF U.K.	All	Brown shrimp	Adult	SW	100 min + 24 h	no	Kuwait	4
MAFF U.K.	All	Limpet	Adult	SW	72 h; 6 h + +	no	Kuwait	4
MAFF U.K.	All	Pogge (fish)	Adult	SW	96 h	no	Kuwait	4

Note: FW = Freshwater; SW = Salt water; Biorem. = Bioremediation agents only; PBCO = Prudhoe Bay Crude Oil.
1 Described in Walker et al., 1993.
2 Sergy et al., 1995.
3 NETAC, 1993.
4 Rycroft et al., 1994.
5 Sergy et al., 1993.

B. Toxicity of Treating Agents

With the notable exceptions of dispersants and shoreline cleaning agents, there is still little published data on the toxicity of oil spill treating agents. One major source is Walker et al. (1993) who solicited and received information on the results of toxicity tests conducted by some of the manufacturers of materials other than dispersants and bioremediation agents. Eighty products were identified through the NCP, product brochures, and literature searches, and questionnaires were sent out asking for more specific information. Responses were received for 43, of which 12 declined to offer additional information. Of the 31 positive responses, toxicity information was provided for 17, including 8 shoreline cleaners, 4 solidifiers, 2 herding agents, 2 elasticity modifiers, and one emulsion breaker. The toxicity of these products was assessed with a total of 44 tests involving a cumulative 24 acute aquatic toxicity endpoints.

The results of all available data (screening with both standardized tests and manufacturers own tests) are summarized in Table 40.3, together with data on ranges of toxicities of dispersants, bioremediation agents and several oils and fuels. The materials are listed, from top to bottom, approximately in order from least toxic (solidifiers, top, Table 40.3) to most toxic (bottom, No. 2 and Bunker C fuel oils).

Although there are many problems and hazards in compiling and comparing this mixed assemblage of data, there are some noteworthy patterns. Toxicity of treating agents ranged from 8 ppm (quite toxic) to >500,000 ppm (not toxic) for 48-h LC_{50}s and from 5 to >500,000 ppm for 96-h LC_{50}s. By comparison, fresh crude and fuel oils are more toxic (lower LC_{50}s), but there is considerable overlap. For example, the overall range of 96-h LC_{50}s for a variety of oils and fuels was 0.2 to >1000 ppm (bottom, Table 40.3). Water soluble fractions from both crude and fuel oils were quite toxic with 96-h LC_{50}s ranging from 1.9 to >19.8 ppm, with fuel oils (No. 2 diesel and Bunker C) being generally the most toxic (0.4 to about 10 ppm).

These comparisons suggest that solidifiers, elasticity modifiers, emulsion treating agents, and herding agents are practically nontoxic, with 96-h LC_{50}s considerably greater than 1000 ppm (mg/L)

Table 40.3   Comparison of Acute Toxicities (LC_{50}s) of Oil Spill Treating Agents, Crude Oils, and Fuel Oil

Class	#Products	#Tested	#Tests	#End-points	LC_{50}, ppm[1] 48 h Min	48 h Max	96 h Min	96 h Max
Solidifiers[2]	10	4	8	4	>500,000		>500,000	NG
Elasticity modifiers[2]	2	2	5	4	>18,000	>100,000	>100,000	NG
Emulsion treating Agents[2]	3	1	1	1	ND	ND	>10,000	NG
Herding agents[2]	2	2	6	5	>4,500	>100,000	>1,000	4,500
Shoreline cleaning agents[2,*]	8	8	24	22	44	>10,000	29	>10,000
Shoreline cleaning agents[3]	51	51	51	1	ND		8	>5,600
Oxidizer[2]	1	0	0	0	ND		ND	
Dispersants[4]	2	2	>50	>25	8	20,000	5	42,000
Bioremediation agents[5]	10	10	3	3	>10,000		710	>10,000
Crude oils[6]	NG	NG	54	54	ND		0.2	>1,000
Crude oils OWD[7]	2	2	3	8	48	>80,000	200	>80,000
Crude oils WSF[7]	2	3	3	8	6.6	>19.8	5.5	>19.8
No. 2 fuel oil[6]	NG	NG	39	8	ND		0.4	8
No. 2 fuel oil OWD[7]	2	1	3	8	1.3	200	3.0	93
No. 2 fuel oil WSF[7]	2	1	3	8	0.9	76.9	3.5	6.3
Bunker C fuel, WSF[7]	1	1	3	8	0.9	4.4	1.7	3.5

[1] Nominal for treating agents including dispersants. [2] From data in Walker et al., 1993. [3] Fingas et al., 1995. [4] From Rycroft et al., 1994. [5] From Sergy et al., 1993. [6] Rice et al., 1977. [7] Anderson et al., 1974. * One reported NOEC = 3.3 and EC_{50} = 54 ppm. NG = Not Given. ND = No Data.

and mostly above 10,000 ppm, well above the toxicity range of oils and fuels. However, the ranges of oil and fuel toxicity overlap considerably with the ranges of toxicities for remaining classes of treating agents including shoreline cleaning agents (LC_{50}s 29 to 10,000 ppm), bioremediation agents (710 to >10,000 ppm), and two dispersants (5 to 42,000 ppm). Some of these agents had LC_{50}s well within the range of dissolved or dispersed fractions of crude and fuel oils (i.e., 1 to 100 ppm). Thus, as classes, shoreline treating agents, bioremediation agents, and dispersants were more toxic than the solidifiers, the elasticity modifier, emulsion treating agents, and herding agents, but generally less toxic than oils. However, in some cases, they had toxicity similar to that of oil.

Obviously, some shoreline cleaning agents are almost as toxic as crude and fuel oils, and their approval and use should be carefully considered. Bioremediation treating agents are also a class of recent current interest because of major developmental efforts during the 1990s. Additional data on the toxicity of bioremediation treating agents have been collected as part of an EPA evaluation of bioremediation testing protocols (NETAC, 1993), but results have not been published. In addition, there is some information from one product, the oleophilic fertilizer, Inipol. Inipol toxicity was determined with and without weathered Prudhoe Bay crude oil using six acute and two chronic estimator marine bioassays (EPA, 1991). Nominal LC_{50}s for Inipol alone for a pandalid shrimp and three marine fish ranged from 100 to 1500 ppm, but were lower for bivalve larvae and mysids (15 to 23 ppm). Mixed with weathered crude oil, shrimp and fish LC_{50}s indicated somewhat less toxicity (92 to 6000 ppm), but inspection of the data also suggests toxicity to the three small invertebrates increased (8.5 to 55 ppm). Finally, in the only known application of mutagenicity testing for oil spill treating agents, it was reported as part of the EPA study that Inipol was not mutagenic.

There are many problems with these (Table 40.3) comparisons, not the least of which is a lack of information on the actual methods and conditions of the tests, including the extent to which exposure concentrations have been confirmed, use of too many different and probably incomparable species and endpoints, etc. These are discussed in the closing comments of this paper.

C. Field Monitoring

Microscale toxicity tests are only just beginning to play a role in assessing toxicity of oil spill treating agents under actual field conditions, and this is where their role may become most important. Using small aquaria set up at an incident command post, Shigenaka et al. (1995) used three native tropical intertidal invertebrates to observe for possible toxic effects of the shoreline cleaner Corexit® 9580 to improve washing effectiveness during a major 1993 oil spill in San Juan, Puerto Rico. Essentially, no effects attributable to the agent were observed when animals were exposed to oil and Corexit-contaminated rinse water. During the response to the March 1989 Exxon Valdez oil spill in Alaska, Claxton et al. (1993) used an Ames *Salmonella* test to record trends and differences in mutagenicity among oiled plots treated with bioremediation products — namely, oleophilic and water-soluble nutrients. During three months of shoreline treatment and monitoring the authors observed continuous decline in mutagenicity, but at the end of the summer of 1989, treated plots appeared to be more mutagenic than untreated plots. This conclusion has not been tested statistically. Coehlo et al. (1995) collected water samples under dispersant-treated and untreated oil slicks in the North Sea but have yet to report the results. Using data supplied by EPA, Mearns et al. (1993) reported water toxicity to mysids in one of two marsh plots treated with a bioremediation product at a 1990 oil spill in Galveston Bay, Texas. Water from untreated plots was not toxic, and it was speculated that copper in a micronutrient mix in the treating agent mixture was high enough, upon dilution, to exceed water quality criteria. Using five different small-scale bioassays, Mearns et al. (1995) monitored and compared rates of toxicity reduction among replicate oiled, nutrient-treated, and nutrient plus oil-degrading bacteria plots, during a major 14-week oil bioremediation project on a sandy beach in Delaware (see also complementary study by Venosa et al., 1996). The tests included both pore water and sediment-phase bioassays including Microtox®, a grass shrimp embryo bioassay, and a 10-day sediment amphipod bioassay. All but one of the tests indicated that nutrient treatment neither accelerated nor slowed loss of toxicity as the oil weathered and degraded. However, based on results using the Fisher and Foss (1993) grass shrimp (*Paleomonetes pugio*) embryo bioassay, pore water from one nutrient and bacteria-treated plot remained toxic for many weeks. In addition, using a 10-day sediment (solid phase) amphipod bioassay, all oiled plots, regardless of treatment, remained highly toxic for at least 14 weeks. Finally, during field trials of a complementary countermeasure (*in situ* burning), Daykin et al. (1994) reported low or no toxicity of subsurface waters under either burned or unburned oil using five marine toxicity tests.

V. FUTURE DIRECTIONS

A. The Responder's View

Response to an oil spill is an organized rapid process allowing little or no room for experimentation. As oil weathers and moves, the windows of opportunity for various counter measures quickly move from one to the next (response niches, Walker et al., 1995). An opportunity to use a particular method (*in situ* burning or dispersants, for example) can appear quickly, then disappear a few hours later if the weather changes or the oil emulsifies. Even once stranded on the shore, cleanup opportunities change drastically as oil thickens or increases in viscosity or pour-point temperature (requiring hotter water to wash). Thus, the fast pace of a spill response leads many authorities to reject untested treatment proposals, and apparently toxic or ineffective products, and, instead, encourages vendors to participate in pre-approving the use of new materials.

Thus the issue of immediate concern is establishment of the process of pre-approving new spill response products. Resource trustees, who must be consulted as part of any pre-approval process, will approve use of products that are effective and safe, but will reject outright application of treating agents with inadequate documentation of toxicity and/or environmental effects. The role of microscale toxicity testing is implicit in this process, mainly because almost all freshwater and marine toxicity testing can be accomplished with small-scale (laboratory) bioassays. Based on existing data, we appear able to advise on approval of some classes of spill treating agents. For example, dispersants and to some extent shoreline cleaning agents have been evaluated and some pre-approved. At first glance, solidifiers, herders, and emulsifiers, by virtue of insolubility, do not appear to be very toxic and could also be pre-approved. Alternatively, toxicity data on some classes of products — namely, bioremediation agents in the U.S. — have yet to be published or the results subjected to peer review. We really don't know where they stand. As demonstrated by EPA (1991), one agent (Inipol) had a range of toxicities comparable to or greater than the weathered oil which it was used to treat.

When selecting a product for a given response, it may be appropriate to consider effective products even though they are more toxic than others. Decisions to use or not to use a product must be weighed in light of what would happen if the treating agent is not used. For example, if treatment is a one-time event and exposure will be short-lived (hours) or the product can be removed after treatment, then it may be useful to use such a product for oil removal, i.e., to clean a marsh prior to a migratory bird arrival. In other words, it is not always the case that responders must reject products solely on the basis of toxicity.

Finally, approving authorities must also consider the extent to which use of a treating agent reduces response damage in sensitive areas. For example, use of even a very nontoxic product may be counter-productive if its application results in trampling marshes. The goal is to balance the scales in favor of minimizing impact and maximizing recovery and restoration.

B. The Toxicologist's View

Oils, fuels, and oil spill treating agents are complex, often incomparable materials that continue to pose great challenges to marine and aquatic toxicologists. During a spill response, all these materials can occur together in several phases or states (surface slicks, tar balls, dispersed droplets, dissolved materials, aerosols, etc.), and the dominant phases can shift with time. Thus the greatest toxicological problem is not so much the selection of species to be employed in toxicity testing, or even the scale of the toxicity testing system, but rather with the properties of the test materials themselves. The issues of exposure have been and continue to be addressed time and again in the literature. The need for their resolution is finally being acknowledged and acted on (Aurand, 1995a and 1995b; Aurand and Coelho, 1995; Betton, 1994; Bragin et al., 1994; Kucklick and Walker, 1995; Mearns, 1995; NRC, 1989; Pace et al., 1995; Rice et al., 1977; and Sergy et al., 1995). To emphasize the challenge, consider the following example. With regard to solidifiers, all exposure concentrations reported to date are nominal, not measured. It may be the case that a particular "insoluble" solidifier produces a very low nominal toxicity, i.e., 500,000 ppm. However, if as little as 0.0001 or 0.001% of that product is insoluble, organisms responding with an LC_{50} of 500,000 ppm nominal may in fact be experiencing product concentrations on the order of 0.5 to 5.0 ppm, which are comparable to concentrations of oil in the water. In any case, as suggested by many of the authors cited above, the extent to which nominal vs. measured concentration is a concern needs to be addressed in continuing use of microscale toxicity tests.

A second issue is exposure time. If a product is to be applied once, and exposure is short, isn't it appropriate to judge toxicity on the basis of exposure times considerably shorter than 96 h? As noted above, the problem of one-time "pulse" exposure is now being directly addressed for dispersants and dispersed oil (Aurand, 1995a and b; Aurand and Coelho, 1995) where exposure is only on the order of minutes to hours. Perhaps consideration of pulse exposure regimes should be given to other classes of spill treating agents as well. Alternatively, if a product is to be applied over a long period of time (weeks), measures of chronic toxicity should be considered.

Once exposure conditions are agreed upon, selection of test species and endpoints is indeed important. Species sensitivity to oil spill treating agents may vary over several orders of magnitude. However, we have little information to judge which are most and least sensitive since there have been no studies comparing species except for dispersants (Walker et al., 1995a). In addition, there is continuing debate about sensitivity of a very popular tool, Microtox, yet it has been used successfully to screen and monitor dispersants (Clayton et al., 1995; Dunn et al., 1994).

C. The Manufacturer's View

Vendors and offerers of new oil spill treating agents are often shocked and upset when authorities reject their well-intended products and proposals. Spill response testing appears to be done long after product development and production and during the final stages of pre-approval. In addition, it is apparent from vendor brochures that, with notable exceptions (dispersants, some shoreline cleaners), products for open use on oil spills are derived from existing materials designed for other purposes (such as drilling or oil production and transport). Accordingly, it would not be inappropriate to propose a process through which low-cost toxicological and effectiveness testing procedures are incorporated as part of product development, not just product pre-approval. The goal would be to carry on development of those products that are clearly less toxic than oil and cease development of those that are not. Microscale toxicity tests such as those listed for Canada and the U.S. EPA in Table 40.2, could easily be applied by manufacturers during the product development phase. Although this adds cost to product development, in the long run it could save time and money for both the producers and users of oil spill treating agents.

However, this does not solve all the problems. As is evident in Table 40.2, manufacturers and vendors are faced with different final-product testing protocols in different countries. This may result in a larger collective body of interesting and modest-quality toxicity data, but it is expensive and may deter development and alienate some manufacturers. Therefore, both manufacturers and response authorities need to help each other to develop international consensus on toxicity testing.

Manufacturers should be clear on product application rates over time and space and, if necessary, offer evidence that it is relatively safe with repeated applications. Finally, manufacturers should offer as much information as possible about the actual forms and concentrations of specific substances in their final product.

D. The Integrated View

In the author's opinion the ideal integrated approach would be an internationally accepted multitiered testing process. During product development, potentially useful materials could be screened for toxicity using one or two reliable freshwater and/or marine bioassays. To avoid surprises in the future, it can be argued that a sensitive species and endpoint should be used at the beginning of product development and screening. In any case, once toxic products are sorted out from nontoxic (possible criterion: product is no more toxic than unweathered oil), the better ones (as also measured by standard laboratory effectiveness tests as per Fingas et al., 1995) could go to the next tier to check on bioaccumulation, chronic toxicity and perhaps mutagenicity. These also can be relatively inexpensive microscale tests (e.g., bioaccumulation in juvenile mussels, *Daphnia* life cycle test, and Ames *Salmonella* mutagenicity bioassays). At this stage, consideration could be given to use regionally appropriate species such as were described long ago by Becker et al. (1973). Those products that look best should then be moved to the field and tested on small-scale experimental spill plots for final effectiveness, toxicity, and a little ecological effects testing (as in Delaware, Venosa et al., 1996 and Mearns et al., 1995). This puts the final onus on state and federal authorities to permit such testing or to offer large-scale facilities, such as mesocosms.

The bottom line is that to make substantial improvement in the development of new response technologies, responders, manufacturers, and toxicologists must cooperate with each other, both nationally and internationally.

ACKNOWLEDGMENTS

The author wishes to thank Dr. Jean Snider, NOAA, and Dr. Kenneth Lee, Fisheries and Oceans, Canada, for encouraging this review, and several anonymous reviewers. The extensive information resources of the NOAA Hazardous Materials Response and Assessments Division Information Center, and its director, John Kaperick, are especially appreciated.

REFERENCES

Anderson, J.W., J.M. Neff, B.A. Cox, H.E. Tatem and G.M. Hightower. 1974. Characteristics of dispersions and water-soluble extracts of crude and refined oils and their toxicity to estuarine crustaceans and fish, *Marine Biology,* 27, 75-88.

Anon. 1994a. The testing of oil spill dispersants under FEPA II. MAFF SOP No. BEG 43. Ministry of Agriculture, Fisheries and Food, Directorate of Fisheries Research, Lowestoft, UK.

Anon. 1994b. Standard procedure for testing oil bioremediation products and toxicity. MAFF SOP No. BEG 45. Ministry of Agriculture, Fisheries and Food, Directorate of Fisheries Research, Lowestoft, UK.

Aurand, D. 1995a. A research program to facilitate resolution of ecological issues affecting the use of dispersants in marine oil spill response. p. 172-190 In P. Lane (ed). The use of chemicals in oil spill response. ASTM STP 1252. American Society for Testing and Materials, Philadelphia, PA.

Aurand, D. 1995b. The application of ecological risk assessment principles to dispersant use planning. p. 253-261 In International Maritime Organization, Second International Oil Spill Research and Development Forum, 23-26 May 1995, London. International Maritime Organization, London, UK.

Aurand, D. and G. Coelho. 1995. Using toxicity data in oil spill response planning. 21 — In Scientific and Environmental Associates, Inc., Eds., Workshop Proceedings, The Use of Chemical Countermeasures Product Data for Oil Spill Planning and Response. Volume 2. Hazardous Materials Response and Assessment Division, National Oceanic and Atmospheric Administration, Seattle, WA.

Becker, C.D., J.A. Lichatowich, M.J. Schneider and J.A. Strand. 1973. Regional survey of marine biota for bioassay standardization of oil and oil dispersant chemicals. API Publication No. 4167. American Petroleum Institute, Washington DC, 2 pp + Append.

Betton, C.I. 1994. Chapter 10: Oils and hydrocarbons. p. 244-263 In P. Calow, Ed., Handbook of ecotoxicology. Volume 2. Blackwell Scientific Publications, London, U.K.

Bragin, G.E., J.R. Clark and C.B. Pace. 1994. Comparison of physically and chemically dispersed crude oil toxicity under continuous and spiked scenarios. MSRC Technical Report Series 94-015. Marine Spill Response Corporation, Washington, DC, 45 pp.

Claxton, L.D., V.S. Houck, R. Williams and F. Kremer. 1991. Effect of bioremediation on the mutagenicity of oil spilled in Prince William Sound, Alaska, *Chemosphere,* 23(5):643-650.

Clayton, J.R., Jr., B.C. Stransky, M.J. Schwartz, D.C. Lees, J. Michel, B. J. Snyder and A. C. Adkins. 1995. Development of protocols for testing cleaning effectiveness and toxicity of shoreline cleaning agents (SCA's) in the field. Volumes 1 and 2. MSRC Technical Report Series 95-020.1 and 95-020.2. Marine Spill Response Corporation, Washington, DC

Coelho, G.M., G.E. Bragin, D.V. Aurand, J.R. Clark and D.A. Wright. 1995. Field and laboratory investigation of the toxicity of physically and chemically dispersed oil. p. 1117-1131 In Proceedings, 18th Arctic and Marine Oilspill Program (AMOP) Technical Seminar, June 14-16, 1995, Edmonton, Alberta, Canada, Environment Canada, Ottawa.

Daykin, M., G. Sergy, D. Aurand, G. Shigenaka, Z. Wang and A. Tang. 1994. Aquatic toxicity resulting from *in situ* burning of oil-on-water. p. 1165-1193 In Proceedings, 17th Arctic and Marine Oilspill Program (AMOP) Technical Seminar, June 8-10, 1994, Vancouver, BC. Environment Canada, Ottawa.

Dunn, R.M., D.J. Mitchell and R.P.J. Swannell. 1994. Potential use of the Microtox (TM) assay as an indicator of the toxicity of dispersed oil. MSRC Technical Report Series 94-018. Marine Spill Response Corporation, Washington, DC, 18 pp.

Fingas, M.F., D.A. Kyle, N. Laroche, B. Fieldhouse, G. Sergy and G. Stoodly. 1995. The effectiveness testing of oil-spill-treating agents. p. 286-296 In P. Lane, Ed., The Use of Chemicals in Oil Spill Response. ASTM STP 1252, American Society for Testing and Materials, Philadelphia, PA, 340 pp.

Fisher, W. S. and S. S. Foss. 1993. A simple test for toxicity of Number 2 fuel oil and oil dispersants to embryos of grass shrimp, *Palaemonetes pugio, Marine Pollution Bulletin,* 26(7):385-391.

Hoff, R., G. Shigenaka, R. Yender and D. Payton. 1994. Chemistry and environmental effects of the shoreline cleaner PES-51TM. HazMat Report 94-2. Hazardous Materials Response and Assessment Division, National Oceanic and Atmospheric Administration, Seattle, WA. March 1994. 22 pp.

Jones, J.R.E. 1964. Fish and river pollution. Butterworth and Co., Ltd., London.

Kucklick, J.H. and A.H. Walker. 1995. Laboratory aquatic toxicity testing of chemical treating agents. 3 — In Scientific and Environmental Associates, Inc., Eds., Workshop Proceedings, The Use of Chemical Countermeasures Product Data for Oil Spill Planning and Response. Volume 2. Hazardous Materials Response and Assessment Division, National Oceanic and Atmospheric Administration, Seattle, WA.

Mearns, A.J. 1995. Low solubility materials. pp. 49-57 in Scientific and Environmental Associates, Eds., Workshop Proceedings, The Use of Chemical Countermeasures Product Data for Oil Spill Planning and Response. April 4-6, 1995, Leesburg, VA. Volume 2. Hazardous Materials Response and Assessment Division, National Oceanic and Atmospheric Administration, Seattle, WA. 293 pp

Mearns, A.J. in press. Exxon Valdez shoreline treatment and operations: implications for response, assessment, monitoring and research. In S. Rice and B. Wright (eds), Proceedings of the Exxon Valdez Oil Spill Symposium, Anchorage, Alaska, February, 1993. American Fisheries Society, Bethesda, MD, in press.

Mearns, A.J., P. Roques and C.B. Henry, Jr. 1993. Measuring efficacy of bioremediation of oil spills: Monitoring, observations and lessons from the Apex oil spill experience. p. 335-343 In 1993 Oil Spill Conference. American Petroleum Institute, Washington, DC

Mearns, A.J., J. Doe, W. Fisher, R. Hoff, K. Lee, R. Siron, C. Mueller and A. Venosa. 1995. Toxicity trends during oil spill bioremediation experiment. p. 1153-1145 In Proceedings, 18th Arctic and Marine Oilspill Program (AMOP) Technical Seminar. June 14-16, 1995. Environment Canada, Ottawa.

Mendelsshon, I.A., Q. Lin, K. Debusschere, S. Penlanmd, C.B. Henry, E.B. Overton, R. J. Portier, M.M. Walsh and N.N. Rabalais. 1995. The development of bioremediation for oil spill cleanup in coastal wetlands: product impacts and bioremediation potential. p. 97-100 In 1995 Oil Spill Conference. American Petroleum Institute, Washington, DC

National Research Council (NRC) (U.S.). 1989. Using oil spill dispersants on the sea. National Academy Press, Washington, DC. 335 pp.

National Environmental Technology Applications Center (NETAC). 1993. Evaluation methods manual oil spill response bioremediation agents. Prepared for the U.S. Environmental Protection Agency Bioremediation Action Committee. National Environmental Technology Applications Center, University of Pittsburgh Applied Research Center, Pittsburgh, PA. 89 pp.

Pace, C.B. and J.R. Clark. 1993. Evaluation of a toxicity test method for dispersant screening in California. MSRC Technical Report Series 93-028. Marine Spill Response Corporation, Washington, DC 34 pp.

Pace, C.B., J.R. Clark, and G.E. Bragin. 1995. Comparing crude oil toxicity under standard and environmentally realistic exposures. p. 611-614 In 1995 Oil Spill Conference. American Petroleum Institute, Washington, DC

Pauwels, S.J. and J.R. Clark. 1993. Overview of international oil spill dispersant toxicity testing requirements. 803-804 In 1993 Oil Spill Conference. American Petroleum Institute, Washington, DC

Rice, S.D., J.W. Short and J.F. Karinen. 1977. Comparative oil toxicity and comparative animal sensitivity. 77-94 In D.A. Wolfe, Ed., Fate and Effects of Petroleum Hydrocarbons in Marine Ecosystems and Organisms. Proceedings of a Symposium, November 10-12, 1976, Olympic Hotel, Seattle, WA. Pergamon Press, New York. 478 pp.

Rycroft, R.J., P. Matthiessen and J.E. Portmann. 1994. MAFF review of the UK oil dispersant testing and approval scheme. Ministry of Agriculture, Fisheries and Food, Directorate of Fisheries Research, Lowestoft, UK. 87 pp.

Scientific and Environmental Associates, Inc., Eds. 1995. Workshop Proceedings: The use of chemical countermeasure product data for oil spill planning and response, April 4-6, 1995, Leesburg, VA. Volumes 1 and 2. Hazardous Materials Response and Assessment Division, National Oceanic and Atmospheric Administration, Seattle, WA. 83 and 293 pp.

Sergy, G., S. Blenkinsopp, D. Westlake, J. Foght and D. McLeay. 1993. The development of laboratory methods for assessing the efficacy and toxicity of oil spill bioremediation agents. p. 355-365 In Proceedings, 16th Arctic and Marine Oilspill Program (AMOP) Technical Seminar, June 7-9, 1993. Environment Canada, Ottawa.

Sergy, G., S. Blenkinsopp, J.R. Harper, B. Humphey and E. Owens. 1995. Recent and emerging Canadian studies addressing oil-on-shorelines issues. p. 388-395 In International Maritime Organization, Second International Oil Spill Research and Development Forum, 23-26 May 1995, London. International Maritime Organization, London, UK.

Shigenaka, G., V.P. Vicente, M. A. McGehee and C.B. Henry, Jr. 1995. Biological effects monitoring during an operational application of Corexit 9580. p. 177-184 In 1995 Oil Spill Conference. American Petroleum Institute, Washington, DC.

Singer, M.M., D.L. Smalheer, R.S. Tjeerdema and M. Martin. 1991. Effects of spiked exposure to an oil dispersant on the early life-stages of two marine species, *Environmental Toxicology and Chemistry,* 10:1367-1374.

Thomas, G.F., G.N. Moore and L. Tyrance-Strickland. 1995. The National Contingency Plan product schedule database. p. 611-614 In 1995 Oil Spill Conference. American Petroleum Institute, Washington, DC

U.S. E.P.A. 1991. Alaska oil spill bioremediation project Science Advisory Board draft report. EPA/600/9-91/046a. United States Environmental Protection Agency, Office of Research and Development, Gulf Breeze, FL. 572 pp.

Venosa, A.D., M.T. Suidan, B.A. Wrenn, K.L. Strohmeier, J.R. Haines, B. L. Eberhart, D. King and E. Holder. 1996. Bioremediation of an experimental oil spill on the shoreline of Delaware Bay, *Environmental Science and Technology,* in press.

Walker, A.H., J. Michel, G. Canevari, J. Kucklick, D. Scholz, C.A. Benson, E. Overton and B. Shane. 1993. Chemical oil spill treating agents: Herding agents, emulsion treating agents, solidifiers, elasticity modifiers, shoreline cleaning agents, shoreline pretreatment agents and oxidation agents. MSRC Technical Report Series 93-015. Marine Spill Response Corporation, Washington DC. 328 pp.

Walker, A.H., J.H. Kucklick, J. Michel, D.K. Scholz and T. Reilly. 1995a. Chemical treating agents: response niches and research and development needs. p. 211-217 In 1995 Oil Spill Conference. American Petroleum Institute, Washington, DC.

Wolfe, D.A., M.J. Hameedi, J.A. Galt, G. Watabayashi, J. Short, C. O'Clair, S. Rice, J. Michel, J.R. Payne, J. Braddock, S. Hanna and D. Sale. 1994. The fate of the oil spilled from the Exxon Valdez, *Environmental Science and Technology,* 28(13): 561A-568A.

CHAPTER 41

The Development, Validation, and Analysis of *Salmonella* Mutagenicity Test Methods for Environmental Situations

Larry D. Claxton

CONTENTS

I. INTRODUCTION

Many benefits of any modern, developing society are accompanied by certain hazards. Numerous types of activities have the potential to affect the nation's water resources. Energy sources, such as electrical power plants, use solid or liquid organic fuels and emit a variety of by-products. Even the pollution emitted into the air can contaminate surface waters. Most manufacturing processes create and discharge some contaminants into the environment. In addition, some environmentally applied chemicals and substances used by individuals (e.g., pesticides, food preservatives, and cancer chemotherapy agents) have been shown to have toxic effects. Researchers and public health specialists have, for many years, contemplated what effect such a plethora of substances may have on public health.[1] Since most environmental exposures to most of these products are chronic, low-level exposures, the greatest concerns generally revolve around carcinogenicity, mutagenicity, and teratogenicity.

Identifying the hazardous components in our environment has been a major effort for the past two decades.[1] Because of increased use of anthropogenic materials, concern about the presence of carcinogenic compounds in water has increased.[2] Some studies correlate the presence of known pollutants (e.g., disinfection by-products in drinking water) with increased incidences of neoplastic diseases.[3-5] Genotoxic compounds can also have adverse effects on ecosystems (e.g., neoplastic diseases in fish).[2,6] In other cases, human populations are exposed in varying degrees to numerous complex environmental mixtures, some of which contain literally thousands of compounds. The use of epidemiology and long-term animal testing has proven inefficient in identifying toxic mixtures and compounds that are either emitted into or generated by environmental processes.[7] Because the vast majority of compounds pose no health threat at environmental concentrations, it is necessary to have rapid, relatively inexpensive tests that can be used to not only identify environmental toxicants but also for comparative assessments and priority setting. It is well documented that mutations play an important role in the cancer process; therefore, short-term genotoxicity tests can be used to identify potential carcinogenic hazards. In the areas of mutagenesis and carcinogenesis, the most extensively used assay is the *Salmonella* mutagenicity assay.[8] Indeed, the most-used genotoxicity assay for water samples has been the *Salmonella* assay.[8,9]

Mutagenicity tests using bacteria have been available for approximately 40 years. In 1951, Demarec et al. used an *Escherichia coli* reverse mutation assay to test 31 chemicals.[10] In 1958, Szybalski[11] tested 431 substances for mutagenicity using a somewhat different *E. coli* system. Much later (1971), Ames and Yamasaki[12] published a mutagen detection system using histidine-requiring mutants of *Salmonella typhimurium*. The results of these early uses of mutation tests created little concern about the presence of environmental mutagens and showed no support for the mutation theory of cancer. The *Salmonella* assay (and later many other *in vitro* tests) was significantly improved after Malling[13,14] used a mammalian metabolizing system to activate dimethylnitrosamine to its mutagenic metabolite. It is now known that many carcinogenesis-initiating compounds must be metabolized before they interact with protooncogenes. Ames[15,16] showed a qualitative correlation between *Salmonella* mutagenicity test results and known carcinogenicity test results for a relatively large number of compounds. Claxton et al.[17] and Ashby and Tennant[18] later showed that the extent of correlation was dependent upon chemical class and therefore mechanisms of metabolism and mutation. Although a number of papers have debated this correlation between mutagenicity tests and potential carcinogenicity, the consensus today is that based on the study of mechanistic actions the *Salmonella* mutagenicity test is a good tool to identify and do baseline comparative assessments of potential genotoxic carcinogens.

Included in the advantages of bacterial *in vitro* assays is the fact that the bacteria can be "customized" and engineered to enhance their ability to detect mutagens and to alter their metabolic capabilities. Also, the technical performance of the assay can be modified to deal with different types and available amounts of substances. Initially, strains were developed to detect different types of gene mutations,[12] thereby increasing the sensitivity of the tester strains. Many others have examined, discussed, and suggested modifications of the standard *Salmonella* mutagenicity assay.[19-38] No other mutagenesis protocol has gained as much general acceptance. A discussion of the *Salmonella* mutagenicity test procedures, special uses, and limitations is the aim of this chapter.

II. BASIC THEORY AND PROTOCOL

The original test procedure[39] detects reverse mutations that occur in histidine-requiring strains of *S. typhimurium*. Indicator (or tester) strains are either spontaneous or induced mutants of the parental strain LT-2. LT-2 will grow on histidine-free medium, but the mutant indicator strains will not grow on histidine-free medium. In the bioassays for mutagenicity, these histidine-requiring mutant strains undergo a reverse mutation to prototropy (wild type). The mutants (often called *revertants*) can be detected on a minimal nutrient medium that either lacks or is deficient in histidine. Since more than one type of gene mutation can be induced in bacterial DNA, more than one indicator strain had to be selected. Two broad classes of indicator strains exist: one class consists of those strains that undergo primarily a base-pair substitution to revert to prototropy; the other class consists of those strains that respond to frameshift mutagens. The use of this simplified classification aids general understanding; however, this is not fully descriptive of mutational mechanisms.

Besides the histidine-locus mutation, the most commonly used strains also have additional mutations. The *rfa* mutation creates a lipopolysaccharide deficiency in the bacterial cell wall, thus increasing the cell's permeability to large molecules. The *uvr*B mutation was incorporated into the frequently used strains to increase their susceptibility to several classes of mutagens by impairing genetic repair. In addition, the pKM101 R-factor plasmid, a modification found in some strains, appears to increase the indicator strain's sensitivity by participating in a type of error-prone repair. Investigators have incorporated other changes into one or more tester strains for special purposes. For example, some available strains have either reduced or enhanced abilities to metabolize nitroarenes;[40] some vary in DNA repair capabilities;[39] and some have been engineered to have specific mammalian metabolizing enzyme abilities.[41-43] The modified strains fulfill special needs when performing complex mixture research and when trying to delineate a chemical's mechanism of action.

The most commonly used protocol for bacterial mutation testing is called the *plate incorporation method*. Parts of this protocol and the principles of it are contained in all the modifications that will be discussed later. With this type of protocol, the bacteria, the substance to be tested, and any exogenous metabolic activation system used are mixed in a melted, soft-agar overlay and then poured onto a minimal media plate. The media consists of a salt solution, glucose, agar, biotin, a trace of histidine, and water. The added trace of histidine allows the bacteria to replicate while exposed to the substance being tested. This trace histidine causes a cloudy background growth (*lawn*) of bacteria upon the plate, but the concentration is not enough to produce individual colonies. Revertants, however, will grow and produce distinct countable colonies.

One should keep in mind that using an agar medium places limits on the test. First, because not all substances are soluble in this medium, the test substances may not be made available to the bacteria. Second, both the variety and amount of solvent material used are limited by the nature of an agar medium and by the toxicity of the solvent to the bacteria. Third, some compounds volatilize out of the medium very quickly. Fourth, although preparing the medium seems simple, small changes in preparation can have major effects. For example, if the medium is too cool when the biotin is added, uneven mixing of the solutions may occur and bacterial survival will differ among plates. Cautious handling of the agar medium cannot be overemphasized.

Knowledge concerning the substance to be tested and the solvent used for the test substance can alleviate many potential problems. Solubility plays a major role in developing a testing strategy. Although a variety of solvents (including water, acetone, and ethanol) can be used and although dimethylsulfoxide (DMSO) is the most commonly used solvent for substances to be tested, aqueous solvents (i.e., water, buffer, and saline) are the most preferred. Most solvents, however, are somewhat toxic to bacteria. One must consider the solubility of the substance in the solvent, what is likely to happen when the solvent plus substance is added to the water-based medium (e.g., precipitation), the effect of pH, the toxicity of both the solvent and the substance to the tester strains, solvent and test substance interaction, and the volatility of each component.

Exogenous metabolism is usually provided by a mammalian tissue homogenate added to the agar overlay before pouring it onto the plate. The most commonly used activation system is the supernatant of an Aroclor-induced, rat liver homogenate (produced by centrifugation at $9000 \times g$)

called *S9*. The Aroclor induces the mixed function oxygenase (MFO) system of enzymes which in turn increases the amount of metabolites produced. For promutagens, this increases the number of revertant colonies per amount of substance placed into the test. Because S9 contains microsomal plus cytosol enzymes, this simple method provides an effective exogenous metabolic activation system. Among the many variations that can be introduced in the exogenous activation portion of the assay are changes in the strain or species of the animal, the tissue used, type of induction, type of activation fraction (e.g., microsomes, whole cells), and the methods of application and exposure. Compounds that produce a mutagenic response in the assay without adding a mammalian metabolic activating system are generally called *direct-acting mutagens*. This is an operational definition only, because bacterial metabolism may have to occur before the compound is mutagenic within the test system. Compounds that require a mammalian enzyme system to be present before detecting any mutagenicity are usually called *indirect mutagens* or *promutagens*. Mutagens identified by *in vitro* tests, therefore, fall into three categories: 1) direct-acting mutagens, 2) endogenously activated mutagens (metabolism provided by the test cells), and 3) exogenously activated mutagens. Also, note that endogenous and exogenous metabolism can either activate or deactivate the compound.

To conduct the plate incorporation assay, *Salmonella* broth culture and a specific dose of the sample are added to a melted agar overlay in a small tube. When exogenous activation is used, the activation mixture is also added to the overlay tube. The mixture is gently vortexed, poured onto a minimal media plate as a smooth even layer, and allowed to harden. These plates are then incubated in darkness at 37°C for 48 to 72 h. This process is repeated for each replicate for each dose tested and for each control. When testing a single compound with five tester strains both with and without exogenous activation using three replicates at each of five doses, 150 total test plates would be used for the test substance. In addition, 30 to 60 plates would typically be needed for spontaneous, positive, and sterility controls.

After incubation, plates are examined for background growth, contamination, and numbers of colonies. The background growth or lawn, due to the trace histidine added to the media, causes the plates to look cloudy when held up to a light source. Clearing of this background indicates that the sample or solvent is toxic to the bacteria. An overgrowth indicates the presence of histidine or histidine-like substitutes. Colonies should be white, evenly growing (though rough edged), and of moderate size. Colonies that grow on the surface rather than within the overlay will be slightly larger. Colonies can also exhibit the color of any highly colored agent tested. All other colony types should be considered as contaminants and noted as such.

After the plates are examined, the number of colonies per plate is counted and recorded. Normally, counts range from zero to 2000 although greater colony counts have been reported. Most electronic counters cannot count accurately above 1500, and the reliability of normal hand counting decreases rapidly for counts exceeding 500.[44] All data counts and additional information must be carefully recorded. Data summary and evaluation will be discussed in another section.

Correct microbiological techniques are needed at every step of the assay. For example, attention must be paid to sterility, especially that of the sample, S9, and media. Tests preferably are done in biological safety hoods that provide a sterile environment for the test while providing protection for the worker. Working under yellow light prevents photoactivation and photodeactivation of any test substance. Also, the best quality incubators should be used.

III. SPECIALIZED TESTING PROTOCOLS

The plate incorporation test protocol is sufficient for screening many substances. However, full advantage of rapid microbial techniques is best achieved by knowing when to use the variety of alternative procedures available. Two highly useful modifications of the test protocol for water-related samples are the microsuspension assay and the spiral plating assay.

A. The Microsuspension and Preincubation Assays

The microsuspension assay[33] is a modification of the preincubation test.[45] For these assays, one incubates the test substance in an overlay tube (or well) with only the bacteria and, if applicable, a metabolic activation mixture for a predetermined period. The preincubation mixture and the overlay agar are mixed and then poured onto a minimal media plate. Several advantages occur with this technique. Some compounds that are negative in the normal plate test (e.g., nitrosamines) are positive (mutagenic) with these protocols. Also, a portion of the preincubation mixture can be plated onto a complete media plate enabling an index of survival to be calculated. The primary differences between the preincubation and microsuspension assays are differences in the volumes and amounts of materials used. These tests can also be done in sealed vials when there is a need to test volatile chemicals. The main drawback to the tests is the significant increase in labor involved.

B. The Spiral Plating Assay

The spiral *Salmonella* assay [46] is an automated assay that generates dose-response data from a continuous concentration gradient on a single agar plate. The spiral plater (Spiral Biotech, Inc., Bethesda, MD) deposits the bacteria, compound, and activation mixtures in an Archimedes spiral onto the surface of a rotating agar plate. An automated scanner counts the colonies along the path of the spiral and correlates the colony count with the applied dose. A complete dose response can be seen when using as little as 50 μL of solution. This assay greatly reduces the amount of sample needed and the amounts of materials and supplies. Its primary drawbacks are the decreased numbers of bacteria exposed at doses equivalent to the plate test and the cost of the equipment. The assay has proven useful not only for routine screening but also for a number of specialized purposes. It is especially useful for the bioassay of complex mixtures[47] and for doing interaction studies.

Table 41.1 provides a brief summary of other available *Salmonella* mutagenicity protocols.

IV. CRITERIA FOR ADEQUATE TESTING

Individual substances vary not only in genotoxicity but also in solubility, mode of action, and toxicity to the indicator organisms. All of these factors must be considered when planning mutagenesis experiments. Initial testing should be designed to give the investigator definite information about toxicity, solubility, and any inherent technical problems with the substance being tested. Although many investigators do a toxicity range-finding assay before performing the mutagenicity assay, it is generally preferable to do a preliminary mutagenicity assay so that information about potential mutagenicity can be incorporated into the design of the more definitive tests.

Although assays are designed on a case-by-case basis, some general guidelines have emerged within the literature. There are a number of generalized guidelines and reviews available.[24,27,48-64]

The reader should also be aware that DNA sequence analysis of mutants recovered from bacterial mutagenicity tests can now be done in an efficient and highly reproducible manner.[65-67] Not only does this sequence analysis confirm the presence of induced mutants, but it also provides information about how these substances produce the induced mutations.

V. SUMMARIZING AND ANALYZING MUTAGENICITY DATA

For many years, negative (nonmutagenic) and positive (mutagenic) results were determined according to a general consensus rule frequently called the *twofold rule*. This title, however, overstates its simplicity. The rule states that for a substance to be considered mutagenic or positive,

Table 41.1   Useful Alternative *Salmonella* Mutagenicity Test Protocols

Primary Modification	Use	Advantage (A:)/Disadvantage (D:)
Spot test[12]		
Single dose to center of plate	Very rapid screening	A: easy, small amounts of test substance needed D: very qualitative, diffusion/toxicity problems
Fluctuation Test[78]		
Liquid medium fluctuation test done in small tubes or wells	Simple, rapid, can use mammalian cell activation system	A: rapid, easy to quantitate D: solvent toxicity, highly dependent on pH
Suspension Test[79]		
Forerunner of the liquid preincubation test	Quantitative for mutation and survival	A: very quantitative assay, detects classes of chemical missed in plate assay; D: labor intensive
TLC Plate Assay[80]		
Modified plate assay poured over TLC plate to detect fractionated mixtures	To aid in the identification of mutagens contained in complex mixtures	A: simple, little sample needed; D: highly dependent on TLC fractionation method and diffusibility of compound
The Prival Method[81]		
Modified plate incorporation conditions to provide reductive metabolism	To test substances, e.g., azo dyes, that require anaerobic reduction to be mutagenic	A: easier than true anaerobic metabolism, good for dyes; D: more costly, time consuming than plate assay
Methods for Testing Gases and Volatile Organic Liquids[82,83,84]		
Plate tests for volatile compounds and gases done in Tedlar bags, glass desiccators, and special chambers	To test volatile compounds and gases	A: normal plate test not effective with volatile materials; D: dosing may require special equipment

the data for a substance must fulfill two criteria. First, the average number of revertants per plate when plotted against the doses tested must produce a regularly increasing dose-response curve. Secondly, at one or more doses, the revertants per plate must be greater than double the spontaneous count for that strain and testing conditions. This rule was and is adequate for most substances, because most bacterial mutagens give dramatic responses in one or more tester strains.

The twofold rule may not serve an investigator in at least two circumstances. Some substances give repeatable dose-response results without a doubling of the spontaneous count. (This can occur, for example, when bacterial toxicity masks mutagenicity or when the compounds precipitate out of solution at the higher doses.) This rule also provides only a qualitative determination of mutagenicity. When doing mechanistic research or comparative mutagenicity studies, there is a need to summarize the mutagenicity with a quantitative value. These needs sparked the use of several types of analysis[68,69] and statistical models.[23,50,70-74] As expected, each approach has definite strengths and weaknesses. Some protocols also may not lend themselves to quantitative evaluation. When using statistical models, the investigator must understand the model's limitations and must decide whether or not any summary values generated are biologically plausible.

In recent years, some of these models have become available for use on personal computers.[75-77] The software made available by Claxton et al.,[75] besides providing four different statistical models, provides the means to record other experimental information, quality control information, and relevant notes. Use of such systems helps to ensure data integrity and completeness and a certain rigor in data analysis; however, the information retained by these means and their interpretation should be done by knowledgeable, experienced investigators.

VI. USE OF THE *SALMONELLA* MUTAGENICITY ASSAY FOR AQUATIC TOXICOLOGY

Although the primary purpose of this chapter is to provide a basic understanding of *Salmonella* mutagenicity assays, those interested in aquatic toxicology can gain a deeper understanding of the assay and its uses by examining the use of the assay in their specialty. Table 41.2 provides the reader a listing of key manuscripts in which the authors employed the assay in water or water-related research. The table also helps the reader to sort out the type of information included in the paper by noting (when given) the type of study, the sources of the samples used, the type of contamination seen in field studies, and the type of treatment, if any, which was applied to the water. Many of the studies were simply the testing of chemicals previously identified in some type of water sample (e.g., MX). There are, however, a significant number of studies in which samples from actual field sites are returned to a laboratory for analysis. Most of the "Lab" studies in Table 41.2, refer to studies in which field-like conditions are simulated in a laboratory (e.g., a microcosm study), and the extracted components are then examined for mutagenicity. Because they provide the details of the methods, a number of papers that describe the development of methods are also listed. Not only have water samples (drinking water, wastewater, and natural waters such as rivers) been examined, but treated waters, effluents, sediments, and extracts of biological materials have also undergone bioassay. Although the listing is not exhaustive, it should provide the reader with a useful resource.

VII. SUMMARY

The *Salmonella typhimurium* mammalian activation assay for mutagenicity is a rapid, reliable, bacterial test for demonstrating whether or not a substance can interact with genetic material to cause a permanent mutation. The assay uses selected histidine-deficient strains which upon treatment with certain types of mutagens revert back to prototropy, produce histidine, and grow on a selective media. A single test can be completed (planned, executed, and analyzed) in 4 or 5 days at modest cost and can be modified in several ways to explore various factors affecting mutagenicity. The assay can be used to screen individual compounds and simple or complex mixtures. Although quantitative interpretation is best done with proven statistical models, the qualitative interpretation of results is relatively easy and reliable for most mutagens. Ample utility of the test has been demonstrated, and the use of bacterial mutagenicity for screening mutagens and potential carcinogens will continue as a broad and useful field of toxicology and aquatic toxicology.

ACKNOWLEDGMENTS

I would like to thank the EPA management (Dr. Lawrence W. Reiter and Dr. Hal Zenick) that has for many years encouraged and rewarded my efforts in this area of toxicology. I also want to thank those co-workers and collaborators who have made significant contributions to the area of microbial mutagenicity testing. This list includes my current co-workers V.S. Houk, T.J. Hughes, S. Warren, J. Creason, M.J. Kohan, and S.E. George. There are many others too numerous to mention. Special thanks to Carolyn Fowler for assistance in preparing this manuscript.

Table 41.2   Reference Literature Illustrating Use of the *Salmonella* Mutagenicity Assay in Aquatic Toxicology

First Author, Year[a]	Ref.[b]	Study Type[c]	Source[d]	Contamination[e]	Treatment[f]
DeMarini, 82	85	Bioassay	DW, NW		
Kool, 82	86	Field	DW		
Maruoka, 82	87	Field, Fraction	NW		
Scully, 82	88	Compound			
Spanggord, 82	89	Compound	WW		
Whong, 82	90	Bioassay			
Zoeteman, 82	91	Lab/Field	Treated NW	BP	Cl, CD, O, UV
Harrington, 83	92	Lab, Bioassay	DW		
Kamra, 83	93	Field	Pulp mill effluent	Cl-BP	Cl
Maruoka, 83	94	Lab	NW	Cl-BP	Cl
Meier, 83	95	Lab	Humic acid chlorination	Cl-BP	Cl
Sato, 83	96	Field, Fraction.	NW Sed.		
Kowbel, 84	97	Lab	Soil Fulvic Acid		O/Cl
Backlund, 85	98	Field/Lab Treat	Humic water		Cl, O, CD, O/C, Cl/CD
Fawell, 85	99	Review	DW		
Nikunen, 85	100	Field	Effluents, sed., fish	Petrochemical effluents	
Meier, 85	101	Compound, Lab	Humic acid solutions		Cl
Sato, 85	102	Field, Fraction	Sed.		
Maruoka, 86	103	Field, Fraction	NW		
Vartiainen, 86	104	Field	DW		Cl
Hopke, 87	105	Field	WW/land		
Meier, 87	106	Compound			
Thomas, 87	107	Compound			
Kool, 88	108	Field	DW	Cl-ORG, OTHER	
Kronberg, 88	109	Field/Lab, Compound		Cl-ORG	Cl
Liimatainen, 88	110	Field	C-DW	Cl-BP	Cl
Matsumoto, 88	111	Compound			
Meier, 88	9	Review			
Urano, 88	112	TD(S)	DW		
Varma, 88	113	Compound Mixture			
Wigilius, 88	114	Field/Lab	DW	Cl-BP	Cl
Wilcox, 88	115	Field	C-DW	Cl-BP	Cl
Backlund, 89	116	Compound, Field, Lab	NW, DW	Cl-Org	Cl, O/AC, Cg
Kira, 89	117	Field, Lab	NW, Biota		
Kool, 89	118	Field, TD(S)	Soils		
Kronberg, 89	119	Compound			
Sayato, 89	120	Lab	Art.		Cl, Cl/O
Waters, 89	121	TD (B, S), Field	WW		
Metcalfe, 90	122	Field, TD(B)	Sed.		
Sakamoto, 90	123	TD(B, S, C), Field	NW		
Sayato, 90	124	Field, TD(S)	NW		
Tikkanen, 90	125	Chemical			
Vartiainen, 90	126	Lab, Field TD(B)	DW		None, Cl
Matsuda, 91	127	Lab TD(F)			O
Ankley, 92	128	Field, Chemical	Sed., NW		
Fracasso, 92	129	Field	WW		
Rodriguez, 92	130	Field, TD(B, F)	Biota		
Marvin, 93	131	TD(F), Field	Sed		
Onodera, 93	132	TD(S)	C-W		
Sayato, 93	133	TD(S, F)	DW		

Table 41.2 (continued)   Reference Literature Illustrating Use of the *Salmonella* Mutagenicity Assay in Aquatic Toxicology

First Author, Year[a]	Ref.[b]	Study Type[c]	Source[d]	Contamination[e]	Treatment[f]
Ottaviani, 93	134	Field, TD(F)	WW		
Rodriguez, 93	135	Field	NW, Biota		
Shibuya, 93	136	TD(S)	DW		
Vargas, 93	137	Field	NW	Petroch.	
Abe, 94	138	Compound			
Koivusalo, 94	139	Field, Epi	DW		
Kusamran, 94	140	Field	NW		
LeCurieux, 94	141	Compound, TD(B)			
DeFlora, 95	142	TD(B)	Biota		
Filipic, 95	143	Field, TD(B, F)	NW, WW		
Rehana, 95	144	Field	NW	Pesticides	
Tahara, 95	145	Lab, Compound			
Rehana, 96	146	Field	NW	Pesticides	

[a] First author of publication and year of publication (Sorted by year, then author).

[b] Number for publication in this chapter's reference list.

[c] Type of study: Lab, laboratory study; Field, Field samples used; Compound, one or more compounds associated with pollution of water bioassayed; WW, wastewater/effluents bioassayed; Fraction., Fractionation of complex mixture sample before bioassay; TD, test development or comparison of the following kinds: (B, bioassay; C, chemical analysis; F, fractionation methods; S, sample preparation); Epi, done in conjunction with epidemiology study.

[d] Type of sample(s) used: DW, drinking water; NW, natural water (e.g., river or lake water); Sed., sediments; WW, wastewaters and effluents; C, chlorinated; biota, tissues/extracts of living materials (e.g., mussels); Art., artificial mixtures.

[e] Contamination known: BP, by-products of treatment; Cl-BP, chlorinated by-products; Cl-ORG, chlorinated organics; Petroch., petrochemical products or by-products

[f] Treatment of water before testing: AC, activated carbon; Cg, coagulation; Cl, chlorination; CD, chlorine dioxide; O, ozone; (Note: commas separate individual sample treatments while a "/" shows multiple treatments of one sample).

REFERENCES

1. Johnson, B.L., Hazardous waste: nature, extent, effects, and societal responses, in *Ecotoxicity and Human Health*, de Serres, F.J. and Bloom, A.D., Eds., CRC Press, Boca Raton, FL, 1995, 1.

2. Stahl, R.G., Jr., The genetic toxicology of organic compounds in natural waters and wastewaters, *Ecotoxicol. Environ. Safety*, 22(1), 94, 1991.

3. Morris, R. D., Audet, A-M., Angelillo, I. F., Chalmers, T. C., and Mosteller, F., Chlorination, chlorination by-products, and cancer: A meta-analysis, *American Journal of Public Health*, 82, 955, 1992.

4. Koivusalo, M., Jaakkola, J., Vartiainen, T., Hakulinen, T., Karjalainen, S., Pukkala, E., and Tuomisto, J., Drinking water mutagenicity and gastrointestinal and urinary tract cancers: An ecological study in Finland, *American Journal of Public Health*, 84(8), 1223, 1994.

5. Griffith, J. and Riggan, W., Cancer mortality in U.S. counties with hazardous waste sites and ground water pollution, *Archives of Environmental Health*, 44, 69, 1989.

6. Mix, M., Cancerous diseases in aquatic animals and their association with environmental pollutants, *Marine Environmental Research*, 20, 1, 1986.

7. Claxton, L.D., Houk, V.S., and George, S.E., Integration of complex mixture toxicity and microbiological analyses for environmental remediation research, in *Ecotoxicity and Human Health*, de Serres, F.J. and Bloom, A.D., Eds., CRC Press, Boca Raton, FL, 1995, 87.

8. Houk, V., The genotoxicity of industrial wastes and effluents, *Mutation Research*, 277, 91, 1992.

9. Meier, J.R., Genotoxic activity of organic chemicals in drinking water, *Mutation Research*, 196, 211, 1988.

10. Demarec, M., Bertani, G., and Flint, J., A survey of chemicals for mutagenic action of *E. coli*, *American Naturalist*, 85, 119, 1951.

11. Szybalski, W., Special microbiological systems, II, Observations on chemical mutagenesis in micro-organisms, *Annuals of the New York Academy of Science*, 76, 475, 1958.
12. Ames, B.N. and Yamasaki, E., The detection of chemical mutagens with enteric bacteria, in *Chemical Mutagens: Principles and Methods for Their Detection*, Hollander, A., Ed., New York, Plenum Press, 1971, Chap. 9.
13. Malling, H., Dimethylnitrosamine: formation of mutagenic compounds by interaction of mouse liver microsomes, *Mutation Research*, 13, 425, 1971.
14. Malling, H.V. and Frantz, C.N., *In vitro* versus *in vivo* metabolic activation of mutagens, *Environmental Health Perspectives*, 6, 71, 1973.
15. McCann, J., Choi, E., Yamasaki, E., and Ames, B., Detection of carcinogens as mutagens in the Salmonella/microsome test: Assay of 300 chemicals, *Proc. Natl. Acad. Sci., USA*, 72, 5135, 1975.
16. Ames, B., Durston, W., Yamasaki, E., and Lee, F., Carcinogens are mutagens, *Proc. Natl. Acad. Sci., USA*, 72, 979, 1973.
17. Claxton, L.D., Stead, A.G., and Walsh, D., An analysis by chemical class of *Salmonella* mutagenicity tests as predictors of animal carcinogenicity, *Mutation Research*, 205(1-4), 197, 1988.
18. Ashby, J. and Tennant, R., Definitive relationships among chemical structure, carcinogenicity and mutagenicity for 301 chemicals tested by the U.S. NTP, *Mutation Research* 257, 229, 1991.
19. Booth, S., Welch, A., and Colin, A., Some factors affecting mutant numbers in the *Salmonella* microsome assay, *Carcinogenesis*, 1, 911, 1980.
20. Arimoto, S., Negishi, K., and Hayatsu, H., A modification of the Ames test procedure: accelerated growth of the His+ revertants, *Mutation Research*, 91(4-5), 407, 1981.
21. Anderson, D. and McGregor, D., The effect of solvents upon the yield of revertants in the *Salmonella*/activation mutagenicity assay, *Carcinogenesis*, 1, 363, 1980.
22. Belser, W.L., Jr., Shaffer, S.D., Bliss, R.D., Hynds, P.M., Yamamoto, L., Pitts, J.N., Jr., and Winer, J.A., A standardized procedure for quantification of the Ames *Salmonella*/mammalian-microsome mutagenicity test, *Environ. Mutagen.*, 3(2), 123, 1981.
23. Cheli, C., DeFrancesco, D., Petrullo, L.A., McCoy, E.C., and Rosenkranz, H.S., The *Salmonella* mutagenicity assay: reproducibility, *Mutation Research*, 74(2), 145, 1980.
24. Claxton, L.D., Douglas, G., Krewski, D., Lewtas, J., Matsushita, H., and Rosenkranz, H., Overview, conclusions, and recommendations of the IPCS collaborative study on complex mixtures, *Mutation Research*, 276(1-2), 61, 1992.
25. Claxton, L.D., The utility of bacterial mutagenesis testing in the characterization of mobile source emissions: a review, *Dev. Toxicol. Environ. Sci.*, 10, 69, 1982.
26. Chen, C., Speck, W., and Rosenkranz, H., Mutagenicity testing with *Salmonella typhimurium* strains, II, The effect of unusual phenotypes on the mutagenic response, *Mutation Research*, 28, 31, 1975.
27. de Serres, F.J. and Shelby, M.D., The *Salmonella* mutagenicity assay: recommendations, *Science*, 203(4380), 563, 1979.
28. Dunkel, V.C., Collaborative studies on the *Salmonella*/microsome mutagenicity assay, *J. Assoc. Off. Anal. Chem.*, 62(4), 874, 1979.
29. Dunkel, V.C., Zeiger, E., Brusick, D., McCoy, E., McGregor, D., Mortelmans, K., Rosenkranz, H.S., and Simmon, V.F., Reproducibility of microbial mutagenicity assays: I. Tests with *Salmonella typhimurium* and *Escherichia coli* using a standardized protocol, *Environ. Mutagen.*, 2, 1, 1984.
30. Greim, H., Goggelmann, W., Summer, K.H., and Wolff, T., Mutagenicity testing with *Salmonella* microsome test, *Arch. Toxicol.*, 46(1-2), 31, 1980.
31. Grafe, A., Lorenz, R., and Vollmar, J., Testing the mutagenic potency of chemical substances in a linear host mediated assay, *Mutation Research*, 31, 205, 1975.
32. Grafe, A., Mattern, I.E., and Green, M., A European collaborative study of the Ames assay. I. Results and general interpretation, *Mutation Research*, 85(6), 391, 1981.
33. Kado, N.Y., Langley, D., and Eisenstadt, E., A simple modification of the *Salmonella* liquid-incubation assay. Increased sensitivity for detecting mutagens in human urine, *Mutation Research*, 121(1), 25, 1983.
34. Matsumoto, T., Yoshida, D., and Mizusaki, S., Enhancing effect of harman on mutagenicity in *Salmonella*, *Mutation Research*, 56, 85, 1977.
35. Myers, L.E., Adams, N.H., Hughes, T.J., Williams, L.R., and Claxton, L.D., An interlaboratory study of an EPA/Ames/*Salmonella* test protocol, *Mutation Research*, 182(3), 121, 1987.

36. Prival, M.J., The *Salmonella* mutagenicity assay: promises and problems, *Ann. N.Y. Acad. Sci.*, 407, 154, 1983.

37. Shahin, M. and von Borstel, R., Comparisons of mutation induction in reversion systems of *Saccharomyces cerevisiae* and *Salmonella typhimurium, Mutation Research*, 53, 1, 1978.

38. Thompson, E.D. and Melampy, P.J., An examination of the quantitative suspension assay for mutagenesis with strains of *Salmonella typhimurium, Environ. Mutagen.*, 3(4), 453, 1981.

39. Ames, B.N., McCann, J., and Yamasaki, E., Methods for detecting carcinogens and mutagens with the *Salmonella*/mammalian-microsome mutagenicity test, *Mutation Research*, 31(6), 347, 1975.

40. Rosenkranz, H.S. and Poirier, L.A., Evaluation of the mutagenicity and DNA-modifying activity of carcinogens and noncarcinogens in microbial systems, *J. Natl. Cancer. Inst.*, 62(4), 873, 1979.

41. Thier, R., Taylor, J.B., Pemble, S.E., Humphreys, W.G., Persmark, M., Ketterer, B., and Guengerich, F.P., Expression of mammalian glutathione S-transferase 5-5 in *Salmonella typhimurium* TA1535 leads to base-pair mutations upon exposure to dihalomethanes, *Proceedings of the National Academy of Sciences, USA*, 90, 8576, 1993.

42. Simula, T.P., Glancey, M.J., and Wolf, C.R., Human glutathione S-transferase-expressing *Salmonella typhimurium* tester strains to study the activation/detoxification of mutagenic compounds: studies with halogenated compounds, aromatic amines and aflatoxin B_1, *Carcinogenesis*, 14(7), 1371, 1993.

43. Simula, T.P., Glancey, M.J., Soderlund, E.J., Dybing, E., and Wolf, C.R., Increased mutagenicity of 1,2-dibromo-3-chloropropane and tris(2,3-dibromopropyl)phosphate in *Salmonella* TA100 expressing human glutathione S-transferases, *Carcinogenesis*, 14(11), 2303, 1993.

44. Claxton, L., Toney, S., Perry, E., and King, L., Assessing the effect of colony counting methods and genetic drift on Ames bioassay results, *Environmental Mutagenesis*, 6, 331, 1984.

45. Yahagi, T., Degawa, Y., Seino, Y., Matsushima, T., Nagao, M., Sugimura, T., and Hashimoto, Y., Mutagenicity of carcinogenic azo dyes and their metabolites, *Cancer Letters,* 1, 91, 1975.

46. Houk, V., Schalkowsky, S., and Claxton, L., Development and validation of the spiral *Salmonella* assay: An automated approach to bacterial mutagenicity testing, *Mutation Research*, 223, 49, 1989.

47. Houk, V.S., Early, G., and Claxton, L.D., Use of the spiral *Salmonella* assay to detect the mutagenicity of complex environmental mixtures, *Environ. Mol. Mutagen.*, 17(2), 112, 1991.

48. Claxton, L., Allen, J., Auletta, A., Mortelmans, K., Nestman, E., and Zeiger, E., A guide for the *Salmonella typhimurium*/mammalian microsome tests for bacterial mutagenicity, *Mutation Research*, 189, 83, 1987.

49. Adler, I.D., A review of the coordinated research effort on the comparison of test systems for the detection of mutagenic effects, sponsored by the E.E.C, *Mutation Research*, 74(2), 77, 1980.

50. Agnese, G., Risso, D., and De, F.-S., Statistical evaluation of inter- and intra-laboratory variations of the Ames test, as related to the genetic stability of *Salmonella* tester strains, *Mutation Research*, 130(1), 27, 1984.

51. Amlacher, E., Short-term tests in screening programs of environmental chemical carcinogens, *Exp. Pathol.*, 22(4), 187, 1982.

52. Auletta, A.E., Dearfield, K.L., and Cimino, M.C., Mutagenicity test schemes and guidelines: U.S. EPA Office of Pollution Prevention and Toxics and Office of Pesticide Programs, *Environ. Mol. Mutagen.*, 21(1), 38, 1993.

53. Bartsch, H., Malaveille, C., Camus, A.M., Martel, P.-G., Brun, G., Hautefeuille, A., Sabadie, N., Barbin, A., Kuroki, T., Drevon, C., Piccoli, C., and Montesano, R., Bacterial and mammalian mutagenicity tests: validation and comparative studies on 180 chemicals, *IARC Sci. Publ.*, 27, 179, 1980.

54. Clayson, D.B., International Commission for Protection Against Environmental Mutagens and Carcinogens. ICPEMC publication No. 17. Can a mechanistic rationale be provided for non-genotoxic carcinogens identified in rodent bioassays?, *Mutation Research*, 221(1), 53, 1989.

55. Gatehouse, D., Critical features of bacterial mutation assays, *Mutagenesis*, 2(5), 397, 1987.

56. Gatehouse, D., Haworth, S., Cebula, T., Gocke, E., Kier, L., Matsushima, T., Melcion, C., Nohmi, T., Ohta, T., Venitt, S., et al., Recommendations for the performance of bacterial mutation assays, *Mutation Research*, 312(3), 217, 1994.

57. Goggelmann, W., Grafe, A., Vollmar, J., Baumeister, M., Kramer, P.J., and Pool, B.L., Criteria for the standardization of *Salmonella* mutagenicity tests: results of a collaborative study. IV. Relationship between the number of his- bacteria plated and number of his+ revertants scored in the *Salmonella* mutagenicity test, *Teratogenesis Carcinog. Mutagen.*, 3(2), 205, 1983.

58. ICPEMC: International Commission for Protection against Environmental Mutagens and Carcinogens (ICPEMC). Committee 2, Final Report. Mutagenesis testing as an approach to carcinogenesis, *Mutation Research*, 99(1), 73, 1982.

59. McCann, J., Horn, L., and Kaldor, J., An evaluation of *Salmonella* (Ames) test data in the published literature: application of statistical procedures and analysis of mutagenic potency, *Mutation Research*, 134(1), 1, 1984.

60. Palmer, G.H., The Ames test: a look at the *Salmonella*/mammalian microsome mutagenicity test, *Vet. Hum. Toxicol.*, 22(1), 23, 1980.

61. Purchase, I.F., Longstaff, E., Ashby, J., Styles, J.A., Anderson, D., Lefevre, P.A., and Westwood, F.R., An evaluation of 6 short-term tests for detecting organic chemical carcinogens, *Br. J. Cancer*, 37(6), 873, 1978.

62. Purchase, I.F., Validation of tests for carcinogenicity, *IARC Sci. Publ.*, 27, 343, 1980.

63. Purchase, I.F., International Commission for Protection against Environmental Mutagens and Carcinogens. ICPEMC working paper 2/6. An appraisal of predictive tests for carcinogenicity, *Mutation Research*, 99(1), 53, 1982.

64. Wilcox, P., Wedd, D.J., and Gatehouse, D., Report of the Association of British Pharmaceutical Industries Collaborative Study Group. Collaborative study to evaluate the inter/intra laboratory reproducibility and phenotypic stability of *Salmonella typhimurium* TA97a and TA102, *Mutagenesis*, 8(2), 93, 1993.

65. Fuscoe, J.C., Wu, R., Shen, N., Healy, S., and Felton, J., Base change analysis of revertants of the *hisD3052* allele in *Salmonella typhimurium*, *Mutation Research*, 201, 241, 1988.

66. Cebula, T. and Koch, W., Use of polymerase chain reaction (PCR), direct sequencing, and DNA colony hybridization in the characterization of *Salmonella typhimurium* revertants, *Environmental Molecular Mutagenesis*, 14(Suppl. 15), 34, 1989.

67. Bell, D.A., Levine, J.G., and DeMarini, D.D., DNA sequence analysis of revertants of the *hisD3052* allele of *Salmonella typhimurium* TA98 using the polymerase chain reaction and direct sequencing: Application to 1-nitropyrene-induced revertants, *Mutation Research*, 252, 35, 1991.

68. Weinstein, D. and Lewinson, T., A statistical treatment of the Ames mutagenicity assay, *Mutation Research*, 51, 433, 1978.

69. Chu, K., Patel, K., Lin, A., Tarone, R., Linhart, M., and Dunkel, V., Evaluating statistical analyses and reproducibility of microbial mutagenicity assays, *Mutation Research*, 85, 119, 1981.

70. Cologne, J. and Breslow, N., Nonparametric regression analysis of data from the Ames mutagenicity assay, *Environmental Health Perspectives*, 102(Supplement 1), 61, 1994.

71. Krewski, D., Leroux, B., Bleuer, S., and Broekhoven, L., Modeling the Ames *Salmonella*/microsome assay, *Biometrics*, 49, 499, 1993.

72. Margolin, B., Kaplan, N., and Zeiger, E., Statistical analyses of the Ames *Salmonella*/microsome test, *Proc. Natl. Acad. Sci., USA*, 78(6), 3779, 1981.

73. Myers, L., Sexton, N., Southerland, L., and Wolff, T., Regression analysis of Ames test data, *Environmental Mutagenesis*, 3, 575, 1981.

74. Stead, A.G., Hasselblad, V., Creason, J.P., and Claxton, L., Modeling the Ames test, *Mutation Research*, 85(1), 13, 1981.

75. Claxton, L.D., Creason, J., Nader, J.A., Poteat, W., and Orr, J.D., GeneTox manager for bacterial mutagenicity assays: a personal computer and minicomputer system, *Mutation Research*, 342, 87, 1995.

76. Leroux, B. and Krewski, D., AMESFIT: a microcomputer program for fitting linear-exponential dose-response models in the Ames *Salmonella* assay, *Environmental Molecular Mutagenesis*, 22, 78, 1993.

77. Myers, L., Adams, N., Rao, T., Shah, B., and Williams, L., Microcomputer software for data management and statistical analysis of the Ames/*Salmonella* test, in *Statistics in Toxicology*, Krewski, D. and Franklin, C., Eds., New York, Gordon and Breach, 1991, 265.

78. Green, M.H., Bridges, B.A., Rogers, A.M., Horspool, G., Muriel, W.J., Bridges, J.W., and Fry, J.R., Mutagen screening by a simplified bacterial fluctuation test: use of microsomal preparations and whole liver cells for metabolic activation, *Mutation Research*, 48(3-4), 287, 1977.

79. Frantz, C.N. and Malling, H.V., The quantitative microsomal mutagenesis assay method, *Mutation Research*, 31, 365, 1975.

80. Houk, V.S. and Claxton, L.D., Screening complex hazardous wastes for mutagenic activity using a modified version of the μLC/*Salmonella* assay, *Mutation Research*, 169, 81, 1986.

81. Prival, M.J. and Mitchell, V.D., Analysis of a method for testing azo dyes for mutagenic activity in *Salmonella typhimurium* in the presence of flavin mononucleotide and hamster liver S9, *Mutation Research*, 97(2), 103, 1982.

82. Claxton, L., Assessment of bacterial bioassay methods for volatile and semi-volatile compounds and mixtures, *Environment International*, 11, 375, 1985.

83. Hughes, T., Simmons, D., Monteith, L., and Claxton, L., Vaporization technique to measure mutagenic activity of volatile organic chemicals in the Ames/*Salmonella* assay, *Environmental Mutagenesis*, 9, 421, 1987.

84. Barber, E.D., Donish, W.H., and Mueller, K.R., A procedure for the quantitative measurement of the mutagenicity of volatile liquids in the Ames *Salmonella*/microsome assay, *Mutation Research*, 90(1), 31, 1981.

85. DeMarini, D.M., Plewa, M.J., and Brockman, H.E., The use of four short-term tests to evaluate the mutagenicity of municipal water, *J. Toxicology and Environ. Health*, 9(1), 127, 1982.

86. Kool, H.J., van Kreijl, C.F., de Greef, E., and van Kranen, H.J., Presence, introduction and removal of mutagenic activity during the preparation of drinking water in the Netherlands, *Environmental Health Perspectives*, 46, 207, 1982.

87. Maruoka, S. and Yamanaka, S., Mutagenicity in *Salmonella typhimurium* strains of XAD-2-ether extract, recovered from Katsura River water in Kyoto City, and its fractions, *Mutation Res.*, 102(1), 13.

88. Scully, F.E. and Bempong, M.A., Organic N-chloramines: chemistry and toxicology, *Environmental Health Perspectives*, 46, 111, 1982

89. Spanggord, R.J., Mortelmans, K.E., Griffin, A.F., and Simmon, V.F., Mutagenicity in *Salmonella typhimurium* and structure-activity relationships of wastewater components emanating from the manufacture of trinitrotoluene, *Environ. Mutagenesis* 4(2), 163, 1982.

90. Whong, W.Z., Stewart, J.D., Adamo, D.C., and Ong, T., Mutagenic detection of complex environmental mixtures using the *Salmonella* arabinose-resistant assay system, *Mutation Res.*, 120(1), 13, 1983.

91. Zoeteman, B.C., Hrubec, J., de Greef, E., and Kool, H.J., Mutagenic activity associated with byproducts of drinking water disinfection by chlorine, chlorine dioxide, ozone, and UV-radiation, *Environmental Health Perspectives*, 46, 197, 1982.

92. Harrington, T.R., Nestmann, E.R., and Kowbel, D.J., Suitability of the modified fluctuation assay for evaluating the mutagenicity of unconcentrated drinking water, *Mutation Res.*, 120 (2-3), 97, 1983.

93. Kamra, O.P., Nestmann, E.R., Douglas, G.R., Kowbel, D.J., and Harrington, T.R., Genotoxic activity of pulp mill effluent in *Salmonella* and *Saccharomyces cerevisiae* assays, *Mutation Res.*, 118, 269, 1983.

94. Maruoka, S. and Yamanaka, S., Mutagenic potential of laboratory chlorinated river water, *Science Total Environ.*, 29, 143, 1983.

95. Meier, J.R., Lingg, R.D., and Bull, R.J., Formation of mutagens following chlorination of humic acid. A model for mutagen formation during drinking water treatment, *Mutation Res.*, 118, 25, 1983.

96. Sato, T., Momma, T., Ose, Y., Ishikawa, T., and Kato, K., Mutagenicity of Nagara river sediment, *Mutation Res.*, 118, 257, 1983.

97. Kowbel, D.J., Malaiyandi, M., Paramasigamani, V., and Nestman, E.R., Chlorination of ozonated soil fulvic acid: Mutagenicity studies in *Salmonella*, *Science Total Environ.*, 37, 171, 1984.

98. Backlund, P., Kronberg, L., Pensar, G., and Tikkanen, L., Mutagenic activity in humic water and alum flocculated humic water treated with alternative disinfectants, *Science Total Environ.*, 47, 257, 1985.

99. Fawell, J.K. and Fielding, M., Identification and assessment of hazardous compounds in drinking water, *Science Total Environ.*, 47, 317, 1985.

100. Nikunen, E., Toxic impact of effluents from petrochemical industry, *Ecotoxicology Environmental Safety*, 9, 84, 1985.

101. Meier, J.R., Ringhand, H.P., Coleman, W.E., Munch, J.W., Streicher, R.P., Kaylor, W.H., and Schenck, K.M., Identification of mutagenic compounds formed during chlorination of humic acid, *Mutation Res.*, 157, 111, 1985.

102. Sato, T., Kato, K., Ose, Y., Nagase, H., and Ishikawa, T., Nitroarenes in Suimon River sediment, *Mutation Res.*, 157, 135, 1985.

103. Maruoka, S., Yamanaka, S., and Yamamoto, Y., Isolation of mutagenic components by high-performance liquid chromatography from XAD extract of water from the Nishitakase River, Kyoto City, Japan, *Science Total Environ.*, 57, 29, 1986.

104. Vartiainen, T. and Liimatainen, A., High levels of mutagenic activity in chlorinated drinking water in Finland, *Mutation Res.*, 169, 29, 1986.

105. Hopke, P.K., Plewa, M.J., and Stapleton, P., Reduction of mutagenicity of municipal wastewaters by land treatment, *Science Total Environ.*, 66, 193, 1987.

106. Meier, J.R., Blazak, W.F., and Knohl, R.B., Mutagenic and clastogenic properties of 3-chloro-4-(dichloromethyl)-5-hydroxy-2(5H)-furanone: a potent bacterial mutagen in drinking water, *Environ. Molecular Mutagenesis*, 10, 411, 1987.

107. Thomas, E.L., Jefferson, M.M., Bennett, J.J., and Learn, D.B., Mutagenic activity of chloramines, *Mutation Res.*, 188, 35, 1987.

108. Kool, H.J. and van Kreyl, C.F., Mutagenic activity of drinking water prepared from groundwater: A survey of ten cities in The Netherlands, *Science Total Environ.*, 77, 51, 1988.

109. Kronberg, L. and Vartiainen, T., Ames mutagenicity and concentration of the strong mutagen 3-chloro-4-(dichloromethyl)-5-hydroxy-2(5H)-furanone and its genometric isomer E-2-chloro-3-(dichloromethyl)-4-oxo-butenoic acid in chlorine-treated tap waters, *Mutation Res.*, 206, 177, 1988.

110. Liimatainen, A., Muller, D., Vartiainen, T., Jahn, F., Kleeberg, U., Klinger, W., and Hanninen, O.A.D., Chlorinated drinking water is mutagenic and causes 3-methylcholanthrene type induction of hepatic monooxygenase, *Toxicology*, 51,281, 1988.

111. Matsumoto, M., Ando, M., and Ohta, Y., Mutagenicity of monochlorodibenzofurans detected in the environment, *Toxicology Letters*, 40, 21, 1988.

112. Urano, K., Haga, N., and Emoto, F., Method for evaluating mutagenicity of water: II. Conditions for applying new sample preparation method to the Ames test, *Science Total Environ.*, 74, 191, 1988.

113. Varma, M.M., Ampy, F.R., Verma, K., and Talbot, W.W., *In vitro* mutagenicity of water contaminants in complex mixtures, *J. Applied Toxicology*, 8, 243, 1988.

114. Wigilius, B., Boren, H., Grimvall, A., Carlberg, G.E., Hagen, I., and Brogger, A., Impact of bleached kraft mill effluents on drinking water quality, *Science Total Environ.*, 74, 75, 1988.

115. Wilcox, P., Williamson, S., Lodge, D.C., and Bootman, J., Concentrated drinking water extracts, which cause bacterial mutation and chromosome damage in CHO cells, do not induce sex-linked recessive lethal mutations in *Drosophila, Mutagenesis*, 3, 381, 1988.

116. Backlund, P., Wondergem, E., Voogd, K., and de Jong, A., Mutagenic activity and presence of the strong mutagen 3-chloro-4-(dichloromethyl)-5-hydroxy-2(5H)-furanone (MX) in chlorinated raw and drinking waters in The Netherlands, *Science Total Environment*, 84, 273, 1989.

117. Kira, S., Hayatsu, H., and Ogata, M., Detection of mutagenicity in mussels and their ambient water, *Bulletin Environ. Contam. Toxicol.* 43, 583, 1989.

118. Kool, H.J., van Kreyl, C.F., and Persad, S., Mutagenic activity in groundwater in relation to mobilization of organic mutagens in soil, *Science Total Environ.*, 84, 185, 1989.

119. Kronberg, L. and Christman, R.F., Chemistry of mutagenic by-products of water chlorination, *Science Total Environ.*, 81, 219, 1989.

120. Sayato, Y., Nakamuro, K., and Ueno, H., Mutagenicity on chlorination of products formed by ozonation of naphthoresorcinol in water, *Mutation Res.*, 226, 151, 1989.

121. Waters, L.C., Schenley, R.I., Owen, B.A., Walsh, P.J., Hsie, A.W., Jolley, R.L., Buchman, M.V., and Condie, L.W., Biotesting of wastewater: A comparative study using *Salmonella* and CHO assay systems, *Environ. Molec. Mutagenesis*, 14, 254.

122. Metcalfe, C.D., Batch, G.C., Cairns, V.W., Fitzsimons, J.D., and Dunn, B.P., Carcinogenic and genotoxic activity of extracts from contaminated sediments in western Lake Ontario, *Science Total Environment*, 94, 125, 1990.

123. Sakamoto, H. and Hayatsu, H., A simple method for monitoring mutagenicity of river water: Mutagens in Yodo river system, Kyoto-Osaka, *Bulletin Environ. Contam. Toxicol.*, 44, 521, 1990.

124. Sayato, Y., Nakamuro, K., Ueno, H., and Goto, R., Mutagenicity of adsorbates to a copper-phthalocyanine derivative recovered from municipal river water, *Mutation Res.*, 242, 313, 1990.

125. Tikkanen, L. and Kronberg, L., Genotoxic effects of various chlorinated butenoic acids identified in chlorinated drinking water, *Mutation Res.*, 240, 109, 1990.

126. Vartiainen, T. and Lampelo, S., Effects of human placental S9 and induced rat liver S9 on the mutagenicity of drinking waters processed from humus-rich surface waters, *Environ. Molecular Mutagenesis,* 15, 198, 1990.

127. Matsuda, H., Ose, Y., Nagase, H., Sato, T., Kito, H., and Sumida, K., Mutagenicity of the components of ozonated humic substance, *Science Total Environ.,* 103, 129, 1991.

128. Ankle, G.T., Lodge, K., Call, D.J., Balcer, M.D., Brooke, L.T., Cook, P.M., Kreis, R.G., Jr., Carlson, A.R., Johnson, R.D., and Niemi, G.J., Integrated assessment of contaminated sediments in the lower Fox River and Green Bay, Wisconsin, *Ecotoxicol. Environ. Safety,* 23, 46, 1992.

129. Fracasso, M.E., Leone, R., Brunello, F., Monastra, C., Tezza, F., and Storti, P.V., Mutagenic activity in wastewater concentrates from dye plants, *Mutation Res.,* 298, 91, 1992.

130. Rodriguez-Ariza, A., Abril, N., Navas, J.I., Dorado, G., Lopez-Barca, J., Pueyo, C., Metal, mutagenicity, and biochemical studies on bivalve mollusks from Spanish coasts, *Environ. Molecular Mutagenesis,* 19, 112, 1992.

131. Marvi, B.D., Allan, L., McCarry, B.E., and Bryant, D.W., Chemico/biological investigation of contaminated sediment from the Hamilton Harbour area of western Lake Ontario, *Environ. Molecular Mutagenesis,* 22, 61, 1993.

132. Onodera, S., Nagatsuka, A., Rokuhara, T., Asakura, T., Hirayama, N., and Suzuki, S., Re-evaluation of solid-phase adsorption and desorption techniques for isolation of trace organic pollutants from chlorinated water, *J. Chromatogr.,* 642, 185, 1993.

133. Sayato, Y., Nakamuro, K., Ueno, H., and Goto, R., Identification of polycyclic aromatic hydrocarbons in mutagenic adsorbates to a copper-phthalocyanine derivative recovered from municipal river water, *Mutation Res.,* 300, 207, 1993.

134. Ottaviani, M., Crebelli, R., Fuselli, S., La Rocca, C., and Baldassarri, L.T., Chemical and mutagenic evaluation of sludge from a large wastewater treatment plant, *Ecotoxicol. Environ. Safety,* 26, 18, 1993.

135. Rodriguez-Ariza, A., Martinez-Lara, E., Pascual, P., Pedrajas, J.R., Abril, N., Dorado, G., Toribio, F., Barcena, J.A., Peinado, J., and Pueyo, C., Biochemical and genetic indices of marine pollution in Spanish littoral, *Science Total Environ.,* Suppl. 1, 109, 1993.

136. Shibuya, N., Ohta, T., Nakadaira, H., Mano, H., Endoh, K., and Yamamoto, M., Mutagenicity of activated carbon adsorbate of drinking water in the Ames assay, Tohoku, *J. Exp. Med.,* 17, 89, 1993.

137. Vargas, V.M., Motta, V.E., and Henriques, J.A., Mutagenic activity detected by the Ames test in river water under the influence of petrochemical industries, *Mutation Res.,* 319, 31, 1993.

138. Abe, A. and Urano, K., Influence of chemicals commonly found in a water environment on the *Salmonella* mutagenicity test, *Science Total Environment,* 153, 169, 1994.

139. Koivusalo, M.T., Jaakkola, J.J., and Vartiainen, T., Drinking water mutagenicity in past exposure assessment of the studies on drinking water and cancer: application and evaluation in Finland, *Environ. Res.,* 64, 90, 1994.

140. Kusamran, W.R., Wakabayashi, K., Oguri, A., Tepsuwan, A., Nagao, M., and Sugimura, T., Mutagenicities of Bangkok and Tokyo river waters, *Mutation Res.,* 325, 99, 1994.

141. Le Curieux, F., Marzin, D., and Erb, F., Study of the genotoxicity of five chlorinated propanones using the SOS Chromotest, the Ames-fluctuation test and the newt micronucleus test, *Mutation Res.,* 341, 1, 1994.

142. De Flora, S., Bagnasco, M., Bennicelli, C., Camoirano, A., Bojnemirski, A., and Kurelec, B., Biotransformation of genotoxic agents in marine sponges: Mechanisms and modulation, *Mutagenesis,* 10, 357, 1995.

143. Filipic, M., Mutagenicity and toxicity of water extracts from the Sora river area, *Mutation Res.,* 342, 1, 1995.

144. Rehana, Z., Malik, A., and Ahmad, M., Mutagenic activity of the Ganges water with special reference to the pesticide pollution in the river between Kachla to Kannauj (U.P.), India, *Mutation Res.,* 343, 2, 1995.

145. Tahara, I., Kataoka, K., Kinouchi, T., and Ohnishi, Y., Stability of 1-nitropyrene and 1,6-dinitropyrene in environmental water samples and soil suspensions, *Mutation Res.,* 343, 109, 1995.

146. Rehana, Z., Malik, A., and Ahmad, M., Genotoxicity of the Ganges water at Narora (U.P.), India, *Mutation Res.,* 367, 187, 1996.

CHAPTER 42

Investigating the Sources and Fate of Genotoxic Substances in Aquatic Ecosystems with the SOS Chromotest

Paul A. White and Chantale Côté

CONTENTS

I. INTRODUCTION

The pioneering works of Muller[1] and Auerbach and Robson[2] established that genes, the basic unit of inheritance, could be altered by exposure to both physical and chemical agents. At the present time, it is generally appreciated that exposure to genotoxic pollutants can result in a wide range of pathological conditions including immune system dysfunction, heritable genetic disorders, and cancer.[3-6] Many researchers have investigated the genotoxicity of industrial waste materials, particularly those that are directly discharged into aquatic systems (see Houk[7] for review). In addition, many researchers have detected genotoxic substances in surface waters (see Stahl[8] and DeFlora et al.[9] for reviews) and sediments[10-16] collected from areas that are known to receive industrial and municipal discharges. Studies of plants and animals that inhabit such areas have revealed a wide range of genotoxic and carcinogenic effects. For example, Klekowski and Levins,[17]

Klekowski and Berger,[18] and Ravindran and Ravindran[19] documented mutagenic and clastogenic effects on aquatic plants. Dixon[20] and Nacci et al.[21] documented genotoxic and clastogenic effects on bivalve mollusks. DiGiulio,[22] Prein et al.,[23] and Alink et al.[24] documented genotoxic effects on fish. Finally, many researchers[25-30] have detected an abnormally high frequency of idiopathic lesions, including neoplasia, in fish and invertebrates that inhabit areas contaminated with industrial and municipal wastes.

Reduction or elimination of exposure to genotoxic substances requires efficient and effective tools for the detection of genotoxicity and mutagenicity. This need has led to the development of a wide range of genotoxicity, mutagenicity, and clastogenicity bioassays. Assays that are commonly used can be divided into three major categories: the *in vitro* assays, the *in vivo* assays, and the *in vivo–in vitro* assays. The *in vivo* assays measure direct mutagenic effect on multicellular animals. While essential for human risk assessment, they are generally less sensitive and more expensive than the *in vitro* assays.[31] Popular *in vivo* tests involve exposure of fruit flies (*Drosophila melanogaster*) (e.g., sex-linked recessive lethal test) or mice (e.g., mouse dominant lethal test and mouse specific-locus test). The *in vivo–in vitro* assays involve exposure *in vivo* followed by tissue removal and effect quantification *in vitro* (e.g., chromosomal aberrations in rat bone marrow, micronuclei in immature mouse erythrocytes and unscheduled DNA synthesis in rat hepatocytes).

The *in vitro* assays usually involve exposure of prokaryotic or eukaryotic cell cultures in a liquid suspension. For the most part, these tests are less expensive and more convenient than their *in vivo* or *in vitro–in vivo* counterparts. They can be subdivided primarily into microbial assays and mammalian assays, with the microbial assays being further subdivided into the prokaryotic (i.e., bacteria) and the eukaryotic (i.e., yeast). The short-term bacterial assays are without doubt the most popular tools for rapid genotoxicity assessment. Several of these involve the reversion of specially constructed auxotrophs (e.g., *his*, *trp*) to prototrophy. Of these, the single most popular test is the *Salmonella*/mammalian microsome assay or Ames test.[32-34] It was this assay that provided the first convincing evidence that many carcinogens are in fact mutagenic.[35-36]

Several short-term bacterial assays such as the Microscreen phage induction assay,[37] the SOS/*umu* assay,[38] and the SOS Chromotest[39] employ the error-prone DNA repair pathway of *Escherichia coli* (i.e., the SOS response) for the detection of DNA damaging agents (see Walker[40] for a review of the SOS response in *E. coli*). The remainder of this chapter will focus on the SOS Chromotest and its use in determining the sources and fate of genotoxic substances in aquatic systems. Further information on other *in vitro* or *in vivo* bioassays mentioned above can be found in Li and Loretz,[31] Kilbey et al.,[41] and/or other chapters (e.g., 8, 12, 16, 24, 41) of this text.

A. Microscale Testing for Genotoxicity Assessment

The advantages of bacterial mutagenicity and genotoxicity assays are obvious. Bacteria and all higher organisms use DNA as the macromolecule for storage of genetic information. The process of DNA replication and repair is more completely understood in bacteria than more complex eukaryotic cells. In addition, bacteria are easy to manipulate, have conveniently rapid growth rates (≈ 2 to 3 doublings hour^{-1}), and can easily be cultured and exposed in microscale containers (<0.5 mL). However, the use of bacteria is not without drawbacks. For example, some substances are not genotoxic until they are *metabolically activated* by oxidative enzymes that are not present in bacteria. These substances are known as pro-genotoxins. In mammals, pro-genotoxins are transformed *in vivo* into potent electrophiles that can readily react with, and damage, DNA. The desire to augment the utility of bacterial assays has lead to the development of *in vitro* activation mixtures (e.g., mammalian liver microsomes or S9 fraction, see Miller and Miller[42] and Malling[43] for historical insight). In addition, differences in the structural organization of bacterial and mammalian chromosomes do not permit bacterial assays to be used for the detection of clastogens.

The aforementioned ability to culture bacteria in small containers has led to the development of microscale versions of some short-term bacterial assays for genotoxicity and mutagenicity. These microscale assays have a conveniently small sample requirement and are capable of processing many samples in a single working day. High throughput has caused microtiter versions of short-term bacterial tests to become popular, cost-effective means for (geno)toxicity assessment.[44] Microscale assays that have appeared in the recent literature include the *Salmonella* microscale fluctuation test,[45-47] the microtiter SOS Chromotest,[48-49] and the luminescent *Salmonella*/microsome assay.[50]

The microtiter SOS Chromotest has several advantages that make it an attractive choice for genotoxicity assessment. In addition to possessing the attributes of other microbial assays, the test system has the following additional advantages:

- Survival of the tester strain is not required
- Sample sterility is not usually required
- Bacteriostatic effects can be simultaneously monitored[51]
- The response pathway responds to a wide range of DNA damage scenarios[52-53]
- The test can easily accept biological samples such as tissue extracts and biological fluids (histidine contamination of biological samples can hamper the performance of *Salmonella* assay[48,54-58])
- The microtiter version of the assay is readily amenable to automation[49,59-60]

The high sample throughput afforded by a combination of microscale testing and automation is particularly important for environmental monitoring and bioassay-directed fractionation. Effective, environmental monitoring must include a wide range of samples in order to encompass both temporal and spatial variability.[61] In addition, *bioassay-directed fractionation* schemes generally require that large numbers of samples be examined in an effort to isolate and identify putative genotoxins in complex environmental samples.[62-63] Bioassay-directed fractionation techniques have been successfully used to investigate mutagens in several complex samples such as urban air particulate matter,[64-65] diesel particulate matter,[66-67] contaminated sediments,[10,12] drinking water,[68-69] and industrial effluents.[70-71]

II. TESTING ENVIRONMENTAL COMPLEX MIXTURES WITH THE SOS CHROMOTEST

Over the past 25 years the Ames/*Salmonella* assay and other short-term, microbial genotoxicity and mutagenicity assays have been successfully used to identify mutagenic substances such as dyes and toners,[72-73] pesticides,[74-75] food additives,[76-77] and drugs.[78-79] Appropriate measures were subsequently instituted to reduce or eliminate human exposure. Despite these successes, chemical-specific analyses ignore the fact that organisms outside the laboratory are exposed to complex mixtures of both known and unknown substances. Genetic toxicologists were among the first environmental toxicologists to appreciate this fact and employ bioassays to investigate complex environmental samples such as urban air particulates, cigarette smoke condensate, and automobile exhaust.[80-82]

The effective use of any bacterial genotoxicity assay, including the SOS Chromotest, for the examination of complex environmental samples requires that several technical challenges be overcome. Genotoxic substances present in complex liquid samples (e.g., surface water, groundwater, urine, wastewater) are frequently present at very low concentrations. As a result, unconcentrated liquid samples frequently elicit responses that are close to the detection limit of the assay system.[48,83-85] Many researchers have employed concentration and/or extraction procedures to overcome this problem.[11,56-57,86-89] For the most part, solid samples are physically incompatible with aqueous assays such as the SOS Chromotest. These samples are often extracted with organic solvents prior to genotoxicity testing.[11,16,90-95] Although concentration and extraction procedures circumvent detection limit problems and provide samples that are generally compatible with the aqueous assay medium, they can introduce bias through loss of volatile substances and chemical alteration of sample components. A wide range of procedures for the extraction and concentration of genotoxins from both solid and liquid samples are available. Few have been rigorously validated for use with bioassays, and all are accompanied by assumptions and limitations. For a more complete discussion, the reader is referred to U.S. EPA,[96] Hunchak and Suffet,[97] Tabor and Loper,[98] and Montreuil et al.[99]

A. Complex Mixtures: Challenges for the SOS Chromotest

As previously noted, the SOS Chromotest is a short-term bacterial genotoxicity assay that employs the SOS response pathway of *Escherichia coli* for the detection of DNA damaging agents. The cornerstone of the assay is a specially constructed strain of *E. coli*, known as PQ37, in which the expression of β-galactosidase is under the express regulatory control of the SOS response system (see Huisman and D'Ari[100] and Quillardet and Hofnung[51] for strain construction and genotype details). The assay involves incubation of PQ37 in aqueous suspension with a suspected compound or complex mixture. Postexposure β-galactosidase activity reflects the genotoxicity of the tested sample. Alkaline phosphatase activity (constitutive in PQ37) provides an indirect measure of bacteriostatic and/or bacteriocidal effects.[39,51] In order to detect both direct-acting substances and substances that require metabolic activation (pro-genotoxins), samples are usually tested both in the presence and the absence of a post-mitochondrial supernatant (S9 fraction) obtained from Arochlor 1254 induced rats.

Approximately one quarter of the SOS Chromotest publications have examined complex environmental samples such as bodily fluids and excreta, foodstuffs, industrial wastes, surface waters and sediments and airborne particulate matter. Table 42.1 provides a summary of these studies. The complex samples listed present several technical challenges for the SOS Chromotest. For example, organic extracts and oily samples are often not easily mixed with the aqueous assay medium.[87,94] Some researchers have attempted to use nonionic detergents to overcome this problem.[94,105,106] Turbid and colored extracts interfere with the spectrophotometric measurements required to assess the level of SOS response induction.[11,86,87,107,112] This problem is frequently overcome by postexposure centrifugation to isolate exposed cells.[58,87,126] Alternatively, some researchers recommend initial optical density readings taken immediately after the addition of colorimetric reagents required for the measurement of β-galactosidase and alkaline phosphatase activity.[84,86,92,93,112,121] However, while initial optical density measurements can correct for the color or turbidity of the tested substance or extract, it cannot correct for interference with enzyme activity. Several researchers have demonstrated that some substances (e.g., $AlCl_3$, landfill leachates, fecal extracts) can alter the activity of the enzymes measured in the SOS Chromotest.[107,118,127] Postexposure inhibition of enzyme activity by sample components will result in ambiguous results. To avoid these problems, several of the aforementioned researchers[58,87,126] prefer centrifugation and sample removal over color correction via initial optical density measurements.

In addition to protocol modifications that permit the SOS Chromotest to be used for the assessment of complex environmental extracts, the authors listed in Table 42.1 have introduced additional protocol modifications. The majority of these concern culture methods, cell concentration, exposure times, S9 concentration, and the composition of the cell lysis/enzyme assay reagents. The net result is a wide range of protocols for the analysis of complex environmental samples, few of which have been validated in any way. White et al.[49] recently published a rapid and effective, microplate version of the assay for the analysis of complex mixtures. They manipulated cell density, exposure time, substrate conversion time, the composition of the cell lysis/enzyme assay reagent, and the concentration of S9 in order to maximize the response to six pure substances (4-nitroquinoline-*N*-oxide, *N*-methyl-*N*'-nitro-*N*-nitrosoguanidine, mitomycin C, captan, 2-naphthylamine and *N*-nitrosodime-

Table 42.1   Summary of Published Studies which Used the SOS Chromotest to Investigate the Genotoxicity of Complex Mixtures

Sample Studied	Results Obtained	Source
1) Urine and Urine Concentrates		
Unconcentrated urine.	Positive response for subjects exposed to toxic mineral oils.	48
XAD-2[†] extract.	No positive response.	57
XAD-2 and C_{18}[§] SEP-PAK extracts.	Positive response for patients receiving antineoplastic drugs (+S9 where required).	56
XAD-2 extract.	Positive response (+S9 and −S9) for patients receiving coal tar treatment.	86
2) Airborne Particulates		
DCM[‡] and acetone extracts of samples from Paris area.	Several positive responses. No response in presence of S9.	91
DCM extracts of samples from Berlin area.	Several positive responses. Higher activity with S9.	101
DCM extracts of samples from Gangzhou, China.	Several positive responses without S9.	102
3) Industrial Wastes and Effluents		
Unconcentrated effluents from 7 industries.	Several positive responses. Algal activation documented.	103
Several waste samples.	Investigated the effect of detergents on the response of complex mixtures.	104
Unidentified industrial waste.	Effect of complex waste on 4-NQO response.	105
Solid waste from semiconductor manufacture.	Several positive responses (PQ37 and PQ243) both with and without S9.	106
Aqueous soil and waste leachates.	Positive responses without S9.	107
DCM extract of aluminum plasma etching waste.	Several positive responses without S9. S9 addition reduced response.	94
XAD-8, XAD-4 extracts of bleached Kraft mill effluent.	Positive response on several fractions (both with and without S9).	108
Kraft pulp spent liquors (pulp and paper waste).	Several positive responses without S9. No response for hardwood pulps.	109
Petroleum refinery effluents (10-fold concentrate).	Positive response for one sample.	110
Unconcentrated effluents from 48 industries.	37 of the 48 samples investigated invoked a significant, positive response.	85
DCM extracts of final effluents from 42 industries.	Samples of varying potency. S9 frequently caused reduction in potency.	87
DCM extracts of effluent suspended particulates.	High potency positive responses (relative to aqueous filtrate).	90
4) Surface Waters, Sediment, and Soils		
Surface water and aqueous sediment extracts.	Several positive responses on surface waters and sediments (with S9).	83
DMSO extracts of marine sediments.	Several positive responses both with and without S9.	111
DMSO extracts of river sediment.	Several positive responses with S9.	112
DCM extracts of surface water and sediments.	Positive responses on water, suspended and bottom sediments (both +S9 and −S9).	11
Unaltered river sediments.	Direct sediment test procedure (DSTP).	113
DCM and cyclohexane extracts of contaminated soils.	Several positive responses both with and without S9.	93
C_{18} extracts of groundwater.	Weak, but detectable responses at several sites (without S9).	80
Raw surface waters and sediment extracts (aqueous and DMSO).	Several positive responses on water and bottom sediment extracts without S9.	84
Aqueous extract of river sediment.	No evidence of genotoxicity.	114
DCM extracts of suspended particulates and sediments.	Suspended particulates — positive without S9. Sediments — frequently require S9.	16
5) Miscellaneous Samples		
Orange juice.	Detected aflatoxin B_1 genotoxicity in juice.	115
Cow's milk.	Aflatoxin M_1 not detected in milk extract.	116

Table 42.1 (continued)   Summary of Published Studies which Used the SOS Chromotest to Investigate the Genotoxicity of Complex Mixtures

Sample Studied	Results Obtained	Source
Various complex mixtures.	Discussed effects of extraction solvents on test results.	117
Aqueous extract of feces.	No clear positive response. Exogenous catalytic activity documented.	58
Aqueous extract of feces.	No positive response.	118
XAD-2 extract of red wine.	Weak positive responses both with and without S9.	119
Refined hard-wood smoke flavor.	No positive response with or without S9.	120
Aqueous and organic extracts of preserved foods.	Several weak, positive responses without S9.	92
Ethyl acetate extract of feces.	Positive responses without S9.	121
Aqueous extract of medicinal plant.	Positive response on SOS spot test.	122
DCM extracts of SRM 1650 and SRM 1649.	Strong, positive response to SRM 1650. Questionable response to SRM 1649.	59
Nitrosated red wine.	Toxic effect. Genotoxicity results inconclusive.	123
CHL◊ extracts of molds in uranium mine samples.	Positive responses, particularly with S9.	95
DCM extracts of melted snow.	Several positive samples. Most required S9 activation.	89
DCM extracts of bivalve mollusks (*Mya arenaria*).	Several positive responses without S9. Evidence of genotoxin bioaccumulation.	124
DCM extracts of freshwater macroinvertebrates and fish.	Several positive responses without S9. Evidence for bio-diminution.	125

† Nonpolar polystyrene resin (Amberlite®).
§ Nonpolar, bonded-phase extraction cartridge.
‡ Dichloromethane.
◊ Chloroform.
From White, P. A., Rasmussen, J. B. and Blaise, C., *Mutat. Res.*, 360, 51, 1996. With permission.

thylamine), and complex extracts of three standard reference materials (coal tar, diesel particulate matter, and urban dust/organics). The validated method was subsequently used to examine the genotoxicity of organic extracts from a wide range of complex samples including industrial effluents, river sediments and tissues of aquatic biota (see White[128] for a complete discussion).

B. A Semiautomated, Microplate Version of the SOS Chromotest for Complex Environmental Samples

The original SOS Chromotest protocol is carried out in 16 × 100 mm (≈15 mL) disposable glass culture tubes.[51] The aforementioned microsuspension methods[48-49] (<0.3 mL) permits greater throughput and increased efficiency. Moreover, the method of White et al.[49] employs an automated laboratory workstation (Biomek™ 1000 Automated Laboratory Workstation, Beckman Instruments, Palo Alto, CA) to perform all required liquid transfers and serial dilutions. The resulting semiautomated method is accurate, precise, and provides a daily throughput of more than 50 samples.

The details of the semiautomated, microplate method used to examine the samples described in this chapter will not be described here. Details of the method and its validation can be found in White et al.[49] Briefly, a 16-h overnight culture (37°C) of *E. coli* PQ37 is diluted in fresh growth medium to an optical density (600 nm) of 0.05 to 0.075 (6 to 8 × 10^6 CFU mL^{-1}). The sample (or solvent blank), dissolved in DMSO, is then added to an aliquot of culture (plus S9 if desired) and serially diluted to provide for a total of six to eight, twofold serial dilutions. Twelve control wells receive PQ37 culture and a solvent blank. Each extract is tested in duplicate or triplicate at the highest possible concentration range. The final background concentration of DMSO is always 5% v/v. Final S9 concentration is usually 1% v/v. 4-Nitroquinoline-*N*-oxide and 2-aminoanthracene are routinely used as positive controls.

Microplates are incubated for 3 h at 37°C and subsequently centrifuged at 2000 × g for 20 min. The supernatant is removed and the bacterial pellet resuspended in 100 µL of 200 mM (Tris[hydroxymethyl])-aminomethane pH 7.5 and 100 µL of cell lysis/enzyme assay reagent. Plates are then mixed for 60 seconds and incubated (37°C) for 60 min after which *initial* optical density readings are taken. Values recorded at 620 nm measure the activity of β-galactosidase, those recorded at 405 nm measure the activity of alkaline phosphatase. Final optical density readings are taken 120 min after cell resuspension and lysis. Calculation of net optical density values (120 min value–60 min value) allows the contribution of any remaining colored substances to be accounted for prior to analysis of the results.

Induction of the SOS response pathway at a sample concentration C, denoted R(C), is the ratio of net (120 min–60 min) β-galactosidase activity to net alkaline phosphatase activity. To account for the background induction of the *sul*A gene, a normalized SOS response induction factor is calculated as IF = R(C)/R(0), where R(0) is the ratio of β-galactosidase to alkaline phosphatase activity averaged over the 12 controls which received blank solvent. The original SOS Chromotest protocol[51] recommends that only induction factor values above 1.5 be considered as *significant* positives. Although some researchers also require evidence of a dose-related response,[59,92,107,108,129] most routinely use the 1.5 cutoff recommended by Quillardet and Hofnung.[51] However, White et al.[49] demonstrated that upper 95% confidence limits of DMSO controls are usually (98%) less than 1.20. Only one out of 293 microplates yielded an upper confidence limit greater than 1.30. As a result, White et al.[49] recommend that normalized SOS induction factors that exceed the upper 95% confidence limit of the control (≈1.2 to 1.3) be regarded as significant positives. Samples that yield IF values that exceed the upper confidence limit of the control at a minimum of three concentrations are denoted genotoxic. Those that yield IF values that exceed the upper confidence limit of the control at one or two concentrations only are denoted marginal.

Three genotoxicity parameters are routinely calculated for each significant, positive response. The genotoxic potency of a sample is assessed via its SOS Response Inducing Potency or SOSIP.[51] The SOSIP (also called SRIP by Langevin et al.[11]) is the initial slope of the concentration-response curve, expressed as SOS induction factor units per unit of concentration (e.g., µg per assay mL, nmoles per assay mL, etc.). The minimum detectable genotoxic concentration or MDGC[11] is the concentration of sample, inferred from the fitted concentration-response curve, required to produce an induction factor that equals the upper 95% confidence limit of the control.

When concentration response curves are distinctly nonlinear, exhibiting zero-order kinetics at high sample concentration, the concentration response curves are analyzed by iterative, nonlinear regression employing a least squares loss function.[130] In such cases, the SOSIP and MDGC are calculated according to Langevin et al.[11] When the range of concentrations tested does not result in a full hyperbola, the SRIP and MDGC are determined by ordinary, least squares linear regression.[130] Figure 42.1 provides examples of concentration response relationships for two industrial effluent extracts examined by White et al.[49] Both samples invoke a significant, positive SOS response and approach zero-order kinetics at high concentrations.

III. ENVIRONMENTAL MONITORING — SOME PROGRESS TO DATE

As part of the St. Lawrence Action Plan (Environment Canada), the SOS Chromotest was employed to investigate the sources and fate of genotoxic contaminants in the St. Lawrence and Saguenay rivers (Canada). Sources examined include a wide range of industrial and municipal facilities that discharge aqueous wastes into the St. Lawrence system. Samples collected from the receiving system itself include suspended and sedimented particulate matter, as well as tissues of a wide range of indigenous biota. The examination of a wide range of samples representing different compartments of both the waste stream and the receiving system, permitted comparative analyses and subsequent general statements about sources and fate. Although the material presented in this

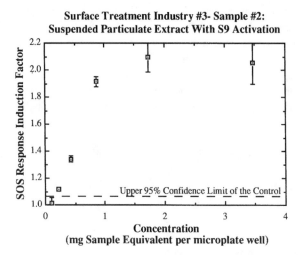

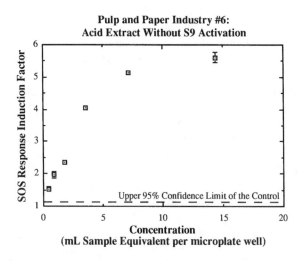

Figure 42.1   Examples of SOS Chromotest concentration-response plots. Each plot shows mean SOS Induction Factor as a function of concentration in mg or mL of equivalent starting material per microplate well. Examples shown are from the industrial effluents results of White et al.[87] Particulate material and aqueous filtrates were extracted by direct partitioning into dichloromethane. Filtrates were extracted following pH adjustment. Particulate material was obtained via glass fiber filtration. Upper panel) Suspended particulate material obtained from a metal surface finishing facility that produces parts for high-performance aircraft engines. SOSIP = 0.25 IF Units per mg sample equivalent per assay mL (microplate well = 0.2 mL). MDGC = 0.96 mg sample equivalent per assay mL. Upper 95% confidence limit of control = 1.092. Lower panel) Acid (pH = 2) extract of aqueous filtrate from a kraft pulp and paper facility. SOSIP = 0.16 IF Units per mL sample equivalent per assay mL (microplate well = 0.2 mL). MDGC = 0.21 mL sample equivalent per assay mL. Upper 95% confidence limit of control = 1.19. In both figures error bars represent one standard error of the mean (n = 2). Extracts of laboratory grade water were used as a matrix control. Confidence limits on solvent controls were calculated according to Welsch et al.[131]

section represents only a fraction of the results obtained to date, it serves to illustrate the manner in which the SOS Chromotest can be used to investigate the sources and ecological behavior of genotoxins in aquatic systems. In all cases appropriate quality control and quality assurance procedures (instrument calibration, extraction of matrix blanks and check samples, etc.) were employed to ensure that the generated data met the standards required to address the intended research questions. A complete overview of the results can be found in White.[128]

A. Genotoxic Hazards of Industrial and Municipal Wastewaters

Figure 42.1 illustrated the SOS Chromotest results for two industrial effluent extracts. In total, White et al.[87] measured the genotoxicity of organic extracts from the final effluents of 42 industries located along the St. Lawrence and Saguenay rivers. Three dichloromethane extracts were prepared from each examined effluent: a suspended particulate matter extract, an aqueous alkaline extract (pH = 12), and an aqueous acidic extract (pH = 2). The results obtained revealed a significant effect of industry type and sample type (solids, acid, and alkaline) on SOS genotoxic potency (as SOSIP in IF units per mg sample equivalent per assay mL). For example, inorganic and organic chemical production industries frequently provided high potency samples. In contrast, municipal wastewater treatment facilities frequently provided low potency samples. Figure 42.2A illustrates the geometric mean, total genotoxic potency for the eight industrial types examined. When genotoxicity is expressed as total SOS genotoxicity per liter of effluent, the results reveal a fourfold variation in potency, with municipal wastewater samples yielding the lowest potency samples.

Although Figure 42.2A illustrates the variation in the genotoxic potency of a variety of industrial and municipal wastewaters, the potential hazard to the downstream ecosystem will be determined by the likelihood that the measured toxicity will actually be realized.[132] Actual measurement of post-emission hazard is difficult. However, since hazard or relative risk increases with increasing exposure, potential hazard can be estimated by combining SOS Chromotest results with daily, volumetric emission rates. The product of genotoxic potency (IF units per L) and mean daily discharge (L per day) provides a daily genotoxic loading value. Figure 42.2B illustrates the geometric mean, total genotoxic loadings of the eight industry categories examined. For simplicity, genotoxic loading values have been converted to their BaP (benzo(a)pyrene) genotoxic equivalence and are expressed as g BaP equivalents per day. In contrast to the results shown in Figure 42.2A, these results indicate that daily genotoxic loadings from municipal wastewaters far exceed those of other industries. Subsequent mass balance calculations indicated that domestic wastewater emissions can account for over 85% of the observed surface water genotoxicity levels in the St. Lawrence river near Montréal.[133]

B. Monitoring of Genotoxins in the Aquatic Environment

The semiautomated, microplate version of the SOS Chromotest can also be used to rapidly examine environmental samples collected from the downstream receiving system. Subsequent to the examination of industrial discharges, the SOS Chromotest was used to investigate the presence of genotoxic substances in the water column, sedimented particulate matter, and the tissues of aquatic biota.

1. Particulate Material

White et al.[90] investigated the sorptive properties of organic genotoxins in industrial effluents. The results revealed that in many instances a substantial fraction (up to 99.8%) of the emitted genotoxicity is associated with particulate material. Subsequent work employed the SOS Chromotest to investigate the presence of particle-associated genotoxins in the receiving system. The investigation involved examinations of organic extracts of both suspended and sedimented particulate matter. The results obtained indicate that while genotoxic substances are associated with both sedimented and suspended particulate matter, the putative genotoxins are likely very different. Suspended particulate matter yielded extracts that are more potent in the absence of S9 activation, however, the reverse is true for bottom sediments (see Figure 42.3). This pattern was consistent across the 21 industrial sites examined. Since bottom sediments contain older material removed from the water column,[134-135] this result, illustrated in Figure 42.3, suggests that direct-acting substances are fairly readily degraded in the downstream ecosystem. This may not be surprising

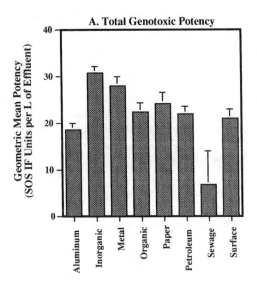

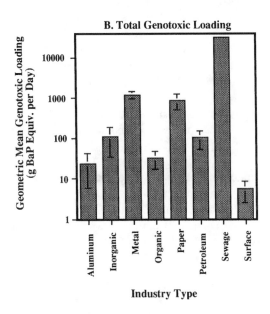

Figure 42.2   Geometric mean genotoxic potency and daily genotoxic loading of dichloromethane extractable substances in industrial effluents. Results are based on examinations of final effluents from 42 industries.[87] *Total* indicates that the values presented are the sum of measurements for the different effluent fractions examined. All potency values are expressed as IF units per L equivalent of effluent (per assay mL). Total suspended solid concentrations (dry mg per L) were used to convert the genotoxic potency of suspended solid samples to IF units per equivalent L of effluent. A) Geometric mean total genotoxic potency of the eight industrial effluent categories examined by White et al.[87] B) Geometric mean total, daily genotoxic loading of the eight industrial effluent categories examined by White et al.[87] The mean genotoxic potency of BaP on the SOS Chromotest (2.52 IF units per μg per assay mL. Mean value based on 40 measurements.) was used to convert daily loading rates from IF units per day to g BaP equivalents per day. In both plots error bars are one standard error of the mean. Where error bars are not provided they were too small to be shown.

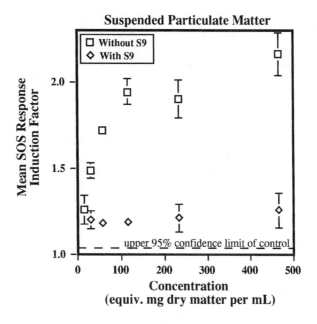

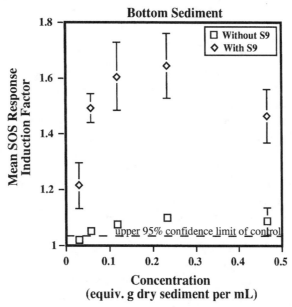

Figure 42.3 SOS Chromotest results for extracts of suspended and sedimented particulate matter from Jonquière, Québec. The site is located on the Saguenay river downstream from two pulp and paper facilities and one large aluminum refinery/inorganic chemical production complex. Suspended particulate material from ≈60L of surface water was collected by glass fiber filtration. Bottom sediments were collected with a 225 cm² Ekman dredge. A) Suspended sediment results. SOSIP without S9 = 17.1 IF units per g dry wt. per assay mL, MDGC without S9 = 7.5 mg dry wt. per mL. SOSIP with S9 = 0.6 IF units per g dry wt. per assay mL, MDGC with S9 = 13.6 mg dry wt. per mL. B) Bottom sediment results. SOSIP without S9 = 0.98 IF units per g dry wt. per assay mL, MDGC without S9 = 15 mg dry wt. per mL. SOSIP with S9 = 21.3 IF units per g dry wt. per assay mL, MDGC with S9 = 2 mg dry wt. per mL. Induction factor values are means of duplicate samples. Error bars are one standard error of the mean. Where error bars are not shown they were smaller than the plotting symbol. Control samples collected from areas that receive no direct industrial input failed to elicit a positive response.

since direct-acting substances are often unstable electrophiles that will readily react with any available nucleophile.[136] In contrast, progenotoxins that predominate in the bottom sediments appear to be less readily degraded in the downstream environment. Many other researchers have also noted that sediments tend to accumulate substances that require S9 activation.[10-12,15,137-140]

Further analyses revealed that the genotoxicity of both suspended and sedimented particulate material is empirically related to the regional, weighted average potency of the previously examined industrial discharges. The results obtained indicate that genotoxins adsorbed to particulate matter in the water column are a diluted version of those adsorbed to particulate material in the effluents. Moreover, progenotoxins in bottom sediments appear to originate in the dissolved, base/neutral, and suspended solid fractions of the examined industrial effluents. A detailed overview of the sediment results for the 21 sites examined can be found in White et al.[16]

2. Biota

Many researchers have employed chemical analyses to investigate the accumulation of geno-toxic, mutagenic, and carcinogenic substances by aquatic biota that inhabit regions contaminated with industrial and municipal wastes.[141-144] However, few have employed bioassays for such inves-tigations. The advantages of bioassays is that they do not require any *a priori* assumptions about the physical–chemical properties of the putative genotoxins, and they are capable of integrating the effects of all mixture components. In addition, as previously noted, microscale assays permit large numbers of samples to be examined in a cost-effective, efficient manner. Despite these advantages, and the fact that aquatic biota are capable of maintaining tissue concentrations of organic pollutants that are as much as 10^5 times higher than water concentrations,[145] few researchers have used microbioassays to detect genotoxins or mutagens in biota tissues. Table 42.2 summarizes the published research efforts that have employed microbioassays to detect mutagens or genotoxins in the tissues of aquatic biota. The majority of the listed studies employed the *Salmonella*/mam-malian microsome assay for mutagenicity detection. Although this assay has been successfully used, most of the researchers listed[54,141,146,148,150] have described problems caused by the presence of histidine in biota extracts. Histidine hampers assay performance and complicates the quantifi-cation of prototrophic reversions attributable to mutagens in the sample.[54,141,147]

For reasons already discussed, the SOS Chromotest is particularly well suited to the examination of biota tissue extracts. The assay was used to examine organic extracts of a variety of fish and macroinvertebrate tissue collected from 18 previously investigated, industrial sites. The results obtained indicate that tissue genotoxin concentrations (expressed as μg benzo(*a*)pyrene equivalents per g wet tissue) are related to lipid content, trophic level, and site contamination. Figure 42.4 illustrates the relationship between tissue genotoxicity and lipid content for both fish and macro-invertebrates. Linear regression analyses revealed a significant effect of both lipid content and biota type. The following model relates log tissue concentrations of BaP equivalents (μg per g) to log lipid content (% wet wt.) and biota type (fish/invert). ($r^2 = 0.28$, n = 116, F Ratio = 21.5, $p < 0.0001$):

$$\frac{\text{Log BaP Equivalents}}{(\mu g \text{ per g wet tissue})} = \begin{bmatrix} -1.15 & \text{Fish} & \mathbf{A} \\ -0.65 & \text{Inverts} & \mathbf{B} \end{bmatrix} + \left[1.12 \pm 0.18 \times \text{Log Lipid} \left(\% \text{ wet wt.} \right) \right] \quad (42.1)$$

All effects are significant at $p < 0.02$. The value associated with the slope term is the standard error of the estimate. Post-hoc contrasts of regression coefficients[130] revealed a significant difference between invertebrate contamination and fish contamination. Coefficients accompanied by the same letter are not significantly different at $\alpha = 0.05$. This result indicates that predicted invertebrate levels will be about half an order of magnitude higher than fish levels. In addition, further analyses of the fish data revealed that for a given lipid content, the more piscivorous species, such as walleye

Table 42.2   Summary of Published Research that Used Short-Term, Microbial Bioassays to Investigate the Presence of Genotoxins or Mutagens in Tissues of Marine and Freshwater Biota

Biota Examined	Extraction Procedure	Bioassay Employed	Result	Source
Mussel (*Mytilus edulis*) tissues	Ethanol extract	Fluctuation assay with several strains of *E.coli*, *S. typhimurium,* and *S. cerevisiae*	Several + responses (some >10-fold control) on samples from industrialized areas	146
Mussel (*Mytilus edulis*) tissues	Ethanol extract	Fluctuation assay with several strains of *S. typhimurium* and *E. coli*	Several + responses on samples from industrialized areas	147
Mussel (*Mytilus edulis*) tissues	Ethanol and nitric/perchloric acid extract	Fluctuation assay with several strains of *E.coli*, *S. typhimurium,* and *S. cerevisiae*	Several + responses (some >10-fold control) on samples from industrialized areas	148
Oysters (*Crassostrea virginica*), Sea squirts (*Mogulla sp.*), Shrimps (*Peneaus duorarum*)	2-propanol extraction followed by XAD[‡] adsorption	Plate incorporation assay with *S. typhimurium* TA98, TA100, TA1535, TA1537, and TA1538	Some marginal responses on Sea squirt (residential bay) and shrimp (market purchased) extracts	54
Sea trout (*Nibea mitsukurii*) liver	Ether/methanol Soxhlet extract	Plate incorporation assay with *S. typhimurium* (TA98, TA100 + TA1537) and *B. subtilus*	Marginal responses on samples collected near pulp mill discharge	149
Mussels (*Mytilus* sp., market purchased)	Ethanol/Soxhlet extract	Plate incorporation assay with *S. typhimurium* (TA 98 and TA100), *Saccharomyces cerevisiae,* and *Schizosaccharomyces pombes*	Several + responses on samples from heavily polluted areas (Venice lagoon)	150
Shrimps (*Peneaus semisulcatus*), Clams (*Circenita callipyga*)	Various methods explored	Fluctuation assay with *S. typhimurium* TA98 and *E. coli* WP$_2$	Strong + response with ethanol extract	151
Mussels (*Mytilus edulis*), Limpets (*Patella vulgata*)	Nitric acid extract	Fluctuation assay with *S. typhimurium* TA98 and TA100	Strong + response following oil spill	152
Mussel tissue (unspecified)	Acetone/blue cotton extract	Plate incorporation assay with *S. typhimurium* TA98	Strong + response (>10-fold control) in industrialized area	153
Various marine mollusks (*Chamelea gallina, Ruditapes decussatus, Crassostrea gigas*)	Ethanol extraction followed by XAD adsorption	Plate incorporation assay with 4 strains of *S. typhimurium* and 5 strains of *E. coli*	Several + responses suggesting polar, oxidative mutagens	154
Zebra mussels (*Dreissina polymorpha*)	Dichloromethane extraction followed by alumina clean-up	Plate incorporation assay with *S. typhimurium* TA100 and YG1029	Strong + response at PAH contaminated sites (+S9 only)	141
Soft-shell clams (*Mya arenaria*) and blue mussels (*Mytilus edulis*)	Dichloromethane extract followed by GPC[§] clean-up.	Microplate version of the SOS Chromotest (*E. coli* PQ37)	Strong + response for PAH contaminated samples	124
Fish (e.g., *Perca* sp., *Esox* sp.) and macroinvertebrates (e.g., insect nymphs, gastropods)	Dichloromethane extract followed by GPC clean-up.	Microplate version of the SOS Chromotest (*E. coli* PQ37)	Positive responses for both fish and invertebrate samples. Evidence for bio-diminution.	125

‡ Nonpolar polystyrene resin (Amberlite®)
§ Gel permeation chromatography
From White, P. A., Blaise, C. and Rasmussen, J. B., *Mutat. Res.*, 1996, in press. With permission.

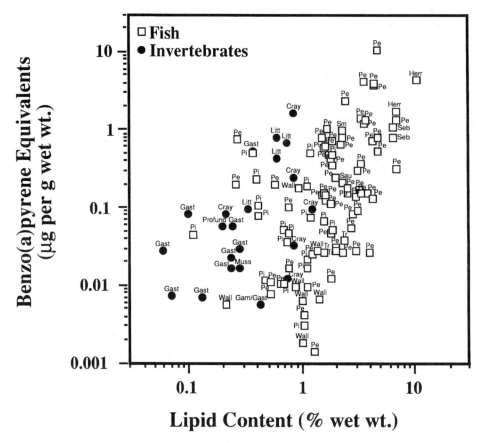

Figure 42.4   The effect of lipid content (% wet wt.) on genotoxin tissue concentration (as µg BaP equivalents per g wet tissue). All tissues were extracted with dichloromethane as described by White et al.[125] Prior to genotoxicity analyses, lipids and other interfering macromolecules were removed via gel permeation chromatography. The mean genotoxic potency of benzo(a)pyrene was used to values from IF per g wet tissue to µg BaP equivalents per g wet tissue. Fish species examined include yellow perch (*Perca flavescens*), northern pike (*Esox lucius*), walleye (*Stizostedion vitreum*), sauger (*Stizostedion canadense*), brook trout (*Salvelinus fontinalis*), Atlantic herring (*Clupea harengus*), rainbow smelt (*Osmerus mordax*), and deepwater redfish (*Sebastes mentella*). Examined macroinvertebrates include a wide variety of gastropods, insect nymphs and larvae, crustaceans, and oligochaetes. Little attempt was made to separate the invertebrate samples into their respective taxonomic groups. Where quantities permitted, samples were separated into gastropods, amphipods (*Gammarus* spp.), decapods (primarily *Orconectes* spp.), bivalve mollusks (Unionidae), and miscellaneous riparian invertebrates. Value labels indicate the type of biota sampled. Pe = Perch, Pi = Pike, Herr = Herring, Seb = Sebastes, Wall = Walleye, Sau = Sauger, Tr = Trout, Sm = Smelt, Cray = crayfish, Gast = gastropods, Gam/Gast = Gammarus/gastropods, Litt = mixed riparian invertebrates, Profundal = mixed benthic sled sample, Muss = mussels. Matrix blanks (fish and invertebrate tissues) were collected from several sites that do not receive any direct industrial input. With the exception of one fish tissue sample, matrix blanks failed to elicit a positive response.

and pike, generally have lower levels of genotoxins in their tissues than omnivorous species such as perch. Thus, in contrast to the biomagnification of persistent organochlorines,[155] the results indicate that genotoxins are biodiminished with increasing trophic level.

While additional chemical analyses did not support PAHs as the putative genotoxins, the results do indicate that the accumulated genotoxins have properties that are similar to those of PAHs. Observed tissue-to-sediment ratios for genotoxicity are between 0.07 and 1.25, and generally less than 1.0. Similar values (<0.1) can be calculated from the zebra mussel and sediment *Salmonella* mutagenicity data published by Marvin et al.[10,141] These values are consistent with PAH-like degradability as opposed to the persistence of organochlorines. For example, fish-to-sediment ratios for persistent, chlorinated hydrocarbons such as DDT and PCBs are in the 10^2 to 10^3 range.[156] In contrast, PAHs generally exhibit biota-to-sediment ratios of 0.01 to 0.1.[157-159] For a complete discussion, the reader is referred to White et al.[125]

IV. SUMMARY AND CONCLUSIONS

Ecotoxicology is the study of the fate and effects of contaminants on ecosystems.[160] As such, it approaches the problem of aquatic habitat contamination from a very broad perspective. The tendency of genetic toxicology to focus on sample potency, and the chemistry of particularly potent samples, is a noteworthy shortcoming. For example, very few investigations of industrial discharges attempt to calculate daily loading rates, the value required to assess potential downstream hazard. In addition, despite numerous examinations of water, sediment, and biota, researchers rarely comment on ecotoxicological phenomena such as bioaccumulation, biomagnification, and persistence. The lack of broad, ecotoxicological investigations in genetic toxicology likely stems from the difficulty in processing the large number of samples required to investigate ecotoxicological phenomena in a cost-effective manner. In addition, methodological difficulties often hamper the use of certain bioassays and complicate examinations of some environmental matrices (e.g., biota tissues).

The purpose of this chapter was to demonstrate that the SOS Chromotest, particularly the semiautomated, microplate version described by White et al.,[49] can efficiently examine the samples required to permit general statements about the sources and ecological behavior of genotoxic, organic contaminants in aquatic systems. The results shown confirm that the assay can be used to examine a wide range of samples including industrial and municipal wastewaters, suspended particulate matter, bottom sediments, and biota tissues. Moreover, the results can be integrated in an effort to provide information about post-emission fate and ecological behavior. For example, Figure 42.5 presents a synthesis of the industrial effluent results of White et al.[87,90] and the subsequent analyses of downstream suspended and sedimented particulate matter.[16] The figure illustrates likely post-emission pathways of industrial genotoxins sorbed to particulate material. Further investigations can investigate how mechanical, chemical, and biological processes that control particle movement and chemical–particle interactions effect the ultimate fate and effects of sorbed genotoxins.

Figure 42.6 presents a synthesis of the industrial effluent results of White et al.[87,90] and the subsequent analyses of downstream biota samples.[125] Although statistical analyses did not reveal any significant trend, the results suggest that sites which experience higher regional genotoxic loads from industrial and municipal wastewaters yield biota samples that are less contaminated. Similar results were obtained by Rodriguez-Ariza et al.[154] in their examinations of bivalve mollusks from the coast of Spain. Rodriguez-Ariza et al. attribute lower tissue mutagenicity levels at more polluted sites to increased levels of enzymes (e.g., glutathione-S-transferase) required for detoxification and excretion of contaminants. Further research will be required to determine if the observed fish and invertebrate tissue burdens are related to the organism's competence for metabolism and excretion of accumulated contaminants.

In conclusion, this chapter has demonstrated that the semiautomated microplate of the SOS Chromotest can be used to examine a wide variety of samples, the results of which can be integrated to provide information about the sources and fate of genotoxins in the environment. The efficacy, simplicity, and low cost of the assay make it well suited for environmental monitoring of aquatic systems. In addition, although no attempt was made to isolate and identify putative genotoxins present in the complex extracts examined, the SOS Chromotest should be well suited to subsequent bioassay-directed chemical analyses.

Fate of Industrial Genotoxins in the Downstream Environment

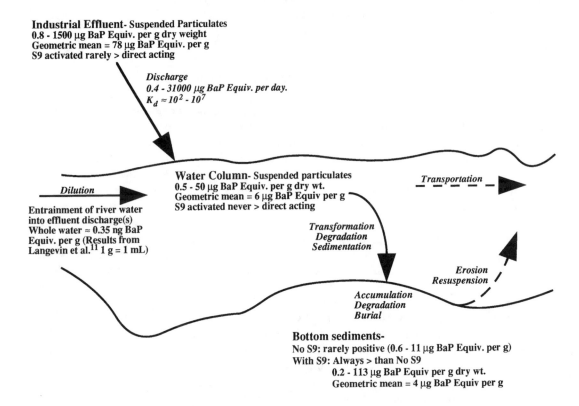

Figure 42.5   Summary of SOS Chromotest results for industrial effluents and riverine particulate matter. Particle-bound genotoxins released into the receiving system are initially diluted with entrained *fresh* water. Diluted suspended particulate matter is approximately an order of magnitude less potent than that found in industrial effluents. Following dilution, sorbed genotoxins can be removed from the water column by degradation, transportation, and sedimentation. The results obtained suggest that once sedimented, direct acting substances, which predominate in the water column, are transformed into progenotoxic substances that require S9 activation. Progenotoxic substances in bottom sediments can accumulate or be lost through burial and further degradation. Comparisons of SPM and BSED total potency values indicate that annual losses are several percent. Annual or seasonal fluctuations in current may cause erosion and resuspension of particle-bound genotoxins. Dashed arrows indicate pathways that were not investigated.

V. FUTURE PROSPECTS

It will certainly prove interesting to investigate the effects of accumulating SOS genotoxins. Although several researchers have established quantitative relationships between SOS response induction in *E. coli* and mutation mutagenesis in *Salmonella*,[161-162] it is not immediately apparent whether such SOS genotoxins will have mutagenic and/or genotoxic effects on higher organisms such as macroinvertebrates and fish. In addition, the consequences of these supposed mutagenic effects are not obvious. Induction of somatic cell mutations can result in neoplastic lesions in exposed organisms.[3] However, in an ecological context, genotoxic effects on somatic cells may be of little consequence, particularly if they become clinically visible long after sexual maturity. Typical latency periods for the formation of malignant tumors results in losses from a population that are most dramatic for the older age classes (e.g., see *Ictalurus nebulosus* tumor frequency data of Maccubbin

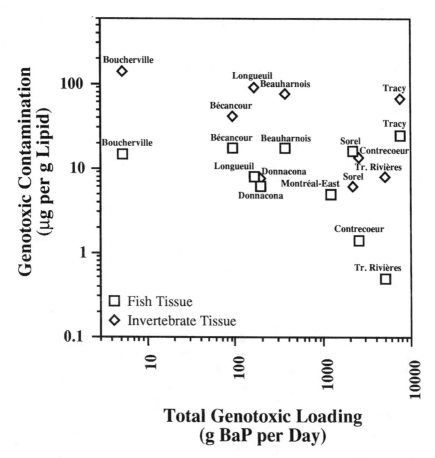

Figure 42.6   The effect of total, regional genotoxic loading from industrial and municipal sources on the observed genotoxic contamination of aquatic biota in the St. Lawrence river. Biota genotoxicity data are from White et al.[125] (see Figure 42.4). Loading values are from White et al.[87] (see Figure 42.2). Where several observations were available, biota contamination values are arithmetic means. Number of observations ranged from 1 to 9. Value labels refer to industrialized regions between Cornwall, Ontario, and Québec City, Québec.

and Ersing[163]). Genotoxic effects on germ cells are likely to be more consequential, with alterations of the germ cell mutation rate effecting the dynamics of entire populations. Recent publications have begun to discuss the population level effects of mutagenic pollutants. The computer simulations of Lynch et al.[164] indicate that a doubling of the deleterious mutation rate will be expected to reduce the longevity of a population by about 50%, with small populations being particularly vulnerable to extinction via "mutational meltdown." Moreover, deleterious mutations introduced into a population will remain in the gene pool long after the pollutant source is extinguished.

ACKNOWLEDGMENTS

The authors would like to thank the research and administrative personnel of the St. Lawrence Center (Environment Canada) in Montreal. Without their support and assistance the research projects discussed in this chapter would not have been possible. The projects discussed were funded by the St. Lawrence Center, the Natural Sciences and Engineering Research Council of Canada, La Fondation Canadienne d'Aide à la Recherche and the World Wildlife Fund (Canada). PAW would like to thank Prof. J. B. Rasmussen for supervision and creative inspiration.

REFERENCES

1. Muller, H. J., Artificial transmutation of the gene, *Science*, 66, 84, 1927.
2. Auerbach, C. and Robson, J. M., Chemical production of mutations, *Nature*, 157, 302, 1946.
3. Kurelec, B., The genotoxic disease syndrome, *Mar. Environ. Res.*, 35, 341, 1993.
4. Dellarco, V. L., Erickson, R. P., Lewis, S. E., and Shelby, M. D., Mutagenesis and human genetic disease: an introduction, *Environ. Molec. Mutagen.*, 25(suppl. 26), 2, 1995.
5. Metcalfe, C. D., Balch, G. C., Cairns, V. W., Fitzsimons, J. D., and Dunn, B. P., Carcinogenic and genotoxic activity of extracts from contaminated sediments in Western Lake Ontario, *Sci. Tot. Environ.*, 94, 125, 1990.
6. Aust, A. E., Mutations and cancer, in *Genetic Toxicology*, Li, A.P. and Heflich, R.H., Eds., CRC Press, Boston, 1991, 94.
7. Houk, V. S., The genotoxicity of industrial wastes and effluents — a review, *Mutat. Res.*, 277, 91, 1992.
8. Stahl, R. G., The genetic toxicology of organic compounds in natural waters and wastewaters, *Ecotox. Environ. Safety*, 22, 94, 1991.
9. De Flora, S., Bagnasco, M., and Zanacchi, P., Genotoxic, carcinogenic and teratogenic hazards in the marine environment, with special reference to the Mediterranean Sea, *Mutat. Res.*, 258, 285, 1991.
10. Marvin, C. H., Allan, L., McCarry, B. E., and Bryant, D. W., Chemico/biological investigation of contaminated sediment from the Hamilton Harbour region of western Lake Ontario, *Environ. Mol. Mutagen.*, 22, 61, 1993.
11. Langevin, R., Rasmussen, J. B., Sloterdijk, H. H., and Blaise, C., Genotoxicity in water and sediment extracts from the St. Lawrence river, using the SOS Chromotest, *Wat. Res.*, 26, 419, 1992.
12. Ho, K. T. and Quinn, J. G., Bioassay-directed fractionation of organic contaminants in an estuarine sediment using the new mutagenic bioassay, Mutatox™, *Environ. Toxicol. Chem.*, 12, 823, 1993.
13. West, W. R., Smith, P. A., Booth, G. M., and Lee, M. L., Isolation and detection of genotoxic components in a black river sediment, *Environ. Sci. Technol.*, 22, 224, 1988.
14. Grifoll, M., Solanas, A. M., and Bayona, J. M., Characterization of genotoxic components in sediments by mass spectrometric techniques combined with *Salmonella*/microsome test, *Arch. Environ. Contam. Toxicol.*, 19, 75, 1990.
15. Balch, G. C., Metcalfe, C. D. and Huestis, S. Y., Identification of potential fish carcinogens in sediment from Hamilton Harbour, Ontario, Canada, *Environ. Toxicol. Chem.*, 14, 79, 1995.
16. White, P. A., Rasmussen, J. B., and Blaise, C., Genotoxic substances in the St. Lawrence system I: Industrial genotoxins sorbed to particulate matter in the St. Lawrence, St. Maurice and Saguenay rivers, *Environ. Toxicol. Chem.*, in press.
17. Klekowski, E. and Levins, D. E., Mutagens in a river heavily polluted with paper recycling wastes: results of field and laboratory mutagen assays, *Environ. Mutagen.*, 1, 209, 1979.
18. Klekowski, E. J. and Berger, B. B., Chromosome mutations in a fern population growing in a polluted environment: a bioassay for mutagens in aquatic environments, *Amer. J. Bot.* 63, 239, 1976.
19. Ravindran, P. N. and Ravindran, S., Cytological irregularities induced by water polluted with factory effluents: a preliminary report, *Cytologia*, 43, 565, 1978.
20. Dixon, D. R., Aneuploidy in mussel embryos (*Mytilus edulis* L.) originating from a polluted dock, *Mar. Biol. Lett.*, 3, 155, 1982.
21. Nacci, D., Nelson, S., Nelson, W., and Jackim, E., Application of the DNA alkaline unwinding assay to detect strand breaks in marine bivalves, *Mar. Environ. Res.*, 33, 83, 1992.
22. DiGiulio, R. T., Habig, C., and Gallagher, E. P., Effects of Black Rock Harbor sediments on indices of biotransformation, oxidative stress, and DNA integrity in channel catfish, *Aquat. Toxicol.*, 26, 1, 1993.
23. Prein, A. E., Thie, G. M., Alink, G. M., Koeman, J. H., and Poels, C. L. M., Cytogenetic changes in fish exposed to water of the River Rhine, *Sci. Total Environ.*, 9, 287, 1978.
24. Alink, G. M., Frederix-Wolters, E. M. H., van der Gaag, M. A., van der Kerkhoff, J. F. J., and Poels, C. L. M., Induction of sister-chromatid exchanges in fish exposed to Rhine water, *Mutat. Res.*, 78, 369, 1980.
25. Black, J. J. and Baumann, P. C., Carcinogens and cancers in freshwater fishes, *Environ. Health Perspect.*, 90, 27, 1991.

26. Vogelbein, W. K., Fournie, J. W., Van Veld, P. A., and Huggett, R.J., Hepatic neoplasms in the mummichog *Fundulus heteroclitus* from a creosote contaminated site, *Cancer Res.*, 50, 5978, 1990.
27. Mix, M. C., Cancerous diseases in aquatic animals and their association with environmental pollutants: a critical literature review, *Mar. Environ. Res.*, 20, 1, 1986.
28. Mix, M. C., Shellfish diseases in relation to toxic chemicals, *Aquat. Toxicol.*, 11, 29, 1988.
29. Krahn, M. M., Rhodes, L. D., Myers, M. S., Moore, L. K., MacLeod, W. D., and Malins, D.C., Associations between metabolites of aromatic compounds in bile and the occurrence of hepatic lesions in English Sole (*Parophrys vetulus*) from Puget Sound, Washington, *Arch. Environ. Contam. Toxicol.*, 15, 61, 1986.
30. Malins, D. C., McCain, B. B., Brown, D. W., Chan, S., Myers, M. S., Landahl, J. T., Prohaska, P. G., Friedman, A. J., Rhodes, L. D., Burrows, D. G., Gronlund, W. D., and Hodgins, H. O., Chemical pollutants in sediments and diseases of bottom-dwelling fish in Puget Sound, Washington, *Environ. Sci. Technol.*, 18, 705, 1984.
31. Li, A. P. and Loretz, L. J., Assays for genetic toxicity, in *Genetic Toxicology*, Li, A. and Heflich, R., Eds., CRC Press, Boston, 1991, 119.
32. Ames, B., Durston, W., Yamasaki, E., and Lee, F., An improved bacterial system for the detection and classification of mutagens and carcinogens, *Proc. Natl. Acad. Sci. USA*, 70, 2281, 1973.
33. Ames, B. N., McCann, J., and Yamasaki, E., Methods for detecting carcinogens and mutagens with the *Salmonella*/mammalian-microsome mutagenicity test, *Mutat. Res.*, 31, 347, 1975.
34. Maron, D. and Ames, B. N., Revised methods for the *Salmonella* mutagenicity test, *Mutat. Res.*, 113, 173, 1983.
35. Ames, B., Durston, W., Yamasaki, E., and Lee, F., Carcinogens are mutagens: a simple test system combining liver homogenates for activation and bacteria for detection, *Proc. Natl. Acad. Sci. USA*, 70, 2281, 1973.
36. McCann, J. and Ames, B., Detection of carcinogens as mutagens in the *Salmonella*/microsome tests: assay of 300 chemicals: discussion, *Proc. Natl. Acad. Sci. USA*, 73, 950, 1976.
37. Rossman, T. G., Molina, M., and Meyer, L. W., The genetic toxicology of metal compounds: I. induction of λ prophage in *E. coli* WP2$_s$(λ), *Environ. Mutagen.*, 6, 59, 1984.
38. Oda, Y., Nakamura, S., Oki, I., Kato, T., and Shinagawa, H., Evaluation of the new system (*umu*-test) for the detection of environmental mutagens and carcinogens, *Mutat. Res.*, 147, 219, 1985.
39. Quillardet, P., Huisman, O., D'Ari, R., and Hofnung, M., SOS Chromotest, a direct assay of induction of an SOS function in *Escherichia coli* K-12 to measure genotoxicity, *Proc. Natl. Acad. Sci. USA*, 79, 5971, 1982.
40. Walker, G. C., Mutagenesis and inducible responses to deoxyribonucleic acid damage in *E. coli*. *Microbiol. Rev.*, 48, 60, 1984.
41. Kilbey, B. J., Legator, M., Nichols, W., and Ramel, C., *Handbook of Mutagenicity Test Procedures*, Elsevier Science Publishing Co., New York, 1984.
42. Miller, J. A. and Miller, E. C., Mechanisms of chemical carcinogenesis: Nature of proximate carcinogens and interactions with macromolecules, *Pharmacol. Rev.*, 18, 805, 1966.
43. Malling, H. V., Dimethylnitrosamine: formation of mutagenic compounds by interaction with mouse liver microsomes, *Mutat. Res.*, 13, 425, 1971.
44. Blaise, C., Microbiotests in aquatic ecotoxicology: characteristics, utility, and prospects, *Environ. Toxicol. Water Qual.*, 6, 145, 1991.
45. Gatehouse, D., Detection of mutagenic derivatives of cyclophosphamide and a variety of other mutagens in a 'microtiter®' fluctuation test, without microsomal activation, *Mutat. Res.*, 53, 289, 1978.
46. Hubbard, S. A., Green, M. H. L, Gatehouse, D., and Bridges, J. W., The fluctuation test in bacteria, in *Handbook of Mutagenicity Test Procedures*, Kilbey, B. J., Legator, M., Nichols, W., and Ramel, C., Eds., Elsevier Science Publishing Co., New York, 1984, 142.
47. Kado, N. Y., Guirguis, G. N., Flessel, C. P., Chan, R. C., Chang, K., and Wesolowski, J. J., Mutagenicity of fine (0.5μm) airborne particles: diurnal variation in community air determined by a *Salmonella* micro pre-incubation (microsuspension) procedure, *Environ. Mutagen.*, 8, 53, 1986.
48. Fish, F., Lampert, I., Halachmi, A., Riesenfeld, G., and Herzberg, M., The SOS Chromotest kit, a rapid method for the detection of genotoxicity, *Toxic. Assess.*, 2:135, 1987.
49. White, P. A., Rasmussen, J. B., and Blaise, C., A semiautomated, microplate version of the SOS Chromotest for the analysis of complex environmental extracts, *Mutat. Res.*, 360, 51, 1996.

50. Côté, C., Blaise, C., Delisle, C. E., Meighan, E. A., and Hansen, P. D., A miniaturized Ames mutagenicity assay employing bioluminescent strain of *Salmonella typhimurium*, *Mutat. Res.*, 345, 137, 1995.

51. Quillardet, P. and Hofnung, M., The SOS Chromotest, a colorimetric bacterial assay for genotoxins: procedures, *Mutat. Res.*, 147, 65, 1985.

52. Elespuru, R., Responses to DNA damage in bacteria and mammalian cells, *Environ. Mol. Mutagen.*, 10, 97, 1987.

53. Houk, V. and deMarini, D., Use of the microscreen phage-induction assay to assess the genotoxicity of 14 hazardous industrial wastes, *Environ. Mol. Mutagen.*, 11, 13, 1988.

54. Sparks, T., Baylis, J., and Chang, C., Comparison of mutagen accumulation in three estuarine species using the *Salmonella*/Microsome activation system, *Mutat. Res.*, 85, 133, 1981.

55. Parry, J., The detection of mutagens in the tissues of marine organisms exposed to environmental pollutants, in *Mutagenicity Testing in Environmental Pollution Control*, Zimmermann, F. and Taylor-Mayer, R., Eds., Ellis Horwood Ltd., Chichester, 1985, 105.

56. Kohn, A., Donchin, M., Jacobs, J., Horn, Y., Gibor, Y., Fish, F., Riesenfeld, G., Kirenberg, F., Lampert, I., Halachmi, A., and Herzberg, M., Detection of genotoxic substances in cancer patients receiving antineoplastic drugs, *Ann. NY Acad. Sci.*, 534, 776, 1988.

57. DeMeo, M., Miribel, V., Botta, A., Laget, M., and Dumenil, G., Applicability of the SOS Chromotest to detect urinary mutagenicity caused by smoking, *Mutagenesis*, 3, 277, 1988.

58. Bosworth, D. and Venitt, S., Testing human faecal extracts for genotoxic activity with the SOS Chromotest: the importance of controlling for faecal enzyme activity, *Mutagenesis*, 1, 143, 1986.

59. Nylund, L., Hakala, E., and Sorsa, M., Application of a semiautomated SOS Chromotest for measuring genotoxicities of complex environmental mixtures containing polycyclic aromatic hydrocarbons, *Mutat. Res.*, 276, 125, 1992.

60. Janz, S., Wolff, G., and Storch, H., SOS Chromotest, a qualitative short-term bacterial assay for the detection of genotoxic compounds in an automated version adapted to the Bioscreen Analyzer System®, *Zentralbl. Mikrobiol.*, 143, 645, 1988.

61. Lewtas, J., Environmental monitoring using genetic bioassays, in *Genetic Toxicology*, Li, A. and Heflich, R., Eds., CRC Press, Boston, 1991, 359.

62. Schuetzle, D. and Lewtas, J., Bioassay-directed chemical analysis in environmental research, *Anal. Chem.*, 58, 1060A, 1986.

63. Alfheim, I., Bjørseth, A., and Møller, M., Characterizations of microbial mutagens in complex samples-methodology and application, *CRC Crit. Rev. Environ. Control*, 14, 91, 1984.

64. Sicherer-Roetman, A., Ramal, M., Voogd, C. E., and Bloeman, H. J., The fractionation of extracts of ambient particulate matter for mutagenicity testing, *Atmos. Environ.*, 22, 2803, 1988.

65. Nishioka, M. G., Howard, C. C., Contos, D. A., Ball, L. M., and Lewtas, J., Detection of hydroxylated nitro aromatic and hydroxylated nitro polycyclic aromatic compounds in an ambient air particulate extract using bioassay-directed fractionation, *Environ. Sci. Technol.*, 22, 908, 1988.

66. Saleem, I., Pero, A., Zator, R., Schuetzle, D., and Riley, R., Ames assay chromatograms and the identification of mutagens in diesel particle extracts, *Environ. Sci. Technol.*, 18, 375, 1984.

67. Pedersen, T. C., Biologically active nitro-PAH compounds in extracts of diesel exhaust particulates, in *Mobile Source Emissions Including Polycyclic Organic Species*, Rondia, D., Ed., D. Reidel Publishing Co., Dordrecht, 1983, 227.

68. Horth, H., Crathorne, B., Gwilliam, R. D., Palmer, C. P., Stanley, J. A., and Thomas, M. J., Techniques for the fractionation and identification of mutagens produced by water chlorination, in *Organic Pollutants in Water. Sampling, Analysis and Toxicity Testing*, Suffet, I. H. and Malaiyandi, M., Eds., Advances in Chemistry Series #214. American Chemical Society, Washington, DC, 1987, 659.

69. Kronberg, L., Holmbom, B., Reunanen, M., and Tikkanen, L., Identification and quantification of the Ames mutagenic compound 3-chloro-4-(dichloromethyl)-5-hydroxy-2(5*H*)-furanone and of its geometric isomer (*E*)-2-chloro-3-(dichloromethyl)-4-oxobutenoic acid in chlorine-treated humic water and drinking water extracts, *Environ. Sci. Technol.*, 22, 1097, 1988.

70. Nestmann, E. R., Lee, E. G-H., Mueller, J. C., and Douglas, G. R., Mutagenicity of resin acids identified in pulp and paper mill effluents using the *Salmonella*/Mammalian-Microsome assay, *Environ. Mutagen.*, 1, 361, 1979.

71. Holmbom, B., Voss, R. H., Mortimer, R. D., and Wong, A., Fractionation, isolation and characterization of Ames mutagenic compounds in Kraft chlorination effluents, *Environ. Sci. Technol.*, 18, 333, 1984.

72. Ames, B. N., Kammen, H. O., and Yamasaki, E., Hair dyes are mutagenic: Identification of a variety of mutagenic ingredients, *Proc. Natl. Acad. Sci. USA*, 72, 2423, 1975.
73. Rosenkranz, H. S., McCoy, E. C., Sanders, D. R., Butler, M., Kiriazides, D. K., and Mermelstein, R., Nitropyrenes: Isolation, identification and reduction of mutagenic impurities in carbon black and toners, *Science*, 209, 1039, 1980.
74. Shirasu, Y., Moriya, M., Kato, K., Furuhashi, A., and Kada, T., Mutagenicity screening of pesticides in the microbial system, *Mutat. Res.*, 40, 19, 1976.
75. Simmon, V.F., Mitchell, A. D., and Jorgenson, T. A., *Evaluation of Selected Pesticides as Chemical Mutagens. In vitro and in vivo Studies*, EPA Technical Report EPA-600/1-77-028. Health Effects Research Laboratory, U.S. Environmental Protection Agency, Research Triangle Park, NC, 1977.
76. Joner, P., Dahle, H., Aune, T., and Dybing, E., Mutagenicity of nitrovin — a nitrofuran feed additive, *Mutat. Res.*, 48, 313, 1977.
77. Sugimura, T. and Sato, S., Mutagens-carcinogens in foods, *Cancer Res.*, 43, 2415s, 1983.
78. Dayan, J., Crajer, M., and Deguingand, S., Mutagenic activity of 4 active-principle forms of pharmaceutical drugs, *Mutat. Res.*, 102, 1, 1982.
79. Rosenkranz, H. S. and Speck, W. T., Mutagenicity of metronidazole: Activation by mammalian liver microsomes, *Biochem. Biophys. Res. Comm.*, 66, 520, 1975.
80. Chrisp, C. E. and Fischer, G. L., Mutagenicity of airborne particles, *Mutat. Res.*, 76, 143, 1980.
81. Kier, L. D., Yamasaki, E., and Ames, B. N., Detection of mutagenic activity in cigarette smoke condensates, *Proc. Natl. Acad. Sci. USA*, 71, 4159, 1974.
82. Wang, Y. Y., Rappaport, S. M., Sawyer, R. F., Talcott, R. E., and Wei, E. T., Direct-acting mutagens in automobile exhaust, *Cancer Lett.*, 5, 39, 1978.
83. Dutka, B. J., Jones, K., Xu, H., Kwan, K. K., and McInnis, R., Priority site selection for degraded areas in the aquatic environment, *Water Poll. Res. J. Canada*, 22, 326, 1987.
84. Wong, P. T. S., Chau, Y. K., Ali, N., and Whittle, D. M., Biochemical and genotoxic effects in the vicinity of a pulp mill discharge, *Environ. Toxicol. Water Qual.*, 9, 59, 1994.
85. Legault, R., Blaise, C., Trottier, S., and White, P., Detecting the genotoxic activity of industrial effluents with the SOS Chromotest microtitration procedure, *Environ. Toxicol. Water Qual.*, 11, 151, 1996.
86. Venier, P., Montini, R., Zordan, M., Clonfero, E., Paleologo, M., and Levis, A., Induction of SOS response in *Escherichia coli* strain PQ37 by 16 chemical compounds and human urine extracts, *Mutagenesis*, 4, 51, 1989.
87. White P. A., Rasmussen, J. B., and Blaise, C., Comparing the presence, potency and potential hazard of organic compounds extracted from a broad range of industrial effluents, *Environ. Molec. Mutagen.*, 27, 116, 1996.
88. Pfeil, R. M., Venkat, J. A., Plimmer, J. R., Sham, S., Davis, K., and Nair, P. P., Qualitative assessment of groundwater quality using a biological indicator: some preliminary observations, *Arch. Environ. Contam. Toxicol.*, 26, 201, 1994.
89. White, P. A., Rasmussen, J. B., and Blaise, C., Genotoxicity of snow in the Montréal metropolitan area, *Water, Air, Soil Pollut.*, 83, 315, 1995.
90. White, P. A., Rasmussen, J. B., and Blaise, C., Sorption of organic genotoxins to particulate matter in industrial effluents, *Environ. Molec. Mutagen.*, 27, 140, 1996.
91. Courtois, Y., Min, S., Lachenal, C., Jacquot-Deschamps, J., Callais, F., and Festy, B., Genotoxicity of organic extracts from atmospheric particles, *Ann. NY Acad. Sci.*, 534, 724, 1988.
92. Poirier, S., Bouvier, G., Malaveille, C., Ohshima, H., Shao, Y., Hubert, A., Zeng, Y., deThé, G., and Bartsch, H., Volatile nitrosamine levels and genotoxicity of food samples from high-risk areas for nasopharyngeal carcinoma before and after nitrosation, *Int. J. Cancer*, 44, 1088, 1989.
93. McDaniels, A., Reyes, A. L., Wymer, L. J., Rankin, C. C., and Stelma Jr., G. N., Genotoxic activity detected in soils from a hazardous waste site by the Ames test and an SOS colorimetric test, *Environ. Molec. Mutagen.*, 22, 115, 1993.
94. Raabe, F., Janz, S., Wolff, G., Merten, H., Landrock, A., Birkenfeld, T., and Herzschuh, R., Genotoxicity assessment of waste products of aluminum plasma etching with the SOS Chromotest, *Mutat. Res.*, 300, 99, 1993.
95. Srám, R., Dobiás, L., Rössner, P., Veselá, D., Vesely, D., Rakusová, R., and Rericha, V., Monitoring genotoxic exposure in uranium mines, *Environ. Health Perspect.*, 101, 155, 1993.

96. USEPA (United States Environmental Protection Agency), *Guidelines for Preparing Environmental and Waste Samples for Mutagenicity (Ames) Testing: Interim Procedures and Panel Meeting Proceedings*, EPA/600/4-85/058, Environmental Monitoring Systems Laboratory, Office of Research and Development, Las Vegas, NV, 1985.

97. Hunchak, K. and Suffet, I. H., Analysis of acetone-hexane artifacts produced in the Soxhlet extraction of solid environmental samples, *J. Chrom.*, 392, 185, 1987.

98. Tabor, M. W. and Loper, J. C., New methods for the isolation of mutagenic components of organic residuals in sludges, *Adv. Chem.*, 214, 675, 1987.

99. Montreuil, C. N., Ball, J. C., Gorse, R. A., and Young, W., Solvent extraction efficiencies of mutagenic components from diesel particles, *Mutat. Res.*, 282, 89, 1992.

100. Huisman, O. and D'Ari, R., An inducible DNA replication-cell division coupling mechanism in *E. coli, Nature*, 290, 797, 1981.

101. Schleibinger, H., Leberl, C., and Rueden, H., Nitrated polycyclic aromatic hydrocarbons (nitro-PAH) in suspended particulate matter. II: comparison of the mutagenic activity of nitro-PAH in the Ames-, the SOS-repair and the SCE-assay, *Int. J. Hyg. Environ. Med.*, 188, 421, 1989.

102. Qian, L. M., Wang, W. X., and Wang, W. T., Genotoxicities of organic extracts of airborne particulates in SOS Chromotest and Ames test, *Environ. Molec. Mutagen.*, 23(Suppl 23), 55, 1994.

103. Harwood, M., Blaise, C., and Couture, P., Algal interactions with the genotoxic activity of selected chemicals and complex liquid samples, *Aquat. Toxicol.*, 14, 263, 1989.

104. Raabe, F., Janz, S., Wolff, G., Birkenfeld, T., and Storch, H., Einfluß der lösungsvermittlung durch liposomen und n-tenside auf die nachweisbarkeit der gentoxischen aktivät von komplexen wasser-löslichen substanzgemischen im SOS-Chromotest, *Zeit. Gesell. Umwelt-Mutation (GUM)*, 1, 14, 1989.

105. Janz, S., Raabe, F., and Wolff, G., Application of an alternative approach for using the SOS Chromotest to screen complex indeterminate mixtures, *Zentralbl. Mikrobiol.*, 145, 177, 1990.

106. Braun, R., Hüttner, E., Merten, H., and Raabe, F., Genotoxicity studies in semiconductor industry. 1. *In vitro* mutagenicity and genotoxicity studies of waste samples resulting from plasma etching, *J. Toxicol. Environ. Health*, 39, 309, 1993.

107. Hoflack, J. C., Férard, J. F., Vasseur, P., and Blaise, C., An attempt to improve the SOS Chromotest responses, *J. Appl. Toxicol.*, 13, 315, 1993.

108. Rao, S. S., Burnison, B. K., Rokosh, D., and Taylor, C. M., Mutagenicity and toxicity of pulp mill effluent, *Chemosphere*, 28, 1859, 1994.

109. Nylund, L., Rosenburg, C., Jäppinen, P., and Vainio, H., Genotoxicity of kraft spent liquors from different types of chlorination procedures, *Mutat. Res.*, 320, 165, 1994.

110. Sherry, J., Scott, B., Nagy, E., and Dutka, B., Investigations of the sublethal effects of some petroleum refinery effluents, *J. Aquat. Ecosystem Health*, 3, 129, 1994.

111. Xu, H., Dutka, B. J., and Kwan, K. K., Genotoxicity studies on sediments using a modified SOS Chromotest, *Toxic. Assess.*, 2, 79, 1987.

112. Lan, Q., Dickman, M., and Alvarez, M., Evidence of genotoxic substances in the Niagara river watershed, *Environ. Toxicol. Water Qual.*, 6, 1, 1991.

113. Kwan, K. K. and Dutka, B. J., A novel bioassay approach: direct application of the Toxi-Chromotest and the SOS Chromotest to sediments, *Environ. Toxicol. Water Qual.*, 7, 49, 1992.

114. Naikai, Z. and Ahlf, W., Preliminary genotoxicity assessment of sediment extract from Le An river, *China Environ. Sci.*, 5, 169, 1994.

115. Riesenfeld, G., Kirsch, I., and Weissman, S. H., Detection and quantification of aflatoxin B_1 in orange juice, *Food Add. Contam.*, 2, 253, 1985.

116. Fremy, J. M. and Quillardet, P., The 'carry-over' of aflatoxin into milk of cows fed ammoniated rations: use of an HPLC method and a genotoxicity test for determining milk safety, *Food Add. Contam.*, 2, 201, 1985.

117. Carver, J. H. and Machado, M. L., Modifying the SOS Chromotest for complex mixtures: solvent effects, *Environ. Molec. Mutagen.*, 8(Suppl. 6), 16, 1986.

118. Venitt, S. and Bosworth, D., Further studies on the detection of mutagenic and genotoxic activity in human faeces: aerobic and anaerobic fluctuation tests with *S. typhimurium* and *E. coli* and the SOS Chromotest, *Mutagenesis*, 1, 49, 1986.

119. Rueff, J., Laires, A., Borba, H., Chaveca, T., Gomes, M., and Halpern, M., Genetic toxicology of flavonoids, the role of metabolic conditions in the induction of reverse mutation, SOS functions and sister-chromatid exchanges, *Mutagenesis*, 1, 179, 1986.

120. Jenek, J., Quillardet, P., Chomiak, D., Borys, A., and Hofnung, M., Absence of genotoxic activity of refined smoke flavor (RSF) in two bacterial short-term tests, *Mutat. Res.*, 204, 235, 1988.

121. Nair, P., Shami, S., Sainz, E., Menon, M., Jerabek, L. B., Jones, D. Y., Judd, J. T., Campbell, W. S., Schiffman, M. H., Taylor, P. R., Schatzkin, A., Guidry, C., and Brown, C. C, Influence of dietary fat on fecal mutagenicity in premenopausal women, *Int. J. Cancer*, 46, 374, 1990.

122. Vargas, V. M. F., Motta, V. E. P., Leitao, A. C., and Henriques, J. A. P., Mutagenic and genotoxic effects of aqueous extracts of *Achyrocline satureoides* in prokaryotic organisms, *Mutat. Res.*, 240, 13, 1990.

123. Laires, A., Gaspar, J., Borba, H., Proença, M., Monteiro, M., and Rueff, J., Genotoxicity of nitrosated red wine and of the nitrosable phenolic compounds present in wine: tyramine, quercetin and malvidine-3-glucoside, *Fd. Chem. Toxicol.*, 31, 989, 1993.

124. White, P. A., Blaise, C., and Rasmussen, J. B., Detection of genotoxic substances in bivalve mollusks from the Saguenay Fjord (Canada), using the SOS Chromotest, *Mutat. Res.*, in press.

125. White, P. A., Rasmussen, J. B., and Blaise, C., Genotoxic substances in the St. Lawrence system II: Extracts of fish and macroinvertebrates from the St. Lawrence and Saguenay rivers (Canada), *Environ. Toxicol. Chem.*, in press.

126. Legault, Richard, Ecotoxicology and Environmental Chemistry, The St. Lawrence Center, Environment Canada, 105 McGill St., Montréal, Québec. H2Y 2E7. Personal communication.

127. Olivier, P. and Marzin, D., Study of the genotoxic potential of 48 inorganic derivatives with the SOS Chromotest, *Mutat. Res.*, 189, 263, 1987.

128. White, P. A., *Detection, Discharge and Ecological Behavior of Genotoxic Organic Contaminants in the St. Lawrence and Saguenay Rivers*, Ph.D. Thesis, McGill University, Montreal, Canada, 1996.

129. Legault, R., Blaise, C., Rokosh, D., and Chong-Kit, R., Comparative assessment of the SOS Chromotest kit and the mutatox test with the *Salmonella* plate incorporation (Ames test) and fluctuation tests for screening genotoxic agents, *Environ. Toxicol. Water Qual.*, 9, 45, 1994.

130. SAS Institute Inc., *SAS/STAT™ User's Guide, Release 6.03 Edition*, SAS Institute Inc., Cary, NC, 1988.

131. Welsch, A. H., Peterson, A. T., and Altmann, S. A., The fallacy of averages, *Amer. Nat.*, 132, 277, 1988.

132. Paasivirta, J., *Chemical Ecotoxicology*, CRC Press/Lewis Publishers, Boca Raton, FL, 1991.

133. White, P. A. and Rasmussen, J. B., The genotoxic hazards of domestic wastes in surface waters, *Mutat. Res.*, in press.

134. Knezovich, J. P., Harrison, F. L., and Wilhelm, R. G., The bioavailability of sediment-sorbed organic chemicals: a review, *Water Air Soil Pollut.*, 32, 233, 1987.

135. Wetzel, R. G. and Likens, G. E., *Limnological Analyses*, Springer-Verlag, New York, NY, 1991.

136. Albertini, R. J. and Robison, S. H., Human population monitoring, in *Genetic Toxicology*, Li, A. P. and Heflich, R. H., Eds., CRC Press, Boca Raton, FL, 1991, 143.

137. Durant, J. L., Hemond, H. F., and Thilly, W. G., Determination of mutagenicity in sediments of the Aberjona watershed using human lymphoblast and *Salmonella typhimurium* mutation assays, *Environ. Sci. Technol.*, 26, 599, 1992.

138. Grifoll, M., Solanas, A. M., Parés, R., Centellas, V., Bayona, J. M., and Albaigés, J., Assessment of mutagenic activity of coastal sediments off Barcelona, *Tox. Assess.*, 3, 315, 1988.

139. Hashizume, T., Ueda, K., Tokutsu, S., Hanawa, I., and Kinae, N., Monitoring of mutagens in river and marine sediments by *Salmonella*/Microsome assay combined with blue cotton method, *Bull. Environ. Contam. Toxicol.*, 49, 497, 1992.

140. Suzuki, J., Sadamasu, T., and Suzuki, S., Mutagenic activity of organic matter in an urban river sediment, *Environ. Pollut. (A)*, 29, 91, 1982.

141. Marvin, C. H., McCarry, B. E., and Bryant, D. W., Determination and genotoxicity of polycyclic aromatic hydrocarbons isolated from *Dreissina polymorpha* (Zebra mussels) sampled from Hamilton harbor, *J. Great Lakes Res.*, 20, 523, 1994.

142. Mix, M. C. and Schaffer, R. L., Benzo(a)pyrene concentrations in mussels (*Mytilus edulis*) from Yaquina Bay, Oregon during June 1976-May 1978, *Bull. Environ. Contam. Toxicol.*, 23, 677, 1979.

143. Dunn, B. P. and Stich, H. F., The use of mussels in estimating benzo(a)pyrene contamination of the marine environment, *Proc. Soc. Exper. Bio. Med.*, 150, 49, 1975.

144. Andelman, J. B. and Suess, M. J., Polynuclear aromatic hydrocarbons in the water environment, *Bull. Wld. Hlth. Org.*, 43, 479, 1970.
145. Veith, G. D., deFoe, D. L., and Bergstedt, B. J., Measuring and estimating the bioconcentration factor of chemicals in fish, *J. Fish. Res. Bd. Can.*, 36, 1040, 1979.
146. Parry, J. M., Tweats, D. J., and Al-Mossawi, M. A. J., Monitoring the marine environment for mutagens, *Nature*, 264, 538, 1976.
147. Parry, J. M. and Al-Mossawi, M. A. J., The detection of mutagenic chemicals in the tissue of the mussel *Mytilus edulis*, *Environ. Pollut.*, 19, 175, 1979.
148. Parry, J. M., Barnes, W., and Kadhim, M., The detection of mutagens in marine waters, in *Progress in Environmental Mutagenesis*, Alacevic, M., Ed., Elsevier/North-Holland Biomedical Press, Amsterdam, 1980, 227.
149. Kinae, N., Hashizume, T., Makita, T., Tomita, I., Kimura, I., and Kanamori, H., Studies on the toxicity of pulp and paper mill effluents- II. Mutagenicity of the extracts of the liver from the Spotted Sea Trout (*Nibea mitsukurii*), *Wat. Res.*, 15, 25, 1981.
150. Frezza, D., Pegoraro, B., and Prescuittini, B., A marine host-mediated assay for the detection of mutagenic compounds in polluted sea waters, *Mutat. Res.*, 104, 215, 1982.
151. Al-Mossawi, M. A. J., Kadri, M., Salama, M., and Salem, A., The efficiency of various solvents in the extraction of chemical mutagens from living tissues: a comparative study, *Mutat. Res.*, 104, 43, 1982.
152. Kadhim, M. and Parry, J. M., The detection of mutagenic chemicals in the tissues of shellfish exposed to oil pollution, *Mutat. Res.*, 136, 93, 1984.
153. Kira, S., Hayatsu, H., and Ogata, M., Detection of mutagenicity in mussels and their ambient water, *Bull. Environ. Contam. Toxicol.*, 43, 583, 1989.
154. Rodriguez-Ariza, A., Abril, N., Navas, J. I., Dorado, G., Lopez-Barea, J., and Pueyo, C., Metal, mutagenicity, and biochemical studies on bivalve mollusks from Spanish coasts, *Environ. Molec. Mutagen.*, 19, 112, 1992.
155. Rasmussen, J. B., Rowan, D. J., Lean, D. R. S., and Carey, J. H., Food chain structure in Ontario lakes determines PCB levels in Lake Trout (*Salvelinus namaycush*) and other pelagic fish. *Can. J. Fish. Aquat. Sci.*, 47, 2030, 1990.
156. Connor, M. S., Fish/sediment concentration ratios for organic compounds, *Environ. Sci. Technol.*, 18, 31, 1984.
157. Eadie, B. J., Faust, W., Gardner, W. S., and Nalepa, T., Polycyclic aromatic hydrocarbons in sediments and associated benthos in Lake Erie, *Chemosphere*, 11, 185, 1982.
158. Varanasi, U., Reichert, W. L., Stein, J. E., Brown, D. W., and Sanborn, H. R., Bioavailability and biotransformation of aromatic hydrocarbons in benthic organisms exposed to sediment from an urban estuary, *Environ. Sci. Technol.*, 19, 836, 1985.
159. Black, J. J., Har, T. F., and Evans, E., HPLC studies of PAH pollution in a Michigan trout stream, in *Chemical Analysis and Biological Fate: Polynuclear Aromatic Hydrocarbons*, Cooke, M. and Dennis, A. J., Eds., Batelle Press, Columbus, OH, 1981, 343.
160. Shane, B. S., Introduction to ecotoxicology, in *Basic Environmental Toxicology*, Cockerham, L. G. and Shane, B. S., Eds., CRC Press, Boca Raton, FL, 1994, 3.
161. White, P. A. and Rasmussen, J. B., SOS Chromotest results in a broader context. Empirical relationships between genotoxic potency, mutagenic potency and carcinogenic potency, *Environ. Molec. Mutagen.*, 27, 270, 1996.
162. Mersch-Sundermann, V., Schneider, U., Klopman, G. G., and Rosenkranz, H., SOS induction in *Escherichia coli* and *Salmonella* mutagenicity: a comparison using 330 compounds, *Mutagenesis*, 9, 205, 1994.
163. Maccubbin, A. E. and Ersing, N., Tumors in fish from the Detroit river, *Hydrobiol.*, 219, 301, 1991.
164. Lynch, M., Conery, J., and Bürger, R., Mutation accumulation and the extinction of small populations, *Amer. Nat.*, 146, 489, 1995.

CHAPTER 43

The Development and Application of Sediment Toxicity Tests for Regulatory Purposes

Marion G. Nipper

CONTENTS

I. INTRODUCTION

Sediments are recognized as sinks and sources of contaminants in aquatic systems. The effects of sediment contamination on the aquatic biota and on the surrounding environment is cause for concern to scientists, legislators, resource managers, and the community in general. Contaminants can be introduced into aquatic environments and accumulate in sediments by several pathways including disposal of liquid effluents, runoff and leachates carrying chemicals originating from a variety of urban, industrial and agricultural activities, as well as aerial deposition.

Bioaccumulation of sediment contaminants and human health issues are also matters of concern. Bioaccumulation tests are useful for the identification of bioavailable contaminants, but their biological meaning is poorly understood. Tests are usually conducted with relatively insensitive species, which do not seem to be adversely affected by the doses of contaminants present in the surrounding environment. The level of contaminants which triggers biological responses other than accumulation in tissues, therefore causing adverse effects on populations and communities, is generally not known. The accumulation of contaminants in sediments and their bioaccumulation by edible species raises human health issues. While both bioaccumulation by aquatic species and human health problems caused by sediment contamination represent important sediment-related environmental problems, they are not included in the scope of this chapter and will not be discussed further.

Regulatory perspectives for chemical sediment criteria with a view to protection of aquatic biota have been a scientific and legislative issue for at least the last decade.[1] Microscale laboratory toxicity tests have become an essential tool in sediment assessment and monitoring programs. They have the objective of estimating the potential biological impact of a given contaminated sample, and results of such tests are useful for the establishment of acceptable concentrations of contaminants in sediments, i.e., concentrations that are not expected to adversely affect the aquatic biota. Therefore, they are useful tools for the derivation of sediment quality guidelines, and for making pass/fail decisions to determine the need of cleanup and restoration measures, and for the disposal of dredged material (e.g., USACE/EPA[2,3]). Important factors which need to be considered in the planning of microscale sediment toxicity tests with a view to their application within the regulatory framework include: sediment sampling and storage procedures; confounding factors in sediment toxicity testing; routes of exposure, bioavailability and modes of action of contaminants in sediments; test organisms to be used and test endpoints; and realism of laboratory results and their interpretation.

This chapter will discuss the main approaches currently available for sediment monitoring and regulatory actions, their limitations and strengths. Most of those approaches strongly rely on the application of microscale toxicity tests. Subjects discussed include: microscale toxicity tests as tools for the identification of contamination problems and derivation of sediment quality guidelines; description of the main numerical sediment quality guidelines and their possible application in the regulatory framework; the sediment quality triad as a means to integrate biological and chemical data for the assessment of sediment pollution problems; and the New Zealand experience as a case study discussing research directions for the development and application of sediment toxicity tests and of sediment quality guidelines at the national level.

II. APPROACHES FOR THE ESTABLISHMENT OF NUMERICAL SEDIMENT QUALITY GUIDELINES AND CRITERIA

A. Overview

Although extensive scientific databases encompassing chemical contamination and biological effects of polluted sediments have been generated in an attempt to assess sediment quality, to date only five numerical sediment quality criteria for nonionic organic compounds have been proposed by the U.S. EPA.[4-8] Nevertheless, also in North America, scientific data sets have been used to derive a variety of sediment quality guidelines using chemical and biological data for a range of priority pollutants.[9-11]

Bulk sediment chemistry data, due to questions about bioavailability of sediment contaminants, are of limited use in a regulatory context.[12] The experience gained with water quality management provides insights for the management of sediment quality and argues for site-specific and effects-based approaches which take into account factors such as bioavailability and integration of the toxicological effects of multiple chemicals.[13] Chemically based sediment quality guidelines, derived from observed biological effects (sediment toxicity and health of benthic communities), and from scientific findings on the bioavailable fraction of contaminants in sediments, provide numerical values (concentrations) of chemicals in sediments above which adverse effects to the benthic biota are expected to occur or below which they are not expected to occur. Some of the current guidelines and approaches used to derive such numerical values are briefly discussed.

B. Equilibrium Partitioning (EqP)

The rationale of the equilibrium partitioning approach is that safe sediment contaminant levels can be established by ensuring that water quality criteria are met in sediment interstitial water.[14] It attempts to model the tendency of a chemical to move from one environmental compartment to another,[15] i.e., from the solid phase to interstitial water and to the biota. The EqP approach addresses the two main technical issues involved in the establishment of sediment quality criteria, which are the varying bioavailability of chemicals in sediments and the selection of appropriate biological effects concentrations. The final chronic value (FCV) from the U.S. EPA water quality criteria has been proposed as the chemical concentration that would be sufficiently protective of benthic organisms.[16]

Some factors that affect the bioavailability of contaminants in sediments, and are therefore important for the EqP approach, include the role of total organic carbon as a binding factor for nonpolar organic contaminants and of acid volatile sulfides (AVS) for some metals.[16-19] However, there are uncertainties in this field, and the effect of partitioning phases for nonpolar organic chemicals other than to organic carbon, and of the presence of mixtures of organic contaminants in sediments remain to be addressed.[20] Even after organic carbon normalization, the bioavailability of some highly hydrophobic organic compounds can vary by a factor of twenty or more.[21] The AVS concentration in sediments can vary seasonally,[22] especially in the superficial oxic layer of sediments. Therefore, to ensure conservative estimations of contaminant limits in sediments for the protection of the benthic biota, the toxicity caused by sediment-associated metals should be esti-mated when AVS is at its lowest concentrations. Under this condition the AVS-associated metals would be most bioavailable, the highest possible toxic effects would be observed, and proposed numerical criteria would be protective of the benthic biota year-round. Another problem with the AVS normalization is that while it is effective in predicting the bioavailability for some metals (Cd, Zn), its role in the binding of other metals (Hg, Cu, As, Cr) is poorly understood. Another problem with this approach is that it is based on single chemical information and does not take into consideration the effects of mixtures.

The convenience of applying the well developed and accepted water quality criteria is a strong reason for the acceptance of the EqP approach. It has been adopted by some countries, for example, The Netherlands, which apply it for freshwater environments on the basis that no toxicity data are available for sediments. Maximum acceptable risk levels for sediment are derived indirectly from maximum acceptable risk levels for water, using the equilibrium partitioning method.[23]

C. The ER-L/ER-M Approach

Two guideline values are established by this approach, the ER-L (effects range-low) and the ER-M (effects range-median), derived from effect data sets composed of a compilation of results of sediment toxicity tests and benthic community surveys. The ER-L and ER-M represent the 10th percentile and the 50th percentile concentrations of the effect data set, respectively, for a number of chemicals,[9,24,25] defining the concentration ranges which are rarely, occasionally, or frequently linked to adverse biological effects. Microscale toxicity tests had an essential role in the development of these guideline values. For example, for the organic chemicals acenaphthene and phenanthrene, the data set utilized for the derivation of the ER-L and ER-M values was composed mainly by toxicity test results (68 and 63% of the data set, respectively)[25] applying some of the endpoints mentioned in Section III.B (e.g., mortality, emergence, reburial, embryo development and abnormality).

This approach considers bulk contaminant concentrations, not taking into consideration the role sediment properties (i.e., TOC, AVS, etc.) play in controlling contaminant bioavailability.[26] Another feature of this approach is that it does not take into consideration the interactive effects of co-occurring contaminants in the test sediments. Therefore, these guidelines should be used as informal screening tools in environmental assessments, rather than precluding the use of toxicity tests or other measures of biological effects[25] which would provide site-based information.

D. The TEL and PEL Approach — Sediment Quality Guidelines for Canada and Florida

Sediment quality guidelines to be used in Florida[10] and in Canada,[11,27] similar to the ER-L/ER-M approach, also define two values for chemical concentrations: the threshold effect level (TEL) and the probable effects level (PEL). Adverse effects are expected to rarely occur below the TEL, and frequently above the PEL. Values between the TEL and the PEL represent the range of concentrations that could, potentially, be associated with biological adverse effects. The TEL for each chemical represents the geometric mean of the lowest 15th percentile concentration of the effects data set and the 50th percentile concentration of the no-effects data set, and the PEL represents the geometric mean of the 50th percentile concentration of the effects data set and the 85th percentile concentration of the no-effects data set. Also similar to the ER-L/ER-M approach, the TEL and PEL were derived by the use of a large number of toxicity test results. For instance, in the example given by the Canadian Council of Ministers of the Environment (CCME)[11] for cadmium, microscale toxicity tests represented 93% of the summarized data set.

This guideline also does not account for bioavailability, cause-effect mechanisms, or confounding factors, and should be used as a screening tool rather than as a standard value which would preclude the use of biological assessments. As noted by Chapman,[28] they serve as a basis for the identification and delineation of areas and contaminants of concern.

E. Screening Level Concentration (SLC)

The SLC approach, rather than concentrating on microscale toxicity tests, essentially uses field data on the co-occurrence in sediments of benthic infaunal invertebrates and different concentrations of nonpolar organic contaminants. By this procedure, it estimates the highest concentration of a particular nonpolar organic contaminant in sediments (normalized to sediment organic carbon concentration) that can be tolerated by approximately 95% of benthic fauna.[29] Although the SLC approach was originally suggested for use with nonpolar organic chemicals, it can be applied for any chemical contaminant as long as the appropriate normalization procedures are used.[30] A problem is that appropriate normalization procedures are not known for many chemicals of concern, especially ionic organics and several metals.

F. Apparent Effects Threshold (AET)

The AET approach[31] derives sediment quality values based on sediment concentrations of a certain chemical above which statistically significant adverse biological effects are always expected. This is established based on large databases of sediment chemistry, and also makes strong use of toxicity test results, as well as benthic infauna composition, for the establishment of numerical guidelines for a variety of chemicals. The application of the AET approach in Puget Sound generated the establishment of standards for 46 chemicals. Nevertheless, when sediment cleanup is recommended, biological testing may be used to confirm or over-ride the results of comparisons of sediment contaminant concentrations with the AET numerical standards.[32] Like the other approaches discussed above, the AET recommends the use of bulk concentrations of chemicals in sediments (based on sediment dry weight), which do not consider bioavailability, confounding factors, and cause-effect mechanisms.

G. General Comments

Numerical sediment quality guidelines are important tools in comprehensive ecosystem management, and microscale sediment toxicity tests are important tools for the derivation of several of the sediment quality guidelines. However, rather than being intended to preclude the need for site-specific considerations, they are intended to provide the scientific basis to support effective environmental management.[33] These guidelines are useful for screening sediments for potentially adverse levels of contamination. However, they should only be used as the first tier in a hierarchy of steps for evaluating contaminated sediments. If contaminant concentrations in sediment exceed the safe thresholds, the evaluation should proceed to direct biological testing in a site-specific manner, using the methods described in the following section.

III. SEDIMENT TOXICITY TEST SYSTEMS

Sediment toxicity tests consist of the exposure of selected test organisms to contaminated sediments, with a view to estimating lethal or sublethal adverse effects of the contaminants on the benthic biota, and statistically comparing these results with those of a control or reference (uncontaminated) sediment. The development of microscale sediment bioassays around the world has been through three generations: 1st — tests devised for water column and adapted for sediment testing (e.g., embryo development tests and Microtox®*); 2nd — acute solid-phase tests; 3rd — sublethal solid-phase tests.[34] These systems can be applied to a number of phylogenetic groups and involve the exposure of organisms to different fractions of the sediment by the use of different routes of exposure, and the assessment of the adverse effects of toxicants on several different endpoints, as discussed below.

A. Routes of Exposure

Toxicity tests can be conducted with the following fractions of contaminants contained in sediments:

- **Solid phase**: consists of the exposure of benthic organisms to the whole sediment, or of water column species to the overlying water in test vessels containing sediments, to assess the effects of contaminants released from the sediment to the surrounding environment.
- **Elutriate**: consists of exposing water column species to sediment elutriates containing the soluble fraction of contaminants, released to the surrounding water by the elutriation process. This process consists of mixing standard amounts of sediment and noncontaminated water, stirring the mixture for a certain period of time, centrifuging or decanting the supernatant, which is then used for the toxicity test. This method is commonly used to assess the potential impact of dumping dredged material into aquatic environments because it simulates its behavior at disposal sites.

---

* Registered trademark of AZUR Environmental, California, USA.

- **Pore water**: consists of the exposure of water column species to the fraction of contaminants contained in the pore water, which is considered to be the dominant route of exposure for many compounds.[35]
- **Chemical extracts**: consists of exposing water column species to the contaminant fraction obtained by chemical extraction of sediments. This has been suggested as a sediment toxicity test system,[36] and is useful for screening sediments for the presence of toxic chemicals. However, it does not provide a realistic assessment of sediment toxicity to the benthic biota, since the chemical extraction process removes contaminants from the sediment which otherwise might not be bioavailable, although it is possible that in certain cases this process may simulate the digestive function of test organisms.

B. Test Endpoints: Types and Applications

The most common endpoints for toxicity tests are survival (acute tests), and growth, reproduction, behavioral alterations (e.g., emergence, reburial, sediment avoidance), fertilization success, and larval development (short-term and life-cycle chronic tests). Sediment toxicity tests are necessary for a variety of reasons, e.g., the identification of polluted sites for remediation actions, and the ranking of sites based on extent and severity of contamination. These needs have resulted in the standardization of three test methods by the U.S. EPA: 10-d solid-phase survival sediment toxicity tests with freshwater amphipods and midges, and with marine amphipods.[37] The development of standardized protocols for sediment sampling and toxicity testing has resulted in comparable data among different regulatory programs.[32] However, the predominant focus on acute tests means that potential long-term effects are not taken into account.

Sediment test methods should be organized in a hierarchical tiering system in order to promote efficient use of resources and screening of sites.[37] While acute tests can be used as screening tools in the early phase of an assessment hierarchy of highly polluted areas[38] with a view to regulatory actions, it may be necessary to also employ chronic toxicity assessments if contaminants are believed to be present at high enough concentrations to pose a risk to benthic communities, or after remediation has taken place and recovery from an acute situation can be observed. In some countries, low levels of contamination prevail, and rather than restricting the study of pollution effects to acute situations, emphasis has to be given to the development and application of chronic toxicity tests for the protection of the aquatic biota. Chronic test methods and endpoints are more common for liquid samples, and although a number of solid phase tests with sublethal endpoints have been developed, these methods have not been widely used to date.

C. Test Organisms

1. Acute Tests

Sediment toxicity test methods with a variety of organisms have been discussed in several chapters of this book. The most common groups for solid phase tests are amphipods, polychaetes, and bivalves for marine and estuarine sediments, and amphipods, insect larvae, and oligochaetes for freshwater sediments. For pore water, elutriate, or sediment extract tests, cladocerans, phytoplankton, and fish are commonly used for freshwater tests; mysids, copepods, echinoid sperm and embryos, mollusk embryos, phytoplankton, macroalgae gametes, bacteria (Microtox), and fish are commonly used for marine and estuarine tests. There is an extensive bibliography regarding these methods, and in addition to this book, a number of good reviews are also available in the scientific literature.[39-41]

2. Chronic Tests

Chronic sediment tests, with lethal and sublethal endpoints, have been developed in different countries for both marine and freshwater species. The main examples are: marine and freshwater amphipod survival, growth, and reproduction;[42-46] mollusk survival, growth, reburial, and sediment avoidance;[47-49] echinoid growth;[50,51] polychaete survival and growth;[52,53] oligochaete survival, growth, and reproduction;[54,55] and insect growth.[46,56] The application of life-cycle or short-term chronic tests is intended to estimate the long-term impact of sediment contaminants.

Both the application of acute and chronic microscale toxicity tests and the use of a variety of species for such tests permit a thorough assessment of the potential impacts of contaminants to the biota. Whereas the use of organisms with different niches, e.g., deposit vs. filter feeders, ensures the toxicological assessment of contaminants which undergo different chemical and binding processes in the sediment, the use of different phylogenetic groups for toxicity testing is important for the assessment of effects of toxicants with different modes of action. Both the test design (e.g., acute vs. chronic) and the selection of test organisms can strongly affect the outcome of small-scale toxicity tests, therefore reinforcing the importance of a careful experimental design, especially when results of toxicity tests are to be applied with regulatory purposes.

D. Variables in Sediment Toxicity Testing

Some variables involved in the treatment applied to sediments intended for use in toxicity tests, such as sampling, handling, and storage, can strongly influence the bioavailability of contaminants, and consequently, experimental results. A variety of confounding factors can mask or enhance the real toxicity of contaminants in sediment tests.[43,57-60] Some of the most important are the natural sediment characteristics, such as particle-size distribution, organic content, and salinity. Ammonia and sulfide also occur naturally in sediments and can be considered as confounding factors; on the other hand, their concentrations can be considerably increased by anthropogenic inputs and place them in the category of toxicants *per se*, and caution should be exercised in the data interpretation of tests where ammonia and sulfide levels are suspected to be responsible for observed toxic effects. None of the whole-sediment approaches discussed in Section II accounts for effects of noncontaminant-related variables. Whereas the effects of some of these factors (e.g., grain size, salinity) can be estimated and corrected for by using a reference sediment, others (e.g., natural levels of ammonia and sulfide) are difficult to match in different sediments. This can lead to false positive test results, i.e., a "toxic" effect caused by natural sediment features, rather than by the presence of chemical contaminants. The sensitivity of the test species to intrinsic factors of the test sediments should be well established prior to the interpretation of test results with contaminated sediments. The lack of control of some of these variables can have serious implications if test results are used for regulatory purposes. This subject is more extensively discussed by Tay et al.[61] in Chapter 39.

IV. THE SEDIMENT QUALITY TRIAD: AN INTEGRATIVE APPROACH

The integration of toxicity test results with benthic assemblage assessments and contaminant concentrations in sediments was proposed in the mid-'80s, with the development of the concept of a sediment quality triad.[62,63] While chemical analyses provide information on concentrations of chemicals, toxicity tests assess their bioavailability and effects under laboratory conditions, and benthic assemblage studies give information on alterations under *in situ* conditions. The triad approach has the advantage of integrating chemical data with the laboratory tests described above, and with field-obtained biological information. However, it is not clear how the importance of chemical, toxicological, and benthic data should be weighted. The triad differs from the guidelines discussed in Section II in the sense that it does not result in numerical values of acceptable levels of contaminants in sediments, therefore it does not permit a quantitative comparison of sites. Nevertheless, it can provide very good information on individual sediment quality and can be applied in a number of situations, such as to evaluate the environmental effects of an oil platform[64] or of a sewage outfall,[65] or to assess the sediment quality of historically contaminated areas (e.g., Puget Sound,[62] San Francisco Bay,[66] Galveston Bay[67]).

V. DEVELOPMENT OF CHRONIC SEDIMENT TOXICITY TESTS AND REGULATORY ISSUES — THE NEW ZEALAND EXPERIENCE

New Zealand's Resource Management Act (1991) requires that a discharge of a contaminant or water into water shall not cause any significant adverse effect on aquatic life after reasonable mixing. This has stimulated research to evaluate the effects of toxic contaminants and establish techniques for assessing and mitigating adverse effects on the environment. However, no sediment quality guidelines have been developed in the country yet.

Due to the lack of sufficient local information to allow the development of sediment guidelines or criteria, generic guidelines are often considered as representing the best immediately available information to be used on scientifically and on legally-defensible grounds.[68] The ER-L/ER-M approach[9,25] has been used to predict the potential future impact of storm water runoff if present levels of contaminants are not controlled, and the Canadian sediment quality guidelines[11] have been applied to assess the potential impact of contaminants in sediments from a harbor area (Roper, personal communication). While approaches for the derivation of sediment quality guidelines developed in North America are currently applied in New Zealand, the search for ecologically realistic tests with native species and their field validation is one of the main focuses of sediment pollution research in the country, with a view to their application in future regulatory actions. Current emphasis is being given to the search for highly sensitive chronic test methods, rather than acute tests. This is mostly due to two factors: the low levels of contamination encountered in most of the country's water bodies, which did not cause observable effects in acute mortality assays; and the fact that the time when concerns about environmental pollution in New Zealand became an issue coincided with the time when other countries were beginning to realize the importance of the application of chronic toxicity tests for the establishment of environmental quality standards. Sensitive sublethal methods are necessary in New Zealand to identify adverse biological effects at early stages of sediment contamination, so that the source of contaminants can be identified, then leading to preventive actions to avoid further input of contaminants, and allowing the restoration of the area prior to the occurrence of high levels of degradation.

It is recognized that major differences in toxicity test results can be observed due to species response differences, based on exposure routes, life history characteristics, and individual sensitivity patterns.[69] Sediment research directions in New Zealand have led to an investigation of the applicability of test methods and approaches developed overseas, and to the development of sensitive test methods encompassing behavioral and growth endpoints for a number of indigenous species.[44,48,49]

The estuarine amphipod *Chaetocorophium* cf *lucasi*, abundant in the muddy sediments of northern New Zealand estuaries, has been used for the development of 28-d life-cycle tests. It can be easily cultured in the laboratory, and methods to establish effects of sediments on growth have been assessed.[44] Neither survival nor growth were affected by a wide range of particle size distribution, and total organic carbon (TOC) or water content. Only sandy sediment (96.8% sand), with low TOC (0.13%) and water content (24.9%) reduced growth significantly, but did not affect 28-d survival. Survival seemed to be more affected by sediment type (origin) than by particle size or organic matter content.[44]

The bivalve *Macomona liliana*, common in sandy intertidal sediments around New Zealand, has been identified as a potential toxicity test species for impact assessment because of the tendency of juveniles to move out of contaminated sediments, as observed in field trials with chlordane.[70] Behavioral tests with this bivalve showed that burial rate was reduced and sediment avoidance was induced at relatively low levels of some common contaminants (lead, copper, and zinc).[49] However, sediment avoidance was also induced by a muddy reference sediment, and the effect of such confounding factors has to be studied further. Growth tests with *M. liliana* are also currently being developed. Survival and growth after a 28-d exposure were not affected by particle size distributions varying from 8.4 to 96.8% sand, water content from 24.9 to 56.3%, and TOC content from 0.13 to 1.4%.[44]

The feasibility of using the echinoid embryo development test to assess pore water toxicity is currently under evaluation. The sensitivity of the method has been assessed using the sand dollar *Fellaster zelandiae*, which is abundant subtidally around the New Zealand coast.[71] The test method proved to have similar sensitivity to reference toxicants (zinc and SDS) as echinoid embryo development tests used elsewhere.[72] Its sensitivity to sewage treatment plant effluents was similar or higher than that of algal growth (microplate tests), Microtox, and mysid 96-h acute tests.

Whereas the sand dollar test method with pore water has potential to become a useful screening tool to identify toxic sediments, pore water tests should not stand alone and recommendations should be made for the application of further chronic tests with the solid phase using benthic species when toxic effects are caused by the pore water. Results should be integrated with chemical- and field-data, ultimately leading to the establishment of sediment quality guidelines for application in the country.

VI. FUTURE DIRECTIONS

Sediment toxicity tests have progressed from the acute to the chronic toxicity assessment approach. Full life-cycle toxicity tests comprise the most ecologically relevant, laboratory-based endpoints currently available, and the ultimate applicability of these requires linkages between responses in the laboratory and responses of populations in the field.[73] The field validation of laboratory tests has been partially achieved by benthic assemblage assessments, and further linkages between laboratory and field responses can be done by *in situ* toxicity tests with the same species used in the laboratory, which will help in estimating the degree of realism and the ecological relevance of laboratory obtained results. Other aims in the field of sediment ecotoxicology include improving the understanding of the role of individual chemicals in complex mixtures of contaminants, and of processes that control contaminant bioavailability, which would improve the normalization procedures. Regarding numerical sediment quality guidelines, the quantitative structure activity relationship (QSAR) approach and methods that account for bioavailability (e.g., $\sum$PAH model) hold promise for the future.[74]

Biological criteria, based on laboratory and field experiments and on a thorough knowledge of ecological processes will eventually lead to the establishment of the most appropriate sediment quality guidelines and criteria, resulting in regulatory actions which in most cases will neither be under- nor overprotective.

ACKNOWLEDGMENTS

I thank Drs. David Roper, Theodore DeWitt, Stephanie Turner, and Julie Hall, as well as an anonymous and a known referee, Dr. Peter Chapman, for valuable reviews and comments.

REFERENCES

1. Chapman, G. A., Establishing sediment criteria for chemicals — regulatory perspective, *Fate and effects of sediment-bound chemicals in aquatic systems*, Dickson, K. L., Maki, A. W., and Brungs, W. A., Eds., Pergamon Press, New York, 1987, Chap. 23.
2. USACE/USEPA, Management plan report: unconfined open-water disposal of dredged material, Phase II (North and South Puget Sound), PSDDA (Puget Sound Dredged Disposal Analysis) Reports, 89 10 10126, US Army Corps of Engineers and United States Environmental Protection Agency, Region 10, Seattle, WA, 1989.
3. USACE/USEPA, Evaluation of dredged material proposed for ocean disposal, EPA-503/8-91/001, US Army Corps of Engineers and United States Environmental Protection Agency, Washington, DC, 1991.

4. USEPA, Sediment quality criteria for the protection of benthic organisms: fluoranthene, EPA-822-R-93-012, United States Environmental Protection Agency, Washington, DC, 1993.

5. USEPA, Sediment quality criteria for the protection of benthic organisms: acenaphthene, EPA-822-R-93-013, United States Environmental Protection Agency, Washington, DC, 1993.

6. USEPA, Sediment quality criteria for the protection of benthic organisms: phenanthrene, EPA-822-R-93-014, United States Environmental Protection Agency, Washington, DC, 1993.

7. USEPA, Sediment quality criteria for the protection of benthic organisms: dieldrin, EPA-822-R-93-015, United States Environmental Protection Agency, Washington, DC, 1993.

8. USEPA, Sediment quality criteria for the protection of benthic organisms: endrin, EPA-822-R-93-016, United States Environmental Protection Agency, Washington, DC, 1993.

9. Long, E. R., and Morgan, L. G., The potential for biological effects of sediment-sorbed contaminants tested in the National Status and Trends Program, NOAA Technical Memorandum NOS OMA 52, National Oceanic and Atmospheric Administration, Seattle, WA, 1990.

10. MacDonald, D. D., Approach to the assessment of sediment quality in Florida coastal waters, Vol. 1 — Development and evaluation of sediment quality assessment guidelines, Florida Department of Environmental Protection, Office of Water Policy, Tallahassee, 1994.

11. CCME (Canadian Council of Ministers of the Environment), Protocol for the derivation of Canadian sediment quality guidelines for the protection of aquatic life, Report CCME EPC-98E, Ottawa, 1995.

12. Southerland, E., Kravitz, M., and Wall, T., Management framework for contaminated sediments (the U.S.EPA sediment management strategy), *Sediment toxicity assessment*, Burton. G. A. Jr., Ed., Lewis Publishers, Boca Raton, 1992, Chap. 15.

13. Adams, W. J., Kimerle, R. A., and Barnett, J. W. Jr., Sediment quality and aquatic life assessment, *Environ. Sci. Technol.*, 26, 1865, 1992.

14. Pavlou, S. P., The use of the equilibrium partitioning approach in determining safe levels of contaminants in marine sediments, *Fate and effects of sediment-bound chemicals in aquatic systems*, Dickson, K. L., Maki, A. W., and Brungs, W. A., Eds., Pergamon Press, New York, 1987, 389.

15. Shea, D., Developing national sediment quality criteria, *Environ. Sci. Technol.*, 22, 256, 1988.

16. Di Toro, D. M., Zarba, C. S., Hansen, D. J., Berry, W. J., Swartz, R. C., Cowan, C. E., Pavlou, S. P., Allen, H. E., Thomas, N. A., and Paquin, P. R., Technical basis for establishing sediment quality criteria for nonionic organic chemicals using equilibrium partitioning, *Environ. Toxicol. Chem.*, 10, 1541, 1991.

17. Di Toro, D. M., Mahony, J. D., Hansen, D. J., Scott, K. J., Hicks, M. B., Mayr, S. M., and Redmond, M. S., Toxicity of cadmium in sediments: the role of acid volatile sulfide, *Environ. Toxicol. Chem.*, 9, 1487, 1990.

18. Ankley, G. T., Phipps, G. L., Leonard, E. N., Benoit, D. A., Mattson, V. R., Kosian, P. A., Cotter, A. M., Dierkes, J. R., Hansen, D. J., and Mahony, J. D., Acid-volatile sulfide as a factor mediating cadmium and nickel bioavailability in contaminated sediments, *Environ. Toxicol. Chem.*, 10, 299, 1991.

19. Carlson, A. R., Phipps, G. L., Mattson, V. R., Kosian, P. A., and Cotter, A. M., The role of acid-volatile sulfide in determining cadmium bioavailability and toxicity in freshwater sediments, *Environ. Toxicol. Chem.*, 10, 1309, 1991.

20. Hoke, R. A., Ankley, G. T., Cotter, A. M., Goldenstein, T., Kosian, P. A., Phipps, G. L., and Vander-Meiden, F. M., Evaluation of equilibrium partitioning theory for predicting acute toxicity of field-collected sediments contaminated with DDT, DDE and DDD to the amphipod *Hyalella azteca*. *Environ. Toxicol. Chem.*, 13, 157, 1994.

21. Landrum, P., How should numerical sediment quality criteria be used? *Human Ecol. Risk Assess.*, 1, 13, 1995.

22. Leonard, E. N., Mattson, V. R., Benoit, D. A., Hoke, R. A., and Ankley, G. T., Seasonal variation of acid volatile sulfide concentration in sediment cores from three northeastern Minnesota lakes, *Hydrobiologia*, 271, 87, 1993.

23. Van de Meent, D., Aldenberg, T., Canton, J. H., Van Gestel, C. A. M., and Slooff, W., Desire for levels — Background study for the policy document "Setting environmental quality standards for water and soil", Report number 670101 002, National Institute of Public Health and Environmental Protection, Bilthoven, 1990.

24. Long, E. R., Ranges in chemical concentrations in sediments associated with adverse biological effects, *Mar. Poll. Bull.*, 24, 38, 1992.

25. Long, E. R., MacDonald, D. D., Smith, S. L., and Calder, F. D., Incidence of adverse biological effects within ranges of chemical concentrations in marine and estuarine sediments, *Environ. Manag.*, 19, 81, 1995.

26. Chapman, P. M., Invited debate/commentary: how should numerical criteria be used?, *Human Ecol. Risk Assess.*, 1, 1, 1995.

27. Smith, S. L., MacDonald, D. D., Keenleyside, K. A., and Gaudet, C. L., The development and implementation of Canadian Sediment Quality Guidelines, *Development and Progress in Sediment Quality Assessment: Rationale, Challenges, Techniques and Strategies,* Munawar, M. and Dave, G., Eds., SPB Academic Publishing, Amsterdam, 1996, 233.

28. Chapman, P. M., A test of sediment effects concentrations: DDT and PCB in the Southern California Bight, *Environ. Toxicol. Chem.*, 15, 1197, 1996.

29. Neff, J. M., Cornaby, B. W., Vaga, R. M., Gulbransen, T. C., Scanlon, J. A., and Bean, D. J., An evaluation of the Screening Level Concentration Approach for validation of sediment quality criteria for freshwater and saltwater ecosystems, *Aquatic toxicology and hazard assessment: 10th volume, ASTM STP 971*, Adams, W. J., Chapman, G. A., and Landis, W. G., Eds., American Society for Testing and Materials, Philadelphia, 1988, p. 115.

30. Chapman, P. M., Current approaches to developing sediment quality criteria, *Environ. Toxicol. Chem.*, 8, 589, 1989.

31. PTI Environmental Services, Briefing report to the EPA Science Advisory Board: The apparent effects threshold approach. Report prepared for the U.S. Environmental Protection Agency, Region 10, Seattle, WA. PTI Environmental Services, Bellevue, WA, 1988.

32. Ginn, T. C., and Pastorok, R. A., Assessment and management of contaminated sediments in Puget Sound, *Sediment toxicity assessment*, Burton. G. A. Jr., Ed., Lewis Publishers, Boca Raton, 1992, Chap. 16.

33. Gaudet, S. L., Keenleyside, K. A., Kent, R. A., Smith, S. L., and Wong, M. P., How should numerical criteria be used? The Canadian approach, *Human Ecol. Risk Assess.*, 1, 19, 1995.

34. Chapman, P., Swartz, R., Roddie, B., Phelps, H., van der Hurk, P., and Butler, R., An international comparison of sediment toxicity tests in the North Sea, *Mar. Ecol. Prog. Ser.,* 91, 253, 1992.

35. Carr, R.S., Marine and estuarine pore water toxicity testing, *Microscale Testing in Aquatic Toxicology: Advances, Techniques, and Practice*, Wells, P. G., Lee, K., and Blaise, C., Eds., CRC Press, Boca Raton, FL, 1998, Chap. 36.

36. Chapman, P. M., Marine sediment toxicity tests, *Chemical and biological characterization of sludges, sediments, dredge spoils, and drilling muds, ASTM STP 976*, Lichtenberg, J. J., Winer, F. A., Weber, C. I., and Fradkin, L., Eds., American Society for Testing and Materials, Philadelphia, 1988, p. 391.

37. USEPA, EPA's contaminated sediment management strategy, EPA 823-R-94-001, United States Environmental Protection Agency, Office of Water, Washington, DC, 1994.

38. ASTM, Standard guide for designing biological tests with sediments, Designation E 1525-93 *ASTM annual book of standards*, Vol. 11.04, American Society for Testing and Materials, Philadelphia, 1993, p. 1339.

39. Lamberson, J. O., DeWitt, T. H., and Swartz, R. C., Assessment of sediment toxicity to marine benthos, *Sediment toxicity assessment*, Burton. G. A. Jr., Ed., Lewis Publishers, Boca Raton, 1992, Chap. 9.

40. Burton, G. A. Jr., Assessing the toxicity of freshwater sediments, *Environ. Toxicol. Chem.*, 10, 1585, 1991.

41. Burton, G. A. Jr., Nelson, M. K., and Ingersoll, C. G., Freshwater benthic toxicity tests, *Sediment toxicity assessment*, Burton. G. A. Jr., Ed., Lewis Publishers, Boca Raton, 1992, Chap. 10.

42. Borgman, U., and Munawar, M., A new standardized sediment bioassay protocol using the amphipod *Hyalella azteca* (Saussure), *Hydrobiologia*, 188/189, 425, 1989.

43. Nipper, M. G., Greenstein, D. J., and Bay, S. M., Short- and long-term sediment toxicity test methods with the amphipod *Grandidierella japonica, Environ. Toxicol. Chem.*, 8, 1191, 1989.

44. Nipper, M. G., and Roper, D. S., Growth of an amphipod and a bivalve in uncontaminated sediments: implications for chronic toxicity assessments, *Mar. Poll. Bull.*, 31, 424, 1995.

45. DeWitt, T. H., Redmond, M. S., Sewall, J. E., Swartz, R. C., Development of a chronic sediment toxicity test for marine benthic amphipods, Report CBP/TRS 89/93, Newport, OR, 1992.

46. ASTM, Standard guide for conducting sediment toxicity tests with freshwater invertebrates, Designation E 1383-93 ASTM *annual book of standards*, Vol. 11.04, American Society for Testing and Materials, Philadelphia, 1993, p. 1173.

47. Burgess, R. M., and Morrison, G. E., A short-exposure, sublethal, sediment toxicity test using the marine bivalve *Mulinia lateralis*: statistical design and comparative sensitivity, *Environ. Toxicol. Chem., 13*, 571, 1994.

48. Roper, D. S., and Hickey, C. W., Behavioral responses of the marine bivalve *Macomona liliana* exposed to copper- and chlordane-dosed sediments, *Mar. Biol.*, 118, 673, 1994.

49. Roper, D. S., Nipper, M. G., Hickey, C. W., Martin, M. L., and Weatherhead, M. A., Burial, crawling and drifting behavior of the bivalve *Macomona liliana* in response to common sediment contaminants, *Mar. Poll. Bull.*, 31, 471, 1995.

50. Thompson, B. E., Bay, S. M., Anderson, J. W., Laughlin, J. D., Greenstein, D. J., and Tsukada, D. T., Chronic effects of contaminated sediments on the urchin *Lytechinus pictus, Environ. Toxicol. Chem.*, 8, 629, 1989.

51. Thompson, B. E., Bay, S. M., Greenstein, D. J., and Laughlin, J. D., Sublethal effects of hydrogen sulfide in sediments on the urchin *Lytechinus pictus, Mar. Environ. Res.*, 31, 309, 1991.

52. Johns, D. M., Pastorok, R. A., and Ginn, T. C., A sublethal sediment toxicity test using juvenile *Neanthes* sp. (Polychaeta: Nereidae), *Aquatic toxicology and risk assessment: fourteenth volume, ASTM STP 1124*, Mayes, M. A., and Barron, M. G., Eds., American Society for Testing and Materials, Philadelphia, 1991, p. 280.

53. Dillon, T. M., Moore, D. W., and Reish, D. J., A 28 day sediment bioassay with the marine polychaete, *Nereis (Neanthes) arenaceodentata, Environmental toxicology and risk assessment — third volume, ASTM STP 1218*, Hughes, J. S., Biddinger, G. R., and Mones, E., Eds., American Society for Testing and Materials, Philadelphia, 1995, p. 201.

54. Wiederholm, T., and Dave, G., Toxicity of metal polluted sediments to *Daphnia magna* and *Tubifex tubifex, Hydrobiologia,* 176/177, 411, 1989.

55. Phipps, G. L., Ankley, G. T., Benoit, D. A., and Mattson, V. R., Use of the aquatic oligochaete *Lumbriculus variegatus* for assessing the toxicity and bioaccumulation of sediment-associated contaminants, *Environ. Toxicol. Chem.*, 12, 269, 1993.

56. Day, K. E., Kirby, R. S., and Reynoldson, T. B., Sexual dimorphism in *Chironomus riparius* (Meigen): impact on interpretation of growth in whole-sediment toxicity tests, *Environ. Toxicol. Chem.*, 13, 35, 1994.

57. DeWitt, T. H., Ditsworth, G. R., and Swartz, R. C., Effects of natural sediment features on survival of the phoxocephalid amphipod, *Rhepoxinius abronius, Mar. Environ. Res.*, 25, 99, 1988.

58. Ankley, G. T., Katko, A., and Arthur, J. W., Identification of ammonia as an important sediment-associated toxicant in the lower Fox River and Green Bay, Wisconsin, *Environ. Toxicol. Chem.*, 9, 313, 1990.

59. Ankley, G. T., Benoit, D. A., Balogh, J. C., Reynoldson, T. B., Day, K. E., and Hoke, R. A., Evaluation of potential confounding factors in sediment toxicity tests with three freshwater benthic invertebrates, *Environ. Toxicol. Chem.*, 13, 627, 1994.

60. Suedel, B. C., and Rodgers, J. H. Jr., Responses of *Hyalella azteca* and *Chironomus tentans* to particle-size distribution and organic matter content of formulated and natural freshwater sediments, *Environ. Toxicol. Chem.*, 13, 1639, 1994.

61. Tay, K. L., Doe, K. G., and MacDonald, A. J., The influence of particle size, ammonia and sulfide on toxicity of dredged materials for ocean disposal, *Microscale Testing in Aquatic Toxicology: Advances, Techniques, and Practice*, Wells, P. G., Lee, K., and Blaise, C., Eds., CRC Press, Boca Raton, FL, 1998, Chap. 39.

62. Long, E. R., and Chapman, P. R., A sediment quality triad: measures of sediment contamination, toxicity and infaunal community composition in Puget Sound, *Mar. Poll. Bull.*, 16, 405, 1985.

63. Chapman, P. M., Sediment quality criteria from the sediment quality triad: an example. *Environ. Toxicol. Chem.*, 5, 957, 1986.

64. Chapman, P. R., Power, E. A., Dexter, R. N., and Andersen, H. B., Evaluation of effects associated with an oil platform, using the sediment quality triad, *Environ. Toxicol. Chem.,* 10, 407, 1991.

65. Chapman, P., Arthur, A., Paine, M., and Taylor, L., Sediment studies provide key information on the need to treat sewage discharged to sea by a major Canadian city, *Water Sci. Technol.*, 28, 255, 1993.

66. Chapman, P. M., Dexter, R. N., and Long. E. R., Synoptic measure of sediment contamination, toxicity and infaunal community composition (the Sediment Quality Triad) in San Francisco Bay, *Mar. Ecol. Prog. Ser.*, 37, 75, 1987.

67. Carr, R. S., Chapman, D. S., Howard, C. L., and Biedenbach, J. M., Sediment quality triad assessment survey of the Galveston Bay, Texas system, *Ecotoxicology*, 5, 1, 1996.
68. Morrisey, D. J., Williamson, R. B., and Roper, D. S., Letter to the editor-How should numerical criteria be used?, *Human Ecol. Risk Assess.*, 1(2), i, 1995.
69. Power, E. A., Munkittrick, K. R., and Chapman, P. M., An ecological impact assessment framework for decision-making related to sediment quality, *Aquatic Toxicology and Risk Assessment: Fourteenth Volume, ASTM STP 1124*, Mayes, M. A., and Barron, M. G., Eds., American Society for Testing and Materials, Philadelphia, 1991, p. 48.
70. Pridmore, R. D., Thrush, S. F., Wilcock, R. J., Smith, T. J., Hewitt, J. E., and Cummings, V. J., Effect of the organochlorine pesticide technical chlordane on the population structure of suspension and deposit feeding bivalves, *Mar. Ecol. Prog. Ser.*, 76, 261, 1991.
71. McKnight, D., An outline distribution of the New Zealand shelf fauna. Benthos survey, station list, and distribution of the Echinoidea. New Zealand department of scientific and industrial research, Bulletin 195, New Zealand Oceanographic Institute, Memoir no 47, 1969, 91p.
72. Nipper, M. G., Roper, D. S., Martin, M. L., and Williams, E. K., Marine toxicity tests development with a New Zealand echinoid, *Second SETAC World Congress, Abstracts Book*, Vancouver, 1995, 315.
73. DeWitt, T. H., Swartz, R. C., Word, J. Q., Scott, K. J., and Chapman, P. M., From death to extinction: interpreting existing and forthcoming amphipod sediment toxicity tests, *Second SETAC World Congress, Abstracts Book*, Vancouver, BC, 1995, 39.
74. Swartz, R. C., Schults, D. W., Ozretich, R. J., Lamberson, J. O., Cole, F. A., DeWitt, T. H., Redmond, M. S., and Ferraro, S. P., $\sum$PAH: a model to predict the toxicity of polynuclear aromatic hydrocarbon mixtures in field-collected sediments, *Environ. Toxicol. Chem.*, 14, 1977, 1995.

PART FOUR

Conclusions and Future Directions

CHAPTER 44

Microscale Testing in Aquatic Toxicology: Conclusions and Future Directions

Kenneth Lee, Peter G. Wells, and Christian Blaise

This book has brought together the knowledge and experience of a number of specialists currently engaged in various areas of microscale aquatic toxicity testing, covering a wide range of organisms. As such, it is directly useful to those with a scientific interest in one of the most exciting research areas of aquatic toxicology, to resource and environmental managers with the mandate to monitor and protect aquatic environments, and to graduate students and new practitioners of aquatic toxicology.

The use of microscale aquatic tests is expected to remain very broad-based and invaluable for chemical screening, pollution prevention and control, and environmental impact assessment. The evolution of different tests, supporting methodologies and applications will involve, among others, the fields of biotechnology, remote sensing and monitoring, and environmental conservation and protection (Figure 44.1). Within this framework, future developments and applications of microscale testing are briefly discussed below.

While quantitative and mechanistic ecotoxicology have traditionally been partners working in isolation, recent advances in biotechnology and small-scale techniques have resulted in a new generation of highly specific bioassays capable of providing first-hand information of cause-effect relationships (e.g., Chapters 11 and 12). These novel "exposure-effects biomarker bioassays" can directly identify the specific toxicants of primary concern in many point source and field situations. As more genomes of model experimental organisms are described (from yeasts and bacteria to fish and amphibians), such focused bioassays can only proliferate.

The development and use of commercialized microscale aquatic products (self-contained, user friendly bioanalytical kits) are expected to continue, particularly for use as easily transportable screening tools for field studies and for the monitoring of liquid and solid matrices (e.g., Chapter 14). With greater reliance on laboratories in the private sector to conduct studies for regulatory compliance, there clearly is a growing need for commercial "turn key" biotests. In parallel, to deal with an ever-increasing demand for assessment of large numbers of varied and complex samples in the laboratory and the demand for a reduction in operating costs, more tests will have to become miniaturized, automated and computer-interfaced to address sample throughput requirements (e.g., Chapters 18 and 20). Simplicity, sensitivity of response and good predictability of toxic potential remain the sought-after features of rapid microscale assays. With the recognition of the need for multitrophic level tests, a greater coverage of phyla and organisms of global significance is inevitable and highly desirable.

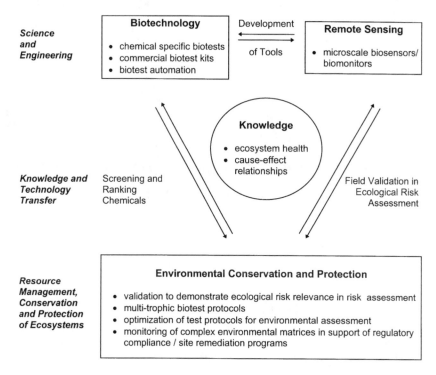

Figure 44.1 Microscale aquatic toxicity tests — future developments and applications.

The diversity of applications with small-scale toxicity assays is only limited by the imagination and creativity of researchers, chemical engineers, technical operators, and environmental conservation and protection specialists. The most important and promising practical applications for the future include:

- Continued studies in microcosms, mesocosms, and in open field environments to demonstrate the ecological relevance of aquatic microscale biotests, i.e., to validate the ability of single tests or suites of tests to predict environmental effects of contaminants. Most recently, the prediction of effects includes estimating the potential of disease to occur in biota including humans (e.g., the *Hydra* teratogenicity assay of Chapter 24, which relates to results of mammalian assays), not solely knowledge of the dose-response relationships.
- Use in Toxicity Identification Evaluations (TIEs) for the screening and study of urban and industrial effluents and other emissions at source. These studies can be coupled with fractionation techniques to identify the principal toxicants of concern in complex mixtures.
- Refinement of multitrophic microscale test batteries (e.g., Chapter 37). The choice and application of microscale biotests will be optimized for use in specific studies and chemical screening matrices. The optimization process will identify the required number of assays for specific purposes (i.e., achieving the objectives with the minimum rather than maximum numbers of assays) and additional complementary data requirements for sound interpretation of study results. Techniques are also being refined to include the use of endemic organisms with high sensitivity to contaminants to ensure the ecological relevance of the overall assessments and chosen endpoints using small-scale assays (e.g., Chapter 15).
- Applications to assess toxicity and hazards in other compartments of the biosphere, besides aquatic, such as soil, ice, snow, and air. This is envisaged as a growth area of ecotoxicological technology, due to the obvious linkages with human health concerns (e.g., air and soil quality), and to the need to prove success and identify operational endpoints following remedial activity at contaminated sites (e.g., use of soil assays). Atmospheric ecotoxicology will rapidly evolve with the appropriate adaptation of microbiotests.

- In living resource management, microscale biotests can be used to address fisheries/aquaculture-pollution-environmental quality issues. It is important to have detailed knowledge of the sensitivity of young life stages of commercially important species to contaminants and other stressors, alone and in combination. Microscale assays facilitate the rapid testing of many experimental treatments, and efficient analysis of large data sets.
- Technology transfer of microbiotest techniques to the world's developing countries is needed. On the basis of simplicity and low-cost, an increasing number of microscale biotests are ideal tools for countries in need of pragmatic and inexpensive environmental monitoring, and on-site testing and control of industrial discharges.

In terms of environmental protection, microscale aquatic toxicity tests have become a principal component of monitoring, assessment, regulatory testing, and site remediation tasks. Their application will reduce and replace our reliance on traditional methods based on the use of whole organism assays with juvenile and adult plants, invertebrates, fish, and amphibians, whether wild-caught or cultured. As shown by numerous chapters in this book, many microscale assays are currently available as ideal alternative tests.

For the foreseeable future, with the development of on-line biosensors, a major application of microscale aquatic toxicity tests will be to predict, prevent, control and monitor chemical emissions at source. At the same time, they will be in use in many other ways, as illustrated by the contributions to this book. There will be a continual interplay between the development of new tools, their role in basic toxicology and ecotoxicology research, their application such as in screening and ranking chemical toxicities, and the field validation of predicted effects in ecological risk assessment (Figure 44.1). Microscale testing in aquatic toxicology has a bright future, with many individuals and disciplines charting the still-early course!

GLOSSARY

Glossary*

Acute test: A study in which the organism is observed over a short period of time, usually one week or less.

A/D ratio: A range derived from the adult NOAEL and LOAEL divided by the developmental toxicity NOAEL and LOAEL.

Algal test battery: A microplate bioassay conducted with several unialgal species at a time. Depending on experimental configuration, one or more microplates may be required to accommodate the algal species in the test wells.

Algaltoxkit: 72-h algal growth inhibition microbiotest, based on the green alga *Selenastrum capricornutum* (also *Raphidocelis subcapitata*), and which makes use of algal cells de-immobilized from "algal beads."

Amphipod: Small, shrimp-like crustacean. Aquatic forms commonly live on the bottom, feeding on algae and detritus, and serving as food for other species.

Aneugenic event: Alteration of a diploid chromosomal set arising from chromosome elimination during anaphase.

Antigen: An antigen is a substance which induces the synthesis of specific antibodies after contact with the vertebrate immune system, e.g., injection into the vertebrate body. Antigens may be whole organisms, e.g., bacteria, or individual molecules of high molecular weight, e.g., proteins, polysaccharides or nucleic acids, but also synthetic polymers. Only parts of the antigen interact directly with the antibodies; they are called epitopes. Most antigens have several epitopes.

Antibody: Antibodies are glycoproteins of the immunoglobulin (Ig) class. They are synthesized by vertebrates (e.g., rabbits, mice, sheep, chickens, or humans) after contact of the immune system with a foreign compound of high molecular mass. The antibody binds selectively to this compound. Most mammals have five classes of antibodies, of which the immunoglobulin G (IgG) is the main class. The majority of immunoassays are based on IgG. The antibody molecule incorporates two identical light and two identical heavy polypeptide chains, covalently linked by interchain disulfide bonds in a Y-shaped molecule. The two binding sites are formed by the variable regions of the heavy (V_H) and light chains (V_L) at the amino terminal end of the molecule. The binding sites interact with the epitope of the antigen or with haptens.

---

* Selected by the contributors as some of the terms important for a general understanding of their field(s) of microscale testing in aquatic toxicology.

Artificial substrate: A material introduced into a natural system to provide a uniform substrate that indigenous organisms can colonize.

Assessment endpoint: An effect criterion by which toxicity is estimated (e.g., mortality, growth, reproduction).

ATP: Adenosine triphosphate. a primary energy carrier in living cells.

Autogamy: A sexual process in ciliated protozoa during which an individual ciliate undergoes self-fertilization.

Axenic: Without any other species of organisms present, including bacteria.

Bioconvection: Circulation or movement of aquatic microorganisms that can be modeled using the physical or convective processes driven by heat.

Bioluminescence: The production of light by living organisms. The result of an enzyme (luciferase)-catalyzed, oxygen-dependent biochemical reaction, in which a substrate (tetradecanal in the case of the marine bacterium *Vibrio fischeri*) is oxidized to an energized form. Upon return to the unenergized ground state, light is emitted. In marine bacteria, luminescence components are coded for by the *lux* genes.

Chemical extracts: One of the possible substrates to be used in microscale sediment toxicity tests. It consists of the fraction of contaminants in the sediment which can be extracted by traditional chemical procedures, e.g., by the use of solvents or acids.

Chemosensory: Any sensory process stimulated by chemical means.

Chemotaxis: Movement away from or toward a chemical stimulus.

Chorion: The outermost layer surrounding an embryo.

Chromosomal mutation: Chromosomal changes visible under the light microscope.

Chronic test: A study in which the organism is observed over a long period of time which may include a substantial portion of its life cycle.

Clastogenic event: Chromosome breakage.

Clone: A collection of cells arising from the division of a single cell, all of which are therefore genetically identical.

Conjugation: A sexual process in ciliated protozoa during which two ciliates temporarily fuse to exchange gametic nuclei.

Control sediment: "Clean" sediment which does not contain concentrations of contaminants that could impact the growth, survival, or behavior of the test organisms. Control sediment might be a natural sediment from an uncontaminated site, or formulated (reconstituted) sediment. This sediment must contain no added test material or substance, and must enable acceptable (i.e., ≥80%)

survival of the test organisms during the test. The use of control sediment provides a basis for interpreting data derived from toxicity test using test sediment(s), and also provides a base sediment for spiking procedures.

DNA damage: A collective description for different changes in DNA that do not refer to normal cell rhythms; for instance single- or double-stranded breakage, e.g., missing purine and pyrimidine bases.

Daphtoxkit: 24- to 48-h invertebrate microbiotest, based on the cladoceran crustaceans *Daphnia magna* or *Daphnia pulex*, hatched from ephippia.

Developmental selectivity: The ability to interfere with *in utero* development at exposure levels lower than are needed to disrupt maternal homeostasis.

Elutriate: One of the possible substrates to be used in microscale sediment toxicity tests. The elutriation process consists of mixing standard amounts of sediment and noncontaminated water, stirring the mixture for a certain period of time, centrifuging or decanting the supernatant, and then using it for the toxicity test. The elutriate contains the soluble fraction of contaminants from sediments.

Energy metabolism: All of the physical and chemical processes in an organism by which energy is made available for its functions.

Enzyme-linked immunosorbent assay (ELISA) and enzyme immunoassay (EIA): ELISA stands for the most common type of immunoassay, the solid-phase enzyme immunoassay. Originally the term ELISA was used for enzyme immunoassays with immobilized antigens or haptens, which are detected by an enzyme-labeled antibody. EIAs are immunoassays which use an enzyme-labeled hapten as a tracer.

Exogenous activation: An externally supplied metabolic activation system applied to an *in vitro* bioassay, for example, the adding of a mammalian liver homogenate to a bacterial bioassay for mutagenicity.

FETAX assay: A developmental toxicity assay employing frog embryos.

Gene-mutation: Changes of DNA at molecular level not visible using a light microscope, e.g., point mutation, deletion, and insertion in larger DNA portions.

Genome: Total genetic material of a cell and or an individual.

Genomutation: Changes in the chromosome number, e.g., Sister Chromatid Exchange (SCE) assay.

Genotoxic: The property associated with substances that are capable of damaging the genetic material of an exposed organism.

Genotoxicity: General toxicity upon the genome (genetic material).

Genotoxicity Test: A test system to demonstrate genotoxic reaction, e.g., DNA-damage and DNA-repair.

Genotoxin: A DNA-damaging substance.

Grain size: This is the measure of frequency and size distribution of the desegregated mineral particles comprising the sediment. Size distributions are commonly reported on the Wentworth Scale, which classifies particles as follows:

NAME	GRADE LIMITS	
	mm	Phi
Coarse sand	1.000–0.500	+1.0
Medium sand	0.500–0.250	+2.0
Fine sand	0.250–0.125	+3.0
Very fine sand	0.125–0.062	+4.0
Silt	0.062–0.004	+8.0
Clay	< 0.004	>8.0

Phi = $-\log_2$ particle diameter in mm.

Hapten: is a low-molecular-weight substance, e.g., a pesticide, that is able to bind to antibodies, although it is too small to induce an immune response by itself. Nevertheless, it may elicit a strong response if covalently bound to a carrier, a high-molecular-weight immunogenic protein or synthetic polymer.

Hydra Assay: A developmental toxicity assay designed to detect selective developmental toxicants.

Hydra attenuata: A freshwater coelenterate (Cnidarian).

***In vivo* test**: Toxicity or genotoxicity test using whole animals or plants.

Induction: The influence exerted by an embryonic organizer or evocator on the differentiation of an adjacent tissue or cell type.

Infauna: Organisms (generally invertebrates and microbes) living within the bottom sediments of water bodies.

Insect growth regulator: A class of insecticides which adversely interferes with the normal growth and development of insects.

K2 values: The coefficient of utilization of assimilated energy for growth; analogous to net growth efficiency.

LC_{50} or LC50: A statistically calculated concentration that is expected to be lethal to 50% of the test organisms. The LC_{50} is always expressed as a time-dependent value (e.g., 24 h, 48 h, 96 h), that is, the concentration estimated to be lethal to 50% of the test organisms after 24, 48, or 96 h of exposure.

LOAEL: The lowest observed adverse effect level in the lowest concentration of a material used in a toxicity test that has a statistically significant adverse effect on the exposed population of test organisms compared with the controls.

LOEC: The lowest-observed-effect concentration. This is the lowest concentration of a test material to which organisms are exposed, that causes adverse effects on the organism, effects which are detected by the observer and are statistically significant. For example, the LOEC might be the lowest concentration at which fertilization success differed significantly from that in the control. LOEC is generally reserved for sublethal effects. Also called the LOAEL.

Library: Within a genetic context, a collection of clones, each of which contains a single copy of a reporter gene at a unique chromosomal site.

Lipopolysaccharide: Complex lipid structure containing unusual sugars and fatty acids found in many gram-negative bacteria, and constituting the chemical structure of the outer layer.

Luciferase: An enzyme from firefly lanterns which acts on its substrate, luciferine, to produce light emission in the expense of ATP.

Lugol's iodine solution: 20g potassium iodide (KI) and 10 g iodine crystals dissolved in 200 mL distilled water to which 20 mL glacial acetic acid has been added.

Margin of safety (MOS): animal NOAEL of a guideline compliant developmental toxicity (teratology) study divided by anticipated human exposure level.

Measurement endpoint: The numerical expression of a particular assessment endpoint or effect criterion (e.g., IC_{50}, NOEC, LOEC).

Metabolic activation: An enzymatic biotransformation in which a chemical is made more toxic.

Metamorphosis: A change in shape or structure of an immature animal to an adult.

Microcosm: A small model of an ecosystem that incorporates some, but not all, of the parts and interactions of the natural system of interest.

Micronucleus: Small intracytoplasmic chromosomal masses, separate from the nucleus and arising from a clastogenic or aneugenic event.

Microtox® Test System: *In vitro* toxicity tests that use bioluminescent bacteria to detect toxic substances in water.

Monospecific test: A microplate bioassay conducted with one algal species (e.g., *S. capricornutum*) at a time.

Mutagen: An agent which causes production of a mutation.

Mutagenic: The property associated with substances that are capable of inducing changes in the base pair sequence of an exposed organism's genetic material.

Mutation: A change in the character of a gene which is perpetuated in subsequent divisions of the cell in which it occurs.

Reverse mutation: Description of a "second" mutation occurring in a mutant cell or organism that "reverts" the cell or organism to prototrophy (or wild type phenotype).

NOAEL: See no-observed-effect-concentration.

NOEC: The no-observed-effect concentration. This is the highest concentration of a test material to which organisms are exposed, that does not cause any observed and statistically significant adverse effect on the organism. For example, the NOEC might be the highest tested concentration at which an observed variable such as fertilization success did not differ significantly from that in the control.

O:N ratio: A metabolic indicator of substrate utilization derived from the ratio of oxygen atoms consumed to ammonia atoms excreted.

Phytotoxicity: The potential of any material (physical, biological, chemical) to cause adverse effects toward vegetal systems.

Plasmid: A small circular DNA, replicating independently of the chromosome. Plasmids often confer antibiotic resistance, and are widely used for the introduction of genes or DNA constructs into bacterial cells.

Polychaete: Member of the class Polychaeta, phylum Annelida, typically segmented marine worms bearing numerous setae.

Pore water: It consists of the liquid fraction of soil or sediments, i.e., the water contained among sediment grains which can be extracted by a variety of methods, e.g., centrifuging or squeezing. One of the possible substrates to be used in microscale sediment toxicity tests.

Power (of a toxicity test): The probability of detecting a difference or effect of a given size if it is present.

Power analysis: A determination of the ability of a statistical test to reject a false hypothesis.

Pro-genotoxin: A genotoxic substance that is benign until metabolically activated by mammalian enzyme systems.

Promoter: A DNA region involved in, and necessary for, initiation of mRNA transcription. It includes the RNA polymerase binding site, the startpoint of transcription, and other potential sites at which regulatory proteins may bind.

Promutagen: A chemical or substance that has mutagenic activity only after it is metabolized.

Protozoans: Animal-like unicellular eukaryotic organisms, usually those that move by means of flagella, cilia, or pseudopods. Members of the Kingdom Protoctista.

Reference sediment: A field-collected sample of presumably "clean" sediment, selected for properties (e.g., particle size, compactness, total organic content) representing sediment conditions that closely match those of the sample(s) of test sediment except for the degree of chemical contaminants. It is often selected from a site uninfluenced by the source(s) of contamination but within the general vicinity of the sites where samples of test sediment are collected.

Reporter gene: A known DNA sequence whose protein product is easily assayed. Such genes, minus their regulatory sequences, can be inserted within target genes whose protein products are unknown or not easily assayed, thus allowing measurement of the expression of the target gene.

Rotoxkit: 24-h invertebrate microbiotest, based on the rotifer *Brachionus calyciflorus*, hatched from cysts.

S9: The 9000 g supernatant of mammalian liver homogenate. S9 contains the microsomal and cytoplasmic enzymes required for *in vitro* activation of pro-genotoxins.

Sediment: Material, such as sand or mud, suspended in or settling to the bottom of a liquid. Sediment input to a body of water comes from natural sources, such as erosion of soils and weathering of rock, or as the result of anthropogenic activities. Certain contaminants tend to collect on and adhere to sediment particles.

Solid phase: One of the possible substrates to be used in microscale sediment toxicity tests. It consists of the whole sediment sample, collected either by corers or by grab samplers.

SOS Response: A complex regulatory network in *Escherichia coli* that is induced by DNA damage. Also known as the error-prone repair pathway, the SOS network regulates the expression of several (>17) unlinked genes.

Spionid: Member of the family Spionidae of the class Polychaeta, phylum Annelida, typically bearing a pair of elongate palps, segments bearing setae of two types including long slender setae (capillary) and short hooded hooks.

SPMD: Semipermeable Polymeric Membrane Device, used as a passive *in situ* sampling tool to monitor the bioavailability of lipophilic chemical contaminants in water, soil, and air.

Sublethal: Detrimental to the organism, but below the level that directly causes death within the test period.

Tandem toxicity testing: Qualitative and quantitative testing approach to determine acute toxicity and genotoxicity of environmental chemical contaminants.

Taxonomic richness: The number of taxa in a biological community.

Teratogen: An agent which causes abnormal development.

Teratology: The branch of embryology concerned with the production, development, anatomy, and classification of abnormal development.

Standard Developmental Teratology Test: Groups of pregnant animals are exposed to a range of test chemical concentrations spanning the maternal and developmental NOAEL through a high dose, altering maternal homeostasis but producing no more than 10% maternal deaths.

Thamnotoxkit: 24-h invertebrate microbiotest, based on the anostracan crustacean *Thamnocephalus platyurus*, hatched from cysts.

Tolerance distribution: a description of the relative sensitivity of organisms to a material; the number of taxa first adversely affected by a toxicant plotted over increments in concentration of the toxicant.

Toxicity: The inherent potential or capacity of a substance to cause adverse effect(s) on living organisms. The effect(s) could be lethal or sublethal, direct or indirect.

(Developmental) toxicity: production of any of the four cardinal signs of disrupted *in utero* development: live terata, dead conceptus, *in utero* growth retardation, decrement of anticipated postnatal function.

Toxicity Identification Evaluation (TIE): Describes a systematic sample pre-treatment (e.g., pH adjustment, filtration, aeration, addition of zeolite, etc.) followed by tests for toxicity. This evaluation is used to identify the agent that is primarily responsible for toxicity in a complex mixture or sediment. The toxicity test can be lethal or sublethal.

Standard developmental toxicity testing: a study involving exposure of pregnant animals (rats, rabbits, mice) to a series of exposure intensities throughout the period of organogenesis.

Toxicity Test: The means by which the toxicity of a chemical or other test material is determined. Measures the degree of response produced by exposure to a specific level of stimulus (e.g., a concentration of chemical, sample of sediment), either: (1) the proportions of organisms affected (quantal), or (2) the degree of effect shown (graded or quantitative).

Toxkit: Generic name for any microbiotest which is dependent on stock culturing, and departs from dormant or immobilized stages of selected test species.

Tracer: A tracer is a substance that produces the signal for measurements in the immunoassay. In the most frequently used format with immobilized antibodies, labeled haptens or antigens are added as a tracer. For a radio immunoassay (RIA) a radioactive compound is used, for the enzyme immunoassay (EIA) an enzyme labeled tracer, and for the fluorescence immunoassay (FIA) a fluorescence labeled tracer is employed.

Transcription: The first stage of gene expression, whereby the DNA sequence of the gene is copied to an RNA intermediate, which will determine the protein ultimately produced via expression of the gene.

Trochophores: free-swimming larval stage of many polychaetes, used in 96-hour tests.

***umuC*:** The induction of the gene *umuC* is part of the specific response of the bacterial cell to DNA-damage. *UmuC* is the acronym for *U*V *mu*tagenesis and *c*hemical repair gene.

***umuC* operon:** The *umuC* operon is a regulatory sequence for the induction of the *umuC* gene.

Xenobiotic: A substance foreign to an environment; a man-made substance or manufactured chemical.

Note: The reader is also directed to the glossary in Rand, G-M., Ed., *Fundamentals of Aquatic Toxicology: Effects, Environmental Fate and Risk Assessment*, 2nd Edition, Taylor and Francis, Washington, D.C., 1995, 1125 p.; to the *Dictionary of Toxicology*, E. Hodgson, R.B. Mailman, and J.E. Chambers, Stockton Press, USA, 1997, 500 p. (approx.), in press; and to the *Standard Methods* published by the American Society for Testing and Materials.

INDEX

A

Acartia tonsa, 438
Acenaphthylene, 208, 216
2-Acetamidofluorine, 208
Acetylcholinesterase, 91, 145
Acetylcholinesterase inhibition test, 78, 79, 85
Acid
  phosphatase, 164
  volatile sulfides (AVS), 633
Acineta, 540
Actin microfilaments, 262
Acute sediment toxicity test, 460
Adenosine triphosphate (ATP), 31, 153
Adsorbable organic halides (AOX), 56
Aeromonas
  hydrophila, 551
  liquefaciens, 214
AET, see Apparent effects threshold
AHH, see Aryl hydrocarbon hydroxylase
AhR, see Ah receptor
Ah receptor (AhR), 55
Alachlor, 17
Alanine aminopeptidase, 81, 82, 87, 220
Alaninpeptidases, 83
Alcohol dehydrogenase, 164
Aldehyde dehydrogenase inhibition test, 78, 79
Aldrin, 208
Algal test battery (ATB), 276
Algaltoxkit, 311–320
  complexity and variability of algal toxicity tests,
    311
  development of Algaltoxkit microbiotest, 313–316
  development of microbiotests with algae, 312
  development of stock-culture-free algal
    microbiotest, 313
  first validation of Algaltoxkit microbiotest,
    316–317
  potential and limitations, 318–319
  precision of Algaltoxkit microbiotest, 317–318
Alkaline agarose gel electrophoresis, 492
Alkylbenzene sulfonates, 49
Ambystoma mexicanum, 508

American Public Health Association (APHA), 333
American Society for Testing and Materials (ASTM),
  5, 424, 428
American Type Culture Collection (ATCC), 292
2-Aminoanthracene, 208
Ammonia excretion rate, 469
Ampelisca abdita, 65, 453, 561
Ampelisca spp., 452
Amphiascus tenuiremis, 531
Amphibians, genotoxicity tests in, 507–519
  amphibian micronucleus test, 508–516
    principle of micronucleus test, 508–509
    studies *in situ*, 512–516
    studies with X-rays and chemicals, 509–512
  in vitro and *in vivo* genotoxicity tests, 507–508
  practicality of amphibian micronucleus test,
    516–517
Amphipod sediment toxicity tests, 451–463
  advances and future directions, 455–459
    ammonia and hydrogen sulfide toxicity, 456
    full life-cycle chronic tests, 456–458
    further miniaturization, 455–456
    integrated usage of amphipod assays, 459
    mechanisms of toxicity in amphipod assays, 459
    spiked sediment controls in amphipod assays,
     456
  historical perspective, 452–453
  techniques and practice, 453–455
    acute tests, 453–454
    sublethal tests, 454–455
Amphipod toxicity, 455
Amphiporeia virginiana, 225, 454, 564, 565
Ampicillin, 240
Amylase, 145, 220
Anabaena, 292, 297, 304
  cylindrica, 274, 275
  flos-aquae, 290, 293
  spp., 293
  variabilis, 295
Analysis of variance (ANOVA), 429, 431
Ankistrodesmus, 297
Ankistrodesmus spp., 293
Ankyra judayi, 293